T0206188

Lecture Notes in Computer Science 577

Edited by G. Goos and J. Hartmanis

A. Finkel M. Jantzen (Eds.)

STACS 92

9th Annual Symposium on Theoretical Aspects of Computer Science
Cachan, France, February 13–15, 1992
Proceedings

Springer-Verlag
Berlin Heidelberg New York
London Paris Tokyo
Hong Kong Barcelona
Budapest

Series Editors

Gerhard Goos
Universität Karlsruhe
Postfach 69 80
Vincenz-Priessnitz-Straße 1
W-7500 Karlsruhe, FRG

Juris Hartmanis
Department of Computer Science
Cornell University
5148 Upson Hall
Ithaca, NY 14853, USA

Volume Editors

Alain Finkel
ENS Cachan, LIFAC
61, Avenue du President Wilson, 94235 Cachan, France

Matthias Jantzen
Fachbereich Informaik, Universität Hamburg
Rothenbaumchaussee 67/69, W-2000 Hamburg 13, FRG

CR Subject Classification (1991): F, G.1–2, C.2, I.1, I.3.5, I.2.6, C.5, E.3

ISBN 3-540-55210-3 Springer-Verlag Berlin Heidelberg New York
ISBN 0-387-55210-3 Springer-Verlag New York Berlin Heidelberg

Printed in Germany

Typesetting: Camera ready by author
Printing and binding: Druckhaus Beltz, Hemsbach/Bergstr.
45/3140-543210 - Printed on acid-free paper

Foreword

The annual Symposium on Theoretical Aspects of Computer Science (STACS) is held each year, alternatively in France and Germany. STACS is organized jointly by the Special Interest Group for Fundamental Computer Science of the Association Française des Sciences et Technologies de l'Information et des Systèmes (AFCET) and the Special Interest Group for Theoretical Computer Science of the Gesellschaft für Informatik (GI).

STACS 92, the ninth of this series, was held in Cachan, February 13-15. It was preceded by symposia at Hamburg (1991), Rouen (1990), Paderborn (1989), Bordeaux (1988), Passau (1987) Orsay (1986), Saarbrücken (1985), and Paris (1984); the proceedings of these symposia are published Lecture Notes in Computer Science Volumes 480, 415, 349, 294, 247, 210, 182 and 166 respectively.

The large number of 179 submitted papers, their scientifc quality and relevance for the symposium once again proved the importance of STACS for many areas of theoretical computer science.

The time schedule of the symposium allowed for acceptance of only 44 of the submitted papers and made parallel sessions unavoidable. Therefore, the selection was very difficult and many good papers had to be rejected. Each submitted paper was sent to four Program Committee members, who were often assisted by other competent referees. The members of the Program Committee are as follows:

<u>H. Alt</u> (Berlin)	<u>M. Jantzen</u> (Hamburg,co-chairman)
<u>A. Arnold</u> (Bordeaux)	<u>A. Pnueli</u> (Rehovot)
<u>D. Beauquier</u> (Paris)	<u>D. Seese</u> (Duisburg)
<u>A. Bertoni</u> (Milan)	G. Turan (Chicago)
M. Cosnard (Lyon)	P. Vitanyi (Amsterdam)
<u>A. Finkel</u> (Cachan,chairman)	<u>I. Wegener</u> (Dortmund)

Those who attended the meeting are underlined. All evaluations arrived on time for this selection meeting.

We would like to express our gratitude to all members of the Program Committee, in particular to those who made the final selection, and to all referees who assisted them. The list of the referees is as complete as we can achieve, and we apologize for any possible omissions or errors. Moreover we would like to thank all those who submitted papers to this symposium. STACS 92 offered three invited talks which opened each day of the symposium:

M.C. Gaudel (Univ. Paris-Orsay): Structuring and Modularizing Algebraic Specifications: the PLUSS Specification Language, Evolutions and Perspectives
T. Hagerup (MPI Informatik): The Log-Star Revolution
I. SIMON (Univ.de Sao Paulo): Compression and Entropy

A number of software systems were presented which showed the possibilities of applying theoretical results to software construction as well as providing a help for doing research.

Except for systems, the papers in this volume are printed in the order of presentation and thus grouped into sessions, most of which are thematic.

We gratefully acknowledge all the institutions and corporations who have supported this conference:

Aérospatiale, Bull, C^3, CNET, CNRS, Dassault, DRED, DRET, ENS Cachan, IBM, INRIA, MACIF, Matra, MRT, PRC-Programmation, SLIGOS, Université de Picardie, Université Paris-Orsay.

Last but not least, we wish to express our gratitude to the members of the Organizing Committee:

P. Estraillier (Paris), A. Finkel (Cachan), A. Petit (Orsay), L. Petrucci (Evry), B. Rozoy (Orsay), V. Ségaut (AFCET, Paris)

Paris, December 1991

Alain Finkel
Matthias Jantzen

List of Referees

ABITEBOUL	A
AKL	S
ALTHÖFER	I
APT	L
ARIKATI	S
ASVELD	P
AURENHAMMER	F
AVENHAUS	J
BADLAENDER	H
BADOUEL	E
BAETEN	J
BALBO	G
BALCAZAR	J
BENZAKEN	V
BERARD	B
BERMOND	J.c.
BERSTEL	J
BEST	E
BETREMA	J
BIDOIT	M
BLOEMER	J
BLOK	W
BOASSON	L
BODLAENDER	H
BOISSONAT	J.d.
BOL	R
BONGIOVANNI	G
BONIZZONI	P
BOUCHERON	M
BOUCHITTE	V
BOUGE	L
BRANDENBURG	F
BRUSCHI	D
BRUSHI	M
BRUYERE	V
BRZOZOWSKI	J
BUHRMAN	H
BURRIS	S
CAMPADELLI	P
CARDONE	F
CARSTENSEN	H
CASPI	P
CASTERAN	P
CEGIELSKI	P
CERIN	C
CHOFFRUT	C
CHRETIENNE	P
COMON	M
CORI	R
COSTANTINI	S
COURCELLE	M
CREPEAU	M
CRESCENZI	P
CROCHEMORE	M
CULIK II	K
D'ANTONA	O
DAMIANI	M
DASSOW	J
DE ANTONELLIS	M
DE CINDIO	F
DE ROUGEMONT	M
DEBRAY	S
DELEST	M
DELPORTE	C
DEN AKHER	R
DEVIENNE	P
DI JANNI	M
DIAZ	J
DIEKERT	V
DONATELLI	S
DOUCET	A
DUCHAMP	G
DURIEUX	J
ENJALBERT	P
ESPARZA	J
FAGLIA	G
FANCHON	J
FAUCONNIER	H
FERREIRA	M
FISHER	P
FORMANN	M
FORTNOW	M
FOURNIER	J.c.
FRAIGNIAUD	P
FROUGNY	C
GABARRO	J
GASTALDO	M
GASTIN	P
GASTINGER	S
GAUDEL	M
GAUDIOT	Jl
GAVALDA	R
GIRE	F
GOLDWURM	M
GOUYOU-BEAUCHAMPS	D
GRAMMATIKAKIS	M
GRANDJEAN	E
GRIGORIEFF	S
GRUSKA	J
GUERRA	G
GUESSARIAN	I
GUSGEN	M
HABEL	M
HABER	F
HANSMANN	W
HASTAD	M
HAUSCHILDT	D
HEERING	J
HEMACHANDRA	L
HERRMANN	E
HOENE	A
HOFFMANN	F
HOHBERG	W
HOOGEBOOM	H
IBARRA	O
JOHNEN	C
JWAINSKY	M
KANT	G
KAPOULAS	G
KENNETH	R
KENYON	C
KHACHIYAN	M
KIRSIG	B
KLEIN	R
KLOKS	T
KOBLER	J
KONIG	J.Cl.
KRANAKIS	E

KRATSCH	D
KREOWSKI	H
KRIEGEL	K
KUCHEROV	G
KUDLEK	M
KUIJK	F
LAMERSDORF	W
LANGE	K
LANZARONE	G
LATTEUX	M
LAVALLEE	Y
LAVAULT	C
LE VERGE	H
LEHMANN	D
LENGANER	T
LESCANNE	P
LITOVSKY	I
LOMBARDI	H
LUEBER-WULF	D
LUO	J
MACKENTHUN	R
MANACHER	G
MANFRED	K
MARRE	B
MARTELLI	A
MASSAZZA	P
MAURI	G
MAZOYER	J
MEHLHORN	K
MEINEL	C
MICHAUX	C
MIGLIOLI	P
MILSTED	K
MOLDT	D
MOREL	P
NEUMANN	B
NICOLAI	F
NILSSON	B
NIVAT	M
ORNAGHI	M
PAPADIMITRIOU	C
PAUL	W
PAULIN	C
PELADEAU	P
PELC	A
PENAUD	J
PERRIN	D
PERSIANO	G
PETIT	A
PETRESCHI	R
PETRUCCI	L
PICARONY	C
PIGHIZZINI	G
PIPERNO	A
POCCHIOLA	M
PODELSKI	A
POMELLO	L
POTTHOFF	A
PUEL	L
RAJSKI	J
RAUZY	A
RAYNAL	M
REGAN	K
REICHEL	B
REISCHUCK	R
REISCHUK	R
REUS	B
RICHTER	M
ROBERT	Y
ROBSON	M
RODENBURG	P
ROMAN	J
ROSSI	G
ROZOY	B
SABADINI	N
SALLE	P
SAMI	Y
SANTHA	M
SATTLER-KLEIN	A
SCHNOESELEN	P
SCHONFELD	M
SCHONING	U
SCHRYVER	M
SCHWARZKOPF	O
SEEBOLD	P
SENIZERGUES	G
SIMONE	C
SLOWINSKI	K
SNIR	M
SOLITRO	U
SONG	S
SOS	T
STRAUBING	H
SYGORA	O
TEL	G
THIBON	J.y.
THOMAS	W
TIER	C
TISON	M
TORAN	J
TORENDIET	L
TRILLING	L
TROMP	J
TRYSTRAM	G
TURAN	G
UKKONEN	E
VALK	R
VALKEMA	E
VAN EMDE BOAS	P
VAN LEEUWEN	J
VATAN	F
VAUQUELIN	B
VELDHORST	M
VIDAL-NAQUET	G
VISHKIN	U
VRTO	I
WAACK	M
WAGNER	K
WAGNER	F
WAINSKY	I
WEIL	P
WELZL	E
WERNISCH	L
WOLFERS	B
YAMINA	S
YVINEC	M
ZIELONKA	W

Contents

Invited Lecture

Parallel Algorithms

Logic and Semantics

Parallel Algorithms 2

Computational Geometry

Automata and Languages

Structural Complexity 1

Computational Geometry and Learning Theory

Invited Lecture

Complexity and Communication

Structural Complexity 2

Distributed Systems

Complexity

Algorithms 1

Cryptography

Algorithms 2

VLSI

Invited Lecture

Words and Rewriting

Algorithms 3

Systems

INVITED LECTURE

Structuring and Modularizing Algebraic Specifications: the PLUSS specification language, evolutions and perspectives

Marie-Claude Gaudel
LRI, Université de Paris-Sud et CNRS (UA 410)
Bâtiment 490
91405 Orsay cedex
France

Abstract

The formal specification of abstract data types using axioms was proposed about fifteen years ago by several authors. It has become the basis of the so-called algebraic specification methods. Numerous algebraic specification languages have been designed and experimented and several specification languages make use of the algebraic approach to specify data types.

This paper reports the experience of the author in the design of the PLUSS specification language and in using it for various case studies: PLUSS has evolved a lot since its beginning and will still evolve. A rationale for its past evolution and a discussion of some possible future changes is given. Moreover, some lessons learned from the case studies on the advantages and drawbacks of algebraic specifications are reported.

1- Introduction

The formal specification of abstract data types using axioms was proposed fifteen years ago by several authors [Zil 74, Gut 75, ADJ 76]. It has become the basis of the so-called algebraic specification methods. Numerous algebraic specification languages have been designed and experimented: CLEAR [BG 80], ACT_ONE [EFH 83], PLUSS [Gau 85], OBJ [FGJM 85], Extended-ML [ST 86], etc, and several specification languages make use of the algebraic approach to specify data types, e. g. LOTOS, LARCH [GH 86], RSL [NHWG 88] (for a comprehensive survey, see [BKLOS 91].)

This paper discusses the past and future evolution of the PLUSS specification language. The aim of PLUSS is to evolve: it has a sort of life-cycle: a version of the language is tested against several case studies, and these tests serve as bases for a revision of the language, and so on. The discussion focuses on the problems of structuring, modularizing, and, to a less extent, parameterizing specifications.

2 - PLUSS-84: rationale for different kinds of modules

2.1 - The first case studies

The very origin of the PLUSS specification language comes from an experiment, carried out at INRIA, in formally specifying a PASCAL compiler for the PERLUETTE compiler generator [Des 83]: it made very clear that such a large specification must be structured in an elaborated way, and that the current linguistic support was not sufficient from this point of view. This problem was not new, since it was already discussed in [BG 77] and a kernel of primitives for structuring algebraic specifications, ASL, was proposed in [SW 83].

The next step[1] was based on a case study in the area of telephony: the idea was to experiment the ASL primitives or some combinations of them to write the case study, and to define a specification language. However, the definition of good structuring constructs was only one of the requirements for the language: another requirement was to provide a way to formally specify the so-called "defence mechanisms" of the telephone switching system, namely error detection and recovery; a last important requirement was to provide a method for constructing ADA programs from specifications.

The error detection and recovery problem will not be discussed here; the solution we adopted for this project was published in [B2G3 85]; further works on this topic are presented in [BBC 86] and [BLG 91].

The preliminary version of PLUSS was very close to ASL: the semantics of a specification was a pair <signature, class of algebras> (it has remained unchanged until now); specifications were built by composition of modules, mainly *enrichment* , possibly with *renaming* , and *visibility control* ; modules can be also obtained by instanciation of *parameterized modules*; these modules being parameterized by specifications, the requirements on arguments were specified as *parameter modules*. The semantics of PLUSS was (and is still) a combination of initial and loose semantics: implicitly the semantics is loose, but it is possible to state for some specification that the class of algebras must be restricted to those algebras isomorphic to the initial one.

---

[1]which took place in the Laboratoires de Marcoussis

As usual, we call the algebras of the class associated to a specification SP the models of SP. We note *Sig*(SP) and *Mod*(SP) the signature and the class of the models of SP.

We were very confident in the use of parameterization for structuring large specifications. However, when working on the case study, it turned out that it was not obvious at all to decide what was parameterized and by what, at least at the beginning of the specification process: parameters had an unpleasant tendency to proliferate and even to become parameterized... The reason of this phenomena was that it is tempting, at this stage, to parameterize the piece of specification one is writing by other pieces of the specification which are not yet fully specified. It is clearly a biased use of parameterization since there will be only one instantiation. Moreover this instantiation has good chances to be by a specification which is also parameterized, and so on. The conclusion was that some specific constructs were needed in the specification language to allow the use of some parts of the specification explicitly stated as "not yet fully specified", without recourse to parameterization.

The program construction method associated with the specification language is not described here: it is a rather elaborated one, based on the decomposition of abstract data types; its principles are given in [Gui 84] and its application to the case study is described in [B2G3 85]. For this discussion, it is sufficient to know that each standard specification module was transformed into an ADA package, the so-called "specification part" of the package being generated from the signature of the specification module. It worked rather well on small examples, but when Michel Bidoit and Brigitte Biebow applied it to the whole case study, they noticed that in some cases the decomposition process did not match the structure of the specification [BB 83]. They solved the problem in introducing in the method some feed-back mechanism aiming at keeping the correspondence between the structures of the specification and the program. We were not satisfied with this solution; we gave to this problem the name "structure clash" and, in some sense, we are still struggling with it[2].

---

[2]It is interesting to note that this phenomena is both well-known and very rarely mentioned in the literature: may be because it occurs on large examples only...

2.2 Overview of the language

PLUSS-84 is presented in [Gau 85]. There is a more detailed report, available from the author, which was presented and discussed during an Alvey Workshop on Formal Specifications in Swindon in October 1984. This presentation of the language deliberately skips those aspects which are not related to the structure of the specifications and the decomposition into modules.

The main novelty was the introduction of two kinds of specification modules: *drafts* and *specs* . *drafts* can be combined by the *enrich* construct; *specs* can be combined by the *use* contruct; a *draft* (or a set of *drafts*) can be converted into a *spec* using the *from* construct; specs can be parameterized, but drafts are not parameterizable. This last point was a methodological decision: drafts were not supposed to be used and reused, but mainly to be enriched and to evolve; eventually they should be converted into a spec which can be parameterized.
The models of a specification which has a *draft* as main module are the class of all algebras satisfying the axioms of the module and some hierarchical constraints; the models of a specification which has a *spec* as main module is the class of all algebras finitely generated w.r.t. some generators, satisfying the axioms of the module and some stronger hierarchical constraints.

The hierarchical constraint associated with the *enrich* construct is the weakest possible, namely consistency between the properties of the enriching and the enriched modules: it means that the models are those satifying both the additional properties and the enriched specification. Let D0 be a draft, let $\Delta D1 = (\Delta S1, \Delta\Sigma 1, \Delta Ax1)$ be some new definitions of sorts, operations and axioms[3], and let D1 be:

draft D1 **enrich** D0
  Δ D1
end D1

We have

$$Mod(D1) =_{def} \{ M \in Alg(\Sigma 1) \mid M \models \Delta Ax1 \text{ and } M|_{\Sigma 0} \in Mod(D0) \}$$

where $\Sigma 1 = Sig(D1)$ and $\Sigma 0 = Sig(D0)$, $M|_{\Sigma 0}$ is the $\Sigma 0$-restrict of the $\Sigma 1$-algebra M (the notations are the usual ones, see [Gau 91].)
Note that $Mod(D0) \supseteq Mod(D1)|_{\Sigma 0}$ and this inclusion is usually strict: some models of D0 may not be extensible into models of D1. A classical example is given

---

[3]There was also a notion of union of drafts in the language, thus multiple enrichments were possible; multiple uses, and mixed enrichments and uses were also allowed.

in figure 1: among the models of STUPID_SET, there are the natural numbers with zero and successor operations; this model cannot be extended into a model of LESS_STUPID which must give an interpretation of ∈, since it makes no distinction between sets containing different elements.

```
draft STUPID_SET use NAT
    sort Set
    operations
        ∅ : -> Set
        insert _ in _ : Nat x Set -> Set
        count : Set -> Nat
    axioms
        count (∅) = 0
        count ( insert n in s) = count (s) +1
    with n:Nat, s:Set
end STUPID_SET

draft LESS_STUPID enrich STUPID_SET use BOOL
    operations
        _∈_ : Nat x Set -> Bool
    axioms
        n ∈ ∅ =false
        n ∈ (insert m in s ) = equal(n,m) or (n ∈ s)
    with n,m:Nat, s:Set
end LESS_STUPID
```

Figure 1

The hierarchical constraints associated with the *use* construct are stronger: *use* is always applied to a *spec* ; the semantics was designed to allow the decomposition into *spec* modules to be kept by the development; it implies that *spec* modules can be implemented in an independent way. Thus the semantics of *use* must state that using a *spec* does not change its semantics, thus its class of models. Let S0 be a spec, let ΔS1 and ΔD2 be some definitions of sorts, operations and axioms, and let S1 and D2 be:

spec S1 **use** S0
 ΔS1
end S1

draft D2 **use** S0
 ΔD2
end D2

Let us note

$$\mathit{Mod}'(S1) =_{def} \{ M \in \mathit{Gen}(\Sigma 1) \mid M \models \Delta Ax1 \text{ and } M|_{\Sigma 0} \in \mathit{Mod}(S0) \}$$

$$\mathit{Mod}'(D2) =_{def} \{ M \in \mathit{Alg}(\Sigma 2) \mid M \models \Delta Ax2 \text{ and } M|_{\Sigma 0} \in \mathit{Mod}(S0) \}$$

We have

$$\mathit{Mod}(S1) =_{def} \mathit{Mod}'(S1) \text{ if } \mathit{Mod}'(S1)|_{\Sigma 0} = \mathit{Mod}(S0), \varnothing \text{ otherwise}$$

$$\mathit{Mod}(D2) =_{def} \mathit{Mod}'(D2) \text{ if } \mathit{Mod}'(D2)|_{\Sigma 0} = \mathit{Mod}(S0), \varnothing \text{ otherwise}$$

As shown in figure 2, it is possible to transform STUPID_SET or LESS_STUPID into specs (and possibly to introduce new operations). However, it is inconsistent to use the SSET spec to write the LESS_STUPID draft.

```
spec SSET from STUPID_SET
        gen ∅, insert_in_
end SSET
spec SLST from LESS_STUPID
        gen ∅, insert_in_
        operations
                remove _ from _ : Nat x Set -> Set
        axioms
                n ∈ (remove m from s ) = (not equal(n,m)) and (n ∈ s)
                count (remove m from s ) = count (s)
        with n,m:Nat, s:Set
end SLST
```

Figure 2

The rationale of this version of the language was to make expressible both a clear separation of concerns in the structure of the specification via the use of *drafts*, and the decomposition of the specified system into modules via the use of *specs*, allowing independent developments from each module. However, this last point is not solved in a completely satisfactory way.

As soon as a system is specified as a directed acyclic graph of *spec* modules, related by the *use* construct, one would like to say that the choices of the programs to be associated with the specification modules are completely independent, the only requirement being that each program is a correct implementation of the corresponding specification module. Moreover, such an approach for modularizing specification should make easier the reuse of existing software modules. However, the above framework was not sufficient to deal with these problems: the semantics was only given for DAG of modules, i.e. for modularized specifications, in terms of the semantics of used subspecifications; there was no semantic definition for modules. Such a definition was given in [GM 88] in order to define a *reusability relation* between algebraic specification modules; we summarize in part 3 the revised version given in Thierry Moineau's Thesis [Moi 91]; another slightly different definition of the semantics of specification modules is the basis of the stratified-loose semantics [Bid 87] which is used in the current version of PLUSS [Bid 89] presented in part 4. In part 5 we discuss the differences of these approaches.

3- Modular implementations of structured specifications [Moi 91]

3.1 Hierarchical specifications

In order to explicit the difference between structure and modularity at the specification level, we consider here hierarchical specifications which can be: the empty specification $\emptyset$, with an empty signature and no axioms; an enrichment $SP0 + \Delta SP$ where SP0 is a hierarchical specification and ΔSP is a specification module; or an union $SP1 \cup SP2$ of two hierarchical specifications.
With any hierarchical specification, is associated its (flat) presentation as in [EFH 83], and a class of hierarchical models which is a subclass of the finitely generated models of this presentation.

Given a hierarchical specification SP and its presentation PR, the class $\mathcal{HMOD}$ (SP) of the hierarchical models of SP is defined as:

- if $SP = \emptyset$, $\mathcal{HMOD}(SP) =_{def} \{\mathbb{1}\}$, where $\mathbb{1}$ is the unique algebra of empty signature;
- if $SP = \emptyset + \Delta SP$, we call it a basic specification and
$$\mathcal{HMOD}(SP) =_{def} \{ I \in \mathcal{Gen}(PR) \mid I \text{ is initial in } \mathcal{Gen}(PR) \}$$
- if $SP = SP0 + \Delta SP$, with $SP0 \neq \emptyset$,
$$\mathcal{HMOD}(SP) =_{def} \{ M \in \mathcal{Gen}(SP) \mid M|_{\Sigma 0} \in \mathcal{HMOD}(SP0) \}$$
- if $SP = SP1 \cup SP2$, with SP1 and $SP2 \neq \emptyset$,
$$\mathcal{HMOD}(SP) =_{def} \{ M \in \mathcal{Gen}(SP) \mid M|_{\Sigma 1} \in \mathcal{HMOD}(SP1) \text{ and } M|_{\Sigma 2} \in \mathcal{HMOD}(SP2) \}$$

A specification SP is said *hierarchically consistent* if and only if $\mathcal{HMOD}(SP) \neq \emptyset$.

In order to avoid name clashes and some causes of inconsistency, we assume that the specifications are *well-structured* [Bid 89], i.e.: a module is the main module of only one subspecification[4]; modules have disjoint signatures (the signature of a module consists of the sorts and operations defined in the module); sorts and their generators are defined in the same module.

3.2 Semantics of specification modules: modular consistency

Let us now give a semantic definition of a specification module: given a specification SP = SP0 + ΔSP, a *realization* of ΔSP *above* SP0 is a mapping $\mathcal{R}$: $\mathcal{HMOD}$ (SP0) -> $\mathcal{HMOD}$ (SP) such that:

$$\forall M0 \in \mathcal{HMOD}(SP0), \mathcal{R}(M0)|_{\Sigma 0} = M0$$

We note $\mathcal{R}eal_{SP0}(\Delta SP)$ the class of all the realizations of ΔSP above SP0.

The notions of realization of specification modules and of hierarchical models of specification are consistent, namely [Moi 91]:

$$\mathcal{HMOD}(SP) = \{ \mathcal{R}(M0) \mid M0 \in \mathcal{HMOD}(SP0) \text{ and } \mathcal{R} \in \mathcal{R}eal_{SP0}(\Delta SP) \}$$

A specification module ΔSP is said *realizable above SP0* if and only if :

$$\mathcal{R}eal_{SP0}(\Delta SP) \neq \emptyset.$$

An example of a specification module which is not realizable is the LESS_STUPID module of figure 1.

A hierarchical specification is said *modularly consistent* if and only if it is hierarchically consistent, well-structured, and all its modules are realizable.

Modular consistency is clearly a much stronger property than hierarchical consistency: in fact, these two properties are complementary. At the beginning of the specification process, one requires hierarchical consistency, of course, and modular consistency is too constraining; when starting the refinements of the specification into programs, hierarchical consistency is not sufficient and modular consistency is highly desirable.

Given a modularly consistent specification SP, which is a composition of a set of modules $\{\Delta SP_i\}$, we note $\lfloor \Delta SP_i/SP$ the unique subspecification of SP such that $\lfloor \Delta SP_i/SP + \Delta SP_i$ is a subspecification of SP. A *modular implementation* of SP is a mapping which gives for each ΔSP_i a realization above $\lfloor \Delta SP_i/SP$. Let us note $\mathcal{MI}$ such a mapping.

---

[4]SPs is a subspecification of SP if and only if : SP = SPs, or SP = SP0 + ΔSP and SPs is a subspecification of SP0, or SP = SP1 ∪ SP2 and SPs is a subspecification of SP1 or SP2 (possibly of both).

Given a subspecification SPs of SP, $\mathcal{MI}|_{SPs}$ is the restriction of $\mathcal{MI}$ to the modules of SPs: it is a modular implementation of SPs.

$\mathcal{MI}$ defines a model of SP which is called its *resulting model* , M[$\mathcal{MI}$] :

- if SP = Ø, M[$\mathcal{MI}$] $=_{def}$ $\mathbb{1}$
- if SP = SP0 + ΔSP, M[$\mathcal{MI}$] $=_{def}$ $\mathcal{MI}$ (ΔSP) (M[$\mathcal{MI}|_{SP0}$])
- if SP = SP1 ∪ SP2, M[$\mathcal{MI}$] $=_{def}$ M[$\mathcal{MI}|_{SP1}$] ⊕ M[$\mathcal{MI}|_{SP1}$], (where ⊕ is the amalgamated sum of algebras.)

In [Moi 81] it is shown that any hierarchical model of a modularly consistent specification is the resulting model of a modular implementation.

This theory is the basis of a formal definition of program components reuse: it makes it possible to define some relations, at the specification level, of reusability of the implementations of a given specification SP for implementing another specification SP' [MG 91].

4- PLUSS-89: a semantics for specification modules

PLUSS-89 was designed by Michel Bidoit and is described in [Bid 89]. Its main difference with respect to the previous version of PLUSS is that it defines some semantics of the specification modules, which follows the so-called stratified-loose approach [Bid 87]. Moreover, there are three kinds of modules: *sketch, draft* and *spec* and all of them can be parameterized.

4.1 The stratified loose semantics of specification modules

The *spec* modules aim at the same goal as in the previous version of PLUSS: they correspond to parts of completed specifications, and they allow independent developments.

Let us consider a *spec* SP, which *uses* a *spec* SP0, i.e.:

spec SP **use** SP0 ΔSP **end** SP

We follow the convention used in the previous parts: SP denotes the whole specification, and ΔSP denotes the module. Σ, Ω and Ax are the signature, generators and axioms of SP, ΔΣ, ΔΩ and ΔAx are the signature, generators and axioms defined in the module ΔSP.

We note $\mathcal{SL}$ (SP) the class of models of SP in the stratified loose approach, and we note $\mathcal{HGEN}$ (SP) the class of Σ-algebras finitely generated w.r.t. Ω , satisfying the axioms Ax, whose Σ0-restricts belong to $\mathcal{SL}$ (SP0).

Following [Bid 87], the semantics of the ΔSP module is defined as the class $\mathcal{F}$ of all the mappings $\mathcal{F}_i$ such that:

- $\mathcal{F}_i$ is a total functor from $\mathcal{SL}$ (SP0) to $\mathcal{HGEN}$ (SP);
- $\mathcal{F}_i$ is a right inverse of the forgetful functor;
- if the class $\mathcal{F}$ is empty ΔSP is said to be inconsistent.

$\mathcal{SL}$ (SP) is the class of all the images of models of $\mathcal{SL}$ (SP0) by the functors of F:

$$\mathcal{SL}\text{ (SP)} =_{\text{def}} \bigcup_{\mathcal{F}_i \in \mathcal{F}} \mathcal{F}_i \left(\mathcal{SL}\text{ (SP0)} \right)$$

In the basic case, i.e. a spec module which uses no other spec, the semantics is the initial one.

The main difference between this approach and the one of [Moi 91] is that the semantics of a module is a class of functors, which preserve morphisms between models, instead of a class of mappings. We discuss some aspects of this difference in part 5.

4.2 Overview of the language and comments on some case studies

As mentioned above, there are three kinds of modules in the current version of PLUSS: *specs* which can be *used* by any kind of other modules; *sketches* which can be *enriched* by other sketches; and *drafts* which can be combined together via the *with* construct.

Sketches are similar to the drafts of the previous version of PLUSS: there is no notion of generators, the models are any algebras satifying the axioms, and the enrichment is as liberal as possible (global consistency must be kept).

Drafts are obtained from sketches (or from skratch) by defining a set of generators among the operations of the signature: the models are finitely generated w.r.t. these generators and any combination of drafts must preserve this property.

The distinction between sketches, drafts and specs points out two major steps in the development of a specification: the passage from sketches to drafts fixes the way it is possible to obtain values of a sort and the passage from drafts to specs fixes the class of models. These two steps were collapsed in the previous version of the language, and it turned out to be inconvenient: sometimes the generators are known very early in the specification process and they can be useful to write more concise specifications or to perform induction proofs.

However, the current version of PLUSS is not completely satisfactory in this respect. The main case study where sketches, drafts and specs have been used is the UGE specification (UGE means the on-board part of an automatic train pilot system) [Dau 91, DM91]. This specification is made of several parts which share a "state" where all the information on the state of the train, the railroad, the next train, etc, is stored. Sketches turned out to be useful to structure all this information: there is a SPEED sketch, a DOOR sketch, etc; however the limitations on the composition of drafts and sketches turned out to be very inconvenient. It is clear that it should be allowed to put together sketches and drafts: there are no semantic reasons to the current restrictions.

The fact that all kinds of modules can be parameterized in PLUSS-89 is in opposition to the choice made in PLUSS-84. This change shows clearly that parameterization at the specification level is not an easy issue: in PLUSS-84 it was sometimes very frustrating to wait until the passage from draft to spec to specify how the specification was parameterized... However, in the various case studies which were performed with the current version of the language, the use of parameterized specifications has not been very convincing, in my opinion: in the Unix file system specification [BGM 89], it was even disappointing since we decided after a while to write a specific version of generic trees when we aimed at reusing an existing one; in the specification of the transit node [MBGHP 89] we made use of existing "primitive" generic specifications such as SET(X), LIST(X), etc but we do not define specific parameterized modules; in the UGE specification we have not used parameterization.

The discussion on parameterization in specifications is currently open: in [SST 90] the authors underline the importance of distinguishing specifications of parameterized programs (i.e. of generic modules) versus parameterized specification of normal programs; in [Bro 91] Manfred Broy mentioned also some doubts on the usefulness of parameterized specifications. Moreover, it is interesting to remark that in the current version of PLUSS sketches and parameters have the same semantics, and instantiation is similar to sketch enrichment: I feel always uncomfortable when different syntactic constructs correspond to the same semantic concept.

5- Module semantics: mappings versus functors

The modularization of specifications introduces very strong constraints on the content of the modules. Figure 1 gives an example of modular inconsistency; more elaborated examples can be found in [GM 88], [Bid 89] and [Moi 91]. [Moi 91] gives sufficient conditions, which are rather restrictive.

The example given in figure 3 is from Thierry Moineau [Moi 91]; it aims at pointing out the difference of the semantics of modules given in part 3 and the stratified loose semantics summarized in part 4.

```
spec SP0 use BOOL                          spec SP  use  SP0
    sort s                                     operations
    generators                                         f: s -> s
        a, b, c: -> s                          axioms
    operations                                         a eq b = true => f(x) =c
        _eq_ : s x s -> Bool                   where x:s
    axioms                                 end SP
        a = c => true = false
        b = c => true = false
        x eq x = true
        a eq c = false
        b eq c = false
        a = b  <=> a eq b = true
    where  x:s
end SP0
```

Figure 3

In SP0, the first and second axioms ensure that a is distinct from c, and b from c; the next axioms ensure that the interpretation of eq returns true if and only if a and b denote the same value, and false otherwise.

The program below corresponds to a correct realization (cf. part 3) of ΔSP:

```
fct f (x:s) returns s
      if eq (a, b) then return (c)
      else case
            eq (x, a) => return (c);
            eq (x, b) => return (b);
            eq (x,c) => return (a);
            endcase
      endif
end f;
```

SP0 has two hierarchical models, A0 and B0 with:

$s^{A0} = \{\alpha, \beta, \gamma\}$ $a^{A0} = \alpha$ $b^{A0} = \beta$ $c^{A0} = \gamma$

$s^{B0} = \{\alpha\beta, \gamma\}$ $a^{B0} = \alpha\beta$ $b^{B0} = \alpha\beta$ $c^{B0} = \gamma$

The program above corresponds to a mapping $\mathcal{R}$: *HMOD* (SP0) -> *HMOD* (SP) with

$\mathcal{R}$ (A0) = A : $s^A = \{\alpha, \beta, \gamma\}$ $a^A = \alpha$ $b^A = \beta$ $c^A = \gamma$

$f^A(\alpha) = \gamma$ $f^A(\beta) = \beta$ $f^A(\gamma) = \alpha$

$\mathcal{R}$ (B0) = B : $s^B = \{\alpha\beta, \gamma\}$ $a^B = \alpha\beta$ $b^B = \alpha\beta$ $c^B = \gamma$

$f^B(\alpha) = \gamma$ $f^B(\beta) = \gamma$ $f^B(\gamma) = \gamma$

$\mathcal{R}$ is not a functor since it does not preserve the morphism ϕ_0 : A0 -> B0 defined by:

$\phi_0(\alpha) = \phi_0(\beta) = \alpha\beta$ and $\phi_0(\gamma) = \gamma$

Thus this realization of ΔSP is not allowed in the framework of the stratified loose semantics.

Conclusion: some hints on the "next" PLUSS

This paper is in no way an announcement of a new version of PLUSS: it summarizes some on-going reflections and some conclusions from experiments with the current version of the language.

The distinction between structuring and modularizing specifications has proved to be very fruitful. The question now is how to allow a flexible use of several kinds of modules, especially at the sketch-draft level where flexibility and ease of use are fundamental.

Concerning modularity, the choice between functors or mappings for the semantics of modules must be studied. It is not clear that functors are necessary, and the additional constraint they introduce could eliminate sensible realization of modules (however, a more convincing example than the one given figure 3 is needed).

One step further would be to allow an even cleaner separation of the specification structure and of the description of the future system architecture, to provide a better way to express this architecture and the correspondences with the specification structure: a specification language should provide some interface between requirement analysis and system design and PLUSS could be improved with this respect.

A last point under consideration is the introduction of some built-in notion of state, in order to allow an easier specification of systems similar to the UGE case study; such an extension of algebraic specifications is proposed in [Dau 92]. However, the important notion of the encapsulation of independent states is not addressed: some of the works presented in [GS 91] could be useful for this point.

References

[ADJ 76] Goguen, J.A., Thatcher, J.W. and Wagner E.G., "An initial algebra approach to the specification, correctness and implementation of abstract data types", in Current Trends in Programming Methodology, Vol.4: Data Structuring, edited by R.T. Yeh, pp.80-149, Prentice-Hall, 1978.

[Bid 87] Bidoit M., "The stratified loose approach : a generalization of initial and loose semantics", in Recent Trends in Data Type Specification, selected papers of the 5th Workshop on Specifications of Abstract Data Types, Gullane, Ecosse, Springer-Verlag L.N.C.S. n° 332, pp. 1-22, Sept. 1987.

[Bid 89] Bidoit M., "PLUSS, un langage pour le développement de spécifications algébriques modulaires", thèse d'Etat, Université de Paris-Sud, Orsay, Mai 1989.

[Bro 91] Broy M., opening lecture of the 6th IEEE International Worshop on Software Specification and Design, Como, Sept. 1991.

[BB 83] Bidoit M. and Biebow B., "Application de la méthode de spécification et de programmation des exceptions et des erreurs à des exemples choisis en téléphonie", rapport final du poste 4 du marché DAII n° 82.35.033, Sept. 1983.

[BBC 86] Bernot G., Bidoit M., Choppy C., "Abstract data types with exception handling : an initial approach based on a distinction between exceptions and errors", Theoretical Computer Science, Vol.46, N°1, Janvier 1986, pp.13-46.

[B2G3 85] Bidoit M., Biebow B., Gaudel M.-C., Gresse C., Guiho G., "Exception handling : formal specification and systematic program construction", I.E.E.E. Transactions on Software Engineering, Vol. SE-11, N°3, March 1985, pp.242-252.

[BG 77] Burstall, R. M. and Goguen, J.A., "Putting theories together to make specifications", in Proceedings of the 5th International Joint Conference on Artificial Intelligence, Cambridge, 1977, pp. 1045-1058.

[BG 80] Burstall, R. M. and Goguen, J.A., "The semantics of CLEAR, a specification language", in Proceedings of the Advanced Course on Abstract Software Specifications, Copenhagen, LNCS n°86, pp.292-332, Springer-Verlag, 1980.

[BGM 89] Bidoit, M., Gaudel, M-C. and Mauboussin, A., "How to make algebraic specifications more understandable ? An experiment with the PLUSS specification language", Science of Computer Programming, 46-1, pp. 1-38, June 1989.

[BKLOS 91] Bidoit M., Kreowski H.-J., Lescanne P., Orejas F. and Sannella D. "Algebraic System Specification and Development: a survey and annotated bibliography", LNCS n° 501, Springer-Verlag, 1991.

[BLG 91] Bernot G. and Le Gall P.,"Exception handling and term labelling",submitted for publication.

[Dau 92] Dauchy P., forthcoming thesis, Université de Paris-Sud, 1992.

[Des 83] Despeyroux-Savonitto J., "An algebraic specification of a Pascal compiler", SIGPLAN notices, vol. 18, n°12, 1983.

[DM 91] Dauchy P., Marre B., "Test data Selection from the Algebraic Specification of a module of an Automatic Subway", European Software Engineering Conference: ESEC'91, Milan, Oct. 1991, LNCS n° 550, pp. 80-100.

[DO 91] Dauchy P., Ozello P., "Experiments with Formal Specifications on MAGGALY", Second International Conference on Applications of Advanced Technologies in Transportation Engineering, Mineapolis, August 1991.

[EFH 83] Ehrig H., Fey W., and Hansen H., "ACT-ONE: an algebraic specification language with two levels of semantics", Technische Universität Berlin, 1983.

[FGJM 87] Futatsugi, K., Goguen, J.A., Jouannaud, J-P. and Meseguer, J., "Principles of OBJ2", in Proceedings of the 12th ACM Symposium on Principles Of Programming Languages, 1987, pp. 51-60.

[Gau 85] Gaudel, M.-C., "Towards Structured Algebraic Specifications", in Esprit'85 Status Report, pp.493-510, North-Holland,1986.

[Gau 91] Gaudel M-C., "Algebraic Specifications", Chapter 22 in Software Engineer's Reference Book, John Mc Dermid ed., Butterworths, 1991.

[Gui 84] Guiho G., Automatic Programming using abstract data types. *Key-note Lecture,* IJCAI 83, Karlsrühe, Aug. 83.

[GB 84] Goguen, J. A. and Burstall, R.M., "Introducing Institutions", in Proceedings of the Logics of Programming Workshop, Carnegie-Mellon University, LNCS n°164, pp. 221-256, Springer-Verlag, 1984.

[GH 86] Guttag, J.V. and Horning, J.J., "Report on the Larch shared language", Science of Computer Programming, vol. 6. pp. 103-134, 1986.

[GM 88] Gaudel, M.-C. and Moineau, Th., "A theory of Software Reusability", in proceedings of European Symposium On Programming: ESOP'88, LNCS n°300, pp. 115-130, Springer-Verlag, 1988.

[Gut 75] Guttag, J.V., "The specification and application to programming of abstract data types", Ph. D. Thesis, University of Toronto, 1975.

[Kap 89] Kaplan S., "Algebraic specification of Concurrent Systems", Theoretical Computer Science, N° 1, Volume 69, pp. 69-115, Dec. 1989.

[Moi 91] Moineau Th., "Réutilisation de Logiciel : une Approche Algébrique, son Application à Ada et les outils associés", Nouvelle thèse, Université Paris-Sud, Orsay, Janvier 1991.

[MBGHP 89] Mauboussin A., Bidoit M., Gaudel M-C., Hagelstein J., Perdrix H., "From an ERAE Requirement Specification to a PLUSS Algebraic Specification: a case study", METEOR Workshop, Algebraic Methods II, Mierlo, Septembre 1989, LNCS n° 490, pp 395-432.

[MG 91] Moineau Th. and Gaudel M-C., "Software reusability through formal specifications", in proceedings of the 1st International Workshop on Software Reusability, Dordmund, July 1991.

[NHWG 88] Nielsen M., Havelund K., Wagner K. and George C., "The RAISE Language, Method and Tools", in Proceedings of VDM'88, LNCS n°328, pp. 376-405, Springer-Verlag,1988.

[SS 91] Saake G. and Sernadas A. eds., "Information Systems—correctness and reusability", Proceedings of Worshop IS-CORE'91, ESPRIT BRA WG 3023, London, Sept. 1991, Informatik-Berichte 91-03, Technische Universität Braunschweig.

[SST 90] Sannella, D.T, Sokolowski S. and Tarlecki, A., "Toward formal development of programs from algebraic specifications: parameterisation revisited", Bericht 6/90, Universität Bremen, April 1990.

[ST 86] Sannella, D.T and Tarlecki, A. "Extended ML : an institution-independent framework for formal program development", In Proceedings Workshop on Category Theory and Computer Programming, Guilford, LNCS n°240, pp.364-389, Springer-Verlag, 1986.

[ST 83] Sannella, D.T. and Wirsing, M., "A kernel language for algebraic specification and implementation", in Proceedings Colloquium on Foundations of Computation Theory, Linkoping, LNCS n°158,pp.413-427,Springer-Verlag, 1983.

[Zil 74] Zilles S.N., "Algebraic specifications of abstract data types", Computation Structures Group memo 119, Laboratory of Computer Science, MIT, 1974.

PARALLEL ALGORITHMS 1

The Parallel Complexity of Tree Embedding Problems
(extended abstract)

Arvind Gupta*
School of Computing Science
Simon Fraser University
Burnaby, Canada, V5A 1S6
email: arvind@cs.sfu.ca

Naomi Nishimura†
Department of Computer Science
University of Waterloo
Waterloo, Ontario, Canada, N2L 3G1
email: nishi@maytag.waterloo.edu

Abstract

The sequential complexity of various tree embedding problems arises in the recent work by Robertson and Seymour on graph minors; here we consider the parallel complexity of such problems. In particular, we present two CREW PRAM algorithms: an $O(n^{4.5})$-processor $O(\log^3 n)$ time randomized algorithm for determining whether there is a topological embedding of one tree in another and an $O(n^{4.5})$-processor $O(\log^3 n \log\log n)$ time randomized algorithm for determining whether or not a tree with a degree constraint is a minor of a general tree. These algorithms are two examples of a general technique that can be used for solving other problems on trees. One by-product of this technique is an $\mathcal{NC}$ reduction of tree problems to matching problems.

1 Introduction

As an outgrowth of the graph minors project, Robertson and Seymour [RSb] have proved the existence of polynomial time sequential algorithms for a number of graph embedding problems. Their main result is to show that in any infinite set of graphs there exist distinct graphs H and G such that H is a minor of G (where H is a *minor* of G if by performing a sequence of edge contractions on a subgraph of G we can obtain a graph isomorphic to H). This result implies that there exists a finite characterization for any family of graphs closed under minors (see [FL88] for a catalog of natural examples of such families). Although we know that such algorithms and characterizations exist, many still remain to be found; the parallel complexity of many of these problems, also deserving of study, is still unknown.

Fundamental to this work is the problem of determining whether or not a graph H is a minor of a graph G. If arbitrary graphs G and H are both part of the input, the graph minor problem is $\mathcal{NP}$-complete [Kar75], even if G and H are trees [Duc91]; we call this the *two-input graph minor problem.* However, for the variant on this problem in which H is fixed, the *one-input graph minor problem*, there is an $O(n^3)$ time sequential algorithm determining whether or not H is a minor of an input graph G [RSa]. It may also be possible

*Research partially performed while the author was a NSERC postdoctoral fellow in the Department of Combinatorics and Optimization, University of Waterloo. Research supported by the Natural Sciences and Engineering Research Council of Canada, the Center for System Sciences and a President's Research Grant, Simon Fraser University.

†Research supported by the Natural Sciences and Engineering Research Council of Canada.

to find polynomial time solutions for two-input graph minor problems in which G and H are restricted to belong to certain classes of graphs more restricted than general trees.

The parallel complexities of both graph minor problems are still unknown. The one-input graph minor problem can be reduced to the problem of finding k disjoint paths in a graph [RSa]. Khuller, Mitchell and Vazirani [KMV89] have developed a fast parallel algorithm for the problem of finding two disjoint paths; the parallel complexity of the disjoint path problem for any $k > 2$ is unknown.

One way of finding solutions for a restricted class of graph minor problems involves the use of the *tree-decomposition* of a graph, a notion used extensively by Robertson and Seymour in their work. Recently, Lagergren [Lag90] has given a fast parallel algorithm for finding the tree-decomposition of a graph when the width of the decomposition is fixed. Combined with results of Bodlaender [Bod88], this implies the existence of fast parallel algorithms for many problems on minor closed graph families that are restricted to having bounded width tree-decompositions.

There has been some study of two-input versions of other embedding problems on trees, for example the problem of deciding if a tree T is isomorphic to a tree T'. For n the sum of the sizes of the trees, Lingas and Karpinski [LK89] and independently Gibbons, Karp, Miller and Soroker [GKMS90] have shown that the problem can be solved in $O(\log^3 n)$ time on a randomized CREW PRAM. Both algorithms proceed by finding all subtrees S' of T' isomorphic to T and combining information using bipartite perfect matching [KUW86, MVV87].

Lingas and Karpinski use a top down approach to solve the problem. The key idea is to find a node v of T such that the removal of the path from the root of T to v splits T into small pieces. The problem is solved for these subpieces and the solutions are pieced together.

Gibbons, Karp, Miller and Soroker apply the tree contraction technique of Miller and Reif [MR85] to T. Each edge e in T is associated with a set of edges $L(e)$ in T'. For $e' \in L(e)$, the subtree of T' consisting of e' and all its descendants contains the subtree of T consisting of e and all its descendants. The problem is solved by maintaining $L(e)$ throughout the stages of tree contraction. The authors also present a logspace reduction from subtree isomorphism to bipartite perfect matching. We observe that the same construction can be used to obtain a reduction from subtree topological embedding to bipartite perfect matching.

In this paper we study the parallel complexity of tree topological embedding and tree minor problems. We decompose a tree from the bottom up using Brent restructuring. As in the results for subtree isomorphism, we combine information using bipartite perfect matching. Our results easily allow us to solve the problem of subtree isomorphism as a special case. Of the three parallel algorithms known for this problem, ours is the closest in spirit to the fastest known sequential algorithm, due to Matula [Mat78].

We observe that the Lingas and Karpinski as well as the Gibbons, Karp, Miller and Soroker techniques can be adapted to solve the topological embedding problem. However, it is not clear that they can be extended to our minor problem. We introduce a new, general technique that can solve the topological embedding problem for arbitrary trees and the minor problem when the degree of T is bounded by $O(\frac{\log n}{\log\log n})$. We show that our technique can also be used to solve subtree isomorphism directly, as well as other problems on trees.

The outline of the remainder of the paper is as follows. In the next section we give definitions and discuss Brent restructuring. Section 3 contains a description of our technique as applied to the solution of the tree topological embedding problem. In Section 4 we discuss other problems for which our technique gives fast algorithms, such as the tree minor problem (when the source tree has degree at most $O(\frac{\log n}{\log\log n})$), the subtree isomorphism problem, and the topological embedding problem for planar planted trees. Finally, in Section 5 our technique is related to other techniques and open problems are presented.

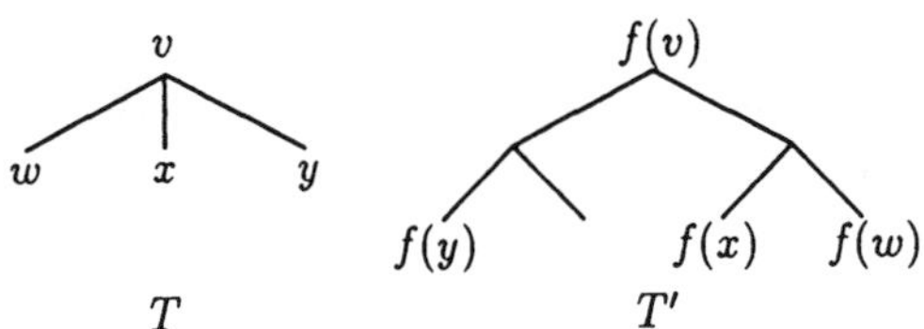

Figure 1: $T \leq_m T'$ but $T \not\leq_e T'$

2 Preliminaries

2.1 Trees

In this paper, all trees are finite and rooted. For T a tree, $V(T)$ and $E(T)$ denote its node and edge set, respectively. The size of T, $|V(T)|$, is denoted by $|T|$ and the root of T is denoted by root(T). Later in this paper we will be considering representations of trees by special types of trees. For clarity, we will refer to *nodes* of the trees T and T' and *vertices* of the representation trees. In these trees, vertices will be labeled by subgraphs of T.

We will distinguish between an arbitrary connected *subgraph* of T, and a *subtree* of T consisting of a node in $V(T)$ and all its descendants. We use T_v to denote the subtree of T rooted at v. At times we will be concerned with subgraphs that arise from removing a subtree from another subtree. For $v \in V(T)$ and $w \in V(T_v)$, the subgraph $T_v \backslash T_w$ denotes the subgraph obtained by removing from T_v all proper descendants of w. In particular, this means that the node $w \in T_v \backslash T_w$. We will call $T_v \backslash T_w$ a *scarred* subtree of T. In this context, we will call v a *scar*, and we will say that T_v *is scarred at* w or, more succinctly, that v is scarred at w.

Much of our work focuses on finding fast parallel algorithms for determining minor and topological embeddings in trees.

Definition: Let T and T' be trees. Then T is *topologically embeddable* in T', $T \leq_e T'$, if there is a one-to-one function $f : V(T) \to V(T')$ such that for any $v, w, x \in V(T)$ the following properties hold. If w is a child of v, then $f(w)$ is a descendant of $f(v)$. If w and x are distinct children of v, then the path from $f(v)$ to $f(w)$ and the path from $f(v)$ to $f(x)$ have exactly the node $f(v)$ in common. Equivalently, we will say that there is a *topological embedding* of T in T'. The topological embedding is a *root-to-root embedding* if $f(\text{root}(T)) = \text{root}(T')$.

Definition: Let T and T' be trees. Then T is a *minor* of T', $T \leq_m T'$, if there is a one-to-one function $f : V(T) \to V(T')$ such that for any $v, w, x \in V(T)$ the following properties hold. If w is a child of v, then $f(w)$ is a descendant of $f(v)$. If w and x are distinct children of v, then the intersection of $T'_{f(w)}$ and $T'_{f(x)}$ is empty. Equivalently, we will say that there is a *minor embedding* of T in T'.

Note that if $T \leq_e T'$, then $T \leq_m T'$. The converse, however, is not necessarily true. For example, every tree is a minor of a sufficiently large binary tree. However, the tree consisting of a root with three children, all leaves, is not topologically embeddable in any binary tree since there can only be two disjoint paths emanating from $f(v)$ (see Figure 1).

Definition: A *planar planted tree* is a tree with an ordering on the children of every internal node.

For T a planar planted tree and v an internal node of T, suppose v has children $c_0, \ldots, c_{k-1}$. We will normally write the children of v as a sequence which specifies the ordering, say $(c_0, \ldots, c_{k-1})$. We can extend the definitions of the topological and minor embeddings to planar planted trees in the natural way.

2.2 Brent Restructuring

In 1974, Brent [Bre74] investigated the problem of converting arbitrary arithmetic formulae to equivalent formulae with small depth and polynomial size. Brent's technique actually gives a general procedure for solving many tree based problems in parallel. The key idea is to recursively partition the tree into subtrees.

The method used by Brent in performing the partitioning forms two different types of subgraphs of the original tree T, namely unscarred and scarred subtrees of T. Lemma 2.1 contains the essential components of the partitioning for the first case and Lemma 2.2 for the second case. Both lemmas, slight generalizations of results by Brent, are stated without proof.

Lemma 2.1. *Let T be a tree with at least two nodes. Then there is a unique node v of T with children $c_1, \ldots, c_k$ such that:*

1. $|T \backslash T_v| \leq \frac{|T|}{2}$, *(or equivalently $|T_v| > \frac{|T|}{2}$); and*
2. $|T_{c_i}| \leq \frac{|T|}{2}$, *for all $1 \leq i \leq k$.*

Lemma 2.2. *Let T be a tree, $|T| > 2$, and ℓ be a leaf of T. Then there is a unique ancestor v of ℓ such that if c is the child of v for which $\ell \in T_c$ then:*

1. $|T \backslash T_v| \leq \frac{|T|}{2}$, *(or equivalently $|T_v| > \frac{|T|}{2}$); and*
2. $|T_c| \leq \frac{|T|}{2}$.

We obtain a partition of the tree into subgraphs by starting with a tree T and recursively applying the two lemmas depending on whether or not a subgraph has a scar. In both Lemma 2.1 and Lemma 2.2, the node v is called the *Brent break* of T. In Lemma 2.2 we can view the leaf ℓ as a place holder for a scar. It is clear that in the partition, each subgraph has at most one scar. In practice, we will view the applications of both Lemma 2.1 and Lemma 2.2 as two-step operations: first T is split into subgraphs $T \backslash T_v$ and T_v and then T_v is split into subtrees $T_{c_1}, \ldots, T_{c_k}$ where $c_1, \ldots, c_k$ are the children of v. The first step will be called a *Brent break* and the second a *child break*; we further distinguish between breaks that are *simple* (where, as in Lemma 2.1, none of the resulting subgraphs are scarred) or *scarred*, as in Lemma 2.2. The set of subgraphs obtained by the application of a Brent break and then a child break form a partition of the original subgraph. We use the term *Brent restructuring* to apply to one Brent break followed by one child break, resulting in a partition of the nodes. It is not difficult to see that $O(\log n)$ recursive applications of the lemmas will result in trees of constant size. For many problems, these can then be processed directly in constant time.

3 The Embedding Algorithm

3.1 Outline of the Algorithm

In this section, we present a randomized parallel algorithm for solving the topological embedding problem on trees. More formally, given trees T and T' (where $n = |T| + |T'|$), we wish to find the set of nodes A in $V(T')$ such that for every node $a \in A$, T can be root-to-root embedded in T'_a. The algorithm relies on the randomized matching algorithm of Mulmuley, Vazirani, and Vazirani [MVV87]; here we present the reduction of topological embedding to matching.

Our algorithm proceeds by finding topological embeddings of larger and larger subgraphs of T in T'. These subgraphs are formed by iterative applications of Brent restructuring. To exploit parallelism, at the outset our algorithm will need access to the subgraphs of T which arise from all the iterations of Brent restructuring. We represent the computation as a tree $\mathcal{B}_T$, the *Brent tree* of T; this idea was first introduced in [Gup85, CGR86]. In $\mathcal{B}_T$, a vertex will correspond to a subgraph of the original tree and an edge between two trees will indicate that the child is derived from the parent by a restructuring step.

After we have created the Brent tree, we complete the computation by processing the Brent tree level by level from the bottom up, where the child of a vertex at level ℓ is at level $\ell + 1$. When a level $\ell + 1$ has been completely processed, we will have determined all the possible locations of T' in which we can topologically embed those subgraphs of T that correspond to vertices at level $\ell + 1$ in $\mathcal{B}_T$. We make use of this information in the processing of level ℓ. Since each level of the Brent tree corresponding to child breaks forms a partition of the nodes of T, the subgraphs to be topologically embedded are disjoint; processing level $\ell - 1$ constitutes solving in parallel a number of independent problems, one associated with each vertex of $\mathcal{B}_T$. The topological embedding information is determined in two steps.. First, we use matching and the information from a level to determine root-to-root embeddings. Then, we propagate this information to the ancestors in T' of the nodes involved in the topological embeddings. We will show that we can set up a matching and update values in constant time, so that the overall running time will be $O(M \log |T|)$, where M is the parallel running time of matching.

3.2 Preprocessing

We first precompute all the ancestors of all nodes in T' and T. For each node v in T, we determine the size of T_v. For each node a in T' and descendant $k \in T'_a$, we determine the unique child of a that is an ancestor of k, which we store in array entry $C(a, k)$. Then, we compute the Brent break of each scarred and each unscarred subtree in T, that is for every node v we compute the Brent break of T_v and for each pair of nodes v and w, we compute the Brent break of $T_v \backslash T_w$. It is not difficult to see that all these computations can be achieved in $O(\log n)$ time using $O(n^2)$ processors by assigning a team of n processors to each node and combining information from each team.

3.3 Creating the Brent Tree

The straight-forward method of creating the Brent tree level by level yields a prohibitively slow parallel algorithm. Instead we begin by considering a graph $\mathcal{G}_T$ containing the tree. The graph $\mathcal{G}_T$ can be generated quickly, but it may contain supplementary vertices. Specifically, the vertex set of the graph consists of all triples (v, w, τ) where $v \in V(T)$, $w \in V(T_v) \cup \epsilon$, and $\tau \in \{B, C\}$. The triple (v, w, τ) represents the tree $T_v \backslash T_w$. The entry τ indicates whether the next step in the restructuring will be a Brent break, if $\tau = B$, or a child break, if $\tau = C$. If $w = \epsilon$, then T_w is the empty tree and the next break applied to T_v will be a simple break (Lemma 2.1); otherwise, the next break applied to $T_v \backslash T_w$ will be a scarred break (Lemma 2.2).

We form the adjacency matrix of $\mathcal{G}_T$ by assigning one processor to each potential edge. For $\alpha, \beta \in V(\mathcal{G}_T)$, let $\alpha = (v_\alpha, w_\alpha, \tau_\alpha)$ and $\beta = (v_\beta, w_\beta, \tau_\beta)$. There is an edge from α to β in $\mathcal{G}_T$ if the tree described by β is one component generated from the tree described by α by the application of a single step of Brent restructuring. Since the breaks alternate between Brent breaks and child breaks, whenever there is an edge from α to β, $\tau_\alpha \neq \tau_\beta$. Each of the following seven sets of additional conditions is sufficient to define an edge; each can be verified in constant time, given the preprocessing stage.

[Simple Child Break] The node v_β is a child of node v_α and $w_\beta = \epsilon$.
[Simple Brent Break] The node $v_\beta = v_\alpha$ and w_β is the Brent break of T_{v_α}.
[Simple Brent Break] The node v_β is the Brent break of T_{v_α} and $w_\beta = \epsilon$.
[Scarred Child Break] The node v_β is a child of v_α, $w_\beta = \epsilon$, and T_{v_β} does not contain w_α.
[Scarred Child Break] The node v_β is a child of v_α and T_{v_β} contains $w_\alpha = w_\beta$.
[Scarred Brent Break] The node $v_\beta = v_\alpha$ and w_β is the Brent break of $T_{v_\alpha} \backslash T_{w_\alpha}$.
[Scarred Brent Break] The node v_β is the Brent break of $T_{v_\alpha} \backslash T_{w_\alpha}$ and $w_\alpha = w_\beta$.

Once we have created $\mathcal{G}_T$, we can extract $\mathcal{B}_T$ by computing the transitive closure of the adjacency matrix of $\mathcal{G}_T$, removing all vertices not reachable from the vertex $(root(T), \epsilon, B)$, and then reversing pointers to point from each node to its parent. The total determination of $\mathcal{B}_T$ can be achieved in $O(\log |T|)$ time using $O(|T|^4)$ processors.

3.4 Processing the Brent Tree

In this section, we consider the processing of a particular level of the Brent tree. To process level ℓ, we determine, for each subgraph of T stored in a vertex at level ℓ, all the locations in T' where the subgraph can be topologically embedded. The processing of a vertex at level ℓ makes use of the information determined in the processing of its children at level $\ell + 1$; Lemma 3.1 makes explicit the relationship between levels.

Lemma 3.1. *Let T be a tree with subgraph S and T' be a tree with subgraph S'. Let v be the root of S and let $w_0, w_1, \ldots, w_{k-1}$ be the children of v. Similarly, let a be the root of S' and let $b_0, b_1, \ldots, b_{h-1}$ be the children of a. If each subgraph rooted at a node w_i can be topologically embedded in a distinct subgraph rooted at a node b_j, then S can be topologically embedded in S'.*

Thus, when processing a level of vertices in $\mathcal{B}_T$ with children formed by Brent breaks, we are combining information about subgraphs formed by a Brent break and storing it as information about the larger, composite subgraph. For a level of vertices with children formed by child breaks, information from subtrees is combined using matching. Since the processing techniques for the two types of levels are similar, we present only the more involved of the two, the technique for processing vertices with children formed by child breaks.

We make use of an array E that holds the topological embedding information pertaining to the current level of the Brent tree. This array is updated after the processing at each level.

At a particular level in the Brent tree, there is at most one subgraph rooted at a particular node, and if the subgraph is scarred, exactly one scar in the subgraph. Consequently, it is not necessary to explicitly store the identity of the scar node for a subgraph at a particular

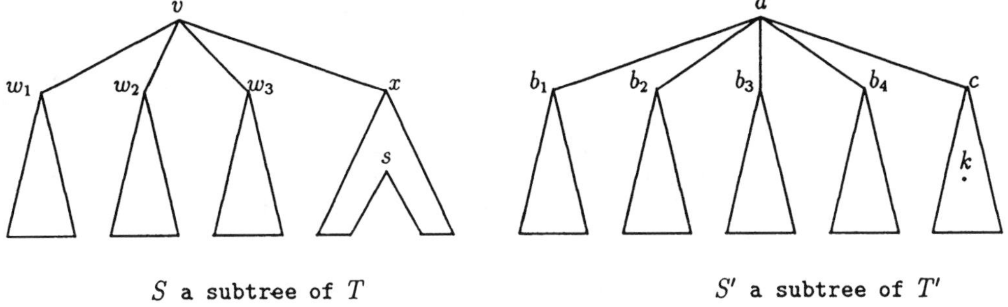

Figure 2: Processing a level of breaks

level; this information is implicit in the Brent tree. For v scarred, with scar s at the current level, the entries of E have the following meanings:

$$E[v(a,k)] = \begin{cases} 2, & \text{if } s \text{ maps to } k \text{ in a root-to-root embedding of } T_v \backslash T_s \text{ in } T'_a; \\ 1, & \text{if } s \text{ maps to } k \text{ in an embedding of } T_v \backslash T_s \text{ in } T'_a; \\ 0, & \text{otherwise.} \end{cases}$$

If the first two conditions are both true, the entry has value 2. For w unscarred, the entries $E[w(a,\epsilon)]$ have analogous meanings with respect to embeddings of T_w in T'_a.

To process a level of vertices, we wish to find for each node-scar pair (v,s) in T and each pair of nodes (a,k) in T', for k a descendant of a, whether or not there is a topological embedding of $T_v \backslash T_s$ in T'_a which maps s to k ($E[v(a,k)] = 1$). There will exist such an embedding either if there is a root-to-root embedding of $T_v \backslash T_s$ in T'_a ($E[v(a,k)] = 2$) or if there is a root-to-root embedding of $T_v \backslash T_s$ in T'_b, for b a descendant of a ($E[v(b,k)] = 2$). We first find all root-to-root embeddings associated with a level (Step 2), and then propagate this information to the ancestors of the nodes in T' involved in the root-to-root embedding (Step 4). Figure 2 illustrates the determination of a particular root-to-root embedding.

1. Copy all of E into a temporary array E'.

2. For each subgraph $T_v \backslash T_s$ in a vertex at the current level, for each a in T', and for each descendant k of a, determine whether there is a root-to-root embedding of $T_v \backslash T_s$ in $T'_a \backslash T'_k$. This can be divided into the following steps:

   (a) For x the child of v that is the ancestor of s, determine all children c of a and all $k \in T'_c$ such that the subgraph $T_x \backslash T_s$ topologically embeds in T'_c with s mapped to a node $k \in T'_c$. This can be determined by reading $E'[x(c,k)]$.

   (b) For each child c determined in Step 2a, for each unscarred child w of v, determine a distinct child b of a, $b \neq c$, such that T_w topologically embeds in T'_b. This can be achieved by setting up a matching problem for each c, as follows:

      i. Create a bipartite graph $G_c = (X, Y, E)$, where the set X corresponds to the unscarred children of v and the set Y corresponds to all the children of a except c. There is an edge from an unscarred child w of v to a child b of a if and only if T_w can be topologically embedded in T'_b. The adjacency matrix for a particular graph G_c is created by a set of *edge processors*, where each processor is assigned to a different potential edge in G_c. For an unscarred child w, a processor assigned to the potential edge (w, b) simply reads the value in $E[w(b, \epsilon)]$ and if the value is nonzero writes a 1 into the appropriate cell in the adjacency matrix, writing a 0 otherwise.

ii. The adjacency matrix for each G_c then forms input to the matching subroutine, the array entry $M[v(a,c)]$ in the $|T| \times |T'| \times |T'|$ array M is set to 1 if and only if there is a matching such that each node in X is covered by an edge in the matching.

3. Update the values of E to reflect the results of the matching. In particular, $E[v(a,k)]$ is assigned a value of 2 if and only if for some child c of a both $E'[x(c,k)] \neq 0$ and $M[v(a,c)] = 1$.

4. If $E[v(b,k)] = 2$, then set $E[v(a,k)]$ to 1 for each ancestor a of b.

The number of processors needed is dominated by the number of edge processors needed to create simultaneously all the adjacency graphs at a particular level. For a fixed vertex of the Brent graph and fixed a, k, and c, we need one processor for each potential edge of G_c, or the product of the number of children of a and the number of children of v. For fixed values of a and k, the unique value of c is contained in array entry $C(a,k)$. By noting that the subgraphs at each level of the Brent graph form a partition of the nodes of T, we can then show that the number of edge processors needed is at most $|T|(|T'|)^3$.

We wish to show that the time complexity for each of the $\log |T|$ levels is constant, not counting the time to determine a matching. It will suffice to consider the complexity of Step 4, since the other steps clearly each take constant time. In Step 4, to broadcast the matching information to the ancestors of each node in T' in a naive way would take $O(\log |T|)$ time. We perform the propagation of information by instead using one processor per piece of information per ancestor, for a total of at most $|T| \cdot |T'|^3$ processors. A processor assigned to ancestor a of b reads information pertaining to b in E and updates the relevant entry for a accordingly. Thus, each level of the tree is processed in constant time, using $O(n \cdot n^3)$ processors. By the best known matching algorithm [MVV87], the total complexity of embedding is $O(\log^3 n)$ using $O(n^{4.5})$ processors on a randomized CREW PRAM.

4 Extensions and Special Cases

The technique of using a Brent graph to structure the upcoming iterations can be used to solve a number of other tree problems. These include a special case of the minor embedding problem, subtree isomorphism, maximum independent set, maximum matching, vertex cover, and decomposition of the target tree into disjoint pattern trees. The minor embedding problem, which we consider below, has no other known solution. Although the other problems in the list can also be solved by using other methods, they demonstrate the robustness of our technique.

As noted earlier, both Lingas and Karpinski [LK89] and Gibbons *et al.* [GKMS90] solve the problem of subtree isomorphism. Here we note that our techniques also solve this problem directly. The essential change to the algorithm of Section 3 is that we only have to worry about root-to-root embeddings. That is, there is no propagation of matching information to the ancestors of a node in T'. Thus, we obtain an $O(\log n)$ time CREW PRAM reduction from subtree isomorphism to bipartite perfect matching.

4.1 The Minor Algorithm

For arbitrary trees T and T', the problem of determining whether or not $T \leq_m T'$ is $\mathcal{NP}$-complete [Duc91]. By making use of our general technique and the relationship between topological and minor embeddings, we obtain a fast parallel randomized algorithm for this problem when T' is arbitrary but the degree of T is in $O(\frac{\log n}{\log\log n})$. In particular we show that there is an $O(\log n \log\log n)$ parallel time reduction from this restricted version of the tree minor problem to bipartite perfect matching on a CREW PRAM.

For the remainder of this section, suppose T and T' are trees and we wish to determine whether or not $T \leq_m T'$, when the degree of T is in $O(\frac{\log n}{\log\log n})$. As in Section 3, we form $\mathcal{B}_T$ and process it level by level. For an unscarred subtree T_v associated with a vertex of $\mathcal{B}_T$, we process the vertex by finding all nodes a of T' such that $T_v \leq_m T'_a$ (the other case, in which the subgraph is scarred, is analogous to this one and hence omitted from further discussion). We call this set of nodes $\mathcal{L}_v$. Since we have processed previous levels of the Brent tree, we have the set $\mathcal{L}_w$ for each child w of v.

In Section 3, to determine whether or not $T_v \leq_e T'_a$, it sufficed to set up a matching, making use of information about the topological embedding of subtrees rooted at children of v in subtrees rooted at children of a. The success of the method relied on the fact that embedding information could be propagated up the tree. In the case of minors, however, this sort of propagation does not occur (recall Figure 1). Instead, we consider the mapping into T' of the tree T_{small} consisting of v and its children. For each such mapping, we then set up a series of matchings, making use of information about the minor relations between the subtrees rooted at the children of v and the subtrees rooted at the nodes of T' to which the leaves of T_{small} (i.e., the children of v) are mapped.

To complete the reduction, we must show how to determine all possible minor embeddings of T_{small} into T'. We will show, in Lemma 4.1, that to determine the minor embedding of one tree in another, it suffices to determine the topological embeddings of a certain class of trees, $\mathcal{U}_n$. Let $\mathcal{U}_n$ be the set of rooted unlabeled trees with $O(\frac{\log n}{\log\log n})$ leaves such that no internal non-root node has degree less than three. It can be shown by induction that $|\mathcal{U}_n| \leq n^{O(1)}$ by first determining a bound on the size of the set of unlabeled rooted trees on n nodes.

To determine how to map T_{small} into T', we make use of the relationship between topological and minor embeddings. We show how to form from any T and T' a class of trees such that for any tree $\bar{T}$ in the class, $\bar{T} \leq_e T'$ implies that $T \leq_m T'$. A tree $\bar{T}$ is formed by substituting certain nodes in T by subtrees.

Definition: Let T be a tree, and let v be an internal node of T that has children $w_0, \ldots, w_{k-1}$. Let O be an ordering of the children of v, say $w_0, \ldots, w_{k-1}$. Let S be a tree with leaves $y_0, \ldots, y_{k-1}$. We can form a new tree T' from T by removing v and in its place substituting S: the parent of v becomes the parent of the root of S, and the children of w_i become the children of y_i, for $0 \leq i < k$. We say that T' is formed from T by *substituting v by S according to O.*

Given arbitrary trees T and T', we define a class of trees, where a tree $\bar{T}$ in the class is defined in the following fashion. First, we will choose for each node v of T a tree $S_v \in \mathcal{U}_n$ and an ordering of children O_v. The tree $\bar{T}$ is then formed from T by substituting v by S_v according to O_v, for each node $v \in T$. The proof from the following lemma appears in the full paper.

Lemma 4.1. *For any trees T and T', and for the class of trees defined as above, if there exists a $\bar{T}$ in the class such that $\bar{T} \leq_e T'$, then $T \leq_m T'$.*

To find a minor embedding of T_{small} in T', it suffices to topologically embed all trees in $\mathcal{U}_n$. We can precompute the Brent tree of each tree in $\mathcal{U}_n$ in $O(\log\log n)$ time on a CREW PRAM, since the number of nodes in such a tree is in $O(\frac{\log n}{\log\log n})$. By extending our results to apply to a special type of embedding in which information is retained at the leaves, an extra factor of $O(\log n)$ to reduce the special embedding to matching yields an $O(\log n(M\log\log n+\log n))$ CREW PRAM algorithm, where M is the parallel time complexity of matching.

4.2 Planar Planted Trees

If the trees T and T' are restricted to being planar planted, the solution to the matching problem has a special form. In particular, if we are trying to embed T_v in T'_a, where v has children $(w_0,\ldots,w_{k-1})$ and a has children $(b_0,\ldots,b_{h-1})$ in the specified order, then if there is an edge in the matching from w_i to b_j, then there can be no edge from $w_{i'}$ to $b_{j'}$, for $i'<i$ and $j'>j$. Intuitively, we insist that the edges in the matching do not cross. The matching for a scarred child break results in a series of pairs of such problems; the method is similar and hence omitted.

We make use of an array of size kh, and assign one processor to each entry. For $0\le j<h$ and $0\le i<k$, the processor assigned to cell $jk+(k-i)$ will write the value i into the cell if there is an edge from w_i to b_j. When all processors have written, the resulting array will consist of h blocks of length k, where the $(j+1)$st block contains, in order from largest to smallest, values i for every w_i to which b_j is connected.

We can specify a matching by selecting occurrences of $0,1,\ldots,k-1$, in order, where selecting an occurrence of i in block $j+1$ signifies matching w_i and b_j. We can search for $0,1,\ldots,k-1$ without paying attention to block boundaries: since we have reversed the order of the i values in each block, it is not possible to select two occurrences from the same block. We append to the ends of the array a dummy start node of value -1 and a dummy end node of value k. We wish to link each occurrence of i to the next occurrence of $i+1$; we can then find the string $012\ldots(k-1)$ by traversing the pointers from the dummy start node to the dummy end node.

We can form the links by computing all nearest largers, using the technique of Berkman *et al.* [BBGSV89], in $O(\log\log n)$ time on an $O(\frac{n}{\log\log n})$-processor COMMON CRCW PRAM. It will now suffice to perform pointer jumping [FW78] on the linked list that is formed to determine whether or not the dummy starter can reach the dummy end, yielding a deterministic COMMON CRCW PRAM algorithm running in $O(\log n)$ time on $O(n^4)$ processors that solves the topological embedding problem for planar planted trees.

5 Conclusions and Open Problems

The key contribution of this paper is a general technique for producing fast parallel algorithms for problems on trees. Our technique is simple and straight-forward, producing reductions between the tree problems and bipartite graph problems. The algorithms that we produce are close in spirit to the natural sequential algorithms for the problems.

We have illustrated the use of our technique in the presentation of an $O(\log n)$ time $O(n^4)$ processor CREW PRAM reduction from the subtree topological embedding problem to bipartite matching. Using the best known algorithm for bipartite matching [MVV87], this results in an $O(\log^3 n)$ time, $O(n^{4.5})$ processor randomized CREW PRAM algorithm. When applied to planar planted trees, this result yields a reduction from the subtree topological embedding problem to a problem with a fast deterministic parallel solution. We have also shown how our techniques can be used to solve the subtree minor embedding problem with the degree of the source tree is at most $O(\frac{\log n}{\log\log n})$. Here we use the technique at two levels within the algorithm. Our technique can also be used to directly solve subtree isomorphism, a problem solved by Lingas and Karpinski [LK89] and Gibbons *et al.* [GKMS90].

There is a need for further investigation into the general class of problems that can be solved in this manner. It remains open whether our degree constraint can be relaxed, for example to $O(\log n)$. In addition, the topological embedding problem is open for more general graphs.

Although the reduction in this paper yields a randomized CREW PRAM algorithm for the restricted version of the tree minor problem, it is possible that the algorithm can be made deterministic. We reduce the tree minor problem to a case of bipartite perfect matching problem in which the degree of the vertices in one part of the partition is at most $O(\frac{\log n}{\log\log n})$. It does not seem that the approach taken by Mulmuley, Vazirani and Vazirani [MVV87] can be modified to yield a deterministic algorithm for perfect matching with this added degree constraint, so a different approach may be necessary.

Acknowledgements

We thank Prabhakar Ragde for bringing to our attention the connections between our work and that of Gibbons, Karp, Miller, and Soroker. We are grateful to the anonymous reviewers from STACS for their useful suggestions, many of which will be incorporated in greater detail in the full version of the paper.

References

[BBGSV89] O. Berkman, D. Breslauer, Z. Galil, B. Schieber, and U. Vishkin, "Highly parallelizable problems," *Proceedings of the 21st Annual ACM Symposium on the Theory of Computing*, pp. 309–319, 1989.

[Bod88] H. Bodlaender, "NC-algorithms for graphs with bounded tree-width," Technical Report RUU-CS-88-4, University of Utrecht, 1988.

[Bre74] R. Brent, "The parallel evaluation of general arithmetic expressions," *Journal of the ACM* **21**, 2, pp. 201–206, 1974.

[CGR86] S. Cook, A. Gupta, and V. Ramachandran, "A fast parallel algorithm for formula evaluation," unpublished manuscript, October 1986.

[Duc91] P. Duchet, "Tree Minors," AMS-IMS-SIAM Joint Summer Research Conference on Graph Minors, 1991.

[FL88] M. Fellows and M. Langston, "Nonconstructive tools for proving polynomial-time decidability," *Journal of the Association for Computing Machinery* **35**, 3, pp. 727–739, July 1988.

[FW78] S. Fortune and J. Wyllie, "Parallelism in Random Access Machines," *Proceedings of the 10th Annual ACM Symposium on the Theory of Computing*, pp. 114–118, 1978.

[GKMS90] P. Gibbons, R. Karp, G. Miller, and D. Soroker, "Subtree subtree isomorphism is in random NC," *Discrete Applied Mathematics* **29**, pp. 35–62, 1990.

[Gup85] A. Gupta, "A fast parallel algorithm for recognition of parenthesis languages," Master's thesis, University of Toronto, 1985.

[Kar75] R. Karp, "On the complexity of combinatorial problems," *Networks* **5**, pp. 45–68, 1975.

[KR88] R. Karp and V. Ramachandran, "Parallel algorithms for shared-memory machines," in *Handbook of Theoretical Computer Science, Volume A*, ed. J. Van Leeuwen, Elsevier, 1990.

[KUW86] R. Karp, E. Upfal, and A. Wigderson, "Constructing a perfect matching is in random NC," *Combinatorica* **6**, 1, pp. 35–48, 1986.

[KMV89] S. Khuller, S. Mitchell, and V. Vazirani, "Processor efficient parallel algorithms for the two disjoint paths problem, and for finding a Kuratowski homeomorph," *Proceedings of the 30th Annual IEEE Symposium on the Foundations of Computer Science*, pp. 300–305, 1989.

[Lag90] J. Lagergren, "Efficient parallel algorithms for tree-decompositions and related problems," *Proceedings of the 31st Annual IEEE Symposium on the Foundations of Computer Science*, pp. 173–181, 1990.

[LK89] A. Lingas and M. Karpinski, "Subtree isomorphism is NC reducible to bipartite perfect matching," *Information Processing Letters* **30** pp. 27–32, 1989.

[Mat78] D. Matula, "Subtree isomorphism in $O(n^{5/2})$," *Annals of Discrete Mathematics* **2**, pp. 91–106, North-Holland, 1978.

[MR85] G. Miller and J. Reif, "Parallel tree contraction and its application," *Proceedings of the 26th Annual IEEE Symposium on the Foundations of Computer Science*, pp. 478–489, 1985.

[MVV87] K. Mulmuley, U. Vazirani, and V. Vazirani, "Matching is as easy as matrix inversion," *Proceedings of the 19th Annual ACM Symposium on the Theory of Computing*, pp. 345–354, 1987.

[RSa] N. Robertson and P. Seymour, "Graph Minors XIII. The disjoint paths problem," in preparation.

[RSb] N. Robertson and P. Seymour, "Graph Minors XV. Wagner's conjecture," in preparation.

[TV85] R. Tarjan and U. Vishkin, "Finding biconnected components and computing tree functions in logarithmic parallel time," *SIAM Journal of Computing* **14**, pp. 862–874, 1985.

A Theory of Strict P-completeness

Anne Condon *

Computer Science Department

University of Wisconsin at Madison

Madison, WI 53706, U.S.A.

Abstract

A serious limitation of the theory of P-completeness is that it fails to distinguish between those P-complete problems that do have *polynomial speedup* on parallel machines from those that don't. We introduce the notion of *strict P-completeness* and develop tools to prove precise limits on the possible speedup obtainable for a number of P-complete problems.

1 Introduction

A major goal of the theory of parallel computation is to understand how much speedup is obtainable in solving a problem on parallel machines over sequential machines. The theory of P-completeness has successfully classified many problems as unlikely to have polylog time algorithms on a parallel machine with a polynomial number of processors. However, the theory fails to distinguish between those P-complete problems that do have significant, polynomial speedup on parallel machines from those that don't. Yet this distinction is extremely important from a practical point of view [16]. In this paper, we refine the theory of P-completeness to obtain *strict* completeness results for a number of P-complete problems.

Kruskal, Rudolph and Snir [10] identified the *speedup* of a problem as a critical parameter of the problem. This is simply the ratio between its sequential running time and its parallel running time on a machine with a polynomial number of processors. A problem has *polynomial speedup* if its speedup is $\Omega(n^\epsilon)$, for some $\epsilon > 0$. Given the practical importance of polynomial speedups, Kruskal et al. [10] introduced a new complexity class, SP ("semi-efficient, polynomial time" or "semi-efficient, parallel") of problems which have

*Work supported by NSF grant number CCR-9100886

polynomial speedup. In earlier work, Simons and Vitter [14] introduced similar complexity classes and showed that versions of the Circuit Value Problem, Ordered Depth First Search and Unification, where the underlying graphs are "dense", have parallel algorithms with running time approximately the square root of the input size. Another example of a problem in SP is that of computing the expected cost of a finite horizon Markov decision process, a well studied problem in optimization theory (we define this in detail in Section 4).

However, other P-complete problems do not seem to have polynomial speedups on PRAMS with a polynomial number of processors. The general Circuit Value Problem, for example, has no known parallel algorithm on a PRAM with a polynomial number of processors that runs in time $O(n^{1-\epsilon})$ for any $\epsilon > 0$. Proving any polynomial lower bound on the parallel complexity of this problem would separate NC from P. Instead, we ask whether the theory of P-completeness can be refined to prove that the Circuit Value Problem does not have polynomial speedup unless *all* problems in P have polynomial speedup.

To address this question, we introduce the notion of *strict P-completeness.* Roughly, a problem is strictly $T(n)$-complete for P if there is a parallel algorithm for the problem with running time that is within a polylog factor of $T(n)$ and moreover, the existence of a parallel algorithm that improves this by a polynomial factor would imply that *all* problems in P (with at least linear running time) have polynomial speedup. We use this notion to investigate limits on the speedups of P-complete problems.

Our first strictly P-complete problem is a restriction of the Circuit Value Problem (CVP) - the *Square* Circuit Value Problem. A square circuit is a synchronous circuit in which the number of gates at every level is equal to the depth of the circuit. We prove that this problem is strictly $n^{1/2}$-complete for P.

We apply this result to obtain a similar result for the nonstationary, finite horizon, Markov decision process problem. This problem is to decide the expected cost of a nonstationary Markov decision process with m states in a given time T. Dynamic programming is a sequential algorithm for this problem and this can be parallelized to obtain polynomial speedup, but this parallel algorithm requires $\Omega(T)$ time. We show that, unless all problems in P have polynomial speedup, any parallel algorithm for the nonstationary Markov decision process problem requires $\Omega(T^{1-\epsilon})$ time on a PRAM with a polynomial number of processors, for all $\epsilon > 0$. We also describe limits on the parallel speedup obtainable for a number of other P-complete problems, including First Fit Bin Packing, Lex First Maximal Independent Set, Ordered Depth First Search and Unification.

We use the RAM and PRAM as our models of sequential and parallel computation, respectively, with a log cost for instructions. The log cost RAM is a model of sequential computation whose cost reflects well the actual cost of solving a problem on a real machine

[2]. Although the PRAM model has been criticized as a practical model of computation [16], it is nevertheless an ideal model for proving completeness results, since if polynomial speedup is not obtainable for a problem on a PRAM, it is also not obtainable on more realistic models of parallel computation. There are many variations of the PRAM model, depending on the way simultaneous access to global memory is resolved. Our results are robust in that they hold for all the standard variations of the PRAM model, such as Exclusive Read, Exclusive Write or Priority Concurrent Read, Concurrent Write (see [9] for definitions of these models).

Our completeness proof for the Square Circuit Value Problem maps a RAM running in time $t(n)$ to a synchronous circuit of width and depth that are within a polylog factor of $t(n)$. Thus, the size of our circuit is significantly smaller than that obtainable from previous results. By combining the reduction of Ladner [11] from a Turing machine computation to the circuit value problem and simulations of RAMs by Turing machines of Cook and Reckhow [2], a RAM running in time $t(n)$ can be mapped to a circuit with depth and width $\Omega(t^2(n))$. Obtaining a synchronous circuit then further squares the size of the circuit (see Greenlaw et al. [7]).

The rest of the paper is organized as follows. In Section 2, we give some background on sequential and parallel models of computation, and define precisely our notion of strict P-completeness. In Section 3, we show how to simulate a RAM by a square circuit whose depth is at most a polylog factor times the running time of the restricted RAM, and prove that the Square Circuit Value Problem is strictly $n^{1/2}$-complete. In Section 4 we extend the results on the Square Circuit Value Problem to obtain limits on the parallel speedups of a number of other P-complete problems.

2 Definitions and Background

In this section we define our models of sequential and parallel computation. We introduce a restricted RAM model in Section 2.1, and describe properties of this model that are needed to prove the main results of the paper. The PRAM model of parallel computation and parallel complexity classes are discussed in Section 2.2. In Section 2.3, we define the notion of strict P-completeness.

2.1 RAMS and Restricted RAMS

Our RAM model is essentially the log cost random access machine of Cook and Reckhow [2]. This model charges for instructions based on the number of *bit operations*. A RAM consists of a program, or finite sequence of instructions, which accesses an unbounded sized memory of cells $R_0, R_1, \ldots$. Each cell can hold an unbounded length integer. The instruction set we use is given in the following table, along with their *costs*, where logs are to the base 2.

Instruction	Cost
$R_i \leftarrow C$	1
$R_i \leftarrow R_j + R_k$	$\log R_j + \log R_k$
$R_i \leftarrow R_j - R_k$	$\log R_j + \log R_k$
$R_i \leftarrow R_{R_j}$	$\log R_j + \log R_{R_j}$
$R_{R_j} \leftarrow R_i$	$\log R_j + \log R_i$
goto L	1
halt	1
if $R_i > 0$ **then** *instruction*	$\log R_i$(+ cost(*instruction*) if $R_i > 0$)
if $R_i = 0$ **then** *instruction*	$\log R_i$(+ cost(*instruction*) if $R_i = 0$)

We assume that the input is stored in contiguous memory cells. We view RAMs as language acceptors and assume that the output bit is stored in R_0. The running time of a RAM on a fixed input is the total cost of the instructions executed on that input. We say a problem, or language, has *sequential running time* $t(n)$ if there is a RAM that, for all inputs of length n, has running time at most $t(n)$. We say that a RAM M *simulates* RAM M' if both M and M' accept the same language.

Cook and Reckhow [2] showed that the largest value computed by M, and hence the largest address referenced by M, is $2^{O(\sqrt{t(n)})}$. Thus, the cells accessed by a RAM running in time $t(n)$ could have addresses of length $\Omega(\sqrt{(t(n))})$. In converting RAMs to circuits, it is convenient to assume that the addresses accessed by a RAM running in time $t(n)$ have length $O(\log t(n))$. In the next lemma, we show how "compact" the address space of a RAM that runs in time $t(n)$, so that all address lengths are $O(\log t(n))$. The cost of this is an additional factor of $O(\log t(n))$ to the running time of the RAM.

Previous work on compacting the space used by a RAM was done by Slot and Van Emde Boas [15]. Using hashing techniques, they showed that if a RAM M uses d distinct addresses, it can be simulated by another RAM M' for which the largest address is numbered $O(d)$. Moreover, if the *space* used by a RAM is defined as the sum over all cells, of the maximum value in every cell, then the space used by M' is within a constant factor of the space used by M. However, the time required by M' may be exponential in the time required by M. In contrast, our simulation guarantees that the time required by M' is within a logarithmic factor of the time required by M, but we pay for this by adding an amount $O(t(n))$ to the space used by M'.

Lemma 1 *A RAM with running time $t(n)$ can be restricted to access only cells whose addresses are $O(t(n))$ on inputs of length n, with only a loss of a factor of $O(\log t(n))$ in its running time.*

Proof: Let M be a RAM with running time $t(n)$. We construct a RAM M' that simulates M, but which uses a table to store the addresses of cells referenced by M, together with the values at those addresses.

The table is organized as a sequence $T_0, T_1, \ldots,$ of balanced binary trees. Each node of a tree is a record consisting of a cell address, the value at that address and pointers to the nodes which are children, if any. The first tree T_0 contains at most two nodes, containing the addresses 0 and 1. For $i \geq 1$, the ith binary tree T_i contains addresses referenced by M between $2^{2^{i-1}}$ and $2^{2^i} - 1$, inclusive. The number $2^{2^i} - 1$ is the *index* of the ith binary tree. For each tree there is also a *tree record*, which contains the index of the tree, a pointer to the root of the tree, if any and a pointer to the next tree record, if any. The trees are built dynamically by M' as it is simulating M.

Suppose the value of cell R_N is needed for an instruction. This is located by M' as follows. First, N is compared with the indices of the binary trees in order, until the first index is found which is $\geq N$. If no such tree exists, new, empty trees and their tree records are created in sequence until one exists. Let i be such that $2^{2^{i-1}} \leq N \leq 2^{2^i} - 1$. Next, the tree T_i is searched for N. If not found, a node with the address N is added to the tree with value 0. Then, the value of N can be retrieved or updated as specified by the instruction.

Space for the trees can be allocated dynamically as needed, so that all space used by the algorithm is stored in contiguous memory cells. The total number of cells needed to store the table is $O(t(n))$, since at most $t(n)$ distinct cells are accessed by M and there are only $O(\log t(n))$ tree records.

The time to maintain this table is as follows. The time to create the empty trees is dominated by the cost of computing the tree indices. Let $2^{2^i} - 1$ be the largest index computed. Then some address, say N, referenced by M is at least $2^{2^{i-1}} + 1$. Since N is computed within time $O(t(n))$ by M, the indices up to $2^{2^{i-1}}$ can be computed in time $O(t(n))$, by repeated additions. The time to then compute 2^{2^i} from $2^{2^{i-1}}$ is $O(2^{2i})$, since, using repeated additions, 2^{i-1} additions are needed, each involving numbers of length $O(2^i)$. Since $\log_2 N \geq 2^{i-1}$, this time is $O(\log^2 N)$. Finally, since from Cook and Reckow [2], $N \leq 2^{O(\sqrt{t(n)})}$, this is $O(t(n))$.

The time to search for the tree in which address N is stored (not counting the time to create new trees, which we have already accounted for) is $O(\log t(n) \log N)$. This is because there are $O(\log t(n))$ trees in the table, and hence N is compared with $O(\log t(n))$ indices, each of length $O(\log N)$. The time to search for N in the tree is $O(\log t(n) \log N)$. This is because the tree is balanced, thus ensuring that N is compared with $O(\log t(n))$ entries; and also all entries in the tree have length $O(\log N)$. Similarly, the time to add N to the tree is $O(\log t(n) \log N)$.

The total running time of M' is the time to create the trees, plus the time to execute all the instructions of M, plus the time to locate the operands of each instruction and the time to update the result of each instruction. The above analysis shows that the time to locate operands or update results in M' is within a factor of $O(\log t(n))$ of the time to do so in M. Hence the running time of M' is $O(t(n) \log t(n))$. □

We say a RAM is *restricted* if for all inputs x, and all k, the kth instruction executed on x (if any) is a function only of k and $|x|$. Note that the **if** statement is considered to be just one instruction. Thus, the flow of control of the program of a restricted RAM is oblivious of the actual input. Our next lemma states that any RAM with running time $t(n)$ can be simulated by a restricted RAM with running time $O(t(n))$. Again, this restriction later simplifies our reduction from a RAM to a circuit in Theorem 4. The proof of this lemma is straightforward.

Lemma 2 *Any RAM with running time $t(n)$ can be simulated by a restricted RAM with running time $O(t(n))$.*

2.2 Parallel RAMs and Complexity Classes

Our model of parallel computation is the PRAM model (Fortune and Wyllie [5]), which consists of a set of RAMs, called *processors*, plus an infinite number of *global* memory cells. An instruction of a processor can access its own memory cells or the global memory cells. There are many variations of the PRAM model, depending on how simultaneous access to a global memory location by more than one processor is handled. However, since the various PRAM models with a polynomial number of processors can simulate each other with at most $O(\log n)$ loss in time, the particular model does not affect our results. See Karp and Ramachandran [9] or Greenlaw, Hoover and Ruzzo [7] for a survey of the various PRAM models.

In what follows, we only consider PRAMs that use a number of processors that is polynomial in the input length. With this assumption, we say that a language has *parallel running time* $T(n)$ if there is a PRAM accepting the language with running time $T(n)$. We say a language L has *parallel speedup* $s(n)$ if there is a PRAM accepting L with running time $t(n)/s(n)$, given that L has sequential running time $t(n)$. L has *polynomial speedup* if it has parallel speedup $\Omega(n^\epsilon)$ for some $\epsilon > 0$. We use the notation $O^*(f(n))$ to denote $O(f(n) \log^{O(1)} n)$.

The class NC is the set of languages with parallel running time $\log^{O(1)} n$. A language L is NC-reducible to language L' if there is an NC-computable function f that is a many-one reduction from L to L'. For an introduction to this class, see [7].

2.3 Strict P-Completeness

In this section we define the notion of strict P-completeness. This is a refinement of the standard definition of P-completeness [7] which is that a problem L in P is P-complete if every problem in P is many-one NC-reducible to L. Roughly, a problem L in P is strictly $T(n)$-complete for P if L has parallel running time within a polylog factor of $T(n)$ and furthermore, if this could be improved by a polynomial factor, then *all* problems in P with at least linear running time have polynomial parallel speedup. We now make this precise.

Let L and L' be problems in P. We say L is *at most $T(n)$-complete with respect to L'* if the following holds. Let $t'(n)$ be any sequential running time for L' where $t'(n) = \Omega(n)$ and let ϵ be any constant > 0. Then there exists a function f such that

- f is an honest, many-one NC reduction from L' to L, and
- $T(|f(x)|) = O(t'(|x|)|x|^{\epsilon})$ for all x.

We say that a problem L is *at most $T(n)$-complete for P* if it is at most $T(n)$-complete with respect to all languages in P. We claim that if L is at most $T(n)$-complete for P, then if the parallel running time of L is a polynomial factor less than $T(n)$, every problem in P with at least linear sequential running time has polynomial speedup.

To see this, suppose that L is at most $T(n)$-complete for P. Let L' be any problem in P with sequential running time $t'(n)$. Let f be a reduction from L' to L satisfying the properties above. Since f is an honest reduction, there is an integer k such that $|f(x)| \geq |x|^{1/k}$, for all x. Now, suppose there is a parallel algorithm A for L that runs in time $O(T(n)n^{-2k\epsilon})$, for some $\epsilon > 0$. Consider the parallel algorithm for L' that on input x, computes $f(x)$ and runs the parallel algorithm A on $f(x)$. This algorithm has running time $O(T(|f(x)|)\ |f(x)|^{-2k\epsilon})$. This is $O(t'(|x|)\ |x|^{-\epsilon})$ since $|f(x)| \geq |x|^{1/k}$ and $T(|f(x)|) = O(t'(|x|)\ |x|^{\epsilon})$. Hence there is a parallel algorithm for L' which achieves speedup $\Omega(n^{\epsilon})$ over the sequential algorithm with running time $t'(n)$. Hence L' has polynomial speedup.

We say L is *strictly $T(n)$-complete for P* (or just *strictly $T(n)$-complete*) if L is at most $T(n)$-complete for P and the parallel running time of L is $O^*(T(n))$. In Theorem 5, we show that the Square CVP is strictly $n^{1/2}$-complete. The next lemma is useful in extending Theorem 5 to bound the parallel speedups of other problems.

Lemma 3 *Suppose L is at most $T(n)$-complete and that there is an honest NC reduction f from L to L'. Let $s(n)$ be the maximum of $|f(x)|$ over instances x of L of size n. Suppose $T(n)$ and $s(n)$ are non-decreasing functions. Then, L' is at most $T'(n) = T(s^{-1}(n))$-complete.*

3 The Square Circuit Value Problem

The Circuit Value Problem, CVP, denotes the set of all Boolean circuits together with an input to the circuit, such that the circuit evaluates to 1 on the input. This problem is well known to be complete for P [11] and in fact many restrictions of the CVP are also P-complete. Greenlaw, Hoover and Ruzzo [7] proved that the CVP is P-complete even when accepted instances must be monotone, alternating, synchronous and with fan-in and fan-out 2. Here, monotone means that the gates are all **and** or **or** gates, alternating means that on any path of the circuit starting at an input gate, the gates alternate between **and** and **or** gates and synchronous means that the inputs for level i of the circuit are all outputs of level $i - 1$.

We now define a *square* circuit to be one with all of the above properties and in addition, the number of gates at every level equals the depth. The *Square CVP* is the subset of CVP where the circuits are square.

The next theorem shows how to simulate RAMs with running time $t(n)$ by square circuits with depth $O^*(t(n))$. Related work on simulating PRAMs by circuits was done by Stockmeyer and Vishkin [17], who showed that parallel time and number of processors for Concurrent Read, Concurrent Write PRAMS correspond to depth and size for unbounded fan-in circuits, where the time-depth correspondence is to within a constant factor and the processor-size correspondence is to within a polynomial. Our proof differs from theirs in that we are simulating a RAM rather than a PRAM and want a precise bound the size of the circuit in terms of the running time of the RAM. Furthermore, our circuit have bounded fan-in and are synchronous.

Theorem 4 *Any RAM that runs in time $t(n) = \Omega(n)$ is simulated by a family of square circuits of depth $O^*(t(n))$.*

Proof: Given a RAM with running time $t(n)$, let M be a restricted RAM that accepts the same language and has the following properties. There is a number $T = O(t(n) \log t(n))$ such that on inputs of length n, (i) T is an upper bound on the maximum running time of M, (ii) Only cells $R_0 \ldots R_T$ are accessed during a computation and (iii) the total sum of the lengths of the non-zero cell values at any step of a computation is at most T. Such a machine M exists, by Lemmas 1 and 2, and since M is a log cost RAM.

We construct a circuit that simulates M on inputs of length n. We describe the construction of the circuit in stages. We first describe how to construct a circuit of the correct size and depth, and then describe the transformations needed to make the circuit square.

High level description. The circuit consists of T layers. The kth layer corresponds to the kth instruction of M that is executed on inputs of length n. Since M is restricted, this depends only on n and k. The output gates of the kth layer output the values of the cells $R_0 \ldots, R_T$ after the kth instruction is executed and are inputs to the $(k+1)$st layer. In addition, the $(k+1)$st layer may also have constant 0,1 inputs. One simple scheme would be to allow T outputs gates per cell, at each layer. However, this would require the circuit to have width $\Omega(t^2(n))$, which is too big.

We use T tuples of output gates, one per bit value of the cells. Each tuple contains $O(\log T)$ gates to record the bit position, $O(\log T)$ gates to record the cell number and 1 gate to record the value of the bit. Tuples with all gates equal to 0 do not correspond to any bit value. Call these tuples the *cell tuples.*

A layer consists of a circuit for an instruction, preceded by *select* circuits that select the correct inputs for the instruction and an *update* circuit that updates the T edge tuples, based on the result of the instruction. Each layer is of size $O^*(t(n))$ and depth $\log^{O(1)} t(n)$. The details can be found in the full paper.

Achieving monotonicity, alternation, restricted fan-in and synchronicity. We now have a circuit of the desired depth and size. To complete the proof, we show how this circuit can be transformed into a square circuit. The construction of Goldschlager [6] can be used to make the circuit monotone. This construction doubles the size of the circuit and does not increase the depth. Techniques of Greenlaw et al. [7] can be used to ensure that the circuit is alternating and that each gate has fan-in and fan-out 2. These techniques can be applied to each layer independently.

Greenlaw et al. [7] also describe how to make the resulting circuit synchronous, but their method at least squares the size of the circuit. We show that for the layered circuit of the above form, their technique can be modified to obtain a square, synchronous circuit of depth $O^*(t(n))$. There are two main steps. First, each layer is converted into a synchronous circuit of depth $O(\log^{O(1)} t(n))$ and size $O^*(t(n))$. This transformation of the layers increases the width of the layers, and hence duplicates the inputs and outputs of each layer. A problem caused by this is that the set of outputs of one layer may not equal the set of inputs to the next layer. The second main step is to connect the synchronized layers. This is done by transforming in a synchronous fashion the set of outputs of one layer so that it equals the set of inputs of the next layer. Again, details can be found in the full paper. □

From the above proof, the following theorem is not hard to prove.

Theorem 5 *The Square CVP is strictly $n^{1/2}$-complete.*

4 Completeness Results for Other Problems

We now apply Theorem 5 to prove strict completeness results for other problems. We first consider the problem of computing the expected cost of a Markov Decision Process, a central and well-studied problem in optimization theory. P-completeness results for Markov decision processes were obtained by Papadimitriou and Tsitsiklis [12]; we strengthen their results in the case of nonstationary finite horizon processes. We also describe a number of problems that are at most $n^{1/2}$-complete for P, but which are not known to be strictly P-complete.

A Markov decision process consists of a set S of m states, including an initial state, s_0. There is a finite set D of decisions and a probability transition function $p(s, s', d, t)$ which gives the probability of going from state s to s' at time t, given that decision d is made. The cost function $c(s, d, t)$ gives the cost of decision d at time t if the process is in state s. We consider the problem of minimizing the expected cost over T steps, starting at the initial state. The decision version of this problem, the *Nonstationary Finite Horizon Markov Decision Process Problem (NMDP)* is defined as follows: given an instance (S, s_0, D, p, c, T), is the expected cost equal to 0?

The problem can be solved sequentially using dynamic programming (Howard [8]). Roughly, the algorithm fills in a table of size $T \times m$ row by row, where entry (t, s) of the table is the minimum expected cost when t is the finite horizon and s is the initial state. The bottleneck in parallelizing this algorithm for NMDP seems to be that the rows of the table must be computed in sequential order. We can show that under the assumption that $T = \Theta(m)$, so that the input size $n = \Theta(m^2T)$, NMDP is strictly $n^{1/3}$-complete for P. Informally, this implies that if there is a parallel algorithm for the NMDP with running time $O(T^{1-\epsilon})$ for some $\epsilon > 0$, then all problems in P have polynomial speedup.

Theorem 6 *The Nonstationary Finite Horizon Markov Decision Process Problem where $T = \Theta(m)$ is strictly $n^{1/3}$-complete.*

We next describe upper bounds on the parallel speedup obtainable for many other P-complete problems, starting with the unrestricted Circuit Value Problem. The proof of the following corollary is immediate from Theorem 5.

Corollary 7 *The CVP is at most $n^{1/2}$-complete for P, even when the circuits are restricted to be monotone, alternating and synchronous, with fan-in and fan-out 2.*

However, it is an open problem whether CVP is strictly $n^{1/2}$-complete, since on the one hand, there is no known $O^*(n^{1/2})$ parallel algorithm with a polynomial number of processors for the CVP and on the other hand, it is not clear that a RAM which runs in time $O(t(n))$ can be simulated by a circuit family of size $O(t(n))$. Examples of other problems that are at most $\sqrt{n}$-complete are given in the following list. The completeness follows from Lemma 3 and the reductions cited below.

- *First Fit Decreasing Bin Packing (FFDBP).* Anderson, Mayr and Warmuth [1] gave an *NC*-reduction from the Monotone CVP with fan-in and fan-out 2 to FFDBP. The reduction maps a circuit with n gates to an instance of FFDBP with $O(n)$ items whose weights that can be represented using $O(\log n)$ bits.
- *Ordered Depth First Search (ODFS).* Reif [13] gave an *NC*-reduction from NOR CVP to ODFS that maps a circuit with n gates to a graph of size $O(n)$. There is a simple reduction from the Square CVP to NOR CVP that preserves the size of the circuit to within a constant factor [7].
- *Unification*: Dwork, Kanellakis and Mitchell [4] reduced the Monotone CVP to Unification, mapping a circuit of size n to an instance of Unification of size $O(n)$.
- *Lex First Maximal Independent Set (LFMIS)*: Cook [3] gave a linear-sized reduction from the Monotone CVP to LFMIS.

Finally, the *Lex First Maximal Clique (LFMC)* is an example of a problem for which there is a $O^*(n^{1/2})$ parallel algorithm, and the problem is at most $n^{1/4}$-complete.

5 Conclusions

We have introduced the notion of strict P-completeness, which refines previous work on P-completeness. We proved that the Square Circuit Value Problem and the Nonstationary Finite Horizon Markov Decision Process Problem are strictly $n^{1/2}$-complete and $n^{1/3}$-complete, respectively. In doing this, we obtained improved reductions from the RAM model to the circuit model of computation. We also obtained limits on the possible speedups for many other P-complete problems; in particular, showing that the CVP is at most $n^{1/2}$-complete.

An interesting open problem is whether the CVP is strictly $n^{1/2}$-complete, strictly n-complete, or somewhere in between. A related question is whether there are ways to prove limits on the parallel speedup that can be obtained for the CVP problem, other than by direct reduction from the RAM model. Finally, we have not severely restricted the number of processors in our PRAM model, since we allow any polynomial number of processors. Can our results be strengthened to obtain sharper limits if there is a stricter bound on the number of processors available?

References

[1] Anderson, R., E. Mayr and M. Warmuth. Parallel Approximation Schemes for Bin Packing, *Information and Computation*, 82(3):262-277, 1989.

[2] Cook, S. A. and R. A. Reckhow. Time bounded random access machines, *Journal of Computer and System Sciences*, 7(4):354-375, 1973.

[3] Cook, S. A. A taxonomy of problems with fast parallel algorithms, *Information and Control*, 64(1-3):2-22,1985.

[4] Dwork, C., P. C. Kanellakis and J. C. Mitchell. On the sequential nature of unification, *Journal on Logic Programming*, 1:35-50,1984.

[5] Fortune, S. and J. Wyllie, Parallelism in random access machines. In 10th Annual ACM Symposium on Theory of Computing, 114-118, 1978.

[6] Goldschlager, L. M. The monotone and planar circuit value problems are log space complete for P, *SIGACT News*, 9(2):25-29,1977.

[7] Greenlaw, R., H. J. Hoover and W. L. Ruzzo. A Compendium of Problems Complete for P, Department of Computing Science University of Alberta Technical Report TR 90-36, March 1991.

[8] Howard, R. A., *Dynamic Programming and Markov Processes*, MIT Press, Cambridge, 1960.

[9] Karp, R. M. and V. Ramachandran. A Survey of Parallel Algorithms for Shared-Memory Machines, *Handbook of Theoretical Computer Science*, North-Holland.

[10] Kruskal, C. P., L. Rudolph and M. Snir. A complexity theory of efficient parallel algorithms, *Theoretical Computer Science*, 71(1):95-132.

[11] Ladner, R. E., The circuit value problem is log space complete for P. *SIGACT News*, 7(1):18-20, 1975.

[12] Papadimitriou, C. H. and J. N. Tsitsiklis. The complexity of Markov decision processes, *Mathematics of Operations Research*, 12(3):441-450, 1987.

[13] Reif, J. Depth first search is inherently sequential, *Information Processing Letters*, 20(5):229-234, 1985.

[14] Simons and Vitter. New classes for parallel complexity: a study of Unification and Other Complete Problems for P. *IEEE Transactions on Computers*, 35(5):403-418, 1986.

[15] Slot, C. and P. Van Emde Boas. The problem of space invariance for sequential machines, Information and Computation, 77::93-122, 1988.

[16] Snyder, L. Type architectures, shared memory, and the corollary of modest potential, *Annual Review of Computer Science* 1:289-317, 1986.

[17] Stockmeyer, L. and U. Vishkin. Simulation of Parallel Random Access Machines by Circuits, *SIAM Journal on Computing*, 14:862-874, 1985.

Fast and Optimal Simulations between CRCW PRAMs

TORBEN HAGERUP

Max-Planck-Institut für Informatik
W–6600 Saarbrücken, Germany

We describe new simulations between different variants of the CRCW PRAM. On a MINIMUM PRAM, the minimum value is written in the event of a write conflict, whereas concurrent writing to the same cell is without effect on a TOLERANT PRAM. Compared to other variants of the CRCW PRAM, the MINIMUM PRAM is very strong, and the TOLERANT PRAM is very weak. In fact, the two variants are near opposite ends of the spectrum of CRCW PRAM variants commonly considered. We show that one step of a (randomized) MINIMUM PRAM with n processors can be simulated with high probability in $O((\log^* n)^3)$ time on a randomized TOLERANT PRAM with $O(n/(\log^* n)^3)$ processors. The simulation is optimal, in the sense that the product of its slowdown and the number of simulating processors is within a constant factor of the number of simulated processors. It subsumes most previous work on randomized simulations between n-processor CRCW PRAMs with infinite memories.

1 Introduction

A *CRCW PRAM* is a synchronous parallel machine with processors numbered $1, 2, \ldots$ and with a global memory that supports simultaneous access to a single cell by arbitrary sets of processors. Whereas the meaning of simultaneous reading is clear, researchers have studied several *variants* of the CRCW PRAM, each distinguished by a different *write conflict resolution rule*. The following write conflict resolution rules and corresponding variants are relevant to the present paper:

FETCH-AND-ADD (Gottlieb *et al.*, 1983): The effect of concurrent writing is as if the processors involved access the memory cell in question sequentially in some order, each processor adding its value to the contents of the cell after first reading the previous contents;

MINIMUM (Cole and Vishkin, 1986): The smallest value that some processor attempts to store in a given cell in a given step gets stored in the cell;

PRIORITY (Goldschlager, 1982): If two or more processors attempt to write to a given cell in a given step, then the lowest-numbered processor among them succeeds and writes its value;

ARBITRARY (Shiloach and Vishkin, 1982): If two or more processors attempt to write to a given cell in a given step, then one of them succeeds, but there is no rule assumed to govern the selection of the successful processor;

COMMON (Shiloach and Vishkin, 1981): All processors writing to a given cell in a given step must be writing the same value, which then gets stored in the cell;

COLLISION (Fich *et al.*, 1988b): If two or more processors attempt to write to a given cell in a given step, then a special collision symbol gets stored in the cell;

COLLISION$^+$ (Chlebus *et al.*, 1988): If the processors attempting to write to a given cell in a given step all attempt to write the same value, then that value gets stored in the cell; if at least two values differ, a collision symbol gets stored in the cell;

Supported in part by the Deutsche Forschungsgemeinschaft, SFB 124, TP B2, VLSI Entwurfsmethoden und Parallelität, and in part by the ESPRIT II Basic Research Actions Program of the EC under contract No. 3075 (project ALCOM).

TOLERANT (Grolmusz and Ragde, 1987): If two or more processors attempt to write to a given cell in a given step, then the contents of that cell do not change;

ROBUST (Hagerup and Radzik, 1990): A cell subjected to concurrent writing afterwards contains a well-defined but unknown value.

A CRCW PRAM working according to the PRIORITY (ARBITRARY, etc.) rule will be called a PRIORITY (ARBITRARY, etc.). When no explicit processor bounds are mentioned in a statement about a particular simulation, we always intend the number of processors of the simulated machine to be denoted by n.

A machine M_2 simulates another machine M_1 with ***slowdown*** T if at most T steps of M_2 are needed to simulate a single step of M_1. As has been done before, we use "$A \leq B$", where A and B are names of variants, to indicate that an n-processor machine of type B can simulate an n-processor machine of type A with constant slowdown. The following hierarchy is easy to establish.

ROBUST $\leq$ TOLERANT $\leq$ COLLISION $\leq$ COLLISION$^+$ $\leq$ ARBITRARY $\leq$ PRIORITY $\leq$ MINIMUM.

Simulations of strong variants of the CRCW PRAM (i.e., variants near the right end of the above chain) on weaker variants have been the object of intensive study in later years (Kučera, 1982, 1983; Fich *et al.*, 1988a, 1988b; Chlebus *et al.*, 1988, 1989; Boppana, 1989; Ragde, 1989; Hagerup and Radzik, 1990; Matias and Vishkin, 1990a, 1990b; Gil, 1990). In the interest of brevity, we limit the discussion below to randomized simulations without an increase in the number of processors (i.e., the simulating machine is allowed no more processors than the simulated machine).

If we are concerned only with the slowdown of a simulation, the best previous results are: $O(\log n)$ for the simulation of any of the above variants on any other variant (or even on a bounded-degree processor network), $O(\log\log n)$ for the simulation of MINIMUM on TOLERANT (Gil, 1990), and $O(\log\log n)$ for the simulation of PRIORITY on ROBUST (Hagerup and Radzik, 1990). Speed is not everything, however, and the simulations mentioned so far use n processors and nonconstant time to simulate one step of an n-processor machine, although the sequential cost of this computation is only $O(n)$. A desirable goal, as in other areas of parallel computation, is clearly to attain ***optimality***, which in the present case means to obtain ***optimal simulations***, for which the product of slowdown and number of simulating processors is within a constant factor of the number of simulated processors. The only previous optimal simulations known to the author were given by Matias and Vishkin (1990a; revised and corrected: 1990b), who showed that MINIMUM and FETCH-AND-ADD can be simulated optimally on ARBITRARY with expected slowdown $O(\log n)$ (by a convention introduced above, n denotes the number of processors of the simulated MINIMUM or FETCH-AND-ADD PRAM; since the simulation is optimal, the simulating ARBITRARY PRAM has $O(n/\log n)$ processors). This result assumes that a prime larger than the size of the memory of the simulated machine is given for free; following previous authors, we shall ignore the difficulty of obtaining such a prime. We are not aware of any lower bounds on the slowdown of randomized simulations between variants of the CRCW PRAM.

For some of the simulations discussed above, the bounds on the slowdown are bounds on the expected value of this parameter only. The probability that an actual run of the simulation exceeds these bounds, which we (over-dramatizing slightly) call the ***error probability***, may therefore be nonnegligible. More satisfactory simulations have bounds on the error probability that depend on the number n of simulated processors and converge rapidly towards zero as n increases. Our simulations are of this kind.

A final problem with which the simulation of (Hagerup and Radzik, 1990) is afflicted is the so-called "memory blowup": A machine with n processors and m global memory cells is simulated by a machine with as many as $\Theta(nm)$ global memory cells. Ideally, we would like the simulation to use no more memory than the m cells of the simulated machine. The simulations of (Matias and Vishkin, 1990a, 1990b) and (Gil, 1990) come close to this ideal, using $O(n+m)$ global memory cells.

The present paper contributes essentially two new simulations, both of which are optimal. One is of ARBITRARY on TOLERANT, the other one of MINIMUM on COLLISION$^+$. In both cases, the slowdown is $O((\log^* n)^2)$ with high probability. The simulations can be combined to yield an optimal simulation of MINIMUM on TOLERANT whose slowdown is $O((\log^* n)^3)$ with high probability. In terms of speed, our simulations are a long way better than all previous results. For PRAMs with infinite memories, they subsume all previous results except one algorithm of (Matias and Vishkin, 1990a, 1990b), which allows the FETCH-AND-ADD PRAM to be simulated, and one algorithm of (Hagerup and Radzik, 1990), which runs on the weaker ROBUST PRAM. As described in Section 6, one of our simulations is easily turned into an optimal simulation of FETCH-AND-ADD on COLLISION$^+$ whose slowdown is $O(\log n)$ with high probability, so that the simulations of (Matias and Vishkin, 1990a, 1990b) can be considered as subsumed as well. Our first algorithm exhibits a direct trade-off between space and reliability. The minimum space needed to simulate an n-processor PRAM with m global memory cells is $\Theta(n + m(\log n)^2)$. The second simulation needs $O(n + m)$ space.

Subsequently to or concurrently with the work reported here, still faster randomized simulations have been devised (Gil, Matias and Vishkin, 1991; Hagerup, 1992). The best of these (Hagerup, 1992) is an optimal simulation with slowdown $O(\log^* n)$ of a MINIMUM PRAM with n processors and m global memory cells on a TOLERANT PRAM with $O(n/\log^* n)$ processors and $O(n+m)$ global memory cells. It uses tools developed in the present paper.

2 Preliminaries

An *allocated* PRAM (Bhatt *et al.*, 1991) is identical to a standard PRAM, except that the set of processor numbers, called the *index set* of the allocated PRAM, is not necessarily of the form $\{1, \ldots, q\}$, where $q \in \mathbb{N}$ (each processor still has an integer processor number). The allocated PRAM allows us to model a situation in which some processors of a standard PRAM appear "inactive" or "dead", presumably because they are engaged in an unrelated computation.

The relations in Lemma 2.1 and Lemma 2.2 below are known as Chernoff bounds and Azuma's inequality, respectively; for proofs, see (Hagerup and Rüb, 1990) and (McDiarmid, 1989).

Lemma 2.1: For every binomially distributed random variable S,
(a) $\Pr(S \geq 2E(S)) \leq e^{-E(S)/3}$;
(b) $\Pr(S \leq E(S)/2) \leq e^{-E(S)/8}$;
(c) For every $z \geq 6E(S)$, $\Pr(S \geq z) \leq 2^{-z}$.

Lemma 2.2: Let $n \in \mathbb{N}$, let $Z_1, \ldots, Z_n$ be independent random variables with finite ranges, and let S be an arbitrary real function of $Z_1, \ldots, Z_n$ with $E(S) \geq 0$. Assume that S changes by at most c in response to an arbitrary change in a single Z_i. Then for every $z \geq 2E(S)$,

$$\Pr(S \geq z) \leq e^{-z^2/(8c^2 n)}.$$

Definition (Matias and Vishkin, 1991; Hagerup, 1991a): For $n \in \mathbb{N}$ and $d, s \in \mathbb{R}$, the *compaction problem* of size n and with parameters (d, s) is the following: Given n bits $x_1, \ldots, x_n$ with $\sum_{j=1}^{n} x_j \leq d$, compute n nonnegative integers $y_1, \ldots, y_n$ such that
(1) For $j = 1, \ldots, n$, $x_j = 0 \Leftrightarrow y_j = 0$;
(2) For $1 \leq i < j \leq n$, if $x_i \neq 0$, then $y_i \neq y_j$;
(3) $\max\{y_j : 1 \leq j \leq n\} = O(s)$.

A natural interpretation of the compaction problem is that $(x_1, \ldots, x_n)$ is a bit vector representation of a set of at most d *active elements*, scattered over an array of size n, the task being to place the active elements in an array of size $O(s)$. We choose our terminology accordingly.

Lemma 2.3 (Hagerup, 1991a): There is a constant $\epsilon > 0$ such that for all $d \geq 1$ and all τ with $\log^* d \leq \tau \leq n$, compaction problems of size n and with parameters (d, d) can be solved on a TOLERANT PRAM using $O(\tau)$ time, $O(n/\tau)$ processors and $O(n)$ space with probability at least $1 - 2^{-n^\epsilon}$.

Lemma 2.4 (Hagerup, 1991b): There is a constant $\epsilon > 0$ such that for all τ with $\log^* n \leq \tau \leq n$, random permutations of $\{1, \ldots, n\}$ can be drawn from the uniform distribution over the set of all permutations of $\{1, \ldots, n\}$ on a TOLERANT PRAM using $O(\tau)$ time, $O(n/\tau)$ processors and $O(n)$ space with probability at least $1 - 2^{-n^\epsilon}$.

When stating that a random sample Y is drawn from some set V with associated probability p, what we intend is that each element of V is included in Y with probability p and independently of all other elements of V. In all cases of interest to us, Chernoff bound (a) shows that the probability that the size of the resulting sample exceeds $2p|V|$ can be ignored, and we shall tacitly do so.

A standard observation is that for a simulation between two variants of the CRCW PRAM, it suffices to consider a step of the simulated machine in which each processor attempts to write to a global memory cell. We therefore consider the input to a simulation of an n-processor PRAM with m global memory cells to be a sequence $(\xi_1, \theta_1), \ldots, (\xi_n, \theta_n)$ of n integer pairs, where $1 \le \xi_j \le m$, for $1 \le j \le n$, and where (ξ_j, θ_j) means that the simulated processor with processor number j attempts to write the value θ_j to the global memory cell with address ξ_j. The task of the simulation is to modify the global memory in accordance with the input and with the write conflict resolution rule of the simulated machine. We shall often identify each physical processor of the simulated machine with a *virtual* processor of the simulating machine.

For the remainder of the paper, fix an input $(\xi_1, \theta_1), \ldots, (\xi_n, \theta_n)$ and let $\Xi = \{\xi_1, \ldots, \xi_n\}$ be the set of used addresses. For each $\xi \in \Xi$, let $\mathcal{Q}(\xi)$ be the set of processors attempting to write to address ξ, and let $W(\xi) = \{j : 1 \le j \le n \text{ and } \xi_j = \xi\}$ be the set of their processor numbers. Each set of the form $W(\xi)$, where $\xi \in \Xi$, will be called a *color class*. Each color class $W(\xi)$ is associated with a separate problem, that of writing the correct value to the global memory cell with address ξ. It is natural to let this problem be solved by the processors in $\mathcal{Q}(\xi)$, i.e., as on an allocated PRAM with index set $W(\xi)$. In a simulation of ARBITRARY, the task of the allocated PRAM is simply to distinguish some (arbitrary) processor, which can then carry out the actual writing. In different contexts, similar problems are known as "leader election" or "selection resolution". In a simulation of MINIMUM, the task of the allocated PRAM is to write the value $\min\{\theta_j : j \in W(\xi)\}$ in a designated output cell.

3 A generic algorithm

The two simulations described in this paper both use instances of a generic algorithm described in this section. Loosely speaking, the generic algorithm inputs a collection of problems of varying sizes, each of which can be solved in $O(\log^* n)$ expected time with a number of processors equal to its size, and manages to solve all problems in the collection in $O((\log^* n)^2)$ time with high probability. In the interest of readability, we have kept the theorem below rather vague (e.g., we employ without definition the notion of a "problem" and of "solving a problem"). The careful reader may fill in the missing details or choose to treat Theorem 3.1 as a meta-theorem, to be specialized and checked in each application.

Theorem 3.1: For every constant $\delta < 1/2$, there is a constant $\epsilon > 0$ such that the following holds for every variant A of the CRCW PRAM with TOLERANT $\le A$. Let $\mathcal{P} = \{P_u : u \in U\}$ be a set of problems (actually, of problem instances), where $U \subseteq \mathbb{N}$ is some set, such that each $P \in \mathcal{P}$ can be associated with an *index set* $I(P) \subseteq \{1, \ldots, n\}$. Assume that there is a real number $p > 0$ and an algorithm $\mathcal{A}$ such that for every $P \in \mathcal{P}$, $\mathcal{A}$ can be executed on an allocated PRAM of type A with index set $I(P)$ in $\tau = O(\log^* n)$ time, after which P has been solved with probability at least p, and each processor of the allocated PRAM knows whether P has been solved. Assume further that $\mathcal{A}$ can be slowed down (i.e., the stated properties are preserved if the operations comprising one step of $\mathcal{A}$ are in fact executed over several steps by fewer than the stated number of processors), that multiple instances of $\mathcal{A}$ can execute in parallel without interfering with each other, that $I(P) \cap I(P') = \emptyset$ for all $P, P' \in \mathcal{P}$ with $P \neq P'$, and that $|I(P)| = O(n^\delta)$, for all $P \in \mathcal{P}$. Then, given a vector $x_1, \ldots, x_n$ such that for $j = 1, \ldots, n$,

$$x_j = \begin{cases} u, & \text{if } x_j \in I(P_u); \\ 0, & \text{if } x_j \notin \bigcup_{u \in U} I(P_u), \end{cases}$$

all problems in $\mathcal{P}$ can be solved on a PRAM of type A using $O((\log^* n)^2)$ time and $O(n\tau/(\log^* n)^2)$ processors with probability at least $1 - 2^{-n^\epsilon}$.

Proof: Let $\epsilon = (1 - 2\delta)/6$, call a problem $P \in \mathcal{P}$ *active* if it has not yet been solved, and call $j \in \{1, \ldots, n\}$ active if j belongs to the index set of an active problem. Since $\mathcal{A}$ can be replaced by an algorithm that executes $\mathcal{A}$ a constant number of times, let us assume without loss of generality that

$p = 7/8$. Choose integers $T, v_1, \ldots, v_T$ such that (1) $v_T = \lfloor n^\epsilon \rfloor$, (2) $v_{t-1} = \lceil \log v_t \rceil$, for $t = 2, \ldots, T$, and (3) $v_1 = 1$. This is clearly possible, and $T = O(\log^* n)$. Also take $v_{T+1} = 2^{v_T}$ and consider the following algorithm.

(1) **for** $t := 1$ **to** T **do**
(2)   **begin**
(3)   Allocate v_t (virtual) processors to each active element in $\{1, \ldots, n\}$;
(4)   For each active problem $P \in \mathcal{P}$,
(5)   simulate v_t independent executions of $\mathcal{A}$ on an allocated PRAM with index set $I(P)$;
(6)   **end**;

For $t = 1, \ldots, T$, let *Stage t* be the tth execution of lines (2)–(6), and let N_t (a random variable) be the number of elements that are active at the end of Stage t. Since in Stage t each active element participates in v_t independent executions of $\mathcal{A}$, each of which fails with probability at most 1/8, clearly

$$E(N_t) \le n \cdot 8^{-v_t} = n \cdot 2^{-v_t} \cdot 2^{-2v_t} \le \tfrac{1}{2} n \cdot v_{t+1}^{-2}.$$

In particular (take $t = T$), the probability that some element is active at the end of the algorithm, i.e., that not all problems in $\mathcal{P}$ are solved, is at most $\frac{1}{2} n \cdot 2^{-2\lfloor n^\epsilon \rfloor}$.

For fixed $t \in \{1, \ldots, T-1\}$, since the outcome of a single execution of $\mathcal{A}$ can affect N_t by at most n^δ, an application of Lemma 2.2 gives that $\Pr(N_t > n \cdot v_{t+1}^{-2}) \le e^{-z}$, where

$$z = \frac{n^2 \cdot v_{t+1}^{-4}}{8 \cdot n^{2\delta} \cdot n} = \frac{n^{1-2\delta}}{8 v_{t+1}^4} \ge \frac{n^{1-2\delta-4\epsilon}}{8} = \frac{n^{2\epsilon}}{8}.$$

But if $N_t \le n \cdot v_{t+1}^{-2}$ for $t = 1, \ldots, T-1$, then Lemma 2.3 can be use to place the active elements in an array of size $O(n/v_t^2)$ at the beginning of Stage t, for $t = 1, \ldots, T$. This on the one hand makes the processor allocation in line (3) a trivial task, and on the other hand shows that the algorithm can be executed using a total of

$$O\left(n\tau \sum_{t=1}^{T} v_t^{-1}\right) = O(n\tau)$$

operations. Decreasing ϵ as necessary, we can assume the probability that a given application of Lemma 2.3 fails to be bounded by $2^{-n^{2\epsilon}}$. The claim of the theorem therefore holds with probability at least

$$1 - \tfrac{1}{2} n \cdot 2^{-2\lfloor n^\epsilon \rfloor} - T e^{-n^{2\epsilon}/8} - T \cdot 2^{-n^{2\epsilon}},$$

which for sufficiently large values of n is at least $1 - 2^{-n^\epsilon}$. ∎

4 Simulation of ARBITRARY on TOLERANT

In this section we first show the result below, which misses optimality by a factor of $\Theta(\log^* n)$. Subsequently we describe a preprocessing stage that reduces the initial problem size by a factor of $\Omega(\log^* n)$ and achieves optimality.

Lemma 4.1: There is a constant $\epsilon > 0$ such that one step of a (randomized) n-processor ARBITRARY PRAM can be simulated by $O((\log^* n)^2)$ steps of a randomized TOLERANT PRAM with $O(n/\log^* n)$ processors with probability at least $1 - 2^{-n^\epsilon}$.

By Theorem 3.1 and the discussion towards the end of Section 2, proving Lemma 4.1 boils down to describing an algorithm for leader election on an allocated TOLERANT PRAM with at most n processors that works in $O(\log^* n)$ expected time. Our main tool is that of a *graduated conditional scattering* (GCS). Graduated conditional scatterings were introduced in (Hagerup and Radzik, 1990) and named in (Hagerup, 1991a).

Definition: For $s \in \mathbb{N}$, a *scattering* with range s is a random experiment carried out by a set $\mathcal{Q}$ of processors as follows: Each processor $Q \in \mathcal{Q}$, independently of other processors, chooses a random number Z_Q from the uniform distribution over $\{1,\ldots,s\}$. Two distinct processors $Q,Q' \in \mathcal{Q}$ *collide* if $Z_Q = Z_{Q'}$.

For our purposes, a scattering succeeds if at least one processor does not collide. The following lemma shows that "if there is room for everybody", then a scattering usually succeeds. Although the proof is omitted, a little experimentation should make the lemma highly plausible to the reader.

Lemma 4.2: For all $k,s \in \mathbb{N}$ with $1 \le k \le s$, if k processors carry out a scattering with range s, then the probability that they all collide is at most $1/s$.

Definition: For $r,s \in \mathbb{N}$, a *graduated conditional scattering* (GCS) with range $r \times s$ is a random experiment carried out by a set $\mathcal{Q}$ of processors as follows: Each processor $Q \in \mathcal{Q}$ first, independently of other processors, chooses a random number I_Q with $\Pr(I_Q = i) = 2^{-i}$, for $i = 1,\ldots,r$, and $\Pr(I_Q = 0) = 2^{-r}$. If $I_Q \neq 0$, it then participates in a scattering S_{I_Q} with range s.

Lemma 4.3: There is a constant $C \in \mathbb{N}$ such that for all $k,r,s \in \mathbb{N}$ with $r \ge \log k$ and $s \ge C$, if a set $\mathcal{Q}$ of k processors carries out a GCS with range $r \times s$, then the probability that all processors in $\mathcal{Q}$ collide is at most $1/s$.

Proof: For $i = 1,\ldots,r$, let $R_i = \{Q \in \mathcal{Q} : I_Q = i\}$.

Case 1: $k \le s$. The claim follows immediately from Lemma 4.2, since $1 \le |R_i| \le s$ for at least one $i \in \{1,\ldots,r\}$.

Case 2: $k > s$. Let $c = 3e^4$ and $i = \lfloor \log(ck/s) \rfloor$. Note that if we take $C \ge c$ (recall that $s \ge C$), then $1 \le i \le r$, and $|R_i|$ is binomially distributed with expected value $E(|R_i|) = 2^{-i}k$. By the choice of i, $s/c \le E(|R_i|) \le 2s/c$. Hence by Chernoff bounds (a) and (b),

$$\Pr\left(\frac{s}{2c} \le |R_i| \le \frac{4s}{c}\right) \ge 1 - 2e^{-s/(8c)}.$$

But if $s/(2c) \le |R_i| \le 4s/c$, it follows from Lemma 1 of (Hagerup and Radzik, 1990) that the probability that all processors in R_i collide is at most

$$\left(\frac{3e^3 \cdot 4s/c}{4s}\right)^{s/(2c)} = e^{-s/(2c)}.$$

Altogether, hence, the probability that all processors collide is at most $3e^{-s/(8c)}$, which for sufficiently large values of C is bounded by $1/s$. ∎

It is obvious that a GCS with range $r \times s$ can be executed on an allocated TOLERANT PRAM in constant time and using $O(rs)$ space, in the sense that each processor learns whether it collides in the GCS. Suppose that in our attempt to solve the leader election problem we prune away all colliding processors. By Lemma 4.3, if we choose $r \ge \log n$, then the probability that we inadvertently prune away all processors (causing the election to fail) is fairly small. On the other hand, we are left with at most rs surviving processors, and we can repeat the process with a much smaller value of r. The details are given in the algorithm below, where L denotes a constant to be determined in the subsequent analysis, and where the subroutine *Scatter*(r,s) lets all active processors carry out a GCS with range $r \times s$ and afterwards makes all colliding processors inactive.

```
Let all processors be active;
t := 1;
r_1 := ⌈log n⌉;
Scatter(r_1, L);
while r_t > L do
  begin
    t := t + 1;
    r_t := max{2⌈log r_{t-1}⌉, L};
    Scatter(r_t, r_t);
  end;
```

Provided that $L \geq 6$, the algorithm finishes in $O(\log^* n)$ time, and it is easy to see that the space needed is $O(\log n)$. Say that a call of *Scatter fails* if it reduces the number of active processors from at least 1 to zero, and that the execution of the algorithm as a whole fails if some call of *Scatter* fails. By Lemma 4.3, if $s \geq C$, where C denotes the constant of the same name in that lemma, then a call of $Scatter(r, s)$ executed by at most 2^r active processors fails with probability at most $1/s$. But if $L \geq C$, the stated conditions can be seen to necessarily hold before each call of *Scatter* executed by the algorithm. Hence the failure probability for the whole algorithm is at most

$$\frac{1}{L} + \frac{1}{r_2} + \frac{1}{r_3} + \cdots + \frac{1}{L},$$

which for sufficiently large values of L is bounded by $1/2$. After the execution of the algorithm, at most L^2 processors remain active. Furthermore, each active processor can be associated with a different position in an $L \times L$ array, and if the execution has not failed, the leader election can be completed in constant time in a trivial manner.

This ends the proof of Lemma 4.1, except that in order to comply with the requirements of Theorem 3.1, we still have to describe how to handle color classes of size $\geq n^\delta$, where $\delta = 1/3$, say. Since the preprocessing described in the remainder of this section in fact elects leaders for all such color classes, we refrain from repeating the argument here.

Theorem 4.4: There is a constant $\epsilon > 0$ such that one step of a (randomized) n-processor ARBITRARY PRAM can be simulated by $O((\log^* n)^2)$ steps of a randomized TOLERANT PRAM with $O(n/(\log^* n)^2)$ processors with probability at least $1 - 2^{-n^\epsilon}$.

Proof: Execute the three steps described below. Following the terminology of the proof of Theorem 3.1, we call a color class $W(\xi)$ as well as its elements active until a leader has been elected from $Q(\xi)$. Whenever convenient, we will identify a processor with its processor number.

Step 1: Draw a random sample Y from $\{1, \ldots, n\}$ with associated probability $1/\lfloor n^{1/6} \rfloor$. Use Lemma 2.3 to allocate at least $n^{1/7}$ processors to each element of Y and use the algorithm of (Chlebus *et al.*, 1989, Theorem 2) (essentially, an $n^{1/7}$-way tournament) to compute a leader in each color class represented in Y.

Analysis of Step 1: By Chernoff bound (b), the probability that some color class of size $\geq n^{1/3}$ does not have at least one representative in Y is $2^{-n^{\Omega(1)}}$, i.e., negligible. Hence Step 1 deactivates all color classes of size $\geq n^{1/3}$, as anticipated in the proof of Lemma 4.1.

In the remainder of the proof, a color class is called *small* if its size is at most $s = (\log^* n)^2$; otherwise it is called *large*. Steps 2 and 3 aim to reduce the number of active small and large color classes, respectively.

Step 2: Begin by partitioning the n virtual processors randomly into s groups of $r = n/s$ elements each, with all partitions being equally likely (in order to avoid problems of rounding, first replace n by the nearest larger multiple of s by introducing at most s dummy processors). Then process the groups one by one in s successive time slots of constant length each. Processing a group means letting each processor in the group try to establish itself as the leader of its color class by writing its processor number to the target cell of its color class. Note that this procedure determines a leader in each color class for which some group contains precisely one element of the color class.

Analysis of Step 2: The random partition into groups can be obtained by computing a random permutation π of $\{1, \ldots, n\}$ (Lemma 2.4) and taking the jth group to consist of the processors with processor numbers in the set $\{(j-1)r+1, \ldots, jr\}$, for $j = 1, \ldots, s$. It is convenient to view π as an arrangement of the numbers $1, \ldots, n$ in an $r \times s$ array, with the jth column defining the jth group, for $j = 1, \ldots, s$. In preparation for the next lemma, view the elements of a fixed color class as black balls and the remaining elements of $\{1, \ldots, n\}$ as white balls.

Lemma 4.5: Let $r, s, k \in \mathbb{N}$ and assume that $k \leq s$ and that $r \geq 2$. Define a *tableau* as an $r \times s$ array of k black balls and $rs - k$ white balls and call a tableau *bad* if no column in the tableau contains exactly one black ball. Then a tableau drawn at random from the uniform distribution over all tableaux is bad with probability at most $\frac{1}{s}\left(\frac{r}{r-1}\right)^s$.

Proof: Assume that the black balls are numbered $1, \ldots, k$, and that the white balls are also distinguishable from each other. The total number of tableaux is then $(rs)!$. Associate with each tableau a function from $\{1, \ldots, k\}$ to $\{1, \ldots, s\}$ by mapping i to the number of the column in the tableau containing the black ball numbered i, for $i = 1, \ldots, k$. It is easy to see that a fixed function is associated with at most $r^k(rs-k)!$ tableaux. Lemma 4.2 now implies that the probability of a random tableau being bad is at most

$$\frac{r^k(rs-k)!s^{k-1}}{(rs)!} \le \frac{r^k s^{k-1}}{(rs-k)^k} = \frac{1}{s}\left(\frac{rs}{rs-k}\right)^k \le \frac{1}{s}\left(\frac{rs}{rs-s}\right)^s = \frac{1}{s}\left(\frac{r}{r-1}\right)^s. \quad \blacksquare$$

Lemma 4.5 implies that for sufficiently large values of n, the probability that a fixed element of a small color class remains active after Step 2 is at most $2/(\log^* n)^2$. Hence if N denotes the total number of active elements in small color classes after Step 2, then $E(N) = O(n/(\log^* n)^2)$. We want to conclude from Lemma 2.2 that in fact $N = O(n/(\log^* n)^2)$, except with probability $2^{-n^{\Omega(1)}}$. Note first that since Step 1 deactivates all color classes of size $\ge n^{1/3}$, a change in π that affects the value of π on $O(1)$ elements in its domain affects N by $O(n^{1/3})$. What remains is hence to describe a method for constructing π from $O(n)$ independent random variables, a change in one of which affects the value of π on $O(1)$ elements. Such a method is furnished by the following standard sequential algorithm for computing random permutations described, e.g., in (Hagerup, 1991b): Let $Z_2, \ldots, Z_n$ be independent random variables with Z_i uniformly distributed over $\{1, \ldots, i\}$, for $i = 2, \ldots, n$, and compute the output permutation π as $\pi_{2,Z_2} \circ \cdots \circ \pi_{n,Z_n}$, where $\pi_{i,j}$ denotes the permutation of $\{1, \ldots, n\}$ that interchanges i and j while leaving all other elements fixed, for $i, j \in \{1, \ldots, n\}$. It is easy to verify that a change in a single Z_i changes the value of π on at most 3 elements in its domain.

Step 3: Draw a random sample Y from $\{1, \ldots, n\}$ with associated probability $1/\log^* n$. Compact Y into an array of size $O(n/\log^* n)$ by means of Lemma 2.3 and use Lemma 4.1 to elect a leader in each color class represented in Y.

Analysis of Step 3: By Chernoff bound (b), the probability that a large color class does not have a representative in Y is $2^{-\Omega(\log^* n)}$. Hence the expected number of active elements in large color classes after Step 3 is $O(n/\log^* n)$. By an easy application of Lemma 2.2, the actual number of such elements is also $O(n/\log^* n)$, except with probability $2^{-n^{\Omega(1)}}$.

Altogether, the analysis of Steps 2 and 3 shows that except with probability $2^{-n^{\Omega(1)}}$, the total number of elements remaining active after Step 3 is $O(n/\log^* n)$. At this point the simulation can be completed by another appeal to Lemma 4.1. $\blacksquare$

Theorem 4.4 is a special case of Theorem 4.6 below, which takes the space used by the simulation into account.

Theorem 4.6: There is a constant $\epsilon > 0$ such that for all s with $\log n \le s \le n^\epsilon$, one step of a (randomized) n-processor ARBITRARY PRAM with m global memory cells can be simulated by $O((\log^* n)^2)$ steps of a randomized TOLERANT PRAM using $O(n/(\log^* n)^2)$ processors and $O(n + ms \log n)$ space with probability at least $1 - 2^{-s}$.

Proof: Assume s to be an integer. We trade reliability for economy of space in the simulation described above by modifying it in two places. First, the main algorithm, which is an instance of the algorithm of Theorem 3.1, is made to operate in its last stage with $v_T = s$ instead of $v_T = \lfloor n^\epsilon \rfloor$. The probability that some element is not deactivated by the algorithm is then at most $\frac{1}{2}n \cdot 2^{-2s} \le \frac{1}{2} \cdot 2^{-s}$, and the space needed is $O(n + ms \log n)$.

Second, Step 1 of the preprocessing is modified as follows: When at least $n^{1/7}$ auxiliary processors have been allocated to each element of the sample Y, the auxiliary processors of each color class carry out a GCS with parameters $\lceil \log n \rceil \times Ks$, for a suitable constant $K \in \mathbb{N}$, and the tournament algorithm of (Chlebus *et al.*, 1989) is applied only to those elements of Y whose auxiliary processors do not all collide in their GCS. Relations derived in the proof of Lemma 4.3 show that for K chosen sufficiently large and for $\epsilon < 1/7$, the probability that all participating processors collide in some GCS is negligible. $\blacksquare$

5 Simulation of MINIMUM on COLLISION$^+$

Our simulation of MINIMUM on COLLISION$^+$ is little more than an assembly of previously known results. We list some of these results.

For our purposes, a ***perfect hash function*** for a sequence $x_1, \ldots, x_n$ of n integers is an injective function from $\{x_1, \ldots, x_n\}$ to $\{1, \ldots, s\}$, where $s = O(n)$, that can be stored in $O(n)$ space and evaluated in constant time by a single processor.

Lemma 5.1 (Bast and Hagerup, 1991): There is a constant $\epsilon > 0$ such that a perfect hash function for a sequence of n integers can be constructed on a COLLISION$^+$ PRAM using $O(\log^* n)$ time, $O(n)$ operations and $O(n)$ space with probability at least $1 - 2^{-n^\epsilon}$.

Given a sequence $x_1, \ldots, x_n$, say that a permutation $\pi_1, \ldots, \pi_n$ of $1, \ldots, n$ is ***duplicate-grouping*** for $x_1, \ldots, x_n$ if for all $i, j, k \in \mathbb{N}$, if $1 \le i \le j \le k \le n$ and $x_{\pi_i} = x_{\pi_k}$, then $x_{\pi_j} = x_{\pi_i}$. A ***padded representation*** of size $s \ge n$ of a sequence $x_1, \ldots, x_n$ is a vector of size s whose non-*nil* elements, taken in order, precisely form the sequence $x_1, \ldots, x_n$. Here *nil* is a special value that cannot occur as an element of a sequence. A padded representation is hence the same as a usual representation, except that we allow intervening "detectable garbage". In the language of (Hagerup, 1990), we do not insist that the representation be compact.

Lemma 5.2 (Hagerup, 1991a): There is a constant $\epsilon > 0$ such that the following problem can be solved on a COLLISION$^+$ PRAM using $O((\log^* n)^2)$ time, $O(n)$ operations and $O(n)$ space with probability at least $1 - 2^{-n^\epsilon}$: Given a sequence $x_1, \ldots, x_n$ of n integers in the range $1 \,.\,.\, n$, compute a padded representation of size $O(n)$ of a duplicate-grouping permutation for $x_1, \ldots, x_n$.

Definition: The ***all nearest zero bit*** problem of size n is, given a bit vector A of size n, to mark each position in A with the position of the nearest zero in A to its left, if any.

Lemma 5.3 (Berkman and Vishkin, 1989; Ragde, 1990): All nearest zero bit problems of size n can be solved on a COMMON or TOLERANT PRAM using $O(\log^* n)$ time, $O(n)$ operations and $O(n)$ space.

Lemma 5.4 (Ragde, 1990, Theorem 1): For all $d \ge 0$, compaction problems of size n and with parameters $(d, d^4 \log n)$ can be solved on a (deterministic) COMMON PRAM using constant time, n processors and $O(n)$ space.

Lemma 5.5 (Shiloach and Vishkin, 1981, Corollary 2): For every fixed $\epsilon > 0$, the minimum of n elements can be computed on a COMMON or TOLERANT PRAM using constant time, $O(n^{1+\epsilon})$ processors and $O(n)$ space.

Theorem 5.6: For every fixed $\epsilon > 0$, the minimum of n elements can be computed on a COMMON or TOLERANT PRAM using constant time, n processors and $O(n)$ space with probability at least $1 - 2^{-n^{1/4-\epsilon}}$.

Proof (inspired by (Reischuk, 1985, Theorem 1)): For the sake of simplicity, we give only the proof for the COMMON PRAM (this is all that we need in the sequel). Without loss of generality assume that the input consists of n pairwise distinct elements $x_1, \ldots, x_n$ and that $\epsilon \le 1/4$. Let $M_1 = \{1, \ldots, n\}$ and execute $T = \lceil 2 + 3/\epsilon \rceil$ ***stages***. Stage t, for $t = 1, \ldots, T$, consists of the following:
Step 1: Draw a random sample Y_t from M_t with associated probability $p_t = 1/\max\{\lfloor n^{3/4-(t-2)\epsilon/4} \rfloor, 1\}$.
Step 2: Compute $m_t = \min_{j \in Y_t} x_j$ and let $M_{t+1} = \{j \in M_t : x_j \le m_t\}$.
Since $p_T = 1$, m_T is the desired result, and all parts of the algorithm other than the computation of $m_1, \ldots, m_T$ can be executed in constant time. For $t = 1, \ldots, T$, call Stage t ***good*** if $|Y_t| \le 12(n^{1/4-\epsilon/4} + 1)$ and $|M_{t+1}| \le \lceil n^{1-t\epsilon/4} \rceil$. If Stage t is good and n is sufficiently large, then $|Y_t|^4 \log n \le n^{1-\epsilon/2}$. In this case the elements of Y_t can be placed in an array of size at most $n^{1-\epsilon/2}$ using Lemma 5.4, and their minimum can be computed in constant time using Lemma 5.5. What remains is to bound the probability that some stage is not good.

For $t = 1, \ldots, T-1$, $|M_{t+1}| > \lceil n^{1-t\epsilon/4} \rceil$ only if none of the indices of the $\lceil n^{1-t\epsilon/4} \rceil$ smallest elements among $x_1, \ldots, x_n$ are included in $Y_1 \cup \cdots \cup Y_t$, the probability of which is at most

$$(1 - p_t)^{n^{1-t\epsilon/4}} \le e^{-p_t n^{1-t\epsilon/4}} \le 2^{-n^{(t-2)\epsilon/4-3/4+1-t\epsilon/4}} = 2^{-n^{1/4-\epsilon/2}}.$$

Furthermore, for fixed M_t with $|M_t| \leq \lceil n^{1-(t-1)\epsilon/4} \rceil$, $|Y_t|$ is binomially distributed with expected value

$$E(|Y_t|) \leq p_t n^{1-(t-1)\epsilon/4} + 1 \leq 2n^{1/4-\epsilon/4} + 1,$$

and Chernoff bound (c) implies that $|Y_t| > 12(n^{1/4-\epsilon/4} + 1)$ with probability at most $2^{-12n^{1/4-\epsilon/4}}$. Hence the probability that some stage is not good is at most $2T \cdot 2^{-n^{1/4-\epsilon/2}}$, which for sufficiently large values of n is bounded by $2^{-n^{1/4-\epsilon}}$. ∎

We are now ready for the main result of this section.

Theorem 5.7: There is a constant $\epsilon > 0$ such that one step of a (randomized) n-processor MINIMUM PRAM with m global memory cells can be simulated by $O((\log^* n)^2)$ steps of a randomized COLLISION$^+$ PRAM with $O(n/(\log^* n)^2)$ processors and $O(n+m)$ global memory cells with probability at least $1-2^{-n^\epsilon}$.

Proof: Recall that the input is a sequence $(\xi_1, \theta_1), \ldots, (\xi_n, \theta_n)$, where (ξ_j, θ_j) means that the jth processor attempts to write the value θ_j to the global memory cell numbered ξ_j. Execute the following algorithm:

Step 1: Use Lemma 5.1 to compute integers $y_1, \ldots, y_n$ in the range $1..s$, for some $s = O(n)$, such that for all $i, j \in \{1, \ldots, n\}$, we have $y_i = y_j \Leftrightarrow \xi_i = \xi_j$.

Step 2: Use Theorem 5.2 to compute a padded representation of size $O(n)$ of a permutation $\pi_1, \ldots, \pi_n$ of $1, \ldots, n$ that is duplicate-grouping for $y_1, \ldots, y_n$ and, by implication, for $\xi_1, \ldots, \xi_n$.

Step 3: Compute a padded representation A of the sequence $(\xi_{\pi_1}, \theta_{\pi_1}), \ldots, (\xi_{\pi_n}, \theta_{\pi_n})$. This is a trivial data routing based on the outcome of Step 2.

After Step 3, the write requests pertaining to a given cell, i.e., the pairs (ξ_j, θ_j) with a given first component, occur together in A in a contiguous segment, separated only by *nil* entries. More precisely, for each $\xi \in \Xi$, define the ***segment*** of ξ as the set $\{i, \ldots, j\}$, where i and j are the first and the last position in A containing an element with first component ξ, respectively. Using Lemma 5.3, each position in A can be marked with the segment to which it belongs, if any. We now appeal to Theorem 3.1, whereby the collection $\mathcal{P}$ contains one problem for each $\xi \in \Xi$. The problem associated with $\xi \in \Xi$ is to compute the minimum second component occurring in the segment of ξ, with *nil* entries acting as neutral elements (i.e., as if they had second components of ∞), and its index set is precisely the segment of ξ. The algorithm $\mathcal{A}$ that solves each problem in constant expected time is furnished by Theorem 5.6. Note that since the index set of each problem is a set of consecutive integers, the distinction between the allocated PRAM "provided" by Theorem 3.1 and the standard PRAM required by Theorem 5.6 can be ignored in the present case, which is precisely the reason for having Steps 1–3. Again problems with index sets of size $\geq n^{1/3}$ require special attention. However, since such problems are solved in the first try with overwhelming probability, according to Theorem 5.6, they can be removed in a preprocessing step (this is not actually necessary).

The space needed is $O(n+m)$ plus what is needed by the applications of Theorem 5.6 in the algorithm of Theorem 3.1. Since the time and number of operations required by these applications are $O(\log^* n)$ and $O(n)$, respectively, Lemma 5.1 implies that they can be executed in $O(n)$ space. ∎

6 Additional simulation results

A straightforward combination of Theorems 4.4 and 5.7 yields an optimal simulation of MINIMUM on TOLERANT with slowdown $O((\log^* n)^4)$. A closer look, however, reveals that the only parts of the algorithm of Theorem 5.7 that need $\omega(\log^* n)$ time are compactions as in Lemma 2.3 and related operations. Since these can be executed directly on TOLERANT, we have

Theorem 6.1: There is a constant $\epsilon > 0$ such that one step of a (randomized) n-processor MINIMUM PRAM can be simulated by $O((\log^* n)^3)$ steps of a randomized TOLERANT PRAM with $O(n/(\log^* n)^3)$ processors with probability at least $1 - 2^{-n^\epsilon}$.

Having created segments in Step 3 of the algorithm of Theorem 5.7, we chose to determine the minimum second component present within each segment. We could just as well execute some other computation locally within each segment. For instance, if we choose instead to carry out a prefix summation of the second components, with *nil* entries acting as zero, the result is an optimal simulation of FETCH-AND-ADD on COLLISION$^+$ whose slowdown is $O(\log n)$ with high probability, improving the result of (Matias and Vishkin, 1990a, 1990b). If the integers appearing in memory cells are of size polynomial in n (a frequent special case), the prefix sums can be computed optimally in $O(\log n/\log\log n)$ time (Cole and Vishkin, 1989), yielding a corresponding optimal simulation of FETCH-AND-ADD on COLLISION$^+$ with slowdown $O(\log n/\log\log n)$. As a final example, let RANDOM be a write conflict resolution rule that stipulates that the successful processor in each write conflict is drawn at random from the uniform distribution over the set of writing processors. The corresponding RANDOM PRAM was previously considered, e.g., by Martel and Gusfield (1989) and by Gil (1990). Using Lemma 2.4, we obtain an optimal simulation of RANDOM on COLLISION$^+$ with slowdown $O((\log^* n)^2)$.

References

BAST, H., AND HAGERUP, T. (1991), Fast and Reliable Parallel Hashing, manuscript. A preliminary version appears in Proc. 3rd Annual ACM Symposium on Parallel Algorithms and Architectures, pp. 50–61.

BERKMAN, O., AND VISHKIN, U. (1989), Recursive *-Tree Parallel Data-Structure, *in* Proc. 30th Annual Symposium on Foundations of Computer Science, pp. 196–202.

BHATT, P. C. P., DIKS, K., HAGERUP, T., PRASAD, V. C., RADZIK, T., AND SAXENA, S. (1991), Improved Deterministic Parallel Integer Sorting, *Inform. and Comput.* **94**, pp. 29–47.

BOPPANA, R. B. (1989), Optimal Separations between Concurrent-Write Parallel Machines, *in* Proc. 21st Annual ACM Symposium on Theory of Computing, pp. 320–326.

CHLEBUS, B. S., DIKS, K., HAGERUP, T., AND RADZIK, T. (1988), Efficient Simulations between Concurrent-Read Concurrent-Write PRAM Models, *in* Proc. 13th Symposium on Mathematical Foundations of Computer Science, Springer Lecture Notes in Computer Science, Vol. 324, pp. 231–239.

CHLEBUS, B. S., DIKS, K., HAGERUP, T., AND RADZIK, T. (1989), New Simulations between CRCW PRAMs, *in* Proc. 7th International Conference on Fundamentals of Computation Theory, Springer Lecture Notes in Computer Science, Vol. 380, pp. 95–104.

COLE, R., AND VISHKIN, U. (1986), Deterministic Coin Tossing and Accelerating Cascades: Micro and Macro Techniques for Designing Parallel Algorithms, *in* Proc. 18th Annual ACM Symposium on Theory of Computing, pp. 206–219.

COLE, R., AND VISHKIN, U. (1989), Faster Optimal Parallel Prefix Sums and List Ranking, *Inform. and Comput.* **81**, pp. 334–352.

FICH, F. E., RAGDE, P., AND WIGDERSON, A. (1988a), Simulations Among Concurrent-Write PRAMs, *Algorithmica* **3**, pp. 43–51.

FICH, F. E., RAGDE, P., AND WIGDERSON, A. (1988b), Relations Between Concurrent-Write Models of Parallel Computation, *SIAM J. Comput.* **17**, pp. 606–627.

GIL, J. (1990), Lower Bounds and Algorithms for Hashing and Parallel Processing, Ph. D. Thesis, The Hebrew University, Jerusalem.

GIL, J., MATIAS, Y., AND VISHKIN, U. (1991), Towards a Theory of Nearly Constant Time Parallel Algorithms, *in* Proc. 32nd Annual Symposium on Foundations of Computer Science, pp. 698–710.

GOLDSCHLAGER, L. M. (1982), A Universal Interconnection Pattern for Parallel Computers, *J. ACM* **29**, pp. 1073–1086.

GOTTLIEB, A., GRISHMAN, R., KRUSKAL, C. P., MCAULIFFE, K. P., RUDOLPH, L., AND SNIR, M. (1983), The NYU Ultracomputer — Designing an MIMD Shared Memory Parallel Computer, *IEEE Trans. Comp.* **32**, pp. 175–189.

GROLMUSZ, V., AND RAGDE, P. (1987), Incomparability in Parallel Computation, *in* Proc. 28th Annual Symposium on Foundations of Computer Science, pp. 89–98.

HAGERUP, T. (1990), Optimal Parallel Algorithms on Planar Graphs, *Inform. and Comput.* **84**, pp. 71–96.

HAGERUP, T. (1991a), Fast Parallel Space Allocation, Estimation and Integer Sorting, Tech. Rep. no. MPI-I-91-106, Max-Planck-Institut für Informatik, Saarbrücken.

HAGERUP, T. (1991b), Fast Parallel Generation of Random Permutations, *in* Proc. 18th International Colloquium on Automata, Languages and Programming, Springer Lecture Notes in Computer Science, Vol. 510, pp. 405–416.

HAGERUP, T. (1992), The Log-Star Revolution, these proceedings.

HAGERUP, T., AND RADZIK, T. (1990), Every Robust CRCW PRAM Can Efficiently Simulate a PRIORITY PRAM, *in* Proc. 2nd Annual ACM Symposium on Parallel Algorithms and Architectures, pp. 117–124.

HAGERUP, T., AND RÜB, C. (1990), A Guided Tour of Chernoff Bounds, *Inform. Proc. Lett.* **33**, pp. 305–308.

KUČERA, L. (1982), Parallel Computation and Conflicts in Memory Access, *Inform. Proc. Lett.* **14**, pp. 93–96.

KUČERA, L. (1983), Erratum and Addendum to: Parallel Computation and Conflicts in Memory Access, *Inform. Proc. Lett.* **17**, p. 107.

MARTEL, C. U., AND GUSFIELD, D. (1989), A Fast Parallel Quicksort Algorithm, *Inform. Proc. Lett.* **30**, pp. 97–102.

MATIAS, Y., AND VISHKIN, U. (1990a), On Parallel Hashing and Integer Sorting, *in* Proc. 17th International Colloquium on Automata, Languages and Programming, Springer Lecture Notes in Computer Science, Vol. 443, pp. 729–743.

MATIAS, Y., AND VISHKIN, U. (1990b), On Parallel Hashing and Integer Sorting, Tech. Rep. no. UMIACS-TR-90-13.1 (revised version), University of Maryland, College Park.

MATIAS, Y., AND VISHKIN, U. (1991), Converting High Probability into Nearly-Constant Time — with Applications to Parallel Hashing, *in* Proc. 23rd Annual ACM Symposium on Theory of Computing, pp. 307–316.

MCDIARMID, C. (1989), On the Method of Bounded Differences, *in* Surveys in Combinatorics, 1989, ed. J. Siemons, London Math. Soc. Lecture Note Series 141, Cambridge University Press, pp. 148–188.

RAGDE, P. (1989), Processor-Time Tradeoffs in PRAM Simulations, *J. Comp. Sys. Sci.*, to appear.

RAGDE, P. (1990), The Parallel Simplicity of Compaction and Chaining, *in* Proc. 17th International Colloquium on Automata, Languages and Programming, Springer Lecture Notes in Computer Science, Vol. 443, pp. 744–751.

REISCHUK, R. (1985), Probabilistic Parallel Algorithms for Sorting and Selection, *SIAM J. Comput.* **14**, pp. 396–409.

SHILOACH, Y., AND VISHKIN, U. (1981), Finding the Maximum, Merging, and Sorting in a Parallel Computation Model, *J. Alg.* **2**, pp. 88–102.

SHILOACH, Y., AND VISHKIN, U. (1982), An $O(\log n)$ Parallel Connectivity Algorithm, *J. Alg.* **3**, pp. 57–67.

LOGIC AND SEMANTICS

Suitability of the Propositional Temporal Logic to Express Properties of Real-Time Systems

Eric Nassor[1,2]
nassor@aar.alcatel-alsthom.fr
[1]L R I
U.R.A. CNRS 410, 91405 Orsay
France

Guy Vidal-Naquet[2]
vidalnaq@aar.alcatel-alsthom.fr
[2]Alcatel Alsthom Recherche
Route de Nozay, 91460 Marcoussis
France

Abstract

We claim in this paper that the Propositional Temporal Logic (PTL) is an adequate logic to specify a reactive system. To prove this affirmation, we present abbreviations which allow the expression of the most important real-time properties and some other important requirements. We show that they are easy to use and that the induced complexity cost is the same as the complexity of the other real-time logics. Finally, we prove that many real-time logics already published have exactly the same expressive power as PTL.

1 Introduction

We claim in this paper that the Propositional Temporal Logic (PTL) is an adequate type of logic for expressing the properties of real-time systems. Although in the last few years classical temporal logic has been used successfully to specify qualitative properties of reactive systems, it is judged generally too restrictive for the analysis of real-time systems, whereas these systems are often critical and crucially need a formal approach. Indeed in these systems, the qualitative requirements are no longer sufficient, they are complemented by quantitative requirements. For example the responsiveness property (an environment stimulus p must be followed by a system reaction q) is replaced by a timed responsiveness property which imposes a bound on time interval between p and q. Many authors have proposed new logics to express these quantitative requirements. These real-time temporal logics use a different model from the Temporal Logic: the timed state sequences [7, 13, 11, 1, 8]. Although the suitability of real-time temporal logics as specification languages has often been shown, most of these previous attempts remain ad hoc with little regard to complexity and expressiveness questions. We show in this paper that these logics have no more expressive power than the classical PTL which, with our abbreviations, can easily express the quantitative requirements.

In Section 2 we recall the definition of the propositional temporal logic. In Section 3 we present abbreviations added to the classical temporal logic which allow us to express quantitative requirements and therefore show that the temporal logic needs no extension to express real-time properties. In Section 4 we show the usage of our abbreviations with some examples: these abbreviations facilitate the expression of the most useful real-time properties. Section 5 gives the complexity of the satisfiability question and of the model-checking problem for our new operators: it remains in Expspace, which is the best that has been achieved for a real-time logic. Section 6 compares our results with some previously proposed real-time temporal logics: many have been shown to be undecidable [2]. We prove that both real-time logics which are decidable, MTL (the Metric Temporal Logic [8, 2]) and TPTL (the Timed Propositional Temporal Logic [1]), are not extensions of PTL, they have exactly the same expressive power. In Section 7 we show some limitations of PTL for real-time systems and some possible extensions, in particular XCTL (the eXplicit Clock Temporal Logic [5]) is shown to be an extension of PTL and TPTL contrary to the claim of its authors. Finally we conclude in Section 8.

2 The Propositional Temporal Logic

We recall here the definition of Propositional Linear Temporal Logic (PTL or L(F,X,U) in the notation of [4]) as given for example in [15].

Syntax.

PTL formulas are built from

1. a set $\mathcal{P}$ of atomic propositions: $p_1, p_2 \ldots$
2. boolean connectives: $\wedge$ and $\neg$
3. temporal operators: $\bigcirc$(next) and U(until).

The formation rules of the formulas are:

1. an atomic proposition $p \in \mathcal{P}$ is a formula
2. if f_1 and f_2 are formulas, so are: $f_1 \wedge f_2, \neg f_1, \bigcirc f_1, f_1 \mathrm{U} f_2$

We also use the following abbreviations:

1. $f_1 \vee f_2 \equiv \neg(\neg f_1 \wedge \neg f_2)$
2. $\Diamond f_1 \equiv True \mathrm{U} f_1$ (sometime)
3. $\Box f_1 \equiv \neg \Diamond \neg f_1$ (always)

Semantics

A structure for a PTL formula is a state sequence, this is a set $\sigma = (\mathcal{P}, S, \pi, N, s_0)$ where

- $\mathcal{P}$ is the set of the atomic propositions
- S is a finite or enumerable set of states
- $\pi : (S \rightarrow 2^{\mathcal{P}})$ assigns truth values to the atomic propositions of the language in each state.
- $N : (S \rightarrow S)$ is a total successor function which for each state gives a unique next state
- s_0 : is a distinguished element of S, this is the initial state of the state sequence.

In the following definitions $N^i(s)$ denote the ith successor of the state s.
For a structure σ and a state $s \in S$ we have:

$$\begin{array}{rll} \sigma, s \models p & \text{iff} & p \in \pi(s) \\ \sigma, s \models f_1 \wedge f_2 & \text{iff} & \sigma, s \models f_1 \text{ and } \sigma, s \models f_2 \\ \sigma, s \models \neg f & \text{iff} & \text{not } \sigma, s \models f \\ \sigma, s \models \bigcirc f & \text{iff} & \sigma, N(s) \models f \\ \sigma, s \models f_1 \mathrm{U} f_2 & \text{iff} & \exists i \geq 0, \left\{ \begin{array}{l} \sigma, N^i(s) \models f_2 \\ \forall j (0 \leq j < i \text{ implies } \sigma, N^j(s) \models f_1) \end{array} \right. \end{array}$$

From these definitions, the meaning of the abbreviations can be deduced:

$$\begin{array}{rll} \sigma, s \models f_1 \vee f_2 & \text{iff} & \sigma, s \models f_1 \text{ or } \sigma, s \models f_2 \\ \sigma, s \models \Diamond f & \text{iff} & \exists i \geq 0, \sigma, N^i(s) \models f \\ \sigma, s \models \Box f & \text{iff} & \forall i \geq 0, \sigma, N^i(s) \models f \end{array}$$

Some authors have argued that it is easy to see that this definition of the temporal logic says nothing about finite interval of time and only deals with certain states (Next) or with semi-infinite interval of time, and therefore this shows that real-time properties are not expressible. We show in the following section that this feeling is false: based on the Until operator we give abbreviations which permit an easy characterization of a finite interval of time.

3 The new abbreviations

Definition 1

We define the family of operators B_k for all $k \in \mathbb{N}^*$ to be the following abbreviations:

$$\begin{array}{rcl} f_1 \mathrm{B}_1 f_2 & \equiv & \neg(\neg f_1 \mathrm{U} f_2) \\ \forall k > 1, f_1 \mathrm{B}_k f_2 & \equiv & (f_1 \mathrm{B}_1 f_2) \vee [(\neg f_1 \wedge \neg f_2) \mathrm{U} (f_2 \wedge (f_1 \vee \bigcirc (f_1 \mathrm{B}_{k-1} f_2)))] \end{array}$$

This operator has the following informal meaning: $f_1 \mathrm{B}_k f_2$ means that f_1 must be true before the kth state where f_2 is true, if this k^{th} state does not exist, the formula is true. f_2 behaves like a clock: its truth occurrence marks the sequencing of events. This approach has the advantage that any atomic proposition (and any formula) can describe a clock. This is exactly what is claimed by authors of synchronous languages [12].

The formal meaning is given by the following theorem:

Theorem 1

$$\sigma, s \models f_1 B_k f_2 \textit{ iff } \forall i_1, \ldots, i_k$$
$$\left\{ \begin{array}{l} 0 \leq i_1 < \cdots < i_k \\ \sigma, N^{i_1}(s) \models f_2, \ldots, \sigma, N^{i_k}(s) \models f_2 \end{array} \right. \quad \textit{implies } \exists j, 0 \leq j < i_k, \sigma, N^j(s) \models f_1$$

Definition 2

In the same way, we define a second family of operators ${}_k\mathrm{B}$ for all $k \in \mathbb{N}^*$:

$$\forall k > 0, f_1\ {}_k\mathrm{B} f_2 \equiv \neg(f_2 \mathrm{B}_k f_1) \vee \Box \neg f_2$$

This operator has the following informal meaning: the formula $f_1\ {}_k\mathrm{B}\ f_2$ means that f_1 is true k times before f_2 is true or f_2 is never true. And the formal meaning is given by the following theorem:

Theorem 2

$$\sigma, s \models f_1\ {}_kB\ f_2 \textit{ iff}$$

$$\forall i \geq 0,\quad \sigma, N^i(s) \models f_2 \textit{ implies } \exists j_1, \ldots, j_k \begin{cases} 0 \leq j_1 < \cdots < j_k \leq i \\ \sigma, N^{j_1}(s) \models f_1, \ldots, \sigma, N^{j_k}(s) \models f_1 \end{cases}$$

Definition 3

Next we define a third family of operators: At_k for all $k \in \mathbb{N}^*$:

$$\begin{aligned} f_1 \mathrm{At}_1 f_2 &\equiv \neg[f_2 \mathrm{B}_1 (f_2 \wedge f_1)] \\ \forall k > 1, f_1 \mathrm{At}_k f_2 &\equiv \neg[f_2 \mathrm{B}_1 (f_2 \wedge \bigcirc (f_1 \mathrm{At}_{k-1} f_2))] \end{aligned}$$

And $f_1 \mathrm{At}_k f_2$ means that at the kth state where f_2 is true, f_1 is true. This is exactly what is given in the formal meaning.

Theorem 3

$$\sigma, s \models f_1 At_k f_2 \textit{ iff}$$

$$\exists j_1, \ldots, j_k, \begin{cases} 0 \leq j_1 < \cdots < j_k \\ \sigma, N^{j_1}(s) \models f_2, \ldots, \sigma, N^{j_k}(s) \models f_2 \\ \forall j < j_k, j \neq j_1, \ldots, j \neq j_k \textit{ implies } \sigma, N^j(s) \not\models f_2 \\ \sigma, N^{j_k}(s) \models f_1 \end{cases}$$

The following section shows how these operators can be used to specify in an easy way the most important real-time properties. They will be used next to prove that some real-time logics have the same expressive power than PTL.

4 Usage

A real-time logic

We now show that the three families of operators, which have been defined in the previous section, are sufficient to express most of the real-time properties, and therefore PTL is expressive enough to specify a reactive system. Two important classes of real-time properties are the *bounded invariance* and the *bounded response* properties [6]. They have the following meaning:

- a *bounded response* property asserts that something will happen within a certain amount of time. A typical application is to state an upper bound on the termination of a computation: if p is the first state of the computation and q the last state of the computation we say that the computation must end before a certain number u of events e (the tick of a clock) have arrived with the following formula:

$$\Box(p \rightarrow q\mathrm{B}_u e)$$

- a *bounded invariance* property asserts that something will hold continuously for a given amount of time. A typical application is to state a lower bound on a system computation: the computation (which starts in state p and ends in state q) cannot end before a certain number l of events e (the tick of the clock) have arrived.

$$\Box(p \rightarrow e_l\mathrm{B}q)$$

These two properties may be combined in order to assert that something must happen during an interval of time. A typical application is to state an upper bound and a lower bound for the termination of a computation: the computation must end after a certain number l and before a certain number u of events e.

$$\Box(p \rightarrow ((q\mathrm{B}_{u-l}e)\mathrm{At}_l e))$$

A multiform time

Our new abbreviations allow the expression of real-time properties as easily as the real-time logics, but it also allows the expression of other properties which are classically difficult to express.

A specification stating that at the third time the button (b) is pressed, the system must respond in less than two seconds (s) can be expressed by:

$$\Box(p \rightarrow ((q\mathrm{B}_2 s)\mathrm{At}_3 b))$$

This property explicitly shows that different clocks may be used in the formulas, and we claim (as already described by authors of synchronous languages [12]) that at a suitable level of abstraction the only notion which is necessary for a real-time system is the notion of a reaction to an external event. Real-time can be considered as an external event like the others, without any privileged role, and all events can be viewed as defining a time. It encourages a more abstract point of view in the specification of a system.

For instance in a railway regulation system, there are events which come from beacons along the track, from wheel revolutions, from a timer. One can consider all these events in a uniform way, and therefore time can be counted in wheel revolutions as well as in milliseconds.

Example

Let us take the classical example of the mutual exclusion [10]. Consider a program consisting of two processes P_1 and P_2 which need to coordinate their entries to critical sections in their code. The program for each process P_i is partitioned into three sections N_i, T_i and C_i. N_i represents the non-critical activity of the process, T_i represents the *trying* section where the process engages a protocol to access the critical section, and C_i represents the critical section itself.

The basic requirement of the mutual exclusion algorithm is that P_1 and P_2 never access simultaneously their critical section. This requirement is expressed by the following formula:

$$\Box\neg(in\_C_1 \wedge in\_C_2)$$

where in_C_i is a control predicate expressing the fact that P_i is currently executing the section C_i.

But such a specification can easily be implemented by a program that does not allow any of the processes to ever access its critical section. Therefore we must add to the specifications the requirements that each process which try to enter its critical section will eventually succeed. This property can be specified by the following formula:

$$\Box(in\_T_i \rightarrow \Diamond in\_C_i)$$

The previous formulas are not sufficient for a real-time system. A typical property of such a system states that the process P_i will enter its critical section in a bounded amount of time after its request. The following formula express that P_i will enter its critical section at worst 10 milliseconds after its request:

$$\Box(in\_T_i \rightarrow in\_C_i \ \mathrm{B}_{10} \ ms)$$

Such a bounded response property can not be true generally if we have not a fairness property stating that a process can not enter its critical section many times if the other process is waiting. Let us call $entry\_in\_C_i$ the predicate expressing that P_i is going in its critical section:

$$entry\_in\_C_i \equiv in\_T_i \wedge \bigcirc in\_C_i$$

The following formula express that P_2 will not enter two times its critical section during the waiting of P_1:

$$\Box(in\_T_1 \rightarrow in\_C_1 \ \mathrm{B}_2 \ entry\_in\_C_2)$$

Another useful property states that there is a maximal rate for the requests. The following formula expresses that a process P_i in its critical section will not try to reenter its critical section for at least 2 milliseconds:

$$\Box(in\_C_i \rightarrow ms \ {}_2\mathrm{B} \ in\_T_i)$$

We have seen that the abbreviations allow to express in an easy way complex real-time properties. But this facility has a cost in complexity. This is shown in the following section.

5 Complexity

The three families of operators which have been introduced are only abbreviations, therefore the expressibility and the decidability of the logic are kept but the complexity is not the same as in PTL.

Theorem 4 *The satisfiability and the model-checking problem are EXPSPACE.*

Each operator that we have added can be expanded in pure PTL logic. The expansion of $f_1 \mathrm{B}_k f_2$ (or $f_1 \ {}_k\mathrm{B} f_2$ or $f_1 \mathrm{At}_k f_2$) where f_1 and f_2 are pure PTL formula, will take ck characters to be written where c is a constant. Therefore the expansion of a formula ϕ take a length in $O(N.K)$, where N is the number of connectives, quantifiers, and grammar operators in ϕ, and K is the product of all constants (the subscripts) occurring in ϕ. The length L of the formula ϕ is $O(N + \log K)$ because of the binary encoding of the constants.

For instance the previous formula stating that a process P_i will not try to reenter its critical section for at least 2 milliseconds:

$$\Box(in\_C_i \rightarrow ms\ {}_2\mathrm{B}\ in\_T_i)$$

can be expanded in pure PTL logic in the following way:

$$\Box(in\_C_i \rightarrow \neg(\neg(\neg in\_T_i \mathrm{U} ms) \vee [(\neg in\_T_i \wedge \neg ms) \mathrm{U} (ms \wedge (in\_T_i \vee \bigcirc \neg(\neg in\_T_i \mathrm{U} ms)))]) \vee \Box \neg in\_T_i)$$

The determination of the satisfiability of a PTL formula (that means to determine whether there is a model such that the formula is true) is a Pspace problem and an algorithm exists which takes a time exponential in the length of the formula. Hence, the satisfiability problem for our new operators is in Expspace, and an algorithm exists which takes an exponential time in $O(N.K)$: that means a doubly exponential time in the length of the timing constants, and an exponential time in the number of connectives, quantifiers, and grammar operators in the formula.

In the same way, the model-checking problem of a PTL formula and a state sequence (that means to determine whether the state sequence is a model for the formula) is a Pspace problem, and an algorithm exists which takes an exponential time in the length of the formula and a linear time in the length of the state sequence. Hence, the model-checking problem for our new operators is in Expspace, and an algorithm exists which takes an exponential time in $O(N.K)$ and a linear time in the length of the state sequence.

This result is not surprising because it was proved in [2] that PTL with only one abbreviation like $\bigcirc_k$ is Expspace, because of the binary encoding of the constants. This result explains why for all the real-time logics which have been proposed, the model-checking problem takes at least double exponential time: this is the case for MTL, TPTL, XCTL (for this last one, the satisfiability problem is undecidable). In the following section we give some more explanations on the relationships between PTL and the other real-time logics.

6 Comparison with real-time logics

First, we recall the model of the real-time logics: the timed state sequences, from which we build a 1-to-n relation to the classical state sequences. Then we recall the syntax and semantic of the Metric Temporal Logic (MTL) and we show that MTL and PTL have the same expressive power. We end the section by comparing PTL with the Timed Propositional Temporal Logic (TPTL).

6.1 The timed state sequences

To obtain a notion of time in temporal logics many authors have added a time in the model of state sequences. A timed state sequence is a set $\sigma_t = (\mathcal{P}, S, \pi, N, s_0, T)$ where:

- $\mathcal{P}$ is the set of the atomic propositions
- S is a finite or enumerable set of states.
- $\pi : (S \rightarrow 2^{\mathcal{P}})$ assigns truth values to the atomic propositions of the language in each state.

- $N : (S \rightarrow S)$ is a total successor function which for each state give a unique next state.
- s_0 : is a distinguished element of S: the initial state of the sequence.
- $T : (S \rightarrow \mathbb{N})$ assigns a time value to each state such that $T(s) \leq T(N(s))$.

We call Σ_T the set of the timed state sequences. This model of timed state sequences is used by many authors of real-time logics, for instance it is used in [2, 5] to define the MTL, TPTL, and XCTL logics.

We now present a relation between timed state sequences and classical state sequences. First we need to add a proposition (called *tick*) to the set $\mathcal{P}$ of all the propositions and we call Σ the set of all the state sequences on $\mathcal{P} \cup \{tick\}$ such that *tick* is infinitely often false. The proposition *tick* marks the tick of the clock and therefore the time: the time of a state is the number of states where *tick* is true from the first state of the sequence. Therefore a timed state sequence, is only a state sequence where we are interested by some states (states where *tick* is false) and the number of ticks from the first state.

Let us call $\mathcal{R}$ the following relation between the timed state sequences and the state sequences:

$$\mathcal{R} \subset \Sigma_T \times \Sigma / (\sigma_t, \sigma) \in \mathcal{R} \Leftrightarrow \begin{cases} \sigma = (\mathcal{P} \cup \{tick\}, S, \pi, N, s_0) \\ \sigma_t = (\mathcal{P}, S_t, \pi_t, N_t, s_t, T) \\ S_t = \{s \in S / tick \notin \pi(s)\} \\ \forall s \in S_t, \pi_t(s) = \pi(s) \\ \forall s \in S_t, N_t(s) = N^i(s) \text{ with } i = min\{j / tick \notin \pi(N^j(s))\} \\ s_t = N^i(s_0) \text{ with } i = min\{j / tick \notin \pi(N^j(s_0))\} \\ \forall s \in S_t, s = N^i(s_0), T(s) = card\{j / j < i, tick \in \pi(N^j(s_0))\} \end{cases}$$

$\mathcal{R}$ is a 1-to-n relation: that means that each state sequence (such that *tick* is infinitely often false) is in relation with a unique timed state sequence, and each timed state sequence is in relation with several state sequences. This relation allows to compare the expressive power of the different logics, because with this relation a formula of any logic characterizes a subset of the state sequences set.

6.2 The Metric Temporal Logic MTL

Several authors have tried to adapt temporal logic to reason about real-time properties by interpreting its modalities as bounded operators. Such constant bounds can be found for example in [8, 2] with the name of Metric Temporal Logic (MTL) and in [13] with the name of Quantized Temporal Logic (QTL). For example the typical bounded response time property "Every p-state is followed by a q-state before 5 time-units" will be written as:

$$\Box(p \rightarrow \Diamond_{<5} q)$$

Syntax

MTL formulas are built from

1. a set $\mathcal{P}$ of atomic propositions: $p_1, p_2 \ldots$
2. boolean connectives: $\wedge$ and $\neg$
3. temporal operators: $\bigcirc_{\sim c}$ and $\mathrm{U}_{\sim c}$ where $\sim$ means $<, >$ or $=$ and c is a positive integer.

The formations rules of the formulas are:

1. an atomic proposition $p \in \mathcal{P}$ is a formula
2. if f_1 and f_2 are formulas, so are: $f_1 \wedge f_2, \neg f_1, \bigcirc_{\sim c} f_1, f_1 \mathrm{U}_{\sim c} f_2$

We also use the following abbreviations:

1. $f_1 \vee f_2 \equiv \neg(\neg f_1 \wedge \neg f_2)$
2. $\Diamond_{\sim c} f_1 \equiv True \mathrm{U}_{\sim c} f_1$
3. $\Box_{\sim c} f_1 \equiv \neg \Diamond_{\sim c} \neg f_1$

Semantics

For a timed state sequence σ_t and a state $s \in S_t$ we have:

$$
\begin{array}{rcl}
\sigma_t, s \models_{MTL} p & \text{iff} & p \in \pi(s) \\
\sigma_t, s \models_{MTL} f_1 \wedge f_2 & \text{iff} & \sigma_t, s \models_{MTL} f_1 \text{ and } \sigma_t, s \models_{MTL} f_2 \\
\sigma_t, s \models_{MTL} \neg f & \text{iff} & \text{not } \sigma_t, s \models_{MTL} f \\
\sigma_t, s \models_{MTL} \bigcirc_{\sim c} f & \text{iff} & \sigma_t, N(s) \models_{MTL} f \text{ and } T(N(s)) \sim T(s) + c \\
\sigma_t, s \models_{MTL} f_1 \mathrm{U}_{\sim c} f_2 & \text{iff} & \exists i \geq 0, \left\{ \begin{array}{l} \sigma_t, N^i(s) \models_{MTL} f_2 \\ T(N^i(s)) \sim T(s) + c \\ \forall j (0 \leq j < i \text{ implies } \sigma_t, N^j(s) \models_{MTL} f_1) \end{array} \right.
\end{array}
$$

Expressiveness

Theorem 5 *MTL and PTL have the same expressiveness.*

We show that each formula of MTL, characterizes a set of timed states sequences and therefore a set of state sequences (through the $\mathcal{R}$ relation). This set is characterized by a PTL formula, which is obtained from the initial MTL formula by a translation of the MTL operators. We denote $\mathcal{T}$ this translation which is defined in the following way:

1. $\mathcal{T}(f \mathrm{U}_{<c} g) = [(\mathcal{T}(f) \vee tick) \mathrm{U} (\mathcal{T}(g) \wedge \neg tick)] \wedge [(\mathcal{T}(g) \wedge \neg tick) \mathrm{B}_c \, tick]$
2. $\mathcal{T}(f \mathrm{U}_{=c} g) = [tick \,_c\mathrm{B}\, (\neg \mathcal{T}(f) \wedge \neg tick)] \wedge [\mathcal{T}(f \mathrm{U}_{<2}\, g) \mathrm{At}_c \, tick]$
3. $\mathcal{T}(f \mathrm{U}_{>c} g) = [tick \,_{c+1}\mathrm{B}\, (\neg \mathcal{T}(f) \wedge \neg tick)] \wedge [((\mathcal{T}(f) \vee tick) \mathrm{U} (\mathcal{T}(g) \wedge \neg tick)) \mathrm{At}_{c+1} \, tick]$
4. $\mathcal{T}(\bigcirc_{<c} f) = \bigcirc([(\mathcal{T}(f) \wedge \neg tick) \mathrm{B}_c \, tick] \wedge [tick \mathrm{U} (\mathcal{T}(f) \wedge \neg tick)])$
5. $\mathcal{T}(\bigcirc_{>c} f) = \bigcirc([tick \,_c\mathrm{B} (\mathcal{T}(f) \wedge \neg tick)] \wedge [tick \mathrm{U} (\mathcal{T}(f) \wedge \neg tick)])$
6. $\mathcal{T}(\bigcirc_{=c} f) = \mathcal{T}(\bigcirc_{<c+1} f) \wedge \mathcal{T}(\bigcirc_{>c-1} f)$

For the other operators and the atomic propositions the translation keeps the same formula:

1. $\mathcal{T}(p) = p$
2. $\mathcal{T}(f \wedge g) = \mathcal{T}(f) \wedge \mathcal{T}(g)$

3. $\mathcal{T}(\neg f) = \neg\mathcal{T}(f)$

More formally, if we denote by $\mathcal{L}_{MTL}$ the language of all the MTL formulas, we show that

$$\forall\phi \in \mathcal{L}_{MTL}, \forall\sigma_t \in \Sigma_t, \forall\sigma \in \Sigma/(\sigma_t, \sigma) \in \mathcal{R}, \forall s \in S_t, \sigma_t, s \models_{MTL} \phi \Leftrightarrow \sigma, s \models \mathcal{T}(\phi)$$

Therefore MTL is not more expressive that PTL. The inverse relation is easy to see: MTL is an extension of PTL.

6.3 The Timed Propositional Temporal Logic

The Timed Propositional Temporal Logic (TPTL) has been proposed in [1]. This logic is obtained from PTL by adding a time quantifier "x" that binds the associated variable with the current time. For instance, the typical bounded response time property that "Every p-state is followed by a q-state within time 5" can be stated as:

$$\Box_x(p \rightarrow \Diamond_y(q \wedge (y \leq x + 5)))$$

[2] gives the demonstration that TPTL has the same expressiveness that MTL if we consider only the initiated state sequences (i.e. the date of the first state is 0). Therefore TPTL has the same expressiveness that PTL.

Most of the real-time logics which have been proposed, are not more expressive than PTL, however some properties of real-time systems cannot be expressed by PTL (neither by TPTL or MTL), as shown in the following section.

7 PTL limitations and extensions

Some properties cannot be expressed by PTL, but are expressible by other logics.

The best known extension to the propositional temporal logic is the Extended Temporal Logic [15]. It gives ETL the power of the monadic second order logic on $(\mathbb{N},<)$, whereas PTL has only the power of the monadic first order logic on $(\mathbb{N},<)$. ETL offers the possibility to count the occurrence of events. There is no contradiction with our results: we can see that the property that p is true in all the even states is not expressible in PTL even if the new abbreviations are added. Surprisingly ETL has exactly the same complexity as PTL. Therefore our abbreviations can be added to ETL to allow an easier expression of the real-time properties with no cost in complexity (the complexity will be the same as the one given in Section 5).

The previous extension of PTL has no cost in complexity, but it is not the case for most of the other extensions.

The Explicit Clock Temporal Logic (XCTL) is presented in [5]. The previous demonstration of the equivalence of TPTL and PTL shows that contrary to what the authors of the above mentioned paper claimed, XCTL and TPTL are not incomparable. They claimed that XCTL cannot express the following property: if the operation α does not take more than 5 time units then the execution time of the complete program will not take more than 1000 time units, but this property is expressible in PTL:

$$\Box(p \rightarrow q\mathrm{B}_5 tick) \rightarrow \Box(u \rightarrow v\mathrm{B}_{1000} tick)$$

and therefore it is expressible in XCTL (we will not give the complete formula in strict PTL, it would take more than 20,000 characters). Hence XCTL is not incomparable, it has a higher expressive power than PTL and TPTL. A typical instance of a property that XCTL can express and that PTL cannot is: "The computation time of the process α (which starts in state p and ends in state q) is always higher than the computation time of the process β (which starts in state q and ends in state r)"

$$\Box(p\wedge(x=T) \rightarrow \Box(q\wedge(y=T) \rightarrow \Box(r\wedge(z=T) \rightarrow [(z-y)<(y-x)])))$$

This property is not expressible because we cannot use expressions like $z-y$. But this extension, obtained by introducing operations on numbers, is too powerful: the satisfiability problem is undecidable for XCTL.

An important limitation of PTL is that all the timing constraints must be integers. Therefore we cannot have accurate timing constraints. This is why some authors have proposed logics which use the set of real numbers to model the time [3, 9]. Unfortunately, these logics have been shown to be undecidable [2].

8 Conclusion

To specify a real-time system we need to express real-time properties: that means quantitative requirements. We have shown in this article that we do not need a new logic: the Propositional Linear Temporal Logic (PTL) which is the classical temporal logic is expressive enough. We have introduced some abbreviations to express in an easy way the quantitative requirements, and we have shown with some examples, how these operators can be used. Most of the extensions to PTL which have been previously proposed are either undecidable or give no more expressive power. If some operators of the other logics are important for easing the expression of the properties, they can be replaced by abbreviations in the syntax of PTL. This gives PTL the same clearness as the other logics without any modification of the associated tools (e.g. model-checker). The ease of usage has a cost in complexity but the induced complexity is no worse than the best real-time logics. That explains why we claim that the Propositional Temporal Logic is a good candidate for the specification of real-time systems.

Acknowledgments

We wish to thank the referees for their critical comments which led to improvements over the earlier version. We also benefited from numerous discussions with Frédéric Boniol.

References

[1] R. Alur and T. A. Henzinger *A really temporal logic* IEEE Symp. on Found. of Comp. Sci., pages 164–169, 1989.

[2] R. Alur and T. A. Henzinger *Real-time logics: complexity and expressiveness* IEEE Logic in Comp. Sci., pages 390–401, 1990.

[3] R. Alur, C. Courcoubetis and D. Dill *Model-checking for real-time systems* IEEE Logic in Comp. Sci., pages 414–425, 1990.

[4] E. A. Emerson and J. Y. Halpern *"Sometimes" and "not never" revisited: on branching versus linear time temporal logic* Journal of the ACM, 33(1):151–178, January 1986.

[5] E. Harel, O. Lichtenstein, and A. Pnueli *Explicit clock temporal logic* IEEE Logic in Comp. Sci., pages 402–413, 1990.

[6] T. A. Henzinger, Z. Manna and A. Pnueli *Temporal proof methodologies for real-time systems* ACM Princ. of Distr. Comp., pages 353–366, 1990.

[7] F. Jahanian and A. K. Mok *Safety analysis of timing properties in real-time systems* IEEE Trans. on Software Engineering, 12(9):890–904, September 1986.

[8] R. Koymans *Specifying real-time properties with metric temporal logic* Journal of Real-Time Systems, 1(2):255–299, 1990.

[9] H. Lewis *A logic of concrete time intervals* IEEE Logic in Comp. Sci., pages 414–425, 1990.

[10] Z. Manna and A. Pnueli *A hierarchy of temporal properties* ACM Princ. of Distr. Comp., pages 353–366, 1990.

[11] J. S. Ostroff *Real-time temporal logic decision procedures* IEEE Real Time Syst. Symp., pages 92–101, 1989.

[12] D. Pilaud and N. Halbwachs *From a synchronous declarative language to a temporal logic dealing with multiform time* Formal Techniques in Real-Time and Fault-Tolerant Systems 1988, LNCS 331, pages 99–110, 1988.

[13] A. Pnueli and E. Harel *Applications of temporal logic to the specification of real-time systems* Formal Techniques in Real-Time and Fault-Tolerant Systems 1988, LNCS 331, pages 84–98, 1988.

[14] A. P. Sistla and E. M. Clarke *The complexity of propositional linear temporal logics* Journal of the ACM, 32(3):733–749, July 1985.

[15] P. Wolper *Temporal logic can be more expressive* Information and Control, (56):72–99, 1983.

Axiomatizations of Backtracking*

Michel Billaud
LaBRI† - Université Bordeaux I
351, Cours de la Libération
33405 Talence Cedex (France)

billaud@geocub.greco-prog.fr

Abstract. *Goal schemes* are terms built from a set of variables (representing goals) and the control structures $\{false,\ true,\ or,\ and\}$ to which we give a sequential *à la Prolog* interpretation. We study equivalence relations induced by some interesting classes of *elementary goals*. We prove that, when goals are allowed to produce side-effects, or when they are restricted to have *finite* behaviours, there are finite complete axiomatizations that can be used to decide the equivalence of goal schemes. We conjecture that there is no such finite axiomatization in the case of *general pure requests*.

Keywords: Prolog, semantics of programming languages, foundations of logic programming, depth-first search, backtracking, equivalence of program schemes, algebraic semantics.

Introduction

We consider here the language $\mathcal{GS}$ of *goal schemes*, namely the set of finite terms built from a countable set X of *goal variables* and the signature $\mathcal{F} = \{false, true, or, and\}$. We will follow a common abuse: these names denote the control structures of sequential Prolog, not their logical ancestors.

This paper deals the *equivalence* of such goal schemes. For example, the following *equivalence laws* are familiar to Prolog programmers:

$$\begin{array}{rclrcl} true\ and\ A & \approx & A & A\ and\ true & \approx & A \\ false\ or\ A & \approx & A & (A\ and\ B)\ and\ C & \approx & A\ and\ (B\ and\ C\ ) \\ A\ or\ false & \approx & A & (A\ or\ B)\ or\ C & \approx & A\ or\ (B\ or\ C) \end{array}$$

Obviously, if we combine these identities, we will obtain only valid identities. But, conversely, can we deduce all valid equalities from a finite (and preferably small) set of axioms?

What is *equivalence of goal schemes*?

Let us start with a set $\mathcal{G}_0$ of *elementary goals*, from which we build the set $\mathcal{G}$ of goals using the control structures of $\mathcal{F}$. We also suppose we have a formal interpreter to tell us the result of a computation.

There is a natural *observational equivalence* relation on $\mathcal{G}$: we write $g_1 \sim g_2$ if these goals can be substituted for each other in any compound goal g without making any difference in the result; in other words they show the same behaviour in all possible contexts.

Now we can define precisely the equivalence of two goal schemes p and q in $\mathcal{GS}$: we will write $p \approx q$ if for each substitution $\sigma : X \to \mathcal{G}$ one has $\sigma(p) \sim \sigma(q)$.

Several questions arise:

- Which kind of *elementary goals* do we want to study? Are we interested in side-effects (like the *write* predicate of Prolog)? Do we restrict the goals to always have a finite behaviour (e.g. databases

*This work has been partially supported by the GRECO de Programmation (METHEOL project)

†Laboratoire Bordelais de Recherche en Informatique, Unité Associée 1304 du C.N.R.S.

queries) ? For example the identity

$$A\ and\ false\ \approx false$$

is valid as long as we consider goals with finite behaviours which don't produce side-effects, but it doesn't hold for $A = write(foo)$, or when A produces infinitely many solutions.

- Is there a *constructive* definition of $\sim$? Many authors have presented *operational* semantics for Prolog (see for example [1, 7]), but it is far better to have a *denotational* semantics [12, 2, 1, 10, 11, 4, 5, 6, 14]. In that case, we have a semantic function S from $\mathcal{G}$ to some set $\mathcal{M}$ of *meanings*, such that $S[g_1] = S[g_2] \Rightarrow g_1 \sim g_2$ (*compositionality*), and conversely $g_1 \sim g_2 \Rightarrow S[g_1] = S[g_2]$ (*full abstraction*). Also, we will require $\mathcal{M}$ to be an algebra of the same type $\mathcal{F}$, and S to be an homomorphism.[1]

- Is there *constructive* characterization of the set $Id_{\mathcal{M}}(X)$ of all valid identities?[2] It would be nice to provide a *finite axiomatization* Σ for $\approx$, that is, a finite set of identities enjoying both consistency and completeness. Consistency means that every identity in the *deductive closure* $D(\Sigma)$[3] is valid, that is $D(\Sigma) \subseteq Id_{\mathcal{M}}(X)$: Σ tells only the truth about $\approx$. *Completeness* is achieved when all valid identities are provable from the axioms, i.e. $Id_{\mathcal{M}} \subseteq D(\Sigma)$: Σ tells *all* the truth. For example, the axioms given below don't constitute a complete axiomatization, because the following valid identities

$$\begin{array}{rcl} false\ and\ A & \approx & false \\ (A\ or\ B)\ and\ C & \approx & (A\ and\ C)\ or\ (B\ and\ C) \end{array}$$

  cannot be deduced from them.

- Can Σ be used as a *decision procedure* ? Suppose we can *orient* the equations in order to obtain a noetherian and confluent relation $\rightarrow_\Sigma$, then the equivalence of two goals schemes could be tested by a simple comparison of their normal forms.

There is now abundant litterature about operational and denotational semantics of Prolog, with or without *cut*, input-output and database predicates, but curiously the two last points have never been adressed (as far as we know), despite their similarity with the classical theory of *program schemes.*

The paper is organized as follows. Section 1 presents a formal framework for the study of backtracking-oriented computation mechanisms. In Section 2 we give a axiomatization for backtracking in the presence of side-effects, and prove its completeness. Section 3 is devoted to the study of goals without side-effects: in the case of pure finite requests we give a complete axiomatization, and we conjecture that there is no such finite axiomatization in the general case of pure requests.

1 Semantics of Backtracking

Backtracking is essentially a technique to find several answers for a given question. For example, if we define the following procedure in Prolog:

```
member(X,[X|L]).
member(X,[Y|L]) :- member(X,L).
```

the evaluation of the goal `member(X,[a,b,c])` (if `X` is free) returns a *stream* of 3 solutions, namely `X=a`, `X=b`, and `X=c` in this order.

Streams slightly generalize lists, in the sense that an interpreter can also produce infinite streams of solutions, or start an improductive endless loop.

Here we will consider streams made of two kinds of elements: *states* (which represent substitutions in Prolog, and *events* (if we are interested in studying side-effects). For example the effect of the goal `member(X,[a,b]), write(X)` is to first write an a on some output device, then to succeed with `X=a`. After that, backtracking will provide another 2 elements in the result stream: a b is written, and the goal succeeds with `X=b`.

[1] That is: $\mathcal{M}$ is equipped with a set of 4 operations $\{tt, ff, \oplus, \otimes\}$ such that $S[false] = ff$, $S[true] = tt$, and for all $g_1, g_2 \in \mathcal{G}$ one has $S[g_1\ or\ g_2] = S[g_1] \oplus S[g_2]$ and $S[g_1\ and\ g_2] = S[g_1] \otimes S[g_2]$. In this case S is defined from its restriction on $\mathcal{G}_0$.

[2] $p \approx q \in Id_{\mathcal{M}}(X)$ is often written $\mathcal{M} \models p \approx q$.

[3] $D(\Sigma)$ is the set of identities which can be *formally proved* using axioms from Σ and simple "algebraic manipulations" (replacement, substitution, etc. See [8] for formal definitions about Universal Algebra). One also writes $\Sigma \vdash p \approx q \Longleftrightarrow p \approx q \in D(\Sigma)$.

1.1 Streams as Meanings

In this technical part we provide a formal definition of streams, and two basic operations on them.

Definition 1.1 *Let $\mathcal{S}$ and $\mathcal{E}$ be (distinct) sets of states and events respectively. "$\bullet$" is a binary constructor, ϵ represents the* empty *stream, $\perp$ is the* undefined *stream (endless loop). The set $Str_\perp(\mathcal{S},\mathcal{E})$ of* finite approximations of streams *is defined inductively:*

$$st \in Str_\perp(\mathcal{S},\mathcal{E}) \Leftrightarrow \begin{cases} st = \perp \\ st = \epsilon \\ st = s \bullet st' \\ st = e \bullet st' \end{cases}$$

$$\text{where } s \in \mathcal{S}, e \in \mathcal{E} \text{ and } st' \in Str_\perp(\mathcal{S},\mathcal{E})$$

Then, we obtain the set $Str_\perp^\omega(\mathcal{S},\mathcal{E})$ of *streams* by ω-completion of $Str_\perp(\mathcal{S},\mathcal{E})$ (see [13, 9]).

Definition 1.2 *$Str_\perp^\omega(\mathcal{S},\mathcal{E})$ is the set of* limits *of ascending chains on $Str_\perp(\mathcal{S},\mathcal{E})$.*

Example: the infinite stream $s \bullet s \bullet s \ldots$ is the limit of the chain $<\perp, s\bullet\perp, s \bullet s \bullet \perp, \ldots>$.

Remark : As a benefit of ω-completion, continuous functions on streams are defined by their restrictions on *finite* approximations :

Definition 1.3 *The* concatenation $\diamond$ *of two streams returns another stream:*

$$\begin{array}{rcll} \perp & \diamond & st'' & = \perp \\ \epsilon & \diamond & st'' & = st'' \\ (s \bullet st') & \diamond & st'' & = s \bullet (st' \diamond st'') \\ (e \bullet st') & \diamond & st'' & = e \bullet (st' \diamond st'') \end{array}$$

$$\forall s \in \mathcal{S}, e \in \mathcal{E} \text{ and } st', st'' \in Str_\perp(\mathcal{S},\mathcal{E})$$

Definition 1.4 *The* extension *operation (hat) turns any function $f : \mathcal{S} \to Str_\perp^\omega(\mathcal{S},\mathcal{E})$ into a function $\widehat{f} : Str_\perp^\omega(\mathcal{S},\mathcal{E}) \to Str_\perp^\omega(\mathcal{S},\mathcal{E})$ defined as follows:*

$$\begin{array}{rcl} \widehat{f}(\perp) & = & \perp \\ \widehat{f}(\epsilon) & = & \epsilon \\ \widehat{f}(s \bullet st') & = & f(s) \diamond \widehat{f}(st') \\ \widehat{f}(e \bullet st') & = & e \bullet \widehat{f}(st') \end{array}$$

$$\forall s \in \mathcal{S}, e \in \mathcal{E} \text{ and } st' \in Str_\perp(\mathcal{S},\mathcal{E})$$

1.2 Modelling Prolog

We already know that *states* in $\mathcal{S}$ represent the *substitutions* of Prolog. The set $\mathcal{G}_0$ of elementary goals can be divided in 2 subsets: *user-defined* and *predefined* goals.

A user-defined goal is a kind of "procedure call"; during the evaluation it is replaced by the corresponding "procedure body". Such procedure bodies are obtained from the Prolog code by a *normalization* of the head of clauses, making all unifications explicits, *and*-ing the goals in the clauses and then *or*-ing the results. For example the `member` procedure turns into:

$$\begin{array}{rcl} \texttt{member(X,Z)} & :- & \texttt{unify(Z,[X|L])} \\ & or & (\texttt{unify(Z,[Y|L])}\ and\ \texttt{member(X,L)}). \end{array}$$

We model the "procedure call" mecanism by a function $reduce : \mathcal{G}_0 \times \mathcal{S} \to \mathcal{G} \times \mathcal{S}$ which, to each user-defined goal (for example `member(a, [a,b,c])`) called in a given state (i.e. a substitution) s, associates an instance of the corresponding procedure body (here `unify([a,b,c],[a|L])` *or* `unify([a,b,c],[Y|L])` *and* `member(a,L)`) to be evaluated in a new state (basically s augmented with two new free variables `Y` and `L`).

The `unify` predefined predicate is given the same treatment: given two terms t_1, t_2 and a substitution s, one has: $reduce(\texttt{unify}(t_1,t_2),s) = (true, mgu(t_1,t_2))$ if the terms have a most general unifier, or $reduce(\texttt{unify}(t_1,t_2),s) = (false, s)$ if they have not.

Output predicates constitute another kind of elementary goals. their behaviour is represented by another function $output : \mathcal{G}_\prime \times \mathcal{S} \rightarrow \mathcal{E}$ that tells what will be written on the "output tape" when the goal is evaluated. For example `write(X)` produces an a when `X` is bound to a in the current state, and succeeds.

To distinguish *output events* from *states*, we will prefix the former with a claim mark "!". For example the evaluation of `write(a)` in the state s results in the stream $!a.s.\epsilon$.

Input predicates can also be modelled in a similar framework, see [5, 6], but for simplicity we wont consider them here.

1.3 Denotational Semantics

Theorem 1.1 *The semantic function $\mathcal{S}$ associated with the standard sequential strategy is the least solution of the functional system:*

$$
\begin{array}{llcll}
\mathcal{S}[\ false\ ]s & = & \epsilon & & \\
\mathcal{S}[\ true\ ]s & = & s \bullet \epsilon & & \\
\mathcal{S}[\ g_1 \text{ or } g_2\ ]s & = & \mathcal{S}[g_1]s \diamond \mathcal{S}[g_2]s & & \\
\mathcal{S}[\ g_1 \text{ and } g_2\ ]s & = & (\widehat{\mathcal{S}[g_2]} \circ \mathcal{S}[g_1])(s) & & \\
\mathcal{S}[\ g_e\ ]s & = & !e \bullet s \bullet \epsilon & & \text{if } output(g_e, s) = !e \\
\mathcal{S}[\ g\ ]s & = & \mathcal{S}[g']s' & & \text{if } reduce(g, s) = (g', s')
\end{array}
$$

for all $g_1, g_2, g' \in \mathcal{G}, g_e, g \in \mathcal{G}_0, s, s' \in \mathcal{S}$ such that g_e has output side-effects, and g has not.

The first four equations clearly show the nature of the algebra $\mathcal{M}$ of meanings: it is made of functions from $\mathcal{S}$ to $Str^{\omega}_{\perp}(\mathcal{S},\mathcal{E})$, and they reveal the homomorphism between control structures and operations on $\mathcal{M}$:

$$
\begin{array}{lcl}
\mathrm{ff} & = & \lambda s \cdot \epsilon \\
\mathrm{tt} & = & \lambda s \cdot s \bullet \epsilon \\
\oplus & = & \lambda f, g \cdot \lambda s \cdot f(s) \diamond g(s) \\
\otimes & = & \lambda f, g \cdot \widehat{g} \circ f
\end{array}
$$

2 An axiomatization for goals with side-effects

In this section we suppose that the goals in $\mathcal{G}_0$ can have side-effects. We will propose a rewriting relation $\rightarrow$ on $\mathcal{GS}$, and prove it to be consistent with $\approx$, noetherian, confluent (so goal schemes have unique normal forms) and also complete, so it can be used as an effective decision procedure for the equivalence of goal schemes. The completeness proof relies on a "grammar" for normal forms, and the choice of an *ad hoc* interpretation.

2.1 A rewriting system

Definition 2.1 *Let $\rightarrow$ be the relation $\mathcal{GS}$ defined by the following system $\mathcal{R}$:*

$$
\begin{array}{rclcll}
A & or & false & \rightarrow & A & R_1 \\
false & or & A & \rightarrow & A & R_2 \\
A & and & true & \rightarrow & A & R_3 \\
true & and & A & \rightarrow & A & R_4 \\
false & and & A & \rightarrow & false & R_5 \\
(A \text{ and } B) & and & C & \rightarrow & A \text{ and } (B \text{ and } C) & R_6 \\
(A \text{ or } B) & or & C & \rightarrow & A \text{ or } (B \text{ or } C) & R_7 \\
(A \text{ or } B) & and & C & \rightarrow & (A \text{ and } C) \text{ or } (B \text{ and } C) & R_8
\end{array}
$$

and we define Σ as the set of equations generated by $\mathcal{R}$: $\Sigma = \{p_i \approx q_i \mid p_i \rightarrow q_i \in \mathcal{R}\}$.

Lemma 2.1 *Σ is sound w.r.t. $\approx$.*

Proof. It suffices to show that all identities in Σ are valid, that is the equations

$$\begin{array}{rclcl} \alpha & \oplus & \mathrm{ff} & = & \alpha \\ \alpha & \otimes & \mathrm{tt} & = & \alpha \\ \mathrm{ff} & \otimes & \alpha & = & \mathrm{tt} \\ (\alpha\oplus\beta) & \oplus & \gamma & = & \alpha\oplus(\beta\oplus\gamma) \end{array} \qquad \begin{array}{rclcl} \mathrm{ff} & \oplus & \alpha & = & \alpha \\ \mathrm{tt} & \otimes & \alpha & = & \alpha \\ (\alpha\otimes\beta) & \otimes & \gamma & = & \alpha\otimes(\beta\otimes\gamma) \\ (\alpha\oplus\beta) & \otimes & \gamma & = & (\alpha\otimes\gamma)\oplus(\beta\otimes\gamma) \end{array}$$

hold for all $\alpha, \beta, \gamma \in \mathcal{M}$. □

2.2 Normal Forms

We will now prove that goals schemes have a unique normal form, and also give an inductive definition of the set $\mathcal{NF}$ of normal forms.

Lemma 2.2 $\rightarrow$ *is noetherian.*

Proof. The weighting function $\tau : \mathcal{GS} \rightarrow \mathbb{N}_1$ defined by $\forall x \in X,\ \forall A, B \in \mathcal{GS}$:

$$\begin{array}{rcl} \tau(false) = \tau(true) = \tau(x) & = & 1 \\ \tau(A\ or\ B) & = & 2\cdot\tau(A) + \tau(B) + 1 \\ \tau(A\ and\ B) & = & (1+\tau(A))\cdot\tau(B) \end{array}$$

has the property $A \rightarrow B \Rightarrow \tau(A) > \tau(B)\ \forall A, B \in \mathcal{GS}$, thus there is no infinite rewriting chain. □

Lemma 2.3 $\rightarrow$ *is locally confluent.*

Proof. We use the classical "critical pairs" method. We study all the possible ways to superpose pairs of left-hand sides[4]. For example the superposition of R_1 and R_8 results in the term $(A\ or\ false)\ and\ C$ which can be reduced into $P_1 = A\ and\ C$ (by R_1), and $P_2 = (A\ and\ C)\ or\ (false\ and\ C)$ (by R_8). Then we check that the members of that "critical pair" reduce to a common term. Here $P_2 \rightarrow_{R6} (A\ and\ C)\ or\ false \rightarrow_{R1} A\ and\ C = P_1$. □

Theorem 2.1 *Every goal scheme has a unique normal form.*

Proof. $\rightarrow$ is noetherian and locally confluent, thus it is confluent. □
Definition. We define two sets $\mathcal{NF}_{or}, \mathcal{NF}_{and}$ by induction:

$$\begin{array}{rllcccccc} \mathcal{NF}_{or} = & \{l\ or\ r \mid & l \in & \{true\} & \cup & X & \cup & & \mathcal{NF}_{and}, \\ & & r \in & \{true\} & \cup & X & \cup & \mathcal{NF}_{or} \cup & \mathcal{NF}_{and}\} \\ \mathcal{NF}_{and} = & \{l\ and\ r \mid & l \in & & & X, & & & \\ & & r \in & \{false\} & \cup & X & \cup & \mathcal{NF}_{or} \cup & \mathcal{NF}_{and}\} \end{array}$$

Theorem 2.2 $\mathcal{NF} = \{false, true\} \cup X \cup \mathcal{NF}_{or} \cup \mathcal{NF}_{and}$

Proof. Let $A = \{false, true\} \cup X \cup \mathcal{NF}_{or} \cup \mathcal{NF}_{and}$. $A \subseteq \mathcal{NF}$ can be proved by induction on the structure of terms; at each step we check that (by construction) no rule can be applied at the root of the term. In the other direction, we study the set $E = \mathcal{NF} - A$. If E is not empty, it contains a t which has no proper subterm in E. Thus one has $t = l\ or\ r$ (or $t = l\ and\ r$), with $l, r \in \mathcal{NF} - E$. A study by cases on l, r quickly reveals the contradiction. Thus E is empty. □

2.3 Completeness

We prove here the *completeness* of Σ w.r.t. $\approx$, that is every $p \approx q$ in $Id_{\mathcal{M}}(X)$ is a syntactic consequence of Σ. To do that, it suffices to show that two distinct goals schemes in normal form show distinct behaviours in a well-chosen interpretation. That is $\forall p, q \in \mathcal{NF}(p \neq q)$ there is a substitution $\sigma : X \rightarrow \mathcal{G}$ and a semantic function S such that $S[\sigma(p)] \neq S[\sigma(q)]$. Here we choose:

- $\mathcal{S} = \{s\}$

[4] The proof has actually been performed by a small program which discovered 17 such superpositions.

- $\mathcal{G}_0 = \{g_0, g_1, g_2, \ldots\}$
- $\sigma(x_i) = g_i$, $\forall i \in \mathbb{N}$ (we suppose $X = \{x_0, x_1, \ldots\}$)
- $\mathcal{E} = \{!a_0, !b_0, !a_1, !b_1, a_2, !b_2, \ldots\}$
- $S[g_i]s = !a_i \bullet s \bullet !b_i \bullet \epsilon$, $\forall i \in \mathbb{N}$.

For convenience, we will define $T(g) = S[\sigma(g)]s$ for all $g \in GS$. Intuitively, $T(g)$ returns a "debugging trace" which shows in which order "calls" and "returns" of the elementary goals happen during a computation.

Definition 2.2 *Let W be the least subset of $Str^{\omega}_{\perp}(S, \mathcal{E})$ such that*

$$W = \{\epsilon, s\} \cup \{\ !a_i \bullet w \diamond (!b_i \bullet w') \mid \forall i \in \mathbb{N}, \forall w, w' \in W\}$$

The reader will notice the strong resemblance of W with some kind of *Motzkin words*. That's not really surprising, as we are clearly trying to code *computation trees* by words.

Lemma 2.4 *For all $w, w' \in W$, one has: $w \diamond w' \in W$ and $w[s/w'] \in W$.*

Proof. Obvious. □

Definition 2.3 *$w \in W$ is* prime *if $\forall w', w'' \in W, w = w' \diamond w''$ implies $w' = \epsilon$ or $w'' = \epsilon$.*

Lemma 2.5 *Every $w \in W$ has a unique decomposition under one of these forms:*

- $w = \epsilon$
- $w = s \bullet \epsilon$
- $w = !a_i \bullet w' \diamond (!b_i \bullet \epsilon)$ *with $w' \in W$*
- $w = w' \diamond w''$ *with $w', w'' \in W - \{\epsilon\}$, and w' is prime.*

Proof. Trivial. □

Lemma 2.6 *W is the set of streams generated by the interpretation of goal schemes: $W = T(\mathcal{GS})$.*

Proof. The inclusion $T(\mathcal{GS}) \subseteq W$ is a straightforward consequence of lemma 2.4. In the other direction, we lead a proof by induction on the structure of W. Let us suppose that $w \in W$, then :

- if $w = \epsilon$, then $w = T(false)$,
- if $w = s \bullet \epsilon$, then $w = T(true)$,
- if $w = !a_i \bullet w' \diamond (!b_i \bullet w'')$, then (by induction hypothesis) there exist $g', g'' \in \mathcal{GS}$ such that $T(g') = w'$ and $T(g'') = w''$. So $w = T((x_i \ and \ g') \ or \ g'')$.

□

Lemma 2.7

$$\begin{aligned} T(\mathcal{NF}_{or}) &= \{w \diamond w' \mid w, w' \in W - \{\epsilon\}\} \\ T(\mathcal{NF}_{and}) &= \{!a_i \bullet w \diamond (!b_i \bullet \epsilon) \mid w \in W - \{\epsilon, s \bullet \epsilon\}\ \} \end{aligned}$$

Proof. It is only a refinement of the previous proof. □

Lemma 2.8 *T is a bijection from $\mathcal{NF}$ to W.*

Proof. Stems from the decomposition lemma 2.5:

- if $w = \epsilon$, then $T^{-1}(w) = false$;
- if $w = s \bullet \epsilon$, then $T^{-1}(w) = true$;
- if $w = !a_i \bullet w' \diamond (!b_i \bullet \epsilon)$ with $w' \in W$ then $T^{-1}(w) = x_i \ and \ T^{-1}(w')$.
- if $w = w' \diamond w''$ with w' prime and $w'' \in W - \{\epsilon\}$, then one has: $T^{-1}(w) = T^{-1}(w') \ or \ T^{-1}(w'')$;

□

So we are now able to conclude:

Theorem 2.3 *Σ is an axiomatization of $\approx$, and $\mathcal{R}$ is a decision procedure for the equivalence of goal schemes.*

3 Axiomatizations for "pure requests"

We now suppose that $\mathcal{E} = \emptyset$, that is we are interested in identities about goals ***without side effects***. As the set of possible elementary goals is a subset of the previous one, identities shown in the previous section are still valid, but we also have some new ones, for example

$$(A \text{ and } false) \text{ or } (B \text{ and } false) \approx (B \text{ and } false) \text{ or } (A \text{ and } false)$$

which cannot be deduced from the previous axiomatization Σ.

In this section, we will first restrict ourselves to goals with *finite behaviours*. In this case, we are able to give an axiomatization Σ' and a decision procedure.

Then, we will discuss the case of possibly infinite behaviours: we will show that no confluent and terminating rewriting system can be used as a decision procedure for the equivalence of goal schemes. As a conjecture, we propose an infinite axiomatization.

3.1 Finite Pure Requests

We suppose here that goals always have a finite behaviour, that is an evaluation always produces a *finite list* of solutions, and stops after a finite number of steps. In other words, the program explores a *finite* search tree, which is the case for many applications of backtracking, and also when elementary goals represent (terminating) database queries, and so on. So applications certainly do not lack.

As we are considering finite behaviours only, the relevant algebra of meanings $\mathcal{M}'$ will be the set of functions from $\mathcal{S}$ to the set of finite streams terminated by ϵ, which is isomorphic to $\mathcal{S}^*$.

3.1.1 A rewriting system

Definition 3.1 *We introduce now a new rule R_9 : A and false $\rightarrow$ false, and the identity E_9 : A and false $\approx$ false. Let $\mathcal{R}' = \mathcal{R} \cup \{R_9\}$, and $\Sigma' = \Sigma \cup \{E_9\}$.*

Lemma 3.1 *Σ' is sound w.r.t. $\approx$.*

Proof. As $\mathcal{M}'$ is a subset of $\mathcal{M}$, all identities of Σ are still valid, and we have shown. E_9 is also valid because in $\mathcal{M}'$ we have the property $\widehat{ff} = \lambda w \cdot \epsilon$ (proof by induction on $w \in \mathcal{S}^*$). □

3.1.2 Normal forms

Lemma 3.2 *The rewriting system $\mathcal{R}'$ is terminating and confluent, so each goal scheme has a unique normal form.*

Proof. The same termination criterium also works for $\mathcal{R}'$, and R_9 doesn't introduce divergent critical pairs. □

Definition 3.2 *Let $\mathcal{NF}'$ be the set of goal schemes in normal form.*

Theorem 3.1 *$\mathcal{NF}' = \{false, true\} \cup X \cup \mathcal{NF}'_{or} \cup \mathcal{NF}'_{and}$, where :*

$$\begin{array}{rllllllll}
\mathcal{NF}'_{or} = & \{l \text{ or } r \mid & l \in & \{true\} & \cup & X & \cup & & \mathcal{NF}'_{and}, \\
 & & r \in & \{true\} & \cup & X & \cup & \mathcal{NF}'_{or} \cup & \mathcal{NF}'_{and}\} \\
\mathcal{NF}'_{and} = & \{l \text{ and } r \mid & l \in & & & X, & & & \\
 & & r \in & & & X & \cup & \mathcal{NF}'_{or} \cup & \mathcal{NF}'_{and}\}
\end{array}$$

Proof. Similar to the previous case (lemma 2.2). □

3.1.3 Completeness

The canvas of the proof is still the same, but we now have to choose another interpretation, precisely because we can no more rely on side-effects to "trace" what happens during an evaluation, and that makes the thing a bit more difficult. Nevertheless, the intuitive idea is similar up to a certain extent: we consider elementary goals g_i which produce streams of two solutions corresponding to "calls" and "returns". We will use states themselves to "memorize" what happened during the computation.

Let $I = \{a_0, b_0, a_1, b_1, \ldots\}$. We choose the following interpretation:

- $\mathcal{E} = \emptyset$
- $\mathcal{S} = I^*$; [5]
- $\mathcal{G}_0 = \{g_0, g_1, \ldots\}$
- $\sigma(x_i) = g_i,\ \forall i \in \mathbb{N}$
- $S[g_i]s = (s.a_i) \bullet (s.b_i) \bullet \epsilon,\ \forall i \in \mathbb{N}$.

Definition 3.3 *Let $\odot : \mathcal{S} \times \mathcal{S}^* \to \mathcal{S}$ be the "prefixing" operation such that $\forall s, s' \in \mathcal{S}, \forall w \in \mathcal{S}^*$:*

$$\begin{aligned} s \odot \epsilon &= \epsilon \\ s \odot (s' \bullet w) &= (s.s') \bullet (s \odot w) \end{aligned}$$

Lemma 3.3 *For all $s, s' \in \mathcal{S}$ and $w, w' \in \mathcal{S}^*$:*

$$\begin{aligned} (s.s') \odot w &= s \odot (s' \odot w) \\ s \odot (w \diamond w') &= (s \odot w) \diamond (s \odot w') \end{aligned}$$

Proof. By induction on w. □

Lemma 3.4 *We define $T : \mathcal{GS} \to \mathcal{S}^*$ by $T(g) = S[\sigma(g)]$e. Then $\forall s \in \mathcal{S},\ \forall g \in \mathcal{GS}$ we have:*

$$S[\sigma(g)]s = s \odot T(g)$$

Proof. Induction on the structure of g. The only difficulty is for the "*and*" case, which makes an intensive use of the previous lemma. □

Definition 3.4 *Let W be the least subset of $\mathcal{S}^*$ such that:*

$$w \in W \Leftrightarrow \begin{cases} w = \epsilon & \\ w = \mathrm{e} \bullet w' & \textit{with } w' \in W \\ w = (a_i \odot w') \diamond (b_i \odot w') \diamond w'' & \textit{with } i \in \mathbb{N};\ w', w'' \in W \end{cases}$$

Lemma 3.5 *W is closed under concatenation, that is $w, w' \in W \Rightarrow w \diamond w' \in W$.*

Proof. By induction on the left argument w. □

Lemma 3.6 $W = T(\mathcal{GS})$

Proof. In order to prove that $T(\mathcal{GS}) \subseteq W$, we use the fact that $T(\mathcal{GS}) = T(\mathcal{NF}')$, so we can lead an proof by induction on the structure of goals *in normal form*. Let $g \in \mathcal{NF}'$:

- if $g = false$, then $T(g) = \epsilon$;
- if $g = true$, then $T(g) = \mathrm{e} \bullet \epsilon$;
- if $g = g'$ *or* g'', then $T(g) = T(g') \diamond T(g'')$. By induction hypothesis, $T(g')$ and $T(g'')$ are in W, which is closed by $\diamond$;
- if $g = x_i$ *and* g', then $T(g) = \widehat{S[g']}(a_i \bullet b_i \bullet \epsilon) = (S[g']a_i) \diamond (S[g']b_i) \diamond \epsilon = (a_i \odot T(g')) \diamond (b_i \odot T(g')) \diamond \epsilon$;

so $g \in \mathcal{G} \Rightarrow T(g) \in W$.

Conversely, for each $w \in W$, there exists a goal scheme g such that $T(g) = w$. Proof by induction on the structure of W:

- if $w = \epsilon$, then $w = T(false)$;
- if $w = \mathrm{e} \bullet \epsilon$, then $w = T(true)$;

[5] To avoid confusions we will represent the "empty state" by "e", and the "concatenation" of states by a small dot "."

- if $w = (a_i \odot w') \diamond (b_i \odot w') \diamond w''$ with $w', w'' \in W$, then there exist $g', g'' \in \mathcal{GS}$ such that $w' = T(g'), w'' = T(g'')$, and then $w = T((x_i \text{ and } g') \text{ or } g''($.

□

Lemma 3.7 *T is a bijection from $\mathcal{NF}$ to W'. Thus Σ'' is complete.*

Proof. Also based on the existence of a unique decomposition for every $w \in W'$ under one of these forms:

- $w = \epsilon$, and then $T^{-1}(w) = false$;
- $w = e \bullet \epsilon$, and then $T^{-1}(w) = true$;
- $w = (a_i \odot w') \diamond (b_i \odot w')$ with $w' \in W'$ and then $T^{-1}(w) = x_i \text{ and } T^{-1}(w')$;
- $w = w' \diamond w''$ with $w', w'' \in W' - \{\epsilon\}$, and w' is prime (same definition as in the previous section) : then one has: $T^{-1}(w) = T^{-1}(w') \text{ or } T^{-1}(w'')$.

□

Theorem 3.2 *Σ' is an axiomatization of $\approx$, and $\mathcal{R}'$ is a decision procedure for the equivalence of goal schemes in the case of "finite requests" without side-effects.*

3.2 Dealing with infinite behaviours

Let us now relax the constraint about infinite behaviours. The relevant algebra of meanings $\mathcal{M}''$ is the set of functions from $\mathcal{S}$ to $Str^{\omega}_{\perp}(\mathcal{S}, \emptyset)$, so one has obviously $\mathcal{M}' \subset \mathcal{M}'' \subset \mathcal{M}$, and consequently $Id_{\mathcal{M}'}(X) \supset Id_{\mathcal{M}''}(X) \supset Id_{\mathcal{M}}(X)$. All inclusions are strict, because:

$$\begin{array}{rcl} A \text{ and } false \approx false & \in & Id_{\mathcal{M}'}(X) - Id_{\mathcal{M}''}(X) \\ (A \text{ and } false) \text{ or } (A \text{ and } false) \approx (A \text{ and } false) & \in & Id_{\mathcal{M}''}(X) - Id_{\mathcal{M}}(X) \end{array}$$

3.2.1 Some facts about infinite behaviours

The definition of an *infinite behaviour* directly stems *a contrario* from that of *finite behaviour*:

Definition 3.5 *A goal g has an infinite behaviour when evaluated in the state s if one of the two following conditions occurs:*

1. *There exists $s_1, s_2, \ldots s_n \in \mathcal{S}$ such that $\mathcal{S}[g]s = s_1 \bullet s_2 \bullet \ldots \bullet s_n \bullet \perp$;*
2. *There exists an infinite family $s_1, s_2, s_2, \ldots$ of states such that $\mathcal{S}[g]s = s_1 \bullet s_2 \bullet \ldots$.*

Definition 3.6 *Let $s \in \mathcal{S}$. $Ib(s) = \{g \in \mathcal{G} \mid \mathcal{S}[g \text{ and } false]s = \perp\}$.*

Lemma 3.8 *$g \notin Ib(s) \Leftrightarrow \mathcal{S}[g \text{ and } false]s = \epsilon$*

Proof. Obvious, as there are only 3 kinds of streams. □

Lemma 3.9 *Infinite behaviours are "left absorbent": $\forall s \in \mathcal{S}, g, g' \in \mathcal{G}$:*

$$g \in Ib(s) \Rightarrow \mathcal{S}[g \text{ or } g']s = \mathcal{S}[g]s$$

Proof. Stems from the properties of partial or infinite streams. □

Lemma 3.10 *$\forall s \in \mathcal{S}, \forall g, g' \in \mathcal{G}$*

- *$g \in Ib(s) \Rightarrow g \text{ or } g' \in Ib(s)$*
- *$g \in Ib(s) \Rightarrow g' \text{ or } g \in Ib(s)$*
- *$g \in Ib(s) \Rightarrow g \text{ and } g' \in Ib(s)$*

- g *or* $g' \in Ib(s) \Rightarrow g \in Ib(s)$ *or* $g' \in Ib(s)$

Proof. Obvious. □

Lemma 3.11 $(A \text{ or } B) \text{ and } false \approx (B \text{ or } A) \text{ and } false$

Proof. Straightforward consequence of lemmas 3.8 and 3.10. □

Theorem 3.3 *There is no confluent and terminating rewriting system to decide the equivalence of goal schemes in the case of possibly infinite requests without side-effects.*

Proof. Let us suppose there is such a rewriting system $\mathcal{R}''$. Then the goal scheme $((A \text{ or } B) \text{ and } false)$ by $\mathcal{R}''$ has a unique normal form $nf[A, B]$. Obviously, the normal form $nf[B, A]$ of $((B \text{ or } A) \text{ and } false)$ is the term obtained by turning A's into B's and vice-versa in $nf[A, B]$. From lemma 3.11, one has $nf[A, B] = nf[B, A]$. Consequently, there is no occurrences of the variables A and B in $nf[A, B]$.

This obviously reveals a contradiction, because the value of $S[nf[(g \text{ or } g' \text{ and } false)]s$ clearly depends on g and g': it is equal to $\perp$ if $(g \text{ or } g') \in Ib(s)$, ϵ otherwise. □

3.2.2 A tentative axiomatization

As a conjecture, we propose the following tentative infinite axiomatization Σ'':

$$
\begin{array}{rclcll}
A & or & false & \approx & A & E_1\\
false & or & A & \approx & A & E_2\\
A & and & true & \approx & A & E_3\\
true & and & A & \approx & A & E_4\\
false & and & A & \approx & false & E_5\\
(A \text{ and } B) & and & C & \approx & A \text{ and } (B \text{ and } C) & E_6\\
(A \text{ or } B) & or & C & \approx & A \text{ or } (B \text{ or } C) & E_7\\
(A \text{ or } B) & and & C & \approx & (A \text{ and } C) \text{ or } (B \text{ and } C) & E_8\\
(A \text{ or } B) & and & false & \approx & (B \text{ or } A) \text{ and } false & E_9\\
A \text{ and } (B \text{ or } C) & and & false & \approx & [(A \text{ and } B\ ) \text{ or } (A \text{ and } C)] \text{ and } false & E_{10}\\
l_i \text{ or } C & or & r_i & \approx & l_i \text{ or } C \text{ or } r'_i & E_{11,i}
\end{array}
$$

where the family $E_{11,i}$ of equations is defined inductively as:

$$
\begin{array}{ll}
l_0 = A \text{ and } B & l_i = X_i \text{ and } (Y_i \text{ or } l_{i-1} \text{ or } Z_i)\\
r_0 = A \text{ and } false & r_i = X_i \text{ and } (T_i \text{ or } r_{i-1} \text{ or } Z_i)\\
r'_0 = false & r_i = X_i \text{ and } (T_i \text{ or } r'_{i-1} \text{ or } Z_i)
\end{array}
$$

(Here $A, B, C, X_i, Y_i, Z_i, T_i$ denote variables, and l_i, r_i, r'_i represent goals schemes being built inductively).

Lemma 3.12 *Σ'' is sound w.r.t. $\approx$.*

Proof.

- Equations E_1 to E_8 need no further explanations, because they were in $Id_{\mathcal{M}}(X)$ so they are also valid for $\mathcal{M}''$.
- The "commutation equation" E_9 stems from lemma 3.11.
- The "distribution equation" E_{10} is given a proof by case. Let $a, b, c \in \mathcal{G}$, and $s \in \mathcal{S}$.
  - if $a \in Ib(s)$ then a *and* $(b$ *or* $c)$, a *and* b, a *and* c, $(a$ *and* $b)$ *or* $(a$ *and* $c)$ also belong to $Ib(s)$, because of lemma 3.10. Consequently

    $$S[a \text{ and } (b \text{ or } c) \text{ and } false](s) = \perp = S[((a \text{ and } b) \text{ or } (a \text{ and } c)) \text{and } false](s)$$

  - if $a \notin Ib(s)$, it's easy to see that a *and* $(b$ *or* $c) \in Ib(s)$ iff there is a state s' in $S[a](s)$ such that b *or* $c \in Ib(s')$, which is equivalent to say that $b \in Ib(s')$ or $c \in Ib(s')$, that is a *and* $b \in Ib(s)$ or a *and* $c \in Ib(s)$.

- Let us have a look at the first "cancellation equation":

$$E_{11,0} : (A \text{ and } B) \text{ or } C \text{ or } (A \text{ and } false) \approx (A \text{ and } B) \text{ or } C \text{ or } false$$

that we can obviously reduce (using R_1) to

$$(A \text{ and } B) \text{ or } C \text{ or } (A \text{ and } false) \approx (A \text{ and } B) \text{ or } C$$

Let $a, b, c \in \mathcal{G}$, and $s \in \mathcal{S}$. One obviously has:

$$\begin{array}{lcl} \mathcal{S}[(a \text{ and } b) \text{ or } c \text{ or } (a \text{ and } false)](s) & = & \mathcal{S}[a \text{ and } b](s) \diamond \mathcal{S}[c](s) \diamond \mathcal{S}[a \text{ and } false](s) \\ \mathcal{S}[(a \text{ and } b) \text{ or } c](s) & = & \mathcal{S}[a \text{ and } b](s) \diamond \mathcal{S}[c](s) \end{array}$$

  - if $a \in Ib(s)$ then $a \text{ and } b \in Ib(s)$, so by lemma 3.9:

$$\mathcal{S}[(a \text{ and } b) \text{ or } c \text{ or } (a \text{ and } false)](s) = \mathcal{S}[(a \text{ and } b) \text{ or } c](s)$$

  - if $a \notin Ib(s)$ then $\mathcal{S}[a \text{ and } false](s) = \epsilon$. Thus (lemma 3.8) we have

$$\mathcal{S}[(a \text{ and } b) \text{ or } c \text{ or } (a \text{ and } false)](s) = \mathcal{S}[(a \text{ and } b) \text{ or } c](s) \diamond \epsilon = \mathcal{S}[(a \text{ and } b) \text{ or } c](s)$$

- The other "cancellation equations" hold for the same reasons, as they state that one can replace *A and false* by *false* in some context if we have already evaluated some *A* in a similar context.

□

As there are no normal forms for goal schemes, we cannot use the same technique of the previous sections to prove the completeness of Σ''. We leave it as an open question.

4 Conclusion

The main contributions of this paper are :

- We have shown that the problem of "goal schemes equivalence" can be easily expressed in terms of identities in an algebra of meanings.
- In the most general case (goals with side-effects) we have given a 8 rules axiomatization for the signature $\{false, true, or, and\}$, we have proved its completeness, and shown that it could be used as a decision procedure.
- We have given the same treatment to the case of "finite requests" without side-effects, for which we found a 9 rules axiomatization.
- As a conjecture, we propose an infinite axiomatization for the case of finite or infinite requests without side-effects. It is shown to be sound, but its completeness remains an open question.

This work is to be continued in several directions:

- First, solve the question we left open. Also, we will have to find a decision procedure for the equivalence of goal schemes, as rewriting based reasoning doesn't work in this case.
- Consider the same question for other classical operators: *cut, negation by failure, if-then-else* etc. We conjecture that there is no *finite* axiomatization in these cases.

References

[1] B. Arbab, D.M. Berry, *Operational and Denotational Semantics of Prolog,* J. Logic Programming 1987, vol. 4, 309-329.

[2] M. Billaud, *Formalisation des Structures de Contrôle de Prolog,* Thèse de Troisième Cycle, Université de Bordeaux I, 1985.

[3] M. Billaud, *Prolog Control Structures, a Formalization and its Applications,* in : K.Fuchi and M. Nivat, eds., Programming for Future Generation Computers (North-Holland, Amsterdam, 1988) 57-73.

[4] M. Billaud, *Simple Operational and Denotational Semantics for Prolog with Cut,* Theoretical Computer Science 71 (North-Holland, 1990) 193-208.

[5] M. Billaud, *Une Sémantique Dénotationnelle Simple pour Prolog avec Prédicats d'Entrées-Sorties,* in: Proc. GULP'91, 6to Convegno sulla Programmazione Logica, Pisa (1991), 161-175.

[6] M. Billaud, *Operational and Denotational Semantics for Prolog with Input-Output Predicates,* INFORMATICA'91, Grenoble.

[7] E. Börger, *A logical Operational Semantics of Full Prolog,* IBM Technical Report IKBS 117 (1990)

[8] S. Burris, H.P. Sankappanavar, *A Course in Universal Algebra,* Graduate Texts in Mathematics 78 (Springer-Verlag, 1978).

[9] B. Courcelle and M. Nivat, *Algebraic Families of Interpretations,* in : Proc. 17th Symp. on Foundations of Computer Science, Houston (1976) also available as LABORIA report 189.

[10] S. Debray and P. Mishra, *Denotational and Operational Semantics for Prolog,* in : Journal of Logic Programming (1988) vol. 5, 61-91.

[11] E. De Vink, *Comparative Semantics for Prolog with Cut,* Report IR-166, Vrije Universiteit, Amsterdam (1988), also in Science of Computer Programming 1989-90, vol. 13, 239-264, North-Holland.

[12] N.D. Jones and H. Mycroft, *Stepwise Development of Operational and Denotational Semantics for Prolog,* in : Proc. 1984 Internat. Symp. on Logic Programming (1984) 281-288.

[13] G. Markowski and B. Rosen, *Bases for chain-complete posets,* in : Proc. 16th Symp. on Foundations of Computer Science, Berkeley (1975) 34-47.

[14] N.D. North, *A denotational definition of Prolog,* National Physical Laboratory Report DITS 106/88 (1988).

[15] S. Rossi, *Semantica denotazionale e operazionale per il Prolog (equivalenza e applicazioni),* Thesi di Laurea, Università degli Studi di Padova (1990).

Joining $k-$ and $l-$ recognizable sets of natural numbers

ROGER VILLEMAIRE

Département de mathématiques et d'informatique
U.Q.A.M., C.P. 8888 succ. A, Montréal (Québec), CANADA H3C 3P8

Summary. We show that the first order theory of $< \mathbb{N}, +, V_k, V_l >$, where $V_r : \mathbb{N}\backslash\{0\} \to \mathbb{N}$ is the function which sends x to $V_r(x)$, the greatest power of r which divides x and k, l are multiplicatively independent (i.e. they have no common power) is undecidable. Actually we prove that multiplication is definable in $< \mathbb{N}, +, V_k, V_l >$. This shows that the theorem of Büchi cannot be generalized to a class containing all k- and all l-recognizable sets.

Introduction. As J.R. Büchi showed (see section 3.), a subset of $\mathbb{N}^n$ represented in base k is recognizable on the alphabet $\{0, 1, \ldots, k-1\}^n$ if and only if it is definable in the first-order theory of $< \mathbb{N}, +, V_k >$, where $V_k(x)$ is the greatest power of k which divides x. This shows that the class of k-recognizable subsets of $\mathbb{N}^n$ ($n \in \mathbb{N}$) is closed under intersection, complementation and projection. Hence a set is in the smallest class containing all k-recognizable sets and closed under intersection, complementation and projection if and only if it is definable in $< \mathbb{N}, +, V_k >$.

A. Joyal asked to which extend it could be possible to generalize the above result joining k- and l-automata. I proved that if one takes the smallest class closed under intersection, complementation and projection which contains all k- and all l-recognizable subsets of $\mathbb{N}^n$ ($n \in \mathbb{N}$) (hence the definable subsets of $< \mathbb{N}, +, V_k, V_l >$), then it contains multiplication. Therefore there is no machine specializing Turing machines by which exactly the sets in this class are recognized. Hence one cannot hope to generalize Büchi's theorem in this way.

In the first three sections we give definitions an results about automata, recognition and logic. In section 4. we reduce the main theorem to some technical result, which we prove in the last section.

1.Automata. Let Σ be an *alphabet*, i.e. a finite set. Σ^* will denote the set of *words* of finite length on Σ containing the *empty* word λ formed of no symbol. Any subset L of Σ will be called a *language* on the alphabet Σ.

DEFINITION.*Let Σ be an alphabet. A* Σ-automata $\mathcal{A}$ *is a quadruplet* (Q, q_0, Γ, T) *where*

Q is a finite set, called the set of states,

q_0 *is an element of* Q, *called the* initial state,
Γ *is a subset of* Q *called the set of* final states
and finally T *is a function of* $Q \times \Sigma$ *to* Q, *called the* transition function.

The transition function T can be extended to a function $T^* : Q \times \Sigma^* \to Q$ in the following way:

$$T^*(q, \sigma) = T(q, \sigma) \text{ for } \sigma \in \Sigma$$
$$T^*(q, \alpha\sigma) = T(T^*(q, \alpha), \sigma) \text{ for } \alpha \in \Sigma^* \text{ and } \sigma \in \Sigma$$

Furthermore we have the following definitions.

DEFINITION. *A word* $\alpha \in \Sigma^*$ *is said to be* accepted *by the* Σ-automata (Q, q_0, Γ, T) *if* $T^*(q_0, \alpha) \in \Gamma$.

DEFINITION. *A language* L *on* Σ *is said to be* Σ-recognizable *if there exists a* Σ-*automata such that the set of words accepted by this automata is exactly* L.

2.Recognition over IN. Let Σ_k be the alphabet $\{0, 1, \ldots, k-1\}$. For $n \in \mathbb{N}$ let $[n]_k$ be the word on Σ_k which is the inverse representation of n in base k, i.e. if $n = \Sigma_{i=0}^{s} \lambda_i k^i$ with $\lambda_i \in \{0, \ldots, k-1\}$, then $[n]_k = \lambda_0 \cdots \lambda_s$.

It is also possible to represent tuples of natural numbers by words on $(\Sigma_k^n)^*$ in the following way. Let $(m_1, \ldots, m_n) \in \mathbb{N}^n$. Add on the right of each $[m_i]_k$ the minimal number of 0 in order to make them all of the same length and call these words ω_i. Let $\omega_i = \lambda_{i1} \cdots \lambda_{is}$ where $\lambda_{ij} \in \Sigma_k$. We represent $(m_1, \ldots, m_n)$ by the word $(\lambda_{11}, \lambda_{21}, \ldots, \lambda_{n1})\ (\lambda_{12}, \lambda_{22}, \ldots, \lambda_{n2}) \cdots (\lambda_{1s}, \lambda_{2s}, \ldots, \lambda_{ns}) \in (\Sigma_k^n)^*$.

DEFINITION. *We say that a set* $X \subseteq \mathbb{N}^n$ *is* k-recognizable *if it is* Σ_k^n*-recognizable.*

3.Büchi's Theorem. Let $P_k(x)$ be the predicate (i.e. subset) on IN defined by "x is a power of k". Let also as we said before $V_k : \mathbb{N} \setminus \{0\} \to \mathbb{N}$ be the function which sends x to $V_k(x)$, the greatest power of k which divides x.

In [2,Theorem 9] Büchi states that a subset of $\mathbb{N}^n$ is k-recognizable if and only if it is definable in the first-order structure $< \mathbb{N}, +, P_k >$, i.e. defined by formulas built up from $=, +, P_k$ using $\wedge$ ("and"), $\neg$ ("not"), $\exists$ ("there exists a natural number such that ..."). Unfortunately, as remarked by McNaughton in [7], the proof is incorrect. Furthermore the statement has been disproved by Semenov in [11, Corollary 4]. Thanks to the work of Bruyère [1], we know that the ideas of Büchi can be used to show the following theorem. (See [1] for a proof among the lines of Büchi's, [8] for a different proof or also [14]).

THEOREM 3.1. **Büchi's Theorem** *A set* $X \subseteq \mathbb{N}^n$ *is* k*-recognizable if and only if it is definable in the first order structure* $< \mathbb{N}, +, V_k >$.

There is another version of Büchi's Theorem in terms of weak monadic logic. Before we speak of it, let us give a useful definition and lemma.

DEFINITION.*Let* $X_{k,j}(x,y)$ *denote the relation* "x is a power of k and the corresponding digit in the representation of y in basis k is j", *for* $k \in \mathbb{N}$ *and* $j \in \{0,1,\ldots,k-1\}$. *We have the following result.*

LEMMA3.2.*The relation* $X_{k,j}(x,y)$ *is definable in* $< \mathbb{N},+,V_k >$ *for* $j = 1,\ldots,k-1$.

PROOF:$X_{k,j}(x,y)$ is defined by the formula

$$V_k(x) = x \wedge \exists z,t\,[z < x \wedge V_k(t) > x \wedge y = z + jx + t] \vee \exists z[z < x \wedge y = z + jx]$$

Here jy represents $\underbrace{y + \cdots + y}_{j-\text{times}}$ which is a term in the language.

This holds since $V_k(x) = x$ is equivalent to x being a power of k and furthermore $z < x$ and $V_k(t) > x$ means that z has 0 as coefficient for all powers of k greater or equal to x and t has coefficient 0 for all powers of k smaller or equal to x.

Usually Büchi's Theorem is stated using the weak monadic theory of $< \mathbb{N}, S >$, where S is the successor function on $\mathbb{N}$. The weak monadic theory of $< \mathbb{N}, S >$ is the extension of first order logic by allowing also the use of the *weak monadic quantifiers* $\forall X$ and $\exists X$, which are interpreted as "*for all finite subsets of* $\mathbb{N}$" and "*there exists a finite subset of* $\mathbb{N}$" respectively. We will write WM $< \mathbb{N}, S >$ for this structure. Hence usually Büchi's Theorem is stated a follows.

THEOREM3.3.**Büchi's Theorem monadic version** *A set* $X \subseteq \mathbb{N}^n$ *is 2-recognizable if and only if it is definable in the weak monadic structure WM* $< \mathbb{N}, S >$.

Let us show that this second form of Büchi's Theorem is equivalent to the first one for $k = 2$. We will give an bi-interpretation of WM $< \mathbb{N}, S >$ in $< \mathbb{N}, +, V_2 >$. First of all, any formula $\varphi(X_1,\ldots,X_s,x_1,\ldots,x_t)$ of WM $< \mathbb{N}, S >$, where $X_1,\ldots,X_s$ are monadic variables (i.e. they represent finite sets) and $x_1,\ldots,x_t$ are first-order variables, is equivalent to a formula with no first-order variables, since one can replace an element by a singleton containing it. Let us now show that there is a bijection η between the subsets of $\mathbb{N}$ and the natural numbers such that for any formula $\varphi(X_1,\ldots,X_s)$, there exists a formula φ^* with the property that $\varphi(X_1,\ldots,X_s)$ holds in WM $< \mathbb{N}, S >$ if and only if $\varphi^*(\eta(X_1),\ldots,\eta(X_s))$ holds in $< \mathbb{N}, +, V_2 >$. And furthermore that for any formula $\psi(x_1,\ldots,x_s)$ there exists a formula ψ^* such that $\psi(x_1,\ldots,x_s)$ holds in $< \mathbb{N}, +, V_2 >$ if and only if $\psi^*(\eta^{-1}(x_1),\ldots,\eta^{-1}(x_s))$ holds in WM $< \mathbb{N}, S >$.

Define $\eta(X) = \sum_{i \in X} 2^i$ and let φ be a formula in the language of WM $< \mathbb{N}, S >$. Replace in it $S(n)$ by $2^n + 2^n$, $X(n)$ (i.e. $x \in X$) by $X_{2,1}(2^n, x)$ and $\exists X$, $\forall X$ by $\exists x$, $\forall x$ and call this new formula φ^*. It is easy to show that the above property holds.

Conversely starting from ψ a formula in the language of $< \mathbb{N}, +, V_2 >$, replace in it $x + y = z$ by

$$\exists R[X(0) \wedge Y(0) \leftrightarrow R(S(0))] \wedge$$

$\forall x(R(S(x)) \leftrightarrow$ " at least two of $X(x)$, $Y(x)$, $R(x)$ hold ")$\wedge$
$\forall x(Z(x) \leftrightarrow$ " only one or all three of $X(x)$, $Y(x)$, $R(x)$ hold ")

In this formula R stands for the " carry over " in the addition of $\sum_{i \in X} 2^i$ and $\sum_{i \in Y} 2^i$. This formula can be easily expressed in the language of WM $< \mathbb{N}, S >$.

Finally replace $V_2(x) = y$ by " Y is a singleton contained in X and for all $x \in \mathbb{N}$ smaller than the element of Y, $X(x)$ does not hold ".

Here also this can easily be expressed by a formula of WM $< \mathbb{N}, S >$ as soon as one note that $x < y$ for x, y natural numbers is equivalent of " every finite subset containing y and closed under the inverse of S must contain x ". The formula ψ^* so obtained has now the required property as one can easily check.

Note that in the translation of φ into φ^*, if we replace 2^n by k^n and $X_{2,1}$ by $X_{k,1}$ we get and interpretation of WM $< \mathbb{N}, S >$ in $< \mathbb{N}, +, V_k >$. We will use this fact later.

4. A question of A.Joyal

DEFINITION. *Two natural numbers k, l are said to be* multiplicatively dependent *if there exists natural numbers n, m such that $k^n = l^m$.*

We have the following facts.

• If k and l are multiplicatively dependent then any set $X \subseteq \mathbb{N}$ which is k-recognizable is also l-recognizable (see [4, Corollary 3.7]).

• A set $X \subseteq \mathbb{N}$ which is a union of a finite set with finitely many arithmetic progressions is k-recognizable for any $k \in \mathbb{N}$ (see [4, Proposition 3.4)

• For k, l multiplicatively independent a set which is k- and l-recognizable is a finite union of a finite set with finitely many arithmetic progressions, hence it is m-recognizable for any m (see [3] or also [6] and [9]).

Therefore for k, l multiplicatively independent the class of k- and the class of l-recognizable sets of natural numbers are as far apart as they can be. This is quite unfortunate from a computational point of view, since recognition depends on the basis. A. Joyal asked if we can find a concept of "machine" and of "recognition" extending k-recognition and l-recognition for k, l multiplicatively independent.

Let $\mathcal{K}$ be the smallest class containing all k-recognizable and all l-recognizable subsets of $\mathbb{N}^n$ (for $n \in \mathbb{N}$) and closed under intersection, complementation and projection. We show that $\mathcal{K}$ contains all the arithmetical hierachy (i.e. the closure of the class of recursive relations under projection and complement), hence that there is no machine model specializing Turing machines by which exactly the sets in $\mathcal{K}$ are recognized. More precisely we show the following.

THEOREM4.1. *The structures* $< \mathbb{N}, +, V_k, V_l >$ *for* k, l *multiplicatively independent and* $< \mathbb{N}, +, \cdot >$ *are inter-definable (i.e. multiplication can be defined in terms of* V_k, V_l *and* V_k, V_l *can be defined in terms of* $\cdot$, $+$*).*

PROOF: Since any recursive function is definable in $< \mathbb{N}, +, \cdot >$ and V_k,V_l are recursive, it follows that V_k,V_l are definable in $< \mathbb{N}, +, \cdot >$. This settles one direction.

For the other direction let $k^{\mathbb{N}}$ be the set of powers of k. We will first show the following results.

LEMMA4.2. *For any* k, l *multiplicatively independent there exists a strictly increasing function* $h : k^{\mathbb{N}} \to k^{\mathbb{N}}$ *definable in* $< \mathbb{N}, +, V_k, V_l >$ *such that the following condition holds.*

(*) $h(k \cdot x) > k \cdot (h(x))$ *for infinitely many* $x \in k^{\mathbb{N}}$ *and furthermore there exists a* $d \in \mathbb{N}$ *such that for any consecutive power of* k, k^n,k^m *satisfying the above inequality* $m - n \leq d$.

We will give the proof of Lemma4.2 in the last section.

COROLLARY4.3. *Let* $h : k^{\mathbb{N}} \to k^{\mathbb{N}}$ *be a strictly increasing function satisfying* (*). *The multiplication of powers of* k *is definable in* $< \mathbb{N}, +, V_k, h >$.

PROOF: We first need to extend the interpretation of WM $< \mathbb{N}, S >$ in $< \mathbb{N}, +, V_k >$ we gave in section 3. to an interpretation of WM $< \mathbb{N}, S, h^* >$ in $< \mathbb{N}, +, V_k, h >$, where $h^* : k^{\mathbb{N}} \to k^{\mathbb{N}}$ is defined by $h(k^n) = k^{h^*(n)}$. This is easily obtain by replacing h^* by h. Note now that addition is WM $< \mathbb{N}, S, h^* >$ will be interpreted as the multiplication of powers of k in $< \mathbb{N}, +, V_k, h >$.

Using this interpretation it is sufficient to show the following.

LEMMA4.4. *Let* $h^* : \mathbb{N} \to \mathbb{N}$ *be a strictly increasing function such that* $h^*(S(x)) > S(h^*(x))$ *for infinitely many* $x \in \mathbb{N}$. *Suppose furthermore that there exists a* $d \in \mathbb{N}$ *such that for any consecutive natural numbers* x, y *satisfying the above inequality* $x - y \leq d$. *Then the addition of natural numbers is definable in WM* $< \mathbb{N}, S >$.

PROOF: This lemma is a slight generalization of the result [13, Theorem 2] of W. Thomas. We will follow the proof of Thomas modifing it to prove the above lemma. The technique is due to C.C. Elgot and M.O. Rabin and the interested reader should have a look at their nice paper [5].

The first important fact is to notice that if we can quantify over finite binary relations over $\mathbb{N}$ then we can define addition in the following way.

$$\forall E \subseteq \mathbb{N} \times \mathbb{N}\ [(x,y) \in E \wedge \forall u,v\ (u,v) \in E \to (u+1, v-1) \in E] \to (z,0) \in E$$

The above formula holds if and only if $x + y = z$, since the part in bracket means that E contains $\{(x,y),(x+1,y-1),\ldots,(x+y,0)\}$. Hence if each finite binary relation satisfying this condition contains $(z,0)$, we must have that $z = x+y$.

Therefore we want to show that there is in WM $< \mathbb{N}, S, h^* >$ a formula $F(X,x,y)$, such that for any finite binary relation $E \subseteq \mathbb{N} \times \mathbb{N}$ there is $X^E \subseteq \mathbb{N}$ for which $F(X^E,x,y)$ holds in WM $< \mathbb{N}, S, h^* >$ if and only if $(x,y) \in E$.

Suppose we can define in WM $< \mathbb{N}, S, h^* >$ disjoint sets K_i which union equals $\mathbb{N}$ and one-to-one functions $f_i : \mathbb{N} \to K_i$, $i = 1,\ldots,d$ such that $f_i^{-1}(x)$ is infinite for all $x \in \mathbb{N}$. We will show that this implies the existence of an $F(X,x,y)$ with the above property.

Let us see how we can define quantification over finite binary relations in WM $< \mathbb{N}, S, f_i, K_i; i = 1,\ldots,d >$. We need the following definitions.

- $Nxt(X,x,y) = X(x) \wedge X(y) \wedge x < y \wedge \forall z[x < z < y \to \neg X(z)]$.

Hence $Nxt(X,x,y)$ means that x and y are in X and that y is the successor of x in this set, i.e. there is no element of X between x and y.

- $Od(X,x) = X(x) \wedge \exists Y\ Y(x) \wedge$ " Y contains the smallest element of X " $\wedge$ $\forall y,z,t[Nxt(X,y,z) \wedge Nxt(X,z,t) \wedge Y(y) \to \neg Y(z) \wedge Y(t)]$.

This can be written as a first-order formula and it means that x is in X and that there is a odd number of elements in this set which are smaller or equal to x.

We can now define $F(X,x,y)$ by the following formula.

$\exists u,v[X(u) \wedge X(v) \wedge Od(X,u) \wedge Nxt(X,u,v) \wedge \bigwedge_{i=1}^{d}(K_i(x) \to f_i(u) = x) \wedge \bigwedge_{i=1}^{d}(K_i(y) \to f_i(v) = y)]$.

To see that this formula has the required property, let $E = \{(x_1,y_1),\ldots,(x_k,y_k)\}$ be an arbitrary finite relation on $\mathbb{N}$. Let x_i be in $K_{\alpha(i)}$ and y_i be in $K_{\beta(i)}$. Take n_1 to be such that $f_{\alpha(1)}(n_1) = x_1$, choose $n_2 > n_1$ such that $f_{\beta(1)}(n_2) = y_1$; this is possible since $f_{\beta_1}^{-1}(y_1)$ is infinite. Choose $n_3 > n_2$ such that $f_{\alpha(2)}(n_3) = x_2$ and so on up to n_{2k}. Let $X^E = \{n_1,\ldots,n_{2k}\}$. Then $F(X^E,x,y)$ holds if and only if $(x,y) \in E$.

Therefore the last thing to show is that we can define such f_i and K_i. Let $K_i = \{x \in \mathbb{N}; S(h^*(x)) \notin \mathrm{Im}h^*\}$, $K_2 = \{x \in \mathbb{N}; x \notin K_1$ and $S^{(2)}(h^*(x)) \notin \mathrm{Im}h^*\}$, ..., $K_d = \{x \in \mathbb{N}; x \notin K_1,\ldots,x \notin K_{d-1}$ and $S^{(d)}(h^*(x)) \notin \mathrm{Im}h^*\}$ (here $S(i)$ is the iteration of the function S). Furthermore let $f_i : \mathbb{N} \to K_i$ be defined in the following way. Let y_i be the first element of K_i.

$$f_i(x) = \begin{cases} y & \text{if } x = h^{*(m)}(S^{(i)}(h^*(y))) \text{ for some } m \in \mathbb{N} \text{ and some } y \in K_i, \\ y_i & \text{otherwise.} \end{cases}$$

It is clear by definition that the sets K_i are disjoint and by $(*)$ that their union is $\mathbb{N}$. Let us show that the functions f_i are well defined. Suppose that $h^{*(m)}(S^{(i)}(h^*(y))) = h^{*(m')}(S^{(i)}(h^*(y')))$ for some m, $m' \in \mathbb{N}$ and y, $y' \in K_i$. Since h^* is one-to-one it follows that $h^{*(m-m')}(S^{(i)}(h^*(y))) = S^{(i)}(h^*(y'))$ (we can suppose without loss of generality that $m > m'$). Since $S^{(i)}(h^*(y'))$ is not in $\mathrm{Im}h^*$

by definition of K_i, we must have that $m = m'$. Hence $S^{(i)}(h^*(y)) = S^{(i)}(h^*(y'))$ and $h^*(y) = h^*(y')$, therefore $y = y'$.

Furthermore the sets $f_i^{-1}(x) = \{y; y = h^{*(m)}(S^{(i)}(h^*(y))), m = 0, 1, \ldots\}$ is infinite for every $x \in \mathbb{N}$ and $i = 1, \ldots, d$. Finally, $f_i(x)$ is definable in WM $< \mathbb{N}, S, h^* >$ by the following formula.

$$\forall X[X(x) \wedge \forall z[X(h(z)) \rightarrow X(z)] \rightarrow X(S^{(i)}(h^*(y)))].$$

This holds since the formula says that any X which contains x and is closed under the inverse of h^* must contain $S^{(i)}(h^*(y))$. This completes the proof.

Let α, β be two words on Σ_k. The concatenation $\alpha^\wedge\beta$ is the word obtained by the letters of α followed by the letters of β. For n, m in $\mathbb{N}$ we will write $n^\wedge m$ for the natural number which corresponds to the concatenation $[n]_k^\wedge[m]_k$. More precisely if $l(n)$ is the length of $[n]_k$ then $n^\wedge m = n + k^{l(n)} \cdot m$.

We now have the following result.

LEMMA4.5. *The concatenation in base k is definable in $< \mathbb{N}, +, V_k, h >$.*

PROOF: $z = x^\wedge y$ holds if and only if

$\exists u(V_k(u) = u \wedge u > x \wedge$"$u$ is the smallest natural number with this property"

$\wedge \forall t[t < u \wedge V_k(t) = t \rightarrow \bigwedge_{j=0}^{k-1}(X_{kj}(t, z) \leftrightarrow X_{kj}(t, x))]$

$\wedge \forall t[V_k(t) = t \rightarrow \bigwedge_{j=0}^{k-1}(X_{kj}(t \cdot u, z) \leftrightarrow X_{kj}(t, y))])$

Since $t \cdot u$ is a product of powers of k this formula define the concatenation in $< \mathbb{N}, +, V_k, h >$ (by Lemma4.3).

By the following result of J.W. Thatcher [12] it follows that any recursive function is definable in $< \mathbb{N}, +, V_k, V_l >$.

LEMMA4.6. *Any recursive function is definable in $< \mathbb{N}, +, ^\wedge >$ where $^\wedge$ is the concatenation in base k.*

PROOF: See [12, Theorem 2] and also the footnote on page 183 in the same paper.

COROLLARY4.7. *The multiplication of natural numbers is definable in $< \mathbb{N}, +, V_k, ^\wedge >$, hence by Lemma4.2 and 4.5 it is definable in $< \mathbb{N}, +, V_k, V_l >$.*

The only thing which remains to be shown is Lemma4.2.

5.Definability in $< \mathbb{N}, +, V_k, V_l >$. We will now prove that for any k, l multiplicatively independent there exists a strictly increasing definable function satisfying $(*)$, this will give a proof to Lemma4.2.

We will often use the following fact: For k, n in $\mathbb{N}$, k and k^n are multiplicatively dependent, hence any set which is k-recognizable is k^n-recognizable, and vice versa (see [4,Corollary 3.7]). This means that V_k is definable in $< \mathbb{N}, +, V_{k^n} >$ and V_{k^n} is definable in $< \mathbb{N}, +, V_k >$, therefore we can consider $< \mathbb{N}, +, V_k >$ and $< \mathbb{N}, +, V_{k^n} >$ to be the same.

For the remainder of this section we will fix the following: k and l are multiplicatively independent natural numbers and Supp(x) will be the set of prime divisors of x. Furthermore $k = p_1^{\alpha_1} \cdots p_m^{\alpha_m}$ and $l = p_1^{\beta_1} \cdots p_n^{\beta_n}$ where the p_i are prime numbers.

We will consider three cases.

Case 1) Suppose Supp$(k) \not\subset$ Supp(l) and Supp$(l) \not\subset$ Supp(k). We can suppose without loss of generality that $k > l$ since we can replace k by one of its multiple. In this case we can easily define the multiplication of a power of k with a power of l.

LEMMA5.1.*Let $g : k^{\mathbf{N}} \times l^{\mathbf{N}} \to \mathbb{N}$ be the multiplication i.e. $g(x, y) = x \cdot y$. The function g is definable in $< \mathbb{N}, +, V_k, V_l >$.*

PROOF:The function $g(x, y) = z$ is defined be the formula saying "z is the smallest natural number such that $V_k(z) = x$ and $V_l(z) = y$".

LEMMA5.2.*Let $f : k^{\mathbf{N}} \to l^{\mathbf{N}}$ be such that $f(x)$ is the smallest power of l greater than x. The function f is strictly increasing and definable in $< \mathbb{N}, +, V_k, V_l >$.*

PROOF:We will show that for any two powers of k there is a power of l inbetween. Take k^r and let l^s be the greatest power of l smaller than k^r. Then $l^{s+1} > k^r$ and furthermore $l^{s+1} < k^r l < k^{r+1}$ since $l < k$ by hypothesis. Hence $k^r < l^{s+1} < k^{r+1}$. Since f is obviously definable in $< \mathbb{N}, +, V_k, V_l >$, this completes the proof.

LEMMA5.3.*Let $u : \mathbb{N} \to k^{\mathbf{N}}$ be such that $u(x)$ is the greatest power of k smaller than x. The function u is definable in $< \mathbb{N}, +, V_k >$.*

PROOF:Obvious.

LEMMA5.4.*Let $h : k^{\mathbf{N}} \to k^{\mathbf{N}}$ be such that $h(x) = u(g(x, f(x)))$. Then $h(x) = x^2$ (i.e. for powers of k only) and the function h is strictly increasing.*

PROOF:By definition $u(g(x, f(x))) = u(x \cdot f(x)) = x \cdot u(f(x))$, for $x \in k^{\mathbf{N}}$. By definition of f, we have $u(f(x)) = x$, hence $u(g(x, g(x))) = x \cdot x$. Therefore $h(x) = x^2$

and h is strictly increasing. Since h is obviously definable in $< \mathbb{N}, +, V_k, V_l >$, this completes the proof.

CONCLUSION.*Here $h(k \cdot x) = k^2 \cdot x^2 > k \cdot x^2 = k \cdot (h(x))$ for all $x \in k^{\mathbb{N}}$, hence $(*)$ holds.*

Case 2) Suppose $\mathrm{Supp}(l) \subset \mathrm{Supp}(k)$ and for any $p_i, p_j \in \mathrm{Supp}(l)$, $\frac{\alpha_i}{\beta_i} = \frac{\alpha_j}{\beta_j} = \frac{\alpha}{\beta}$, where α, $\beta \in \mathbb{N}$. Hence $k^\beta = l^\alpha u$, $u \neq 1$, $\alpha \neq 0$, $(l, u) = 1$. Since k, k^β and l, l^α are multiplicatively dependent we can replace k^β by k and $l^\alpha$by l and assume that $k = lu$.

LEMMA5.5.*Let $f : k^{\mathbb{N}} \to l^{\mathbb{N}}$ be as in Case 1. The function f is strictly increasing. Furthermore there exists a $d \in \mathbb{N} \setminus \{0\}$ such that $f(xk^d) \geq f(x)l^{(d+1)}$. As before this function is definable in $< \mathbb{N}, +, V_k, V_l >$.*

PROOF:It follows as in the proof of Lemma 5.2 that f is stricly increasing.

For the second claim take d to be the smallest natural number such that $u^d > l$. Since $f(x)$ is the smallest power of l greater than x, we have that $\frac{f(x)}{l} < x$. Hence $\frac{f(x)k^d}{l} < xk^d$, so $f(x)l^d = \frac{f(x)l^d l}{l} \leq \frac{f(x)l^d u^d}{l} < xk^d$. Therefore $f(xk^d) > f(x)l^d$.

LEMMA5.6.*Let $g' : l^{\mathbb{N}} \to k^{\mathbb{N}}$ be the function which sends l^m to k^m. The function g' is multiplicative (i.e. $g'(x \cdot y) = g'(x) \cdot g'(y)$) and strictly increasing. Furthermore g' is definable in $< \mathbb{N}, +, V_k, V_l >$.*

PROOF:Since $(l, u) = 1$ it follows that $V_l(k^n) = V_l(l^n u^n) = V_l(l^n) = l^n$. Hence we can define $g'(x) = y$ by the formula $V_l(y) = x$. The remaining of the proof is obvious.

LEMMA5.7.*Let $h = g' \circ f : k^{\mathbb{N}} \to k^{\mathbb{N}}$. The function h is strictly increasing. Furthermore for all x in $k^{\mathbb{N}}$, $h(k^d \cdot x) > k^d \cdot h(x)$. (The d is the one of Lemma 5.5). Finally h is definable in $< \mathbb{N}, +, V_k, V_l >$.*

PROOF:Since f and g' are strictly increasing by Lemma 5.5 and 5.6 respectively it follows that h is strictly increasing.

Let us show that for all $x \in k^{\mathbb{N}}$, $h(k^d \cdot x) > k^d \cdot h(x)$. This is the same as showing that for all $x \in k^{\mathbb{N}}$ $g'(f(xk^d)) > g'(f(x))k^d$. Since by Lemma 5.5 we have that for any x in $k^{\mathbb{N}}$ $f(xk^d) \geq f(x)l^{(d+1)}$, it follows (applying Lemma 5.6) that $g'(f(xk^d)) \geq g'(f(x)l^{(d+1)})$. Furthermore by Lemma 5.6 $g'(f(xk^d)) \geq g'(f(x))g'(l^{(d+1)})$ hence $g'(f(xk^d)) > g'(f(x))k^d$ since g' is increasing by Lemma 5.6. Therefore $h(k^d \cdot x) > k^d \cdot h(x)$.

Finally it follows by Lemma 5.5 and 5.6 that h is definable in$< \mathbb{N}, +, V_k, V_l >$.

CONCLUSION.*By Lemma 5.11 we know that $h(k^d \cdot x) > k^d h(x)$ for all $x \in k^{\mathbf{N}}$. Since h is strictly increasing $h(k \cdot x) \geq k \cdot h(x)$. Therefore it follows from $h(k^d \cdot x) > k^d \cdot h(x))$ that for some $y \in \{x, k \cdot x, \ldots, k^d \cdot x\}$, $h(k \cdot y) > k \cdot h(y)$. Hence $(*)$ holds.*

Case 3) Let Supp$(l) \subset$ Supp (k) and for some p_i, p_j, $\frac{\alpha_i}{\beta_i} < \frac{\alpha_j}{\beta_j}$. Hence $m < n$ with as before $k = p_1^{\alpha_1} \cdots p_n^{\alpha_n}$ and $l = p_1^{\beta_1} \cdots p_m^{\beta_m}$. We can suppose without loss of generality that $\frac{\alpha_1}{\beta_1} = \min\{\frac{\alpha_i}{\beta_i}; i = 1, \ldots, m\}$ and $\frac{\alpha_m}{\beta_m} = \max\{\frac{\alpha_i}{\beta_i}; i = 1, \ldots m\}$. Furthermore since k^{β_1},k and l^{α_1}, l are multiplicatively dependent we can suppose without loss of generality that $\frac{\alpha_1}{\beta_1} = 1$, hence $\frac{\alpha_m}{\beta_m} > 1$.

LEMMA5.8.*Let $f' : k^{\mathbf{N}} \to l^{\mathbf{N}}$ be the function which sends k^r to l^s, where $s = \lceil \frac{r\alpha_m}{\beta_m} \rceil$ (the smallest natural number greater or equal to $\frac{r\alpha_m}{\beta_m}$). The function f' is definable in $< I\!N, +, V_k, V_l >$ and strictly increasing.*

PROOF:We will show that $f'(x) = y$ can be defined by the formula " y is the smallest power of l such that $\forall u[V_l(u) \geq y \to V_k(u) \geq x]$".

Let $x = k^r$ and $y = l^s$. Take $u = p_1^{\gamma_1} \cdots p_n^{\gamma_n}$ some natural number, where some γ_i can be zero. There is no loss of generality in assuming that u in the above formula is of this form since any prime factor different of $p_1, \ldots, p_n$ would not change the value of $V_l(u)$ and $V_k(u)$.

Now $V_k(u) = V_k(p_1^{\gamma_1} \cdots p_n^{\gamma_n}) = k^{\min\{\lfloor \frac{\gamma_i}{\alpha_i} \rfloor; i=1,\ldots,n\}}$ and in the same way $V_l(u) = l^{\min\{\lfloor \frac{\gamma_i}{\beta_i} \rfloor; i=1,\ldots,m\}}$. Hence $V_l(u) \geq y \to V_k(u) \geq x$ is equivalent to "$\min\{\lfloor \frac{\gamma_i}{\beta_i} \rfloor; i = 1, \ldots, m\} \geq s$ implies that $\min\{\lfloor \frac{\gamma_i}{\alpha_i} \rfloor; i = 1, \ldots, n\} \geq r$". Furthermore this holds exactly if "for all i, $\gamma_i \geq s\beta_i$" implies "for all i, $\gamma_i \geq r\alpha_i$". Therefore $\forall u[V_l(u) \geq y \to V_k(u) \geq x]$ holds if and only if $r\alpha_i \leq s\beta_i$ for all i. Hence " y is the smallest power of l such that $\forall u[V_l(u) \geq y \to V_k(u) \geq x]$" if and only if $s = \lceil r\frac{\alpha_m}{\beta_m} \rceil$. The function f' is strictly increasing since $\frac{\alpha m}{\beta_m} > 1$. This completes the proof.

LEMMA5.9.*Let $g'' : l^{\mathbf{N}} \to k^{\mathbf{N}}$ be the function which sends l^r to k^r. The function g'' is definable in $< I\!N, +, V_k, V_l >$ and strictly increasing.*

PROOF:Since $\frac{\beta_1}{\alpha_1} = 1$, we can argue as in the proof of Lemma 5.8 to show that $g''(x) = y$ can be defined by the formula " y is the smallest power of k such that $\forall u V_k(u) \geq y \to V_l(u) \geq x$". The function is strictly increasing by definition.

LEMMA5.10.*Let $h = g'' \circ f'$. The function h is strictly increasing and $h(k^r) = k^s$, with $s = \lceil \frac{r\alpha_m}{\beta_m} \rceil$.*

PROOF:This is obvious from Lemma5.9.

CONCLUSION. *Here for any* $x = k^{u\cdot\beta_m - 1}$, $1 < u \in \mathbb{N}$ *we have that* $h(k \cdot x) = h(k^{u\cdot\beta_m}) = k^{u\cdot\alpha_m} = k \cdot k^{u\cdot\alpha_m - 1} = k \cdot k^{\lceil u\cdot\alpha_m - 1\rceil} \geq k \cdot k^{\lceil u\cdot\alpha_m - \frac{\alpha_m}{\beta_m}\rceil} = k \cdot k^{\lceil (u\cdot\beta_m - 1)\cdot\frac{\alpha_m}{\beta_m}\rceil}$ $= k \cdot h(x)$. *Hence* $(*)$ *holds with* $d = \beta_m$.

Acknowledgments. Many thanks go to D. Niwiński for pointing out to me that my original proof could be shorten by the use of Thomas' result. I would like to thank Professor André Joyal for financial support and many fruitful discussions and also Professor Pierre Leroux for financial support. Many thanks to Christian Michaux for many discussions and for sending me material on automata and logic. I would like to thank the Laboratoire de combinatoire et d'informatique-mathématique de l'Université du Québec à Montréal for its hospitality. This work has been done with the financial support of La Fondation de l'UQAM, the author holding the J.A. de Sève post-doctoral scholarship. Finally many thanks go to the referees for suggesting improvements and for pointing out to me the existence of related material, especially [12].

References

[1] V. Bruyère, Entiers et automates finis, U.E. Mons (mémoire de licence en mathématiques) 1984-85

[2] J.R. Büchi, Weak second-order arithmetic and finite automata, Z. Math. Logik Grundlagen Math. 6 (1960), pp 66-92

[3] A. Cobham, On the Base-Dependence of Sets of Numbers Recognizable by Finite-Automata, Math. Systems Theory 3, 1969, pp 186-192.

[4] S. Eilenberg, Automata, Languages and Machines, Academic Press 1974.

[5] C.C. Elgot, M.O. Rabin, Decidability and undecidability of extensions of second (first) order theory of (generalized) successor, J.S.L. vol. 31 (2) 1966, pp 169-181.

[6] G. Hansel, A propos d'un théorème de Cobham, in: D. Perrin, ed., Actes de la Fête des Mots, Greco de Programmation, CNRS, Rouen (1982).

[7] R. McNaughton, Review of [2], J. Symbolic Logic 28 (1963), pp 100-102.

[8] C. Michaux, F. Point, Les ensembles k-reconnaissables sont définissables dans $< \mathbb{N}, +, V_k >$, C.R. Acad. Sc. Paris t. 303, Série I, no 19, 1986, p.???

[9] D. Perrin, Finite Automata, in: J. van Leeuwen, Handbook of Theoretical Computer Science, Elsevier 1990.

[10] W.V. Quine. Concatenation as a basis for arithmetic. J.S.L. (1946) vol. 11 pp 105-114.

[11] A.L. Semenov, On certain extensions of the arithmetic of addition of natural numbers, Math. USSR. Izvestiya, vol 15 (1980), 2, p.401-418

[12] J.W. Thatcher, Decision Problems for multiple successor arithmetics, J.S.L. (1966) vol. 11 pp 182-190.

[13] W. Thomas, A note on undecidable extensions of monadic second order successor arithmetic, Arch. math. Logik 17 (1975), pp 43-44.

[14] R. Villemaire, $< \mathbb{N}, +, V_k, V_l >$ is undecidable. (preprint).

PARALLEL ALGORITHMS 2

On the Performance of Networks with Multiple Busses

Friedhelm Meyer auf der Heide[1)]
Hieu Thien Pham[2)]
Fachbereich Mathematik/Informatik and Heinz–Nixdorf–Institut
Universität–GH Paderborn
Warburger Str. 100
W–4790 Paderborn
Germany

Abstract

We address the following questions:

1) To which extend can the computation power of parallel processor networks be increased by using busses, i.e. by providing broadcast facilities in the networks?
2) To which extend can shared memory cells of PRAMs be replaced by links? (For this question, note that a shared memory cell can be viewed as a global bus.)

We show upper and lower bounds for computing associative operations such as ADDITION or MAXIMUM on networks with busses. Our bounds are based on simple graph theoretical properties of the networks.

As to question 1, these results demonstrate that busses can increase the performance of networks with large diameter. For example, computing MAXIMUM on a d-dimensional mesh with N processors needs time $\Theta(\sqrt[d]{N})$ without busses, but only time $\Theta\left(\sqrt[d+1]{\frac{N}{m}} + \log\log N\right)$ with m CRCW-busses.

As to question 2, these results demonstrate that the storage requirement of optimal PRAM algorithms can be reduced by adding a network with small diameter. For example, an N-processor CRCW-PRAM with an underlying binary tree network needs $m \approx \frac{N}{\operatorname{polylog} N}$ (i.e. m with $\log m = \log N - \Theta(\log\log N)$) shared memory cells to compute MAXIMUM in optimal time $\Theta(\log\log N)$ whereas, without links, $\Theta(N)$ shared memory cells are necessary.

We further consider a very simple, easy to realize class of networks with busses, namely planar networks with planar busses. We describe a planar system of EREW-busses for square meshes on which associative operations can be performed in optimal time $\Theta(\log N)$ — compared to $\Theta(\sqrt{N})$ without busses. On the other hand, we prove that SORTING on planar networks cannot be sped up by the use of additional planar busses.

---

1) Partially supported by DFG-Grants Me 872/1–4 and Di 412/2–1
2) Supported by DFG-Grant Me 872/1–4

1 Introduction

Communication in large parallel computers is usually realized using links or busses. Links connect pairs of processors by communication channels. Busses connect an arbitrary number of processors and work like a broadcast medium, i.e. in each computation round, only one value can be written on the bus, which then can be read by all members of the bus.

In this paper, we consider parallel machines which communicate both by busses and by links. Thus, a machine M is specified by a set $\mathcal{P} = \{P_1, P_2, \ldots, P_N\}$ of N processors which are connected via links from a set E to a bounded degree network $\mathcal{G} = (\mathcal{P}, E)$, and which in addition contain m busses B_1,B_2,...,B_m. A bus B_i is specified by the set of processors connected to it. The main questions addressed in this paper are:

1. *To which extend can computations of parallel processor networks be sped up by the use of busses?*

2. *To which extend can shared memory cells of a PRAM be replaced by links? (Note that a shared memory cell can be viewed as a global bus as well.)*

Previous Work. Almost all previous work considers d-dimensional meshes with additional busses. For d-dimensional networks with m global busses — "global bus" means that all processors are connected to the bus — Bokhari [Bok84] (for $d = 2$, $m = 1$) and Aggarwal [Agg86] show that computing an associative operation on N elements distributed over the N processors can be done in time $O\left(\sqrt[d+1]{\frac{N}{m}} + \log N\right)$. Note that without busses, it needs time $\Theta(\sqrt[d]{N})$. ($\sqrt[d]{N}$ is the diameter of the mesh.)

In [Agg86], the algorithm for MAXIMUM is shown to be optimal within a constant factor for "conservative flow algorithms". Such algorithms may only communicate input variables (via links or busses) but no more complicated information.

Bar-Noy and Peleg consider meshes with a bus on each row and each column, and show the interesting phenomenon that the square mesh with N processors is not optimal for computing associative operations. It becomes faster when the mesh has edge-lengths $N^{3/8}$ and $N^{5/8}$ [BP89].

Stout presents algorithms on meshes with busses on rows and columns for several problems [Sto86].

New Results. For a bus, we can consider several types of read/write modes as they are well known for PRAMs. We prove upper and lower bounds for the most common modes: EREW (exclusive read, exclusive write), CREW (concurrent read, exclusive write), and CRCW (concurrent read, concurrent write) with the PRIORITY-write conflict resolution. For definitions of such rules see e.g. [Weg87].

We first consider networks with m global busses computing associative operations as ADDITION, MULTIPLICATION or, most important, MAXIMUM on N variables x_1,x_2,...,x_N which can take values from $\mathbb{N}$. We assume that initially each processor knows one variable, and finally, P_1 knows the result. As at most N busses can be used per step, we assume w.l.o.g. that $m \leq N$.

For a network $\mathcal{G} = (\mathcal{P}, E)$, $t \geq 0$, a *t-dominating set* is a subset $S \subseteq \mathcal{P}$ such that the t-environments of the nodes from S cover $\mathcal{P}$. A *t-net* is a subset $S \subseteq \mathcal{P}$ such that the t-environments of the nodes from S are disjoint. Let $L_{\mathcal{G}}(t)$ denote the cardinality of a smallest t-dominating set in $\mathcal{G}$, and $l_{\mathcal{G}}(t)$ the cardinality of a largest t-net in $\mathcal{G}$.

We only state the results for computing the MAXIMUM and comment on the other associative operations. We prove:

- *Computing MAXIMUM needs time* $O\left(\min_{t>0}\left\{t + \frac{L_{\mathcal{G}}(t)}{m}\right\} + h(N)\right)$ *on a bounded degree network* $\mathcal{G}$ *with* m *global busses, where* $h(N) = \log N$ *for EREW-, CREW-busses and* $h(N) = \log\log N$ *for CRCW-busses.*

For this result, we use a very simple algorithm with only elementary internal computation and only short message lengths. The following lower bound holds for the case that the processors have arbitrary computation power, and that messages can have arbitrary lengths.

- *Computing MAXIMUM needs time* $\Omega\left(\min_{t>0}\left\{t + \frac{l_{\mathcal{G}}(t)}{m}\right\} + h(N)\right)$ *on a network* $\mathcal{G}$ *with* m *global busses, where* $h(N)$ *is as above.*

For other associative operations as ADDITION, the same results hold, but $h(N)$ equals $\log N$ even for CRCW-busses. For comments on Boolean operations, see Section 6. The results above give matching upper and lower bounds for many networks $\mathcal{G}$. For example,

- In d-dimensional meshes with N processors, $L_{\mathcal{G}}(t)$ and $l_{\mathcal{G}}(t)$ are in $\Theta\left(\frac{N}{t^d}\right)$. Therefore, MAXIMUM requires time $\Theta\left(\sqrt[d+1]{\frac{N}{m}} + \log N\right)$ for EREW-, CREW-busses and $\Theta\left(\sqrt[d+1]{\frac{N}{m}} + \log\log N\right)$ for CRCW-busses. In contrast, MAXIMUM takes time $\Theta(\sqrt[d]{N})$ in meshes without busses.
- In N-processor bounded degree tree networks, Butterfly networks, cube-connected cycles networks etc., $L_{\mathcal{G}}(t)$ and $l_{\mathcal{G}}(t)$ are in $\Theta\left(\frac{N}{c^t}\right)$ for some constant $c > 1$. Therefore, MAXIMUM takes time $\Theta\left(\log\frac{N}{m} + \log\log N\right)$ on such networks with m global CRCW-busses whereas it needs time $\Theta(\log N)$ in these networks without busses. This yields that $m = \frac{N}{\text{polylog} N}$ (i.e. m with $\log m = \log N - \Theta(\log\log N)$) shared memory cells of an N-processor CRCW-PRAM are sufficient for computing MAXIMUM in optimal time $\Theta(\log\log N)$ if the processors of the PRAM have in addition links as the above networks. Note that $\Theta(N)$ shared memory cells are necessary without links.

Our lower and upper bounds generalize those from [Bok84], [Agg86] for meshes with busses mentioned above to arbitrary bounded degree networks. More important, our lower bound works for general algorithms, not only for conservative flow algorithms. This answers an open question posed in [Agg86] and generalizes the answer to arbitrary networks with multiple busses.

In the second part of the paper, we restrict ourselves to networks with busses which have a simple layout in the plane, namely *planar networks with planar busses.* A system of m busses $B_1, B_2, \ldots, B_m$ is *planar* if there are edge-disjoint trees $T_1, T_2, \ldots, T_m$, where T_i connects the nodes (processors) from B_i, such that the T_i's are edge-disjoint, and such that the union of the T_i's forms a planar graph. For realistic reason, we assume that each processor is joined to at most a constant number of busses. We show:

- *There is a system of planar EREW-busses on the square mesh with N processors on which associative operations can be executed in optimal time $\Theta(\log N)$. The bus system is very simple: the busses have size 4 and are disjoint.*
- *SORTING on any planar network with planar CRCW-busses is as slow as on square mesh without busses: it needs time $\Omega(\sqrt{N})$.*

The lower bound is shown by a simple application of the Planar Separator Theorem and holds for conservative flow algorithms as described above. Allowing arbitrary internal computation power and arbitrary lengths of messages, SORTING would become as easy as computing associative operations, i.e. we would obtain $\Theta(\log N)$ run time for SORTING on planar networks with planar EREW-busses.

Part of this material, specialized to meshes, is contained in the Diploma Thesis of the second author [Pha90].

The paper is organized as follows: Section 2 describes the algorithm for associative operations on networks with global busses, Section 3 contains the corresponding lower bound. Section 4 describes the algorithm for associative operations on meshes with planar busses, Section 5 the lower bound for SORTING on planar networks with planar busses. Section 6 contains some remarks and open problems.

2 An Algorithm for Networks with Global Busses

The computation model considered in this section is any bounded degree network $\mathcal{G} = (\mathcal{P}, E)$ with N processors and m global busses for which any read/write conflict resolution can be assumed. As at most N busses can be used per step, we assume w.l.o.g. that $m \leq N$.

Definition 2.1 Let $\mathcal{G} = (\mathcal{P}, E)$ be a graph. For $t \geq 0$, a *t-dominating set* in $\mathcal{G}$ is a subset $S \subseteq \mathcal{P}$ such that the t-environments of the nodes from S cover $\mathcal{P}$. (The t-environment of a node v of $\mathcal{G}$ consists of all nodes that can be reached from v via a path of length at most t.) Let $L_{\mathcal{G}}(t)$ denote the cardinality of a smallest t-dominating set in $\mathcal{G}$.

Theorem 2.2 *Any associative operation can be computed on a bounded degree network $\mathcal{G} = (\mathcal{P}, E)$ with m global busses in time*

$$O\left(\min_{t>0}\left\{t + \frac{L_{\mathcal{G}}(t)}{m}\right\} + h(N)\right),$$

where $h(N) = \log N$ except for the case of computing MAXIMUM with CRCW-busses. In this case, $h(N) = \log\log N$.

Proof Consider a network $\mathcal{G} = (\mathcal{P}, E)$ with m global busses. Fix $t > 0$. Let S be a t-dominating set of $\mathcal{G}$ with minimum cardinality $L_{\mathcal{G}}(t)$. For $P \in S$, let $U(P)$ be a subset of the t-environment of P such that the sets $U(P)$, $P \in S$, form a disjoint partition of $\mathcal{P}$. Consider the following algorithm for ADDITION on $\mathcal{G}$. (Note that it works for all other associative operations as well.)

Algorithm 2.3

Phase 1: For each processor $P \in S$, the processors from $U(P)$ use their links to add up their input values and store the result in P.
Comment: Now we have to add up the $|S|$ subsums stored in the processors from S.

Phase 2: While there are $k > m$ values to be added up: execute $\min\{m, k-m\}$ additions using the busses.

Phase 3: For the remaining m values, use a PRAM algorithm for adding up m values with N processors and m shared memory cells.

Obviously, the above algorithm computes the sum of the values. Phase 1 takes time $O(t)$. Phase 2 takes time $O\left(\frac{|S|}{m}\right) = O\left(\frac{L_{\mathcal{G}}(t)}{m}\right)$, because each pass of the while loop takes constant time and $\left\lceil\frac{|S|}{m}\right\rceil - 1$ passes are executed. It is straight forward to see that step 3 takes time $O(\log m)$ with EREW-busses. It is shown in [SV81] that time $O\left(\max\left\{1, \log\log m - \log\log\frac{N}{m}\right\}\right) = O(\log\log N)$ suffices for CRCW-PRAMs with N processors and m shared memory cells, thus for our machine with m CRCW-busses. Therefore the algorithm needs time $O\left(t + \frac{L_{\mathcal{G}}(t)}{m} + h(N)\right)$. The theorem follows by choosing t such that the term $\left(t + \frac{L_{\mathcal{G}}(t)}{m}\right)$ is minimized. (Note that $\max\left\{1, \log\log m - \log\log\frac{N}{m}\right\}$ is of lower order than $\log\log N$ only if m is very small. But in this case the overall run time bound is governed by the term $\min_{t>0}\left\{t + \frac{L_{\mathcal{G}}(t)}{m}\right\}$.) □

3 A Lower Bound for Networks with Global Busses

In this section, we prove the lower bound for computing MAXIMUM on a network with N processors and m global busses. Recall that we assume $m \leq N$. For this lower bound, we allow a very strong computation model. A computation round consists of

- *a computation step:* every processor P executes an arbitrarily complex computation, depending on its internal configuration. In particular, it computes values $v(P)$, $w(P)$ and $r(P)$. $v(P)$ is the value to be written. $w(P)$ and $r(P)$ specify links or busses incident to P used for writing and reading, respectively.
- *a write step:* P writes $v(P)$ on bus or link $w(P)$.
- *a read step:* P reads the value written on bus or link $r(P)$.

Definition 3.1 Let $\mathcal{G} = (\mathcal{P}, E)$ be a graph. For $t \geq 0$, a *t-net* in $\mathcal{G}$ is a subset $S \subseteq \mathcal{P}$ such that the t-environments of the nodes from S are disjoint. Let $l_{\mathcal{G}}(t)$ denote the cardinality of a largest t-net in $\mathcal{G}$.

Theorem 3.2 *Computing MAXIMUM on a network $\mathcal{G} = (\mathcal{P}, E)$ with m global busses requires time*

$$\Omega\left(\min_{t>0}\left\{t + \frac{l_{\mathcal{G}}(t)}{m}\right\} + h(N)\right),$$

where $h(N) = \log N$ for EREW-, CREW-busses and $h(N) = \log\log N$ for CRCW-busses.

Remark: The same result can (much easier) be obtained for other associative operations on the integers as ADDITION or MULTIPLICATION. In these cases, $h(N) = \log N$ even holds for CRCW-busses.

Proof of Theorem 3.2 Let A be an optimal algorithm for MAXIMUM on our network with busses. W.l.o.g., it needs the same number T of steps for all inputs. We prove

(a) $T = \Omega\big(h(N)\big)$ and

(b) $T \geq \frac{l_{\mathcal{G}}(T)}{m}$.

As (b) implies that $T = \Omega\left(T + \frac{l_{\mathcal{G}}(T)}{m}\right)$, and $T + \frac{l_{\mathcal{G}}(T)}{m} = \Omega\left(\min_{t>0}\left\{t + \frac{l_{\mathcal{G}}(t)}{m}\right\}\right)$, (a) and (b) imply the theorem.

- ad (a): $T = \Omega\big(h(N)\big)$ is shown for EREW-, CREW-PRAMs in [CDR86] and for CRCW-PRAMs in [FMW87]. Clearly, these lower bounds for PRAMs also hold for our models.

- ad (b): The proof for $T \geq \frac{l_{\mathcal{G}}(T)}{m}$ is inspired by the lower bounds in [FMW87]. It uses an adversary argument. First, we fix a T-net $V_1 \subseteq \mathcal{P}$ in $\mathcal{G}$ of maximum cardinality $l_{\mathcal{G}}(T)$ and fix the values of input variables $x_i \notin V_1$ to 1.

  Now, intuitively, during the whole algorithm a processor P can learn at most one input variable $x_i \in V_1$ which is transported to P only via links, i.e. without using busses. The reason is that, by definition of V_1, no processor has more than one processor from V_1 in its T-environment. Thus, each further $x_j \in V_1$ learned by P has to use a bus on its way to P. Our adversary now fixes the value of every input variable which some time is sent via a bus, reducing V_1 more and more.

  In order to make this argument formal we have to make the computation oblivious, i.e. make the computation pattern ("which processor uses which link or bus at time t") independent of the input. For this purpose, we use a simple application of the Pidgeon Hole Principle.

  For $V \subseteq X := \{x_1, x_2, \ldots, x_N\}$, $S \subseteq \mathbb{N}$, $i \in \{1, 2, \ldots, N\}$ let $L^i = S$ for $x_i \in V$, $L^i = \{a_i\} \subseteq \mathbb{N}$ for $x_i \in X \setminus V$, such that each $a_i < \min S$, $A = \{a_i \mid x_i \in X \setminus V\}$. The variables in V are the *living variables*. Then $L(V, S, A) := \times_{i=1}^{N} L^i$.

Lemma 3.3 *For each $t \geq 1$, there is (V_t, S_t, A_t) as above, such that for inputs from $L_t = L(V_t, S_t, A_t)$, the following holds before step t:*

(i) *Each processor only knows at most one living variable.*

(ii) *For all processors P, all $i \in V_t$: if P knows x_i, then P did not write on a bus since it got known x_i.*

(iii) $|S_t| = \infty$.

(iv) $|V_t| \geq l_{\mathcal{G}}(T) - m(t-1)$.

(v) $S_t \subseteq S_{t-1}$, $V_t \subseteq V_{t-1}$.

Before we prove the lemma, we conclude (b) from it. For inputs from L_T the processor P_1 (which has to get the result) only knows about one living variable after T steps. Because of the structure of L_T, it only can know the correct maximum, if $|V_T| = 1$. Thus, by (iv), $1 \geq l_{\mathcal{G}}(T) - m(T-1)$, which implies (b). □

Proof of Lemma 3.3 - We proceed by induction on t.

- $t = 1$: Let $V_1 \subseteq \mathcal{P}$ be a T-net in $\mathcal{G}$ with maximum cardinality $l_{\mathcal{G}}(T)$. Fix each $x_i \notin V_1$ to 1, i.e. $a_i = 1$. Let $S_1 := \mathbb{N} \setminus \{1\}$. The resulting set $L_1 = L(V_1, S_1, A_1)$ obviously fulfills the lemma for $t = 1$.
- $t > 1$: Let $L_{t-1} = L(V_{t-1}, S_{t-1}, A_{t-1})$ fulfill the lemma for $t-1$.
  We first restrict L_{t-1}, such that the communication in step $t-1$ becomes oblivious. Consider P_1. It knows at most one living variable, say x. Thus $w(P_1)$, $r(P_1)$ (addresses for writing and for reading) solely depend on the value of x. As this function f that maps S_{t-1} to $(w(P_1), r(P_1))$ has finite range, there is an infinite subset $R_1 \subseteq S_{t-1}$, such that f is constant on R_1 (Pidgeon Hole Principle). Restrict S_{t-1} to R_1 and go on in the same way with $P_2, P_3, \ldots, P_N$, restricting R_1 to R_2, R_2 to R_3 and so on. As a result, we obtain an infinite $R_N \subseteq S_{t-1}$ such that, for inputs from $L' = L(V_{t-1}, R_N, A_{t-1})$, each processor has fixed links/busses which are used for write/read in step $t-1$, i.e. for inputs from L', the communication in step $t-1$ is oblivious.
  Now fix all those input variables x_i to $c := \min R_N$ which are known by a processor that writes on some bus in step $t-1$. Add c to A_{t-1} to get A_t, remove these x_i's from V_{t-1} to get V_t, remove c from R_N to get S_t.
  Clearly, $L_t = (V_t, S_t, A_t)$ fulfills (iii), (iv), (v). The above rule for restricting V_t ensures that also (ii) holds. The reason why (i) holds is as follows:
  By construction, a processor knowing a living variable x since step $t' < t$ has never written on a bus from time t' on, because otherwise this variable x would become a constant. Thus, a processor only can learn a living variable, if it is communicated to him only via links. Thus, P can only learn living variables from its T-environment during the whole algorithm. As already V_1, and therefore also V_t form a T-net, no processor has more than one living variable in its T-environment, which implies (i).

□

4 An Algorithm for Square Meshes with Planar Busses

In this section, we demonstrate how to construct a very simple planar system of EREW-busses for a $(\sqrt{N} \times \sqrt{N})$ mesh, such that we can execute any associative operation in optimal time $\Theta(\log N)$. In our system, each bus contains only four processors and each processor is member of at most one bus.

Definition 4.1 A system of m busses $B_1, B_2, \ldots, B_m$ is *planar* if there are edge-disjoint trees $T_1, T_2, \ldots, T_m$, where T_i connects the nodes (processors) from B_i, such that the T_i's are edge-disjoint, and such that the union of the T_i's forms a planar graph.

Theorem 4.2 *There is a planar system of EREW-busses on a $(\sqrt{N} \times \sqrt{N})$-mesh such that associative operations can be executed in optimal time $\Theta(\log N)$.*

Proof The lower bound holds because it is even true for EREW-PRAMs. For the upper bound, we first specify the construction of the planar busses.
Let $k, L \in \mathbb{N}$, $k \leq L$. We define a (L,k)-bus system for a $(2^k L \times 2^k L)$-mesh as follows:

- $k = 1$: a $(L,1)$-bus system contains one bus that connects the four nodes in the upper left corners of the four $(L \times L)$-blocks the $(2L \times 2L)$-mesh consists of.
- $k > 1$: a (L,k)-bus system consists of the busses of the four $(L, k-1)$-bus systems recursively defined in the four $(2^{k-1}L \times 2^{k-1}L)$-meshes, the $(2^k L \times 2^k L)$-mesh consists of, and one further bus B.
  B connects the four nodes which are the kth nodes in the diagonals of the lower right $(L \times L)$-blocks in the four $(2^{k-1}L \times 2^{k-1}L)$-meshes. (Note that $k \leq L$, i.e. these nodes are well defined.)

It is easy to check that the above bus system is planar when the busses are drawn in the ⊔-shape as depicted in Figure 1.

Now let $L := \min\{s \in \mathbb{N} \mid 2^s s \geq \sqrt{N}\}$. For simplicity, we assume that $\sqrt{N} = 2^k L$ for some $k \leq L$, $k \in \mathbb{N}$, i.e. $L \approx k \approx \log N$. Then the planar bus system for the $(\sqrt{N} \times \sqrt{N})$-mesh is the (L,k)-bus system. For an example, see Figure 1.

Now we describe recursively an algorithm for ADDITION for $(2^k L \times 2^k L)$-meshes with a (L,k)-bus system. (Note that in the same way arbitrary associative operations can be performed.) We assume that the result appears in the kth node of the diagonal of the lower right $(L \times L)$-block in the $(2^k L \times 2^k L)$-mesh.

- $k = 1$: Use the links to compute the sum of the elements in each of the four $(L \times L)$-blocks in the $(2L \times 2L)$-mesh and store the results in the upper left nodes. Use the bus to add these four subsums. (Note that the bus connects these four nodes.)
- $k > 1$: Compute recursively the subsums of the four $(2^{k-1}L \times 2^{k-1}L)$-submeshes, the $(2^k L \times 2^k L)$-mesh consists of, and store them in the $(k-1)^{th}$ node of the diagonal of the lower right $(L \times L)$-block of each $(2^{k-1}L \times 2^{k-1}L)$-submesh.
  Use the links to move the four subsums to the kth node in the respective diagonal and compute their sum using the bus connecting these nodes.

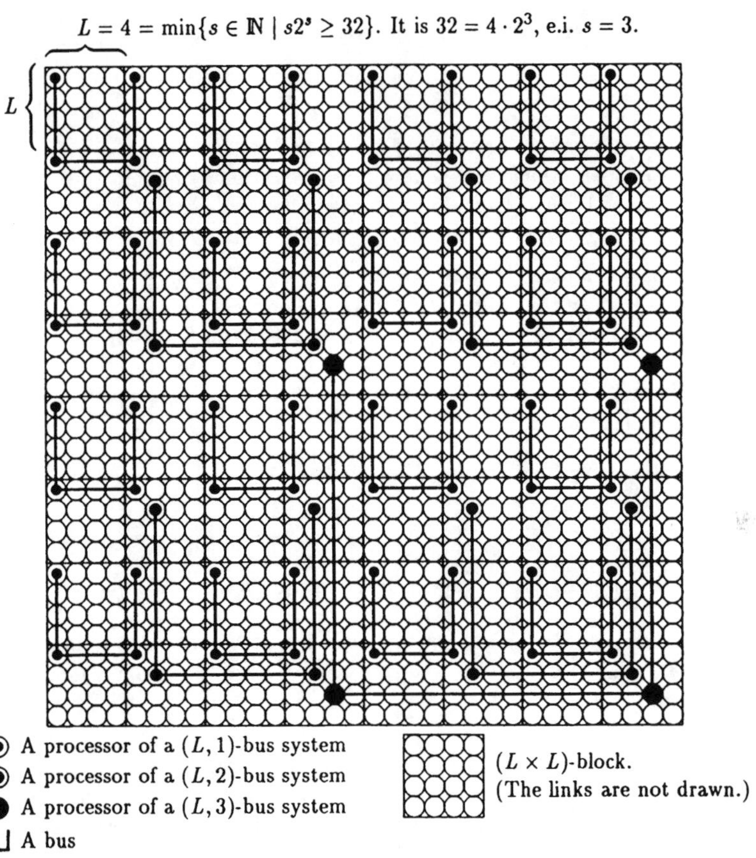

Fig. 1: A (32×32)-mesh with planar busses

Clearly, our algorithm is correct. Let $T(k)$ denote the time needed for a $(2^k L \times 2^k L)$-mesh, then we obtain:

$$\begin{aligned} T(1) &= 2(L-1)+3 \\ T(k) &= T(k-1)+5 \qquad \text{for } k > 1. \end{aligned}$$

Thus, we get $T(k) = 2L + 5k - 4$. As we have chosen $L \approx k \approx \log N$, we obtain the desired $O(\log N)$ time bound. □

5 A Lower Bound for Planar Networks with Planar Busses

In the previous section, we saw that a very simple planar system of EREW-busses can help to execute associative operations on a square mesh in optimal time. In this section, we show that SORTING on any planar network with planar CRCW-busses is as slow as on square mesh without busses, assuming that every processor is joined to at most c busses, for a constant c.

It is well known that it is possible to sort on a $(\sqrt{N} \times \sqrt{N})$ mesh without busses in $\Theta(\sqrt{N})$ steps [SS89]. We now show that SORTING on any planar network with any planar system of CRCW-busses needs $\Omega(\sqrt{N})$ steps, using the Planar Separator Theorem [Meh84]:

Theorem 5.1 (Planar Separator Theorem) *Let $\mathcal{G} = (V, E)$ be a planar graph with $|V| = N$. Then there is a partition X,S,Y of V such that*

(i) $|X| \leq \frac{2N}{3}$, $|Y| \leq \frac{2N}{3}$

(ii) $|S| \leq 4\sqrt{N}$

(iii) *S separates X from Y, i.e. no edge connects a node from X to a node from Y.*

Theorem 5.2 *Let $\mathcal{G} = (\mathcal{P}, E)$ be a planar network with N processors containing in addition a planar system of CRCW-busses. Then every conservative flow algorithm for SORTING on $\mathcal{G}$ needs $\Omega(\sqrt{N})$ steps.*

Proof Let $B_1, B_2, \ldots, B_m$ be the m planar busses. Note that we here do not restrict the number m of busses but only their structure. Let $\mathcal{G}' = (\mathcal{P}, E')$ be the planar graph consisting of the edges of the T_i's as described in Definition 4.1. Theorem 5.1 now guarantees that we find a partition X, S, Y of $\mathcal{P}$ relative to $\mathcal{G}$ and another partition X', S', Y' of $\mathcal{P}$ relative to $\mathcal{G}'$ which fulfill the following inequalities:

$$\begin{aligned} |S|, |S'| &\leq 4\sqrt{N} \\ \frac{N}{3} - 4\sqrt{N} \leq |X|, |Y|, |X'|, |Y'| &\leq \frac{2N}{3}. \end{aligned}$$

W.l.o.g., we assume $|X'| \leq |Y'|$ and $|X' \cap Y| \leq |X' \cap X|$, which imply $|X'| \leq \frac{N}{2}$ and $|X' \cap Y| \leq \frac{N}{4}$. Thus, we may conclude:

$$|X' \cap X| \geq \frac{\frac{N}{3} - 8\sqrt{N}}{2} = \frac{N}{6} - 4\sqrt{N} \quad \text{and}$$

$$|Y \cap Y'| = |Y| - |X' \cap Y| - |S' \cap Y| \geq \left(\frac{N}{3} - 4\sqrt{N}\right) - \frac{N}{4} - 4\sqrt{N} = \frac{N}{12} - 8\sqrt{N}.$$

The above inequalities yield:

- There are at most $|S|$ links that connect the processors from $V := Y \cap Y'$ to the processors from $U := X \cap X'$. Thus, in one step at most $|S| \leq 4\sqrt{N}$ input variables can be sent from V to U via links.

- As $V \subseteq Y'$ and $U \subseteq X'$, every bus that contains both a processor from V and a processor from U must contain a processor from S', too. Since every processor is joined to at most c busses, there are at most $c \cdot |S'|$ busses that connect the processors from V to the processors from U. Therefore, in one step at most $c \cdot |S'| \leq 4c\sqrt{N}$ input variables can be sent from V to U via busses.

Totally, no more than $O(\sqrt{N})$ input variables can be sent in one step from V to U. Thus, it takes $\Omega(\sqrt{N})$ steps for SORTING on $\mathcal{G}$ if at least $\frac{N}{12} - 8\sqrt{N}$ input values from V have to be sent to U. □

6 Conclusion

In this paper, we have demonstrated that the use of busses can improve the performance of networks, in particular of networks with not too small diameter. Several questions remain open:

1. Can busses help for computing other than associative functions? For example, how many (global) busses are necessary to allow SORTING (with realistic constraints on internal computation power and length of messages) in time less than $\Omega(\sqrt{N})$ on a 2-dimensional mesh? Can lower bound methods for PRAMs (e.g. [MW87]) be adapted?

2. Can we get tight bounds for computing Boolean associative operations like OR, AND or PARITY? Note that OR and AND can be done with one CRCW-bus in constant time. On the other hand, a CREW-PRAM needs time $\Theta(\log N)$, see [Weg87]. How much time does a network with m CREW-busses need? Can techniques for PRAMs be adapted? PARITY needs time $\Theta\left(\frac{\log N}{\log\log N}\right)$ on CRCW-PRAM, see [Weg87], can such proof techniques be used to get lower bounds in our computation models?

References

[Agg86] A. Aggarwal. Optimal bounds for finding maximum on array of processors with k global buses. *IEEE Trans. on Comp.*, C-35:62–64, Jan. 1986.

[Bok84] S. H. Bokhari. Finding maximum on an array processor with a global bus. *IEEE Trans. on Comp.*, C-33:133–139, Feb. 1984.

[BP89] A. Bar-Noy and D. Peleg. Square meshes are not always optimal. In *ACM Symp. on Parallel Algorithms and Architectures*, pages 138–147, 1989.

[CDR86] S. Cook, C. Dwork, and R. Reischuk. Upper and lower bounds for parallel random access machines without simultaneous writes. *SIAM Journal on Comp.*, 15(1):87–97, 1986.

[FMW87] F. Fich, F. Meyer auf der Heide, and A. Wigderson. Lower bounds for parallel random access machines with unbounded shared memory. In F. P. Preparata, editor, *Advances in Computing Research*, volume 4 of *Parallel and Distributed Computing*, 1987.

[Meh84] K. Mehlhorn. *Graph Algorithms and NP-Completeness*, volume 2 of *Data Structures and Algorithms. EATCS Monographs on Theoretical Computer Science.* Springer Berlin, 1984.

[MW87] F. Meyer auf der Heide and A. Wigderson. The complexity of parallel sorting. *SIAM Journal on Comp.*, 16(1):100–107, 1987.

[Pha90] H. T. Pham. Fundamental algorithms for networks with busses (in German). *Diploma Thesis, Universität Dortmund*, Dec. 1990.

[SS89] I. D. Scherson and S. Sen. Parallel sorting in two-dimensional VLSI models of computation. *IEEE Trans. on Comp.*, 38:238–249, Feb. 1989.

[Sto86] Q. F. Stout. Meshes with multiple buses. In 27th *IEEE Symp. on Foundations of Comp. Science*, pages 264–273, 1986.

[SV81] Y. Shiloach and U. Vishkin. Finding the maximum, merging, and sorting in a parallel computation model. *Journal of Algorithms*, 2:88–102, 1981.

[Weg87] I. Wegener. *The Complexity of Boolean Functions.* Series in Computer Sience. Wiley-Teubner, 1987.

Efficient Algorithms for Solving Systems of Linear Equations and Path Problems *

(Extended Abstract)

Venkatesh Radhakrishnan, Harry B. Hunt III and Richard E. Stearns
University at Albany, SUNY

Abstract

Efficient algorithms are presented for solving systems of linear equations defined on and for solving path problems [11] for treewidth k graphs [20] and for α-near-planar graphs [22]. These algorithms include the following:

1. $O(nk^2)$ and $O(n^{3/2})$ time algorithms for solving a system of linear equations and for solving the single source shortest path problem,

2. $O(n^2k)$ and $O(n^2 \log n)$ time algorithms for computing A^{-1} where A is an $n \times n$ matrix over a field or for computing A^* where A is an $n \times n$ matrix over a closed semiring, and

3. $O(n^2k)$ and $O(n^2 \log n)$ time algorithms for the all pairs shortest path problems.

One corollary of these results is that the single source and all pairs shortest path problems are solvable in $O(n)$ and $O(n^2)$ steps, respectively, for any of the decomposable graph classes in [5].

Key words: Algorithms and data structures, Mathematics of computation, Systems of Linear Equations, Path Problems, Gaussian Elimination.

1 Introduction

Let F be a field; and let R be a closed semiring [4, 15, 11]. Let A be the $n \times n$ matrix $(a_{i,j})$ over F or R; and let G_A be the directed graph (V_A, E_A), where $V_A = \{x_i | 1 \leq i \leq n\}$ and $E_A = \{(x_i, x_j) | a_{i,j} \neq 0\}$. Let $\mathcal{L}_{A,b}$ be a system of linear equations on F or on R of the form[1]

$$x_i = a_{i,1}x_1 + a_{i,2}x_2 + \cdots + a_{i,n}x_n + b_i \; (1 \leq i \leq n)$$

A number of researchers [4, 2, 19, 7, 17, 23, 15, 11, 18] have studied the complexity of the problems **P1, P2** and **P3** below and the use of these problems in solving path problems for graphs. These path problems include both the **single-source** and the **all-pairs shortest path problems** (denoted by sssp and apsp respectively) [2, 11, 18], all of the path problems in [11] and all of the path expression problems in [23].

P1 : Solve $\mathcal{L}_{A,b}$.

*Supported in part by NSF Grant CCR 89-03319. Address: Department of Computer Science, SUNY at Albany, Albany, NY 12222,USA.

[1] For fields, the equations can be of the equivalent form

$$a_{i,1}x_1 + a_{i,2}x_2 + \cdots + a_{i,n}x_n = b_i \; (1 \leq i \leq n)$$

P2 : Compute the matrix A^{-1}, if A^{-1} exists.

P3 : Compute the matrix A^*, when A is a matrix over a closed semiring R.

Rose [19] and Bunch et al. [7] have shown how certain properties of the graph G_A can be used to characterize the amount of fill-in that occurs in solving $\mathcal{L}_{A,b}$ by means of Gaussian elimination and, equivalently in computing the LDU-decomposition [14] of the matrix A. (Fill-ins are the non-zeros created in the factoring process where A contains zeros.) Lipton et al. [17] and Pan et al. [18] have shown how the ideas of [19] and [7] and the planar separator theorem of [16] can be used to solve problem **P1** in $O(n^{3/2})$ operations and to solve problems **P2** and **P3** in $O(n^2 \log n)$ operations. Here, we show how the ideas of [19] and [7] can be used to solve problems **P1**, **P2** and **P3**, when

(1) the graph G_A is presented, together with one of its tree-decompositions with treewidth k [20], and

(2) the graph G_A is presented, together with one of its planar layouts with $\leq \alpha \cdot |V_A|$ crossovers of edges [22].

Throughout this paper by a "treewidth k graph", we mean a graph presented together with one of its tree decompositions of treewidth k; and by an "α-near-planar graph", we mean a graph with vertex set V presented together with one of its planar layouts with $\leq \alpha \cdot |V|$ crossovers of edges.

A synopsis of our results appears in Table 1. Exactly analogous results hold for the shortest path problems in [11] and the path expression problems in [23]. In the remainder of this section, we discuss the importance of our results and how they extend and complement the known results in the literature. Finally, an overview of the rest of the paper is presented.

Table 1

Problem	*Treewidth k graphs*	*α-Near-planar graphs*
Systems of equations over a field	$O(nk^2)$	$O(n^{3/2})$
Systems of equations over a closed semiring	$O(nk^2)$	$O(n^{3/2})$
sssp	$O(nk^2)$	$O(n^{3/2})$
apsp	$O(n^2k)$	$O(n^2 \log n)$
A^{-1}	$O(n^2k)$	$O(n^2 \log n)$
A^*	$O(n^2k)$	$O(n^2 \log n)$

Our results for treewidth k graphs extend both

1. the well-known $O(nk^2)$ algorithm for the problem **P1** when the matrix A is a bandwidth k matrix, and

2. the $O(n)$ and $O(n^2)$ algorithms for the sssp and apsp problems, for series-parallel graphs G in [13].

In both cases, the representation of A or of G_A assumed in the literature can be converted into a treewidth k graph for G_A in linear time. For bandwidth l matrices, the corresponding treewidth k is l itself. Thus, our results for problems **P1** when G_A is a treewidth k graph are asymptotically of the same order as the known results for bandwidth k matrices. On the other hand, many naturally occuring $n \times n$ matrices of high bandwidth yield low treewidth graphs. For example, the $n \times n$ ϵ-matrices in [11] have bandwidth $n-1$ but result in treewidth 2 graphs. Our results for treewidth k graphs also show that the sssp and apsp problems are solvable in $O(n)$ and $O(n^2)$ time for each of the recursively-defined graph classes studied in [5], when graphs are presented along with their recursive derivations as assumed in both [5] and [13]. These graph classes include rooted trees, series-parallel graphs, ProtoHalin graphs and bandwidth k graphs. Many other natural families of graphs like outerplanar, chordal graphs with maximum clique size k and circular arc graphs with maximum clique size k have constantly bounded treewidth. Finally, most work to date on the complexity of graph problems for treewidth k graphs [3, 6, 5, 12] has considered **NP**- or **#P**- hard problems and has resulted in algorithms for these problems that are often linear in n (the number of vertices in the graph) but are exponential in k (the treewidth of the corresponding tree decomposition). In contrast, our algorithms for path problems from [11] including the sssp and apsp problems for treewidth k graphs are linear or quadratic in both n and k.

Our results for α-near-planar graphs are also of the same asymptotic order as the known results from [17, 18] for planar graphs. Our motivation for considering α-near-planar graphs is the following:
1.Planarity of graphs is unstable, in the sense that the addition of a single edge can destroy planarity.
2. Except for path problems for planar graphs and certain grid graphs, planar systems of linear equations are unnatural. For example, the complete graph on five vertices is not a subgraph of any planar graph. This imposes very tight constraints on the matrix A for the graph G_A to be planar. On the other hand, adding a single edge to an α-near-planar yields at most an $(\alpha+1)$-near-planar graph. Also, for all $\alpha > 0$, α-near-planar graphs can have arbitrarily large cliques as subgraphs.
3.There are several different areas which yield naturally α-near-planar graphs that are not planar. These include wide-area communication networks (eg. ARPA network in [21] is 0.2-near-planar, NSFNET backbone in [8] is 0.4-near-planar) and digital circuits laid out on chips.
4. The almost-planar graphs in [17] are α-near-planar for some $\alpha > 0$ but not conversely (since α-near-planar graphs allow arbitrarily large cliques).

Lastly we note two important properties of our results for α-near-planar graphs. First, since α-near-planar graphs for any $\alpha > 0$ are not closed under subgraphs, the results of [17] or [18] do not directly apply to them. Second, the proofs of our results actually yield $O((n+c)^{3/2})$ and $O(n(n+c)\log(n+c))$ algorithms for the problems considered here for n vertex graphs layed out in the plane with $\leq c$ crossovers of edges.

The rest of the paper is organized as follows. Section 2 provides definitions of some important concepts which are used in the rest of the paper. Section 3 contains algorithms for solving systems of linear equations over a field, evaluation of the determinant and the inverse. Section 4 discusses algorithms for path and path expression problems, for computing A^* and for solving systems of linear equations over a closed semiring.

2 Preliminaries

We present the definitions of tree decomposition from [20] and of planar layout with crossovers from [22].

Definition 2.1 *Let G=(V,E) be a graph. A tree-decomposition of G is a pair* $(\{X_i|i \in I\}, T = (I,F))$, *where* $\{X_i|i \in I\}$ *is a family of subsets of V, T=(I,F) is a tree with the following properties:*

1. $\bigcup_{i\in I} X_i = V$
2. *For every edge e=(v,w)∈E, there is a subset* $X_i, i \in I$ *with* $v \in X_i$ *and* $w \in X_i$,
3. *For all i,j,k∈ I, if j lies on the path in T from i to k, then* $X_i \bigcap X_k \subseteq X_j$

The **treewidth** *of a tree-decomposition* $(\{X_i|i \in I\}, T)$ *is* $\max_{i\in I}|X_i| - 1$.

It is easily seen that if there is a tree-decomposition with treewidth k, then there is a tree-decomposition of treewidth k such that $|I| \leq |V| - k + 1$. In this paper, we only assume that $|I| = O(n)$.

Definition 2.2 *Let $G = (V, E)$ be a graph. A planar layout with crossovers for G is a planar graph $G' = (V', E')$ together with a set C of crossover nodes and a function $f : E' \to E$ such that*

1. $C \bigcap V = \phi$
2. $V' = V \bigcup C$
3. *Each node of C has degree 4.*
4. *For all (a, b) in E, $\{e \in E' | f(e) = (a, b)\}$ is the set of edges on a simple path from a to b in G' involving no other nodes of V.*

Let $\alpha > 0$, an α-near-planar graph is a graph with vertex set V which is presented together with one of its planar layouts with $\leq \alpha \cdot |V|$ crossovers of edges.

We say that crossover node c is associated with edges e_1 and e_2 of E if and only if there are edges e_1' and e_2' in E' with endpoint c such that $e_1 = f(e_1')$ and $e_2 = f(e_2')$.

3 Solution of a system of linear equations over a field

As in [14], we use LDU decomposition to solve by Gaussian elimination the system of linear equations $Ax = b$. Here A is an $n \times n$ matrix, x is an $n \times 1$ vector of variables and b is an $n \times 1$ vector of constants. The solution process consists of two steps. First, we factor A by means of row operations into $A = LDU$ where L is lower triangular, D is diagonal and U is upper triangular. Second, we solve the simplified systems $Lz = b$, $Dy = z$ and $Ux = y$.

The LDU decomposition is found as follows: Let $A_0 = A = \begin{bmatrix} a_{1,1} & r_1 \\ c_1 & B_1 \end{bmatrix}$,

where r_1 is a $1 \times (n-1)$ vector and c_1 is an $(n-1) \times 1$ vector and B_1 is an $(n-1) \times (n-1)$ matrix.

Then $A_0 = \begin{bmatrix} 1 & 0 \\ c_1/a_{1,1} & I \end{bmatrix} \begin{bmatrix} a_{1,1} & 0 \\ 0 & A_1 \end{bmatrix} \begin{bmatrix} 1 & r_1/a_{1,1} \\ 0 & I \end{bmatrix}$ where $A_1 = B_1 - c_1 r_1 / a_{1,1}$.

This is the system obtained by eliminating x_1; and its graph is the graph obtained by removing vertex v_1 and joining every pair of vertices adjacent to v_1. The LDU decomposition is obtained one column at a time of L, one diagonal element at a time of D and a row at a time of U. The elimination of the variable x_i proceeds as follows

Let $A_{i-1} = \begin{bmatrix} a_{i,i}^{i-1} & r_i \\ c_i & B_i \end{bmatrix}$. Then $A_{i-1} = \begin{bmatrix} 1 & 0 \\ c_i/a_{i,i}^{i-1} & I \end{bmatrix} \begin{bmatrix} a_{i,i}^{i-1} & 0 \\ 0 & A_i \end{bmatrix} \begin{bmatrix} 1 & r_i/a_{i,i}^{i-1} \\ 0 & I \end{bmatrix}$

where $A_i = B_i - c_i r_i / a_{i,i}^{i-1}$.

Corresponding to the factorization $A = LDU$ is the graph $G^* = (V, E^*)$ such that $\{v_i, v_j\} \in E^*$ iff $i > j$ and $l_{i,j}$ is non-zero or $i > j$ and $u_{i,j}$ is non-zero.

For a given order of elimination of variables (henceforth called an elimination order), the fill-in(A) is the set of edges of G^* which are not edges of G_A.

3.1 Systems of equations with treewidth k

Consider a system of equations whose non-zero structure corresponds to a treewidth k graph. We use the fact that a treewidth k graph is a subgraph of a k-tree. Rose [19] observed that a k-tree is a perfect elimination graph; i.e. there is an elimination order in which there is no fill-in.

Lemma 3.1 *If the graph G_1 corresponding to an $n \times n$ matrix A_1 is a subgraph of the graph G_2 corresponding to an $n \times n$ matrix A_2, then fill-in(A_1) $\subseteq$ fill-in(A_2), if the same elimination order is used in both cases.*

We now present an algorithm for treewidth k graphs to renumber the variables in order to obtain $O(nk)$ fill-in.

Algorithm 3.1 (Numbering algorithm for treewidth k graphs.)

The variables are renumbered so as to reduce the fill-in as follows: Let $G = (V, E)$ be the graph corresponding to the non-zero structure of A. Let its tree-decomposition be $(\{X_i | i \in I\}, T = (I, F))$ with $\{X_i | i \in I\}$ a family of subsets of V, and T a tree such that $|X_i| \leq k+1$ for all $i \in I$. i.e. treewidth $= k$.

Traverse the tree T bottom up and number the unnumbered variables at the node $i \in I$, which do not occur in the ancestor of i in T.

Theorem 3.2 *Algorithm 3.1 takes time $O(nk)$ for an n-vertex treewidth k graph and has a fill-in of $O(nk)$ associated with it.*

Proof. There are at $O(n)$ nodes in I each of which takes at most $k+1$ steps to number giving a total time of $O(n(k+1))$. Consider the order of elimination of variables which corresponds to the order in which they are numbered. When a variable is eliminated, i.e. the vertex corresponding to it is removed, the vertices adjacent to it, which are numbered higher get connected to each other by an edge.

Consider the way the numbering is done. A variable is numbered at the highest node of T in which it appears. Hence the variables(vertices) adjacent to it and numbered higher also occur at that node of T. Thus the fill-in edges introduced do not increase the treewidth.

The number of edges in G^* is at most the number of edges in a graph of treewidth k. A graph of treewidth k is a subgraph of a k-tree. The number of edges in a k-tree of n vertices is $k(k-1)/2 + (n-k)k = O(nk)$. Thus the fill-in is $O(nk)$. □

We will now obtain bounds on the running time of the LDU decomposition and the solution of the system of linear equations.

Theorem 3.3 *A system of linear equations $\mathcal{L}_{A,b}$ with treewidth k graph G_A can be solved in time $O(nk^2)$.*

Proof. The number of operations required for the elimination of x_i is the product of the number of non-zero elements in the ith row and the ith column of A_{i-1}. However, in the graph corresponding to A_{i-1}, variables adjacent to x_i (and hence numbered higher than x_i), occur at the highest node of T in which x_i appears. Since the number of variables in X_j for any $j \in I$ is no more than $k+1$, the number of variables adjacent to x_i in A_{i-1} is no more than k. Thus the number of non-zero entries in row i of A_{i-1} and in column i of A_{i-1} is no more than k, so that the number of multiplications is no more than $(k+1)k$ and the number of additions is no more than k^2. Thus the number of operations for each $j \in I$ is $O(k^2)$. The total time to eliminate $n-1$ variables is therefore $O(nk^2)$. The back-substitution takes $O(nk)$ time, since there are $O(nk)$ elements in L and U and $O(n)$ one diagonal element at a time of D and a row at a time of U. The elimination elements in D. Hence, the overall time to solve a linear system of equations over a field given its tree-decomposition is $O(nk^2)$. □

There is a close relationship between the parse trees of decomposable graphs [5] and tree-decompositions. This relationship enables us to show that $O(n)$ and $O(n^2)$ algorithms also exist for the single source and all pairs shortest path problems for each of the classes of decomposable graphs in [5], when graphs are represented by their parse trees. Thus the results of this section and section 4 generalize the results in [13], for single source and all-pairs shortest path problems for series parallel graphs.

3.2 α-Near-planar systems of equations

Next, we solve a system of equations whose non-zero structure corresponds to a α-near-planar graph. We use the fact shown below that the fill-in for α-near-planar graphs is not too much more than the fill-in for planar graphs. This contrasts with our treatment of treewidth bounded systems of equations where we used the fact that the fill-in was contained in the fill-in for k-trees.

Let $Ax = b$ be the system to be solved. Let the graph $G = (V, E)$ correspond to its non-zero structure. Let $G' = (V \bigcup C, E')$ be its α-near-planar layout. Recall that G' is a planar graph.

Apply the planar separator theorem to G' to obtain the sets A', B', C'. C' is a separator for G' such that there are no edges in E' from A' to B'. However if we consider the vertices of V in A' and the vertices of V in B', there may be edges in E between them.

Repeat the following for every crossover in C': Let c be a crossover in C' corresponding to edges (v_1, v_2) and (v_3, v_4) of E. If the two vertices of an edge in E are in A' and B', move one of them to C'.

Consider $v_1 \in A'$ and $v_2 \in B'$ where $v_1, v_2 \in V$. If $(v_1, v_2) \in E$, then there is a path in E' between v_1 and v_2 consisting of no other nodes of V. Since there are no edges of E' between vertices in A' and vertices in B', there must be a path of crossover nodes from v_1 to v_2 passing through C', i.e. there is a crossover $c \in C'$ on this path. However, if $c \in C'$ was a crossover, then one of the vertices of each of the edges crossing over at c would be in C'. i.e. either v_1 or v_2 would be in C'. Therefore there is no edge in E between a vertex in A' and a vertex in B'. Hence the following theorem.

Theorem 3.4 *Let $G = (V, E)$ be presented with a near-planar layout with crossovers $G' = (V \bigcup C, E')$. Let $n = |V|$ and $c = |C|$. Then, a partition of $V \bigcup C$ into sets A', B', C' can be found in $O(|V \bigcup C|)$ time such that*

1. $|A'|, |B'| \leq \frac{2}{3}|V \bigcup C| = \frac{2}{3}(n + c)$
2. $|C'| \leq 6\sqrt{2}\sqrt{|V \bigcup C|} = 6\sqrt{2}\sqrt{n + c}$
3. *A' is a layout of the vertices of $A' \bigcap V$ and B' is a layout of the nodes in $B' \bigcap V$.*
4. *There are no edges in E or in E' between vertices in A' and vertices in B'.*

The separator of Theorem 3.4 guarantees that the number of vertices in the individual parts is smaller than the original graph or that the individual parts have fewer crossovers than the original graph or both. It generalizes the planar separator theorem to graphs laid out in the plane with crossovers. For α-near-planar graphs it gives a $O(\sqrt{n})$ separator theorem. We note again that α-near-planar graphs are not closed under subgraph and can have arbitrarily large cliques.

Algorithm 3.2 (Numbering algorithm for graphs with planar layouts with crossovers.)

Given a graph $G = (V, E)$ along with its layout with crossovers $G' = (V \bigcup C, E')$, we number the vertices of V from $a = 1$ to $b = n$ recursively as follows:

If $|V \bigcup C| \leq n_0$, number the vertices of V arbitrarily.

Otherwise, find sets A', B', C' as described earlier. Let the number of unnumbered vertices of V in A', B', C' be i,j,k respectively. Number the vertices in $C' \bigcap V$ arbitrarily from $b-k+1$ to b. Delete all edges of E with both endpoints in C' (Consequently some edges of E' may have to be deleted). Apply the algorithm recursively to the layout $B' \bigcup C'$ to number the unnumbered variables of $B' \bigcap V$ from $b-k-j+1$ to $b-k$. Apply the algorithm recursively to the layout $A' \bigcup C'$ to number the unnumbered variables of $A' \bigcap V$ from a to $a+i-1$.

Theorem 3.5 *Algorithm 3.2 takes $O(n \log n)$ time to number an n-vertex α-near-planar graph with n vertices.*

Proof. Let the layout of a near-planar graph $G = (V, E)$ be $G' = (V \bigcup C, E')$. Let $|V| = n$, $|V \bigcup C| = n'$. We note that $n' \leq (1+\alpha)n$. The time taken to number the graph can be expressed by the following recurrence relation

$$t(n,n') \leq \begin{cases} c_1(n') + \max\{t(n_1,n_1') + t(n_2,n_2')\} & if\ n' > n_0 \\ c_0 & if\ n' \leq n_0 \end{cases}$$

where $n_1 + n_2 \leq n$

$n' \leq n_1' + n_2' \leq n' + 6\sqrt{2}\sqrt{n'}$

$\frac{1}{3}n' \leq n_i' \leq \frac{2}{3}n' + 6\sqrt{2}\sqrt{n'}$ $1 \leq i \leq 2$

Here n' is the number of nodes in the layout and n is the number of nodes in the graph being considered. The individual layouts for $A' \bigcup C'$ and $B' \bigcup C'$ have sizes n_1' and n_2' respectively. The number of unnumbered vertices of V in $A' \bigcup C'$ and $B' \bigcup C'$ are n_1 and n_2 respectively. The solution to this recurrence relation is $t(n,n') = O(n' \log n')$ which is proved by a technique similar to [17]. □

Theorem 3.6 *The fill-in associated with the numbering of Algorithm 3.2 for α-near-planar graphs is $O(n \log n)$.*

Proof. Suppose that the numbering algorithm is applied to an n-vertex near-planar graph with a layout with c crossovers with l vertices previously numbered. Assume $n' = n + c > n_0$ and let A', B', C' be the vertex partition generated by the algorithm. The maximum number of fill-in edges whose lower numbered endpoint is in C' is

$$|C'|(|C'|-1)/2 + |C'|l$$

The maximum number of vertices of the graph in C' can be no more than $4\sqrt{2}\sqrt{n'}$. Thus the maximum number of fill-in edges whose lower numbered endpoint is in C' is no more than

$$(4\sqrt{2})^2 n'/2 + 4\sqrt{2}l\sqrt{n'}$$

Two vertices v and w are joined by a fill-in edge iff there is a path from v to w through vertices numbered less than both v and w. Thus no fill-in edge joins a vertex in A' with a vertex in B'. Let $f(l,n,n')$ be the maximum number of fill-in edges whose lower numbered endpoint is numbered by the algorithm. We obtain the following recurrence relation

$$f(l,n,n') \leq \begin{cases} n(n-1)/2 & if\ n' \leq n_0 \\ (4\sqrt{2})^2 n'/2 + 4\sqrt{2}l\sqrt{n'} + \max\{f(l_1,n_1,n_1') + f(l_2,n_2,n_2') & otherwise \end{cases}$$

where the maximum is taken over values satisfying

$$l_1 + l_2 \leq l + 8\sqrt{2}\sqrt{n'}$$

$$n \le n_1 + n_2 \le n + 4\sqrt{2}\sqrt{n'}$$
$$n_i \le 2/3n' + 4\sqrt{2}\sqrt{n'} \quad 1 \le i \le 2$$
$$1/3n' \le n_i' \le 2/3n' + 6\sqrt{2}\sqrt{n'}$$
$$n' \le n_1' + n_2' \le n' + 6\sqrt{2}\sqrt{n'}$$

The solution to this recurrence relation is

$$f(l,n,n') = c_1(l+n')\log n' + c_2 l\sqrt{n'}$$

which is proved by a technique similar to [17]. We note that $n' \le (1+\alpha)n$. Hence the result. □

Theorem 3.7 *An system of linear equations $\mathcal{L}_{A,b}$ with α-near-planar graph G_A can be solved in time $O(n^{3/2})$.*

Proof. The number of operations required for the elimination of x_i is the product of the number of non-zero elements in the ith row and the ith column of A_{i-1}. Thus a bound on the number of operations associated with a separator C' generated by one call of the recursive numbering algorithm is

$$\sum_{i=1}^{4\sqrt{2}\sqrt{n'}} (i+l)(i+l) \le c_1 n'^{3/2} + c_2 l n' + c_3 l^2\sqrt{n'} + c_4 n' + c_5 l\sqrt{n}$$

Let $g(l,n,n')$ be the number of operations for performing the elimination when l vertices are already numbered, n variables are to be eliminated and the layout for this system has n' vertices. Then the following recurrence is satisfied

$$g(l,n,n') \le \begin{cases} c_6 n^3 & if\ n' \le n \\ c_1 n'^{3/2} + c_2 l n' + c_3 l^2\sqrt{n'} + c_4 n' + c_5 l\sqrt{n'} + \max\{g(l_1,n_1,n_1') + g(l_2,n_2,n_2')\} & otherwis \end{cases}$$

where the maximum is taken over values satisfying

$$l_1 + l_2 \le l + 8\sqrt{2}\sqrt{n'}$$
$$n \le n_1 + n_2 \le n + 4\sqrt{2}\sqrt{n'}$$
$$n_i \le 2/3n' + 4\sqrt{2}\sqrt{n'} \quad 1 \le i \le 2$$
$$1/3n' \le n_i' \le 2/3n' + 6\sqrt{2}\sqrt{n'}$$
$$n' \le n_1' + n_2' \le n' + 6\sqrt{2}\sqrt{n'}$$

The solution to this recurrence relation is

$$g(l,n,n') = c_7(n')^{3/2} + c_8 l n' + c_9 l^2\sqrt{n'}$$

which is proved by a technique similar to [17]. Thus, since $n' \le (1+\alpha)n$, the total time to eliminate n variables is therefore $O(n^{3/2})$. The substitution takes $O(n\log n)$ time, since there are $O(n\log n)$ elements in L and U and $O(n)$ elements in D. Hence, the overall time to solve a linear system of equations over a field given its layout with c crossovers is $O(n^{3/2})$. □

3.3 Inverse of a matrix

A^{-1} is found by solving n systems of the form $Ax = b$ where b corresponds to an $n \times 1$ vector of zeros except for the ith element which is a 1 which gives the ith column of A^{-1}. The advantage of solving these systems by LDU decomposition is that the elimination is done once and the solution just involves substitution n times.

Corollary 3.8 *The inverse of an $n \times n$ matrix A with treewidth k graph G_A can be found in $O(n^2 k)$.*

Proof. The elimination takes time $O(nk^2)$ and each of the n substitutions take $O(nk)$ time since the fill-in is $O(nk)$. □

Corollary 3.9 *The inverse of an $n \times n$ matrix given the α-near-planar layout of the graph corresponding to it can be found in $O(n^2 \log n)$.*

Proof. The elimination takes time $O(n^{3/2})$ and each of the n substitutions take $O(n \log n)$ time since the fill-in is $O(n \log n)$. □

4 Solution of a system of linear equations over a closed semiring

Following [4], we solve the system of linear equations $\mathcal{L}$ over a closed semiring $\{R, +, \cdot, *, 0, 1\}$ as in [15], of the form

$$x = A \cdot x + b \tag{1}$$

as follows: First, we decompose A into upper and lower triangular matrices U and L over R such that $A^* = U^* \cdot L^*$. Second, we solve the coupled system of equations

$$y = L \cdot y + b \tag{2}$$

$$x = U \cdot x + y \tag{3}$$

The operations used to obtain U and L very closely resemble those used to obtain the LDU decomposition of a matrix A over a field F. This close resemblance enables us to make the two observations below on the relative fill-ins and the operation counts on solving problem **P1** over a field F and over a closed semiring R.

The triangular decomposition is obtained as follows: Writing $M^{(0)} = A$, we compute $M^{(k)}$ successively as

$$m_{i,j}^{(k)} = \begin{cases} m_{i,k}^{(k-1)}(m_{k,k}^{(k-1)})^* & \text{for } k < i \leq n, j = k \\ m_{i,j}^{(k-1)} + m_{i,k}^{(k-1)} \cdot (m_{k,k}^{(k-1)})^* \cdot m_{k,j}^{(k-1)} & \text{for } k < i, j \leq n \\ m_{i,j}^{(k-1)} & \text{otherwise} \end{cases}$$

L is the strictly lower triangular part of $M^{(n)}$ and U is the upper triangular part of $M^{(n)}$

In comparison, the operations performed in an LDU decomposition are as follows:

$$m_{i,j}^{(k)} = \begin{cases} m_{i,k}^{(k-1)}/m_{k,k}^{(k-1)} & \text{for } k < i \leq n, j = k \\ m_{k,j}^{(k-1)}/m_{k,k}^{(k-1)} & \text{for } k < j \leq n, i = k \\ m_{i,j}^{(k-1)} - m_{i,k}^{(k-1)}/m_{k,k}^{(k-1)} \cdot m_{k,j}^{(k-1)} & \text{for } k < i, j \leq n \\ m_{i,j}^{(k-1)} & \text{otherwise} \end{cases}$$

Theorem 4.1 *The fill-in obtained in the triangular decomposition of a matrix over a closed semiring is no more than the fill-in obtained in the LDU decomposition of a matrix A' over a field such that A and A' have the same non-zero structure, provided the same elimination order is used.*

Proof. The fill-in corresponds to the elements of $M^{(n)}$ which are non-zero when the corresponding elements of A are zeros. The operation which contributes to fill-in in the triangular decomposition is the one for the case $k \leq i, j \leq n$, and the fill-in occurs when $m_{i,j}^{(k-1)}$ is zero, $m_{i,k}^{(k-1)}$ is non-zero and $m_{k,j}^{(k-1)}$ is non-zero.

Similarly, the operation which contributes to the fill-in in the LDU decomposition is the one corresponding to the case $k \leq i, j \leq n$, and the fill-in occurs when $m_{i,j}^{(k-1)}$ is zero, $m_{i,k}^{(k-1)}$ is non-zero and $m_{k,j}^{(k-1)}$ is non-zero. Thus the operations in the triangular decomposition have corresponding operations in the LDU decomposition and a fill-in occurs in the triangular decomposition only if a fill-in occured in the LDU decomposition. Hence the theorem. □

Theorem 4.2 *The number of operations required to obtain the triangular decomposition of a matrix over a closed semiring is no more than the fill-in obtained in the LDU decomposition of a the number of operations required for the LDU decomposition of a matrix A' over a field such that A and A' have the same non-zero structure, provided the same elimination order is used.*

Proof. If instead of the fill-in, we are interested in estimating the operation count for obtaining the triangular decomposition, (where the count is only for operations involving non-null elements), we find that a non-trivial operation occurs in the triangular decomposition only if a non-trivial operation occurs in the LDU decomposition. Hence the result. □

The above two theorems enable us to directly translate our results in section 3 for problem **P1** for systems of linear equations over a field to those for problem **P1** for systems of linear equations over a closed semiring as follows.

Corollary 4.3 *A system of linear equations $\mathcal{L}_{A,b}$ over a closed semiring with treewidth k graph G_A can be solved in $O(nk^2)$ time.*

Corollary 4.4 *A system of linear equations $\mathcal{L}_{A,b}$ over a closed semiring with α-near-planar graph G_A can be solved in $O(n^{3/2})$ time.*

Shortest path problems have the closed semiring $(\Re, min, +, \infty, 0)$ associated with them.

The single source shortest path problem corresponds to solving a system of equations of the form $x = A \cdot x + b$ where A corresponds to the distance matrix and b is a vector of zeros except for a 1 in the position corresponding to the source. Thus the corollaries obtained earlier can be used to obtain $O(nk^2)$ and $O(n^{3/2})$ algorithms when the graph is a treewidth k graph or an α-near-planar graph respectively.

The all pairs shortest path problem corresponds to solving n systems of equations of the form $x = A \cdot x + b$ with the same distance matrix A but with different b corresponding to the different sources. Thus apsp can be solved in $O(n^2 k)$ and $O(n^2 \log n)$ when the graph is a treewidth k graph or an α-near-planar graph respectively.

A^* is found by solving n systems of the form $x = A \cdot x + b$ where b corresponds to an $n \times 1$ vector of zeros except for the ith element which is a 1 which gives the ith column of A^{-1}. Hence the time to compute A^* for a matrix A whose corresponding graph is treewidth k or α-near-planar is $O(n^2 k)$ or $O(n^2 \log n)$ respectively.

The results obtained above can also be applied to the path expression problems in [23] and to the path algebra problems in [11] using the appropriate closed semiring to obtain equivalent results.

References

[1] A.Arnborg, D.G.Corneil and A.Proskuworski, "Complexity of finding embeddings in a k tree", SIAM J.Alg. and Discr.Methods 8, 1987, pp.277-284.

[2] A.V.Aho, J.E.Hopcroft, J.D.Ullman, "The Design and Analysis of Computer Algorithms", Addison-Wesley, 1974.

[3] A.Arnborg, J.Lagergren, D.Seese,"Problems easy for decomposable graphs",Proceedings of ICALP 88, Springer-Verlag LNCS 317(1988),38-51

[4] R.C.Backhouse and B.A.Carre,"Regular algebra applied to pathfinding problems", J. Inst. Math. Appl., 15,1974,pp.161-186.

[5] M.W.Bern, E.L.Lawler and A.L.Wong,"Linear-time Computation of Optimal Subgraphs of Decomposable Graphs",J. Alg. 8(1987),pp.216-235.

[6] H.L.Bodlaender, "Dynamic programming on graphs with bounded tree-width", RUU-CS-87-22, Proceedings of ICALP 88, Springer-Verlag LNCS 317(1988),105-118.

[7] J.R.Bunch and D.J.Rose,"Partitioning, tearing, and modification of sparse linear systems",J. Math. Anal. Appl. 48(1974),pp.574-593.

[8] D.E.Comer,"Internetworking with TCP/IP",vol.I, Prentice-Hall,1991.

[9] G.Dahlquist and A.Björck,"Numerical methods",Prentice-Hall,1974.

[10] M.R.Garey and D.S.Johnson, "Computers and Intractability: A Guide to the Theory of NP-Completeness", W.H.Freeman and Company,1979.

[11] M.Gondran and M.Minoux,"Graphs and Algorithms",John Wiley,1984.

[12] Y.Gurevich, L.Stockmeyer, and U.Vishkin, "Solving NP-hard problems on graphs that are almost trees and an application to facility location problems",J. ACM, vol.31,pp.459-473,1984.

[13] R.Hassin and A.Tamir, "Efficient algorithms for optimization and selection on series-parallel graphs", SIAM J. Alg. Disc. Meth., vol 7, pp.379-389, 1986.

[14] I.N.Herstein and D.J.Winter,"Matrix Theory and Linear Algebra",Macmillan,1988.

[15] D.J.Lehmann,"Algebraic structures for transitive closure", Theor. Comp. Sc., vol. 4, pp.59-76, 1977.

[16] R.L.Lipton and R.E.Tarjan, "Applications of a planar separator theorem", SICOMP, vol.9, pp.615-629, 1980.

[17] R.J.Lipton, D.J.Rose and R.E.Tarjan,"Generalized nested dissection", SIAM J. Numer. Analysis 16(2),pp.346-358(1979).

[18] V.Pan and J.Reif,"Parallel nested dissection for path algebra computations", Operation research letters 5(4),pp.177-184,1986.

[19] D.J.Rose,"A graph-theoretic study of the numerical solution of sparse positive definite systems of linear equation",Graph Theory and Computing,R.Read,ed., Academic Press,pp.183-217, 1972.

[20] N.Robertson and P.D.Seymour,"Graph Minors III, algorithmic aspects of tree-width",J. of Algorithms 7,1986,pp.309-322.

[21] M.Schwartz,"Computer-communication network design and analysis",Prentice Hall, 1977.

[22] R.E.Stearns and H.B.Hunt III, "Power indices, structure trees, and easier hard problems", SUNY Albany Tech. Rep. TR 89-21, 1989.

[23] R.E.Tarjan, "A unified approach to path problems", J. ACM 28(3),pp.594-614(1981)

EFFICIENT SUBLINEAR TIME PARALLEL ALGORITHMS FOR DYNAMIC PROGRAMMING AND CONTEXT-FREE RECOGNITION

Lawrence L. Larmore

University of California at Riverside
Dept. of Computer Science, Riverside, USA

Wojciech Rytter

Warsaw University, Institute of Informatics
ul. Banacha 2, 00-913 Warszawa 59, Poland

Abstract

We design a sublinear time parallel algorithm for the computation of the general dynamic programming recurrences. Its total work matches the work of the best known sequential algorithm. It is the first optimal sublinear parallel time algorithm for this problem. Using similar methods we construct also sublinear time parallel algorithms for the recognition of linear, unambiguous and deterministic cfl's. Their total works are $O(T(n)n^{\alpha}\log(n))$, for arbitrarily small constant $\alpha<1$, where $T(n)$ is the work of the best known sequential algorithm for the problem. This reduces substantially the gap between the total work of sublinear time parallel algorithms and that of the best known sequential algorithms.

1. Introduction

The fast parallel computation usually means a polylogarithmic time computation. However the efficiency of such algorithms in the sense of the total work (product of time and number of processors) is not always satisfactory. By a fast parallel computation we mean here a sublinear time computation. Then we show that the efficiency of such fast parallel algorithms for dynamic programming and context-free recognition can be improved considerably (compared with known NC algorithms). The best NC-algorithms for the dynamic programming had total work of the order n^6, see [Ry 89], improved to $n^6/\log^4 n$ in [VHL 90].

There are linear time parallel algorithms time with cubic work, see [Ko 75] and [GKT 79]. An algorithm with $O(n^4)$ total work was given in [HVL 90]. Our algorithm is the first sublinear time algorithm whose total work is also cubic. By an optimal parallel algorithm we mean here a sublinear time algorithm whose total work is of the same order as the work of the best known sequential algorithm for a given problem. The best known total work of the sequential algorithm for the dynamic programming problem is cubic. Hence our algorithm is optimal in the sense defined above. We also construct sublinear time parallel algorithms for the recognition of the main subclasses of cfl's, whose total work, in each case, is close to that of the cost of the best known sequential algorithm .

Denote by $T_D(n)$,$T_U(n)$ and $T_L(n)$ the cost of the best known sequential algorithm for the recognition of, respectively, deterministic, unambiguous and linear cfl's. Then $T_D(n) = O(n)$, $T_U(n) = O(n^2)$ and $T_L(n) = O(n^2)$. The best known total work of polylogarithmic time parallel algorithms for these problems are $O(n^2)$ for the recognition of deterministic cfl's, see [KR 87], [CCMR 90] and [CM 90], $O(n^3)$ for unambiguous cfl's, see [CCMR 90], and $O(n^3)$ for linear cfl's, see [ALMKT 89] and [Ry 88]. Hence the known parallel algorithms for these three subclasses of cfl's are optimal within a linear factor. Our main result is the reduction of this factor to $O(n^\alpha \log(n))$, for arbitrarily small α. The most advanced algorithm is here the one for unambiguous cfl's.

We demonstrate a useful algorithmic technique for the design of sublinear time algorithms. Each of the algorithms presented in the paper is a hybrid, consisting of two distinct phases, the first of which mimics a known polylogarithmic algorithm for the problem, but is executed only for small subproblems, and the second of which mimics a known sequential algorithm, but advances in larger steps.

Our model of the parallel computations is a CRCW PRAM, see for example [GR 88].

2. General context-free recognition and dynamic programming

We start with general context-free recognition and show how it can be extended to the dynamic programming problem. One could start with the dynamic programming and treat the context-free recognition as a special case. However we need our particular algorithm for cfl's in the next section. The recognition problem for context-free languages (cfl's, in short) is: we are given a constant size description of a language L by a context-free grammar G or a pushdown automaton A, and an input string w. Check if $w \in L$. The size of the problem is the length n of the input string w. We start with some abstract algorithms and explain the main ideas using the framework of composition systems. By a composition system we mean a triple $(N,T,\otimes)$, where N is the carrier of the system. $T \subseteq N$ is the set of generators and $\otimes$ is a binary operation $\otimes : N \times N \rightarrow 2^N$. The values of $x \otimes y$ are subsets of N consisting of elements composed of x and y. Denote by Closure(X) the smallest set containing X and closed under $\otimes$. Define the generability problem as the one consisting in computing the set Closure(T). For a given context-free grammar G in Chomsky normal form and input string $w = a_1 \dots a_n$ of length n we define the corresponding composition system $S_{G,w}$. Its elements are triples (A,i,j), where A is a nonterminal and (i,j) corresponds to a substring $a_{i+1} \dots a_j$. The set T of generators corresponds to one step valid derivations of substrings of length one.

$$(B,i,k) \otimes (C,k,j) = \{(A,i,j) : A \rightarrow BC\}.$$

(Write A->BC iff it is a production of the grammar. $->^*$ is the transitive closure of the relation ->.) We refer the reader to [BKR 91] for details. The recognition problem is reduced to the generability problem for the system $S_{G,w}$. We say that a triple (A,i,j) is realizable iff $(A,i,j) \in$ Closure(T).

We have

$$(A,i,j) \text{ is realizable iff } A \rightarrow^* a_{i+1} \dots a_j.$$

We can assume w.l.o.g that we deal with context-free grammars in Chomsky normal form without useless nonterminals. The system $S_{G,w}$ is tree-like, it means that all elements in the derivation tree of any element from the generators are distinct. A useful property of this system is that the number of leaves is the same in all trees generating the same element. For x=(A,i,j) let size of x be $|x|=j-i$.

Let $h=n^{\alpha}$, where $0<\alpha<1$. Assume for simplicity that h divides n. Our aim is to construct an efficient parallel algorithm working in time $n^{1-\alpha}=n/h$ and computing the set Closure(T). The general structure of the algorithm consists of computing in n/h stages n/h pairwise disjoint sets called layers. The union of all these sets is Closure(T). We define

$N(k) = \{ x \in N : (k-1)h < |x| \leq kh \}$ and

$Layer(k) = Closure(T) \cap N(k)$.

For $k \leq n/h$ define also the sets

$Gen(k) = \bigcup_{i \leq k} Layer(i)$; $Twice(k) = Gen(k) \otimes Gen(k)$;

where $X \otimes Y = \{ x \otimes y : x \in X, y \in Y \}$.

Denote by $\text{Easy-closure}_k(Z)$ the set of all elements of N_k which can be generated from the set $Layer(1) \cup Z$ by a generation tree T of the form presented in Fig.1.a. It has the structure similar to the generation trees in linear context-free grammars.

T is a binary tree whose leaves are in Layer(1) except exactly one leave which is in Z. We require that each internal node of T has at least one son which is a leaf.

We require also that the leaf of T which is in Z is at the lowest level and $|x|-|z| < h$ and all internal nodes are in N_k.

Fig.1. A generation tree realizing an "easy generation" and the corresponding path in the graph D_h. $z \in Z$ and $y_i \in Layer(1)$, for each leaf y_i different from z. The internal nodes v are in N_k .

```
Algorithm  GEN;
begin
Compute  Layer_1 = Layer(1);  Δ_1 := Layer_1 ⊗ Layer_1 ;
Gen := Layer_1;   Twice := Δ_1;
for k:=2 to n/h {sequentially} do
     begin  { Gen = Gen(k-1) }
     Layer_k := Easy-closure_k( Twice);
     Δ_k := Layer_k ⊗ Gen ∪ Gen ⊗ Layer_k ∪ Layer_k ⊗ Layer_k ;
     Gen := Gen_{k-1} ∪ Layer_k;  Twice := Twice ∪ Δ_k;
     { Layer_k = Layer(k); Twice = Gen(k) ⊗ Gen(k) }
     end;
if (S,0,n) ∈ Gen then return true; { Gen = Closure(T) }
end algorithm.
```

Lemma 1

a) $Layer(k+1) = Easy\text{-}closure_{k+1}(\ Twice(k))$, for $1 \le k < n/h$.

b) $Gen(k) \otimes Gen(k) = Twice(k-1) \cup \Delta_k$.

Proof.

(a) Assume that $x \in Layer(k+1)$ and examine the structure of a generation tree T of x, see Fig.2.

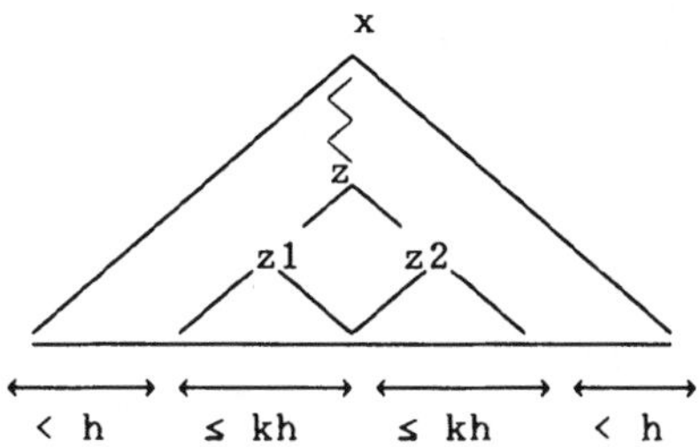

Fig.2. A tree generating an element x of Layer(k+1).$(x,z) \in D_h^*$.

Let z be the deepest element of Layer(k+1) in this tree. Its sons z1, z2 are both in layers preceding the (k+1)-th layer, hence they are in Gen(k). This implies that $z \in Twice(k)$. The tree T has the structure presented in Fig.2.

We have $|x|-|z| \le h$. Consequently $x \in Easy\text{-}closure_{k+1}(Twice(k))$ This completes the proof of point (a). ■

(b) This point follows trivially from equalities

$$\mathrm{Gen}(k) \otimes \mathrm{Gen}(k) = (\mathrm{Gen}(k-1) \cup \mathrm{Layer}_k) \otimes (\mathrm{Gen}(k-1) \cup \mathrm{Layer}_k) = \mathrm{Twice}(k-1) \cup \Delta_k.$$

In the efficient implementation of the algorithm we use a graph-theoretic approach. Define the following graph $D_h(k) = (V,E)$ called the k-th dependency graph, where $V = N(k)$ and

$E = \{ (x,y) : x \in y \otimes z$ or $x \in z \otimes y$ for some element $z \in \mathrm{Layer}(1) \}$.

Denote by D_h the (disjoint) union of all graphs $D_h(k)$. From the point of view of the parallel complexity the crucial point is an efficient precomputation of the transitive closure D_h^* of the acyclic directed graph D_h in time proportional to its depth. Assume that the sets Layer_k, Gen, Twice and Δ_k are represented by boolean arrays.

Lemma 2

(a) Layer(1) can be computed in $O(n^\alpha)$ time with $O(n^{1+2\alpha})$ processors.

(b) The transitive closure D_h^* of D_h can be computed in n^α time with $O(n^{2+3\alpha})$ processors.

(c) The statement "$\Delta_k := \mathrm{Layer}_k \otimes \mathrm{Gen} \cup \mathrm{Gen} \otimes \mathrm{Layer}_k$" can be executed in constant time with $O(n^{2+\alpha})$ processors.

Proof.

(a) The set Layer(1) consists of realizable triples (A,i,j), where $j-i \le n^\alpha$. $|\mathrm{Layer}(1)| \le |N(1)| = O(n^{1+\alpha})$. Let L be a boolean array whose all entries contain initially "false". All realizable elements can be computed in a bottom-up way as follows.

```
for each (A,i,j) ∈ N(1) do in parallel
  L(A,i,j):= (i+1=j) and (A->a_{i+1});
repeat n^{1+α} times
     for each (A,i,j) ∈ N(1), A->BC, i<k<j do in parallel
        if L(B,i,k) and L(C,k,j) then L(A,i,j):= true;
```

After executing this algorithm we have

$\mathrm{Layer}(1) = \{ (A,i,j) \in N(1) : L(A,i,j) = \mathrm{true} \}$;

This algorithm works in time n^α with $O(n^{1+2\alpha})$ processors. We have the same number of triples (i,k,j) such that $i<k<j$ and $j-i < n^\alpha$. Hence this number of processors is enough. The time corresponds to the height of the trees. The height of the trees realizing elements $(A,i,j) \in \mathrm{Layer}(1)$ is $O(n^\alpha)$. This completes the proof of point (a).

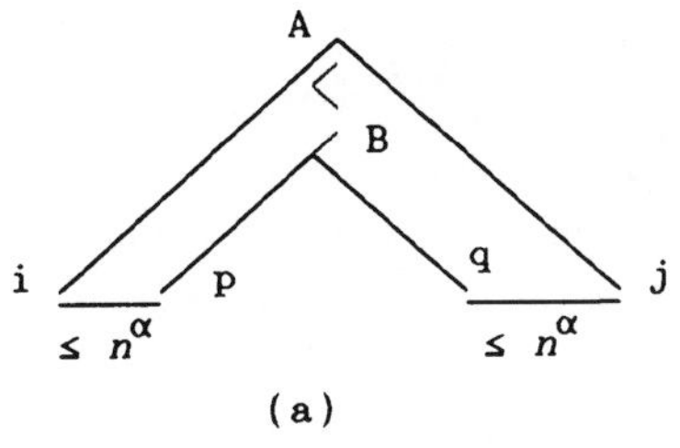

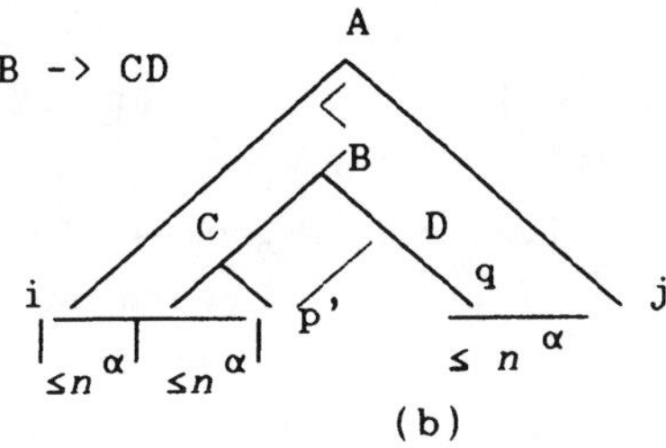

Fig.3. Tree T1 corresponding to an edge ((A,i,j),(B,p,q)) of D_h^* and its extension by a "small" tree corresponding to (C,p,p').

(b) Each edge of D_h^* corresponds to a "thin" partial tree, see Fig.3.a. There are $O(n^{2+2\alpha})$ pairs ((A,i,j),(B,p,q)), where p-i and j-q are of order n^α. Assume that we have already computed the incidency table for D_h. The algorithm for the computation of D_h^* is a kind of parallel bfs and is based on the fact that the length of paths in the graph D_h are bounded by n^α. We compute the transitive closure in n^α phases. Each edge of D_h^*,corresponds to a path in D_h. One phase essentially consists in extending the already computed edges of D_h^* by an edge corresponding to an element in Layer(1), see Fig.3.b. It can be written informally as follows:

for each x=(A,i,j),z=(B,p,q),y=(C,p,p') and z1=(D,p',q) satisfying the situation in Fig.3.b **do in parallel**
  if $D_h^*(x,z)$ **and** $D_h(z,z1)$ **then** $D_h^*(x,z1)$:=**true**;

It is easy to see that $O(n^{2+3\alpha})$ processors are enough here. The time is n^α and the total work is $O(n^{2+4\alpha})$. ■

(c) The following informal statement shows how to compute efficiently $Layer_k \otimes Gen$.

for each A->BC, $(B,i,k) \in Layer_k$ and $(C,k,j) \in Gen$ do in parallel
  add (A,i,j) to $Layer_k \otimes Gen$.

There are $O(n^{1+\alpha})$ triples (B,i,k) in $Layer_k$ and there are O(n) indices j. Hence $O(n^{2+\alpha})$ processors suffice. This ends the proof.■

Theorem 1
Each context-free language can be recognized in $n^{0.75}$ parallel time with $O(n^{2.25})$ processors, or in $n^{1-\alpha}$ time with $O(n^{2+\alpha}+n^{1+4\alpha})$ processors, for $0<\alpha<1$.

Proof

We have $n^{1-\alpha}$ iterations. In one iteration $n^{2+\alpha}$ processors are enough, to compute Δ_k due to Lemma 2.c. The computation of Easy-closure$_k$(Twice) can be done efficiently after precomputing Layer(1) and D_h^*. There are $O(n^{1+3\alpha})$ edges in the graph D_h^* such that $x \in N(k)$, because $|N(k)| = O(n^{1+\alpha})$ and $|x|-|y| = O(n^{\alpha})$.

For each such edge (x,y), in parallel, we add x to Easy-closure if $y \in$ Twice. $O(n^{1+3\alpha})$ processors suffice to execute it. Now the claim follows directly from lemma 1 and Lemma 2.

The recognition problem for general cfl's is closely related to the computation of some dynamic programming problems. The dynamic programming problem is given by a system of recurrence equations:

(*) $\quad cost(i,j) = \min \{ cost(i,k) + cost(k,j) + f(i,k,j) : i<k<j \}$ if $i<j-1$;

$\quad cost(i,i+1) = init(i)$,

where f(i,j,k) and init(i) are functions computable sequentially in O(1) time. The integers i,j are in the range $0 \le i,j \le n$. The size of the problem is n. Various applications and more detailed discussion can be found in [BKR 91]. The dynamic programming problem can be formulated as a problem of computing costs of elements in a weighted composition system. Assume that for each $p,q,r \in N$ we are given the cost f(p,q,r) of applying the composition $q \otimes r$ to generate p in one step from q and r. Assume also that for each initial element $p \in T$ we are given its initial cost init(p). The cost c(T') of the generation tree T' is the sum of the numbers f(p,q,r), taken over all elements p, q, r such that p is a father of q, r, plus the weights init(p) of leaves of T'. Then cost(x) for an element $x \in N$ is the minimum cost of a tree generating x from the set T.

Theorem 2

The dynamic programming recurrences can be computed in $n^{0.75}\log(n)$ parallel time using $O(n^{2.25}/\log(n))$ processors. Total work is cubic.

Proof.

The computation of costs of weighted composition systems can be applied to the dynamic programming problems, see [BKR 91]. In this case the corresponding composition system S consists of elements (i,j), where $0 \le i<j \le n$. The composition is defined as follows:

$(i,k) \otimes (k,j) = \{(i,j)\}$;

The weight of each such composition is f(i,k,j);

$T = \{(i,i+1) : 0 \le i < n\}$.

It is easy to see that the value of cost(i,j) given by recurrences (*) is the same as with respect to the weighted composition system S. We use essentially the same algorithm GEN as for the unweighted composition systems.

The sets N(k), Layer(k), Gen(k) and Twice(k) have the same meaning as for unweighted system in section 2. In the present system S we have: $N(k) = \{ (i,j) : (k-1)n^{\alpha} \le j-i < k\, n^{\alpha}$.

In the new version of the algorithm GEN we compute the auxiliary table cost1(x). Initially each entry of this table is set to infinity. Then the table is successively updated. The condition cosd1(x) = cost(x) reflects the fact that the cost of the element x is correctly computed. The invariants

$\{$ Gen = Gen(k-1) $\}$ and

$\{$ $Layer_k$ = Layer(k); Twice = Gen(k) $\otimes$ Gen(k) $\}$

are replaced now by stronger invariants:

$\{$ Gen = Gen(k-1) and cost1(x) = cost(x) for all $x \in$ Gen $\}$;

$\{$ $Layer_k$ = Layer(k); Twice = Gen(k) $\otimes$ Gen(k) ;

cost1(x) = cost(x) for each $x \in$ ($Layer_k \cup$ Twice) $\}$.

Hence whenever we generate in the algorithm GEN a set $X \subseteq N$ then we have to update the table cost1 to guarantee that the costs of all elements of X are correctly computed. There are basically two places in the algorithm GEN where we generate new sets:

$\Delta_k := Layer_k \otimes Gen \cup Gen \otimes Layer_k \cup Layer_k \otimes Layer_k$;

$Layer_k$:= Easy-closure$_k$(Twice);

In fact the computation of the layers is not needed in the new algorithm (they can be computed trivially). However it helps to understand the behaviour of the algorithm. First we show how to update costs in Δ_k. The implementation is similar to that presented in the proof of Lemma 2.c. The following informal statement shows how to update costs in $Layer_k \otimes$ Gen, assuming that the costs are correctly computed for elements in Gen and $Layer_k$.

```
    for each i<j do in parallel
    begin
c(i,j):= min { cost1(i,s) + cost1(s,j) + f(i,s,j) : (s,j) ∈ N(k)}
cost1(i,j) := min (cost1(i,j); c(i,j)}
    end.
```

The statement above can be executed in log(n) time using $O(n^{2+\alpha}/\log(n))$ processors, due to the fact that $|N(k)| = O(n^{\alpha})$. Similarly we can update costs for elements in sets Gen $\otimes$ $Layer_k$ and $Layer_k \otimes Layer_k$. Denote by $Weighted\text{-}easy\text{-}closure_k(Z)$ the procedure which computes correctly the minimal costs of the generation of elements $x \in N_k$ from the set Layer(1) $\cup$ Z by a generation tree T of the form presented in Fig.1.a. We define the following graph $WD_h(k)$ called the k-th weighted dependency graph. This graph has the same set of nodes and edges as $D_h(k)$. The weight of the edge (x,y) is defined to be the minimum value of (cost(v)+f(x,y,v)) taken over all elements v in Layer(1) such that $x \in y \otimes v$ or $x \in v \otimes y$. Let $WPATH_k(x,y)$ be the minimal weight of a path from x to y in $WD_h(k)$. Using the same reasoning as in Lemma 1.a, it is easy to prove

$$cost(x) = \min \{ WPATH_k(x,y) + cost(y) : y \in Twice(k-1) \}.$$

for each $x \in Layer(k)$ and $0 < k \le n/h$.

The procedure $Weighted\text{-}easy\text{-}closure_k(Twice)$ can be implemented in a similar way as Easy-closure in the proof of Lemma 2.b. The values of $WPATH_k(x,y)$ can be precomputed. The transitive closure of the weighted directed graph $WD_h(k)$ an be precomputed similarly as the transitive closure of D_h in Lemma 2. Then the procedure $Weighted\text{-}easy\text{-}closure_k(Twice)$ can be performed efficiently in parallel. The analysis of the complexity is essentially the same as in the proofs of Lemma 2 and Theorem 1. We omit the technical details. ∎

3. Parallel recognition algorithms for subclasses of cfl's

We use the same algorithm GEN for an efficient recognition of unambiguous cfl's. It can be improved due to the property of unambiguity. We say that the directed graph has the unique path property (u.p.p., in short) iff for each two nodes v1, v2 there is at most one directed path from v1 to v2. The proofs of the three lemmas below are omitted.

Lemma 3 (key lemma)

Assume that the grammar is unambiguous and without useless nonterminals. Then

(a) the graph D_h has the u.p.p.

(b) the sets Δ_k in the algorithm GEN are pairwise disjoint.

Lemma 4

(a) Assume that C is a boolean circuit with n nodes and m edges. Assume also the C has the u.p.p. Then the values of all the nodes of C can be computed in logarithmic time with O(n+m) processors.

(b) Assume that the grammar is unambiguous and the graph D_h is computed. Then Easy-closure$_k$(Twice) can be computed in logarithmic time with $O(n^{1+2\alpha})$ processors.

Theorem 3

Recognition of unambiguous cfl's can be done in $O(n^{1-\alpha}\log(n))$ time with total work $O(\ T_U(n)n^{\alpha}\log(n)\ )$ for $0 < \alpha < 1$.

Proof.

The computation of Easy-closure$_k$(Twice) takes log(n) time with $n^{1+2\alpha}$ processors. The total cost of executing all such operations in $n^{1-\alpha}$ iterations is $O(n^{2+\alpha})$. It is enough to make a similar estimate for the total cost of computing sets Δ_k. The crucial point is that these sets are disjoint, due to Lemma 3.b. Assume that $X \subseteq N$. We maintain the following lists for each two positions i, j and for each nonterminal A (with their lengths):

$\text{Left}(X,A,j) = \{\ i : (A,i,j) \in X\}$, and

$\text{Right}(X,A,i) = \{\ j : A,i,j) \in X\}$.

This allows to execute efficiently the instruction:

$\Delta_k := \text{Layer}_k \otimes \text{Gen}(k-1) \cup \text{Gen}(k-1) \otimes \text{Layer}_k \cup \text{Layer}_k \otimes \text{Layer}_k$;

The sets $\text{Layer}_k \otimes \text{Gen}(k-1)$, $\text{Gen}(k-1) \otimes \text{Layer}_k$ and $\text{Layer}_k \otimes \text{Layer}_k$ are pairwise disjoint, due to the unambiguity. The thesis follows from the following claim: assume we are given the list representations of two sets $X, Y \subseteq \text{Closure}(T)$. Then the list representation of the set $X \otimes Y$ can be computed in log(n) time with $O(|X\otimes Y|+n)$ processors. ■

The proofs of the next theorems use ideas from [KR 88] and [Ry 88].

Theorem 4

Recognition of deterministic cfl's can be done in $O(n^{1-\alpha}\log(n))$ time with total work $O(T_D(n)n^{\alpha})$.

Theorem 5

There exists a parallel algorithm for the recognition of linear cfl's which works in $O(n^{1-\alpha}\log n)$ time with total work $O(n^{2+\alpha})$.

The algorithm from the theorem can be applied to the edit distance, the longest common subsequence and the string shuffling problems.

References

[AKLMT 89] M.Atallah, R.Kosaraju, L.Larmore, G.Miller, Teng, Parallel tree construction, SPAA'89

[BKR 91] L.Banachowski, A.Kreczmar, W.Rytter, Analysis of algorithms and data structures, Chapters 6-7, Addison Wesley (1991)

[CM 90] M.Chytil, B.Monien, Caterpillars and context free languages, STACS'90

[CCMR 90] M.Chytil, M.Crochemore, B.Monien, W.Rytter, On the parallel recognition of unambiguous context free languages, accepted to Theoretical Computer Science

[GR 88] A.Gibbons, W.Rytter, Efficient parallel algorithms, Cambridge University Press (1988)

[GR 89] A.Gibbons, W.Rytter, Optimal parallel algorithm for dynamic expression evaluation and context free recognition, Information and Computation (1989)

[GT 79] L.Guibas, H.Thompson, C.Thompson, Direct VLSI implementation of combinatorial algorithms, Caltech Conf. on VLSI (1979)

[HU 79] J.Hopcroft, J.Ullman, Introduction to automata, languages and computation, Addison Wesley (1979)

[HLV 90] S.Huang, H.Liu, V.Viswanathan, A sublinear parallel algorithm for some dynamic programming problems, Int.Conf. on Parallel Processing 3 (1990) 261-264

[Ko 75] S.R.Kosaraju. Speed of recognition of context-free language by array automata, SIAM J.Comp. 4 (1975) 333-340

[KR 88] P.Klein, J.Reif, Parallel time O(log n) acceptance of deterministic cfl's, SIAM Journal on Comp. 17 (19880, page 484

[Ry 87] W.Rytter, Parallel time O(log n) recognition of unambiguous cfl's, Information and Computation 73 (1987) 315-322

[Ry 88] W.Rytter, On the parallel computation of costs of paths on a grid graph, IPL 29 (1988) 71-74

[Ry 85] W.Rytter, The complexity of two way pushdown automata and recursive programs, Combinatorial algorithms on texts (ed.A.Apostolico, R.Capocelli), Springer-Verlag (1985) 341-356

[Ry 82] W.Rytter, Time complexity of unambiguous path systems, IPL 15:3 (1982)

[VHL 90] V.Viswanathan, S.Huang, H.Liu, Parallel dynamic programming, IEEE Conf. on Parallel Processing (1990) 497-500

COMPUTATIONAL GEOMETRY

A Simplified Technique for Hidden-Line Elimination in Terrains

*Franco P. PREPARATA**   *Jeffrey Scott VITTER*†

Dept. of Computer Science
Brown University
Providence, R. I. 02912–1910
U. S. A.

Abstract

In this paper we give a practical and efficient output-sensitive algorithm for constructing the display of a polyhedral terrain. It runs in $O((d+n)\log^2 n)$ time, where d is the size of the final display. While the asymptotic performance is the same as that of the previously best known algorithm, our implementation is simpler and more practical, because we try to take full advantage of the specific geometrical properties of the terrain. Our main data structure maintains an implicit representation of the convex hull of a set of points that can be dynamically updated in $O(\log^2 n)$ time. It is especially simple and fast in our application since there are no rebalancing operations required in the tree.

1 Introduction

A large number of scenes in graphics applications can be modeled efficiently and effectively by polyhedral terrains. A terrain is a three-dimensional closed polyhedron having the property that, for each location (x, y) in the plane, the values z for which (x, y, z) is on the polyhedron form either the empty set, a single point, or a closed interval.

Reif and Sen [ReS] recently did pioneering work on the generation of displays of terrains. They gave an $O((d+n)\log^2 n)$-time output-sensitive algorithm for the hidden-line elimination of an n-edge terrain whose display consists of d segments. Their technique resorts to general-purpose primitives such as ray shooting and is therefore ponderous and not immediately likely to lead to practical implementation.

The purpose of this paper is essentially methodological, since it shows how to take advantage of the specific nature of the objects involved in order to get a much simpler and, therefore, more practical algorithm, with the same asymptotic complexity as that of Reif and Sen. Our emphasis rests on the viewpoint that the analysis of asymptotic performance does not exhaust

---

*Support was provided in part by National Science Foundation grant CCR–8906419. The author can be reached via electronic mail addressed to franco@cs.brown.edu.

†Support was provided in part by a National Science Foundation Presidential Young Investigator Award CCR–9047466 with matching funds from IBM Corporation, by National Science Foundation grant CCR–9007851, by the U.S. Army Research Office under grant DAAL03–91–G–0035, and by the Office of Naval Research and the Defense Advanced Research Projects Agency under contract N00014–91–J–4052, ARPA order 8225. The author can be reached via electronic mail addressed to jsv@cs.brown.edu.

the objectives of algorithmic design, and that simplicity and ease of implementation are equally important criteria.

In the next section we define the display problem for terrains and discuss useful geometrical features of terrains. In Section 3, we describe our main data structure. It can be viewed as an efficient implicit representation of the lower convex hull of a semidynamic set of points (by which we mean that the universe of points is known beforehand). In Section 4 we give the $O((d+n)\log^2 n)$-time algorithm for terrain display, based upon a $O(\log^2 n)$-time algorithm for dynamically maintaining the implicit representation of the lower hull.

In the Appendix we describe the following result of independent interest: If we remove the semidynamic assumption, we can replace our CBT data structure with a more general balanced tree and get an alternative dynamic data structure for the lower convex hull that is simpler than the one proposed by Overmars and van Leeuwen [OvL]. The new data structure uses only one balanced tree, as opposed to the linear number of balanced trees required in [OvL]. The representation of the lower hull is implicit rather than explicit, but it can support all the applications described in [OvL] with the same performance guarantees.

2 Preliminaries

Let R be a planar subdivision in the (x,y) plane defined by n vertices. We assume for simplicity and with essentially no loss of generality that R has a unique unbounded region. For each open bounded region (simple polygon) r of R, we have an affine function $f_r(x,y)$ of (x,y); for the unbounded region r^* we have $f_{r^*}(x,y)=0$.

A *polyhedral terrain* (or simply *terrain*) τ is a polyhedron given by the following function $z(x,y)$ defined on the domain $\bigcup_{r\in R} r$ of the plane:

$$z(x,y)=\bigcup_{r\in R} f_r(x,y).$$

As stated, z is a function defined only in the interiors of the regions of R. We extend the definition of z to the edges of R, so that if edge e is shared by regions r' and r'', then for each point u of e, we define the following z-coordinate interval:

$$z(u)=\Big[\min\{f_{r'}(u),f_{r''}(u)\},\ \max\{f_{r'}(u),f_{r''}(u)\}\Big]. \tag{1}$$

Frequently, for each point u of e, we have $f_{r'}(u)=f_{r''}(u)$; however, in general, the polyhedron defined by the terrain τ can have as many as $2n$ vertices. Each vertical line intersects τ in a single connected domain, either a point or an interval.

Let $v_0=(x_0,y_0,z_0)$ be the *observation point* of the viewer and let $\mathbf{n}=(n_x,n_y,n_z)$ be the normal to the display plane P. Informally, the observer's eye is placed at v_0 and points in direction $\mathbf{n}$; the terrain can be imagined as being projected onto P. Without loss of generality, we let $n_x=0$, and we assume that R is *regular* [PrS], in that R admits a complete set $\Sigma=\{\sigma_1, \sigma_2,\dots,\sigma_p\}$ of separators monotone with respect to the x-axis. According to standard convention, if $i<j$ then a line parallel to the x-axis and intersecting both σ_i and σ_j intersects σ_i closer to $y=y_0$ than it intersects σ_j. The number p of separators is $O(n)$. The set of separators Σ naturally induces a set of three-dimensional polygonal lines $\Sigma'=\{\sigma'_1,\sigma'_2,\dots,\sigma'_q\}$ such that if two polygonal lines $\sigma'_i,\sigma'_j\in\Sigma'$ project to the same $\sigma\in\Sigma$, then by (1) for $i<j$ we have $z(u_1)\le z(u_2)$ whenever u_1 in σ'_i and u_2 in σ'_j project to the same point of σ. We let $\overline{\sigma_j}$ denote the central projection of $\sigma'_j\in\Sigma'$ to the display plane P, and we define $\overline{\Sigma}=\{\overline{\sigma_1},\overline{\sigma_2},\dots,\overline{\sigma_q}\}$.

As in [ReS], the construction of the display of the terrain τ is done incrementally, by successively processing the polygonal lines of $\overline{\Sigma}$, starting from $\overline{\sigma_1}$ closest to the observation point and proceeding away to $\overline{\sigma_q}$. The construction maintains in the display plane P an x-monotone polygonal chain ρ, called the *silhouette*, that separates the "clear" from the "opaque."

Specifically let $\overline{\sigma_j}$ be the currently processed polygonal line of $\overline{\Sigma}$. The advancing mechanism of the construction intersects $\overline{\sigma_j}$ with ρ. The portion of $\overline{\sigma_j}$ lying below ρ is eliminated; the portion of ρ lying below $\overline{\sigma_j}$ is reported (as part of the display). The upper envelope of $\overline{\sigma_j}$ and ρ forms the new updated silhouette ρ:

$$\rho := \sup(\rho, \overline{\sigma_j}),$$

where sup will also denote a function that implements the silhouette update.

Our approach makes use of several interesting geometric features of the silhouette:

1. The silhouette is an x-monotone chain (in the display plane P) consisting of lower-convex subchains separated by vertices that are the display-images of original vertices of the terrain. By lower-convex, we mean that the subchains are counterclockwise from left to right.

2. Each of the lower-convex subchains introduced above contains at most N segments, where $N = O(n)$ is the number of edges of τ, since each edge contributes at most one segment to each convex chain.

3. The worst-case total number of edges of the silhouette is $O(n\,\alpha(n))$, where $\alpha(n)$ is a functional inverse of Ackermann's function, which follows by the well-known result on the one-sided envelope of a collection of segments [HaS].

4. A chain $\overline{\sigma_j}$ is a concatenation of edges and gaps. (Gaps correspond to edges appearing in previously handled chains.) Thus each time the left extreme p' of an edge of $\overline{\sigma_j}$ belongs to the display (that is, it lies above ρ), then the abscissa of p' is the left extreme of the range of some leaf of T.

3 Silhouette Data Structure

The data structure we use to store the silhouette ρ is an implicit recursive representation of the lower convex hull of the vertices of ρ. From now on, whenever we refer to a point, segment, edge, terrain, etc., we refer to their projections in the display plane P. Since the abscissæ of the vertices of τ are known *a priori*, we order them and denote by X their ordered set. We store them as leaves in a *contracted binary tree* (CBT) [LeP, PVY], which we call T. We recall here for the reader's convenience that the nodes in a CBT T, which we call the ***active nodes***, are a subset of the nodes of a "skeletal" balanced binary tree T^*, whose leaves correspond 1–1 to the members of a fixed ordered universe X (which in our case is the set of abscissæ of the vertices). For simplicity we assume that $|X|$ is a power of 2. More formally, the ***active nodes*** are defined as follows:

1. The root is always active.

2. A leaf is active if it is stored in T and otherwise it is inactive.

3. Any internal nonroot node is active if both of its subtrees contain active nodes.

The number of nodes in the CBT T is clearly bounded by twice the number of leaves in T. The CBT has depth at most $\log |X|$ and no update ever involves rebalancing since the skeletal tree T^* has depth $\log |X|$. Insertions and deletions in T can be done in constant time, when a pointer is given to the element's neighbor; otherwise they require $O(\log |X|)$ time. Moreover, adjacent leaves and adjacent nodes in symmetric order are accessible from one another in constant time. For more details, the reader is referred to [PVY].

We store the silhouette in the following manner: Our universe X is an ordered multiset of abscissæ, one per vertex of τ. (If $k > 1$ vertices have the same abscissa, we order the k copies of the abscissa according to the following ordering of the corresponding vertices: the vertices whose edges have τ to their left are ordered from top to bottom, and are followed by the vertices whose edges have τ to their right, ordered from bottom to top.) Let $\{\lambda_1, \lambda_2, \ldots \lambda_p\}$, $p \leq n$, be the active leaves of the CBT T, each corresponding to a vertex of τ appearing in the silhouette ρ in left-to-right order. With a slight abuse of notation, we use λ_j to represent both the leaf in the CBT and the vertex in τ it corresponds to. With each λ_j, we associate a secondary structure S_{λ_j} (to be discussed below) that stores the convex subchain, denoted $\rho(\lambda_j)$, terminating at the vertex λ_j on the right. Each leaf λ_j, for $j \geq 2$, identifies an interval *range*(λ_j), defined to be $[x(\lambda_{j-1}), x(\lambda_j)]$. The range of an internal node is defined to be the union of the ranges of its leaves.

In each internal node w of T we store two additional items:

1. We store a segment called the *bridge* (denoted *bridge*(w)), which is the common supporting segment of the lower convex hulls of the portions of the silhouette corresponding to the left and right subtrees of w, respectively. (A lower convex hull of a set of points P in the plane is the convex hull of $P \cup \{(0, +\infty)\}$.) If the supporting segment is not well defined (because the concatenation of the nonvertical portions of the two convex hulls is convex), we make the convention that *bridge*(w) is the first segment of the right convex hull. For example, in Figure 1, the bridge spanning the convex subchains $\rho(\lambda_{17})$ and $\rho(\lambda_{20})$ is the first segment ℓ in $\rho(\lambda_{20})$'s convex subchain.

2. We store the abscissa *rightmost*(w) of the rightmost vertex of the subtree rooted at w.

An example of T is pictured in Figure 1a.

The rest of this section describes the general organization of each secondary data structure S_{λ_j}, which stores the convex subchain $\rho(\lambda_j)$ of ρ that terminates at λ_j. The internal vertices of $\rho(\lambda_j)$ are not vertices of τ, but rather they are defined implicitly by intersections of polygonal lines in $\overline{\Sigma}$. Fortunately we can represent $\rho(\lambda_j)$ in such a way that it can itself be stored efficiently in a CBT (which always requires a fixed universe). We denote the CBT by S_{λ_j}.

We set the universe of CBT S_{λ_j} to be the set A of edges in $\overline{\Sigma}$ ordered according to their slope. The convex subchain $\rho(\lambda_j)$ is represented by the edges of A that form it. More specifically, let $e_{i_1}, e_{i_2}, \ldots, e_{i_t}$, where $e_{i_k} = \overline{(p_{k-1}, p_k)}$, be the edges of $\overline{\Sigma}$ that contain the segments of $\rho(\lambda_j)$. The marked leaves of the CBT S_{λ_j} are $e_{i_1}, e_{i_2}, \ldots, e_{i_t}$. Edge e_{i_k} represents the vertex p_k of $\rho(\lambda_j)$, and p_0 is represented implicitly; for $1 \leq k < t$, p_k is the intersection of e_{i_k} and $e_{i_{k+1}}$, and $p_t = \lambda_j$.

We associate the root of each CBT S_{λ_j} with the leaf λ_j of the CBT T. We denote the resulting "combined" tree by $\widehat{T}$. Because $\rho(\lambda_j)$ is a convex subchain, all the bridges in the recursive decomposition of $\rho(\lambda_j)$ (based on the structure of the CBT S_{λ_j}) are simply segments in $\rho(\lambda_j)$ and thus do not need to be stored explicitly. The bridge *bridge*(w) for an internal node w in S_{λ_j} is the segment whose leaf (edge) in S_{λ_j} is the symmetric-order successor of w in S_{λ_j}, which can be found in constant time. A typical CBT S_{λ_j} is pictured in Figure 1b.

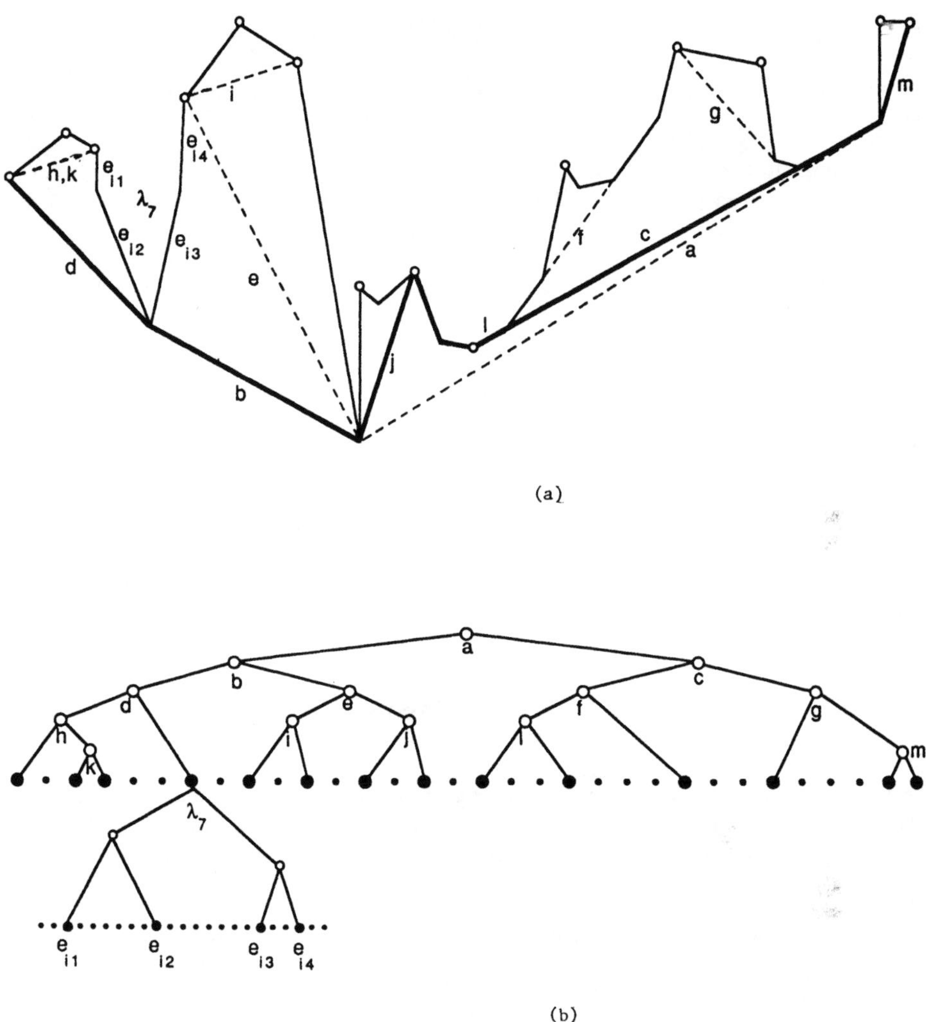

Figure 1: (a) The current silhouette ρ. The lower convex hull of ρ consists of the segments d, b, a, m. The two subhulls whose bridging (by segment a) gives the full hull are pictured in bold lines. All the bridges in the recursive decomposition of the full hull are indicated by dashed lines, except for bridge ℓ, which is also a segment of the silhouette. (b) The combined tree data structure $\hat{T}$ consists of the upper CBT T and the lower CBTs $S_{\lambda_1}, S_{\lambda_2}, \ldots$. Upper CBT T stores the vertices of the silhouette ρ that are display-images of vertices of the terrain τ. Each internal node w in the diagram is labeled by the segment *bridge*(w). For simplicity, the only lower CBT pictured is S_{λ_7}, which stores λ_7's convex subchain $e_{i_1}, e_{i_2}, e_{i_3}, e_{i_4}$.

4 Dynamic Update of the Silhouette

The dynamic update of the silhouette ρ caused by processing polygonal chain $\overline{\sigma_j}$ consists of three tasks:

Task 1 Computation of the intersections of ρ and $\overline{\sigma_j}$. We define an intersection of ρ and $\overline{\sigma_j}$ to be a changeover point in which $\overline{\sigma_j}$ changes from being on or above ρ on one side of the intersection to being strictly below ρ on the other side.

Task 2 Dynamic update of the display: We output to the display the portions of ρ lying below $\sup(\rho, \overline{\sigma_j})$.

Task 3 Dynamic update of the data structure $\hat{T}$: We remove from $\hat{T}$ the portions of ρ lying on or below σ_j, and we insert into $\hat{T}$ the portions of $\overline{\sigma_j}$ lying on or above ρ.

The three tasks can be implemented concurrently left-to-right on σ_j and ρ. Tasks 1 and 2 are handled in Section 4.1 in $O(d' + n' \log^2 n)$ time, where n' is the number of edges of $\overline{\sigma_j}$ and d' is the number of segments output to the display during the processing of $\overline{\sigma_j}$. In Section 4.2 we show that $\hat{T}$ can be updated in $O((d'+1)\log^2 n)$ time. This gives us our main theorem:

Theorem 1 *Hidden-line elimination for terrains can be done without the need for balanced-tree data structures in $O((d+n)\log^2 n)$ time, where d is the number of segments in the resulting display.*

4.1 Computation of the Intersections and Dynamic Update of the Display

We do Tasks 1 and 2 by handling $\overline{\sigma_j}$ edge-by-edge from left to right, as indicated in the following loop. Let $e = \overline{(p', p'')}$ denote the current edge of $\overline{\sigma_j}$ being processed, starting with the point $p \in e$. That is, segment $e' = \overline{(p, p'')}$ is the portion of e currently being processed. We distinguish two cases:

1. *p and the portion of e' immediately to its right are on or above ρ.* In this case we march along the silhouette ρ until either we find an intersection q of e' with ρ or we reach the abscissa of p'' (in which case we set $q := p''$) (Task 1). The portion of ρ lying below $\overline{(p,q)}$ is reported as appearing in the display (Task 2). The above loop continues with $p := q$.

2. *The portion of e' immediately to the right of p is below ρ.* In this case the display does not change in a neighborhood of p (Task 2), and we have to search for the leftmost intersection q of e' with ρ; if none exists, we set $q := p''$ (Task 1). The loop then continues with $p := q$.

   The search for q begins with the node w of T that has the smallest level (leaves of T have level 0) among those whose range contains that of e'. We use the segment $bridge(w)$ and test it against the line $line(e')$ containing e'. We denote the line containing $bridge(w)$ by $line(bridge(w))$. We distinguish three cases, as illustrated in Figure 2:

   (a) *$bridge(w)$ does not extend below $line(e')$.* If $line(bridge(w))$ intersects $line(e')$ to the right (respectively, left) of $bridge(w)$, then w's left subtree (respectively, right subtree) can be excluded from the search. (See Figure 2a.)

   Comment: If, for example, $line(bridge(w))$ intersects $line(e')$ to the right of $bridge(w)$, then all vertices of ρ pertaining to the left child of w are confined to the shaded region in Figure 2a obtained by intersecting the closed halfplane above $line(bridge(w))$ and the closed halfplane to the left of the vertical line $x = x(rightmost(left\_child(w)))$. Clearly none of this region's edges can intersect e'.

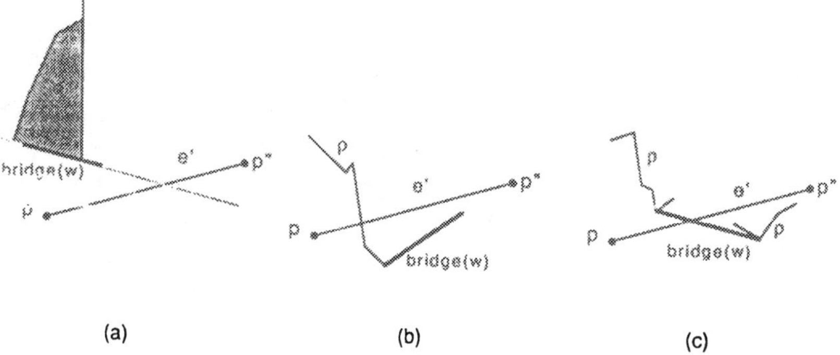

Figure 2: The different cases in the recursive search for the first intersection between edge segment e' and the silhouette ρ.

(b) *bridge*(w) *lies entirely below line*(e'). In this case, w's right subtree can be excluded from the search. (See Figure 2b.)

Comment: By assumption, the part of e' immediately to the right of p is below ρ. Since ρ contains the left endpoint of *bridge*(w), which is below *line*(e'), there must be an intersection between *line*(e') and the portion of ρ pertaining to w's left child.

(c) *bridge*(w) *lies partly on or above line*(e') *and partly below line*(e'). In this case, there is certainly at least one intersection of e' with ρ, but we may not know which subtree of w contains the portion of ρ having the first intersection with e', as in Figure 2c. The search is continued in the left subtree, in accordance with the left-to-right update policy. If no intersection is found in the left subtree, processing continues with the right subtree. The search for q halts whenever an intersection is found.

Finding the intersections of $\overline{\sigma_j}$ with ρ and the dynamic update of the display in Case 1 requires $O(n' + d')$ time, where n' is the number of edges of $\overline{\sigma_j}$ and d' is the number of segments output to the display during the processing of $\overline{\sigma_j}$.

We now show that the time required for Case 2 is $O(n' \log^2 n)$. If e' does not intersect ρ, we traverse a path from node w to a leaf of $\widehat{T}$, each time under Case 2a spending constant time at each node, for a total of $O(\log n)$ time. Suppose instead that e' intersects ρ. We consider a path in $\widehat{T}$ from w to the leaf pertaining to the segment containing the intersection q. In the worst case, each node on this path is a right child and at each such node we launch an unsuccessful search through the left sibling as specified by Case 2c. Clearly there are $O(\log n)$ nodes from which such unsuccessful searches can be taken, each search taking $O(\log n)$ time, before q is found.

4.2 Dynamic Update of $\widehat{T}$

In the algorithm in Section 4.1 for computing intersections between the polygonal line $\overline{\sigma_j}$ being processed and the silhouette ρ, the polygonal line $\overline{\sigma_j}$ alternates between being on or above ρ and being below ρ. The dynamic update of the combined CBT data structure $\widehat{T}$ (Task 3) is concerned with the portion of $\overline{\sigma_j}$ that lies on or above ρ (which corresponds to Case 1 of Tasks 1 and 2).

Let p denote the last processed intersection point of $\overline{\sigma_j}$ with ρ. Immediately after p, $\overline{\sigma_j}$ is on or above ρ. Let $e_1, e_2, \ldots, e_k$, with $e_i = \overline{(p_{i-1}, p_i)}$, denote a maximal subchain of $\overline{\sigma_j}$ such that $p \in e_1$, $p \neq p_0$, and edges $e_2, \ldots, e_{k-1}$ lie completely on or above ρ, as illustrated in Figure 3a

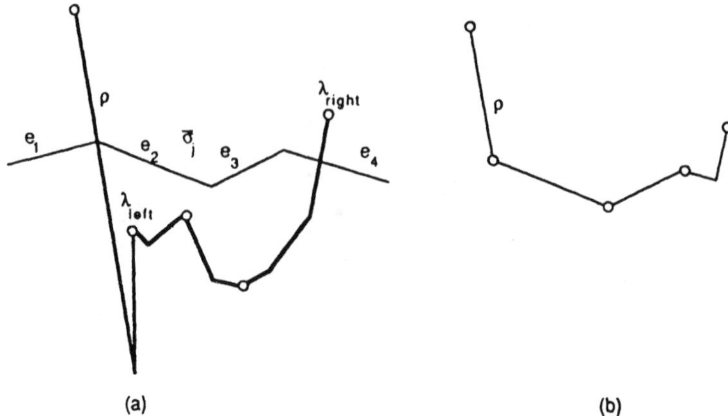

Figure 3: (a) The silhouette before the processing of $\overline{\sigma_j} = e_1, e_2, \ldots, e_4$. (b) The silhouette after the processing.

for the case $k = 4$. Let λ_{left} be the first vertex of τ whose convex subchain $\rho(\lambda_{left})$ intersects e_1 at p, and let λ_{right} be the last vertex of τ whose convex subchain $\rho(\lambda_{right})$ intersects e_k as close to p along $\overline{\sigma_j}$ as possible. We modify T by deleting the leaf λ_{left} and inserting the leaves p_1, p_2, ... p_{k-1}. The convex subchain $\rho(p_i)$, for $2 \leq i \leq k-1$, is initialized to contain the singleton edge e_i. These updates to $\widehat{T}$ require $O(k)$ time. The convex subchain $\rho(p_1)$ is initialized to contain the edges of $\rho(\lambda_{left})$ up to its intersection with e_1, followed by the portion of e_1 after the intersection. Similarly, in the convex subchain $\rho(\lambda_{right})$, the initial portion, up to the intersection with e_k, is deleted and replaced with the portion of e_k up to the intersection. If $k = 1$, then $\rho(\lambda_{right})$ in addition stores the initial part of $\rho(\lambda_{left})$ up to its intersection with e_1. This is illustrated in Figure 3b. These updates involve splits of the CBTs $S_{\lambda_{left}}$ and $S_{\lambda_{right}}$, which, if $k = 1$, are spliced together; this can be done easily in $O(\log n)$ time. The total time for this process is thus $O(\log n + k)$.

We mark the leaves $p_1, \ldots, p_{k-1}, \lambda_{right}$ for later processing. The next portion of the polygonal line $\overline{\sigma_j}$, starting with the intersection of e_k and $\rho(\lambda_{right})$, lies below the silhouette ρ and is processed as in Case 2 in Section 4.1, seeking the subsequent intersection of $\overline{\sigma_j}$ and ρ. The portion of $\overline{\sigma_j}$ that remains after Case 2 is again on or above ρ, so we repeat the above process.

After the above processing of the polygonal line $\overline{\sigma_j}$ is completed, there may be several marked leaves in the CBT T. They represent the vertices of τ whose convex subchains were created or modified as a result of processing $\overline{\sigma_j}$. In the remainder of the dynamic update, we perform a postorder traversal of the portion of T that lies above the marked leaves, in order to update the bridges stored in those nodes.

At each internal node w in T lying on a rootward path from each of the marked leaves (on the way up in the postorder traversal), we compute in $O(\log n)$ time the updated bridge segment *bridge*(w) for node w by processing the bridge information in the nodes below w in T, as follows: We initialize ℓ and r to be the left and right children of w. We use *line*(*bridge*(ℓ)) and *line*(*bridge*(r)) to denote the straight lines containing the segments *bridge*(ℓ) and *bridge*(r). Let λ denote the rightmost leaf *rightmost*(ℓ) in the subtree rooted at ℓ. We carry out the procedure below until ℓ and r are each leaves in $\widehat{T}$. The resulting ℓ and r are made the endpoints of *bridge*(w).

As long as ℓ and r are both internal nodes of $\widehat{T}$, we repeatedly do the actions corresponding to the following two cases:

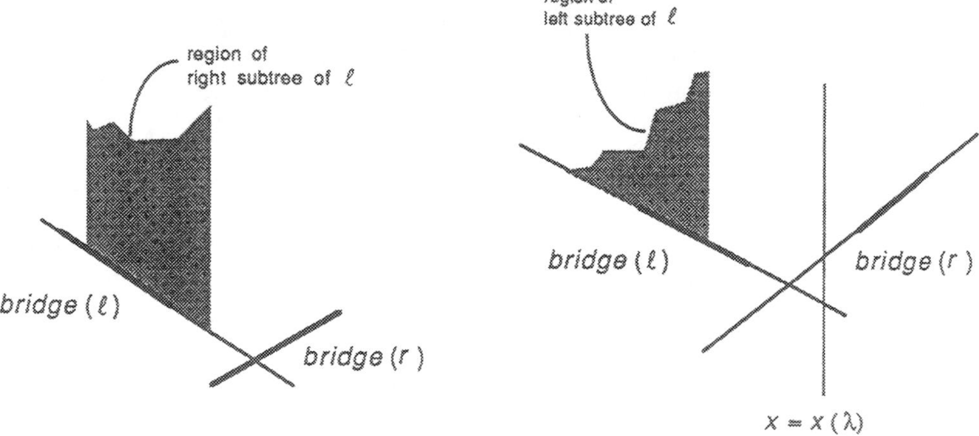

Figure 4: Recursive descent needed to determine *bridge*(w) for a node w in $\widehat{T}$. The bridges for the left child ℓ and the right child r of w are shown. (a) The left anchor point for *bridge*(w) must be in ℓ's left subtree. (b) The left anchor point for *bridge*(w) must be in ℓ's right subtree.

1. *Some portion of bridge*(r) *(respectively, bridge*(ℓ)*) lies below line(bridge*(ℓ)*) (respectively, line(bridge*(r)*))*. We set ℓ to be ℓ's left child (respectively, we set r to be r's right child). (See Figure 4a.)

   Comment: If, for example, some portion of *bridge*(r) lies below *line*(*bridge*(ℓ)), then all vertices of ρ pertaining to the right child of ℓ are confined to the shaded region pictured in Figure 4a, obtained by intersecting the closed halfplane above *line*(*bridge*(ℓ)) and the strip delimited by the two vertical lines $x = x$(left extreme of *bridge*(ℓ)) and $x = x(\lambda)$. Clearly none of this region's vertices can support *bridge*(w).

2. *Both bridge*(ℓ) *and bridge*(r) *lie on or above the line extension of the other.* We denote by *int*(*line*(*bridge*(ℓ)), λ) and *int*(*line*(*bridge*(r)), λ), respectively, the ordinates of the intersections of *line*(*bridge*(ℓ)) and of *line*(*bridge*(r)) with the vertical line $x = x(\lambda)$.

   (a) If *int*(*line*(*bridge*(ℓ)), λ) $\leq$ *int*(*line*(*bridge*(r)), λ), then we set ℓ to be ℓ's right child. (See Figure 4b.)

   (b) If *int*(*line*(*bridge*(ℓ)), λ) $\geq$ *int*(*line*(*bridge*(r)), λ), then we set r to be r's left child. (This case is symmetric to Case 2a.)

   Comment: Since we are negating Case 1, the wedge formed by *line*(*bridge*(ℓ)) and *line*(*bridge*(r)) is lower-convex. Thus if, for example, we have *int*(*line*(*bridge*(ℓ)), λ) $\leq$ *int*(*line*(*bridge*(r)), λ), then the vertices of ρ pertaining to the left child of ℓ are confined to the shaded region pictured in Figure 4b, obtained by intersecting the closed halfplane above *line*(*bridge*(ℓ)) and the closed halfplane to the left of the vertical line $x = x$(right extreme of *bridge*(ℓ)). None of this region's vertices can support *bridge*(w).

When this loop terminates, either ℓ or r is a leaf in the combined CBT $\widehat{T}$. We repeatedly do the following step until ℓ and r are both leaves: If ℓ (respectively, r) is a leaf and lies on or below *line*(*bridge*(r)) (respectively, *line*(*bridge*(ℓ))), then we set r to be r's right child (respectively, ℓ to be ℓ's left child); otherwise we set r to be r's left child (respectively, ℓ to be ℓ's right child).

At this point, both ℓ and r are leaves of $\widehat{T}$. We set *bridge*(w) to be the segment whose endpoints are represented by the leaves ℓ and r.

The time to compute each bridge by the above procedure requires $O(\log n)$ time. The number of bridges to be updated as a result of processing $\overline{\sigma_j}$ is $O(n' \log n)$, where n' is the number of edges in $\overline{\sigma_j}$. Thus the time required for all the bridge updates caused by $\overline{\sigma_j}$ is $O(n' \log^2 n)$. Theorem 1 follows by summing the running times needed for the three tasks in the processing of $\overline{\sigma_j}$.

References

[HaS] S. Hart and M. Sharir, "Nonlinearity of Davenport-Schinzel Sequences and of a Generalized Path Compression Scheme," *Combinatorica* 6 (1986), 151–177.

[LeP] D. T. Lee and F. P. Preparata, "Parallel Batch Planar Point Location on the CCC," *Information Processing Letters* 33 (December 1989), 175–179.

[OvL] M. H. Overmars and J. van Leeuwen, "Maintenance of Configurations in the Plane," *Journal of Computer and System Sciences* 23 (1981), 166–204.

[PrS] F. P. Preparata and M. I. Shamos, *Computational Geometry*, Springer-Verlag, New York, 1985.

[PVY] F. P. Preparata, J. S. Vitter, and M. Yvinec, "Computation of the Axial View of a Set of Isothetic Parallelepipeds," *ACM Transactions on Graphics* 9 (July 1990), 278–300.

[ReS] J. H. Reif and S. Sen, "An Efficient Output-Sensitive Hidden-Surface Removal Algorithm and its Parallelization," *4th Annual ACM Symposium on Computational Geometry* (June 1988), 193–200.

Appendix

In this appendix we describe how to use our lower convex hull data structure of Sections 3 and 4 in a variety of dynamic applications described by Overmars and van Leeuwen [OvL]. In these applications the universe of points is not necessarily known in advance, so we use a balanced tree instead of a CBT to implement L. The points S stored in L are ordered from left to right, and those with the same abscissa are ordered from top to bottom. There is no need for the secondary data structures that we described in Sections 3 and 4; all the points of S are stored as leaves of L. The resulting data structure makes use of a single balanced tree L as opposed to the data structure of [OvL], which uses a linear number of balanced trees; there, each node in the tree has a secondary data structure represented by a balanced tree.

The following theorem shows that the dynamic applications [OvL] can be implemented within the same time and space bounds using our data structure L for the lower convex hull.

Theorem 2 *Our lower convex hull data structure L and the symmetrically defined upper convex hull data structure U support the following operations:*

- *A set of n points in the plane can be "peeled" in $O(n \log^2 n)$ time.*
- *The convex layers of a set of n points in the plane can be determined in $O(n \log^2 n)$ time.*
- *A connecting spiral of a set of n points in the plane can be determined in $O(n \log^2 n)$ time.*
- *A set of points in the plane can be maintained at the cost of $O(\log^2 n)$ time per insertion or deletion so that the query "does p lie on the convex hull of the set of points?" can be answered in $O(\log n)$ time.*
- *Two sets A and B of points in the plane can be maintained at the cost of $O(\log^2 n)$ time per insertion or deletion so separability can be determined in constant time.*

The following theorem is useful for determining the convex layers:

Theorem 3 *Using our lower convex hull data structure L, the lower convex hull of a set of points $S = \{p_1, p_2, \ldots p_n\}$ can be output in $O(h \log \frac{n}{h})$ time, where h is the current size of the lower convex hull of S.*

Let $x1(bridge(w))$ and $x2(bridge(w))$ denote the left and right abscissae, respectively, of segment $bridge(w)$. We prove Theorem 3 by giving the pseudocode that outputs the lower convex hull of the points in left-to-right order. Let λ_1 and λ_n denote the abscissæ of the leftmost and rightmost points; in case of a tie, we choose the bottommost point. The printing of the lower hull is done via the procedure call *print_hull*(root of L, λ_1, λ_n). In the following pseudocode, w is a node of L, and *left* and *right* are abscissæ satisfying *left* $\leq$ *right*.

```
print_hull(w, left, right);
  begin
    if left < x1(bridge(w)) then
      print_hull(left_child(w), left, x1(bridge(w)));
    if [x1(bridge(w)), x2(bridge(w))] ⊆ [left, right] then
      Output bridge(w);
    if x2(bridge(w)) < right then
      print_hull(left_child(w), x2(bridge(w)), right)
  end;
```

The running time is bounded by the path length to the nodes in the tree where the segments of the lower convex hull are stored as bridges. If there are h segments on the lower hull, we can show by convexity arguments that the path length is $O(h \log \frac{n}{h})$. The justification for the pseudocode is given in the following lemma:

Lemma 1 *The segment $bridge(w)$ belongs to the lower convex hull of the set of points S represented in L if and only if*

$$\left(x1(bridge(w)),\ x2(bridge(w))\right) \bigcap \left(x1(bridge(v)),\ x2(bridge(v))\right) = \emptyset \qquad (2)$$

for each ancestor v of w in L.

Proof: ($\Longrightarrow$) Suppose that $bridge(w) = \overline{(p_{left}, p_{right})}$ is part of the lower convex hull. Then the lower hull lies in the halfplane $H^+(w)$ on or above $line(bridge(w))$. For any ancestor v of w, all the points stored in the subtree of L rooted at v, including the endpoints of $bridge(w)$, are contained in $H^+(v)$. Therefore, $line(bridge(v))$ intersects the vertical line $x = x(p_{left})$ (respectively, $x = x(p_{right})$) at point q_{left} (respectively, q_{right}) at or below p_{left} (respectively, p_{right}). If $bridge(w)$ and $bridge(v)$ are colinear, condition (2) must hold, since the bridges can intersect only at their endpoints. Otherwise if the bridges are not colinear and condition (2) does not hold, then the segment $\overline{(q_{left}, q_{right})}$ contains a point of $bridge(v)$ not in $H^+(w)$, which contradicts the assumption that $bridge(w)$ belongs to the lower convex hull.

($\Longleftarrow$) Suppose that $bridge(w) = \overline{(p_{left}, p_{right})}$ does not belong to the lower convex hull. Then there is a point $p \in S$ not in $H^+(w)$. By definition of bridge, p cannot lie directly under $bridge(w)$, but must be either to the left or the right of the vertical strip spanned by $bridge(w)$. Without loss of generality, let us assume that (1) $p \in S$ is to the right of such strip, (2) p is not in $H^+(w)$, and (3) p is the leftmost point in S satisfying (1) and (2). Let v be the lowest common ancestor in L of the leaves for p_{right}, p_{left} and p. By the definition of $bridge(w)$, it follows that node w is in the left subtree of v and the leaf for p is in the right subtree of v. This implies that the left endpoint q of $bridge(v)$ precedes p in S. Assume, for a contradiction, that $x(p_{right}) \leq x(q)$. By definition of p, there is no point in S between p_{right} and p that is below $line(bridge(w))$. Since q is either between p_{right} and p in S or else on or directly above p_{right}, we have $q \in H^+(w)$. Since both q and p are in $H^+(v)$, while $q \in H^+(w)$ and $p \notin H^+(w)$, it follows that $line(bridge(v))$ intersects $line(bridge(w))$ at or to the right of p_{right} in S, and thus $p_{left} \notin H^+(v)$, which contradicts the definition of $bridge(v)$. Therefore, we must have $x(q) < x(p_{right})$, and condition (2) is violated. □

The rest of this section is devoted to proving Theorem 2. The first application of peeling can be implemented easily by observing that the bridge stored at the root of L must be on the lower convex hull. Hence we can "peel" off the two endpoints of the bridge, dynamically update L in $O(\log^2 n)$ time, and continue until no points remain.

For the second application, we need to peel off an entire layer before updating L and U. Theorem 3 can be invoked to output the ith layer in $O(h_i \log n)$ time, where h_i is the size of the ith layer, and $\sum_i h_i = n$. Deleting the points in a layer takes $O(\log^2 n)$ time per point, resulting in a total time bound of $O(n \log^2 n)$.

The third application can be done in a similar way, except that after each layer is removed, the last point on the previous layer is left temporarily in L and U. The first segment on the convex hull starting at that point, which is the desired spiral segment that connects the previous layer to the next, can be found by executing the procedure for Theorem 3 up until the first segment is output, which takes $O(\log n)$ time.

In the fourth application, we can determine whether a point p lies on the convex hull in $O(\log n)$ time by traversing L and U from the leaf for p upward to their respective roots. By Lemma 1, the point p is on the lower (respectively, upper) convex hull if and only if every bridge in L (respectively, in U) encountered either contains p or has an x-range to the left or right of $x(p)$.

For the last application, we use lower and upper convex hull data structures for each of the sets A and B. We can separate A and B if and only if one of the sets has a lower convex hull that does not intersect the upper convex hull of the other. For simplicity let us consider whether the lower convex hull of A (including its interior) intersects the upper convex hull of B (including its interior); the other case is symmetrical. We store the lower convex hull of A in data structure L_A and the upper convex hull of B in data structure U_B.

Let a and b denote the roots of L_A and U_B. We carry out the actions corresponding to the following cases repeatedly until either it is determined that there is or isn't an intersection or else either a or b is a leaf. If a is a leaf, for example, a corresponds to a single segment, and we test whether a's endpoints are on or inside the upper convex hull of B using a procedure like that used for the previous application. We leave the justification to the reader.

1. *The vertical strip lying on or directly above $bridge(a)$ intersects the vertical strip lying on or directly below $bridge(b)$.* In this case there is an intersection.

2. *$bridge(a)$ is on or below $line(bridge(b))$ or $bridge(b)$ is on or above $line(bridge(a))$.* If $bridge(a)$ is to the left (respectively, right) of $bridge(b)$, then we set a to be a's right (respectively, left) child and b to be b's left (respectively, right) child.

3. *Cases 1 and 2 do not hold.*

   (a) If $bridge(a)$ and $bridge(b)$ are parallel, there can be no intersection.

   (b) If $int(line(bridge(a)), line(bridge(b)))$ is to the right (respectively, left) of $bridge(a)$, then we set a to be a's right (respectively, left) child.

   (c) If $int(line(bridge(a)), line(bridge(b)))$ is to the right (respectively, left) of $bridge(b)$, then we set b to be b's right (respectively, left) child.

In each iteration we descend one level in L_A or U_B. Thus we can determine separability in $O(\log n)$ time.

A Competitive Analysis of Nearest Neighbor Based Algorithms for Searching Unknown Scenes

(Preliminary Version)

Bala Kalyanasundaram Kirk Pruhs

Computer Science Department

University of Pittsburgh

Pittsburgh, PA 15260 USA

Abstract: We consider problems involving robot motion planning in an initially unknown scene of obstacles. The two specific problems that we examine are mapping the scene and searching the scene for a recognizable target whose location is unknown. We use competitive analysis as a tool for comparing algorithms. In the case of convex obstacles, we show a tight $\Theta(\min(k, \sqrt{k\alpha}))$ bound on the competitiveness for these problems, where α and k are aspect ratio and number of objects, respectively. This lower bound also holds for randomized algorithms. We derive an almost tight bound on the competitive ratio for the Nearest Neighbor heuristic.

We also propose allowing multiple robots to cooperatively search the scene. For scenes that contain only objects of bounded perimeter and bounded aspect ratio, we show that m robots can achieve a competitive factor of $O(\sqrt{k}/m)$, which is optimal. Thus, the robots can cooperate without interference in searching the scene. For general scenes we show that a competitive factor of $O((\sqrt{k\alpha}\log m)/m)$ is achievable, provided $m = O(\sqrt{k/\alpha})$.

1 Problem Statement

The setting for the problems that we consider is a Euclidean plane sprinkled with k obstacles. Inhabiting this environment is at least one robot that is equipped with a vision sensor. The robot learns about the objects and the environment only through the visual information provided by the sensor. More precisely, the robot is only aware of parts of objects that lie within its line of sight currently, or that lied within its line of sight at some point in the past. The goal of the robot is move through the scene, avoiding obstacles, until it accomplishes some goal. The two major goals we consider are:

Searching: In the searching problem the robot must find the location of some recognizable object, called the *target* in the scene. Imagine that the robot is looking for a McDonalds in an unknown city. The robot may not know the location of the McDonalds, but it can certainly recognize the golden arches once it sees them. More precisely, the robot must move along an obstacle avoiding path P terminating at the target. Alternatively, one might only require the robot to move along an obstacle avoiding path P terminating at point from which some part of the target may be seen. All of our bounds also hold for this variant of searching.

Mapping: In the mapping problem the robot's goal is to determine the location of all of the obstacles. More precisely, the robot must move along an obstacle avoiding path P that has the property that there is an obstacle avoiding straight line path from every point on the boundary of each obstacle to some point in P. As we shall see the difficulty of the problem is unchanged even if the robot is required to see only a part of each object, instead of the entire perimeter.

In each of these problems we want to minimize the length of the robot's path P. Since the robot is learning the environment as it moves, we can view these problems as on-line problems. In this paper we will use competitive analysis to compare different on-line algorithms for these problems. Let OPT be the shortest path that would allow the robot to accomplish its goal. We can assume that OPT is computed by an off-line algorithm that has a bird's eye view of the environment. A robot's algorithm is *c-competitive*, or equivalently has a *competitive factor* of c, if the ratio of the length of P over the length of OPT is at most c.

The problems of searching and mapping have been studied extensively in the robot navigation literature (see [SY] for a current survey). Yet, there has been no previous work using competitive analysis for the problems that we consider. However, there has been some recent investigations, using competitive analysis, into some related problems that are set in unknown scenes, which we now survey.

Papadimitriou and Yannakakis [PY] pioneered the use of competitive analysis in the study of problems involving navigation of a scene using visual information. They consider the problem of finding a short obstacle avoiding path to a target with a *known* location in the plane. Papadimitriou and Yannakakis showed one can achieve a constant competitive factor if every obstacle in the scene has bounded aspect ratio. The *aspect ratio* of a convex object O is defined to be R/r where R is the radius of the smallest circle that circumscribes O and r is the largest circle that inscribes O [BRS]. A scene of convex objects has aspect ratio α if the aspect ratio of each object is at most α. They also showed that no on-line algorithm with a constant competitive factor exists if the objects are convex polygons with unbounded aspect ratio. Blum, Raghavan, and Schieber [BRS] and Karloff, Rabani, and Ravid [KRR] obtained further results in the case of objects with unbounded aspect ratio.

Baeza-Yates, Culberson, and Rawlins [BCR] consider several nice problems that, while not directly related to the problems we consider, have a similar flavor to the searching problem. Deng and Papadimitriou [DP] originally proposed using competitive analysis to study the mapping problem.

2 Summary of Results

The ability of the robot to navigate quickly depends on the type of environment that it inhabits. One factor that makes it more difficult to navigate is the number of objects in the scene. Another property of the scene that makes searching hard is the aspect ratio of the environment [BRS,PY]. We will thus generally measure the competitiveness as a function of k, the number of objects in the scene, and α, the aspect ratio of the scene. In this paper we consider only convex objects to avoid mazelike scenes [BRS]. We use $\mathcal{M}(k,\alpha)$ to denote the expression $min(k, \sqrt{k\alpha})$.

In section 3 we establish a bound of $\Omega(\mathcal{M}(k,\alpha))$ on the competitive factor for searching and mapping problems. This lower bound also holds for randomized algorithms, even if the map of the terrain, sans the target, is known in advance. In section 4, we analyze the competitive factor of one of the simplest and most common heuristics, Nearest Neighbor. For searching Nearest Neighbor's competitive can be as high as $\Omega(2^k)$. For the mapping problem we show that Nearest Neighbor's competitive factor is $\Theta(\alpha^{3/2}\sqrt{k})$. The main reason that Nearest Neighbor is not optimal is that it does not localize it's search around the origin. In section 5, we show that a careful modification to Nearest Neighbor, which we call Bifold Nearest Neighbor (BNN), has an a competitive factor that is $O(\mathcal{M}(k,\alpha)\log k)$. More importantly, the new algorithm is almost as simple as Nearest Neighbor. We finish by giving an algorithm, Bifold Tourist (BT), that has a competitive factor that is $O(\mathcal{M}(k,\alpha)$. Hence BT is asymptotically strongly competitive; That is, it has an optimal competitive factor modulo a multiplicative constant.

In section 6 we propose a generalization of the searching problem that allows multiple robots. The goal for the robots is to cooperatively search the scene for the target. We measure the time, assuming uniform speed, to find the target. We now compare this cost to the time that it would take to travel to the target on the shortest obstacle avoiding path at the same speed. We call this the *Search Party Problem*. Ideally, we would like the robots to cooperate without interference. This can achieved for environments with objects of bounded perimeter and bounded aspect ratio. Thus m robots can find the target after at most $O(\sqrt{k}/m)$ times the optimal off-line time.

Finally, we consider the Search Party Problem for general scenes with convex obstacles. In this case, we show that m robots can achieve a competitive factor or $\Omega((\mathcal{M}(k,\alpha)\log m)/m)$, provided $m = O(\sqrt{k/\alpha})$. Hence, for scenes that contain objects with bounded aspect ratio, $O(\sqrt{k})$ robots can improve the competitive factor to $O(\log k)$.

Independently of our investigation, Deng, Kameda and Papadimitriou [DKP] studied a variant of our version of the mapping problem in the case of arbitrary polygonal obstacles. In the version considered by Deng et al the robot's path must not only see the boundary of each object but also see every point not occupied by an object. Using a frontier based approach, they showed that the interior and the exterior of a polygon can be explored on-line with a constant competitive factor. They could extend this frontier based approach to obtain an $O(k)$-competitive algorithm for mapping a scene of k objects.

We now compare our results to those of Deng, Kameda, and Papadimitriou. Deng et al consider only mapping and not searching. Comparing results for mapping is a little difficult because of the difference in definitions of the property that the robot's path P must satisfy. We chose our definition to allow for a unified treatment of searching and mapping. However, upper bounds translate between the different versions of the mapping problem. Thus, Deng et al's constant competitive algorithm for mapping the boundary of an arbitrary simple polygon is also constant competitive under our definition; Also, our $O(\mathcal{M}(k,\alpha))$ bound for mapping a scene with convex objects can be viewed as a refinement of their $O(k)$ upper bound. While we have a tight $\Omega(\mathcal{M}(k,\alpha))$ bound on the competitiveness for our version of mapping, for the version of Deng et al the best lower bound for the competitive factor that is known is $\Omega(\sqrt{k})$.

3 Lower Bounds

Theorem 3.1 *The competitive factor for any on-line algorithm for searching, and mapping, is* $\Omega(\mathcal{M}(k,\alpha))$.

Proof: Assume first that $\alpha < k$. We then prove an $\Omega(\sqrt{k\alpha})$ bound. We build a scene from the set of obstacles shown in figure 3.1a. For the rest of this proof we will call such a collection a *rectangle*. In figure 3.1a the black regions are obstacle free and are very narrow. The critical fact is that by circling the rectangle the perimeter of each object can be seen, but the black obstacle free square region in the center of the rectangle can not be seen. The width of the rectangle is 1 and the height is α.

Let r be of the order of $\sqrt{k\alpha}$. We then combine these obstacles as shown in figure 3.1b. Each rectangle in figure 3.1b represents the collection of objects from figure 3.1a. There are r/α rows and r columns. Hence the whole picture fits within a $2r$ by $2r$ square region. The number of objects, k, is then on the order of r^2/α. The dashed lines represent lines of sight through the rectangles. One of the rectangles is then modified by adding a small square target to the center obstacle free region. The robot's goal in the search problem is to find this small square. To see this square the robot must visit the center of the rectangle.

The off-line cost in the searching problem is clearly $O(r)$. By walking around the outside of the scene, and by visiting the one modified rectangle the off-line algorithm for the mapping problem can see the perimeters of all the objects while traveling at most a distance of $O(r)$. In the worst case the on-line algorithm must visit the center of each of the r^2/α rectangles, at a cost of $\Theta(\alpha)$ per visit. Thus the total cost to the robot is $\Omega(r^2)$. This gives us a lower bound on the competitiveness to be order of r, which is $\Omega(\sqrt{k\alpha})$.

If $\alpha \geq k$ then $\mathcal{M}(k, \alpha) = k$. We can then repeat the above construction with the aspect ratio $\alpha = k$. ♠

For axis-parallel squares of bounded perimeter one can obtain a lower bound of $\Omega(\sqrt{k})$ for the competitive factor for searching and mapping. This result generalizes to higher dimensions in the following manner.

Theorem 3.2 *The competitive factor for any on-line algorithm for the searching and mapping problem in* $\Re^d$ *is* $\Omega(k^{(d-1)/d})$ *in the case of objects with bounded aspect ratio.*

Theorem 3.3 *The competitive factor for any on-line randomized algorithm against oblivious adversary for the searching and mapping problem in* $\Re^d$ *is* $\Omega(k^{(d-1)/d})$ *in the case of objects with bounded aspect ratio.*

4 Nearest Neighbor

In this section we analyze the competitiveness of the Nearest Neighbor (NN) heuristic. Lumelsky et al [LMS] define NN in the following manner.

Nearest Neighbor: Each time an object is visited it is completely circumnavigated. After an object is circumnavigated, the nearest unvisited obstacle is selected to be visited next. We shall define the distance between two objects O_1 and O_2 to be the length of the shortest obstacle free path connecting any point on O_1 to any point on O_2. Notice that circumnavigating each visited obstacle guarantees that the robot always knows the location of the closest unvisited obstacle. For the searching problem, once NN has seen some part of the target it moves directly to the target and quits. In the Mapping problem NN terminates when it has seen all of the environment.

Theorem 4.1 *The competitive factor for Nearest Neighbor is* $\Omega(\alpha\mathcal{M}(k, \alpha))$ *for the mapping problem.*

Proof Sketch: Assume for now that $\alpha < k$. We construct a scene as shown in figure 4.1b. Each triangle in figure 4.1b represents a collection of objects as shown in figure 4.1a. We call the collection of objects pictured in figure 4.1a, a *triangle*. The scene in figure 4.1b contains α triangles and each triangle contains k/α objects. In figure 4.1a, the obstacle free regions are shown in dark; Furthermore, it is crucial that these obstacle free regions are narrow. Each diamond in figure 4.1a has side length $r\alpha^{3/2}/\sqrt{k}$, and width $r\sqrt{\alpha/k}$. Note that all the objects from each triangle can be seen by walking around the inner circle of radius r. Hence, the optimal off-line cost is r. NN will visit order of k objects with cost of $r\alpha^{3/2}/\sqrt{k}$ per object, for a total cost of $\Omega(r\alpha^{3/2}\sqrt{k})$. Hence we get a competitive factor of $\Omega(\alpha^{3/2}\sqrt{k})$. This is greater than $k\alpha$ if $\alpha \geq k$.

If $\alpha \geq k$ then each triangle consists of a constant number of objects. NN will visit order of k objects at a cost of αr per object. Hence we get a competitive factor of $\Omega(k\alpha)$. ♠

Obstacle avoidance is a major problem faced by the robot. The following definition captures the hindrance posed by an obstacle to the movement of the robot.

Definition 4.1 *We define the resistivity of an object O to be the maximum, over all points x and y in its perimeter, of the ratio of the length of the shortest path from x to y following the boundary of O divided by the length of the line seqment xy.*

We now show that resistivity is equivalent to aspect ratio, modulo a multiplicative constant, for convex objects. Resistivity has the benefit that it extends to nonconvex objects.

Lemma 4.1 *Let O be a convex object with aspect ratio α, resistivity β and area A and perimeter P. Then, $P^2 = O(\alpha A)$ and $\alpha = \Theta(\beta)$.*

Lemma 4.2 *[RSL] Let X be a set of n points in any metric space. Then Nearest Neighbor produces a tour of length at most $O(\log n)$ times the optimal tour.*

Definition 4.2
(a) Let S be the scene under consideration. For every object O in S, we define the point on it first touched by the robot during its navigation as its representative point.
(b) The representative metric space G for a scene S contains exactly those representative points where the distance between any two points (in G) is the shortest obstacle avoiding distance between the corresponding points in S.

Lemma 4.3 *Let S be the scene and let G be the corresponding representative metric space generated by the robot's navigation using the Nearest Neighbor heuristic. The order in which points in G are visited using Nearest Neighbor heuristic is identical to the order in which the corresponding objects are visited in S by the robot using the Nearest Neighbor heuristic (assuming that ties are broken in the same manner).*

Lemma 4.4 *Assume that the scene fits within an r by r box S. Then the total cost to NN is at most $O(r\mathcal{M}(k,\alpha)\log k)$.*

Proof Sketch: First we prove that the cost of circumnavigating objects is $O(r\mathcal{M}(k,\alpha))$. Assume for now that $k \geq \alpha$. The cost of circumnavigating objects is $\sum_{i=1}^{i=k} P_i$ where P_i is the perimeter of ith object. Let A_i be the area of the ith object. We then have that:

$$[\sum_{i=1}^{i=k} P_i]^2 \leq 2k\sum_{i=1}^{i=k} P_i^2 \leq 2k\sum_{i=1}^{i=k} \alpha A_i \leq 2k\alpha r^2$$

Therefore $\sum_{i=1}^{i=k} P_i$ is at most $r\sqrt{2\alpha k} = O(r\mathcal{M}(k,\alpha))$.

Now we consider the case when $k < \alpha$. Due to convexity, an object inside S has a perimeter of at most $4r$. Therefore, the total cost of circumnavigating all of the objects is $4kr = O(r\mathcal{M}(k,\alpha))$.

We bound the cost of moving from an object to another by considering the representative metric space. We first bound the length of the optimal tour in the representative metric space by $O(r\mathcal{M}(k,\alpha))$. Finally, applying Lemma 4.2 and Lemma 4.3 we get the desired result.

In order to simplify our arguments and discussions in Section 5, we bound the optimal tour in the following fashion. Partition S into k boxes of side length $r/\sqrt{k}$. Consider the grid formed by the boxes. We will form a graph using the grid and objects (say set C) that cross some box boundary (or the grid). Ignore the portion of the grid that are hidden by objects in C. Vertices of the induced graph G_1 consist of grid points, object corners, and the intersection of grid edges and the sides of the objects. (See figure 4.4). Also, the edges of G_1 are formed by the grid edges and the sides of the objects.

Now consider a depth first traversal, say T, of G_1. The length of the traversal is at most twice the cumulative length of the edges in G_1. But the cumulative length of the edges in G_1 is at most $r\sqrt{k}$ plus the sum of the perimeter of objects in C, which is at most $O(r\mathcal{M}(k,\alpha))$.

We are now left with objects that lie completely inside some box. Observe that a part of the traversal T takes through portions of the grid that are not covered by objects.

Consider a box B in S. Restrict our attention to representative points inside the box B that are not induced by objects in C. We now form a new edge weighted graph G_2 with these representative points as nodes. Let x and y be representative points in B, and let X and Y be the corresponding objects. There is an edge between representative x and y in G_2 if the shortest obstacle avoiding distance between x and y is of length at most $10r/\sqrt{k}$. Note that if part of X is visible from part of Y then there must be an edge between x and y. Recall that the weight of an edge is the shortest obstacle avoiding distance between the two representative points. Observe that G_2 may be disconnected. For each component of G_2, form a spanning tree. Let the resultant forest be $T(G_2, B)$.

Similarly, construct $T(G_2, B)$'s for each box B in the square. We now connect the components of $T(G_2, B)$ (for all B) in the following fashion. Observe that there is at least one representative point in each component/tree of a $T(G_2, B)$ from which the perimeter of its box B can be reached by an obstacle avoiding path of length $O(r/\sqrt{k})$. Therefore, by adding a single path for each component and taking a union with trees in $T(G_2, B)$ for each B and G_1, we get a global graph G. Notice that G is connected and the representative points of every object are included in the traversal of G. Consider a component D in a forest $T(G_2, B)$. Assume that there are n nodes/representative points in D. Then, the cumulative weight of D including the path leading to the boundary of the box is at most $10nr/\sqrt{k}$. Since there are exactly k objects/representative points, cumulative weight of edges in all of the forests $T(G_2, B)$'s is at most $10kr/\sqrt{k}$. Therefore, cumulative weight of G is at most $10r\sqrt{k} + O(r\mathcal{M}(k,\alpha))$. Observe that a depth first traversal of G gives a traveling salesman tour of the representative points, which is at most $O(r\mathcal{M}(k,\alpha))$. Now applying Lemma 4.2 and Lemma 4.3, we get the desired bound. ♠

Lemma 4.5 *Let $\beta > 2$ be the resistivity of a convex object. The smallest internal angle θ of the convex object satisfies the property $\tan\theta \geq 2/\beta$.*

Theorem 4.2 *The competitive factor for Nearest Neighbor is $O(\alpha\mathcal{M}(k,\alpha)\log k)$ for the mapping.*

Proof Sketch: Applying the previous lemma, it not hard to show that the off-line cost will be $\Omega(r/\alpha)$, where r is the diameter of the scene. We finish by citing lemma 4.4. ♠

On the other hand, we show that Nearest Neighbor performs poorly for the searching problem.

Theorem 4.3 *The competitive factor for NN for searching in the plane is $\Omega(2^k)$.*

Proof Sketch: Assume the robot starts at the origin. Objects are placed at the points $(-1,0)$, $(1,0)$, $(3,0)$, $(7,0)$, ..., $(2^i-1,0)$ ... $(2^{k-1}-1,0)$. A small target is placed behind the obstacle at $(-1,0)$. Assuming that the target is sufficiently small NN will move first to $(1,0)$ then to $(3,0)$ etc., incurring a cost of $\Theta(2^k)$, before returning to the target. ♠

5 Bifold Algorithms

In this section we carefully modify Nearest Neighbor to obtain algorithms with lower competitive factors. First, by using Nearest Neighbor in an exponentially growing sequence of neighborhoods around the origin, we obtain a competitive factor that is asymptotically only $O(\log k)$ away from the optimal. We call this algorithm Bifold Nearest Neighbor (BNN). Finally, we finish by giving an algorithm, Bifold Tourist (BT), that is asymptotically strongly competitive.

Assume that the robot starts at the origin. Let d be half the distance to the closest object, and define $\rho(i) = 2^i d$. Denote by S_i the axis parallel square centered at the origin that has side length $2\rho(i)$.

Bifold Nearest Neighbor: BNN is divided into phases. In the ith phase BNN searches the unvisited obstacles in S_i that are reachable from the origin by a obstacle avoiding path that does not leave S_i. During phase i the robot repeatedly selects the nearest object O with the following properties: (1) O lies at least in part in S_i, (2) O has not be completely circumnavigated yet. If O lies entirely within $S_i - S_{i-1}$ then the perimeter of O is circumnavigated. Otherwise, only the portion of the perimeter that lies completely within S_i is traversed. For the searching problem phase i ends when the algorithm knows of no more unvisited objects in S_i. For the mapping problem BNN first circumnavigates the portion of the perimeter of S_i that is reachable without leaving S_i before ending. Note that some objects in S_i may not be visited if some object slices completely through S_i.

Theorem 5.1 *BNN is $O(\log k \mathcal{M}(k, \alpha))$ competitive for the mapping and searching problems.*

Proof Sketch: By a proof analogous to the proof of Lemma 4.4, the cost to BNN for phase i is $O(\rho(i) \log k_i \mathcal{M}(k_i, \alpha))$. where k_i is the number of objects visited in phase i. We can the prove by induction that: The total cost of BNN at the end of phase i is a most the twice the cost of the ith phase. ♠

Before we describe BT we first consider a simplier scenario in which the robot knows a boundary square for the scene and, k, the number of objects in the scene. We present an algorithm Tourist that is optimal for this scenario.

Tourist: Let the robot start at the center of an r by r square S. Consider the region R of S that is reachable without leaving S. See figure 5.1. Recall that in the proof of Lemma 4.4, the analysis divides the S into boxes of side length $r/\sqrt{k}$. Here, since the robot knows k, it maintains an imaginary grid of width $r/\sqrt{k}$. As in the proof of Lemma 4.4, the robot first performs a depth first traversal T of the graph G_1. Observe that *a priori* knowledge G_1 is not necessary. During this traversal, the robot merely observes the locations of other objects within boxes. Also, it should be remembered that if an object is not completely in the S, then only the portion of the perimeter inside the region is circumnavigated.

Now the robot visits objects that are not in G_1, but are in R. This is accomplished by traversing T once again, while temporarily leaving T to circumnavigate objects inside the boxes in R. Assume that the robot temporarily leaves T at a point p to visit an unvisited object X in a box B. While circumnavigating X, the robot might see portions of new objects (say the set N) that lie inside the box B. The robot then adds N to the current stack of objects that it needs to visit in B. The robot recursively continues visiting objects in the stack, and adding newly seen objects in B to the stack. An object is removed from the stack when it is circumnavigated. Observe the manner in which objects were added to the stack guarantees that the amortized cost of traversing between objects in the stack is at most $O(r/\mathcal{M}(k,\alpha)$ for each object. Once this detour terminates, the robot returns to T at the point p.

Lemma 5.1 *The length of the robot's tour generated by Tourist is* $O(r\ \mathcal{M}(k,\alpha))$.

Proof Sketch: The proof is analogous to the argument found in the proof of Lemma 4.4. However, we do not need to resort to using Lemma 4.2. ♠

The assumption that the robot needs to know k was only for convenience, since it can learn the value of k by using exponential search, without affecting the competitive factor. Then, quite like BNN, BT keeps doubling the size of an imaginary square S_i centered at the starting point. Within a square, BT applies the Tourist search described above.

Theorem 5.2
(a) The cost incurred by BT at the end of phase i is $O(\rho(i)\ \mathcal{M}(k,\alpha))$.
(b) BT is $O(\mathcal{M}(k,\alpha))$ *competitive for the mapping and searching problem in the plane.*

Proof Sketch: Part (a) can be proved by induction on number of phases. Proof of part (b) is analogous to the proof of Theorem 5.1. ♠

6 Search Party Problem

In this section we consider a variant of the search problem in which the on-line algorithm is given several robots, say m, to conduct the search. We assume that the robots move at a constant rate and that we wish to measure the time until the target is found. Hence the optimal off-line cost is the time to traverse the shortest obstacle avoiding path to the target. The on-line algorithms goal is to divide the work as evenly as possible between the robots. This even distribution of work has to be done as quickly as possible.

As before we may assume that the the robots know k, the number of objects, that all the objects lie within an r by r square S, and that the off-line cost is $\Omega(r)$. S is divided up into a square grid G of smaller squares of size $r/\sqrt{k}$ by $r/\sqrt{k}$. Hence there is a total of k grid squares. Denote the grid square in the ith row and j column by $G_{i,j}$, $1 \leq i \leq \sqrt{k}$, $1 \leq j \leq \sqrt{k}$. We first consider an algorithm for scenes with objects of bounded aspect ratio and bounded perimeter.

Algorithm Estimate: The $2\sqrt{k}$ horizontal and vertical grid lines are partitioned evenly among the m robots. Each robot then walks its lines traversing the perimeter of any object that they encounter. The robots then meet and determine the number of objects $C(i,j)$ seen in each $G_{i,j}$. The robots divide the grid squares among themselves so that the apparent amount of work allocated to each robot is roughly equivalent. Each robot is given a set grid squares that are consecutive in row major order. Furthermore if R is the set of grid squares given to some robot then $\sum_{G_{i,j}\in R}((r/\sqrt{k})\min(C(i,j),1)) \leq 2r\sqrt{k}/m$. Each robot then visits its assigned grids in row major order, examining each grid square using Nearest Neighbor.

We need the following lemmas. Let $A(i,j)$ be the actual number of objects in $G_{i,j}$.

Lemma 6.1 *Let S be r by r square that contains k objects of bounded perimeter and aspect ratio. The length of the tour produced by Nearest Neighbor is* $O(r\sqrt{k})$.

Proof Sketch: Assume that $k = 2^i$. We prove it by induction on i where the induction hypothesis is: The cost of NN is at most $cr\sqrt{2^i}$. ♠

Lemma 6.2 *For all i, j, $A(i,j) \leq (C(i,j))^2$.*

Lemma 6.3 *The cost to the robot for algorithm Estimate is $O(r\sqrt{k}/m)$.*

Proof Sketch: By Lemma 6.1, the cost to a robot for $G_{i,j}$ is $O((r/\sqrt{k})\sqrt{A(i,j)})$. By lemma 6.2, $(r/\sqrt{k})\sqrt{A(i,j)} = O((r/\sqrt{k})C(i,j))$. By summing over the $G_{i,j}$ visited by the robot the result follows. ♠

Theorem 6.1 *Estimate has a competitive factor of $\Theta(\sqrt{k}/m)$ in the plane assuming the objects have bounded aspect ratio and bounded perimeter.*

For scenes that consist of objects of unbounded aspect ratio or unbounded perimeter it is not possible to estimate the distribution of the objects as precisely as in lemma 6.1. We now sketch an algorithm, Converge, for general scenes with convex obstacles.

Algorithm Converge: Further assume that the robots know α; It can always be found using a doubling search. Assume, for convenience, that $k \geq \alpha$. Each robot takes $\sqrt{k}/m$ rows of grid squares from G. Each robot then starts working on its grid squares, using the depth first search from the Tourist algorithm in each square. After time $O(r\sqrt{k\alpha}/m)$, half of the robots must be finished. They then split the remaining grid squares evenly amongst themselves and repeat this process recursively.

Theorem 6.2 *Let $m = O(\sqrt{k/\alpha})$.*
(a) The cost/time incurred using algorithm Converge is $O(r\sqrt{k\alpha}\log m/m)$.
(b) The algorithm Converge has a competitive factor that is $O(\sqrt{k\alpha}\log m/m)$.

Proof Sketch: Each robot travels a distance that is $O(r\alpha\log m + r\sqrt{k\alpha}\log m/m)$ units, assuming the robot always moves. Here, the first cost is due to moving between phases and the second cost is due searching. We want the second cost to be the asymptotically dominant cost. Hence, we require that $m = O(\sqrt{k/\alpha})$. ♠

It is interesting to note that similar results can be obtained even if the robots have different speeds.

7 Conclusion and Open Problems

It is interesting to note that at the heart of the searching and mapping problems is a graph traversal game, which we will call *Hide and Seek*. In Hide and Seek both the seeker, say a cat, and the hider, say a mouse, start at the same position in some space containing k havens for the mouse. After the mouse hides in one of k havens the cat searches the possible havens looking for the one hiding the mouse. The cat's goal is to minimize the ratio of the distance that it travels over the shortest path for the mouse. The competitive factor for Hide and Seek in $\Re^d$ is $\Theta(k^{(d-1)/d})$. Hide and Seek seems to also be at the heart of the problem of on-line minimum weight matching in a metric space [KP1].

Recently [KP2] we have derived competitive factors for these problems in terms of average aspect ratio $\bar{\alpha}$. The resultant competitive factor (for most of the problems) can be obtained by replacing α by $\bar{\alpha}$. There is a number of directions in which the results of this paper could be extended. There is $\log k$ gap between the lower and upper bounds of Bifold Nearest Neighbor.

It would also be interesting to consider extending the results for searching and mapping to nonconvex objects. Similarly, one could consider extending the results for the search party problems to a more general class of objects. One could also consider the problem of finding a short path to a *known* destination in the case of multiple robots.

One natural task that one might require of the robot is that it visit and circumnavigate each obstacle. This gives rise to what we call the *Visual Traveling Salesman Problem*. Recently [KP3], we have shown that a constant competitive algorithm exists for this problem.

Acknowledgements : The first author was supported in part by NSF Grant CCR-9009318.

References

[BCR] R. Baeza-Yates, J. Culberson, and G. Rawlins, "Searching with Uncertainty", to appear in *Information and Computation.*

[BRS] A. Blum, P. Raghavan, and B. Schieber, "Navigating in Unfamiliar Geometric Terrain", *STOC*, pp. 494–504, 1991.

[DP] X. Deng, and C. Papadimitriou, "Exploring an Unknown Graph", *FOCS*, pp. 355-361, 1990.

[DKP] X. Deng, Kameda, and C. Papadimitriou, "How to Learn an Unknown Environment", *FOCS*, pp. 298–303, 1991.

[KP1] B. Kalyanasundaram, and K. Pruhs, "On-line Weighted Matching", *SODA*, pp. 234–240, 1991.

[KP2] B. Kalyanasundaram, and K. Pruhs, "A Competitive Analysis of Algorithms for Searching Unknown Scenes", *submitted for publication.*

[KP3] B. Kalyanasundaram, and K. Pruhs, "On-line Construction of Short Traveling Salesman Tours" *submitted for publication.*

[KRR] H. Karloff, Y. Rabani, and Y. Ravid, "Lower Bounds for Randomized k-server and Motion Planning Algorithms", *STOC*, 1991.

[LMS] V. Lumelsky, S. Mukhopadhyay and K. Sun, "Dynamic Path Planning in Sensor-Based Terrain Acquisition", *IEEE Transactions on Robotics and Automation*, vol. 6, pp. 462–472, 1990.

[PY] C. Papadimitriou, and M. Yannakakis, "Shortest Paths Without a Map", *ICALP*, pp. 610–620, 1989.

[RSL] D. Rosenkrantz, R. Stearns, and P. Lewis, "An Analysis of Several Heuristics for the Traveling Salesman Problem", *SIAM J. of Computing*, 6, pp. 563–581, 1977.

[SY] J. Schwartz and C. Yap, *Algorithmic and Geometric Aspects of Robots*, Lawrence Erlbaum, London.

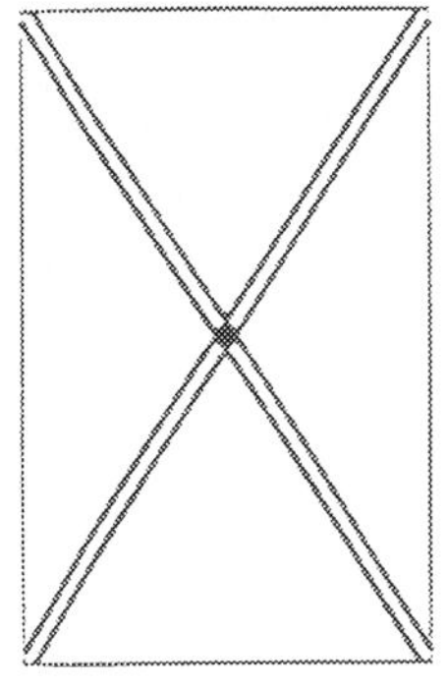

Figure 3.1a

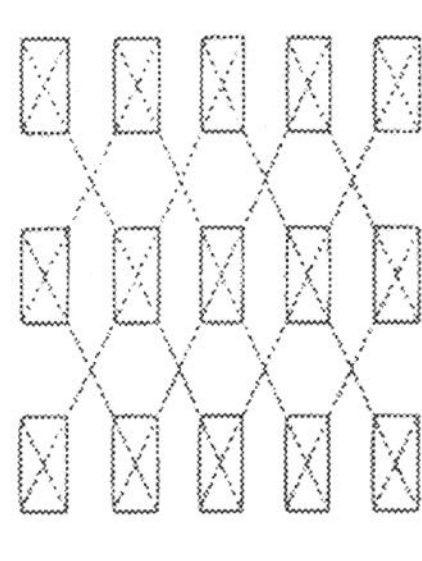

Figure 3.1b

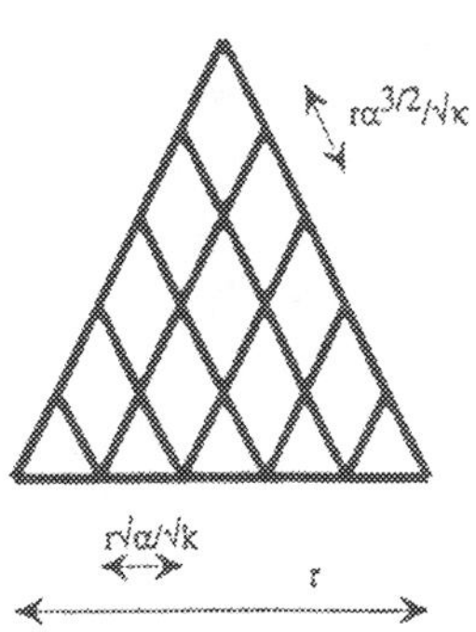

Figure 4.1a

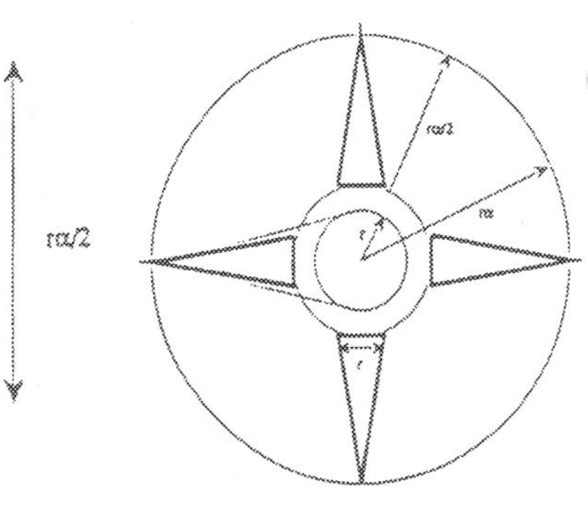

Figure 4.1b

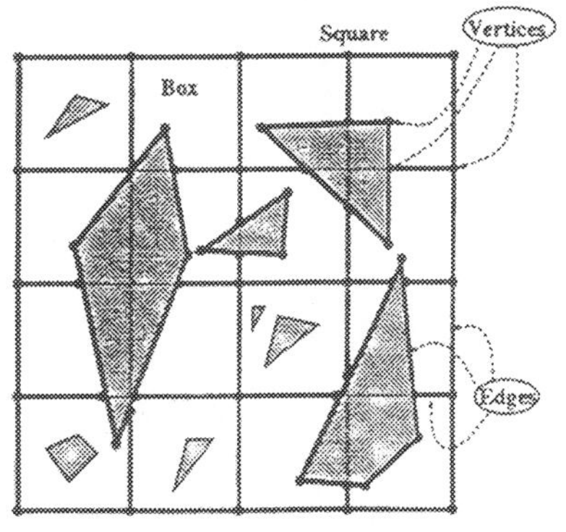

Figure 4.4

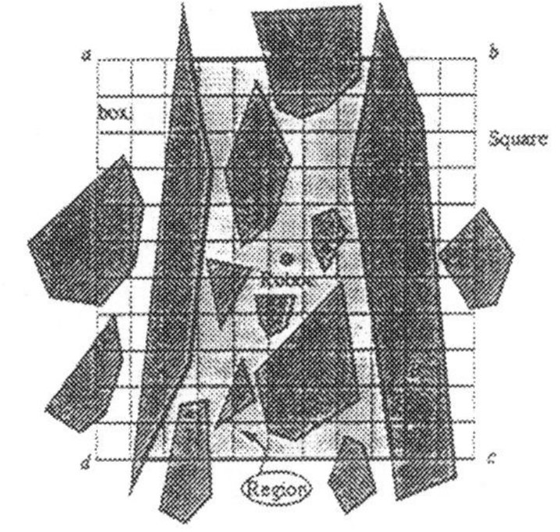

Figure 5.1

AUTOMATA AND LANGUAGES

Equality and disequality constraints on direct subterms in tree automata

Bogaert B. & Tison S.*

LIFL (U.A. 369 CNRS)
University of Lille Flandres-Artois
UFR IEEA - Bâtiment M3
F-59655 Villeneuve d'Ascq cedex (France)
mail : *bogaert*, *tison* *@lifl.lifl.fr*

Keywords: Tree Automata, non linearity.

Abstract: We define an extension of tree automata by adding some tests in the rules. The goal is to handle non linearity. We obtain a family which has good closure and decidability properties and we give some applications.

1 Introduction

Automata, on words or on terms, are an important tool, the decision results they helped to establish are numerous. The main idea of this paper is to add tests in the rules of automata of finite terms, in order to take account of the non-linearity of terms. Let us remind that a term is non-linear if and only if the same variable occurs at least twice in it. All instances of a non-linear term have identical subtrees, obtained by substitution of the same variable. Non linearity is an important property, appearing in many domains, such as logic programming, rewriting, ... Phenomena generated by non-linearity are complex and there is a gap between the"linear case" and the "non-linear case": many properties satisfied in the first case, are lost in the second one (the homomorphic image of a recognizable set is recognizable when the homomorphism is linear, it is not otherwise; the problem of inductive reducibility is polynomial when the left-hand terms are linear [Jouannaud,Kounalis], exponential otherwise [Plaisted, Kapur]). The tools already defined to explicitly manipulate comparisons between terms are rare, axiomatization for algebras of trees [Comon, Maher] is the main one.

In 1981, M. Dauchet and J. Mongy [Mon81] have defined a class of automata, called *Rateg*. In *Rateg* automata, left-hand sides of transitions rules may be non-linear terms with variables at depth more than one. States appear at depth one. For instance $a(q_1(b(x,x)), q_2(c(y,a(x),y)), q_2(b(x,y)), q_3(y))$ is a correct left-hand side of a *Rateg* rule. This rule, in fact, imposes equalities of several subterms. This class, called *Rateg*, is closed under finite intersection and union, but the emptiness is undecidable, as it was proved by J. Mongy. From the order-sorted algebras point of view, bottom up tree automata correspond to a usual signature - states are equivalents to sorts, and transition rules to declarations-. As for *Rateg* automata, they are close to signatures with "term declarations"; this notion was introduced by Schmidt-Schauss [Sch88]: the declarations are of the form t:S, which means that the term t is of sort S; the terms in the declarations may be non linear and of any depth. The corresponding tree language class is strictly included in *Rateg*. Schmidt-Schauss has proved the unification of terms to be undecidable in the general term declaration case and this result can also be obtained as a corollary of Mongy results.

Our goal is to extend the classical definition of automata, while keeping both (good) closure properties (especially under boolean operations) and decision properties (emptiness). Intuitively, the power of *Rateg* come from the ability to overlay equality constraints, and, so, to generate non local tests between subterms (the Post problem can easily be coded with *Rateg*). In order to avoid this phenomenon and to keep good decision properties, two approaches seem to be interesting : either to limit the number of tests, or to restrict limit the form of the tests (more precisely, the position of tested terms). The way we have explored is the second one: the tests used in the automata can only concern "brothers". A rule can use these tests to impose either equalities, or differences, or any combination of equalities and differences beetween brother terms. For instance, $a(q_1(x_1), ...q_n(x_n))$, where x_i are non necessarily

* *This work was supported, in part, by the "PRC Mathématiques et Informatique"*

different variables are correct left hand sides of rules, but we can also impose to some subterms to be distinct, with the general form of rules : $a(q_1(x_1), \ldots, q_n(x_n)) \rightarrow q$ *when* $x_1 = x_2 \wedge x_2 \neq x_3 \wedge x_3 = x_4 \ldots$

We prove here that the corresponding tree language class, $Rec_{\neq}$, is a boolean algebra, closed under (quasi-)alphabetic homomorphisms; it strictly contains the closure of *Rec* recognizable sets of trees under boolean operations and alphabetic tree homomorphisms. The emptiness and finiteness properties are decidable. So, we get an extension of *Rec*, while keeping good decision and closure properties, excepting the closure under inverse morphisms. Most of the constructions and proofs are similar to the standard ones; some of them are more intricate and the "situation" is more complex; for example, the emptiness property becomes NP-hard in $Rec_{\neq}$.

This extension may seem to be a weak extension of tree automata; but already if we consider the "first cousin" case (we can also test equality between first cousins, i.e. each variable is at depth 2), emptiness property is undecidable [Tom91]. So, without restricting the number of allowed tests, it seems dificult to define a stronger extension while keeping good decidability properties: $Rec_{\neq}$ is near the "undecidability border". On the other hand, our automata seem to be close to automata on very simple dags; the main difference is the ability to ensure differences between brothers and this point is very important (particularly to obtain closure under complementation and to define determinization).

Thus, $Rec_{\neq}$ is a real extension of *Rec*. For example, it provides a tool to manipulate some non recognizable tree languages, like (quasi-)alphabetic homomorphic images of recognizable tree languages, or set of normal forms of some rewrite systems; furthermore the good properties of the family allow to use this class in place of *Rec* class to extend some results, as it was done by H.Comon and C.Delor [CoDe91] : in order to obtain solved forms, they give transformation rules for first order formulas in which atoms are either equations or membership constraints (terms must belong to $Rec_{\neq}$ sets of trees); particularly, it provides, in some cases, an algorithm for testing inductive reducibility which is simpler that the general one.

2 About tree automata

2. 1 Rational sets of trees

The set of terms (or trees) over a –finite– ranked alphabet Σ and i variables will be denoted $T(\Sigma)_i$, so $T(\Sigma)_0$ represents the set of ground terms. Σ_i denotes the set of letters of arity i.
The concatenation product of a term $t \in \Sigma_n$ with n terms $t_1, \ldots, t_n$ will be denoted $t(t_1, \ldots, t_n)$. For better readability, $t(t_1)$ will sometimes be denoted $t.t_1$, where $t \in T(\Sigma)_1$. $t(t(t \ldots (t_1) \ldots))$ will also be denoted $t^i.t_1$.

We shall call *REC* the class of sets of finite trees accepted by a finite automaton. This class coincides with the class of rational sets of finite trees.

Several behaviours can be defined for automata. We shall consider only frontier-to-root automata on finite terms.

Definition 2.1 [1]

A frontier-to-root tree automaton is defined by a finite ranked alphabet Σ, a finite set Q of states, a subset F of final states, a set of transition rules $\mathcal{R} \subseteq \bigcup_i \Sigma_i \times Q^{i+1}$.

Transitions : *A transition rule will be denoted*

$$a(q_1(x_1), \ldots, q_n(x_n)) \rightarrow q \quad \text{where } x_1, \ldots, x_n \text{ are distinct variables.}$$

$t \rightarrow t' \iff$
$t = u.a(q_1(u_1), \ldots, q_n(u_n)), t' = u.q.a(u_1, \ldots, u_n)$
for some rule $(a(q_1, \ldots, q_n) \rightarrow q) \in \mathcal{R}$
$u_i \in T(\Sigma)_0$, $u \in T(\Sigma)_1$ *A ground term t is accepted if and only if exists a final state q, such that $t \xrightarrow{*} q(t)$ (we shall simply note $t \xrightarrow{*} q$ and say "t reaches q").*

2. 2 Automata with equality tests

The main idea is to add conditions in rules, in order to handle non-linear terms. The definition of automata rules can be extended as follows :

$$a(q_1(t_1),\ldots,q_n(t_n))[equality\ constraints][disequality\ constraints]\rightarrow q$$

where $t_1,\ldots,t_n$ are terms Equality constraints can be expressed by a finite set of equations which must be satisfied by the instances of occurring variables. Disequality constraints consist in a finite set of disequations, that instances of variables must satisfy.

From this general definition, several kinds of automata may be defined, by fixing 3 parameters :
- depth of the occurrences of variables in the constraints
- existence of equality constraints
- existence of disequality constraints

Class :	*Rec*	$Rec_{\neq}$	"first cousin case"	Rateg
Depth of variables	1	1	2	finite
Equality constraints	No	Yes	Yes	Yes
Disequality constraints	No	Yes	No	No
Boolean closure properties	$\cup, \cap, comp$	$\cup, \cap, comp$	$\cup, \cap$	$\cup, \cap$
Emptiness	decidable	decidable	undecidable	undecidable

Rateg automata have been introduced in 1981 by M. Dauchet and J. Mongy. Equality constraints can be expressed by the non-linearity of left-hand terms, for instance the rule

$$b(q_1(x), q_2(a(x)))\rightarrow q$$

can be applied only if the states q_1 and q_2 appear respectively on the first and the second branch of the "b", and if $second\_subtree\_of\_b = a(first\_subtree\_of\_b)$.

In the Rateg class, which is closed under union and intersection operations, emptiness (satisfiability of a specification) is undecidable. The proof has been established by J.Mongy, as a corollary of a more general result : $\phi_1(R_1)\cap\phi_2(R_2) = \emptyset$ is an undecidable property [Mon 81]. (ϕ_i are tree morphisms and R_i belong to Rec).
The Post correspondence problem can also be coded in the emptiness problem in the Rateg class : the automaton built to code Post correspondence problem uses the overlaying of constraints, which allows, for instance, to impose equalities between any number of subtrees, at any depth of the tree.

Thus, the defined class is too large. Even if we compare "first cousin terms" the emptiness is undecidable [Tom 91]. If we want to get interesting decision properties, we need to impose restrictions on the tests.

3 Comparisons between brother terms

The automata of class $Rec_{\neq}$ impose equality and disequality constraints on variables occuring at depth one. Both equality and disequality constraints will be expressed by a ***constraints expression.***

Definition 3 [1]

> *An atomic constraints expression is either an equation $x_i = x_j$, or an inequality $x_i \neq x_j$ or sign $\perp$ (null constraint), where x_i and x_j denote variables (in the following x_i will always denote the ith son of a node)*
> *A constraints expression is either an atomic expression, or a boolean combination of constraints expressions.*
> *The set of constraints expressions will be denoted CE. CE_n is the set of constraints expressions over at most n variables.*

Definition 3 [2]
A tuple of terms $(t_i)_{1\leq i\leq n}$ satisfies a constraints expression c iff the evaluation of c for the valuation ($\forall i \leq n, x_i = t_i$) is true, when "=" is interpreted as equality of terms, "$\neq$" as its negation, "$\perp$" is interpreted as true, $\wedge$ and $\vee$ as the usual boolean functions.

Definition 3 [3]
An automaton with comparisons between brothers ($REC_{\neq}$ automaton) is given by a finite ranked alphabet Σ, a finite set of states Q, a subset F of final states and a set $\mathcal{R} \subseteq \bigcup_i \Sigma_i \times CE_i \times Q^{i+1}$ of rules – a rule $(a, f, q_1, \ldots, q_n, q)$ will be denoted $(a(q_1, \ldots, q_n)[f] \rightarrow q)$ –.
Transitions :
$t \rightarrow t'$ if and only if
$t = u.a(q_1(t_1), \ldots, q_n(t_n)),\ t' = u.q.a(t_1, \ldots, t_n)$
$\exists f \in CE_n$ ($(t_i)_{1\leq i\leq n}$ satisfies f, $(a(q_1, \ldots, q_n)[f] \rightarrow q) \in \mathcal{R}$)

A tree t is said to be accepted if there exists a final state q, such that $t \xrightarrow{*} q(t)$.

Examples :

• The automaton A_1 defined on the alphabet $\{\underset{0}{\bar{a}}, \underset{1}{a}, \underset{2}{b}\}$, (numbers below letters denote the rank) the unique state $final$, and the rules

$$\bar{a} \rightarrow final, \quad a(final) \rightarrow final, \quad b(final, final)[x_1 = x_2] \rightarrow final$$

accept only well-balanced trees (i.e. : all branches are identical) : for instance $b(a(\bar{a}), a(\bar{a}))$ will be accepted, but not $b(a(\bar{a}), \bar{a}))$

• On the same alphabet, we consider an automaton A_2, its set of states is $\{q, final\}$, and its set of rules :

$$\bar{a} \rightarrow q, \quad a(q) \rightarrow q, \quad b(q, q)[x_1 = x_2] \rightarrow q, \quad b(q, q)[x_1 \neq x_2] \rightarrow final$$

$$a(final) \rightarrow final, \quad b(q_1, q_2) \rightarrow final \ for \ any \ (q_1, q_2) \ s.t. \ q_1 = final \ or \ q_2 = final$$

The recognized set is the set of non well-balanced trees.

4 Transformations of automata

4. 1 Deterministic and complete automata

Definition 4.1 [1]
An automaton A is said to be deterministic (resp. complete) iff, for any $t_1, \ldots, t_n$, for any a in Σ_n, for any $q_1, \ldots, q_n$, there is at most (resp. at least) one rule $(a(q_1, \ldots, q_n)[f] \rightarrow q)$, such that $t_1, \ldots, t_n$ satisfies the constraint expression f.

An equivalent definition of a deterministic automata would be : for any pair of rules $(a(q_1, \ldots, q_n)[f] \rightarrow q)$ and $(a(q_1, \ldots, q_n)[f'] \rightarrow q')$, $f \wedge f'$ cannot be satisfied. This property is obviously decidable.

As usually, when an automaton is deterministic (resp. complete), each tree reaches at most (at least) one state. The completion is easy and similar to the usual case: we have eventually to add a state.

Example
The automata A_1 is not complete. It can be completed by adding the state q and the rules :

$$b(final, final)[x_1 \neq x_2] \rightarrow q$$

$$a(q) \rightarrow q, \quad b(q_1, q_2) \rightarrow q \ for \ any \ (q_1, q_2) \ s.t. \ q_1 = q \ or \ q_2 = q$$

In order to simplify the proofs and the constructions, one condition on the constraints may be added: a constraints expression c over n variables is said to express a *full constraint*, iff for any (i, j) in CE_n, either c implies $x_i = x_j$, or c implies $x_i \neq x_j$. It is then obvious that, for every n, we can restrict CE_n to the subset CE'_n of full constraints expressions such that each n-uple satisfies one and only one constraints expression of CE'_n. For instance if we consider 3 variables, CE'_3 could be

$$\{x_1 = x_2 \wedge x_1 = x_3,\ x_1 = x_2 \wedge x_1 \neq x_3,$$
$$x_1 \neq x_2 \wedge x_1 = x_3,\ x_1 \neq x_2 \wedge x_2 = x_3,\ x_1 \neq x_2 \wedge x_1 \neq x_3 \wedge x_2 \neq x_3\}$$

Clearly, each automaton is effectively equivalent to an automaton with full constraints. We omit here the construction which just consists in transforming the constraint expression of each rule into an equivalent disjunction of full constraints and then in splitting the rule according each member of the disjunction. (This operation increases the number of rules; it is useful in order to simplify the proofs but we could omit it.) When we restrict ourselves to full constraints, the notion of determinism (resp. completeness) is similar to the usual one: for each left-hand side of rule, there is at most (resp. at least) one corresponding rule.

4. 2 Determinization

Proposition 4.2 [1]

For every $\mathcal{F} \in REC_{\neq}$, there exists a complete and deterministic automaton recognizing $\mathcal{F}$

Algorithm

The input $A = (\Sigma, Q, F, \mathcal{R})$ is supposed to be an automaton with full constraints.

$Q_0 \leftarrow \{X | X = \{q | \exists \bar{a} \in \Sigma_0, (\bar{a} \rightarrow q) \in \mathcal{R}\}\}$
$\mathcal{R}_0 \leftarrow \{(\bar{a}[\perp] \rightarrow X) | X = \{q | (a[\perp] \rightarrow q) \in \mathcal{R}\}\}$
repeat :
$\bar{Q}_i \leftarrow \bar{Q}_{i-1} \cup$
$\{X |\ \exists a \in \Sigma_n, \exists (X_i)_{1 \leq i \leq n} \text{ in } Q_{i-1}, \exists d \in CE_n,$
$X = \{q | \exists (q_i)_{1 \leq i \leq n} \text{ in } Q, q_k \in X_k, \quad (a(q_1, \ldots, q_n)[d] \rightarrow q) \in \mathcal{R}\}$
$\}$
$\bar{\mathcal{R}}_i \leftarrow \bar{\mathcal{R}}_{i-1} \cup$
$\{(a(X_1, \ldots, X_n)[d] \rightarrow X) |$
$a \in \Sigma_n, \quad (X_i)_{1 \leq i \leq n} \text{ in } \bar{Q}_{i-1}, \quad d \in CE_n,$
$X = \{q | \exists (q_i)_{1 \leq i \leq n} \text{ in } Q, q_k \in X_k, \quad (a(q_1, \ldots, q_n)[d] \rightarrow q) \in \mathcal{R}\}$
$\}$
until $\bar{Q}_i = \bar{Q}_{i-1}$ (which implies $\bar{\mathcal{R}}_i = \bar{\mathcal{R}}_{i-1}$)
$\bar{Q} \leftarrow \bar{Q}_i$
$\bar{\mathcal{R}} \leftarrow \bar{\mathcal{R}}_i$
$\bar{F} \leftarrow \{X | X \cap F \neq \emptyset\}$

4. 3 Normalization

Definition 4.3 [1]

An automaton is said to be normalized if it is deterministic and if for each rule $(a(q_1, \ldots, q_n)[f] \rightarrow q)$, and for each atomic constraint $x_i = x_j$ or $x_i \neq x_j$ of f, $q_i = q_j$
A normalized automaton is said to be normalized-complete if for each letter a, each n-uple $q_1, \ldots, q_n$, each formula f of CE'_n satisfied by $q_1, \ldots, q_n$, there exists exactly one rule $(a(q_1, \ldots, q_n)[f] \rightarrow q)$.

Proposition 4.3 [2]

For every $\mathcal{F} \in REC_{\neq}$, there exists a normalized-complete automaton recognizing $\mathcal{F}$

The determinism of an automaton ensures that $(t \xrightarrow{*} q, t' \xrightarrow{*} q',\ q \neq q') \ \Rightarrow t \neq t'$.

So, any rule which imposes the equality of branches on which the states are differents, will never be applied and can be suppressed.

5 Closure properties

5. 1 Boolean operations

Proposition 5.1 [1]
$REC_{\neq}$ is a boolean set algebra.

Union : Let $\mathcal{F}, \mathcal{F}'$, two sets of $REC_{\neq}$, recognized by the automata $A = (\Sigma, Q, F, \mathcal{R})$ and $A' = (\Sigma', Q', F', \mathcal{R}')$. We suppose that Q and Q' are disjoints. The automaton $(\Sigma \cup \Sigma', Q \cup Q', F \cup F', \mathcal{R} \cup \mathcal{R}')$ recognizes $\mathcal{F} \cup \mathcal{F}'$.

Complementation : A is supposed to be a normalized-complete and deterministic automaton. Then, for every tree t, there exists a unique state q, reached by t. So, if we consider the set of final states $Q \setminus F$, the accepted set of trees will be $T(\Sigma) \setminus \mathcal{F}$

5. 2 Tree homomorphisms

Proposition 5.2 [1]
$REC_{\neq}$ is closed under alphabetic tree homomorphisms.

The proof is more intricate than in the *REC* case. Indeed, it needs to take into account the cardinality of $h^{-1}(t)$, because non-injective morphisms can erase differences. It is a bit long and tedious; the proof may be found in [Bog 90].

In contrast to the *REC* case, we obtain the following negative result :

Proposition 5.2 [2]
$REC_{\neq}$ is not closed under inverse homomorphisms.

Example
We consider the homomorphism $h : \Sigma = \{\bar{a}, a, a', b\} \rightarrow \Sigma' = \{\bar{a}, a, b\}$ given by

$$h(\bar{a}) = \bar{a}, \quad h(a(x)) = a, \quad h(a'(x)) = a, \quad h(b(x, y)) = b(x, y)$$

The set $\mathcal{F}$ of well-balanced trees (in the same sense as in 3.) of $T(\Sigma')$ belongs to $REC_{\neq}$, but $\mathcal{F}' = h^{-1}(\mathcal{F}) \notin REC_{\neq}$. For any i, $b(a^i.\bar{a}, a'^i.\bar{a}) \in \mathcal{F}'$, so a $REC_{\neq}$ automaton recognizing $\mathcal{F}'$ would contain a rule like $b(q_i, q'_i)[\bot] \rightarrow final$ where $a^i.\bar{a}$ and $a'^i.\bar{a}$ reach q_i and q'_i respectively.
But $b(a^i.\bar{a}, a'^j.\bar{a}) \notin \mathcal{F}'$, for $i \neq j$. Thus for any pair of naturals (i, j), with $i \neq j$, there would exist two distinct states $q_i \neq q_j$ with $a_i.\bar{a}$ reaches q_i and $a^j.\bar{a}$ reaches q_j. This is incompatible with finiteness of Q.

6 Decision problems

6. 1 A decision algorithm for emptiness property

Proposition 6.1 [1]
For any $REC_{\neq}$-automaton A, the property "$\mathcal{F}(A) = \emptyset$" is decidable.

The input of the algorithm we are going to present is supposed to be a deterministic automaton (if it is not the case, the previous determinization algorithm may be applied). max_arity denotes the maximal arity of symbols in Σ.
For any rule $r = (a(q_1, \ldots, q_n)[f] \rightarrow q)$, m^r denotes the number of distinct left-hand side states and $(\tilde{q_1}^r, \ldots, \tilde{q}_{m^r}{}^r)$ denotes these states, in the same order as their first occurrence.

The problem consists in deciding the reachability of each state. Indeed, it needs to calculate the number of trees reaching each state, as explained below:
If we consider a rule r : $(b(q, q)[x_1 \neq x_2] \rightarrow q')$, this rule can be applied if and only if there exists a least two different trees reaching the state q. If the state q' appears only at the right side of the rule r, then it can be reached if and only if q is reached by at least two 'different' trees.
So, we introduce, for each rule r, a calculable function $Prod_r(k_1, \ldots, k_m)$ which represents the number of tree using the rule r at their root, when $\alpha_1, \ldots, \alpha_n$ trees reach $(\tilde{q_1}^r, \ldots, \tilde{q}_{m^r}{}^r)$.
For any state $\tilde{q_i}^r$, occurring α_i times in r, constraints of f impose the existence of at least β_i distinct

trees reaching $\tilde{q}$, with $\beta_i \leq \alpha_i$.
Then, $Prod_r(k_1, \ldots, k_m) = \prod_{i=1}^{m^r}\left(\frac{k_i!}{|k_i-\beta_i|!} \times ppm(k_i, \beta_i)\right)$ with $ppm(x, y) = 1$ *if* $x > y$, 0 *otherwise.*

For example, let r be the rule $b(q(x_1), q(x_2))[x_1 \neq x_2] \rightarrow q'$; then, $Prod_r(1) = 0, Prod_r(2) = 2, Prod_r(3) = 6$.
If there exists i, in $[1, m^r]$, $k_i \geq max\_arity$, then either $Prod_r(k_1, \ldots, k_{m^r}) = 0$ or $Prod_r(k_1, \ldots, k_{m^r}) \geq max\_arity$. This is used for the termination of the algorithm, as it is sufficient to count trees which reach each state, up to the maximal arity of the alphabet.

Algorithm
For every state q, $number_q^0 \leftarrow 0$
$i \leftarrow 1$

Repeat
  For every state q
    $number_q^i \leftarrow 0$
  For every rule r, with right hand state q :
    $number_q^i \leftarrow min(\ max\_arity,\ number_q^i + Prod_r(number_{\tilde{q}_1}^{i-1}, \ldots, number_{\tilde{q}_{m^r}}^{i-1})\ )$
  $i \leftarrow i+1$
until, for every state q, $number_q^i = number_q^{i-1}$
  or $\exists q \in Fin, number_q^i \neq 0$
if $\forall q \in Fin,\ number_q = 0$, then $\mathcal{F}(A) = \emptyset$
  else $\mathcal{F}(A) \neq \emptyset$
fi

Complexity measure
The main loop is executed at most $1 + |Q| \times (max\_arity - 1)$ times. The complexity of each execution only depends on the number of rules. So the complexity of the algorithm is polynomialy bounded by the size of the -deterministic- automaton. For the automaton presented below, the size of accepted trees is greater or equal than $1 + |Q| \times (max\_arity - 1)$.

Example
The alphabet is $\{\underset{0}{\bar{a}}, \underset{1}{a}, \underset{1}{b}, \underset{1}{c}, \underset{3}{@}\}$, the states $\{q_a, q_b, q_c, final\_state\}$ and the rules :

$\bar{a} \rightarrow q_a$, $a(q_c) \rightarrow q_a$, $b(q_a) \rightarrow q_b$, $,c(q_b) \rightarrow q_c$
$@(q_c, q_c, q_c)[x_1 \neq x_2 \wedge x_2 \neq x_3 \wedge x_1 \neq x_3] \rightarrow final\_state$.
Then, one of the smallest accepted trees is $@(bc\bar{a}, bcabc\bar{a}, bcabcabc\bar{a})$. Its height is 9 $(1 + 4 \times 2)$.

6. 2 Complexity of the property

We have seen that the complexity of the property is polynomial for deterministic automata; however, it is NP-hard for non-deterministic automaton, contrarly to the *REC* case:

Proposition 6.2 [1]
The property "$\mathcal{F}(A) = \emptyset$" is NP-hard.

We can prove it by coding the problem of the satisfiability of boolean expressions : "for a given expression $E(x_1, \ldots, x_n)$, is there a valuation satisfying E ?". This problem is known to be NP-complete.
We shall code a boolean expression, and a valuation of its variables, into a tree defined on the alphabet $\{\wedge, \vee, \neg, x_1, \ldots, x_n, true, false\}$. The rank of x_i is 1, the rank of *true* and *false* is 0. For instance, the expression $E(x_1, x_2, x_3) = (x_1 \vee x_2) \wedge (x_2 \vee \neg x_3)$ with the valuation $x_1 = true$, $x_2 = false$, $x_3 = true$ will be represented by
$\wedge(\vee(x_1.true, x_2.false), \vee(x_2.false, \neg(x_3.true)))$

Let us consider a boolean expression E. The expression E if satisfiable if and only if there exists a tree t such that :
- t represents E. This can be controlled by a usual tree automaton. Its size is linearily bounded to the size of the expression E. The state reached by an accepted tree will be denoted *Expr*.
- the valuation encoded by t is correct, i.e. for every variable, each occurrence of it is associated

to the same value. For one variable, the size of an automaton which can control the correctness is polynomialy bounded to the number of variables. We shall call $correct_i$ the state reached by a tree t which represents a correct valuation of x_i.
- the result of the computation equals "true". Computation can be made by an automaton, with the two states $value\_true$, $value\_false$
The non-deterministic automaton obtained by the union of these automata, and the additional rule

$$@(value\_true, Expr, correct_1, \ldots, correct_n)[x_1 = x_2 = \ldots = x_(n+2)] \rightarrow satisfiable$$

(where @ is a symbol of rank $n+2$ and $satisfiable$ the unique final state), accepts a non-empty set if and only if the expression E is satisfiable.

6. 3 Finiteness problem

Proposition 6.3 [1]
For any $REC_{\neq}$-automaton A, the property "$\mathcal{F}(A)$ is finite" is decidable.

The decision algorithm is similar to the one we have built for the emptiness property. The proof uses the following lemma :
The set $\mathcal{F}_q(A)$ of trees reaching q is infinite iff
$\exists q' \in Q, \exists u \in T(\Sigma)_0, \exists v, w \in T(\Sigma)_1$, such that
$\bar{u} \xrightarrow[A]{*} q', v(u) \xrightarrow[A]{*} q', w(v(u)) \xrightarrow[A]{*} q$
This lemma is equivalent to :
$\mathcal{F}_q(A)$ is infinite iff
$\exists q' \in Q, \exists n, n', \exists u \in T(\Sigma)_0, \exists \bar{v} \in T(\Sigma)_n, \exists \bar{w} \in T(\Sigma)_{n'}$, such that
u is not a subterm of $\bar{v}$, $\bar{v}(u, \ldots, u)$ is not a subterm of $\bar{w}$
$u \xrightarrow[A]{*} q' \quad \bar{v}(u, \ldots, u) \xrightarrow[A]{*} q', \quad \bar{w}(\bar{v}(u, \ldots, u), \ldots, \bar{v}(u, \ldots, u)) \xrightarrow[A]{*} q$
We can prove that any element of the string defined by
$u_0 = u$, and $\quad u_i = \bar{v}(u_{i-1}, \ldots, u_{i-1})\_where\_i > 0$
reaches the state q'.

7 Subclasses and characterizations

In this part we are going to study an important subclass of $REC_{\neq}$. It is called $REC_=$. In $REC_=$ the rules can only impose equalities, never differences. We have got some (non effective) characterizations of $REC_{\neq}$-automata recognizing elements of REC, of $REC_=$ or of its boolean closure.

7. 1 The $REC_=$ subclass

Definition 7.1 [1]
An automata with comparisons between brothers is a $REC_=$-automata if the rules uses formulas built only with connectors $\wedge, \vee$ and with the equality of terms predicate.

Properties
Boolean operations : This subclass is closed under union and intersection, but it is not closed under complementation. For instance we can consider the example 2.1.
The set Bal of well-balanced trees (cf. section 3) is accepted by a $REC_=$-automaton (the automaton we have described in 3 is, in fact, a $REC_=$ one). But r $T(\Sigma) \setminus Bal$ does not belong to $REC_=$.
To accept only non well-balanced trees, we need to avoid the application of the main rule when the two subterms are identical. This is possible only if we can impose the presence of distinct subtrees, and if we can use the negation in the formula.

Homomorphisms : $REC_=$ is exactly the image of REC for quasi-alphabetic tree morphisms. So it corresponds to algebras with non-linear signature. $REC_=$ is not closed under inverse homomorphisms (even alphabetic homomorphisms).

7. 2 Characterizations

The following characterizations are not effective, but, as a corollary, we prove that the boolean closure of $REC_=$ is a real subclass of $REC_{\neq}$. We shall consider only complete-normalized automata. The sets of formulas used are called CE'_i. We define a partial order relation, on formulas.

Definition 7.2 [1]

Let f_1 and f_2 be two formulas in CE'_n. $f_1 \leq f_2$ if and only if there exists f'_1, f'_2, respectively equivalent to f_1 and f_2, such that f'_1 can be obtained from f'_2 by replacing equality signs "=" with signs "≠".

$f_1 \leq f_2$ if and only if, for any pair of terms t, t', (f_1 imposes $t = t'$) implies (f_2 imposes $t = t'$)
For instance, $(x_1 = x_2 \wedge x_1 \neq x_4) \leq (x_1 = x_2 \wedge x_1 = x_4) \leq (x_1 = x_2 \wedge x_2 = x_3 \wedge x_3 = x_4)$

Let $x_1, \ldots, x_n$, satisfying f_1. $f_1 \leq f_2$ if and only if there exists a n-uple satisfying f_2, obtained by replacing in $x_1, \ldots, x_n$ each occurrence of a value v_a by a value v_b.

Definition 7.2 [2]

Let A, be a $REC_{\neq}$-automaton. A relation ψ defined on the set of states Q, is said to be compatible with rules if and only if, for any pair of rules $(a(q_1, \ldots, q_n)[f] \rightarrow q)$ and $(a(q'_1, \ldots, q'_n)[f'] \rightarrow q')$ such that $q_i \ \psi \ q'_i$ and $f \leq f'$, then $q \ \psi \ q'$.

Definition 7.2 [3]

For every $REC_{\neq}$-automaton A, the relation denoted ∇_A, defined on the set of states, is the compatible-with-rule closure of the identity relation.

Proposition 7.2 [4]

Let $\mathcal{F}$, be a set of trees
1) $\mathcal{F} \in REC$ if and only if there exists a complete and deterministic $REC_{\neq}$-automaton A, such that $\nabla_A = Identity$
2) $\mathcal{F} \in REC_=$ if and only if there exists a complete and deterministic $REC_{\neq}$-automaton A, such that ∇_A is a cycle-free relation and $(q \in F, q \nabla_A q') \Rightarrow q' \in F$
3) $\mathcal{F}$ belongs to the boolean closure of $REC_=$ if and only if there exists a complete and deterministic $REC_{\neq}$-automaton A, such that ∇_A is a cycle-free relation.

This result can be used to prove that $REC_{\neq}$ is not the boolean closure of $REC_=$, exhibing an element of $REC_{\neq}$ which can't be accepted by an automaton A with a relation ∇_A circuit-free. Such an example is given by the set of trees on $\{\underset{0}{\bar{a}}, \underset{1}{a}, \underset{3}{c}\}$ such that at least one symbol c appears be with two identical sons, different from the third one.
This set belongs to REC_{ne}, but can't be recognized by an automaton A, with A circuit-free, and doesn't belong to the boolean closure of $REC_=$. Intuitively, in the boolean closure of $Rec_=$, we may expres that there is an occurrence of $c(t_1, t_2, t_3)$, with $t_1 \neq t_2$, and one occurrence of $c(u_1, u_2, u_2)$, but we are unable to ensure the coincidence of the two occurrences.

8 Applications

$Rec_{\neq}$ enables us to manipulate some non recognizable tree languages, which could be, for example, defined as (quasi-)alphabetic images of recognizable sets or set of normal forms. For example, it provides a new tool for studying some properties of rewrite systems, in particular -but non anecdotic-cases. We give here some precise examples of applications:

8. 1 Inductive reducibility

For a given rewriting system S, a term t is said to be inductively reducible if every ground instance of it is reducible.
We get a single-exponential decision algorithm in a particular case :
- In t, and in each left-hand term of the system S, the non-linearity (if any) occurs only on brother terms –example : $b(b(x, x), a(y)$–

First, the set *Red* of reducible trees belongs to $REC_{\neq}$. *Red* is accepted by a (non-deterministic) automaton with two "global" states $\{reducible, irreducible\}$, and some "local" (locally useful) states :

for each left-hand term $left$ of S, for each position $p \neq \epsilon$ of $left$, such that the node at position p is a letter a (not a variable), we add the state $local_{left,p}$, and the rule $a(q_1, \ldots, q_n)[f] \rightarrow local_{left,p}$, where $q_i = local_{left,p.i}$, if $t \backslash p.i$ is not a variable, and $q_i = irreducible$, otherwise; f codes the non linearity of variables, if any. For the root of the term $left$, we construct a rule in the same way, but the reached state is $reducible$.
Example :
when $left = b(b(x,x), a(y))$, the rules will be :
$a(irreducible) \rightarrow local_{left,2}$
, $b(irreducible, irreducible)[x_1 = x_2] \rightarrow local_{left,1}$
$b(local_{left,1}, local_{left,2}) \rightarrow reducible$.
The automaton is completed with rules such that if the state *reducible* appears under any letter, with any constraint, then the state *reducible* is reached above the letter.

Second, the set of instances of a term t is also accepted by a $REC_{\neq}$-automaton : the construction looks like the previous one.

$Rec_{\neq}$ is closed under boolean operations, so we can construct an automaton for irreducible instances of t. After determinization, we get an automaton whose size is exponentially bounded to the system size. The emptiness decision algorithm allows to decide whether t is inductively reducible or not.

8. 2 Equational formulæ with membership constraints

The good closure and decidability properties of $REC_{\neq}$ allow to extend some results proved for REC, as described in [CoDe91]. They first have defined a set of transformation rules for first order formulæ whose atoms are either equations beteween terms or "membership constraints" $t \in s$ where s is a regular tree language (a sort). Then, they have extended their results, by considering the $Rec_{\neq}$ family. It can be applied to inductive proofs and provides an algorithm for inductive reducibility in some cases; this algorithm is more general than the preceding one (but its complexity should be greater).

8. 3 Order-sorted calculi with term declarations

In [Sch 88], M.Schmidt-Schauss has studied the notion of order-sorted algebras with explicit term declarations -i.e. declarations of the form t:S-. He proved that unification is then undecidable, which can also be considered as a corollary of Mongy's results [Mon 81]. By using $Rec_{\neq}$, we obtain than unification remains decidable for a large class of term declarations (when the non-linearity (if any) occurs only at depth one) [BoCoTi 91].

9 Conclusion

Excepted for the closure under inverse morphisms, the closure and decision properties of *Rec* are preserved; so, the studied family seems to bring interesting decision algorithms (for instance in a particular case of inductive reducibility, c.f. 5.1), and other algorithmic developements are probably possible. Further results on minimization of $REC_{\neq}$ automata supply an algorithm for checking recognizability (in REC sense) of $REC_{\neq}$ tree languages [BoTi 91] but the decidability of some properties, like membership in $REC_{=}$ or in its boolean closure are open problems.
It would be also interesting to define other extensions of tree automata, for example by limiting the number of tests, without restriction on the positions of compared terms. It would provide a general tool to study, for example, set of normal forms.

Bibliography

[Bog 90] B. Bogaert : " Automates d'arbres avec tests d'égalités."- *PhD Thesis, Université de Lille I. (1990)*

[BoCoTi 91] B.Bogaert, H.Comon & S.Tison : " Order-sorted calculi with term deckarations."- *UNIF'91. (1991)*

[BoTi 91] B.Bogaert, S.Tison : " Minimization of tree automata with tests"- *LIFL Report (to appear) (1991)*

[Com 88] H. Comon : " Unification et Disunification. Théorie et applications."- *PhD Thesis, INPG (Grenoble) (1988)*

[Com 89] H. Comon : " Inductive proofs by specification transformations."- *Proc. Rewriting technics and applications, Chapel-Hill, LNCS 375 (1989)*

[CoDe 91] H. Comon & C. Delor : " Equational formulae with membership constraints."- *Technical report, LRI Paris-Sud Orsay (1991)*

[Cou 89] B. Courcelle : " On recognizable sets and tree automata. Resolutions of equations in algebraic structures."- *Academic Press, Ait-Kaci & M.Nivat ed. (1989)*

[DaTi 90] M. Dauchet & S. Tison : " Réduction de la non-linéarité des morphismes d'arbres."- *Rapport IT 196, LIFL, Université de Lille (1990)*

[JoKo 86] J.P. Jouannaud & E. Kounalis : " Automatic proofs by induction in equationnal theory without constructors."- *Proc. 1st IEEE symp. Logic in Computer Science, Cambridge, Mass. (1986)*

[GéSt 84] F. Gécseg & M. Steinby : " Tree automata."- *Akadémiai Kiadó, Budapest (1984)*

[KaNaZh 85] D. Kapur, P. Narendran, H. Zhang : " On sufficient completeness and related properties of term rewriting systems."- *Research Report TR 87-26, Computer science department, State University of New-York at Albany (1985)*

[Kou 90] E. Kounalis : " Testing for inductive (co-)reductibility."- *Proc CAAP 90 (1990)*

[Mah 88] M.J. Maher : " Complete axiomatizations of the algebras of finite, rationnal and infinite trees."- *Proc 3rd IEEE Symp. Logic in Computer Sc. (1988)*

[Mon 81] J. Mongy : " Transformations de noyaux reconnaissables d'arbres. Forêts RATEG."- *PhD Thesis, Université de Lille I (1981)*

[Oya 87] M. Oyamagushi : " The Church-Roser property for ground term rewriting systems is decidable."- *TCS 49, pp 43-79 (1987)*

[Oya 90] M. Oyamagushi : " The reachability and joinability problems for right-ground term rewriting systems."- *J. Inf. Process (to appear)*

[Tha 73] J.W. Thatcher : " Tree automata : an informal survey."- *Currents in the theory of computing. A.V. Aho ed. Prentice Hall (1973)*

[Sch 88] M. Schmidt-Schauss : " Computational aspects of an order-sorted logic with term declarations"- *PhD Thesis, Univ. Kaiserlautern (1988)*

[Tis 90] S. Tison : " Automates comme outils de décision."- *Mémoire d'Habilitation - Université de Lille (1990)*

[Tom 91] M. Tommasi : " Automates avec tests d'égalités entre cousins germains"- *LIFL IT-report (to appear) (1991)*

Deterministic Regular Languages

Anne Brüggemann-Klein
Institut für Informatik
Universität Freiburg
Rheinstr. 10–12, 7800 Freiburg
Germany

Derick Wood
Department of Computer Science
University of Waterloo
Waterloo, Ontario N2L 3G1
Canada

Abstract

The ISO standard for Standard Generalized Markup Language (SGML) provides a syntactic meta-language for the definition of textual markup systems. In the standard the right hand sides of productions are called *content models* and they are based on regular expressions. The allowable regular expressions are those that are "unambiguous" as defined by the standard. Unfortunately, the standard's use of the term "unambiguous" does not correspond to the two well known notions, since not all regular languages are denoted by "unambiguous" expressions. Furthermore, the standard's definition of "unambiguous" is somewhat vague. Therefore, we provide a precise definition of "unambiguous expressions" and rename them deterministic regular expressions to avoid any confusion. A regular expression E is *deterministic* if the canonical ϵ-free finite automaton M_E recognizing $L(E)$ is deterministic. A regular language is *deterministic* if there is a deterministic expression that denotes it. We give a Kleene-like theorem for deterministic regular languages and we characterize them in terms of the structural properties of the minimal deterministic automata recognizing them. The latter result enables us to decide if a given regular expression denotes a deterministic regular language and, if so, to construct an equivalent deterministic expression.

Classification: Automata and formal languages, esp. formal models in document processing

1 Introduction

Document processing systems like editors, formatters, and retrieval systems deal with many different types of documents, like books, articles, memos, dictionaries, or letters, in addition to user-defined document types or customized versions of "public" types. Recently, the Standard Generalized Markup Language (SGML) [ISO86] has been established as a common platform for the syntactic specification of document types and conforming documents. SGML is an ISO standard and has been endorsed by a number of publishing houses throughout North America and Europe, by the European Community, and by the U.S. Department of Defense.

Document types in SGML are defined by context free grammars that are a mixture of Harrison and Ginsburg's bracketed grammars [GH67] and LaLonde's regular right side grammars [LaL77]. Regular expressions form the right-hand sides of productions, but not all regular expressions are allowed, only those that are "unambiguous" in the sense of Clause 11.2.4.3 of the standard. The intent of the standard is to make it easier for a human to write regular expressions that can be interpreted unambiguously. To achieve this the standard requires each regular expression to be "unambiguous" in the sense that "an element ... that occurs in the document instance must be able to satisfy only one primitive content token without looking ahead in the document instance." In other words, only such regular expressions are valid that permit us to uniquely determine which appearance of a symbol in an expression should match a symbol in an input word without looking beyond that symbol in the input word. This requirement specifies exactly the class of *deterministic expressions* that we investigate here.

An alternative motivation for our study is that the theory of regular languages has become a cornerstone in practical applications involving e.g. specification, pattern matching, and the construction of scanners and parsers. Perhaps the most frequently occurring task is to construct, to a specification in the form of a regular expression, an automaton that can recognize the specified objects. Usually, one first constructs a non-deterministic finite automaton with ϵ-transitions (ϵ-NFA) in time linear in the size of E, eliminates the ϵ-transitions in quadratic time, and finally converts the resulting NFA into a deterministic finite automaton (DFA) [HU79]. The intermediate step can be avoided by directly constructing an ϵ-free automaton [BEGO71,ASU86]. It has been claimed [BS86] that this NFA is the canonical representation because it has a natural connection with the derivatives [Brz64] of the original expression. Since it takes exponential time in the worst case to convert an NFA into a DFA, it is natural to ask for which regular expressions E the canonical NFA M_E is already deterministic. Such expressions are exactly what we have called deterministic above. It can be tested in linear time whether a regular expression is deterministic, and if so, the canonical deterministic automaton can also be constructed in linear time [Bru91].

In this paper, we first give a rigorous definition of *deterministic* regular expressions. Then, we investigate the *deterministic* regular languages, i.e. regular languages that can be denoted by a deterministic expression. As we will see, the deterministic regular languages are a proper subclass of the regular languages; for example, for each $n \geq 1$, the expression $(0+1)^*0(0+1)^n$ denotes a regular language that is not deterministic.

First, we state a Kleene-like theorem for the class of deterministic regular languages. Next, we characterize the deterministic regular languages in terms of the minimal deterministic automata that recognize them. To each regular language L, the minimal deterministic finite automaton M_L recognizing L is uniquely determined. We show that deterministic regular languages L can be symbolized by structural properties of M_L. For a state q of M_L, let the *orbit* $\mathcal{O}(q)$ of q denote the strongly connected component of q, i.e. the states of M_L that can be reached from q and vice versa. Some states of $\mathcal{O}(q)$, called *gates*, connect the orbit to the outside world. Now L is deterministic if and only if all orbits of M_L define deterministic regular languages and if for each orbit $\mathcal{O}(q)$, all gates of $\mathcal{O}(q)$ have identical connections to the outside.

Then we show that any deterministic regular language defined by a DFA M with a single orbit is of the form $v\backslash L^*$, where L is deterministic and $v\backslash L^*$ denotes the set of words w such that vw is in L^* (the Brzozowski derivative of L^* by v). Furthermore, a minimal DFA recognizing L can be constructed from M. Together, these results yield an algorithm that decides, given a DFA M, whether its language is deterministic and, if so, constructs an equivalent deterministic expression. The decision algorithm runs in time quadratic in the size of M, but the corresponding deterministic expression can be exponential in the size of M.

To give an example, for each word w, the language $\Sigma^*w\Sigma^*$ of all words over Σ containing w as a subword is a deterministic regular language.

Most proofs in this paper are just sketches. The complete proofs can be found in the full version [BW91].

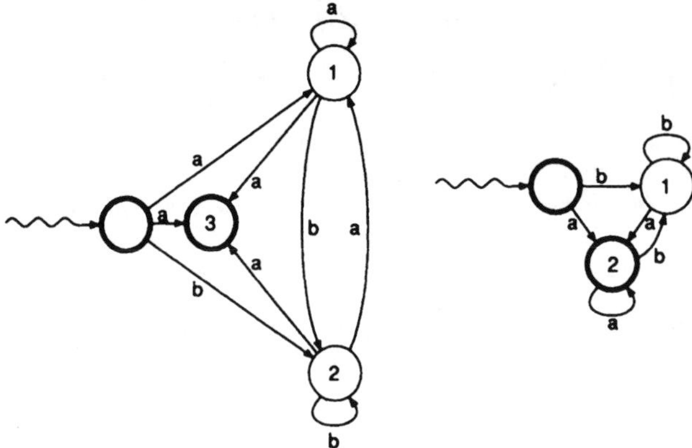

Figure 1 The Glushkov automata corresponding to $(a + b)^*a + \epsilon = (a_1 + b_2)^*a_3 + \epsilon$ and $(b^*a)^* = (b_1^*a_2)^*$.

2 Deterministic regular expressions

The notion of a symbol in a word being satisfied or matched by a symbol in a regular expression has been explained by a number of authors [BEGO71,ASU86,Hen68]. We paraphrase here the description of Hennie [Hen68]. If a word is denoted by an expression, it must be possible to spell out that word by tracing an appropriate "path" through the expression. If we indicate positions in expressions by subscripts, then the word *abba* is denoted by the expression $(a+b)^*a+\epsilon = (a_1+b_2)^*a_3+\epsilon$ because it corresponds to the path that starts at a_1, visits b_2 twice, and finally arrives at a_3. The set of subscripted symbols in an expression E is denoted by $pos(E)$. Of course, the structure of the expression restricts the positions adjacent symbols of a word can be matched with. For instance, if a symbol in a word is matched by a_3 in $(a_1 + b_2)^*a_3 + \epsilon$, then no further symbol of the word can be matched with a symbol in the expression. These restrictions have first been formalized by Glushkov [Glu61].

This description suggests viewing a regular expression E as an automaton M_E whose states correspond to the positions or occurrences of symbols in E and whose transitions connect positions that can be consecutive on a path through E. We call M_E the *Glushkov automaton* of E. Figure 1 shows the two Glushkov automata corresponding to the expressions $(a+b)^*a+\epsilon = (a_1+b_2)^*a_3+\epsilon$ and $(b^*a)^* = (b_1^*a_2)^*$. In addition to the states of M_E that correspond to positions in E, the *true* states, there is one unnamed state in the Glushkov automata of Figure 1 which acts as the initial state.

In general, Glushkov automata are non-deterministic, as $(a+b)^*a+\epsilon = (a_1+b_2)^*a_3+\epsilon$ illustrates. After matching an input symbol a with a_1, a further a can either be matched by a_1 or a_3. Thus, there is a transition on a from a_1 to a_1 and to a_3 in the Glushkov automaton. This example leads naturally to a precise definition of what the SGML standard means by a deterministic expression.

Definition 2.1 A regular expression E is *deterministic* if M_E is deterministic, i.e., if M_E is a DFA. A regular language is *deterministic* if there is some deterministic expression that denotes it.

Figure 1 illustrates that $(a+b)^*a+\epsilon$ is not a deterministic expression. Nevertheless, the language denoted by $(a+b)^*a+\epsilon$ is a deterministic regular language, since it is also denoted by $(b^*a)^*$, which is a deterministic expression.

We define M_E inductively, rather than in terms of the formalism introduced by Glushkov [Glu61] that has been used by a number of authors [BEGO71,ASU86,BS86].

M_E has the form $M_E = (Q_E \dot{\cup} \{q_I\}, \Sigma, \delta_E, q_I, F_E)$, with Q_E comprising the true states corresponding to positions in E, q_I the new initial state, Σ the input alphabet, $\delta_E : (Q_E \cup \{q_I\}) \times \Sigma \longrightarrow 2^{Q_E}$ the transition function, and $F_E \subseteq Q_E \cup \{q_I\}$ the set of final states. To simplify the discussion, we follow the general convention that regular expressions are built from symbols in Σ and the empty string symbol ϵ, but not the empty set symbol $\emptyset$. (Nevertheless, we consider the empty set to be a deterministic regular language.) To *identify* two states in an automaton means to replace them by a new state that has exactly the transitions that both of the old ones had.

Definition 2.2 Given a regular expression E, we define the construction of M_E, illustrated in Figure 2, inductively as follows.

$[E = \epsilon \text{ or } a]$ M_ϵ and M_a are illustrated in Figure 2.

$[E = F + G]$ In M_E the initial states of M_F and M_G are identified. Let

$$Q_E = Q_F \dot{\cup} Q_G, \quad \text{(disjoint union, possibly after renaming states)}$$

$$F_E = F_F \cup F_G,$$

$$\delta_E(q,a) = \begin{cases} \delta_F(q,a) & \text{if } q \in Q_F \\ \delta_G(q,a) & \text{if } q \in Q_G \\ \delta_F(q_I,a) \cup \delta_G(q_I,a) & \text{if } q = q_I. \end{cases}$$

$[E = FG]$ In M_E a copy of the initial state of M_G is identified with each final state of M_F. Let

$$Q_E = Q_F \dot{\cup} Q_G, \quad \text{(disjoint union, possibly after renaming states)}$$

$$F_E = \begin{cases} F_F \cup (F_G \setminus \{q_I\}) & \text{if } q_I \in F_G, \\ F_G & \text{otherwise,} \end{cases}$$

$$\delta_E(q,a) = \begin{cases} \delta_F(q,a) & \text{if } q \in (Q_F \cup \{q_I\}) \setminus F_F, \\ \delta_F(q,a) \cup \delta_G(q_I,a) & \text{if } q \in F_F, \\ \delta_G(q,a) & \text{if } q \in Q_G. \end{cases}$$

$[E = F^*]$ In M_E all transitions from q_I in M_F are added to the final states of M_E. Let

$$Q_E = Q_F, F_E = F_F \cup \{q_I\},$$

$$\delta_E(q,a) = \begin{cases} \delta_F(q,a) \cup \delta_F(q_I,a) & \text{if } q \in F_F, \\ \delta_F(q,a) & \text{otherwise.} \end{cases}$$

Proposition 2.1 *M_E recognizes the language denoted by E.*

Glushkov automata have some peculiar structural properties that are worth investigating.

First, the initial state of M_E has no incoming transitions. This makes the construction correct in the sense of Proposition 2.1. Furthermore, the states directly connected to the initial state correspond exactly to the positions of E that can match the first character of a word of E, and the final states in M_E (besides s_I) correspond exactly to the ones that match the last character of a word of E. Thus, the initial state has transitions to *first positions* in E, and the final states in M_E (besides s_I) are *final positions* of E.

Second, for a subexpression F of E, the structure of M_F is retained in M_E. Each true state of M_F is also a true state of M_E, and all transitions between true states in M_F belong also to M_E. Furthermore, among all positions of F, exactly the final ones are final states of M_E or

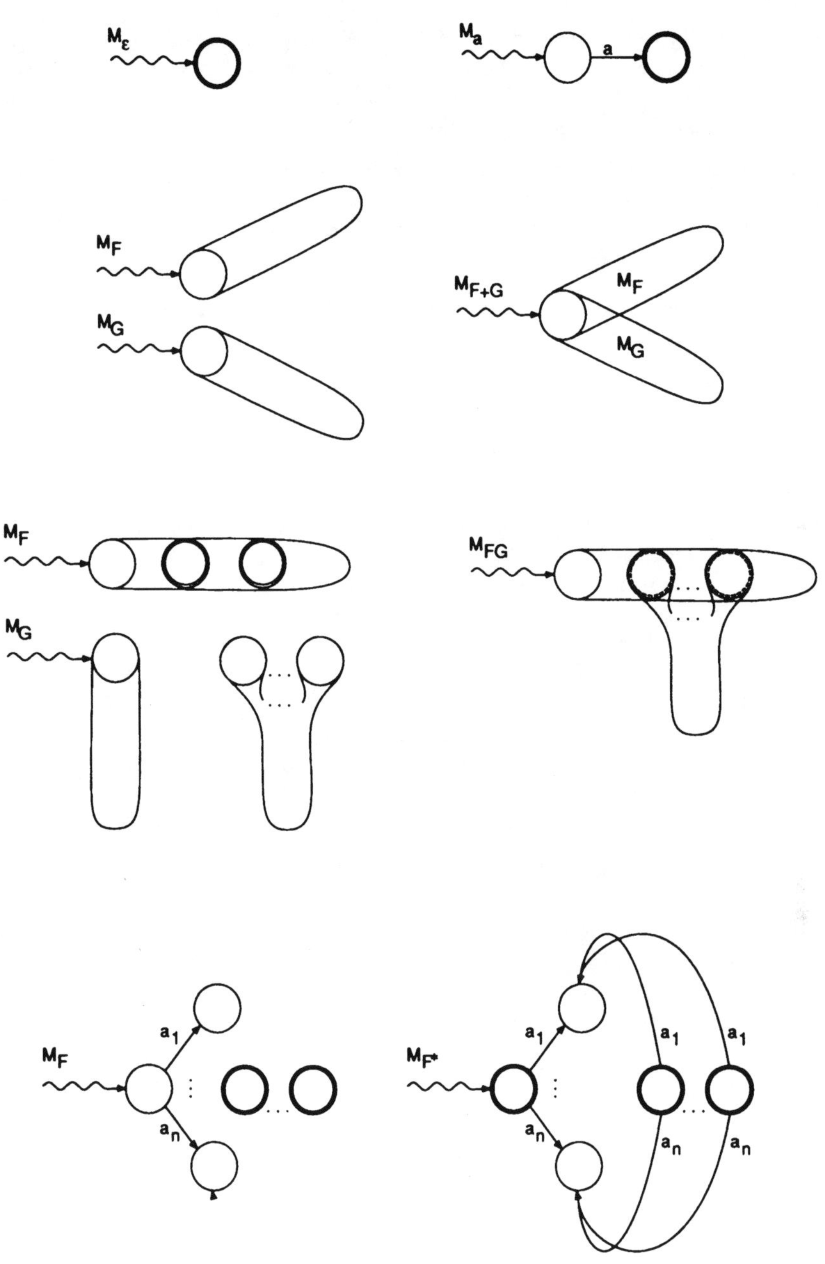

Figure 2 The inductive definition of M_E.

have transitions in M_E to positions outside of F. These conditions are even fulfilled uniformly, meaning that either all or none of the final positions of F are final in E, and that either all or none have a transition in M_E on an $a \in \Sigma$ to a position y of E outside of F. Hence, the final positions of F are an interface of M_F to the surrounding parts of M_E.

Finally, we will be especially interested in maximal starred subexpressions of E, i.e. subexpressions of the form F^* that are not subexpressions of another starred subexpression G^* of E. For such an F^*, any transitions in M_E between positions of F^* are already transitions in M_{F^*}. In this sense, M_{F^*} is closed within M_E.

Proposition 2.2 *Given a regular expression E, we can decide if it is deterministic in time linear in the size of E.*

Proof The Glushkov automaton can be constructed from E in such a way that a *new* transition is introduced at each computation step [Bru91]. As soon as we encounter a transition that makes the automaton under construction non-deterministic, we stop and report E to be non-deterministic. At this point, only time linear in the size of E has been spent. □

Book et al. [BEGO71] have defined a regular expression E to be unambiguous if the Glushkov automaton M_E is unambiguous, i.e. if for each word w there is at most one computation of M_E that accepts w, or, equivalently, at most one path through E that spells out w. They have shown that each regular language can be denoted by an unambiguous regular expression. A deterministic expression E is an unambiguous one where for each word w the corresponding path through E can be computed incrementally from w with just one symbol of look-ahead. Thus, deterministic regular expressions are related to unambiguous ones in the same way that LL(1) grammars are related to unambiguous context-free grammars. This analogy can be made precise: It is possible to translate a regular expression E in a natural way into an equivalent context-free grammar G_E such that E is deterministic if and only if G_E is LL(1).

3 The characterization theorem

We first state without proof a Kleene-like theorem for deterministic regular languages. We then consider the cyclic structure of M_E that we capture in terms of *orbits*. The structure of the orbits is essentially preserved under minimization, and, hence, orbits turn out to be exactly the right tool for characterizing deterministic regular languages.

We begin by defining three functions for languages that can also be adapted to apply to regular expressions. These functions are: *first*(L), the set of symbols that appear as the first symbol of some word in L; *last*(L), the set of symbols that appear as the last symbol of some word in L; and *followlast*(L), the set of symbols that follow a prefix of some word in L, where the prefix is also a word in L. More formally we have:

Definition 3.1 For $L \subseteq \Sigma^*$, let

$$\mathit{first}(L) = \{a \in \Sigma \mid aw \text{ is in } L \text{ for some word } w\},$$

$$\mathit{last}(L) = \{a \in \Sigma \mid wa \text{ is in } L \text{ for some word } w\},$$

$$\mathit{followlast}(L) = \{a \in \Sigma \mid vaw \text{ is in } L, \text{ for some word } v \text{ in } L \setminus \{\epsilon\} \text{ and some word } w\}.$$

Theorem 3.1 *The deterministic regular languages are the smallest class $\mathcal{D}$ of languages that satisfies the following conditions.*

1. *$\emptyset$, ϵ, and $\{a\}$ are in $\mathcal{D}$.*
2. *If $A, B \in \mathcal{D}$ and $\mathit{first}(A) \cap \mathit{first}(B) = \emptyset$, then $A \cup B \in \mathcal{D}$.*

3. *If $A, B \in \mathcal{D}$, $\epsilon \notin A$, and followlast$(A) \cap$ first$(B) = \emptyset$, then $AB \in \mathcal{D}$.*
4. *If $A \in \mathcal{D}$, then $A \setminus \{\epsilon\} \in \mathcal{D}$.*
5. *If $A \in \mathcal{D}$ and followlast$(A) \cap$ first$(A) = \emptyset$, then $A^* \in \mathcal{D}$.*

It is well known that, for each regular language L, the minimum-state deterministic automaton M_L recognizing L is uniquely determined. We argue that, by examining the cyclic structure of M_L, we can decide whether L is deterministic. Furthermore, for a deterministic regular language L, we can construct a deterministic expression for L from M_L.

For each DFA $M = (Q, \Sigma, \delta, q_0, F)$ recognizing L, the equivalence class construction [ASU86] results in a DFA $\overline{M} = (\overline{Q}, \Sigma, \overline{\delta}, [q_0], \overline{F})$ which is isomorphic to M_L.[1] For a deterministic expression E denoting a language L, a minimum-state DFA M_L can be constructed directly from M_E via the equivalence class construction. For a non-deterministic expression E, however, the NFA M_E has first to be converted to a DFA via the subset construction [ASU86]. Thus, we are looking for properties of M_E that are preserved under state minimization, but not under subset construction. We start from the structural properties of M_E noted in Section 2.

Definition 3.2 Let $M = (Q, \Sigma, \delta, q_I, F)$ be an NFA. For $q \in Q$, the strongly connected component of q, i.e. the states of M that can be reached from q and from which q can be reached as well, is called the *orbit* of q and denoted by $\mathcal{O}(q)$. We consider the orbit of q to be *trivial* if $\mathcal{O}(q) = \{q\}$ and there are no transitions from q to itself in M.

Definition 3.3 A state q in an NFA $M = (Q, \Sigma, \delta, q_I, F)$ is called a *gate* of its orbit if either q is a final state or there are $q' \in Q \setminus \mathcal{O}(q)$ and $a \in \Sigma$ with $q \xrightarrow{a} q'$. The NFA M has the *orbit property* if each orbit of M is homogeneous with respect to its gates, i.e. if, for all gates q_1 and q_2 with $\mathcal{O}(q_1) = \mathcal{O}(q_2)$, we have:

- q_1 is a final state if and only if q_2 is a final state.
- $q_1 \xrightarrow{a} q$ if and only if $q_2 \xrightarrow{a} q$, for all $q \in Q \setminus \mathcal{O}(q_1) = Q \setminus \mathcal{O}(q_2)$ and for all $a \in \Sigma$.

In Figure 3, both automata have three orbits, namely $\{1\}$, $\{2,3\}$, and $\{4\}$. The singleton orbits are trivial, and each state is a gate of its orbit. The left automaton fulfills the orbit property, the right one does not.

Now we can formulate a necessary condition for a regular language to be deterministic.

Theorem 3.2 *The minimal DFA M_L recognizing a deterministic regular language L has the orbit property.*

This gives us our first example of a regular language that is not deterministic. The rightmost automaton in Figure 3 is the minimal DFA for the language denoted by $(a+b)^*(ac+bd)$, and it does not have the orbit property. Thus, this language cannot be deterministic.

The proof of Theorem 3.2 is in two steps. First, we describe the orbits of the Glushkov NFA M_E constructed in Definition 2.2 and show that M_E has the orbit property for *all* regular expressions. Next, we show that the orbit property is preserved under state minimization. Thus, if a language L is symbolized by a *deterministic* regular expression E, the minimal automaton $M_L = \overline{M_E}$ has the orbit property.

Lemma 3.3 *Let E be a regular expression and $x \in pos(E)$. If there is no starred subexpression F^* of E with $x \in pos(F)$, then $\mathcal{O}(x) = \{x\}$ and $\mathcal{O}(x)$ is trivial. On the other hand, if F^* is the maximal starred subexpression of E with $x \in pos(F)$, then $\mathcal{O}(x) = pos(F)$, and $\mathcal{O}(x)$ is not trivial. Finally, the orbit of the initial state q_I is trivial.*

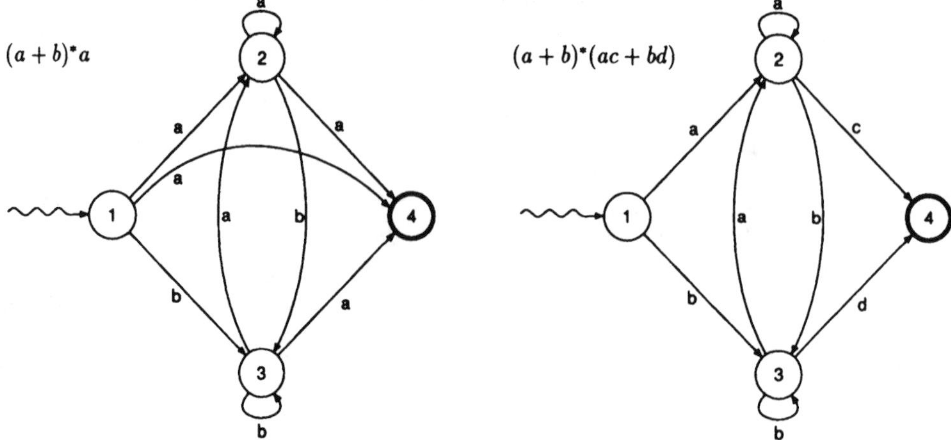

Figure 3 Two NFAs, one fulfills the orbit property, the other one does not.

Lemma 3.4 *Let E be a regular expression. Then,*

1. *M_E has the orbit property and*
2. *if F^* is a maximal starred subexpression of E with $pos(F) \neq \emptyset$, then the last positions of F are the gates of the orbit $pos(F)$.*

Figure 4 shows the Glushkov automaton for $a^*(bca^*)^* = a_1(b_2c_3a_4^*)^*$. The gates of orbit $\{2,3,4\}$ are the states 3 and 4, i.e. the last position of the maximal starred subexpression $(bca^*)^* = (b_2c_3a_4^*)^*$.

Now, consider a *deterministic* automaton M and its minimization $\overline{M}$. If states $p_1, \ldots, p_n$ of M form an orbit in M, then the equivalence classes $[p_1], \ldots, [p_n]$ belong to the same orbit in $\overline{M}$, which may, however, contain further elements. Figure 4 shows the Glushkov automaton M_E for $E = a^*(bca^*)^* = a_1(b_2c_3a_4^*)^*$ and the minimal automaton $\overline{M_E}$. All states of M_E besides state 2 are equivalent, $\{1\}$ is an orbit of M_E, but $[1]$ does not form a complete orbit in $\overline{M_E}$. Nevertheless, the orbit of $[1]$ in $\overline{M_E}$ is completely generated by another orbit of M_E, namely $\{2,3,4\}$. This is a general phenomenon, namely, for each orbit K of $\overline{M}$, there is an orbit C of M that fully generates K, i.e. $K = \{[q] \mid q \in C\}$. C is called a *lift* of K. Now, if a lifted orbit C is homogeneous with respect to its gates, then so is K. Thus, we have the following lemma, which concludes the proof of Theorem 3.2.

Lemma 3.5 *The orbit property is preserved under state minimization.*

The orbit property has gained us a necessary condition for a regular language to be deterministic. Another necessary condition evolves if we examine the orbits themselves in isolation.

Definition 3.4 Let M be a DFA.[1] For $q \in Q_M$, let the *orbit automaton* M_q of q, be the automaton obtained by restricting the state set to $\mathcal{O}(q)$ with initial state q and final states the gates of $\mathcal{O}(q)$ in M. We say the orbit of q is *deterministic* if $L(M_q)$ is deterministic. A regular language L is said to be an *orbit language* if and only if there is a DFA M with a single orbit that recognizes L.

Theorem 3.6 *For each deterministic language L, the minimal DFA recognizing L has only deterministic orbit languages.*

[1]Some minor technicalities are involved here, because the transition functions of M and $\overline{M}$ are only partially defined. The details are in the full version.

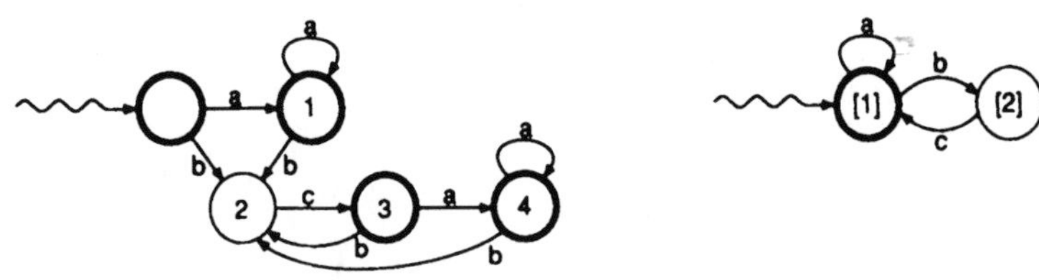

Figure 4 The Glushkov automaton for the expression $a^*(bca^*)^* = a_1(b_2c_3a_4^*)^*$ and its minimization.

For the proof, again we first look at the orbit languages of Glushkov automata and then consider state minimization. Let E be a deterministic expression, and let q be a state of M_E with a non-trivial orbit. Then, q is a position of a maximal starred subexpression F^* of E. Let L be the language of F^*, and let v be a word that leads from the initial state to q in M_{F^*}. Since M_{F^*} is closed within M_E, the orbit language of q in M_E is also the orbit language of q in M_{F^*}, which in turn is

$$v\backslash L := \{w \in \Sigma^* \mid vw \in L\}.$$

The language $v\backslash L$ is known as the *derivative* of L by v [Brz64]. Thus, a non-trivial orbit language of M_E is a derivative of a language denoted by a maximal starred subexpression of E.

The proof of the next proposition is in the full paper.

Proposition 3.7 *The derivative of a deterministic regular language is also deterministic.*

As a corollary, we have:

Lemma 3.8 *Let E be a deterministic regular expression. Then, all orbit languages of M_E are deterministic regular languages.*

Again, this property is preserved under minimization, as can be seen from the next lemma.

Lemma 3.9 *Let M be a DFA and $\overline{M}$ be its reduction. Then, for each state of $\overline{M}$ there is an equivalent state q in M such that*

1. *the orbit of q in M is a lift of the orbit of $[q]$ in $\overline{M}$, and*
2. *the orbit languages of q in M and $[q]$ in $\overline{M}$ are identical.*

This concludes the proof of Theorem 3.6.

The necessary conditions for a minimal DFA to recognize a deterministic regular language as given in Theorems 3.2 and 3.6 are also sufficient:

Theorem 3.10 *Let L be a regular language and M be the minimal DFA recognizing L. Then, L is a deterministic regular language if and only if M has the orbit property and all orbits of M are deterministic.*

Proof We show the implication from right to left by induction on the number of orbits of M. Let $M = (Q, \Sigma, \delta, q_I, F)$ have more than one orbit. Furthermore, let $q_1, \ldots, q_n$ be the distinct states outside $\mathcal{O}(q_I)$ that are reachable in one step from a gate of $\mathcal{O}(q_I)$. All gates of $\mathcal{O}(q_I)$ have an a_i-transition to q_i, and no other outgoing transitions from gates of $\mathcal{O}(q_I)$ to the outside exist. The a_i are pairwise distinct and M_{q_I} has no a_i-transition from a final state.

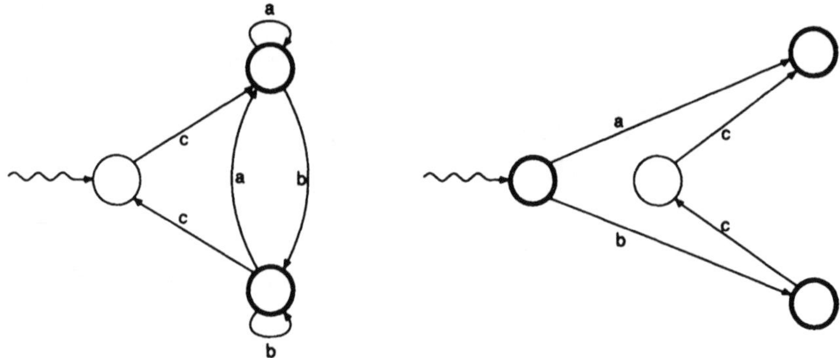

Figure 5 An a, b-consistent DFA and its a, b-cut.

Let M_i be the automaton whose states are the states of M that are reachable from q_i as the initial state. Because M_i has fewer orbits than M, M_i is deterministic. Furthermore,

$$\mathrm{L}(M) = \mathrm{L}(M_{q_I})(a_1\mathrm{L}(M_1) \cup \ldots \cup a_n\mathrm{L}(M_n))$$

or

$$\mathrm{L}(M) = \mathrm{L}(M_{q_I})(a_1\mathrm{L}(M_1) \cup \ldots \cup a_n\mathrm{L}(M_n) \cup \{\epsilon\}),$$

and a deterministic expression for M can be constructed from deterministic expressions for M_{q_I} and $M_1, \ldots, M_n$. □

Theorem 3.10 is the first step of a decision algorithm for deterministic regular languages:

Theorem 3.11 *Given a DFA M, we can decide in time quadratic in the size of M whether the language of M is deterministic. If so, an equivalent deterministic expression can be constructed.*

If a minimal DFA M has the orbit property, then its orbit automata are also minimal. This precludes to apply Theorem 3.10 directly for the orbit automata. On the other hand, we know already that each deterministic orbit language of M has the form $v\backslash\mathrm{L}(F^*)$, where F^* is a deterministic expression. To conclude the proof of Theorem 3.11, we show how a minimal DFA for $\mathrm{L}(F)$ can be constructed from M.

Definition 3.5 A DFA M is *a-consistent*, for $a \in \Sigma$, if there is a state $f_M(a)$ in M such that all final states of M have an a-transition to $f_M(a)$.

Definition 3.6 Let M be a_i-consistent, for $a_i \in \Sigma$, $1 \leq i \leq n$, $n \geq 1$. The *$a_1, \ldots, a_n$-cut* $M(a_1, \ldots, a_n)$ of M is constructed as follows.

1. A new state q_0 is added to M and it is connected to $f_M(a_i)$ with an a_i-transition, for all i.
2. All transitions with a_i from each final state are removed from M.
3. Finally, q_0 is made initial and final.

Figure 5 gives an example of an a, b-consistent DFA and its a, b-cut.

Theorem 3.12 *Let M be a minimal DFA recognizing a language L. Assume that M consists of a single, non-trivial orbit. Let $a_1, \ldots, a_n$ be the elements of Σ for which M is consistent. Then, L is deterministic if and only if*

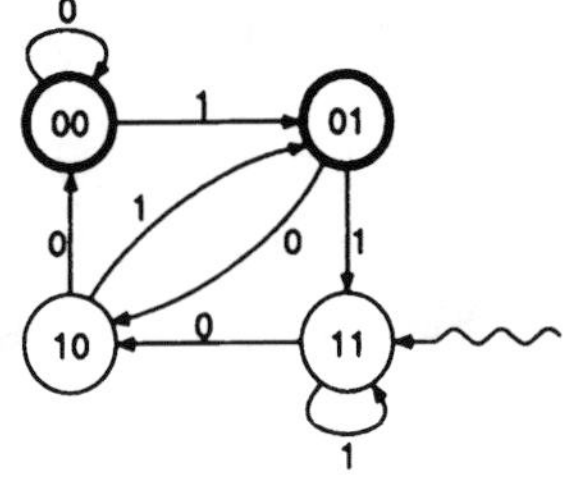

Figure 6 The minimal DFA for $(0+1)^*0(0+1)$.

1. $n \geq 1$.

2. $\mathrm{L}(M(a_1, \ldots, a_n))$ *is deterministic.*

If L *is deterministic and* $a_1, \ldots, a_n \in \Sigma$ *are chosen as above, we can construct a word* $v \in \Sigma^*$ *with*

$$L = \mathrm{L}(M) = v\backslash \mathrm{L}(M(a_1, \ldots, a_n))^*,$$

and a deterministic expression for language L *can be constructed from a deterministic expression for language* $\mathrm{L}(M(a_1, \ldots, a_n))$.

In lieu of a proof, we illustrate Theorem 3.12 with two examples. The language recognized by the left automaton in Figure 5 is deterministic, because the a, b-cut is denoted by the deterministic regular expression $a + b(\epsilon + cc) + \epsilon$. One deterministic expression for the whole language is $c(a + b(\epsilon + cc))^*$.

Figure 6 shows the minimal DFA recognizing $(0+1)^*0(0+1)$. It consists of a single orbit with two gates, 00 and 01, but is neither 0- nor 1-consistent. Thus, $(0+1)^*0(0+1)$ does not denote a deterministic language. The same is true for $(0+1)^*0(0+1)^n$ for each $n \geq 1$.

References

[ASU86] Alfred V. Aho, Ravi Sethi, and Jeffrey D. Ullman. *Compilers: Principles, Techniques, and Tools. Addison-Wesley Series in Computer Science*, Addison-Wesley, Reading, Massachusetts, 1986.

[BEGO71] Ronald Book, Shimon Even, Sheila Greibach, and Gene Ott. Ambiguity in graphs and expressions. *IEEE Transactions on Computers*, C-20(2):149–153, February 1971.

[Bru] Anne Brüggemann-Klein. Regular expressions into finite automata. To appear in the conference proceedings of Latin '92.

[Brz64] Janusz A. Brzozowski. Derivatives of regular expressions. *Journal of the ACM*, 11(4):481–494, October 1964.

[BS86] Gerard Berry and Ravi Sethi. From regular expressions to deterministic automata. *Theoretical Computer Science*, 48:117–126, 1986.

[BW91] Anne Brüggemann-Klein and Derick Wood. On the expressive power of SGML document grammars. In preparation, 1991.

[GH67] Seymour Ginsburg and Michael M. Harrison. Bracketed context-free languages. *Journal of Computer and System Sciences*, 1(1):1–23, March 1967.

[Glu61] V.M. Glushkov. The abstract theory of automata. *Russian Mathematical Surveys*, 16:1–53, 1961.

[Hen68] Frederick C. Hennie. *Finite-State Models for Logical Machines.* John Wiley, New York, 1968.

[HU79] John E. Hopcroft and Jeffrey D. Ullman. *Introduction to Automata Theory, Languages and Computation. Addison-Wesley Series in Computer Science*, Addison-Wesley, Reading, Massachusetts, 1979.

[ISO86] ISO 8879. Information processing—text and office systems—standard generalized markup language (SGML). October 1986. International Organization for Standardization.

[LaL77] Wilf R. LaLonde. Regular right part grammars and their parsers. *Communications of the ACM*, 20(10):731–741, October 1977.

STRUCTURAL COMPLEXITY 1

The Extended Low Hierarchy Is an Infinite Hierarchy [1]

Ming-Jye Sheu   Timothy J. Long [2]
Department of Computer and Information Science
The Ohio State University
Columbus, Ohio 43210, U.S.A.

Abstract

Balcázar, Book, and Schöning introduced the extended low hierarchy based on the Σ-levels of the polynomial-time hierarchy as follows: for $k \geq 1$, level k of the extended low hierarchy is the set $EL_k^{P,\Sigma} = \{A \mid \Sigma_k^P(A) \subseteq \Sigma_{k-1}^P(A \oplus \text{SAT})\}$. Allender and Hemachandra and Long and Sheu introduced refinements of the extended low hierarchy based on the Δ and Θ-levels, respectively, of the polynomial-time hierarchy: for $k \geq 2$, $EL_k^{P,\Delta} = \{A \mid \Delta_k^P(A) \subseteq \Delta_{k-1}^P(A \oplus \text{SAT})\}$ and $EL_k^{P,\Theta} = \{A \mid \Theta_k^P(A) \subseteq \Theta_{k-1}^P(A \oplus \text{SAT})\}$. In this paper we show that the extended low hierarchy is properly infinite by showing, for $k \geq 2$, that $EL_k^{P,\Sigma} \subset EL_{k+1}^{P,\Theta} \subset EL_{k+1}^{P,\Delta} \subset EL_{k+1}^{P,\Sigma}$. Our proofs use the circuit lower bound techniques of Hastad and Ko. As corollaries to our constructions, we obtain, for $k \geq 2$, oracle sets B_k, C_k, and D_k, such that $\text{PH}(B_k) = \Sigma_k^P(B_k) \neq \Delta_k^P(B_k)$, $\text{PH}(C_k) = \Delta_k^P(C_k) \neq \Theta_k^P(C_k)$, and $\text{PH}(D_k) = \Theta_k^P(D_k) \neq \Sigma_{k-1}^P(D_k)$.

1 Introduction

The low and high hierarchies in NP were introduced by Schöning in order to understand the internal structure of NP [Sch83]. The low hierarchy starts at P and grows "upward" towards the NP-complete sets, while the high hierarchy starts with the complete sets in NP and grows "downward" towards P. More formally, set A is in level k of the *low hierarchy* if $\Sigma_k^P(A) = \Sigma_k^P$, and A is in level k of the *high hierarchy* if $\Sigma_{k+1}^P = \Sigma_k^P(A)$. Many interesting sets have been located in these hierarchies: $\leq_T^P$-complete sets and $\leq_T^{SN}$-complete sets for NP form the first two levels of the high hierarchy, while sparse NP sets, NP sets with small circuits, and graph isomorphism have been placed in various levels of the low hierarchy.

These hierarchies were first extended to sets outside of NP by Balcázar, Book and Schöning [BBS86]. In particular, they defined the *extended low hierarchy* as follows: set A is in level k of the extended low hierarchy if $\Sigma_k^P(A) \subseteq \Sigma_{k-1}^P(A \oplus \text{SAT})$. Once again, many interesting classes of sets have been located in the extended low hierarchy such as sparse sets, sets with small circuits, and General Left Cuts of real numbers. Recently, refinements of the extended low hierarchy based on the Δ-classes and Θ-classes of the polynomial-time hierarchy were introduced by Allender and Hemachandra [AH91] and by Long and Sheu [LS91], respectively. Many of the sets previously located in the extended low hierarchy were more carefully and precisely located in new levels of the refined extended low hierarchy.

[1] This work was supported in part by NSF Grant CCR-8909071

[2] The second author is currently a visiting faculty member in the Department of Computer Science at New Mexico State University.

An interesting open question is whether the low hierarchy or the extended low hierarchy consists of an infinite number of distinct levels. This question is of interest for several reasons. If these hierarchies are properly infinite, then they provide a natural formal framework for categorizing and comparing the relative "power and complexity" of sets: sets at higher levels of the low hierarchy are more "complicated" than sets at lower levels. (This is analogous to using the arithmetic hierarchy to categorize sets based on relative degrees of noncomputability.) But more importantly, for the low hierarchy, if it contains at least two distinct levels, then P$\neq$ NP. We also note that it is not even known if the low hierarchy is properly infinite under very natural assumptions, such as the polynomial-time hierarchy consisting of an infinite number of distinct levels.

For the extended low hierarchy, the situation is somewhat different. In proving optimal lower bounds for the extended low hierarchy, Allender and Hemachandra [AH91] constructed a sparse set that is not in $EL_2^{P,\Sigma}$, level two of the extended low hierarchy based on the Σ-levels of the polynomial-time hierarchy. Since it is known that all sparse sets are in $EL_3^{P,\Theta}$, the third level of the extended low hierarchy based on the Θ-levels of the polynomial-time hierarchy, this shows that the extended low hierarchy has at least two distinct levels up through $EL_3^{P,\Theta}$. However, it was not known if the extended low hierarchy has distinct levels beyond $EL_3^{P,\Theta}$.

In this paper we show that the extended low hierarchy is in fact a properly infinite hierarchy. Specifically, we prove that

$$\forall k \geq 2, EL_k^{P,\Sigma} \subset EL_{k+1}^{P,\Theta} \subset EL_{k+1}^{P,\Delta} \subset EL_{k+1}^{P,\Sigma},$$

where $EL_k^{P,\Sigma}$ ($EL_k^{P,\Delta}$, $EL_k^{P,\Theta}$) denotes level k of the extended low hierarchy based on the Σ-levels (Δ-levels, Θ-levels, respectively) of the polynomial-time hierarchy. This is the first example of a low hierarchy in complexity theory that has been shown to be properly infinite.

Our proofs follow the circuit lower bound techniques from Yao [Yao85], Håstad [Has87], and Ko [Ko89]. Of particular interest for us is Ko's result that there are sets relative to which the polynomial-time hierarchy collapses to any finite level:

$$\forall k \geq 1, \exists A\ [\mathrm{PH}(A) = \Sigma_k^P(A) \neq \Sigma_{k-1}^P(A)].$$

In Section 3, we extend Ko's separation results by showing the existence of oracle sets A, B, and C such that for all $k \geq 2$, $\Delta_k^P(A) \neq \Sigma_k^P(A)$, $\Theta_k^P(B) \neq \Delta_k^P(B)$, and $\Sigma_{k-1}^P(C) \neq \Theta_k^P(C)$. In order to obtain our results, we introduce, for the first time, circuit models for the Δ and Θ-classes of relativized polynomial-time hierarchies. In Section 4, we combine the separation results of Section 3 with encoding techniques to construct oracles D_k, E_k, and F_k such that

$$\forall k \geq 1\ \exists D_k\ [\mathrm{PH}(D_k) = \Sigma_k^P(D_k) \not\subseteq \Delta_k^P(D_k \oplus \mathrm{SAT})].$$

$$\forall k \geq 2\ \exists E_k\ [\mathrm{PH}(E_k) = \Delta_k^P(E_k) \not\subseteq \Theta_k^P(E_k \oplus \mathrm{SAT})].$$

$$\forall k \geq 2\ \exists F_k\ [\mathrm{PH}(F_k) = \Theta_k^P(F_k) \not\subseteq \Sigma_{k-1}^P(F_k \oplus \mathrm{SAT})].$$

These results show that every level of the extended low hierarchy is distinct. Immediate corollaries are the existence of oracles G_k, H_k, and I_k, for all $k \geq 2$, such that $\mathrm{PH}(G_k) = \Sigma_k^P(G_k) \neq \Delta_k^P(G_k)$, $\mathrm{PH}(H_k) = \Delta_k^P(H_k) \neq \Theta_k^P(H_k)$, and that $\mathrm{PH}(I_k) = \Theta_k^P(I_k) \neq \Sigma_{k-1}^P(I_k)$. These corollaries answer open questions posed by Ko [Ko89].

2 Preliminaries

The classes of the relativized polynomial-time hierarchy PH(A) are $\{\Sigma_k^P(A), \Pi_k^P(A), \Delta_k^P(A), \Theta_k^P(A) \mid k \geq 0\}$, where

$$\Sigma_0^P(A) = \Pi_0^P(A) = \Delta_0^P(A) = \Theta_0^P(A) = \mathrm{P}(A), \text{ and for } k \geq 0,$$

$$\Sigma_{k+1}^P(A) = \mathrm{NP}(\Sigma_k^P)(A),$$

$$\Pi_{k+1}^P(A) = \text{co-}\Sigma_{k+1}^P(\mathrm{A}),$$

$$\Delta_{k+1}^P(A) = \mathrm{P}(\Sigma_k^P(A)),$$

$$\Theta_{k+1}^P(A) = \Theta(\Sigma_k^P(A)).$$

Note that the Θ operator is defined as follows: for any class $\mathcal{C}$, a set $L \in \Theta(\mathcal{C})$ if and only if there is a deterministic polynomial time oracle machine M and a set $B \in \mathcal{C}$ such that $L = L(M, B)$ and, on any input x, M makes at most $O(\log |x|)$ queries to B. The classes of polynomial-time hierarchy is defined as PH($\emptyset$).

The extended low hierarchy was first defined by Balcázar, Book, and Schöning. Refinements of the extended low hierarchy based on Δ and Θ-levels of the polynomial-time hierarchy were introduced by Allender and Hemachandra and Long and Sheu, respectively. Proposition 2.2 shows the basic relationships between the various hierarchies.

Definition 2.1 *[BBS86, AH91, LS91]*

1. *For each* $k \geq 1$, $EL_k^{P,\Sigma} = \{A \mid \Sigma_k^P(A) \subseteq \Sigma_{k-1}^P(A \oplus \mathrm{SAT})\}$.
2. *For each* $k \geq 2$, $EL_k^{P,\Delta} = \{A \mid \Delta_k^P(A) \subseteq \Delta_{k-1}^P(A \oplus \mathrm{SAT})\}$.
3. *For each* $k \geq 2$, $EL_k^{P,\Theta} = \{A \mid \Theta_k^P(A) \subseteq \Theta_{k-1}^P(A \oplus \mathrm{SAT})\}$.

Proposition 2.2 *[AH91, LS91] For each* $k \geq 1$*:*

1. $EL_k^{P,\Sigma} \subseteq EL_{k+1}^{P,\Theta} \subseteq EL_{k+1}^{P,\Delta} \subseteq EL_{k+1}^{P,\Sigma}$.
2. $EL_1^{P,\Sigma} = EL_2^{P,\Theta} = EL_2^{P,\Delta}$.

Observe the following facts:

(1) For all sets A and for all $k \geq 1$, $\Sigma_k^P(A) \subseteq \Sigma_k^P(A \oplus \mathrm{SAT}) \subseteq \Sigma_{k+1}^P(A)$.

(2) For all sets A and for all $k \geq 1$, $\Delta_k^P(A) \subseteq \Delta_k^P(A \oplus \mathrm{SAT}) \subseteq \Delta_{k+1}^P(A)$.

(3) For all sets A and for all $k \geq 2$, $\Theta_k^P(A) \subseteq \Theta_k^P(A \oplus \mathrm{SAT}) \subseteq \Theta_{k+1}^P(A)$.

It is easy to obtain the following lemma from the above facts.

Lemma 2.3

1. $\forall k \geq 0, \forall A[\Sigma_{k+1}^P(A) \subseteq \Sigma_k^P(A) \Rightarrow A \in EL_{k+1}^{P,\Sigma}]$.
2. $\forall k \geq 1, \forall A[\Delta_{k+1}^P(A) \subseteq \Delta_k^P(A) \Rightarrow A \in EL_{k+1}^{P,\Delta}]$.
3. $\forall k \geq 1, \forall A[\Theta_{k+1}^P(A) \subseteq \Theta_k^P(A) \Rightarrow A \in EL_{k+1}^{P,\Theta}]$.

Now we give some background on circuits and their restrictions. A circuit is a rooted tree, where each non-leaf node is associated with a gate. We use four types of gates: AND, OR, MaxOdd, and XOR in this paper. For a MaxOdd gate, let its input lines from left to right be $b_0, b_1, \ldots, b_n$. A MaxOdd gate outputs 1 if and only if i is odd, where b_i is the rightmost input line with input 1. An XOR gate outputs 1 if and only if the number of its 1-inputs is odd. The fanins of all the gates are unlimited. Each leaf of a circuit is associated with a constant 0, a constant 1, a variable x, or a negated variable $\overline{x}$. The *size* of a circuit C is defined to be the number of gates in C. Given a circuit C with only AND and OR gates, its dual circuit $\hat{C}$ is defined inductively as follows: the dual of a constant or a variable is its negation, and the dual of a circuit C which is an OR (AND) of n children $C_i, 1 \leq i \leq n$, is the AND (OR) of $\hat{C}_i$, $1 \leq i \leq n$.

A circuit computes a function on its variables in the obvious way. Each variable is represented by v_z for some string $z \in \Sigma^*$. Let V be the set of variables in a circuit C. A *restriction* ρ of a circuit C is a mapping from V to $\{0, 1, *\}$. Let $C\lceil_\rho$ denote the circuit C' obtained from C by replacing each variable x with $\rho(x) = 0$ by 0 and each y with $\rho(y) = 1$ by 1. If a restriction ρ maps no variables to $*$, we say ρ is a complete assignment for the variables of C. Also, we say a restriction ρ *completely determines* C if $C\lceil\rho$ computes a constant function 0 or 1. For any set $A \subseteq \Sigma^*$, the natural assignment ρ_A on all variables v_z, $z \in \Sigma^*$ is: $\rho_A(v_z) = 1$ if $z \in A$ and $\rho_A(v_z) = 0$ if $z \notin A$. Random restrictions of circuits were first introduced by Furst, Saxe and Sipser [MFS81]. We use two probability spaces of restrictions $R^+_{q,B}$ and $R^-_{q,B}$ from [Has87], as well as the restriction $g(\rho)$, where $\rho \in R^+_{q,B}$ or $R^-_{q,B}$.

Circuits can be used to describe the computation of oracle machines on input strings. The following proposition is taken from Ko [Ko89].

Proposition 2.4 *[Ko89] Let M be a deterministic oracle machine with runtime $\leq p(n)$, where p is a polynomial. Then, for each x, there is a depth-2 circuit $C = C_{M,x}$ satisfying the following properties:*

1. *C is an* OR *of* AND*s,*
2. *the top fanin of C is $\leq 2^{p(|x|)}$ and the bottom fanin of C is $\leq p(|x|)$, and*
3. *for any set A, $C\lceil_{\rho_A} = 1$ if and only if $M^A(x)$ accepts.*

Intuitively, each AND gate of C represents one of the accepting computation paths of M on input x, where the variables of the AND gate represent the query strings and answers of that path. It is not difficult to see that the above proposition also holds if we require that the circuit C be an AND of ORs.

We call a depth-$(k+1)$ circuit a Σ_k-circuit if it has alternating OR and AND gates, starting with a top OR gate. A $\Sigma_k(m)$-circuit is a Σ_k-circuit such that (1) the number of gates at each level 1 to level $k-1$ is bounded by 2^m, and (2) fanins of level $k+1$ are $\leq m$. A depth-$(k+1)$ circuit is a Π_k-circuit ($\Pi_k(m)$-circuit) if its dual circuit is a Σ_k-circuit ($\Sigma_k(m)$-circuit). Now we can use circuits to represent the predicate $x \in L$, where $L \in \Sigma_k^P(A)$ for some A.

Proposition 2.5 *Let $k \geq 1$. For any set A and any language $L \in \Sigma_k^P(A)$, there is a polynomial q such that for every x, there exists a $\Sigma_k(q(|x|))$-circuit $C_{L,x}$ such that $C_{L,x}\lceil_{\rho_A} = 1$ if and only if $x \in L$ Also, for each variable v_z ($\overline{v_z}$) in $C_{L,x}$, $|z| \leq q(|x|)$.*

Ko [Ko89] introduced a family of functions f_k^n, $n \geq 1, k \geq 1$, computed by special types of circuits. We will use these same functions as well as introducing similar functions that will be used in our separation results in the next section.

Definition 2.6 *[Ko89] For each $n \geq 1$ and $k \geq 1$, a C_k^n circuit is a depth-k circuit such that*

1. *C_k^n has alternating OR and AND gates, with a top OR gate,*
2. *all the fanins of C_k^n are exactly n, except the bottom fanins which are exactly $\sqrt{n}$, and*
3. *each leaf of C_k^n is a unique positive variable.*

We let f_k^n denote the function computed by C_k^n.

Definition 2.7

1. *For each $n \geq 1$ and $k \geq 1$, an H_k^n circuit is a depth-k circuit with a top* MaxOdd *gate having fanin exactly n such that each subcircuit of the top* MaxOdd *gate is a C_{k-1}^n circuit computing a f_{k-1}^n function and such that all the variables in H_k^n are unique. We let h_k^n denote the function computed by H_k^n.*
2. *For each $n \geq 1$ and $k \geq 1$, a D_k^n circuit is a depth-k circuit with a top* XOR *gate having fanin exactly 2 such that each subcircuit of the top* XOR *gate is a C_{k-1}^n circuit computing a f_{k-1}^n function and such that all the variables in D_k^n are unique. We let l_k^n denote the function computed by D_k^n.*

Before ending this section, we state some important lemmas, including the switching lemma on constant-depth circuits from Håstad [Has87].

Lemma 2.8 (Switching Lemma) *[Has87] Let G be an* AND *of* ORs *with bottom fanin $\leq t$ and let $\mathcal{B} = \{B_j\}$ be a partition of the variables in G. Then, for a random restriction ρ from $R_{q,\mathcal{B}}^+$, $Pr(G\lceil_{\rho g(\rho)}$ is not equivalent to a circuit of* OR *of* ANDs *with bottom fanin $\leq s$) is bounded by α^s, where $\alpha < 6qt$. The above probability also holds with $R_{q,\mathcal{B}}^+$ replaced by $R_{q,\mathcal{B}}^-$, or with G being an* OR *of* ANDs *circuit to be converted to an* AND *of* ORs *circuits.*

The importance of the switching lemma is that a depth-$(k+1)$, $k \geq 2$, alternating OR-AND gate circuit with small bottom fanins can be converted to an equivalent depth-k alternating OR-AND gate circuit with small bottom fanins. This "shrinking" effect is done by converting the bottom 2-level OR-of-AND (AND-of-OR) subcircuits to AND-of-OR (OR-of-AND) subcircuits and then collapse the AND (OR) gates at level k and $k-1$ of the circuit. Our next lemma states this shrinking effect.

Lemma 2.9 *For each $k \geq 2$, let $\{C_i\}_{i=1}^n$ be a collection of $\Sigma_k(r)$-circuits and $\Pi_k(r)$-circuits. Let V be the set of the variables in $\{C_i\}_{i=1}^n$ and let $\mathcal{B} = \{B_j\}$ be a partition of the variables in V. q is a number between 0 and 1 and ρ is a random restriction from $R_{q,\mathcal{B}}^+$ or $R_{q,\mathcal{B}}^-$. The probability that every $\Sigma_k(r)$-circuit and every $\Pi_k(r)$-circuit, after applying the random restriction $\rho g(\rho)$, can be converted to a $\Sigma_{k-1}(r)$-circuit and a $\Pi_{k-1}(r)$-circuit is greater than 2/3, if $n \leq 1/3(2\alpha)^r$ where $\alpha < 6qr$.*

Lemmas 2.10 and 2.11 show the effects of the random restriction $\rho \in R_{q,\mathcal{B}}^+$ or $R_{q,\mathcal{B}}^-$ on the circuits that compute f_k^n, h_k^n, and l_k^n. Lemma 2.10 shows that the random restriction $\rho \in R_{q,\mathcal{B}}^+$ (or $R_{q,\mathcal{B}}^-$, depending on whether k is odd or even) transforms a circuit computing f_k^n, (h_{k+1}^n, or l_{k+1}^n), to a circuit computing a function very close to f_{k-1}^n (h_k^n, or l_k^n). Lemma 2.11 is a stronger form of Lemma 2.10.

Lemma 2.10 *For each $k \geq 2$, there exists an integer n_k such that the following hold for all $n \geq n_k$: Let C be a C_k^n circuit computing an f_k^n function (an H_{k+1}^n circuit computing an h_{k+1}^n function , or a D_{k+1}^n circuit computing an l_{k+1}^n function), let $q = n^{-1/3}$, and let ρ be a random restriction from $R_{q,\mathcal{B}}^+$ if k is even or from $R_{q,\mathcal{B}}^-$ if k is odd. The probability that $C\lceil_{\rho g(\rho)}$ contains a subcircuit computing an f_{k-1}^n function (an h_k^n function, or an l_k^n function, respectively) is greater than 2/3.*

Lemma 2.11 *For each $k \geq 2$, let $\{C_i\}_{i=1}^t$ be t C_k^n circuits, each computing a function f_k^n, with pairwise disjoint variables. Let $q = n^{-1/3}$. If $t \leq 2^{\delta n^{1/6}}$ with $\delta < 1$, then for sufficiently large n, there exists a restriction ρ from $R_{q,\mathcal{B}}^+$ if k is even, or from $R_{q,\mathcal{B}}^-$ if k is odd, such that every $C_i\lceil_{\rho g(\rho)}$, $1 \leq i \leq t$, contains a subcircuit that computes an f_{k-1}^n function. This is also holds H_{k+1}^n and D_{k+1}^n circuits and h_{k+1}^n and l_{k+1}^n functions.*

3 Separation results for relativized polynomial-time hierarchies

In this section, we will separate the Σ, Δ, and Θ-classes of relativized polynomial-time hierarchies. The proof will follow circuit lower bound techniques from [Has87] and [Ko89]. The idea of the proofs is to define a family of circuits for Σ-classes (Δ, and Θ-classes, respectively) of the relativized polynomial-time hierarchy. To separate class $\mathcal{C}_1$ from class $\mathcal{C}_2$, for example, we will find a function $\mathcal{F}$ that is computable in $\mathcal{C}_1$ and is not computable by the family of circuits for $\mathcal{C}_2$. This is done by showing that, in order to compute the specific function $\mathcal{F}$, the size of the circuits for $\mathcal{C}_2$ must exceed some double exponential bound on the fanins of the circuit for the function $\mathcal{F}$. To separate Σ_k and Δ_k-classes, we will let the function $\mathcal{F}$ be an f_k^n function for some $n \geq 1$, to separate Δ_k and Θ_k-classes, we will let the function $\mathcal{F}$ be an h_k^n function for some $n \geq 1$, and finally to separate Θ_k-classes and Σ_{k-1}-classes, we will let the function $\mathcal{F}$ be an l_k^n function for some $n \geq 1$.

We now prove the separation of Σ and Δ-classes of the relativized polynomial-time hierarchy. Separations of Δ, Θ-classes and Θ, Σ-classes are done in a similar manner. Our goal is to find, for each $k \geq 2$, a set $L_k^A \in \Sigma_k^P(A)$ such that $L_k^A \notin \Delta_k^P(A)$. By dovetailing the diagonalization for each $k > 0$, the oracle set A can be constructed so that $\Delta_k^P(A) \subset \Sigma_k^P(A)$ for all $k > 0$.

In order to use the circuit lower bound techniques, we need to define circuits for $\Delta_k^P(A)$. Consider a deterministic polynomial time bounded oracle machine M using oracle set B. By Proposition 2.4, there is a two level OR of ANDs circuit $C_{M,x}$ such that $x \in L(M,B)$ if and only if $C_{M,x}\lceil_{\rho_B}= 1$. Each AND gate of $C_{M,x}$ represents an accepting computation path of M on x, relative to some oracle. The fanins (variables) of each AND gate are the query strings in the corresponding computation path. Now if we replace B by $K^{k-1}(A)$, the complete set for $\Sigma_{k-1}^P(A)$, then the language accepted by M is in $\Delta_k^P(A)$. As stated in Section 2, there is a polynomial q and a $\Sigma_{k-1}(q(|z|))$-circuit $C_{K^{k-1},z}$ such that $z \in K^{k-1}(A)$ if and only if $C_{K^{k-1},z}\lceil_{\rho_A}= 1$. We then replace each positive variable v_z of $C_{M,x}$ by $C_{K^{k-1},z}$ if z is of the correct syntactic form, or by 0 if z is not of the correct syntactic form. For each negative variable v_z of $C_{M,x}$, we replace it by the dual circuit of $C_{K^{k-1},z}$ if z is of the correct syntactic form, or by 1 if z is not of the correct syntactic form. The entire circuit consists of $k+2$ levels. Now we formally define, for $k \geq 2$, $\Delta_k(m)$-circuits.

Definition 3.1 *For $k \geq 2$, C is a $\Delta_k(m)$-circuit if and only if*

1. *C has $k+2$ levels, and the top gate is an* OR *with fanin $\leq 2^m$,*
2. *the gates on second level are* AND *gates with fanins $\leq m$,*
3. *each* AND *gate G_j at the second level has a distinct label of m bits, $b_1 \cdots b_m$, and if $b_i = 1$ then the ith subcircuit from the left of G_j is a $\Sigma_{k-1}(m)$-circuit, and if $b_i = 0$ then the ith subcircuit from the left of G_j is a $\Pi_{k-1}(m)$-circuit,*
4. *for any two* AND *gates G_j and G_n, labeled l_j and l_n, at the second level, if the first r bits, $r < m$, of labels l_j and l_n are the same, then the first r subcircuits from the left of G_j and G_n are the same, and*
5. *for any two* AND *gates G_j and G_n, labeled l_j and l_n, at the second level, if the first bit from the left of l_j and l_n, where l_j and l_n differ, is b_r, $r \leq m$, then the rth subcircuits from the left of G_j and G_n are dual circuits.*

The following lemma states that for all $k \geq 2$ and for any arbitrary set A, if $L \in \Delta_k^P(A)$ then there is a Δ_k-circuit C such that $x \in L \Leftrightarrow C\lceil_{\rho_A}= 1$.

Lemma 3.2 *Let $k \geq 2$. For every set $L \in \Delta_k^P(A)$, there is a polynomial q such that for every input string x, there is a corresponding $\Delta_k(q(|x|))$-circuit $C_{L,x}$ such that $x \in L \Leftrightarrow C_{L,x}\lceil_{\rho_A}= 1$.*

Our goal is, for every $k \geq 2$, to find a set A such that $\Sigma_k^P(A)$ properly contains $\Delta_k^P(A)$. First, for fixed $k \geq 1$, define

$$L_k^A = \{x0^{kn} \mid |x| = 2n, (\exists y_1, |y_1| = n)(\forall y_2, |y_2| = n) \cdots (Q_k y_k, |y_k| = n)\\ 0xy_1y_2 \ldots y_k \in A\},$$

where Q_k is $\exists$ if k is odd and Q_k is $\forall$ if k is even. It is clear that $L_A \in \Sigma_k^P(A)$.

Let $M_1, M_2, \cdots$ be an effective enumeration of deterministic polynomial-time bounded oracle machines with running time bounded by polynomials $p_1, p_2, \cdots$, respectively. The construction of set A is by stages. Let $t(n)$ denote the maximum length of strings in A or $\overline{A}$ decided by stage n. Let $D(n)$, the diagonalization area for stage n, be the set $\{x \in \Sigma^* \mid t(n-1)+1 \leq |x| \leq t(n)\}$. At stage n of the construction, we will find a witness 0^m, m is a multiple of $k+2$, such that, $0^m \in L_k^A \Leftrightarrow 0^m \notin L(M_n, K^{k-1}(A))$; that is, at each stage $n > 0$, we want to satisfy the following requirement R_n:

$$R_n : (\exists x = 0^m)(\exists B \subseteq D(n))[0^m \in L_k^B \Leftrightarrow 0^m \notin L(M_n, K^{k-1}(A(n-1) \cup B))],$$

where $A(n-1)$ denotes the strings in A decided prior to stage n. Let C_0 be the obvious k-level circuit for the predicate $0^m \in L_k^B$. C_0 contains a subcircuit that computes an $f_k^{2^{m/(k+2)}}$ function. By Lemma 3.2, there exist a polynomial q and a $\Delta_k(q(m))$-circuit C_n that corresponds to the computation of the machine M_n on input 0^m with oracle $K^{k-1}(A)$. Since the length of query strings to A in the circuit C is bounded by $q(m)$, we let $t(n) = q(m)$. The requirement R_n can then be rewritten as

$$R'_n : (\exists x = 0^m)(\exists B \subseteq D(n))[C_0\lceil_{\rho_B} = 1 \Leftrightarrow C_n\lceil_{\rho_{A'}} = 0],$$

where $A' = A(n-1) \cup B$. Now let $s_k(v)$ be the minimum r such that a $\Delta_k(r)$-circuit can compute an f_k^v function. That requirement R'_n can be satisfied follows from the following lower bound on $s_k(v)$.

Theorem 3.3 *For all $k \geq 2$, for sufficiently large v, $s_k(v) \geq \delta v^{1/5}$, where $\delta = 1/48$.*

Sketch of Proof: The proof is by induction on k. For the base case of $k = 2$, let C_2^v be a circuit computing f_2^v; that is, C_2^v is an OR of v AND's, each having fanin $= \sqrt{v}$. Let C be a $\Delta_2(r)$-circuit, with $r < \delta v^{1/5}$. The proof of the existence of a restriction ρ that $C_2^v\lceil_\rho \neq C\lceil_\rho$ is done by traditional counting argument. Consider C, a $\Delta_2(r)$-circuit. Recall that, for each AND gate g_l at second level of C, g_l is labeled with r bits, $b_{l_1} b_{l_2} \ldots b_{l_r}$, and if $b_{l_i} = 1$ then the ith subcircuit from the left of g_l is a $\Sigma_1(r)$-circuit, and if $b_{l_i} = 0$ then the ith subcircuit from the left of g_l is a $\Pi_1(r)$-circuit. Also, for any other AND gate $g_{l'}$ at the second level with label $b_{l'_1} b_{l'_2} \ldots b_{l'_r}$, where $b_{l_j} = b_{l'_j}$ for $1 \leq j < i$ and $b_{l_i} \neq b_{l'_i}$, the ith-subcircuits of g_l and g'_l are dual circuits. Thus, if we find a restriction ρ that makes the ith subcircuit of g_l output 1, then the ith subcircuit of $g_{l'}$ will output 0. Therefore, the entire subcircuit g'_l at level 2 of C can be replaced by a constant 0. By making assignments to at most $r^2 < (\delta v^{1/5})^2 < v$ variables, we will be able to completely determine the output of C while the output of C_2^v remains undetermined.

Our induction hypothesis is that for some fixed $k' \geq 2$, if C is a $\Delta_{k'}(r)$-circuit with $r < \delta v^{1/5}$, then for sufficiently large v, C does not compute the function $f_{k'}^v$. For the induction step, consider $k = k' + 1 > 2$. Consider a C_k^v circuit that computes an f_k^v function. Let $q = v^{-1/3}$ and let $\mathcal{B} = \{B_j\}$, where B_j is the set of variables of the bottom AND gate H_j of C_k^v. By way of contradiction, assume that there exists a circuit C, that is a $\Delta_k(r)$-circuit with $r < \delta v^{1/5}$, computing the function f_k^v. We will argue that after applying a random restriction ρ from $R^+_{q,B}$, if k is even, or from $R^-_{q,B}$, if k is odd, to C, the resulting circuit $C\lceil_{\rho g(\rho)}$ is a $\Delta_{k-1}(r)$-circuit with $r < \delta v^{1/5}$ that computes the function f_{k-1}^v, which contradicts our induction hypothesis.

To apply the induction hypothesis, we need to verify that (1) with a high probability, $C\lceil_{\rho g(\rho)}$ is equivalent to a $\Delta_{k-1}(r)$-circuit, and (2) with a high probability, $C_k^v\lceil_{\rho g(\rho)}$ contains a subcircuit computing the function f_{k-1}^v. Part (1) is proved by the Lemma 2.9. To apply Lemma 2.9, let us choose $s = t = \delta v^{1/5}$. We have $\alpha < 6qt = 1/8 \cdot v^{-2/15}$. By definition, the total number of $\Sigma_k(r)$-circuits and $\Pi_k(r)$-circuits in a $\Delta_k(r)$-circuit are bounded by $r \cdot 2^r \leq 2^{2r}$. Thus, we only need to show that $2^{2r} \leq 1/(3(2\alpha)^r)$. We have

$$2^{2r} \cdot 3(2\alpha)^r = 3 \cdot (8\alpha)^r \leq 3 \cdot (v^{-2/15})^r \leq 1$$

for sufficiently large v. This proves part (1). Part (2) was proved in Lemma 2.10. This completes the proof of Theorem 3.3. □

To satisfy requirement R'_n, notice that C_n is a $\Delta_k(q(m))$-circuit corresponding to the predicate $0^m \in L(M_n, K^{k-1}(A))$ and that C_0, corresponding to the predicate $0^m \in L_k^A$, contains a subcircuit that computes an $f_k^{2^{m/(k+2)}}$ function. Thus, we need to choose an integer m so large that Theorem 3.3 holds for $s_k(2^{m/(k+2)})$ and such that $q(m) < 1/48 \cdot 2^{m/5(k+2)}$. We now have the desired theorem.

Theorem 3.4 *There exists an oracle A such that for all $k \geq 1$, $\Delta_k^P(A) \subset \Sigma_k^P(A)$.*

Separations of Δ, Θ-classes and Θ, Σ-classes are proved similarly.

Theorem 3.5 *There exist oracles B and C such that for all $k \geq 2$, $\Theta_k^P(B) \subset \Delta_k^P(B)$ and that $\Sigma_{k-1}^P(C) \subset \Theta_k^P(C)$.*

Theorem 3.6 *There exists an oracle D such that for all $k \geq 2$, $\Sigma_{k-1}^P(D) \subset \Theta_k^P(D) \subset \Delta_k^P(D) \subset \Sigma_k^P(D)$.*

4 The extended low hierarchy is an infinite hierarchy

In this section, we will separate $EL_k^{P,\Sigma}$, $EL_k^{P,\Delta}$ and $EL_k^{P,\Theta}$ for each $k \geq 3$ and thus show that the extended low hierarchy contains an infinite number of distinct levels. From Lemma 2.3 and the definition of the extended low hierarchy, we will construct a set A such that the polynomial-time hierarchy relative to A collapses to, for example, level k of the Σ-classes and, at the same time, level k of the Σ-classes is not contained in $\Delta_k^P(A \oplus \mathrm{SAT})$. Thus, A will be in $EL_{k+1}^{P,\Sigma}$ but not in $EL_{k+1}^{P,\Delta}$. The main technical difficulty in the proofs is to avoid potential conflicts between membership of strings in A during the processes of encoding and diagonalization. This is again solved by using circuit lower bound techniques. We start with the separation of $EL_k^{P,\Sigma}$ and $EL_k^{P,\Delta}$.

Theorem 4.1 *For each $k \geq 2$, there exists a recursive set A such that* $\mathrm{PH}(A) = \Sigma_k^P(A) \not\subseteq \Delta_k^P(A \oplus \mathrm{SAT})$.

Sketch of Proof: To collapse $\mathrm{PH}(A)$ to $\Sigma_k^P(A)$, we will "$\Sigma_k^P(A)$-encode" a $\Sigma_{k+1}^P(A)$-complete set into A. At the same time, we separate $\Sigma_k^P(A)$ from $\Delta_k^P(A \oplus \mathrm{SAT})$. For each $k \geq 1$, define the predicate

$$\tau_k(A;x) \equiv (\exists y_1, |y_1| = n) \cdots (Q_k y_k, |y_k| = n)\ 1xy_1y_2\ldots y_k \in A,$$

where $|x| = 2n$. Thus, for each x of length $2n$, the value of $\tau_k(A;x)$ only depends on strings in A of length $(k+2)n+1$. It is easy to see that $\tau_k(A;x)$ is a predicate in $\Sigma_k^P(A)$. In our construction, we will encode the $\Sigma_{k+1}^P(A)$-complete set $K^{k+1}(A)$ using the predicate τ_k. For the separation, we will again use L_k^A from the last section. Also, it is clear that, for any pair of strings x and y, whether $x \in L_k^A$ and $\tau_k(A;y)$ are true depends on the membership in A of different strings.

The construction of A will make sure that for all strings $x \in \Sigma^*$, $x \in K^{k+1}(A) \Leftrightarrow \tau_k(A; x0^{|x|})$. Thus, $\mathrm{PH}(A) = \Sigma_{k+1}^P(A) = \Sigma_k^P(A)$. We will also construct A so that $L_k^A \notin \Delta_k^P(A \oplus \mathrm{SAT})$. The set up of $t(n)$ and $D(n)$ is similar to that in the previous section. Let m be a multiple of $k+2$. At stage n of the construction, we will find an assignment of strings to A and $\overline{A}$ so that $0^m \in L_k^A$ if and only if $0^m \notin L(M_n, K^{k-1}(A \oplus \mathrm{SAT}))$. In stage n, we also "$\Sigma_k^P(A)$-encode" into A strings $x \in K^{k+1}(A)$, $t(n-1)+1 \leq (k+2)\cdot|x|+1 \leq t(n)$, such that $x \in K^{k+1}(A) \Leftrightarrow \tau_k(A;x0^{|x|})$. The encoding is divided into two steps: before and after finding the witness separating $\Sigma_k^P(A)$ and $\Delta_k^P(A \oplus \mathrm{SAT})$. In the first step, we $\Sigma_k^P(A)$-encode into A all the strings $x \in K^{k+1}(A)$, where $t(n-1)+1 \leq (k+2)\cdot|x|+1 < m$. In the second step, we $\Sigma_k^P(A)$-encode into A all the strings $x \in K^{k+1}(A)$, where $m \leq (k+2)\cdot|x|+1 \leq t(n)$.

Let C_x be the circuit corresponding to the predicate $\tau_k(A;x0^{|x|})$ and let C_0 be the circuit corresponding to the predicate $0^m \in L_k^A$. By Lemma 3.2, there is a polynomial q and a corresponding $\Delta_k(q(m))$-circuit C_n such that $C_n\lceil\rho_{A\oplus\mathrm{SAT}} = 1$ if and only if $0^m \in L(M_n, K^{k-1}(A \oplus \mathrm{SAT}))$. We let $t(n) = q(m)$. Note that C_n can be simplified to another $\Delta_k(q(m))$-circuit C_n' by the following replacements: (1) all the variables v_z in the leaves of C_n having the form $z = 1z'$ are replaced by 1 if $z' \in \mathrm{SAT}$ and by 0 if $z' \notin \mathrm{SAT}$, and (2) all the variables v_z of the form $z = 0z'$ are replaced by $v_{z'}$. Thus, for any set A, $C_n\lceil_{\rho_{A\oplus\mathrm{SAT}}} = C_n'\lceil_{\rho_A}$ and $C_n'\lceil_{\rho_A} = 1$ if and only if $x \in L(M_n, K^{k-1}(A \oplus \mathrm{SAT}))$.

The requirement for stage n can be described as the following

$$\begin{aligned} R_n : (\exists B_0, B_1 \subseteq D(n))[&B_0 \cap B_1 = \emptyset,\ C_n'\lceil_{\rho_{B_1,B_0}} \neq *, \text{ and for all } x, \\ &t(n-1) < (k+2)|x|+1 < m, C_x\lceil_{\rho_{B_0,B_1}} = 1 \Leftrightarrow x \in K^{k+1}(A(n-1) \cup B_1), \\ &\text{and } C_0\lceil_{\rho_{B_1,B_0}} = C_x\lceil_{\rho_{B_1,B_0}} = *, \text{ for all } x, m \leq (k+2)|x|+1 \leq t(n)], \end{aligned}$$

where ρ_{B_1,B_0} is the restriction that v_z is 1 if $z \in B_1$ and v_z is 0 if $z \in B_0$. Construction of A at stage n is done by the following steps.

Step 1. For all the strings x, $t(n-1) < (k+2)|x|+1 < m$, assign strings to $A(n)$ and $\overline{A(n)}$ such that $\tau_k(A;x0^{|x|})$ is true if and only if $x \in K^{k+1}(A)$.

Step 2. Replace all the variables in C'_n that are in $A(n)$ by 1 and replace those in $\overline{A(n)}$ by 0. The resulting circuit, let's still call it C'_n, remains a $\Delta_k(q(m))$-circuit.

Step 3. Find a pair of finite sets B_1 and B_0 that satisfy the requirement R_n. Assign variables in B_1 to $A(n)$ and assign variables in B_0 to $\overline{A(n)}$.

Step 4. For all the strings x, $m \leq (k+2)|x|+1 \leq t(n)$, assign remaining free strings to $A(n)$ and $\overline{A(n)}$ such that $\tau_k(A; x0^{|x|})$ is true iff $x \in K^{k+1}(A)$. If there are still remaining free strings of length between $t(n-1)+1$ and $t(n)$, assign them to $\overline{A(n)}$. Go to the next stage.

The key step is Step 3. The following lemma guarantees that the requirement R_n can be satisfied at Step 3.

Lemma 4.2 *For all $k \geq 2$, let $\{C_i\}_{i=1}^t$ be t circuits, each computing a function f_k^v, with pairwise disjoint variables. Let C be a $\Delta_k(r)$-circuit. If $t \leq 2^{\delta v^{1/6}}$ and $r \leq \delta v^{1/5}$, with $\delta = 1/48$, then for sufficiently large v, there exists a restriction ρ such that $C\lceil_\rho \neq *$ and $C_i\lceil_\rho = *$ for all $i = 1, \cdots, t$.*

Sketch of Proof: The proof is by induction on k. For the case of $k = 2$, the proof is similar to the proof of Theorem 3.3, where we found a restriction ρ such that $C\lceil_\rho \neq *$ and $C_k^v\lceil_\rho = *$, where C is an $\Delta_k(r)$-circuit with $r < \delta v^{1/5}$ and C_k^v computes an f_k^v function. The number of strings that we added to the oracle set was at most $r^2 < (\delta v^{1/5})^2 < \sqrt{v}$. Here, every C_i, $1 \leq i \leq t$, is an OR of v AND gates, each of fanin $\sqrt{v}$. Therefore, the same restriction ρ satisfies that $C_i\lceil_\rho = *$, for all i, $1 \leq i \leq t$.

For the induction step, consider for any $k > 2$. We need to verify that (1) with a high probability, $C\lceil_{\rho g(\rho)}$ is equivalent to a $\Delta_{k-1}(r)$-circuit, and (2) with a high probability, $C_i\lceil_{\rho g(\rho)}$ contains a subcircuit computing the function f_{k-1}^v for all $i = 1, \cdots, t$. Part (1) is identical to that in the proof of Theorem 3.3. Part (2) follows from Lemma 2.11. □

It is easy to see that C_0 contains a $C_k^{2^{m/(k+2)}}$ subcircuit computing an $f_k^{2^{m/(k+2)}}$ function. Also, for all x, $m \leq (k+2)|x|+1 \leq t(n)$, C_x contains a subcircuit computing $f_k^{2^{m/(k+2)}}$. There are at most $2^{q(m)}$ many such C_x. To satisfy requirement R_n, we need to choose an integer m so large that $q(m) \leq (1/48)2^{m/6(k+2)}$. Thus, there exists an oracle set A such that $\Sigma_{k+1}^P(A) = \Sigma_k^P(A)$ and $\Sigma_k^P(A) \not\subseteq \Delta_k^P(A \oplus B)$, for each $k \geq 2$. This completes the proof of Theorem 4.1.

Corollary 4.3 *For each $k \geq 2$, $EL_k^{P,\Delta} \subset EL_k^{P,\Sigma}$.*

Corollary 4.4 *For each $k \geq 1$, there exists an oracle set B such that* PSPACE(B) = PH(B) = $\Sigma_k^P(B) \supset \Delta_k^P(B)$.

Proof. Simply replace the complete set $K^{k+1}(B)$ by a complete set for PSPACE(B) and replace SAT by the empty set in the proof of Theorem 4.1. □

The separations of $EL_k^{P,\Delta}$ and $EL_k^{P,\Theta}$ and $EL_k^{P,\Theta}$ and $EL_{k-1}^{P,\Sigma}$ are proved similarly. We have the following theorems and corollaries.

Theorem 4.5 *For each $k \geq 2$, there exist recursive sets C and D such that* $\mathrm{PH}(C) = \Delta_k^P(C) \not\subseteq \Theta_k^P(C \oplus \mathrm{SAT})$ *and that* $\mathrm{PH}(D) = \Theta_k^P(D) \not\subseteq \Sigma_{k-1}^P(D \oplus \mathrm{SAT})$.

Corollary 4.6 *For each $k \geq 2$, $EL_k^{P,\Sigma} \subset EL_{k+1}^{P,\Theta} \subset EL_{k+1}^{P,\Delta} \subset EL_{k+1}^{P,\Sigma}$.*

Corollary 4.7 *For each $k \geq 2$, there exist oracle sets E and F such that* $\mathrm{PSPACE}(E) = \mathrm{PH}(E) = \Delta_k^P(E) \supset \Theta_k^P(E)$ *and that* $\mathrm{PSPACE}(F) = \mathrm{PH}(F) = \Theta_k^P(F) \supset \Sigma_{k-1}^P(F)$.

References

[AH91] E. Allender and L. Hemachandra. Lower bounds for the low hierarchy. *J. ACM*, 1991. to appear.

[BBS86] J. Balcázar, R. Book, and U. Schöning. Sparse sets, lowness, and highness. *SIAM J. Comput.*, 15:739–747, 1986.

[Has87] J. D. Håstad. *Computational limitations for small-depth circuits.* PhD thesis, Massachusetts Institute of Technology, 1987.

[Ko89] K. Ko. Relativized polynomial time hierarchies having exactly k levels. *SIAM J. Comput.*, 18(2):392–408, April 1989.

[LS91] T. Long and M. Sheu. A refinement of the low and high hierarchies. Technical Report OSU-CISRC-2/91-TR6, The Ohio State University, 1991.

[MFS81] M. Furst, J. Saxe, and M. Sipser. Pairty, circuits, and the polynomial-time hierarchy. In *Proc. 22th Annual IEEE Symposium on Foundations of Computer Science*, pages 260–270, 1981.

[Sch83] U. Schöning. A low and a high hierarchy within NP. *J. Comput. System Sci.*, 27:14–28, 1983.

[Yao85] A. Yao. Separating the polynomial-time hierarchy by oracles. In *Proc. 26th IEEE Symp. on Foundations of Computer Science*, pages 1–10, 1985.

Locally Definable Acceptance Types for Polynomial Time Machines

Ulrich Hertrampf
Institut für Informatik
Universität Würzburg
D-8700 Würzburg
Federal Republic of Germany

Abstract. *We introduce m-valued locally definable acceptance types, a new model generalizing the idea of alternating machines and their acceptance behaviour. Roughly, a locally definable acceptance type consists of a set F of functions from $\{0,\ldots,m-1\}^r$ into $\{0,\ldots,m-1\}$, which can appear as labels in a computation tree of a nondeterministic polynomial time machine. The computation tree then uses these functions to evaluate a tree value, and accepts or rejects depending on that value. The case $m = 2$ was (in some different context) investigated by Goldschlager and Parberry [GP86]. In [He91b] a complete classification of the classes (F)P is given, when F consists of only one binary 3-valued function. In the current paper we justify the restriction to the case of one binary function by proving a normal form theorem stating that for every finite acceptance type there exists a finite acceptance type that characterizes the same class, but consists only of one binary function.*

Further we use the normal form theorem to show that the system of characterizable classes is closed under operators like $\exists$, $\forall$, $\bigoplus$, and others. In a similar fashion we show that all levels of boolean hierarchies over characterizable classes are characterizable. As corollaries from these results we obtain characterizations of all levels of the polynomial time hierarchy and the boolean hierarchy over NP*, or more generally Σ_k^p.*

1 Introduction

One of the aims of structural complexity theory is giving closed characterizations for many interesting complexity classes. Especially characterizations via some type of (in most cases nondeterministic) polynomial time machines are very useful for further investigation of such classes.

One possible way of finding such NP-machine characterizations is posing some global condition on the number of accepting paths of such machines. This results in the concept of counting classes, like the ones defined in [BGH90]: a counting class is associated with some polynomial time predicate $Q(x,\alpha)$, and the class consists of all sets, such that there exists a machine with $a(x)$ accepting paths on input x, and $Q(x,a(x))$ is true if and only if x is in the set.

Several other definitions of the term "counting classes" have been given in the literature. For a good overview see [GNW90]. More approaches to characterize complexity classes by polynomial time machines have been developed, like the introduction of oracles, possibly with restricted access (cf. [Wa88]). One common property of these approaches is some global condition on the work of the machines. An external observer has to decide whether the predicate Q is fulfilled, or whether an oracle is really used in a legal way...

In contrast to this, we introduce a local evaluation scheme on the nodes of the computation tree. One can, for instance, think of the nondeterministic steps in the computation as recursive procedure calls. Surprisingly, depending on the given set of acceptance functions, many interesting classes can be characterized that way. Moreover each of these classes can be characterized by only one binary function.

The organization of this paper is as follows: In Section 2 we give the basic definitions, and some first consequences that are used in the later sections. Section 3 shows that all classes characterizable by finite locally definable acceptance types can already be characterized by an acceptance type consisting of only one binary function. In Section 4 we prove several facts which can be exploited to show that many important complexity classes can be characterized by finite acceptance types, like for instance all levels of the polynomial time hierarchy, as well as all levels of the boolean hierarchies over any class that is characterizable itself.

2 Acceptance Types on m values

The usual way to define acceptance of a nondeterministic polynomial time machine is the following: The machine accepts if and only if there exists at least one accepting path. This can be seen as a global condition, but it can be described as a local condition in the computation tree too, in the following manner:

Leaves (i.e. nodes in the computation tree that correspond to final configurations in the computation) take the value 1, if they correspond to accepting configurations, 0 otherwise. Now let inductively from the leaves to the root every node compute its own value as an OR over the values of its successor nodes. That way we assign a value to each node of the computation tree, especially the root node. Let the computation (as a whole) accept, if and only if the root evaluates to 1.

Obviously such an evaluation leads to acceptance if and only if there is at least one accepting path in the computation. Thus we can identify the local OR-computation on computation trees of polynomial depth with the class NP. An obvious generalization of this method, allowing AND and OR-computation in the nodes of a tree leads to the identification of the set $\{\text{AND}, \text{OR}\}$ with the class of sets recognizable by alternating machines in polynomial time, which is proven to be exactly PSPACE in [CKS81].

We further generalize this method in the following definitions:

Definition Let m be an integer, $m \geq 2$. A *function from m-valued logic* is a function from $\{0,\ldots,m-1\}^r$ into $\{0,\ldots,m-1\}$ for some number r (the arity of the function).

Definition An *m-valued locally definable acceptance type* is a set F of functions from m-valued logic. Every locally definable acceptance type F corresponds to a complexity class, denoted by (F)P in the following way: A set belongs to (F)P if there is a polynomial time nondeterministic machine M, and an evaluation scheme for M, fulfilling the following conditions:

- The leaves of the computation tree are evaluated to a constant (the value depending on the state of the machine),
- in every node with only one successor the value is preserved,

- in every node with more than one successor, say r successors, a function from F of arity r is evaluated (the choice of the function depending on the state of the machine) with the variables replaced by the values of the successor nodes (we assume a given order on the successor nodes, for example let the program of the underlying Turing machine be written down sequentially and say, the successor according to the first applicable rule is the leftmost successor, and so on),

- the input is accepted, if and only if the root evaluates to 1.

It is easy to see that for all locally definable acceptance types F the relation $(F)\mathrm{P} \subseteq \mathrm{PSPACE}$ holds. On the other hand we saw that $(\{\mathrm{AND},\mathrm{OR}\})\mathrm{P} = \mathrm{PSPACE}$, and that $(\{\mathrm{OR}\})\mathrm{P} = \mathrm{NP}$. Further it should be clear how to show that $(\{\mathrm{AND}\})\mathrm{P} = \mathrm{coNP}$.

Now we introduce the closure of an m-valued locally definable acceptance type F, denoted by Closure(F). An obvious consequence of this, stated in the subsequent lemma, is the equality of $(F)\mathrm{P}$ with $(\mathrm{Closure}(F))\mathrm{P}$, which can easily be proved by induction.

Definition The closure of an m-valued locally definable acceptance type is defined inductively:

- The constant functions $0,\ldots,m-1$, the identity function, and the functions from F belong to the closure of F.

- If a function f belongs to the closure of F, then all functions f_π obtained from f by permuting the variables belong to the closure of f.

- If the r-ary function f and the functions $g_1,\ldots,g_r$ belong to the closure of F, then also the function obtained by substituting $g_1,\ldots,g_r$ into f belongs to the closure of F.

- If f belongs to the closure of F, then all functions obtained from f by identifying some variables or introducing dummy variables belong to the closure of F.

Lemma 1 *Let F be an m-valued locally definable acceptance type. Then*

$$(F)\mathrm{P} = (\mathrm{Closure}(F))\mathrm{P}.$$

3 A Normal Form

In this section we want to prove that for all finite locally definable acceptance types F there exists one binary function g such that the classes $(F)\mathrm{P}$ and $(\{g\})\mathrm{P}$ coincide. We do this in two steps. First we show that one function (possibly with high arity) can replace a finite number of functions, and in the second step we will prove that a k-ary function can be replaced by a binary function (possibly increasing the number of values).

Lemma 2 *Let F be a finite locally definable acceptance type. Then there exists one function g such that $(F)\mathrm{P} = (\{g\})\mathrm{P}$.*

Proof: By Lemma 1 we can assume that all functions of F have the same number of variables (introducing dummy variables is possible inside the closure of F). So let all functions in F be from $\{0,\ldots,m-1\}^k$ into $\{0,\ldots,m-1\}$. Further let the functions in F be enumerated, say $f_0,\ldots,f_{r-1}$. Define g as follows:

$g : \{0,\ldots,m+r-1\}^{k+1} \mapsto \{0,\ldots,m+r-1\}$
$g(a_1,\ldots,a_k,j) = f_{j-m}(a_1,\ldots,a_k)$ if $m \le j < m+r$ and $a_i < m$ for all i
$g(x_1,\ldots,x_{k+1}) = 0$ else

Clearly every (F)P-computation can be simulated by a $(\{g\})$P-computation. For the converse note that the output value of g is always less than m. Thus an (F)P-machine is always able to check whether a given g-node of a simulated $(\{g\})$P-computation has as rightmost input a value indicating one of the functions $f_0,\ldots,f_{r-1}$, and whether all other inputs are either outputs of an evaluation of function g themselves, or are constant values less than m. If one of these conditions is violated, the simulation goes to a final state with value 0 and this will certainly be the right value for a simulation, since the $(\{g\})$P-machine at this node of the computation tree will evaluate to 0 too. On the other hand, if indeed the rightmost input for the simulated function is a constant value greater than $m-1$, and all other inputs are either constant values less than m or values evaluated by function g themselves, then the (F)P-machine can simulate that step by going to a configuration corresponding to the function indicated in the rightmost input value, applied to the other input values.

This proves the claim of the lemma. ∎

Lemma 3 *Let f be a function from $\{0,\ldots,m-1\}^k$ into $\{0,\ldots,m-1\}$. Then there exists one binary function g such that $(\{f\})$P $= (\{g\}$P.*

Proof: For the following definition of function g we will use the values $0,\ldots,m-1$, as well as vectors of dimension 2 to k of these values, and additionally two special values named * and #. These are $m' = 2 + \sum_{i=1}^{k} m^i$ different values that can eventually be renamed to numbers $0,\ldots,m'-1$ to fit the general definition of our functions.

Define function g as follows (let a_i denote values from $0,\ldots,m-1$, and $[\ldots]$ denote vectors):

$g(a_1,a_2) = [a_1,a_2]$
$g([a_1,\ldots,a_r],a_{r+1}) = [a_1,\ldots,a_r,a_{r+1}]$ $(r < k)$
$g([a_1,\ldots,a_k],*) = f(a_1,\ldots,a_k)$
$g(y,*) = 0$ if y is not a k-dimensional vector
$g(y,z) = \#$ in all other cases.

We first claim that $(\{f\})\text{P} \subseteq (\{g\})\text{P}$. This can be seen by the fact that one step of an $(\{f\})$P-machine like in Figure 3.1 can be simulated by a $(\{g\})$P-machine with a subtree like the one shown in Figure 3.2.

To show conversely that $(\{g\})\text{P} \subseteq (\{f\})\text{P}$ we use Figures 3.1 and 3.2 again. An $(\{f\})$P-machine can check in polynomial time deterministically two steps in advance whether a given configuration of a $(\{g\})$P-machine has a subtree of the form of Figure 3.2. If so, the $(\{f\})$P-machine produces a subtree like the one of Figure 3.1, otherwise it produces a final configuration that evaluates to 0.

Why do we have to check two steps in advance? This is because a wrong form might lead to value # instead of 0 (as can be seen from the value table above). But if it is already ensured that the right successor of the current node evaluates to *, then we really know (!) that the current node with an ill-formed subtree will evaluate to 0.

In fact the simulation can always check two steps in advance, only in the beginning a short initial phase must be performed. But if the root of the whole tree evaluates to #, it certainly does not matter that the simulation evaluates to 0, since both results mean rejection of the input. This proves the lemma. ∎

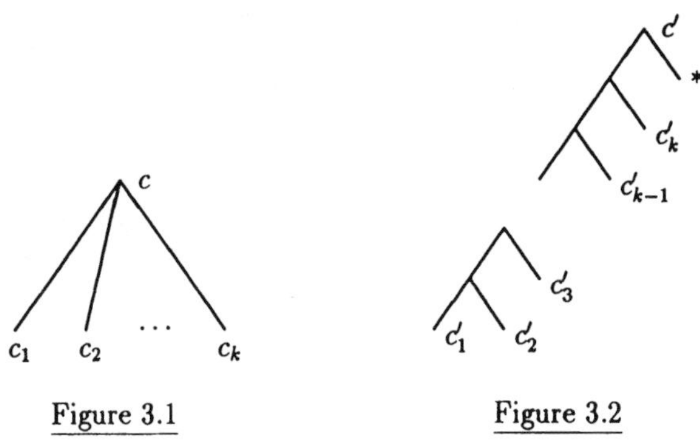

Figure 3.1 Figure 3.2

Combining Lemmas 2 and 3 we obtain our desired normal form theorem:

Theorem 4 *Let F be a finite locally definable acceptance type. Then there exists one binary function g such that* $(F)\mathrm{P} = (\{g\})\mathrm{P}$.

4 Characterization of Important Complexity Classes

In order to show that many very important complexity classes can be characterized by finite locally definable acceptance types, we first want to prove several results that can be viewed as closure properties of the system of characterizable classes. Then the main results can be deduced directly as corollaries from these closure properties.

We start with the closure under complementation:

Lemma 5 *Let F be an m-valued locally definable acceptance type, and let $\mathcal{C} = (F)\mathrm{P}$. Then there exists a constant m' and an m'-valued locally definable acceptance type F' such that co-$\mathcal{C}$ = $(F')\mathrm{P}$.*

Proof: According to theorem 4 let f be a k-valued binary function (some constant k) such that $(F)\mathrm{P} = (\{f\})\mathrm{P}$. Thus f is defined on $\{0,\ldots,k-1\}^2$ with values in $\{0,\ldots,k-1\}$. Extend the domain to $\{0,\ldots,k+2\}^2$ by setting $f(x,y) = 0$ if $x \geq k$ or $y \geq k$. Define a second function $g : \{0,\ldots,k+2\}^2 \mapsto \{0,\ldots,k+2\}$ by

$$g(x,y) = k+1, \quad \text{if } x < k,\ x \neq 1,\ y = k$$
$$g(x,y) = k+2 \quad \text{otherwise.}$$

For the moment assume that the definition of $(\{f,g\})\mathrm{P}$ is by acceptance on root value $k+1$ (instead of value 1) and rejection otherwise. We claim that with this acceptance behaviour $(\{f,g\})\mathrm{P}$ = co-$\mathcal{C}$. Once the claim is proved it is clear that simply renaming the values gives the desired result.

To prove the claim we first show that co-$\mathcal{C} \subseteq (\{f,g\})\mathrm{P}$: Let $L \in \mathcal{C}$. Let M be a nondeterministic machine accepting L as an $(\{f\})\mathrm{P}$-machine. Design a new machine M' as follows:

> On input x, M' initially is in a state augmented by function g. M' splits into two configurations, the first one starting a complete simulation of M on input x, the second one being a final configuration with constant value k.

Obviously M' accepts if and only if M rejects the input. Thus co-$\mathcal{C} \subseteq (\{f,g\})$P.

Now we have to show conversely that $(\{f,g\})$P $\subseteq$ co-$\mathcal{C}$ or, equivalently, that co-$(\{f,g\})$P $\subseteq$ $(\{f\})$P. Let $L \in (\{f,g\})$P, let M be a nondeterministic machine accepting L as an $(\{f,g\})$P-machine. Design a new machine M' as follows:

> On input x, if the initial configuration of M corresponds to function f then accept. (Note that in this case M will certainly not evaluate its root value to $k+1$, and thus will not accept x.)
>
> If the right successor of the initial configuration of M does not correspond to constant value k then accept. (Again M will certainly not accept, because value k can only be reached as a constant value, but g requires right input k to be able and produce value $k+1$.)
>
> If the left successor of the initial configuration of M does not correspond to function f then accept. (The reason for that is similar to the ones given above.)
>
> Now perform an "f-simulation" of the left subtree.

Here, an "f-simulation" means the following:

> If both successors of an f-node in the computation tree are either f-nodes themselves, or constants less than k, then take a normal nondeterministic step augmented by f. Otherwise take on a final state with constant value 0.

One can easily check that the "f-simulation" in M' yields a simulation of a subtree of M with the same value in the subtree's root. Since the initial part ensures that we only perform an "f-simulation" if the top of the tree looks as shown in Figure 4.1, it can be checked that M' accepts if and only if M rejects, since $g(x,k) = k+1$ if and only if $x \neq 1$, given that $x < k$, which is clear in that case, because x is an output value of function f.

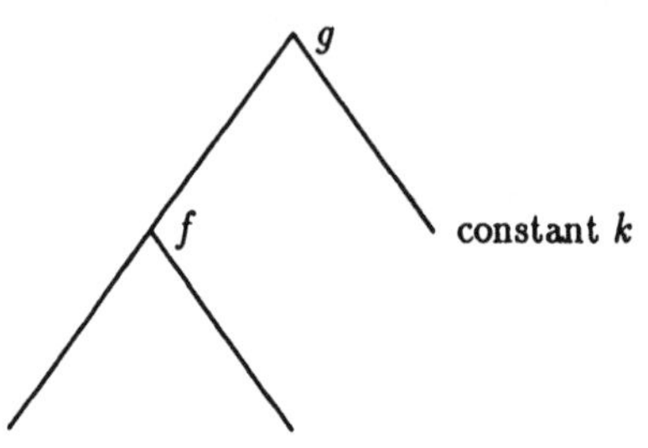

Figure 4.1

That completes the proof of the lemma. ∎

Definition Let $\mathcal{C}$ be a class of languages. Recall that $\exists\mathcal{C}$ is defined the following way:

A set A belongs to $\exists\mathcal{C}$ if and only if there exists a set $B \in \mathcal{C}$ and a polynomial p such that

$$x \in A \iff \exists y : |y| \leq p(|x|) \wedge \langle x,y \rangle \in B.$$

Similar definitions hold for $\forall\mathcal{C}$ and $\oplus\mathcal{C}$, and as a generalization of the latter MOD$_k\mathcal{C}$ ($k \geq 2$). Moreover we define $\mathcal{A}$-$\mathcal{C}$[1] for a finite set $\mathcal{A}$ of nonnegative integers as follows:

A set A belongs to $\mathcal{A}$-$\mathcal{C}$ if and only if there exists a set $B \in \mathcal{C}$ and a polynomial p such that

[1] For the special case $\mathcal{C}$ = P the classes $\mathcal{A}$-P were already defined by Gundermann, Wechsung [GW87], however they called their classes $\mathcal{A}$-NP in contrast to our definition.

$$x \in A \iff \#\{y : |y| \leq p(|x|) \wedge \ <x,y>\in B\} \in \mathcal{A}.$$

Lemma 6 *Let F be a finite m-valued locally definable acceptance type, and let $C = (F)$P. Then there exist finite locally definable acceptance types for all classes $\exists C$, $\forall C$, MOD_kC, and $\mathcal{A}$-C.*

The proof techniques for this lemma are similar to the one of Lemma 5. Refer to [He91a] for a full proof.

In order to capture the intermediate classes of the polynomial time hierarchy, Δ_k^p and Θ_k^p, we need operators like Δ and Θ, which were implicitly defined by Wagner [Wa90] as follows: A set A belongs to ΔC if there exists a set B in C, and a polynomial p such that

- $<x,i+1>\in B \implies <x,i>\in B$ and
- $x \in A \iff \max\{i : i \leq 2^{p(|x|)} \wedge \ <x,i>\in B\}$ is odd.

The definition of Θ is similar, only $i \leq 2^{p(|x|)}$ has to be replaced by $i \leq p(|x|)$. Since for our purposes here, the monotonicity condition is rather unconvenient, we omit it, thus defining new operators Δ' and Θ':

Definition The operators Δ' and Θ' are defined as follows:

A set A belongs to $\Delta' C$ if and only if there exists a set $B \in C$ and a polynomial p such that

$$x \in A \iff \max\{i : i \leq 2^{p(|x|)} \wedge \ <x,i>\in B\} \text{ is odd.}$$

A set A belongs to $\Theta' C$ if and only if there exists a set $B \in C$ and a polynomial p such that

$$x \in A \iff \max\{i : i \leq p(|x|) \wedge \ <x,i>\in B\} \text{ is odd.}$$

It can easily be shown that $\Delta' C = \Delta\exists C$, and $\Theta' C = \Theta\leq^p_{\text{disj}} C$, where the operator $\leq^p_{\text{disj}}$ performs the closure of C under polynomial time disjunctive reductions. In particular we obtain $\Delta'\text{NP} = \Delta\exists\text{NP} = \Delta\text{NP} = \Delta_2^p$ and $\Theta'\text{NP} = \Theta\leq^p_{\text{disj}}\text{NP} = \Theta\text{NP} = \Theta_2^p$, and in general $\Delta'\Sigma_k^p = \Delta_{k+1}^p$ and $\Theta'\Sigma_k^p = \Theta_{k+1}^p$.

Lemma 7 *Let C be characterizable by a finite locally definable acceptance type. Then also $\Delta' C$ and $\Theta' C$ are characterizable by finite locally definable acceptance types.*

We omit the proof here. Refer to [He91a] for details.

Lemma 8 *Let C_1 and C_2 be characterizable by finite locally definable acceptance types. Then also the classes*

$$C_1 \wedge C_2 := \{A \cap B : A \in C_1,\ B \in C_2\}$$
$$C_1 \vee C_2 := \{A \cup B : A \in C_1,\ B \in C_2\}$$

are characterizable by finite locally definable acceptance types.

Proof: We use pairs instead of ordinary numbers in the definition of the new functions, but of course one can finally code such pairs into numbers in a bijective manner. Let $C_1 = (\{f\})$P and $C_2 = (\{g\})$P. Define

$$f'(\,[x,f],\,[y,f]\,) = [\,f(x,y),\,f\,]$$
$$f'(\,[x,z_1],\,[y,z_2]\,) = [\,0,\,f\,] \qquad \text{if } z_1 \neq f \text{ or } z_2 \neq f.$$

Analogously we define g':

$$g'([x,g],[y,g]) = [g(x,y), g]$$
$$g'([x,z_1],[y,z_2]) = [0, g] \quad \text{if } z_1 \neq g \text{ or } z_2 \neq g.$$

Finally let

$$h([1,f],[1,g]) = [1, h]$$
$$h([x,z_1],[y,z_2]) = [0, h] \quad \text{in all other cases.}$$

For the class $(\{f',g',h\})$P we assume acceptance on root value $[1,h]$. Obviously $C_1 \wedge C_2 \subseteq (\{f',g',h\})$P. For the converse note that an $(\{f',g',h\})$P-machine can only accept if the start configuration corresponds to function h with a left subtree whose root corresponds to f' and a right subtree whose root corresponds to g'. Now the claim follows from the fact that an $(\{f\})$P-machine can perform an "f'-simulation" in the sense of the proof of Lemma 5, and similarly a $(\{g\})$P-machine can perform a "g'-simulation". This completes the proof of the claim for $C_1 \wedge C_2$. The claim for $C_1 \vee C_2$ can be deduced by using Lemma 5 and deMorgan's laws. ∎

Combining Lemmas 5 – 8 and applying them to special well known classes leads to the following results:

Theorem 9 *The system of classes characterizable by finite locally definable acceptance types is closed under the unary operations co-, $\exists$, $\forall$, Δ', Θ', MOD_k ($k \geq 2$), $\mathcal{A}$- ($\mathcal{A}$ a finite set of nonnegative integers), and the binary operations $\wedge$, $\vee$.*

Corollary 10 *All levels of the polynomial time hierarchy, consisting of Σ_k^p, Π_k^p, Δ_k^p, Θ_k^p for $k \geq 0$, can be characterized by finite locally definable acceptance types.*

Corollary 11 *All levels of the boolean hierarchies over Σ_k^p, especially all levels of the boolean hierarchy over* NP *can be characterized by finite locally definable acceptance types.*

5 Acknowledgments

Thanks are due to Klaus W. Wagner, Würzburg, and Hans-Jörg Burtschick, Berlin, for several fruitful discussions on the topics of this paper.

References

[BGH90] R. Beigel, J. Gill, U. Hertrampf, *Counting Classes: Thresholds, Parity, Mods, and Fewness*, Proc. 7th Symposium on Theoretical Aspects of Computer Science, LNCS 415 (1990), pp. 49–57.

[CKS81] A.K. Chandra, D.C. Kozen, L.J. Stockmeyer, *Alternation*, J. ACM 28 (1981), pp. 114–133.

[GNW90] T. Gundermann, N.A. Nasser, G. Wechsung, *A Survey on Counting Classes*, 5th Structure in Complexity Theory (IEEE) (1990), pp. 140–153.

[GP86] Leslie M. Goldschlager, Ian Parberry, *On the Construction of Parallel Computers from various Bases of Boolean Functions*, Theoretical Computer Science 43 (1986), pp. 43–58.

[GW87] T. Gundermann, G. Wechsung, *Counting Classes of Finite Acceptance Types*, Computers and Artificial Intelligence 6 (1987), pp. 395–409.

[He91a] Ulrich Hertrampf, *Locally Definable Acceptance Types for Polynomial Time Machines*, Technical Report No. 28, Universität Würzburg, 1991.

[He91b] Ulrich Hertrampf, *Locally Definable Acceptance Types - The Three-Valued Case*, Technical Report No. 29, Universität Würzburg, 1991.

[Wa88] Klaus W. Wagner, *Bounded Query Computations*, Proc. 3rd Structure in Complexity Theory (IEEE) (1988), pp. 260–277.

[Wa90] Klaus W. Wagner, *Bounded Query Classes*, SIAM J. Comput. 19 (1990), pp. 833–846.

The Theory of the Polynomial Many-One Degrees of Recursive Sets is Undecidable

Klaus Ambos-Spies and André Nies
Mathematisches Institut
Universität Heidelberg
Im Neuenheimer Feld 294
D-6900 Heidelberg

The polynomial time bounded versions of many-one (p-m) and Turing (pT) reducibility, introduced by Karp [12] and Cook [8], respectively, are major tools for proving intractability and, more generally, for classifying the complexity of solvable but intractable problems. Ladner [13] started the investigation of the algebraic structure of these reducibility notions and of the degree structures induced by them. The first results on these structures obtained by Ladner and various other authors (see e.g. [1,14,15]) led to the impression that these degree orderings are rather homogeneous structures. Only in the last five years, by using more involved techniques, this impression has been corrected (see e.g. [2,16,17]). For the polynomial Turing reducibility these recent investigations culminated in Shinoda and Slaman's theorem [16] that the theory of the pT-degrees of recursive sets is undecidable (in fact equivalent to true first order arithmetic). Less progress had been made in the investigation of the structure of the p-m-degrees. This is mainly due to the fact that, in contrast to the upper semilattice (u.s.l.) of the pT-degrees, the u.s.l. of p-m-degrees is distributive ([1]). All the recent global results on the theory of the pT-degrees, however, use nondistributive coding schemes, whence these arguments do not carry over to p-m-reducibility. So Shinoda and Slaman [16] raised the question whether the theory of the p-m-degrees of recursive sets is undecidable too.

Here we answer this question affirmatively. We show that the partial ordering $\mathcal{E}^2=(\Sigma_2^0,\subseteq)$ of the Σ_2^0-sets (of the arithmetical hierarchy) under inclusion is elementarily definable with parameters (e.d.p.) in the u.s.l. $\mathcal{R}_{p\text{-}m}$ of the p-m-degrees of recursive sets. By a theorem of Herrmann ([10,11]) this implies that the theory $\mathrm{Th}(\mathcal{R}_{p\text{-}m})$ is undecidable.

A similar reduction scheme has been used recently by the authors and Shore in [5].There the undecidability of the theory of the weak-truth-table degrees of the recursively enumerable sets has been shown by defining the partial ordering of the Σ_3^0-sets in this structure.

The outline of the paper is as follows. In Section 1 we shortly state the model theoretic part needed for the reduction. For more details on this part see [5]. Section 2 contains the algebraic part of the argument. Here we show that for any e.d.p. independent subset A of an upper semi lattice (u.s.l.) P, the partial ordering $(\mathrm{DI}(A),\subseteq)$ of the ideals which are generated by subsets of A and possess exact pairs is e.d.p. in P too provided that the ideal (A) generated by A possesses an exact pair. Moreover, we describe one possible way how to get such an independent set A in distributive upper semi lattices. In Section 3 we show that the recursively presentable subclasses of a recursively presentable class $\mathcal{A}$ are just the Σ_2^0-subclasses of $\mathcal{A}$ (w.r.t. some fixed representation of $\mathcal{A}$). In Section 4 we use this result together with a characterization in [4] of the ideals of p-m-degrees which possess exact pairs to show that for any independent class **A** of p-m-degrees with recursive basis the partial ordering (DI(**A**),$\subseteq$) is isomorphic to the partial ordering (Σ_2^0,$\subseteq$) of the Σ_2^0 sets under inclusion. Finally, in Section 5 we show how we can obtain such a set **A** of p-m-degrees and we conclude from this that the theory Th($\mathcal{R}_{p\text{-}m}$) of the p-m-degrees of recursive sets is undecidable.

The results in Sections 1 to 3 do not refer to p-m-reducibility at all, and the results of Section 4 hold for almost all standard subrecursive reducibilities. Only the final part in Section 5 relies on some specific features of p-m-reducibility. So to obtain undecidability results for other subrecursive reducibility structures, only this part has to be redone. We comment on this further in Section 6.

1. The Reduction Scheme

Rabin has shown that the elementary theory of Boolean pairs is undecidable. This result has been improved by Burris and McKenzie [7] who defined a class $\mathcal{B}$ of Boolean pairs whose elementary theory is hereditarily undecidable. Now any structure in which a class of structures with hereditarily undecidable theory is elementarily definable with parameters (e.d.p.) has a hereditarily undecidable theory again (see [7]). So any theory in which a class of Boolean pairs containing the class $\mathcal{B}$ is elementarily definable with parameters is hereditarily undecidable again. Herrmann [10,11] used this observation to show that, for any number n≥1, the elementary theory of the partial ordering $\mathcal{E}^n$=(Σ_n^0,$\subseteq$) of the Σ_n^0-sets in the arithmetical hierarchy is hereditarily undecidable.

1.1. Theorem (Herrmann). For any n≥1, the elementary theory Th($\mathcal{E}^n$) of the partial ordering $\mathcal{E}^n$=(Σ_n^0,$\subseteq$) of the Σ_n^0-sets under inclusion is hereditarily undecidable.

Again, this undecidability result carries over to structures in which some partial ordering $\mathcal{E}^n$ can be elementarily defined with parameters.

1.2. Corollary. Let **P**=(P;≤) be a partial ordering such that, for some n≥1, the partial ordering $\mathcal{E}^n$ is e.d.p. in **P**. Then the elementary theory Th(**P**) of **P** is (hereditarily) undecidable.

2. Ideals in Upper Semilattices

Throughout this section let **P**=(P;≤,∨,0) be an infinite upper semilattice (u.s.l.) with least element 0, i.e., (P;≤) is a partial ordering in which the supremum a∨b of any two elements a and b of P exists. For a subset A of P let (A) denote the ideal generated by A, i.e.,

$$(A) = \{a \in P : \exists\, n\geq 1\ \exists\, b_1,\ldots,b_n \in A\ (a \leq b_1 \vee \ldots \vee b_n)\}$$

(by convention, (∅)={0}) and let

$$I(A) = \{(B) : B \subseteq A\}$$

be the class of ideals generated by the subsets of A. A subset A of P is called *independent* if

(2.1) $\forall a\in A\ \forall a_1,\dots,a_n\in A-\{a\}\ (a \not\leq a_1\vee\dots\vee a_n)$.

For a countably infinite independent set A the class I(A) is isomorphic to the power set $\mathcal{P}(\omega)$ of the set ω of natural numbers:

2.1. Lemma. Let A be a countably infinite independent subset of P, say $A=\{a_i : i\geq 0\}$, where $a_i\neq a_j$ for $i\neq j$. The function $\Phi: \mathcal{P}(\omega)\rightarrow I(A)$ defined by $\Phi(X) = (\{a_i : i\in X\})$ is an isomorphism from $(\mathcal{P}(\omega);\subseteq)$ onto $(I(A);\subseteq)$.

Proof. Obviously, Φ is onto and preserves the ordering. To show that Φ is one-to-one and that Φ^{-1} preserves the ordering too, it suffices to show that, for any $X,Y\subseteq\omega$ with $\Phi(X)\subseteq\Phi(Y)$, $X\subseteq Y$. For a contradiction assume that $\Phi(X)\subseteq\Phi(Y)$ but $X\not\subseteq Y$, say $i_0\in X-Y$. Then $a_{i_0}\in\Phi(X) = (\{a_i : i\in X\})$ whence, by our assumption that $\Phi(X)\subseteq\Phi(Y)$, $a_{i_0}\in\Phi(Y) = (\{a_i : i\in Y\})$. So there are numbers $j_0,\dots,j_n\in Y$ such that $a_{i_0}\leq a_{j_0}\vee\dots\vee a_{j_n}$. Since, by choice of i_0, $a_{i_0}\notin\{a_{j_0},\dots,a_{j_n}\}$ this contradicts independence of A.

In general ideals are second order objects whence we cannot talk about them in the first order language of $(P;\leq)$. There is an important subclass of ideals, however, which can be elementarily defined with parameters: We call elements a and b of P an *exact pair* for an ideal I of **P** if

$\forall x\in P\ (x\leq a\ \&\ x\leq b \Leftrightarrow x\in I)$.

In this case we write $I=I(a,b)$ and we let

$DI(A) = \{I\in I(A) : I \text{ has an exact pair}\}$

be the class of ideals generated by subsets of A which possess an exact pair. Note that membership in the ideal I(a,b) can be expressed by the first order formula

(2.2) $\phi_\in(x,y,z) \equiv z\leq x\ \&\ z\leq y$,

i.e.

(2.3) $c\in I(a,b) \Leftrightarrow (P;\leq)\models\phi_\in(a,b,c)$.

Similarly, for ideals with exact pairs we can define the inclusion relation in the elementary theory of $(P;\leq)$ as follows. Let

(2.4) $\phi_\leq(x_1,y_1,x_2,y_2) \equiv \forall z\ (\phi_\in(x_1,y_1,z) \Rightarrow \phi_\in(x_2,y_2,z))$.

Then

(2.5) $I(a_1,b_1)\subseteq I(a_2,b_2) \Leftrightarrow (P;\leq)\models\phi_\leq(a_1,b_1,a_2,b_2)$.

The following theorem gives sufficient conditions on independent sets for guaranteeing that $(DI(A),\subseteq)$ is elementarily definable with parameters.

2.2. Theorem. Let A be an independent subset of P such that (A) possesses an exact pair and A is e.d.p. Then $(DI(A),\subseteq)$ is e.d.p.

For the proof of the theorem we need the following lemma.

2.3. Lemma. Let A be an independent subset of P. Then, for any $x\in(A)$, the following are equivalent

(i) $x=0$ or $\exists\, n\geq 1\ \exists\, a_1,\dots,a_n\in A\ (x = a_1\vee\dots\vee a_n)$

(ii) $\forall y{<}x \, \exists a{\in}A \, (a \le x \,\&\, a \not\le y)$

Proof. The implication "(i) ⇒ (ii)" is trivial (and does not require independence of A). For a proof of the nontrivial implication fix $x \in (A)$ for which (ii) holds. Since $x \in (A)$ there are $a_1,...,a_m \in A$ such that $x \le a_1 \vee ... \vee a_m$. So, by independence of A, there is a subset $\{a_{i_1},...,a_{i_n}\}$ of $\{a_1,...,a_m\}$ such that

$\{a : a \in A \,\&\, a \le x\} = \{a_{i_1},...,a_{i_n}\}$.

Now, if $\{a_{i_1},...,a_{i_n}\}$ is empty, then x=0 by (ii). Otherwise, we claim that $x = a_{i_1} \vee ... \vee a_{i_n}$. Obviously $a_{i_1} \vee ... \vee a_{i_n} \le x$. So for a contradiction assume that $a_{i_1} \vee ... \vee a_{i_n} < x$. Then, for $y = a_{i_1} \vee ... \vee a_{i_n}$, $y < x$ but $\{a : a \in A \,\&\, a \le y\} = \{a_{i_1},...,a_{i_n}\} = \{a : a \in A \,\&\, a \le x\}$ contrary to (ii).

Proof of Theorem 2.2. By (2.5) above it suffices to give a formula $\Psi(x,y,z_1,...,z_k)$ s.t. for some parameters $c_1,...,c_k$,

(2.6) $\forall a,b \in P \, (I(a,b) \in DI(A) \Leftrightarrow (P;\le) \models \Psi(a,b,c_1,...,c_k))$.

To describe such a formula Ψ we first note that, for any ideal I, $I \in I(A)$ if and only if

$I \subseteq (A) \,\&\, \forall x \in I\text{-}\{0\} \, \exists n \ge 1 \, \exists a_1,...,a_n \in A \, (x \le a_1 \vee ... \vee a_n \,\&\, a_1 \vee ... \vee a_n \in I)$.

By Lemma 2.3, we can conclude that

(2.7) $I \in I(A) \Leftrightarrow I \subseteq (A) \,\&\, \forall x \in I \, \exists y \in I \, (y \ge x \,\&\, \forall z{<}y \, \exists w \in A \, (w \le y \,\&\, w \not\le z))$

By hypothesis of the theorem there are $u,v \in P$ such that

(2.8) $(A) = I(u,v)$

and there are a first order formula ϕ_A and parameters $w_1, ..., w_k \in P$ s.t., for any $x \in P$,

(2.9) $x \in A \Leftrightarrow (P;\le) \models \phi_A(x,w_1,...,w_k)$.

By replacing (A) by I(u,v), (2.7) yields the following characterization of DI(A):

(2.10) $I(a,b) \in DI(A) \Leftrightarrow$

$$I(a,b) \subseteq I(u,v) \,\&\, \forall x \in I(a,b) \, \exists y \in I(a,b) \, (y \ge x \,\&\, \forall z{<}y \, \exists w \in A \, (w \le y \,\&\, w \not\le z))$$

Using (2.3), (2.5) and (2.9), the right hand side of (2.10) can be converted to the desired formula Ψ.

By Theorem 2.2 and Corollary 1.2, to prove undecidability of the theory of a partial ordering $(P;\le)$ it suffices to find an e.d.p. independent subset A of P s.t. (A) possesses an exact pair and $(DI(A);\subseteq)$ is isomorphic to the partial ordering $(\Sigma^0_n,\subseteq)$ of the Σ^0_n-sets (for some n). For the partial ordering $\mathcal{R}_{p\text{-}m}$ we will show in the next two sections that for certain infinite independent subsets A of $\mathcal{R}_{p\text{-}m}$, namely those with a recursive basis, (A) has an exact pair and $(DI(A);\subseteq) \cong (\Sigma^0_2;\subseteq)$. So it will suffice to find such a set A which will be e.d.p. In the remainder of this section we describe one possible way to define independent sets in a distributive u.s.l.

An upper semi lattice P is *distributive* if

(2.11) $\forall a_1,...,a_n,b \in P \, (b \le a_1 \vee ... \vee a_n \Rightarrow$

$$\exists c_1,...,c_n \in P \, (c_1 \le a_1 \,\&\, ... \,\&\, c_n \le a_n \,\&\, b = c_1 \vee ... \vee c_n).$$

Call $a,b \in P$ a *minimal pair* (for P) if $a,b > 0$ and $a \wedge b = 0$ (where $a \wedge b = c$ means that the infimum of a and b exists and equals c). An element c of P is called a *top* if there is a minimal pair a,b such that $c = a \vee b$. Otherwise, c is a *nontop*.

2.4. Theorem. Let P be distributive and let $A = \{a_i: i\geq 0\}$ be a subset of P s.t., for any $i\neq j$, a_i, a_j is a minimal pair. Then A is independent. If moreover (A) has an exact pair and, for any i, a_i is a nontop then A is e.d.p.

Proof (sketch). By distributivity of P, for any pairwise different elements $a, a_1, ..., a_n$ of A,

$$a \wedge (a_1 \vee \ldots \vee a_n) = 0.$$

So A is independent.
For the remainder of the proof assume that $(A) = I(u,v)$ and that a_i is a nontop for all $i\geq 0$. Then, again by (2.11), the elements of A are just the maximal nontop elements of (A), i.e.,

$$x\in A \Leftrightarrow x \leq u,v \;\&\; x \text{ is a nontop } \&\; \forall\, y\, (x < y \leq u,v \Rightarrow y \text{ is a top}).$$

Obviously, the right hand side of this equivalence can be described by a first order formula $\phi(x,u,v)$ with parameters u and v.

3. Indexings of Recursively Presentable Classes

Let $\Sigma=\{0,1\}$ be the binary alphabet and let Σ^* be the set of finite strings over Σ. To simplify notation we will identify Σ^* with ω, by identifying the nth string z_n of Σ^* (w.r.t. the canonical ordering of Σ^*) with the number n. In the following the term "set" will refer to subsets of ω. Moreover, as usual, we do not distinguish between a set and its characteristic function, i.e., $x\in A$ iff $A(x)=1$ and $x\notin A$ iff $A(x)=0$.

A class $\mathcal{C}$ of recursive subsets of ω is *recursively presentable (r.p.)* if there is a recursive set $U\subseteq\omega\times\omega$ such that $\mathcal{C}=\{U^{(n)} : n\geq 0\}$, where $U^{(n)} = \{x : (n,x)\in U\}$. We call U a *universal set for* or an *indexing of* $\mathcal{C}$. Given an indexing U of $\mathcal{C}$, we say $\mathcal{D}$ is a Σ^0_n*-subclass* of $\mathcal{C}$ (w.r.t. U) if $\mathcal{D} = \{U^{(n)} : n\in H\}$ for some Σ^0_n set H. Finally we say that a class $\mathcal{C}$ is *somewhere closed under finite variants (s.c.f.v.)* if, for some $C\in \mathcal{C}$, any finite variant of C belongs to $\mathcal{C}$ too, i.e.,

$$\forall\, D\, (D =^* C \Rightarrow D\in \mathcal{C}),$$

where $D =^* C$ means that the symmetric difference $(D-C)\cup(C-D)$ of C and D is finite.

The following theorem implies that an s.c.f.v. subclass $\mathcal{D}$ of an r.p. class $\mathcal{C}$ is r.p. if and only if it is a Σ^0_2-subclass of $\mathcal{C}$.

3.1. Theorem (Indexing Theorem). Let $\mathcal{C}$ be an r.p. class and let U be an indexing of $\mathcal{C}$.

(a) Let $\mathcal{D}$ be any Σ^0_2-subclass of $\mathcal{C}$ w.r.t. U and let E_0 be any recursive set. Then $\mathcal{D}\cup\{E : E =^* E_0\}$ is recursively presentable. In particular, if $\mathcal{D}$ is s.c.f.v. then $\mathcal{D}$ itself is recursively presentable.

(b) Let $\mathcal{D}$ be a recursively presentable class. Then $\mathcal{C}\cap\mathcal{D}$ is a Σ^0_2-subclass of $\mathcal{C}$ w.r.t. U.

Proof. (a) Since $\{E : E =^* E_0\}$ is r.p. and since the union of any two r.p. classes is r.p. again, it suffices to show that there is an r.p. class $\mathcal{D}'$ such that

(3.1) $\mathcal{D} \subseteq \mathcal{D}' \subseteq \mathcal{D} \cup \{E : E =^* E_0\}$.

Fix $H \in \Sigma_2^0$ such that $\mathcal{D} = \{U^{(n)} : n \in H\}$. Since H is Σ_2^0, there is a recursive set G such that

(3.2) $x \in H \Leftrightarrow \exists y\ \forall z\ ((x,y,z) \in G)$.

Define $V \subseteq \omega \times \omega$ by

$V(<x,y>,w) =$ **if** $\forall z \leq w$ ($(x,y,z) \in G$) **then** $U(x,w)$ **else** $E_0(w)$ **fi**

where $\lambda x,y.<x,y>$ is a recursive bijection between $\omega \times \omega$ and ω. We claim that V is a universal set for a class $\mathcal{D}'$ satisfying (3.1). Obviously V is recursive. So it suffices to show that

$\mathcal{D} \subseteq \{V^{(n)} : n \geq 0\} \subseteq \mathcal{D} \cup \{E : E =^* E_0\}$.

First take $D \in \mathcal{D}$. Then $D = U^{(x)}$ for some $x \in H$. So, by (3.2), there is a number y such that $(x,y,z) \in G$ for all numbers z. By definition of V this implies $D = V^{(<x,y>)}$. For a proof of the second inclusion, fix $V^{(n)}$ and pick x and y such that $n = <x,y>$. Now if $(x,y,z) \in G$ for all numbers z then, by definition of V and by (3.2), $V^{(n)} = U^{(x)}$ and $x \in H$, whence $V^{(n)} \in \mathcal{D}$. Otherwise, $V^{(n)} =^* E_0$, i.e.,

$V^{(n)} \in \{E : E =^* E_0\}$.

(b) Let V be an indexing for $\mathcal{D}$ and define H by

$$H = \{n \in \omega : \exists m\ \forall x\ (\ U^{(n)}(x) = V^{(m)}(x))\}.$$

Then H is a Σ_2^0 set and $\mathcal{C} \cap \mathcal{D} = \{U^{(n)} : n \in H\}$.

4. Exact Pairs in the p-m Degrees

Recall that a set A is polynomial time many-one (p-m) reducible to a set B ($A \leq_{p\text{-}m} B$), if there is a polynomial time computable function $f: \omega \to \omega$ such that, for any number n, $A(n) = B(f(n))$ (where the time is measured in the length $|z_n| = \log(n)$ of the string z_n representing n). The relation $\leq_{p\text{-}m}$ is a preordering whence

$A =_{p\text{-}m} B \Leftrightarrow A \leq_{p\text{-}m} B \ \&\ B \leq_{p\text{-}m} A$

is an equivalence relation. The equivalence classes are called p-m-degrees, The p-m-degree of a set A is denoted by

$deg_{p\text{-}m}(A) = \{B : B =_{p\text{-}m} A\}$.

We will denote p-m-degrees of recursive sets by lower case boldface letters **a**, **b**, **c**, ... The p-m-reducibility induces a partial ordering on the p-m-degrees by

$\mathbf{a} \leq \mathbf{b} \Leftrightarrow \exists A \in \mathbf{a}\ \exists B \in \mathbf{b}\ (\ A \leq_{p\text{-}m} B\ )$.

This partial ordering has a least element **0**, namely the degree of the polynomial time computable sets. (To avoid trivial pathologies we assume that the empty set and the set ω are equivalent to the other PTIME sets.) Moreover, this partial ordering is an u.s.l., where the supremum of two degrees **a** and **b** is obtained by taking the degree of the effective disjoint union $A \oplus B = \{2n : n \in A\} \cup \{2n+1 : n \in B\}$ of any two members A and B of **a** and **b**, respectively:

$deg_{p\text{-}m}(A) \vee deg_{p\text{-}m}(B) = deg_{p\text{-}m}(A \oplus B)$.

So we may adopt the notation on upper semilattices from Section 2 to the u.s.l. of the p-m-degrees of recursive sets, which in the following will be denoted by

$\mathcal{R}_{p\text{-}m} = (\mathbf{REC}_{p\text{-}m}; \leq, \vee, \mathbf{0})$.

We call a class $\mathbf{A}$ of p-m-degrees of recursive sets *recursively presentable (r.p)* if there is an r.p. class $\mathcal{A}$ of recursive sets such that $\mathbf{A} = \{deg_{p\text{-}m}(C) : C \in \mathcal{A}\}$. If U is an indexing of $\mathcal{A}$ we call it an *indexing* of $\mathbf{A}$ too. We will be interested in a special type of indexings. If U is an indexing of $\mathbf{A}$ such that

$$deg_{p\text{-}m}(U^{(i)}) \neq deg_{p\text{-}m}(U^{(j)})$$

for i≠j then we call U a *recursive basis* of $\mathbf{A}$. An ideal $\mathbf{I}$ in $\mathcal{R}_{p\text{-}m}$ is called a recursively presentable (r.p.) ideal if $\mathbf{I}$ is generated by some r.p. class of p-m-degrees, i.e. $\mathbf{I} = (\mathbf{A})$ for some r.p. $\mathbf{A}$.

The following characterization of ideals in $\mathcal{R}_{p\text{-}m}$ with exact pairs has been given in Ambos-Spies [4]:

4.1. Theorem (Exact Pair Theorem; [4]**).** Let $\mathbf{I}$ be an ideal in $\mathcal{R}_{p\text{-}m}$. The following are equivalent.

(i) $\mathbf{I}$ is a recursively presentable ideal.

(ii) $\{B : deg_{p\text{-}m}(B) \in \mathbf{I}\}$ is r.p.

(ii) $\mathbf{I}$ possesses an exact pair in $\mathcal{R}_{p\text{-}m}$.

The Indexing Theorem and the Exact Pair Theorem imply the following representation theorem for Σ^0_2-sets in $\mathcal{R}_{p\text{-}m}$.

4.2. Corollary (Representation Theorem). Let $\mathbf{A}$ be an independent set of p-m-degrees with recursive basis U. Then $(DI(\mathbf{A});\subseteq) \cong (\Sigma^0_2;\subseteq)$.

Proof. Let $\mathcal{A} = \{U^{(i)} : i \geq 0\}$ and $\mathbf{a}_i = deg_{p\text{-}m}(U^{(i)})$ $(i \geq 0)$. Then, since $\mathbf{A}$ is independent and U is a basis for $\mathbf{A}$, it follows from Lemma 2.1 that the function Φ: $\mathcal{P}(\omega) \rightarrow I(\mathbf{A})$ defined by $\Phi(X) = (\{\mathbf{a}_i : i \in X\})$ is an isomorphism from $(\mathcal{P}(\omega);\subseteq)$ onto $(I(\mathbf{A});\subseteq)$. So it suffices to show that the restriction of Φ on Σ^0_2 is a mapping onto the subclass $DI(\mathbf{A})$ of $I(\mathbf{A})$, i.e., that

(4.1) $\quad X \in \Sigma^0_2 \Rightarrow \Phi(X) \in DI(\mathbf{A})$

and

(4.2) $\quad \mathbf{I} \in DI(\mathbf{A}) \Rightarrow \Phi^{-1}(\mathbf{I}) = \{i : \mathbf{a}_i \in \mathbf{I}\} \in \Sigma^0_2$.

For a proof of (4.1) fix $X \in \Sigma^0_2$. Note that $\Phi(\emptyset) = \{\mathbf{0}\} \in DI(\mathbf{A})$. So w.l.o.g. we may assume that X is nonempty, say $i_0 \in X$. Then the class $\mathcal{D} = \{U^{(i)} : i \in X\}$ is a Σ^0_2-subclass of $\mathcal{A}$ with respect to the indexing U. So, by Theorem 3.1, the class $\mathcal{D}' = \mathcal{D} \cup \{E : E =^* U^{(i_0)}\}$ is r.p. Since p-m-degrees are closed under finite variants and since $U^{(i_0)} \in \mathcal{D}$,

$$\{deg_{p\text{-}m}(D) : D \in \mathcal{D}'\} = \{deg_{p\text{-}m}(D) : D \in \mathcal{D}\} = \{\mathbf{a}_i : i \in X\}.$$

So $\Phi(X) = (\{\mathbf{a}_i : i \in X\}) = (\{deg_{p\text{-}m}(D) : D \in \mathcal{D}'\})$, whence, by definition, $\Phi(X)$ is an r.p. ideal. So it follows with Theorem 4.1 that $\Phi(X)$ possesses an exact pair, i.e., $\Phi(X) \in DI(\mathbf{A})$.

Finally, for a proof of (4.2), assume that $\mathbf{I} \in DI(\mathbf{A})$. Then, by Theorem 4.1, the class

$$\mathcal{D}_{\mathbf{I}} = \{D : deg_{p\text{-}m}(D) \in \mathbf{I}\}$$

is recursively presentable. So, by Theorem 3.1(b), $\mathcal{A} \cap \mathcal{D}_{\mathbf{I}} = \{U^{(i)} : \mathbf{a}_i \in \mathbf{I}\}$ is a Σ^0_2-subclass of $\mathcal{A}$ (with respect to the indexing U), i.e. $\{i : \mathbf{a}_i \in \mathbf{I}\} \in \Sigma^0_2$.

5. Undecidability of the Theory of the p-m-Degrees

To complete the proof that the elementary theory of $\mathcal{R}_{p\text{-}m}$ is undecidable, we next list some results from the literature which will imply that there is a recursively presentable independent subset of $\mathbf{REC}_{p\text{-}m}$ which is e.d.p. in $\mathcal{R}_{p\text{-}m}$.

5.1. Proposition (Ambos-Spies [1]). The u.s.l. $\mathcal{R}_{p\text{-}m}$ is distributive.

5.2. Theorem (Ambos-Spies). Let B be any recursive set such that $B \notin$ PTIME. There is a recursive set C such that

(i) $\quad deg_{p\text{-}m}(B)$ and $deg_{p\text{-}m}(C)$ form a minimal pair

and

(ii) $\quad deg_{p\text{-}m}(C)$ is a nontop.

Moreover, the set C is uniformly recursive in B.

The proof of Theorem 5.2, which combines the minimal pair technique of [4] with the nontop technique of [3], can be found in [1].

5.3. Corollary. There is a recursive set $A \subseteq \omega \times \omega$ such that, for any $i,j \geq 0$ with $i \neq j$,

(5.1) $\quad deg_{p\text{-}m}(A^{(i)})$ and $deg_{p\text{-}m}(A^{(j)})$ form a minimal pair

and

(5.2) $\quad deg_{p\text{-}m}(A^{(i)})$ is a nontop.

Proof. Define the sets $A^{(i)}$ by induction on i as follows: Let $A^{(0)}$ be any recursive set such that $deg_{p\text{-}m}(A^{(0)})$ is a nontop. Such a set exists by Theorem 5.2. Given $A^{(0)},\ldots, A^{(i)}$, let $A^{(i+1)}$ be the set C provided by Theorem 5.2 for $B = A^{(0)} \oplus \ldots \oplus A^{(i)}$. Then, by uniformity of Theorem 5.2, A is recursive while, by part (ii) of the theorem, (5.2) holds. Finally, by part (i) of the theorem, for any i, $deg_{p\text{-}m}(A^{(i+1)})$ and $deg_{p\text{-}m}(A^{(0)} \oplus \ldots \oplus A^{(i)})$ form a minimal pair. Since, for $j \leq i$,

$$A^{(j)} \leq_{p\text{-}m} A^{(0)} \oplus \ldots \oplus A^{(i)},$$

this implies (5.1).

By applying our general observations on definability in upper semilattices and the Representation Theorem to the preceeding result, we obtain the desired definability theorem which implies undecidability.

5.4. Theorem (Definability Theorem). The partial ordering $\mathcal{E}^2 = (\Sigma^0_2; \subseteq)$ of the Σ^0_2 sets under inclusion is elementarily definable with parameters in $\mathcal{R}_{p\text{-}m}$.

Proof. Fix a recursive set $A \subseteq \omega \times \omega$ as in Corollary 5.3. Let $\mathcal{A} = \{A^{(i)} : i \geq 0\}$, $\mathbf{a}_i = deg_{p\text{-}m}(A^{(i)})$ and $\mathbf{A} = \{\mathbf{a}_i : i \geq 0\}$. Then, by (5.1), A is a recursive basis of **A**. So, by the Exact Pair Theorem, (**A**) possesses an exact pair in $\mathcal{R}_{p\text{-}m}$. Since, by Proposition 5.1,
$\mathcal{R}_{p\text{-}m}$ is distributive, we may conclude from (5.1) and (5.2) and from Theorem 2.4 that **A** is

independent and e.d.p. So, by Theorem 2.2, (DI(A),$\subseteq$) is e.d.p. and, by the Representation Theorem, (DI(A);$\subseteq$) $\cong$ (Σ_2^0;$\subseteq$).

5.5. Corollary (Undecidability Theorem). The elementary theory of the polynomial many-one degrees of recursive sets is (hereditarily) undecidable.

Proof. By Corollary 1.2 and Theorem 5.4.

6. Further Results

The Exact Pair Theorem and hence the Representation Theorem holds for most of the standard subrecursive reducibilities. So to prove that the theory of such a reducibility structure is undecidable it suffices to prove the existence of an independent class **A** of degrees which has a recursive basis and which is elementarily definable with parameters in the given structure. This approach can be used to give undecidability results for all of the standard polynomial time reducibilities: The argument of this paper directly carries over to p-1-tt reducibility and, as Ambos-Spies and Yang [6] have shown, to the honest versions of p-m and p-1-tt reducibilities. (Downey, Gasarch and Moses [9] have independently shown that the theory of the honest p-m-degrees of nonrecursive sets is undecidable. Their proof, however, which is adapted from some argument in recursion theory, is based on results on nonrecursive sets and does not give undecidability of the honest p-m-reducibility on the recursive sets.) Furthermore, a simpler than the original undecidability proof for p-T-reducibility can be extracted from Shinoda and Slaman´s paper. The technical main theorem of [16] gives an independent and e.d.p. set of p-T-degrees with a recursive basis and with various additional properties required for the proof given there. By our analysis we do not have to ensure these additional properties which considerably simplifies the proof of the theorem. Ambos-Spies and Yang [6] have adapted this argument to the honest p-Turing degrees.

References

1. K. Ambos-Spies, On the structure of polynomial time degrees, in "STACS 84", Lecture Notes in Computer Science, 166 (1984) 198-208, Springer Verlag.
2. K. Ambos-Spies, On the structure of the polynomial time degrees of recursive sets, Tech. Rep. 206 (1985), Abteilung Informatik, Universität Dortmund.
3. K. Ambos-Spies, An inhomogeneity in the structure of Karp degrees, SIAM Journal on Computing 15 (1986) 958-63.
4. K. Ambos-Spies, Minimal pairs for polynomial time reducibilities, in "Computation Theory and Logic", Lecture Notes in Computer Science, 270 (1987) 1-13, Springer Verlag.
5. K. Ambos-Spies, A. Nies and R.A. Shore, The theory of the recursively enumerable weak truth-table degrees is undecidable, J. Symbolic Logic, to appear.
6. K.Ambos-Spies and D.Yang, Honest polynomial time reducibilities (in preparation).
7. S. Burris and R. McKenzie, Decidability and Boolean representations, Memoirs Amer. Math. Soc. 32, No. 246, 1981.
8. S.A. Cook, The complexity of theorem proving procedures, Proc. Third Annual ACM Symp. on Theory of Comput., 1971, 151-158.

9. R. Downey, W.I. Gasarch and M. Moses, The structure of the honest polynomial m-degrees (to appear).
10. E. Herrmann, Definable Boolean pairs in the lattice of the recursively enumerable sets, in "Proceedings of the Conference on Model Theory, Diedrichshagen, DDR,1983", Seminarberichte Nr. 49, Sektion Mathematik, Humboldt-Universität, Berlin, 1983, 42-67.
11. E. Herrmann, The undecidability of the elementary theory of the lattice of recursively enumerable sets (Abstract), in "Proceedings of the Second Frege Conference, Schwerin, DDR, 1984" (G. Wechsung, ed.), Akademie-Verlag, Berlin 1984, 66-72.
12. R.M. Karp, Reducibility among combinatorial problems, in "Complexity of Computer Computations", Plenum, N.Y., 1972, 85-103.
13. R.E. Ladner, On the structure of polynomial time reducibility, J. ACM 22 (1975), 155-171.
14. L.H. Landweber, R.J. Lipton and E.L. Robertson, On the structure of sets in NP and other complexity classes, Theor. Comput. Sci. 15 (1981) 103-123.
15. K. Mehlhorn, Polynomial and abstract subrecursive classes, J. Comput. System Sci. 12 (1976) 147-178.
16. J. Shinoda and T.A. Slaman, On the structure of the polynomial degrees of the recursive sets (to appear) [Abstract in "Structure in Complexity Theory Third Annual Conference", IEEE Comput. Soc. Press, 1988].
17. R.A. Shore and T.A. Slaman, The p-T-degrees of the recursive sets: lattice embeddings, extensions of embeddings and the two quantifier theory (to appear) [Abstract in "Structure in Complexity Theory Fourth Annual Conference", IEEE Comput. Soc. Press, 1989].

COMPUTATIONAL GEOMETRY AND LEARNING THEORY

A plane-sweep algorithm for finding a closest pair among convex planar objects

Frank Bartling
FB 12, Informatik, Universität - GH - Siegen, Postfach 10 12 40, D - 5900 Siegen, Germany

Klaus Hinrichs
FB 15, Informatik, Westfälische Wilhelms-Universität, Einsteinstr. 62, D - 4400 Münster, Germany

Abstract

Given a set of geometric objects a *closest pair* is a pair of objects whose mutual distance is smallest. We present a plane-sweep algorithm which finds a closest pair with respect to any L_p-metric, $1 \le p \le \infty$, for planar configurations consisting of n (possibly intersecting) compact convex objects such as line segments, circular discs and convex polygons. For configurations of line segments or discs the algorithm runs in asymptotically optimal time O(n log n). For a configuration of n convex m-gons given in a suitable representation it finds a closest pair with respect to the Euclidean metric L_2 in time O(n log(n·m)).

Key words: computational geometry, plane-sweep algorithm, closest-pair problem.

1 Introduction

One of the fundamental problems in computational geometry is that of finding a closest pair among a given set S of geometric objects. The (minimal) distance between two objects $s_1, s_2 \in S$ is defined as $d(s_1, s_2) := \min\{d(p, q): p \in s_1, q \in s_2\}$, where d denotes any L_p-metric ($1 \le p \le \infty$). $\Omega(n \log n)$ is known to be a lower bound for the worst-case time complexity of the closest-pair problem (for n objects) in the algebraic decision tree model of computation.

The closest-pair problem can be considered as a generalization of the 'intersection detection problem': A set S of *compact* objects is pairwise disjoint if and only if the distance of any closest pair is greater than 0. A well known intersection detection algorithm is that given by Shamos and Hoey for configurations of line segments in the plane [SH 76]; this algorithm can also be applied to objects which are 'monotone in the x-direction' [S 85].

For sets of points optimal algorithms are known for solving the closest-pair problem efficiently ([PS 85], [HNS 88]). For sets of line segments and circular arcs in the plane a closest pair with respect to the Euclidean metric can be found in optimal time O(n log n) by first computing the Voronoi diagram and then extracting a closest pair from the Voronoi diagram in linear time. The algorithms ([F 87], [Y 87]) for computing the Voronoi diagram of line segments and circular arcs require that the objects do only intersect at endpoints. Therefore it is necessary to preprocess the configuration by an intersection detection algorithm. Given n simple m-gons (an m-gon is a polygon with up to m edges) in the plane, a closest pair with respect to the Euclidean metric can be found by applying the just described technique to all the n·m edges. The intersection detection algorithm for the n m-gons costs $\Theta(n{\cdot}m{\cdot}\log(n{\cdot}m))$ time ([SH 76]), hence the worst-case time complexity of this algorithm is $\Theta(n{\cdot}m{\cdot}\log(n{\cdot}m))$.

In this paper we describe an easy to implement plane-sweep algorithm 'PSCP' which *directly* computes a closest pair with respect to *any* L_p-metric, $1 \le p \le \infty$, in a set S of n compact *convex* objects in the plane. 'PSCP' combines ideas from the intersection detection algorithm by Shamos and Hoey [SH 76] and from the closest-pair algorithm for configurations of points given by Hinrichs, Nievergelt and Schorn [HNS 88]. It is not necessary to preprocess the configuration by an intersection detection algorithm. For configurations consisting of n line segments, circular discs or convex polygons whose number of edges is treated as a constant 'PSCP' runs in asymptotically optimal time $\Theta(n \log n)$. When S consists of n convex m-gons each of which is given by its vertices in cyclic order, 'PSCP' finds a closest pair with respect to the Euclidean metric in time $O(n \log(n \cdot m))$; this runtime is achieved by employing an optimal $O(\log m)$ algorithm [CD 87] for detecting whether two convex m-gons intersect, and an optimal $O(\log m)$ algorithm ([E 85], [CW 83]) for computing the minimal Euclidean distance between two non-intersecting convex m-gons.

The paper proceeds as follows. In section 2 we give an informal description of the plane-sweep algorithm 'PSCP'. We introduce the relation '$\leq_{Obj}$' on sets of compact convex objects which is a total order when restricted to the objects stored in the y-table during the plane-sweep. In section 3 we show the implementation of 'PSCP'. In section 4 we prove that 'PSCP' computes a correct result, and in section 5 we give a worst-case analysis of 'PSCP'.

2 Plane-sweep applied to the closest pair problem

Our algorithm 'PSCP' is based on the well known plane-sweep method. Plane-sweep's name is derived from the image of sweeping the plane from left to right with a vertical line (front, or cross section), stopping at every transition point (event) of a geometric configuration to update the cross section. All processing is done at this moving front, without any backtracking, with a look-ahead of only one point. The events are contained in the x-queue, the status of the sweep is maintained by the y-table. In the slice between two events, the properties of the geometric configuration detected so far do not change, and therefore the y-table does not have to be updated. The skeleton of a plane-sweep algorithm is as follows:

```
initialize x-queue;
initialize y-table;
while not empty(x-queue) do p := next(x-queue); transition(p) end;
```

The procedure 'transition' is the advancing mechanism of the plane-sweep. It embodies all the work to be done when a new event is encountered; it moves the front from the slice to the left of an event p to the slice immediately to the right of p.

In this paper we apply the plane-sweep principle to solve the closest pair problem in a *set S consisting of n convex compact objects* with respect to any L_p-metric d ($1 \leq p \leq \infty$). For each object $s \in S$ let s[L] denote the smallest and s[R] the largest point in s according to the lexicographic order '$\leq_x$':

$$\forall\, p, q \in E^2,\ p \leq_x q :\Leftrightarrow (p.x < q.x) \text{ or } (p.x = q.x) \text{ and } (p.y \leq q.y).$$

During the sweep 'PSCP' maintains the smallest distance δ of a pair of objects detected so far. As soon as 'PSCP' finds an intersecting pair of objects, i.e. $\delta = 0$, it reports this intersecting pair of objects as a closest pair and stops.

The y-table stores the objects of S intersecting the δ-slice to the left of the sweep line. These are exactly those objects of S which can be closer than δ to any object of S not yet encountered by the sweep line, i.e. lying completely to the right of the sweep line.

The x-queue is initialized with two sets of events: insertion events and deletion events. Insertion events are given by the left end points s[L] for $s \in S$. An object s is inserted into the y-table when the sweep line encounters its left end point s[L]. Deletion events are determined by the right end points s[R] for $s \in S$ and the smallest distance δ of a pair of objects detected so far. An object s is removed from the y-table as soon as it no longer intersects the δ-slice lying to the left of the sweep line, i.e. if the position of the sweep line is to the right of $s[R].x + \delta$. This may happen either if the sweep line proceeds to the right or if the δ-slice shrinks because δ becomes smaller. Since δ can not grow during the sweep the sequence in which deletion events are processed is determined by the right end points s[R] for $s \in S$ only, but does not depend on the current value of δ. In contrast to the conventional plane-sweep scheme the deletion events processed by algorithm 'PSCP' do not necessarily correspond to a predetermined position of the sweep-line: the only thing we know is that when a deletion event for an object s is executed the sweep line is at or to the right of $s[R].x + \delta$.

During initialization of the x-queue we test whether any two objects $s_1, s_2 \in S$, $s_1 \neq s_2$, intersect in their end points: $s_1[L] = s_2[L]$ or $s_1[R] = s_2[R]$. If such a pair of objects is found the algorithm 'PSCP' stops. Therefore in the following we make the

Assumption: $\forall\ s_1, s_2 \in S,\ s_1 \neq s_2$: $s_1[L] \neq s_2[L]$ and $s_1[R] \neq s_2[R]$.

Now the orders in which insertion events and deletion events are processed induce an 'order of insertion $<_{Ins}$' and an 'order of deletion $<_{Del}$' on S: For $s_1, s_2 \in S$, $s_1 <_{Ins} s_2 :\Leftrightarrow s_1[L] <_x s_2[L]$ and $s_1 <_{Del} s_2 :\Leftrightarrow s_1[R] <_x s_2[R]$.

In order to allow an efficient processing of the objects contained in the y-table it is necessary to find a canonical total order '$\leq_{Obj}$' on these objects. Such a total order will be defined in the following. For preparation we need the following definitions:

Definition 2.1: For $s \in S$ let Q(s) be the line segment with end points s[L] and s[R]. Define $HQ(s) := HL(s) \cup Q(s) \cup HR(s)$, where $HL(s) := \{(x, y) \in E^2: y = s[L].y,\ x < s[L].x\}$ and $HR(s) := \{(x, y) \in E^2: y = s[R].y,\ x > s[R].x\}$. For $s \in S$ the set $E^2 \setminus HQ(s)$ consists of two connected components, the upper component UC(s) above HQ(s), and the lower component LC(s) below HQ(s).

For a line segment s we have Q(s) = s; for a circular disc s the set Q(s) is the horizontal diameter of s, and HQ(s) is a horizontal line through the center of s.

Definition 2.2: For $s \in S$ and $\delta > 0$ define

$$D(s, \delta) :=]s[L].x - \delta,\ s[L].x] \times E \text{ and}$$
$$A(s, \delta) := \{s' \in S: s' <_{Ins} s,\ s' \cap D(s, \delta) \neq \emptyset\}.$$

Definition 2.3: Let $s_1, s_2 \in S$, $s_1[L] <_x s_2[L]$. Then we define

$$s_1 <_{Obj} s_2 \ :\Leftrightarrow\ s_2[L] \in UC(s_1) \vee (s_2[L] \in HQ(s_1) \wedge s_1[L].y \leq s_2[L].y)$$
$$s_2 <_{Obj} s_1 \ :\Leftrightarrow\ \neg(s_1 <_{Obj} s_2)$$

For arbitrary objects $s_1, s_2 \in S$ we define

$$s_1 \leq_{Obj} s_2 \ :\Leftrightarrow\ (s_1 = s_2) \vee ((s_1 \neq s_2) \wedge (s_1 <_{Obj} s_2)).$$

From the definition of '$\leq_{Obj}$' we can follow immediately:

Lemma 2.4: The relation '$\leq_{Obj}$' on S is reflexive and antisymmetric.

Observation: The relation '$\leq_{Obj}$' on S is in general not transitive, as the following two examples show. In both cases we have $s_1 <_{Obj} s_2 <_{Obj} s_3 <_{Obj} s_1$.

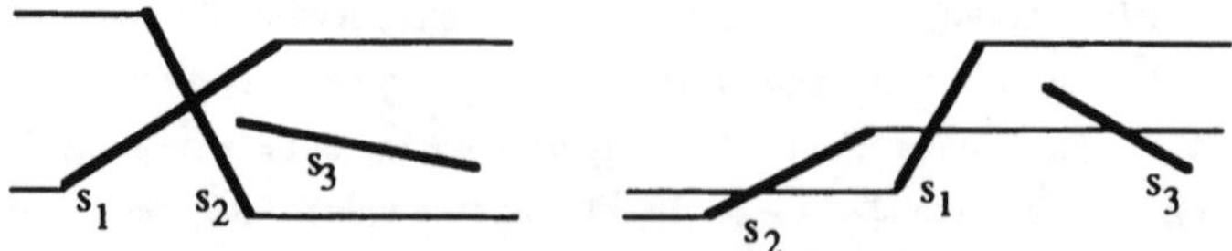

Lemma 2.5: Let $G := \{(x, y): x = a\}$ be a vertical line intersecting two disjoint objects $s_1, s_2 \in S$. Then:

$$\forall\ p_1 \in s_1 \cap G,\ \forall\ p_2 \in s_2 \cap G:\ p_1.y < p_2.y \Leftrightarrow s_1 <_{Obj} s_2.$$

Proof: Consider points $q_1 \in Q(s_1) \cap G$ and $q_2 \in Q(s_2) \cap G$. Since $Q(s_1)$ and $Q(s_2)$ do not intersect we obviously have $q_1.y < q_2.y \Leftrightarrow s_1 <_{Obj} s_2$. Since the sets $s_1 \cap G$ (containing q_1) and $s_2 \cap G$ (containing q_2) are disjoint intervals the assertion follows.

Lemma 2.6: Let $s_1, s_2 \in S$, $\delta > 0$ and $T := Q(s_1) \cap D(s_2, \delta)$. If $T \neq \emptyset$ we have:

(1) $T \subset \{(x, y): y < s_2[L].y\} \Rightarrow s_1 <_{Obj} s_2$.

(2) $T \subset \{(x, y): y > s_2[L].y\} \Rightarrow s_2 <_{Obj} s_1$.

Proof: If $T \subset \{(x, y): y < s_2[L].y\}$ we have $s_2[L] \in UC(s_1)$ and therefore $s_1 <_{Obj} s_2$. If $T \subset \{(x, y): y > s_2[L].y\}$ we have $s_2[L] \in LC(s_1)$ and therefore $s_2 <_{Obj} s_1$.

Lemma 2.7: Let $\delta > 0$, $D :=]x - \delta, x] \times E$ a vertical slice of width δ and $s_1, s_2 \in S$, such that $s_1 \cap D \neq \emptyset$, $s_2 \cap D \neq \emptyset$ and $d(s_1, s_2) \geq \delta$. Then:

$$\forall\ p_1 \in s_1 \cap D,\ \forall\ p_2 \in s_2 \cap D:\ p_1.y < p_2.y \Leftrightarrow s_1 <_{Obj} s_2.$$

Proof: The projections of $s_1 \cap D$ and $s_2 \cap D$ to the y-coordinate are intervals I_1 and I_2; from $d(s_1, s_2) \geq \delta$ it follows that $s_1 \cap s_2 = \emptyset$ and $I_1 \cap I_2 = \emptyset$. We have to show that for any $y_1 \in I_1$ and $y_2 \in I_2$: $y_1 < y_2 \Leftrightarrow s_1 <_{Obj} s_2$. If there exists a vertical line $G := \{(x, y): x = a\}$ with $a \in]x - \delta, x]$ which intersects both s_1 and s_2 this follows from Lemma 2.5. If no such vertical line exists one of the following two cases applies. Case (1) $x - \delta < s_1[R].x < s_2[L].x \leq x$: Since $I_1 \cap I_2 = \emptyset$ we obtain $s_1 <_{Obj} s_2 \Leftrightarrow s_2[L] \in UC(s_1) \Leftrightarrow s_1[R].y < s_2[L].y \Leftrightarrow y_1 < y_2$ for any $y_1 \in I_1$ and $y_2 \in I_2$. Case (2) $x - \delta < s_2[R].x < s_1[L].x \leq x$: Since $I_1 \cap I_2 = \emptyset$ we obtain $s_1 <_{Obj} s_2 \Leftrightarrow s_1[L] \in LC(s_2) \Leftrightarrow s_1[L].y < s_2[R].y \Leftrightarrow y_1 < y_2$ for any $y_1 \in I_1$ and $y_2 \in I_2$.

Lemma 2.8: Let $\delta > 0$, $D :=]x - \delta, x] \times E$ a vertical slice of width δ and $A \subset S$ such that

(a) $\forall\ s \in A$: $s \cap D \neq \emptyset$.

(b) '$\leq_{Obj}$' is a total ordering on A.

(c) $\forall\ s_1, s_2 \in A$, s_1 and s_2 neighbors with respect to '$\leq_{Obj}$': $d(s_1, s_2) \geq \delta$.

Then the following holds:

(1) $\forall\ y \in E$ the set $]x - \delta, x] \times \{y\}$ intersects at most one object from A.

(2) $\forall\ s_1, s_2 \in A,\ \forall\ p_1 \in s_1 \cap D,\ \forall\ p_2 \in s_2 \cap D$: $p_1.y < p_2.y \Leftrightarrow s_1 <_{Obj} s_2$.

(3) For $s' \in S \setminus A$ with $s'[L].x = x$ the relation '$\leq_{Obj}$' is a total ordering on $A \cup \{s'\}$.

Proof: For any $s \in A$ the projection of $s \cap D$ to the y-coordinate is an interval. Now (1) and (2) follow directly from Lemma 2.7. Due to Lemma 2.4 all we have to do to prove (3) is to show that '$\leq_{Obj}$' is transitive on $A \cup \{s'\}$. We split the set A into the subsets $A_1 := \{s \in A: s <_{Obj} s'\}$ and $A_2 := \{s \in A: s' <_{Obj} s\}$. It suffices to show that $s_1 <_{Obj} s_2$ for all $s_1 \in A_1$, $s_2 \in A_2$. Let $s_1 \in A_1$ and $s_2 \in A_2$ and assume $s_2 <_{Obj} s_1$: Consider the sets $T_1 := Q(s_1) \cap D$ and $T_2 := Q(s_2) \cap D$. Since $s' <_{Obj} s_2$ Lemma 2.6 (1) implies $T_2 \cap \{(x, y): y \geq s'[L].y\} \neq \emptyset$. Hence the assumption $s_2 <_{Obj} s_1$ and (2) imply that $T_1 \subset \{(x, y): y > s'[L].y\}$. From Lemma 2.6 (2) we conclude $s' <_{Obj} s_1$, a contradiction.

Invariant of the plane-sweep algorithm: Conditions (a), (b) and (c) of Lemma 2.8 describe the invariant that is maintained during the sweep from left to right. The set 'A' corresponds to the objects stored in the y-table of 'PSCP', x corresponds to a position of the sweep line, and D corresponds to the δ-slice lying to the left of the sweep-line. Condition (a) is maintained by removing objects from the y-table that no longer intersect the δ-slice. Lemma 2.8 (3) guarantees that condition (b), i.e. the total ordering of the objects stored in the y-table, is preserved after insertion of a new object into the y-table. In order to maintain condition (c) of Lemma 2.8 we have to compute distances of objects which become neighbors with respect to '$\leq_{Obj}$' in the y-table and update δ, if necessary. We obtain such new neighbor pairs after inserting a new object into the y-table or after removing an object from the y-table. As soon as we find an intersecting pair of neighbors, δ becomes 0, and 'PSCP' reports this intersecting pair of objects as a closest pair and stops.

In section 4 we will prove that this plane-sweep algorithm 'PSCP' computes a correct closest pair of objects. It seems to be surprising that 'PSCP' finds the correct result by just testing pairs of objects which become neighbors in the y-table with respect to '$\leq_{Obj}$'. In the next section we show a detailed implementation of 'PSCP'.

3 Implementation

We give a sketch of the implementation of the algorithm 'PSCP'.

The function 'distance(s_1, s_2)' yields the distance between the objects s_1 and s_2. If at least one of the two objects is the 'nil' object 'distance' returns '∞'. The variable 'delta' contains the distance between the closest pair of objects 'closest_1' and 'closest_2' found so far. The procedure 'checkDelta' updates the variables closest_1, closest_2 and delta, if the two objects s_1 and s_2 are closer than the closest pair found so far:

```
procedure checkDelta(s1, s2: ConvexObject);
begin
  newDelta := distance(s1, s2);
  if newDelta < delta then delta := newDelta; closest_1 := s1; closest_2 := s2; end
end; { checkDelta }
```

The x-queue is implemented by two sorted arrays 'insertQueue' and 'deleteQueue':

```
var insertQueue, deleteQueue: array[1..maxN] of ConvexObject;
```

'insertQueue' and 'deleteQueue' each store all the objects in S, sorted according to '$\leq_{Ins}$' and '$\leq_{Del}$', respectively. If a pair of objects $s_1, s_2 \in S$ with $s_1[L] = s_2[L]$ or $s_1[R] = s_2[R]$ is detected during the initialization of these arrays then 'PSCP' stops after setting 'closest_1 := s_1', 'closest_2 := s_2' and 'delta := 0'. Two counters i and j keep track of the next insertion and deletion events to be processed. The function 'emptyX' returns 'true' when no more events need to be processed:

```
function emptyX: boolean;
begin return (i > n) and (j > n) end;
```

The procedure 'nextEvent' assumes that 'insertQueue[n+1]' contains a sentinel object with s[L].x = ∞:

```
type Event = (InsEvt, DelEvt);
```

```
procedure nextEvent(var e: Event;  var s: ConvexObject);
begin
  s1 := insertQueue[i];  s2 := deleteQueue[j];
  if  s1[L].x – s2[R].x ≥ delta  then   e := DelEvt;  s := s2;   j := j+1
                                 else   e := InsEvt;  s := s1;   i := i+1     end
end;  { nextEvent }
```

The y-table is implemented as a balanced binary tree. It is empty after initialization. It stores the active objects from S according to the total ordering '$\leq_{Obj}$'. Inserting in and deleting from the y-table is realized by the procedures 'insertY(s)' and 'deleteY(s)'. Successor and predecessor of an object s in the y-table with respect to '$\leq_{Obj}$' are returned by the procedures 'succY(s)' and 'predY(s)'. If no successor or predecessor exists, these procedures return 'nil'.

The initialization part of the plane-sweep is as follows:

```
initialize insertQueue;  i := 1;
initialize deleteQueue;  j := 1;
initialize empty y-table Y;
delta := ∞;
```

The events are processed by the following loop:

```
while  not emptyX and delta ≠ 0  do  nextEvent(e, s);  transition(e, s)  end;
```

The procedure 'transition' encompasses all the work to be done when processing an event:

```
procedure transition(e: Event;  s: ConvexObject);  { Processes a transition event }
begin
  if  evt = InsEvt  then    insertY(s);  checkDelta(s, succY(s));  checkDelta(s, predY(s))
  else  { evt = DelEvt }  checkDelta(succY(s), predY(s));  deleteY(s)  end
end { transition };
```

4 Correctness

In this section we will answer the following questions: 1) Why does 'PSCP' find an intersecting pair of objects in a configuration which includes intersections? (Theorem 4.3) 2) Why does 'PSCP' find a closest pair in a configuration without intersections? (Theorem 4.7)

As before d denotes any L_p-metric ($1 \leq p \leq \infty$).

Definition 4.1: For each object $s \in S$ encountered during the sweep let A(s) denote the set of objects which are contained in the y-table at the moment just before s gets inserted. We call the objects in A(s) *active with respect to s.*

Remark: If δ is the current value of the variable 'delta' just before an object s is inserted into the y-table then δ is greater than 0 since otherwise the algorithm would have stopped. Furthermore we have $A(s) = A(s, \delta)$ (see Definition 2.2).

Lemma 4.2: Let $s_1, s_2 \in S$, $s_1 \cap s_2 \neq \emptyset$, $s_1 <_{Obj} s_2$. Let p be the smallest point with respect to '$\leq_x$' of the set $s_1 \cap s_2$, let p_1 and p_2 be the smallest points with respect to '$\leq_x$' of the sets $Q(s_1) \cap \{(x, y): x = p.x\}$ and $Q(s_2) \cap \{(x, y): x = p.x\}$. Then we have: $p_1.y \leq p.y \leq p_2.y$.

Proof: We prove $p_1.y \le p.y$, the other inequation can be shown in an analogous way. If $s_1[L].x = p.x$ then $p_1 = s_1[L]$ and therefore $p_1.y \le p.y$. Let $s_1[L].x < p.x$ and assume $p.y < p_1.y$. We distinguish three cases. Case (1) $s_2[L].x = p.x$: The definition of p implies $s_2[L].y \le p.y$, together with the assumption we obtain $s_2[L].y < p_1.y$. Therefore $s_2[L] \in LC(s_1)$, i.e. $s_2 <_{Obj} s_1$, and we have a contradiction to $s_1 <_{Obj} s_2$. Case (2) $s_1[L].x < s_2[L].x < p.x$: Since $s_1 <_{Obj} s_2$ we have $s_2[L] \in (UC(s_1) \cup Q(s_1)) \cap \{(x, y): s_1[L].x < x < p.x\}$. Now the assumption $p.y < p_1.y$ implies that the line segment from $s_2[L]$ to p, which is completely contained in s_2, intersects $Q(s_1)$ in a point q with $q.x < p.x$. Since $q \in s_1 \cap s_2$ we obtain a contradiction to the definition of p. Case (3) $s_2[L].x \le s_1[L].x$: Since $s_1 <_{Obj} s_2$ the halfray $\{(x, y): x = s_1[L].x, y \ge s_1[L].y\}$ intersects $Q(s_2)$ in a point p'. Now the assumption $p.y < p_1.y$ implies that the line segment from p to p', which is completely contained in s_2, intersects $Q(s_1)$ in a point q with $q.x < p.x$. Since $q \in s_1 \cap s_2$ we obtain a contradiction to the definition of p.

Theorem 4.3: Let p be an intersection point of the configuration S (i.e. a point in the plane belonging to at least two objects from S), such that for any other intersection point q we have: $p <_x q$. Then algorithm 'PSCP' finds an intersecting pair of objects and stops before any object $s \in S$ with $s[L].x > p.x$ is inserted into the y-table.

Proof: Let $F := \{s \in S: s[L].x \le p.x\}$ and $H := \{s \in S: s[L].x > p.x\}$. Let s_F be the largest element in F and s_H – if $H \ne \emptyset$ – the smallest element in H with respect to '$\le_{Ins}$'. The only interesting case is that no intersecting pair of objects has been found before inserting s_F into the y-table, and s_F does not intersect its neighbors in the y-table. We consider the moment just after processing s_F when the y-table contains exactly the elements of $A(s_F) \cup \{s_F\}$, and we show that the algorithm finds an intersecting pair without processing any further insertion events.

Let $C := \{s \in S: p \in s\}$. Obviously $C \subset A(s_F) \cup \{s_F\}$; hence '$\le_{Obj}$' is a total ordering on C. Let $s_1, s_2 \in C$, $s_1 <_{Obj} s_2$, be two neighbors in C with respect to '$\le_{Obj}$'.

Let $Z := \{s \in A(s_F) \cup \{s_F\}: s_1 <_{Obj} s <_{Obj} s_2\}$. Since no intersection has been found so far, s_1 and s_2 cannot be neighbors with respect to '$\le_{Obj}$' in $A(s_F) \cup \{s_F\}$; therefore $Z \ne \emptyset$. Since s_1 and s_2 are neighbors in C we have $Z \cap C = \emptyset$.

Let $G := \{(x, y): x = p.x\}$; let p_1 and p_2 be the smallest points with respect to '$\le_x$' in $Q(s_1) \cap G$ and $Q(s_2) \cap G$. Lemma 4.2 implies $p_1.y \le p.y \le p_2.y$; split G into the following subsets:

$Q_1 := G \cap \{(x, y): y < p_1.y\}$, $Q_2 := G \cap \{(x, y): p_1.y \le y \le p.y\}$,

$Q_3 := G \cap \{(x, y): p.y < y \le p_2.y\}$, $Q_4 := G \cap \{(x, y): y > p_2.y\}$.

$Q_2 \subset s_1$ and the definition of p imply that no $s \in Z$ intersects $Q_2 \setminus \{p\}$. Since $Z \cap C = \emptyset$, no $s \in Z$ includes p. Thus no $s \in Z$ intersects Q_2. No object $s \in Z$ intersects Q_1 since $s_1 <_{Obj} s$. For objects $s \in Z$ with $Q(s) \cap G \ne \emptyset$ we have $Q(s) \cap G \subset Q_3 \cup Q_4$; since $Q(s) \cap G \subset Q_4$ would imply $s_2 <_{Obj} s$ we have $Q(s) \cap Q_3 \ne \emptyset$; thus for $K := \{s \in Z: Q(s) \cap G \ne \emptyset\}$ we have $K = \{s \in Z: Q(s) \cap Q_3 \ne \emptyset\}$. We now distinguish the cases $K = \emptyset$ and $K \ne \emptyset$.

Case (1): $K = \emptyset$. Then for each $s \in Z$ we have $G \cap s = \emptyset$ and thus $s[R].x < p.x$. This implies $s <_{Del} s_1$ and $s <_{Del} s_2$ for each $s \in Z$. Consider the two cases $H = \emptyset$ and $H \ne \emptyset$.

Case (1.1): $H = \emptyset$. In this case after processing s_F all objects in $A(s_F) \cup \{s_F\}$ have to be deleted from the y-table. The deletion of the last object $s \in Z$ causes s_1 and s_2 to become neighbors in the y-table implying that the intersection of s_1 and s_2 is found. Of course it may happen that not all elements in Z are deleted because of a previous detection of another intersecting pair of objects.

Case (1.2): $H \ne \emptyset$. We will show indirectly that the algorithm detects an intersecting pair and therefore stops before s_H is inserted into the y-table; so assume that s_H gets inserted. At least one object in Z is active with respect to s_H; otherwise the algorithm would have detected an intersection and stopped since s_1 and s_2 would have become neighbors in the y-table by deleting the last object of Z. Since $s <_{Del} s_1$ and $s <_{Del} s_2$ for all $s \in Z$ the objects s_1 and s_2 are active with respect to s_H, too. Let δ denote the current value of variable 'delta' immediately before s_H is inserted into the y-table. The set $A = A(s_H)$ together with $D = D(s_H, \delta)$ satisfies the conditions (a), (b) and (c) of Lemma 2.8. Due to Lemma 2.8 (1), $D(s_H, \delta)$ does not include p, i.e. $p.x \le s_H[L].x - \delta$. From $s[R].x < p.x$ for all $s \in Z$ it follows that none of the objects in Z can be active with respect to s_H, a contradiction.

Case (2): $K \neq \emptyset$. In this case $Q_3 \neq \emptyset$ and therefore $p.y < p_2.y$. Furthermore we have $s_2[L].x < p.x$ since $s_2[L].x = p.x$ would imply $s_2[L] = p_2$ and result in the contradiction $s_2[L].y > p.y$. Hence the triangle $\Delta \subset s_2$ with vertices $s_2[L]$, p and p_2 is not degenerated. For each object $s \in K$ we have $s[L].x = p.x$ since $s[L].x < p.x$ would imply $s \cap (\Delta \setminus G) \neq \emptyset$, a contradiction to the definition of point p. Let s' be the element of K with the property $s'[L].y = \max\{s[L].y : s \in K\}$ and $Z' := \{s \in Z: s' <_{Obj} s <_{Obj} s_2\}$. By construction we have $Z' \cap K = \emptyset$; thus $s[R].x < p.x$ for each $s \in Z'$ implying $s <_{Del} s'$, $s <_{Del} s_1$ and $s <_{Del} s_2$. We again distinguish two cases.

Case (2.1): $H = \emptyset$. We can argue as above if s_1 is replaced by s' and Z by Z'.

Case (2.2): $H \neq \emptyset$. Assume that s_H gets inserted into the y-table. At least one object in Z' is active with respect to s_H; otherwise s' and s_2 would have become neighbors in the y-table after deleting the last object of Z'. Since for all $s \in Z'$ we have $s <_{Del} s_1$ and $s <_{Del} s_2$ also the objects s_1 and s_2 are active with respect to s_H. As above by Lemma 2.8 (1) the slice $D(s_H, \delta)$ does not include p and therefore none of the objects from Z' can be active with respect to s_H, a contradiction.

Lemma 4.4: Let p, p_1, p_2 and q be points in the Euclidean plane E^2 such that

(a) $p_1.y < p_2.y$, $p_1.y \le p.y \le p_2.y$ and $p_1.y \le q.y \le p_2.y$.

(b) $p.x \le \min(p_1.x, p_2.x)$ and $q.x \ge \max(p_1.x, p_2.x)$.

Then the following holds:

(1) $d(q, p_2) \ge d(p_1, p_2)$ and $p.y \le q.y \Rightarrow d(p, p_1) \le q.x - p.x$.

(2) $d(q, p_1) \ge d(p_1, p_2)$ and $p.y \ge q.y \Rightarrow d(p, p_2) \le q.x - p.x$.

Proof: We will prove (1), (2) can be shown similarly. Let $d(q, p_2) \ge d(p_1, p_2)$ and $p.y \le q.y$. The bisector $B(p_1, q) = \{z \in E^2: d(z, p_1) = d(z, q)\}$ of p_1 and q divides the plane into two Voronoi-regions, $V(p_1) = \{z \in E^2: d(z, p_1) < d(z, q)\}$ and $V(q) = \{z \in E^2: d(z, q) < d(z, p_1)\}$. As shown in [L 80], the bisector $B(p_1, q)$ can be treated as a continuous curve which is either non-increasing or non-decreasing ([L 80], Lemma 3).

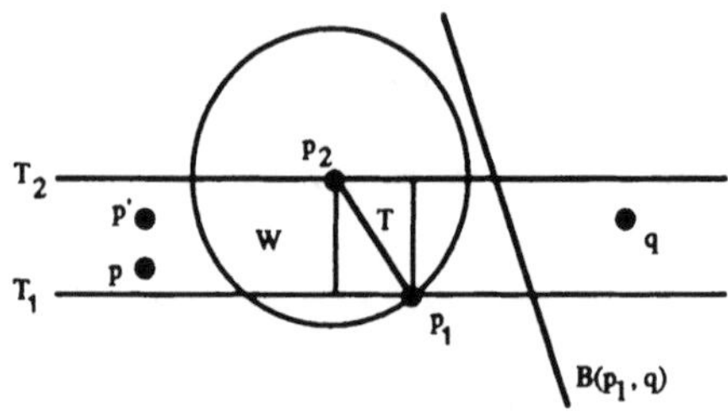

The situation of Lemma 4.4 (1) for the Euclidean metric

Let $T_1 := \{(x, y) \in E^2: y = p_1.y, x \le p_1.x\}$, $T_2 := \{(x, y) \in E^2: y = p_2.y, x \le p_2.x\}$ and T be the line segment having endpoints p_1 and p_2. Let W be the 'wedge' with boundary $T_1 \cup T_2 \cup T$:

$$W = \{ (x,y) \in E^2 : p_1.y \le y \le p_2.y,\ x \le \left(\frac{p_2.x - p_1.x}{p_2.y - p_1.y}\right)(y - p_1.y) + p_1.x \}$$

From $d(q, p_2) \geq d(p_1, p_2)$ it follows $p_2 \in V(p_1) \cup B(p_1, q)$. Since the region $V(p_1)$ is star-shaped with nucleus p_1 ([L 80], Corollary 2), the line segment T is included in $V(p_1) \cup B(p_1, q)$. For all points $z \in T_1$ condition (b) implies $z.x \leq p_1.x \leq q.x$ and therefore $d(z, q) \geq d(z, p_1)$, i.e. $z \in V(p_1) \cup B(p_1, q)$. Since the bisector $B(p_1, q)$ treated as a curve is either nonincreasing or nondecreasing, it cannot intersect the interior of the wedge W. Therefore W is included in $V(p_1) \cup B(p_1, q)$.
The conditions (a) and (b) imply that $p' := (p.x, q.y) \in W$ and therefore $d(p', p_1) \leq d(p', q)$. Since $p.y \leq q.y$ we have $d(p, p_1) \leq d(p', p_1)$; it follows $d(p, p_1) \leq d(p', p_1) \leq d(p', q) = q.x - p.x$.

Lemma 4.5: Let $s_1, s_2 \in S$ such that $s_1 \cap s_2 = \emptyset$, $s_1 <_{Obj} s_2$. Let $p_1 \in s_1$, $p_2 \in s_2$ be points with $d(p_1, p_2) = d(s_1, s_2)$. Then we have $p_1.y \leq p_2.y$.

Proof: If $p_1.x = p_2.x$ the assertion follows directly from Lemma 2.5. We now consider the case $p_1.x < p_2.x$ (the case $p_2.x < p_1.x$ can be handled in an analogous way). Assume $p_1.y > p_2.y$. The choice of p_1 and p_2 now implies that the open rectangle $]p_1.x, p_2.x[\times]p_1.y, p_2.y[$ contains no point of s_1 or s_2. Since s_2 is convex this implies $s_2[L].y \leq p_2.y$. If $s_2[L].x \leq p_1.x$ the vertical line $G := \{(x, y): x = p_1.x\}$ intersects s_2 'below' p_1; Lemma 2.5 now implies $s_2 <_{Obj} s_1$, a contradiction. If $s_2[L].x > p_1.x$ then $s_2[L] \in LC(s_1)$ which also implies $s_2 <_{Obj} s_1$, a contradiction. Hence the assumption $p_1.y > p_2.y$ is wrong, and we obtain $p_1.y \leq p_2.y$.

Notation: For Lemma 4.6 and Theorem 4.7 we introduce the following notations.
Let S be a configuration of *pairwise disjoint* convex compact objects.
$\delta_{min} := \min\{d(s, s'): s, s' \in S, s \neq s'\}$ denotes the smallest distance in S.
Let $s_1, s_2 \in S$, $s_1 <_{Obj} s_2$, be a closest pair and $p_1 \in s_1$, $p_2 \in s_2$ be points with $d(p_1, p_2) = d(s_1, s_2) = \delta_{min}$. By Lemma 4.5 $p_1.y \leq p_2.y$.
Let T denote the line segment with end points p_1 and p_2, $T_1 := \{(x, y) \in E^2: x = p_1.x, y \leq p_1.y\}$ and $T_2 := \{(x, y) \in E^2: x = p_2.x, y \geq p_2.y\}$. The 'chain' K consisting of T_1, T and T_2 splits the plane into two parts: let L denote the 'left' and R the 'right' connected component of $E^2 \setminus K$.
Let $x_1 := \min(p_1.x, p_2.x)$, $x_2 := \max(p_1.x, p_2.x)$ and $U := [x_1, x_2] \times [p_1.y, p_2.y]$.
Let $Z(s_1, s_2) := (HQ(s_2) \cup LC(s_2)) \cap (HQ(s_1) \cup UC(s_1))$.

The following figure illustrates these definitions:

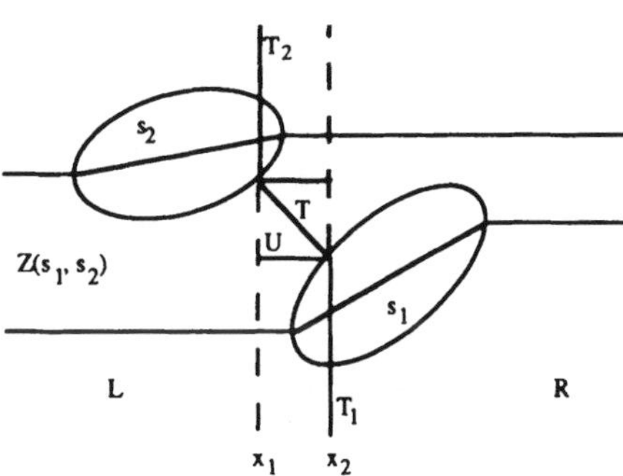

Lemma 4.6: The closest pair s_1, s_2 and the points $p_1 \in s_1$ and $p_2 \in s_2$ with $d(p_1, p_2) = d(s_1, s_2) = \delta_{min}$ can be chosen such that for each $s \in S$ with $s_1 <_{Obj} s <_{Obj} s_2$ one of the following statements hold:

(1) $(s[R] \in Z(s_1, s_2)) \wedge (s[R].x \leq x_1) \wedge (s <_{Del} s_1) \wedge (s <_{Del} s_2)$.

(2) $(s[L] \in Z(s_1, s_2)) \wedge (s[L].x \geq x_2) \wedge (s_1 <_{Ins} s) \wedge (s_2 <_{Ins} s)$.

Sketch of proof: We use the notation as introduced above. Choose the closest pair s_1, s_2 and the points $p_1 \in s_1$, $p_2 \in s_2$ such that no object from $S \setminus \{s_1, s_2\}$ intersects U and neither s_1 nor s_2 intersects the interior of U (for L_p-metrices d where $1 \le p < \infty$ this condition is true for every closest pair of objects s_1, s_2 and every pair of points $p_1 \in s_1$, $p_2 \in s_2$ with $d(p_1, p_2) = \delta_{min}$; for the L_∞-metric try to choose p_1 and p_2 such that $p_1.x = p_2.x$; if this is not possible in the configuration S at hand, then one can choose s_1, s_2 and p_1, p_2 arbitrary).
Let $s \in S$ such that $s_1 <_{Obj} s <_{Obj} s_2$. According to Lemma 2.5 s does neither intersect T_1 nor T_2; due to the definition of p_1 and p_2 s does not intersect T; therefore either $s \subset L$ or $s \subset R$. Then one can show that $s \subset L$ implies statement (1) and $s \subset R$ implies statement (2). In the case $p_1.x = p_2.x$ this is trivial. In the cases $x_1 = p_2.x < p_1.x = x_2$ and $x_1 = p_1.x < p_2.x = x_2$ one mainly has to show by applying Lemma 2.7 that the assumptions $s[L] \in]x_1, x_2[\times E$ and $s[R] \in]x_1, x_2[\times E$ for objects s with $s_1 <_{Obj} s <_{Obj} s_2$ lead to contradictions.

Theorem 4.7: If the objects in S are pairwise disjoint then algorithm 'PSCP' finds a closest pair.

Proof: We use the notation as introduced above. Choose the closest pair s_1, s_2 and the points $p_1 \in s_1$, $p_2 \in s_2$ according to Lemma 4.6. Then the set $A := \{s \in S: s_1 <_{Obj} s <_{Obj} s_2\}$ can be split into the subsets $A_1 := \{s \in A: s <_{Del} s_1, s <_{Del} s_2\}$ and $A_2 := \{s \in A: s_1 <_{Ins} s, s_2 <_{Ins} s\}$. Without loss of generality let $s_2 <_{Ins} s_1$.
If δ denotes the current value of variable 'delta' when s_1 gets inserted into the y-table then by construction $\delta \ge \delta_{min}$. In the case $\delta = \delta_{min}$ a closest pair has already been found and nothing remains to be shown. Therefore let $\delta > \delta_{min}$. Then $s_1[L].x - s_2[R].x \le d(s_1, s_2) = \delta_{min} < \delta$ implies that s_2 is active with respect to s_1. If s_1 becomes a neighbor of s_2 in the y-table when it is inserted, then the algorithm computes their distance and finds a closest pair.
Now assume that $B := \{s \in A(s_1): s_1 <_{Obj} s <_{Obj} s_2\} \neq \emptyset$. Since $B \subset A$ and $s <_{Ins} s_1$ it follows from Lemma 4.6 that $B \subset A_1$ and therefore $A_1 \neq \emptyset$.
Let s" be the largest object in A_1 with respect to '$\le_{Del}$'. We have s" $<_{Del} s_1$ and s" $<_{Del} s_2$ and either s" $\in$ B or $s_1 <_{Ins}$ s"; in any case there exists a moment at which all objects s_1, s_2 and s" are contained in the y-table simultaneously. Because of s" $<_{Del} s_1$ and s" $<_{Del} s_2$ the objects s_1 and s_2 are contained in the y-table at that moment when s" is deleted.
If $A_2 = \emptyset$ then obviously the deletion of s" causes the objects s_1 and s_2 to become neighbors in the y-table, and the algorithm computes their distance and finds a closest pair.
In the following let $A_2 \neq \emptyset$. Let s' be the smallest object from A_2 with respect to '$\le_{Ins}$'. We consider the moment immediately before s' gets inserted into the y-table. Let δ' denote the current value of variable 'delta'; let $\delta' > \delta_{min}$ (if $\delta' = \delta_{min}$ nothing remains to be shown). Now it suffices to show that s" is not active with respect to s'; then s" is deleted from the y-table before s' is inserted, and the deletion of s" causes s_1 and s_2 to become neighbors. Let $C := A \cap A(s')$. Assume $C \neq \emptyset$. We will show that this assumption leads to a contradiction.
The choice of s' implies $C \subset A_1$. Hence, for each $s \in C$ we have $s <_{Del} s_1$ and $s <_{Del} s_2$; now $s_1 <_{Ins} s'$, $s_2 <_{Ins} s'$ and $C \neq \emptyset$ imply $s_1 \in A(s')$ and $s_2 \in A(s')$ and therefore $C \cup \{s_1, s_2\}$ forms a section in A(s'). Hence also the set $C \cup \{s_1, s_2\}$ and the slice $D(s', \delta')$ satisfy conditions (a), (b) and (c) of Lemma 2.8. Let $s \in C$; by Lemma 4.6 we have $s[R] \in Z(s_1, s_2) \cap L$ and thus $\min\{p_1.y, s_1[L].y\} \le s[R].y \le \max\{p_2.y, s_2[L].y\}$; by Lemma 2.8 (1) we obtain $p_1.y < s[R].y < p_2.y$ for each $s \in C$. Therefore $C = \emptyset$ if $p_1.y = p_2.y$. It remains to consider the case $p_1.y < p_2.y$:

Let $q := s'[L]$; By Lemma 4.6 $q \in Z(s_1, s_2) \cap R$. Thus for q.y one of the following three cases occurs: (1): $p_1.y < q.y < p_2.y$; (2): $s_1[R].y \le q.y \le p_1.y$; (3): $s_2[R].y \ge q.y \ge p_2.y$. We handle these cases separately.

Case (1): $p_1.y < q.y < p_2.y$. Let $C_1 := \{s \in C: s[R].y \le q.y\}$ and $C_2 := \{s \in C: s[R].y > q.y\}$. Assume that $C_1 \ne \emptyset$. The points s[R], $s \in C$, are included in D(s', δ'); from Lemma 2.8 (2) it follows that the total ordering on C induced by the values s[R].y, $s \in C$, is equal to the total ordering '$\le_{Obj}$'; therefore s_1 has a neighbor $s^* \in C_1$ at the moment under consideration. Let $p = s^*[R]$; since s* and s_1 are neighbors their mutual distance has been checked; thus $\delta' \le d(s^*, s_1) \le d(p, p_1)$. The points p, p_1, p_2 and q satisfy conditions (a) and (b) of Lemma 4.4. Because of the choice of p_1 and p_2 we have $d(q, p_2) \ge d(p_1, p_2)$. Since $s^* \in C_1$ we have $p.y \le q.y$. Then by Lemma 4.4 (1) we obtain $d(p, p_1) \le q.x - p.x$. Since $\delta' \le q.x - p.x$ the object s* is not active with respect to s', a contradiction. Therefore $C_1 = \emptyset$. By applying Lemma 4.4 (2) one can show analogously that also the assumption $C_2 \ne \emptyset$ leads to a contradiction. Therefore $C = C_1 \cup C_2 = \emptyset$.

Case (2): $s_1[R].y \le q.y \le p_1.y$. Let $s^* \in C$ be the predecessor of s_2 in A(S'). Let $p := s^*[R]$; then $\delta' \le d(s^*, s_2) \le d(p, p_2)$. Let $q' := (q.x, p.y)$; the points p, p_1, p_2 and q' satisfy the conditions (a) and (b) of Lemma 4.4. Let T_1 be the line segment with end points p_1 and $s_1[R]$. Due to convexity $T_1 \subset s_1$; hence the distance between any point of T_1 and q is at least $\delta_{min} = d(p_1, p_2)$. Since $s_1[R].y \le q.y \le p_1.y$ the line segment T_1 includes a point z with $z.y = q.y$; now $q.x - z.x = d(q, z) \ge d(p_1, p_2)$; obviously we have $p_1.x \le z.x$, and therefore it follows $q.x - p_1.x \ge d(p_1, p_2)$, implying $q'.x - p_1.x \ge d(p_1, p_2)$; we conclude $d(q', p_1) \ge d(p_1, p_2)$; now by Lemma 4.4 we obtain $d(p, p_2) \le q'.x - p.x = q.x - p.x$. Hence $\delta' \le q.x - p.x$ which implies s* is not active with respect to s', a contradiction. Therefore $C = \emptyset$.

Case (3): $s_2[R].y \ge q.y \ge p_2.y$. This case can be handled analogously to case (2): consider the successor $s^* \in C$ of s_1 in A(S') and show that it cannot be active with respect to s'. Therefore also in this case $C = \emptyset$.

Since $C = \emptyset$, s" is deleted from the y-table before s' is inserted, and the deletion of s" causes s_1 and s_2 to become neighbors. Therefore the algorithm computes their distance and finds a closest pair.

5 Analysis

Definition 5.1: Let the input configuration S be subset of a specific class C of compact, convex point sets of a certain type in the plane, e.g. the class of line segments or the class of m-gons. We assume that there exist functions g_C and h_C which depend on the underlying class C of objects such that:

- For each $s \in C$ the points s[L] and s[R] can be computed by at most $O(g_C)$ operations.
- For any $s_1 \in C$, $s_2 \in C$ their (minimal) distance $d(s_1, s_2) := \min\{d(p, q): p \in s_1, q \in s_2\}$ with respect to any L_p-metric d ($1 \le p \le \infty$) can be computed by performing at most $O(h_C)$ operations.

Theorem 5.2: The plane-sweep algorithm 'PSCP' solves the closest-pair problem for a configuration S as defined above in time $O(n \cdot \log n + n \cdot (g_C + h_C))$.

Proof: For computing all points s[L] and s[R], $s \in S$, we need $O(n \cdot g_C)$ operations. The initialization of the x-queue, i.e. sorting the objects with respect to '$\le_{Ins}$' and '$\le_{Del}$', can be accomplished in O(n log n) worst-case time. Each comparison between two objects with respect to '$\le_{Obj}$' costs constant time; therefore the cost for all operations performed on the y-table during the plane-sweep, i.e. inserting, deleting objects and finding neighbors, sums up to O(n log n) time in the worst case. 'PSCP' computes the distance of at most $3n - 5$ (for $n \ge 4$) pairs of objects, implying that all these distance computations take $O(n \cdot h_C)$ time. This completes the proof of Theorem 5.2.

From this theorem we obtain immediately

Theorem 5.3: The plane-sweep algorithm 'PSCP' solves the closest-pair problem for a configuration S of n line segments or n circles in time O(n·log n).

Theorem 5.4: Let the input configuration S consist of n convex m-gons each of which is represented by an array containing its vertices given in cyclic order. Then the plane-sweep algorithm 'PSCP' finds a closest pair with respect to the Euclidean metric in time O(n log(n·m)).

Proof: For a convex m-gon s given by its vertices in cyclic order the extreme points with respect to the lexicographic order can be determined in time O(log m) by Fibonacci search [CD 87]. We can detect in optimal time O(log m) whether two convex m-gons intersect [CD 87]. The minimal Euclidean distance between two non-intersecting convex m-gons can be computed in optimal time O(log m) ([E 85], [CW 83]). Thus we have $g_C \in O(\log m)$ and $h_C \in O(\log m)$, and the assertion follows from Theorem 5.2.

6 Open problems

Our proof of correctness in section 4 as well as the proof of Lemma 2.8 depend heavily on the convexity of the objects in S. The question arises whether algorithm 'PSCP' can be generalized to solve the closest-pair problem for less restricted objects, e.g. objects which are 'monotone in the x-direction' [S 85]. Another open question is whether the run time stated in Theorem 5.4 is optimal in the algebraic decision tree model of computation.

References

[CD 87] B. Chazelle, D. P. Dobkin: Intersection of Convex Objects in Two and Three Dimensions, Journal of the ACM, Vol. 34, No. 1 (1987), 1-27.

[CW 83] F. Chin, C. A. Wang: Optimal Algorithms for the Intersection and the Minimum Distance Problems between Planar Polygons, IEEE Trans. Comput. 32 (12), 1203-1207 (1983).

[E 85] H. Edelsbrunner: Computing the Extreme Distances between two Convex Polygons, Journal of Algorithms 6, 213-224 (1985).

[F 87] S. Fortune: A Sweepline Algorithm for Voronoi Diagrams, Algorithmica 2, 153-174 (1987).

[HNS 88] K. Hinrichs, J. Nievergelt, P. Schorn: Plane-Sweep Solves the Closest Pair Problem Elegantly, Information Processing Letters 26, 255-261 (1988).

[L 80] D. T. Lee: Two-Dimensional Voronoi Diagrams in the L_p-metric, Journal of the ACM 27 (4), 604-618 (1980).

[PS 85] F. P. Preparata, M. I. Shamos: Computational Geometry: An Introduction, Springer-Verlag, Berlin, Heidelberg, New York, 1985.

[SH 76] M. I. Shamos, D. Hoey: Geometric Intersection Problems, Proc. 17th Ann. IEEE Symp. on Foundations of Computer Science, 208-215 (1976).

[S 85] M. Sharir: Intersection and Closest-Pair Problems for a Set of Planar Discs, SIAM J. Comput. 14, 448-468 (1985).

[Y 87] C. K. Yap: An O(n log n) Algorithm for the Voronoi Diagram of a Set of Simple Curve Segments, Discrete Comput. Geometry 2, 365-393 (1987).

Linear Approximation of Simple Objects

Jean-Marc Robert* and Godfried Toussaint†

Abstract

Let S be a set of m convex polygons in the plane with a total number of n vertices. Each polygon has a positive weight. This paper presents algorithms to solve the *weighted minmax approximation* and the *weighted minsum approximation* problems. For the first problem, a line minimizing the maximum weighted distance to the polygons can be found in $O(n^2 \log n)$ time and $O(n^2)$ space. The time and space complexities can be reduced to $O(n \log n)$ and $O(n)$, respectively, when the weights are equal. For the second problem, a line minimizing the sum of the weighted distances to the polygons can be found in $O(n^2 \log n)$ time and $O(n)$ space. For both problems, we also obtain similar results for sets of n circles or line segments.

1 Introduction

The problem of approximating sets of points in the plane by lines is encountered in fields as statistical analysis, computer vision, pattern recognition and computer graphics and it is usually referred to as the *linear approximation* or the *linear regression* problem. The problem consists of finding the "best" line approximating a set of points. There are many possibilities for the optimality criterion used. A good survey of algorithms that use a variety of criteria can be found in [II88]. For example, we may want to find a line minimizing the maximum orthogonal distance to the points or minimizing the sum of these distances. In [HIIRY], [DKM84] and [HT88] algorithms solving these problems are presented. A very different parallel strip criterion is used in [Tou85]. Finally, Guibas, Overmars and Robert [GOR91] considered yet another optimality criterion. They presented an algorithm to find a line containing the maximum number of data points.

In some applications, the data points to be approximated are not defined precisely but they are themselves approximated by simple objects such as polygons or circles. Naturally, in such cases, we may still want to find the "best" approximating line. In [O'R81], O'Rourke examined the problem of finding a line consistent with a set of data ranges *i.e.* a line intersecting a set of vertical line segments. This problem has been later generalized to more complex objects and corresponds to the *line transversal* problem. A set of objects has a line transversal if there is a line intersecting each of its members. Atallah and Bajaj [AB87] presented an $O(n \log n)$ time algorithm to determine whether a set of n circles has a line transversal or not. In [EMPRWW82], Edelsbrunner *et al.* described an $O(n \log n)$ time algorithm to compute the description of all the line transversals of a set of n line segments. The optimality of these two algorithms has been later proved by Avis, Robert and Wenger [ARW89]. Edelsbrunner, Guibas and Sharir [EGS89] reduced the line transversal problem for a set of polygons with a total number of n vertices into the problem of finding the envelopes of a linear number of line segments. The *lower envelope* of a set of functions $F = \{f_i : \Re \rightarrow \Re | 1 \leq i \leq n\}$ is the function LE_F defined as $LE_F(x) = \min_i f_i(x)$. The *upper envelope* of F is the function UE_F defined as $UE_F(x) = \max_i f_i(x)$. The envelopes of sets of line segments are defined similarly. An $O(n \log n)$ time algorithm described by Hershberger [Her89] for computing the envelopes of n line segments gives an $O(n \log n)$ time algorithm for

*Dép. d'Informatique et de Recherche Opérationnelle, Université de Montréal, C.P. 6128, Succ. A, Montréal, Qué., Canada, H3C 3J7. This work was done while this author was at McGill University.

†School of Computer Science, McGill University, 3480 University St., Montréal, Qué., Canada, H3A 2A7.

solving the line transversal problem for sets of polygons with a total of n vertices. Finally, Egyed and Wenger [EW89] presented a linear time algorithm to compute the description of all the line transversals of sets of n mutually disjoint translates of a simple convex object.

Such an exact approximating line for a set of objects does not always exist but, in some applications, a line minimizing the maximum orthogonal distance to the objects is sufficient. This problem is called the *minmax approximation* (MMA) problem. We present in the next section an $O(n \log n)$ time and $O(n)$ space algorithm to find a line minimizing the maximum orthogonal distance to sets of polygons with a total number of n vertices and to sets of n circles or line segments. For sets of circles or line segments, this algorithm is optimal. A weighted version of the MMA problem is also considered. This problem is called the *weighted minmax approximation* (WMMA) problem and can be solved in $O(n^2 \log n)$ time and $O(n^2)$ space.

In Section 3, we consider the *weighted minsum approximation* (WMSA) problem. Here, the approximating line must minimize the sum instead of the maximum of the weighted orthogonal distances to the objects. We show how to find such an approximating line in $O(nm \log m)$ time and $O(n)$ space for sets of m polygons with a total of number of n vertices and $O(n^2)$ time and $O(n)$ space for sets of n line segments. The time complexity can be reduced to $O(n^2)$ for sets of n circles.

2 The WMMA Problem for Objects

Let $S = \{\mathcal{O}_1, \mathcal{O}_2, \ldots, \mathcal{O}_m\}$ be a set of compact convex objects in the plane. With each object $\mathcal{O}_i$, we associate a positive weight ω_i. Let $cp(\mathcal{O}_i, l)$ denote a point of the object $\mathcal{O}_i$ at minimum orthogonal distance from the line l and let $\delta(\mathcal{O}_i, l)$ be the orthogonal distance between $cp(\mathcal{O}_i, l)$ and l. Throughtout this paper the notion of distance refers to the Euclidean distance. Let $H^+(l)$ and $H^-(l)$ correspond, respectively, to the upper and the lower closed half-planes delimited by l. A line l defines a partition of S into three disjoint subsets

$$\begin{aligned} S_{up}(l) &= \{\mathcal{O}_i \in S \,|\, \mathcal{O}_i \subseteq H^+(l) \setminus l\}, \\ S_{on}(l) &= \{\mathcal{O}_i \in S \,|\, \mathcal{O}_i \cap l \neq \emptyset\} \text{ and} \\ S_{low}(l) &= \{\mathcal{O}_i \in S \,|\, \mathcal{O}_i \subseteq H^-(l) \setminus l\}. \end{aligned}$$

The following lemma characterizes any optimal solution for the WMMA problem and is a generalization of a lemma proved by Morris and Norback [MN83] for sets of points.

Lemma 2.1 *Let S be a set of compact convex objects in the plane. If S does not have any line transversal, any optimal line l for the WMMA problem is at maximum weighted distance from at least three objects, one on one side of l, say $\mathcal{P}$, and two on the other side, say $\mathcal{A}$ and $\mathcal{B}$, and the orthogonal projection on l of $cp(\mathcal{P}, l)$ lies between the orthogonal projection of $cp(\mathcal{A}, l)$ and $cp(\mathcal{B}, l)$.*

In order to reformulate the previous lemma in a more useful way for the unweighted case, we introduce a few definitions. Let $up\_sup(\mathcal{O}_i, \Theta)$ and $low\_sup(\mathcal{O}_i, \Theta)$ be the upper and the lower supporting line with slope Θ to the object $\mathcal{O}_i$, respectively. Let $UT(S, \Theta)$ be the set of upper supporting lines with slope Θ to the members of S *i.e.* $UT(S, \Theta) = \{up\_sup(\mathcal{O}_i, \Theta) | \mathcal{O}_i \in S\}$. An upper supporting line $up\_sup(\mathcal{O}_i, \Theta)$ is *extreme* for the set S if $up\_sup(\mathcal{O}_i, \Theta)$ is "lower" than any other supporting line in $UT(S, \Theta)$ *i.e.* the closed upper half-plane delimited by $up\_sup(\mathcal{O}_i, \Theta)$ contains all the supporting lines in $UT(S, \Theta)$. The set of lower supporting lines $LT(S, \Theta)$ and the extremal notion for lower supporting lines are defined similarly. Finally, a line is a *common* supporting line for a pair of objects if it is a lower or an upper supporting line for both objects.

The MMA problem for a set of objects is equivalent to finding the *thinnest strip* that intersects every object of the set. A strip is simply an open region delimited by two parallel lines. The orthogonal distance between the two parallel lines delimiting a strip corresponds to the *width* of the strip. By analogy with a line transversal, a *strip transversal* of a set of objects is a strip intersecting all the objects of the set. Thus, the *medial axis* of a thinnest strip transversal of a set of objects corresponds to a line minimizing the maximum orthogonal distance to the objects of the set. The medial axis of a strip is the line parallel and equidistant to the boundaries of the strip. Hence, the previous lemma can be restated as:

Lemma 2.2 *Let S be a set of compact convex objects in the plane. If S has no line transversal, a thinnest strip transversal is delimited by an extreme common supporting line to two objects and an extreme supporting line to a third one.*

This lemma gives a very simple brute-force algorithm for the MMA problem. By enumerating the common supporting lines of every pair of objects, we can find a thinnest strip transversal. To reduce the time complexity of this algorithm, we have to find a way to enumerate only the extreme common supporting lines and to determine efficiently the width of the strip transversal associated with any extreme common supporting line.

In [EMPRWW82], Edelsbrunner *et al.* reduced the line transversal problem for a set of line segments to a double-wedge intersection problem in the dual space. Later, Atallah and Bajaj [AB87] and Edelsbrunner, Guibas and Sharir [EGS89] used a similar transformation to solve the line transversal problem for different kinds of objects. The ideas developed in these papers can be used to solve the thinnest strip transversal problem as well.

The dual transform $\mathcal{D}$ maps a point $p = (x_p, y_p)$ in the primal space to the non-vertical line $\mathcal{D}(p)$ defined by the equation $b = -x_p \cdot a + y_p$ in the dual space and a non-vertical line l defined by the equation $y = a_l \cdot x + b_l$ in the primal space to the point $\mathcal{D}(l) = (a_l, b_l)$ in the dual space [Bro80]. Since $\mathcal{D}$ is a bijection, its inverse $\mathcal{D}^{-1}$ is properly defined. An important porperty of $\mathcal{D}$ is that it preserves the above/on/below relation between points and lines. This dual transformation can be extended for any compact convex object $\mathcal{O}$. In this case, $\mathcal{D}(\mathcal{O})$ corresponds to all the lines intersecting $\mathcal{O}$ in the primal space. More precisely, $\mathcal{D}(\mathcal{O})$ is given by $\bigcup_{l \cap \mathcal{O} \neq \emptyset} \mathcal{D}(l)$. The boundary of $\mathcal{D}(\mathcal{O})$ is decomposed into an upper part *up_bd*$(\mathcal{D}(\mathcal{O}))$ and a lower part *low_bd*$(\mathcal{D}(\mathcal{O}))$ corresponding to the upper and lower supporting lines of $\mathcal{O}$, respectively.

In the dual space, the upper common supporting line of a pair of objects $\mathcal{O}_i$ and $\mathcal{O}_j$ is given by the intersection point of *up_db*$(\mathcal{D}(\mathcal{O}_i)) \cap$ *up_db*$(\mathcal{D}(\mathcal{O}_j))$. An upper common supporting line is extreme if and only if the corresponding point in the dual space is below or on every *up_bd*$(\mathcal{D}(\mathcal{O}_i))$. Thus, it is sufficient to construct the lower envelope of $\{up\_bd(\mathcal{D}(\mathcal{O}_i)) | \mathcal{O}_i \in S\}$ to enumerate the extreme common upper supporting lines.

We are now ready to outline our algorithm solving the MMA problem for compact convex simple objects in the plane.

Algorithm MMA

Input: A family $S = \{\mathcal{O}_i\}$ of convex polygons, line segments or circles.
Output: A line l minimizing the maximum distance to the objects in S.

1. Compute the thinnest vertical strip intersecting the objects.
2. Compute the dual representation of the objects $\mathcal{D}(\mathcal{O}_i)$.
3. Find the lower envelope LE_S of $\{up\_bd(\mathcal{D}(\mathcal{O}_i))\}$ and the upper envelope UE_S of $\{low\_bd(\mathcal{D}(\mathcal{O}_i))\}$.
4. Traverse the envelopes and a the vertex $v = (a_v, b_v)$ of one of the envelopes minimizing the value $(UE_S(a_v) - LE_S(a_v))/\sqrt{a_v^2 + 1}$.

End of the Algorithm

Since the dual transformation $\mathcal{D}$ mapping a point to a line and vice versa is not defined for a vertical line, the case of a vertical strip must be solved separately. In this case, it is sufficient to project all the objects on the x-axis and to find the shortest interval of the x-axis intersecting all the intervals corresponding to the projections of the objects.

The crucial step of Algorithm MMA is Step 3 where the upper and lower envelopes are built. These envelopes depend heavily on the structure of objects in S. In the next theorem, we show how to construct these envelopes for some specific sets of objects and how to traverse them to find an optimal solution for the MMA problem.

Theorem 2.3 *Algorithm MMA correctly solves the MMA problem in $\Theta(n)$ time for a set of n sorted vertical line segments, in $\Theta(n \log n)$ time for a set of n circles or a n line segments and in $O(n \log n)$ time for a set of polygons with a total of n vertices.*

Proof The algorithm is divided into two parts. The construction of the envelopes and their scan. Once the envelopes are built, the structure of the objects is not important anymore. Hence, we begin by presenting how to build the envelopes efficiently and, later, how to process them to find an optimal solution to the MMA problem.

Let S be a sorted set of n vertical line segments. An easy way to build the lower envelope of $\{up\_bd(\mathcal{D}(O_i))\}$ is to compute the lower hull of the upper endpoints of the line segments. This step can be done in linear time since the endpoints are sorted. The edges of the lower hull correspond to the vertices of the lower envelope of $\{up\_bd(\mathcal{D}(O_i))\}$. An edge of the lower hull represents a extreme upper supporting line of a pair of line segments, say O_i and O_j. In the dual space, this common upper supporting line corresponds to the intersection point of $up\_bd(\mathcal{D}(O_i))$ and $up\_bd(\mathcal{D}(O_j))$ which is below or on any other $up\_bd(\mathcal{D}(O_k))$, for $k \neq i, j$. Therefore, this point corresponds to a vertex of the lower envelope. Using a similar argument, we can show that the vertices of the lower hull corresponds to the edges of the lower envelope. The upper envelope can be computed similarly.

For sets of circles, Atallah and Bajaj [AB87] proposed an $O(n \log n)$ time algorithm to compute the linear size envelopes of the dual representation of n circles. For sets of line segments, the envelopes correspond to the envelopes of $2n$ half-lines. The length of the lower envelope of n half-lines unbounded to the right is linear. The same observation is also true for n half-lines unbounded to the left. Finally, the merge of these two envelopes does not increase the length of the resulting envelope by more than a constant factor. Therefore, the length of the envelopes is in $O(n)$ and can be computed in $O(n \log n)$ time by using the divide-and-conquer algorithm presented in [Her89].

For sets of polygons with a total of n vertices, the envelopes of the dual representation of these polygons correspond to the envelopes of $O(n)$ line segments. The length of these envelopes is in $O(n\alpha(n))$. Nevertheless, Hershberger [Her89] presented an $O(n \log n)$ time algorithm to compute such envelopes.

We are now ready to process the envelopes to find a solution to the MMA problem. By Lemma 2.2, each vertex of the envelopes is a candidate for an optimal solution. By scanning the two envelopes from left to right, we can find an optimal solution in time proportional to the length of the envelopes. Since both envelopes are scanned at the same time, the coordinates of the next vertex to be considered and the corresponding point on the other envelope with the same a-coordinate can be determined in constant time. Let a_v be the a-coordinate of a vertex of one of the envelopes. Then, compute the value $(UE_S(a_v) - LE_S(a_v))/\sqrt{a_v^2+1}$ and associate it to the vertex considered. This value corresponds to the width of the strip delimited by the parallel lines $\mathcal{D}^{-1}((a_v, UE_S(a_v)))$ and $\mathcal{D}^{-1}((a_v, LE_S(a_v)))$. A vertex with the minimum value gives an optimal solution to the MMA problem. Note that if this value is less than or equal to zero, the solution corresponds to a line transversal for the set.

The time complexity of the algorithm is determined by the time taken by the construction of the envelopes. The optimality of the algorithm for sets of circles or line segments comes from the lower bounds on the line transversal problem for these objects given in [ARW89]. □

We now consider a constrained version of the MMA problem where the approximating line has to contain the origin. We can show that a line passing through the origin and minimizing the maximum orthogonal distance to the objects must be at maximum distance of at least two objects. Thus, Algorithm MMA can be transformed to solve the constrained problem as well. By projecting the vertices of the envelopes on the a-axis, we divide the axis into intervals. The portions of the envelopes associated to each interval correpond to a line segment or a hyperbolic arc. In both cases, it is possible with analytic methods to find an optimal solution in constant time for any given interval. A closer look at the proofs of the lower bounds on the line transversal problems shows that these lower bounds still apply when the line transversal must contain the origin. Therefore, the solution for sets of circles or line segments is also optimal. The following corollary summarizes these results:

Corollary 2.4 *A line containing the origin and minimizing the maximum distance to a set of objects can be found in $\Theta(n)$ time for a set of n sorted vertical line segments, in $\Theta(n \log n)$ time for a set of n circles or n line segments, in $O(n \log n)$ time for a set of polygons with a total of n vertices.*

We can formulate weighted versions of the two problems considered so far in this section. The weighted distance, denoted $\omega\delta(O,l)$, between an object O and a line l is defined as $\omega \cdot \delta(O,l)$ where ω is the weight associated with O. We begin by solving the constrained problem. The next lemma show that this problem can be reduced to the unweighted case. Lee and Wu [LW86] proved the same result for a set of weighted points.

Lemma 2.5 *The constrained version of the WMMA problem where the approximating line must contain the origin can be reduced to the unweighted version of the same problem.*

Sketch of the proof Let $S = \{O_i\}$ be a set of compact convex objects. For any object $O_i \in S$, let ωO_i be the object defined by the set $\{(\omega_i x, \omega_i y) | (x,y) \in O_i\}$ and let S^* be the set containing all the objects ωO_i. A line containing the origin and minimizing the maximum distance to the objects in S^* minimizes also the maximum weighted distance to the objects in S. □

By combining Corollary 2.4 and Lemma 2.5, we obtain the following result:

Corollary 2.6 *A line containing the origin and minimizing the weighted maximum distance to a set of objects can be found in $\Theta(n)$ time for a set of n sorted vertical line segments, in $\Theta(n \log n)$ time for a set of n circles and for a set of n line segments, in $O(n \log n)$ time for a set of polygons with a total of n vertices.*

For the general problem, Morris and Norback [MN83] showed how to reduce any instance of the WMMA problem for a set of n points to an instance of the constrained MMA problem where the approximating line must contain the origin for a set of $O(n^2)$ points. In [LW86], Lee and Wu used this fact to solve the WMMA problem in $O(n^2 \log n)$ time and $O(n^2)$ space for sets of n points. This reduction can be extended for sets of convex objects.

Lemma 2.7 *The WMMA problem for a set of n convex objects can be reduced to the constrained MMA problem where the approximating line must contain the origin for a set of $O(n^2)$ convex objects.*

Sketch of the proof Let $S = \{O_i\}$ be a set of compact convex objects. For any pair of objects O_i and O_j of S, construct the object ωO_{ij} defined as

$$\left\{ \left(\frac{\omega_i \omega_j}{\omega_i + \omega_j}(x_i - x_j), \frac{\omega_i \omega_j}{\omega_i + \omega_j}(y_i - y_j) \right) \mid (x_i, y_i) \in O_i \text{ and } (x_j, y_j) \in O_j \right\}.$$

Let S^* be the set containing all the objects ωO_{ij}. A line containing the origin and minimizing the maximum distance to the objects in S^* can be translated to obtain a line minimizing the maximum weighted distance to the objects in S. □

Therefore, by combining Corollary 2.4 and Lemma 2.7, we obtain the last result of this section.

Corollary 2.8 *A line minimizing the maximum weighted distance to a set of objects can be found in $O(n^2 \log n)$ time and $O(n^2)$ space for a set of n circles or line segments and for a set of polygons with a total of n vertices.*

3 The WMSA Problem for Objects

Let $S = \{\mathcal{O}_1, \mathcal{O}_2, \ldots, \mathcal{O}_m\}$ be a set of compact convex objects in the plane. With each object $\mathcal{O}_i$, we associate a positive weight ω_i. Let l be a line and $S_{up}(l)$, $S_{on}(l)$ and $S_{low}(l)$ be the partition of S induced by l as defined in the previous section. Define the values $W_{up}(l)$, $X_{up}(l)$ and $Y_{up}(l)$ as follows:

$$\begin{aligned} W_{up}(l) &= \sum_{\mathcal{O}_i \in S_{up}(l)} \omega_i, \\ X_{up}(l) &= \sum_{\mathcal{O}_i \in S_{up}(l)} \omega_i \; x(cp(\mathcal{O}_i, l)) \quad \text{and} \\ Y_{up}(l) &= \sum_{\mathcal{O}_i \in S_{up}(l)} \omega_i \; y(cp(\mathcal{O}_i, l)). \end{aligned}$$

For any point p in the plane, $x(p)$ and $y(p)$ correspond to the x- and y-coordinate of p, respectively. The values $W_{low}(l)$, $X_{low}(l)$ and $Y_{low}(l)$ are defined similarly.

Lemma 3.1 *Let S be a set of compact convex objects in the plane. A line minimizing the sum of the weighted distances to the objects of S must be a supporting line of at least two objects.*

The above lemma gives a simple $O(m^3 \log n)$ time brute-force algorithm to solve the problem for sets of m polygons with a total number of n vertices. For every one of the $O(m^2)$ lines supporting at least two objects, compute in $O(m \log n)$ time the sum of the weighted distances to the polygons and choose the supporting line minimizing the sum. Fortunately, we can obtain a better solution.

For sets of n non-vertical line segments, the WMSA problem can be solved in $O(n^2 \log n)$ time and $O(n)$ space by sweeping with a vertical line the arrangement corresponding to the dual representation of the line segments. This solution can be extended to hold for a set of convex polygons. For sets of n circles, the arrangement can be swept with a topological line instead of a vertical line. This fact reduces the time complexity of the algorithm to $O(n^2)$.

As we pointed out in Section 2, a line supporting two objects corresponds to a point in the intersection of the boundaries of the dual representation of both objects. Hence, the candidates for the optimal approximating lines are given by the vertices of the arrangement of the boundaries of $\mathcal{D}(\mathcal{O}_i)$. This arrangement is denoted $\mathcal{A}(\mathcal{D}(S))$. For sets of line segments, $\mathcal{A}(\mathcal{D}(S))$ represents an arrangement of the $2n$ lines since the dual representation of a non-vertical line segment is a double wedge delimited by two intersecting lines. The algorithm solving the WMSA problem will report all the vertices of $\mathcal{A}(\mathcal{D}(S))$ in such a way that the sums of the weighted distances to the line segments can be evaluated efficiently.

The algorithm may be related to the algorithms reporting the intersections between line segments presented by Shamos and Hoey [SH76]. To implement it, we need two different data structures: a list L allowing the swap of any two elements and the access to any element in constant time and a priority queue Q supporting the Insertion, Deletion and Minimum operations in logarithmic time. These two structures can be realized by an array and a height-balanced search tree as an AVL or a 2-3-tree, respectively.

To initialize the sweeping process, we sort the lines according to their slopes and find a vertical line intersecting $\mathcal{A}(\mathcal{D}(S))$ at the left of all vertices. The leftmost vertex of the arrangement is determined by the intersection two lines with consecutive slopes. At any time, the list L contains the sorted list of the lines intersecting the sweep line ($L[1]$ is the line with the maximum b-coordinate intersection point) and the queue Q contains the candidates for the closest vertex to the right of the sweep line. At the beginning, the order in which the lines cut the sweep line corresponds to the order of the sorted lines. The closest vertex to the right of the sweep line is determined by two adjacent lines in L. Thus, Q is initialized with the intersections of all the pairs of adjacent lines in L and the closest vertex to the right of the sweep line is found using the

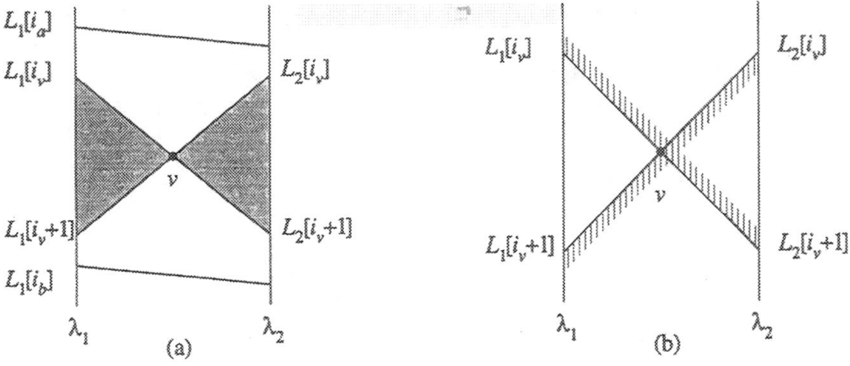

Figure 1: Updates of $L[i]$'s data for line segments

Minimum operation which gives the vertex v with the minimum a-coordinate in Q. When the sweep line crosses v, L and Q must be updated. Let $L[i]$ and $L[j]$ be two lines intersecting the sweep line at v and such that i is minimum and j is maximum. The fact that i and j are extreme implies that $L[i]$ has the smallest and $L[j]$ has the greatest slope of all the lines crossing the sweep line at v. The points determined by $L[i-1] \cap L[i]$, $L[i] \cap L[j]$ and $L[j] \cap L[j+1]$ (when $i = 1$ or $j = n$, the corresponding intersection point is disregarded) are replaced by the points determined by $L[i-1] \cap L[j]$ and $L[i] \cap L[j+1]$ in Q and the sequence $L[i], ..., L[j]$ is reverse in L. The updates can be done in $O(\log n + k)$ time where k represents the number of lines incident to v. Since $\mathcal{A}(\mathcal{D}(S))$ corresponds to a planar graph with $O(n^2)$ vertices, the sum of the numbers of lines incident to the vertices is in $O(n^2)$. Hence, the sweeping part of the algorithm takes $O(n^2 \log n)$ time overall.

To compute efficiently the sums of the weighted distances to the line segments, we associate to each entry $L[i]$ a line $l[i]$ in the primal space such that $\mathcal{D}(l_i)$ represents the intersection between the sweep line and $L[i]$. We also associate to $L[i]$ the sets $S_{up}(l[i])$ and $S_{low}(l[i])$ and the values $X_{up}(l[i])$, $Y_{up}(l[i])$, $W_{up}(l[i])$, $X_{low}(l[i])$, $Y_{low}(l[i])$ and $W_{low}(l[i])$. These values permit to compute the sum of the weighted distances between the line segments and the line $l[i]$ defined by $y = a_i x + b_i$ in constant time. This sum is given by

$$\frac{(Y_{up}(l[i]) - Y_{low}(l[i])) - a_i(X_{up}(l[i]) - X_{low}(l[i])) - b_i(W_{up}(l[i]) - W_{low}(l[i]))}{\sqrt{1+a_i^2}}.$$

Hence, the problem of computing the sums efficiently can be reduced to the problem of maintaining the values of $X_{up}(l[i])$, $Y_{up}(l[i])$, $W_{up}(l[i])$, $X_{low}(l[i])$, $Y_{low}(l[i])$ and $W_{low}(l[i])$ efficiently. We show in the next lemma how to do all these updates in $O(n^2)$ time overall.

Lemma 3.2 *The overall time to compute the sums of the weighted distances between the line segments in S and all the candidate lines is in $O(n^2)$.*

Proof Let L_1 be a copy of L when the sweep line intersects $\mathcal{A}(\mathcal{D}(S))$ at $a = a_1$. Similarly, L_2 is defined at $a = a_2$. The vertical lines at $a = a_1$ and $a = a_2$ are called λ_1 and λ_2, respectively. Let v be the only vertex of $\mathcal{A}(\mathcal{D}(S))$ between λ_1 and λ_2. Thus, L_2 can be obtained from L_1 by switching two adjacent elements, say $L_1[i_v]$ and $L_1[i_v+1]$ and the sum of the weighted distances between the line segments and the line corresponding to $\mathcal{D}^{-1}(v)$ in the primal can be computed in constant time knowing all the data associated to $L_1[i_v]$ and $L_1[i_v+1]$.

Suppose that v is a vertex of a double wedge $\mathcal{D}(\mathcal{O})$ for some $\mathcal{O} \in S$ (see Figure 1(a)), the data associated with $L_2[i_v]$ corresponds to those of $L_1[i_v]$ and the data of $L_2[i_v+1]$ are given by those of $L_1[i_v+1]$. The data for $L[i]$ where $i < i_v$ or $i > i_v+1$ must be also updated when the sweep line crosses v. The endpoint of $\mathcal{O}$ closest to $l_2[i]$ is different from the endpoint closest to $l_1[i]$ and the values $X_{low}(l_2[i])$, $Y_{low}(l_2[i])$, $X_{up}(l_2[i])$ and $Y_{up}(l_2[i])$ change accordingly. We have to add and subtract a value to $X_{low}(l_2[i])$ and $Y_{low}(l_2[i])$ or to $X_{up}(l_2[i])$ and $Y_{up}(l_2[i])$. Hence, for each vertex of a double wedge, the updates can take as much as $O(n)$ time. Fortunately, there are only n such vertices. The sum of the weighted distances to the line segments is not evaluated here, since v does not represent any candidate line.

Now, suppose that v is not a vertex of a double wedge and $L_1[i_v]$ belongs to the lower boundary of a double wedge $\mathcal{D}(\mathcal{O}_1)$ and $L_1[i_v+1]$ belongs to the upper boundary of a double wedge $\mathcal{D}(\mathcal{O}_2)$ (see Figure 1(b)). In this case, $\mathcal{O}_1 \in S_{up}(l_1[i_v+1])$ and $\mathcal{O}_2 \in S_{low}(l_1[i_v])$ when the sweep line intersects $\mathcal{A}(\mathcal{D}(S))$ just before v. When the sweep line crosses v, $S_{up}(l_2[i_v])$ is given by $S_{up}(l_1[i_v+1]) \setminus \{\mathcal{O}_1\}$ and $S_{low}(l_2[i_v])$ by $S_{low}(l_1[i_v+1])$. The values $S_{up}(l_2[i_v+1])$ and $S_{low}(l_2[i_v+1])$ are obtained similarly from $S_{up}(l_1[i_v])$ and $S_{low}(l_1[i_v])$, respectively. Therefore, the values of $X_{low}(l_1[i_v])$, $Y_{low}(l_1[i_v])$, $X_{up}(l_1[i_v+1])$ and $Y_{up}(l_1[i_v+1])$ have to be modified accordingly. These updates and the evaluation of the expression giving the sum of the weighted distances can be done in constant time. The other possibilities for $L_1[i_v]$ and $L_1[i_v+1]$ are treated in the same way. The data associated with $L[i]$, for $i < i_v$ or $i > i_v+1$, does not change when the sweep line crosses v. Therefore, the sum of the weighted distances and the updates of the data needed to compute this sum can be done in constant time.

We should say something about the degeneracies. Suppose that $k > 2$ lines intersect at a vertex v. This vertex can be process in $O(k^2 + tn)$ time (where t represents the number of double wedge vertices coinciding with v) simply by sorting the lines according to their slopes and applying the technique used for the non-degenerate vertices to process the lines in sorted order. For example, the data for $L_2[i_v]$ is obtained by updating the data for $L_1[i_v+k]$ according to the lines $l_1[i_v], ..., l_1[i_{v+1}-1]$. This can be done without altering the overall $O(n^2)$ time. This follows from the fact that the sum of the square of the incidences of the vertices in an arrangement of n lines is in $O(n^2)$ (see [EG89]). □

This lemma can be easily extended for sets of m convex polygons with a total of n vertices. The dual representation of a polygon is delimited by two concave chains. The dual arrangement has complexity $O(mn)$ as each edge of a concave chain can only be intersected by another chain twice. Since there are $2m$ chains and at most $2n$ edges, the total number of vertices is less than or equal to $8mn$. During the sweep of the arrangement, the sweep line intersects $\mathcal{A}(\mathcal{D}(S))$ at only $2m$ points and meets two kinds of vertices: the $O(n)$ vertices of the concave chains and the $O(mn)$ vertices formed by the intersections of concave chains. The first kind of vertex does not represent a candidate for the approximating line and is processed the same way as the double wedge vertices in the case of sets of line segments. Hence, each of these vertices can be processed in $O(m)$ time. The second kind of vertex corresponds to the candidate lines for the WMSA problem and is handled exactly as in the case of sets of line segments. Each of these vertices can be processed in constant time. Therefore, the sums of the weighted distances can be computed in $O(mn)$ time. Finally, since the operations on Q take only $O(\log m)$ time each, the sweep of the arragement takes $O(mn \log m)$ time.

For sets of n circles, it is possible to reduce the time complexity to $O(n^2)$ by sweeping the arrangement with a topological line instead of a vertical line. The center and the radius of a circle are enough to determine the closest distance between the circle and a given candidate line. Hence, only the partition of S induced by l is really needed to compute the sums of the weighted distances efficiently. The dual representation of a circle corresponds to the closed region between the branches of a hyperbola. Even if $\mathcal{A}(\mathcal{D}(S))$ is an arrangement of n hyperbolas, the topological line sweep algorithm of Edelsbrunner and Guibas [EG89] still works in this case [SH89]. Thus, the time complexity of the algorithm solving the WMSA problem can be reduced to $O(n^2)$.

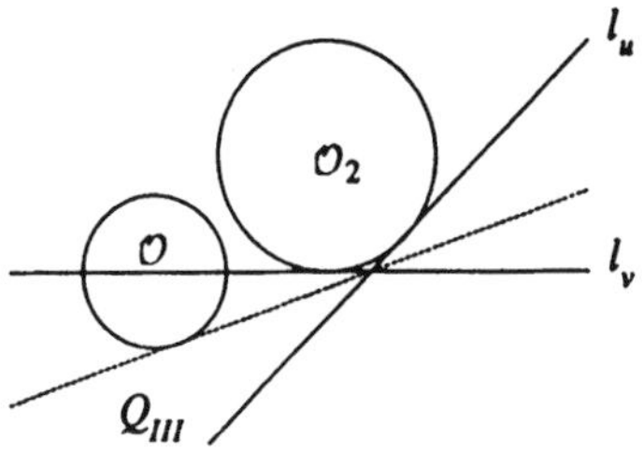

Figure 2: Updates of $L[i]$'s data for circles

For any line l, redefine the values $W_{up}(l)$, $X_{up}(l)$, $Y_{up}(l)$ and define the value $R_{up}(l)$ as follows:

$$\begin{aligned} W_{up}(l) &= \sum_{O_i \in S_{up}(l)} \omega_i, \\ X_{up}(l) &= \sum_{O_i \in S_{up}(l)} \omega_i \, x_i, \\ Y_{up}(l) &= \sum_{O_i \in S_{up}(l)} \omega_i \, y_i \quad \text{and} \\ R_{up}(l) &= \sum_{O_i \in S_{up}(l)} \omega_i \, r_i. \end{aligned}$$

The circle O_i is determined by its center $c_i = (x_i, y_i)$ and its radius r_i. The values $W_{low}(l)$, $X_{low}(l)$, $Y_{low}(l)$ and $R_{low}(l)$ are defined similarly.

The sum of the weighted distances between the circles and a line l, defined by $y = ax + b$, is given by

$$\frac{(Y_{up}(l) - Y_{low}(l)) - a(X_{up}(l) - X_{low}(l)) - b(W_{up}(l) - W_{low}(l))}{\sqrt{1+a^2}} - (R_{up}(l) + R_{low}(l)).$$

During the sweep of $\mathcal{A}(\mathcal{D}(S))$, the values $W_{low}(l)$, $X_{low}(l)$, $Y_{low}(l)$, $R_{low}(l)$, $W_{up}(l)$, $X_{up}(l)$, $Y_{up}(l)$ and $R_{up}(l)$ are maintained for each "active" vertex of $\mathcal{A}(\mathcal{D}(S))$. A vertex is active if it has been visited and it is adjacent to a non-visited vertex.

Let v be a non-visited vertex adjacent to an active vertex u and let l_v and l_u be the corresponding lines in the primal space. The values associated to v can be computed from the ones associated to u in constant time, if there is no degeneracy. Suppose that u is determined by the intersection of the boundaries of $\mathcal{D}(O_1)$ and $\mathcal{D}(O_2)$ and v by the ones of $\mathcal{D}(O_2)$ and $\mathcal{D}(O_3)$. Without loss of generality, we can assume that l_u and l_v are lower supporting lines of O_2 (see Figure 2). The symmetric difference of $S_{up}(l_u)$ and $S_{up}(l_v)$, denoted $S_{up}(l_u)\Delta S_{up}(l_v)$, is a subset of $\{O_1, O_3\}$. Consider the partition of the plane induced by l_u and l_v. Let $O \in S_{up}(l_u)\setminus\{O_3\}$. Suppose that $O \notin S_{up}(l_v)$. Then, O must intersect the quadrant Q_{III} without intersecting l_u and a lower supporting line to O which passes through the intersection point of l_u and l_v must exist. This implies that the vertices u and v are not adjacent in $\mathcal{A}(\mathcal{D}(S))$. Hence, $O \in S_{up}(l_v)$ and $S_{up}(l_u) \setminus \{O_3\} \subseteq S_{up}(l_v)$. Similarly, $S_{up}(l_v) \setminus \{O_1\} \subseteq S_{up}(l_u)$. From these facts, $S_{up}(l_u)\Delta S_{up}(l_v) \subseteq \{O_1, O_3\}$ and $S_{up}(l_v)$ can be obtained from $S_{up}(l_u)$ in constant time by removing O_3 if it is in $S_{up}(l_u)$ and by adding O_1 if it is in $S_{up}(l_v)$. Similarly, $S_{low}(l_v)$ can be computed from $S_{low}(l_u)$. Naturally, the values associated to $S_{up}(l_v)$ and $S_{low}(l_v)$ can also computed in constant time. The degeneracies can be allowed without altering the overall time complexity. We now outline the algorithm that solves the WMSA problem.

Algorithm WMSA

Input: A family $S = \{O_i\}$ of convex polygons or circles and their corresponding weights.
Output: A line l minimizing the sum of the weighted distances to the objects.

1. Dualize the objects in S.
2. Sweep the dual arrangement with a line (topological or vertical).
3. When the sweep line meets a new vertex v
   (a) Let l_v be the line corresponding to $\mathcal{D}^{-1}(v)$.
   (b) Update $S_{up}(l_v)$ and $S_{low}(l_v)$.
   (c) Compute $X_{up}(l_v)$, $Y_{up}(l_v)$, $W_{up}(l_v)$, $X_{low}(l_v)$, $Y_{low}(l_v)$ and $W_{low}(l_v)$ (also $R_{up}(l_v)$ and $R_{low}(l_v)$, for circles).
   (d) Determine the sum of the weighted distances between the objects and l_v.
4. Output the line minimizing the sum of the weighted distances to the objects in Step 3c.

End of the Algorithm

The correctness of Algorithm WMSA follows from Lemmas 3.1 and 3.2 and the correctness of the line sweep method. Therefore, we have obtained the following result:

Theorem 3.3 *Algorithm WMSA solves the WMSA problem in $O(n^2 \log n)$ time and $O(n)$ space for a set of n line segments, in $O(mn \log m)$ time and $O(n)$ space for a set of m polygons with a total of n vertices, in $O(n^2)$ time and $O(n)$ space for a set of n circles.*

Finally, we can show that a line passing through the origin and minimizing the sum of the weighted distances to a set of n objects must be a supporting line to one of the objects. This fact reduces the number of candidate lines for the approximating line to $O(n)$. By processing these lines in increasing order of their slope, it is possible to evaluate all the sums of the weighted distances to the objects very efficiently. We outline the algorithm solving the constrained WMSA problem where the approximating line must contain the origin.

Algorithm OWMSA

Input: A family $S = \{O_i\}$ of convex polygons or circles and their corresponding weights.
Output: A line l containing the origin and minimizing the sum of the weighted distances to the objects.

1. For each object, find the lines containing the origin and supporting the object.
2. Let $t_1, ..., t_k$ be the list of the supporting lines sorted in increasing order of their slope.
3. Let t_0 be the y-axis. Determine the partition of S induced by t_0 and compute $X_{up}(t_0)$, $Y_{up}(t_0)$, $X_{low}(t_0)$ and $Y_{low}(t_0)$ (the upper and the lower regions correspond to the left and the right regions, respectively).
4. For all supporting lines t_i do
   (a) Compute $X_{up}(t_i)$, $Y_{up}(t_i)$, $X_{low}(t_i)$ and $Y_{low}(t_i)$ (also $R_{up}(t_i)$ and $R_{low}(t_i)$, for circles).
   (b) Determine the sum of the weighted distances between the objects and t_i.
5. Output the supporting line minimizing the sum of the weighted distances to the objects.

End of the Algorithm

The correctness of this algorithm relies only on the fact that a line containing the origin and minimizing the sum of the weighted distances to the objects must support at least one object. Its time complexity depends on the time needed to compute the supporting lines in Step 1 and to update the data used to determine the sum in Step 4(b). For sets of n circles, the $O(n)$ supporting lines can be computed and sorted in $O(n \log n)$ time. By processing the supporting lines by increasing order of their slopes, the updates in Step 4(b) can be done in $O(n)$ time overall. As we saw for the general WMSA problem, the partition of S induced by a candidate line and the values associated to it are sufficient to compute the sum of the distances between the circles and the line. Since the supporting lines are processed in increasing order of their slopes, any circle can be transfered from one subset to another subset of the partition at most three times. This fact implies that the overall time to do the updates is in $O(n)$. For sets of n line segments and sets of m polygons with a total of n vertices, the supporting lines can be computed and sorted in $O(n \log n)$ time and in $O(n + m \log m)$ time, respectively. As we remarked earlier any object is transfered from one subset to another subset of the partion at most three times. Unfortunately, the partition of S and the values associated to it are not sufficient. Between two consecutive candidate lines, the vertex of a polygon or the endpoint of a line segment closest to the candidate lines may change without the object being transfered from one subset to another subset of the partition. Thus, we have to keep track of which vertex or endpoint of any object is closest to the candidate lines at any given moment. To do this, we simply have to compute the lines extending the edges of the polygons or extending the line segments and to sort them according to their slopes. For sets of n line segments, this can be done in $O(n \log n)$ time. For sets of m polygons with a total of n vertices, this can be done in $O(n \log m)$ time by using merge sort. Thus, the events in Step 4 will be the supporting lines *and* the lines extending the edges of the polygons or the line segments. Nevertheless, each step can be done in constant time. Therefore, the updates can be done in $O(n)$ time overall once the sort of the lines has been done. Therefore, we have just established the following theorem:

Theorem 3.4 *Algorithm OWMSA finds a line containing the origin and minimizing the sum of the weighted distances in $O(n \log n)$ time and $O(n)$ space for a set of n circles or line segments, in $O(n \log m)$ time and $O(n)$ space for a set of m polygons with a total of n vertices.*

4 Conclusion

We have presented optimal algorithms solving the MMA problem and the constrained WMMA problem where the approximating line has to contain the origin for sets of circles or line segments. The main open problem of this paper is to obtain an optimal algorithm for the general WMMA problem. There is a big gap between the $\Omega(n \log n)$ time lower bound and the $O(n^2 \log n)$ time solution we have obtained for sets of n circles or line segments. We should remark that for sets of n points in the plane, this problem can be solved optimally in $O(n \log n)$ time [HIIRY].

For the WMSA problem, the $O(n^2)$ time complexity of our solution for sets of n circles matches the time complexity of the algorithm presented in [HIIRY] for sets of n points in the plane. For sets of n line segments, we may want to remove the extra $\log n$ factor.

Finally, the approximation problems presented in this paper can be formulated in higher dimension. In [HIIRY], $O(n^{\lfloor (d+1)/2 \rfloor})$ time and $O(n^d)$ time algorithms solving the WMMA problem and the WMSA problem, respectively, for sets of n points in $\boldsymbol{E}^d$ have been presented. By extending the technique used for solving the MMA problem, we have obtained an $O(n^2)$ time algorithm to solve the MMA problem for sets of n vertical line segments. For sets of n general line segments, the time complexity of our solution is $O(n^4)$ time. These solutions are based on the fact that an optimal solution is determined by four line segments (generalization of Lemma 2.1 in $\boldsymbol{E}^3$) and the $O(n^2)$ time algorithm to compute the envelopes of n half-planes [EGS89]. On the other hand, the techniques used to solve the WMSA problem in the plane do not seem to generalize to higher dimensions.

References

[AB87] M. Atallah and C. Bajaj. Efficient algorithms for common transversals. *Inf. Proc. Letters*, 25:87–91, 1987.

[ARW89] D. Avis, J.-M. Robert, and R. Wenger. Lower bounds for line stabbing. *Inf. Proc. Letters*, 33:59–62, 1989.

[Bro80] K.Q. Brown. *Geometric transforms for fast geometric algorithms.* PhD thesis, Carnegie-Mellon University, Pittsburg, PA, 1980.

[DKM84] N.N. Doroshko, M.M. Korneenko, and N.N. Metelskij. Optimal placement of a line relative to a planar point system. *Institute of Math., Byelorussian Academy of Sciences*, 208:1–19, 1984.

[EG89] H. Edelsbrunner and L.J. Guibas. Topologically sweeping an arrangement. *J. Comp. Syst. Sc.*, 38:165–194, 1989.

[EGS89] H. Edelsbrunner, L.J. Guibas, and M. Sharir. The upper envelope of piecewise linear functions: algorithms and applications. *Disc. Comp. Geom.*, 4:311–336, 1989.

[EMPRWW82] H. Edelsbrunner, H.A. Maurer, F.P. Preparata, A.L. Rosenberg, E. Welzl, and D. Wood. Stabbing line segments. *BIT*, 22:274–281, 1982.

[EW89] P. Egyed and R. Wenger. Stabbing pairwise disjoint translates in linear time. In *Proc. of the 5th Annual ACM Symp. on Comp. Geom.*, pages 364–369, 1989.

[GOR91] L. J. Guibas, M. Overmars, and J.-M. Robert. The exact fitting problem for points. In *Proc. of the 3rd Can. Conf. on Comp. Geom.*, pages 171–174, 1991.

[Her89] J. Hershberger. Finding the upper envelope of n line segments in $O(n \log n)$ time. *Inf. Proc. Letters*, 33:169–174, 1989.

[HIIRY] M.E. Houle, H. Imai, K. Imai, J.-M. Robert, and P. Yamamoto. Orthogonal weighted linear L_1 and L_∞ approximation and applications. To appear in *Disc. App. Math.*

[HT88] M.E. Houle and G.T. Toussaint. Computing the width of set. *IEEE Trans. Patt. Anal. Mach. Intell.*, 10:761–765, 1988.

[II88] H. Imai and M. Iri. Polygonal approximations of a curve. In G.T. Toussaint, editor, *Computational Morphology*, pages 71–86. North-Holland, 1988.

[LW86] D.T. Lee and Y.F. Wu. Geometric complexity of some location problems. *Algorithmica*, 1:193–212, 1986.

[MN83] J.G. Morris and J.P. Norback. Linear facility location – solving extensions of the basic problem. *Eur. J. Oper. Res.*, 12:90–94, 1983.

[O'R81] J. O'Rourke. An on-line algorithm for fitting straight lines between data ranges. *Comm. of the ACM*, 24:574–578, 1981.

[SH76] M.I. Shamos and D. Hoey. Geometric intersection problems. In *Proc. of the 15th IEEE Found. of Comp. Science*, pages 208–215, 1976.

[SH89] J. Snoeyink and J. Hershberger. Sweeping arrangements of curves. In *Proc. of the 5th Annual ACM Symp. on Comp. Geom.*, pages 354–363, 1989.

[Tou85] G.T. Toussaint. On the complexity of approximating polygonal curves in the plane. In *Proc. IASTED Symp. on Robotics & Automation*, pages 59–62, 1985.

Language Learning without Overgeneralization*

Shyam Kapur[†]
Gianfranco Bilardi[‡]

Abstract

Language learnability is investigated in the Gold paradigm of inductive inference from positive data. Angluin gave a characterization of learnable families in this framework. Here, learnability of families of recursive languages is studied when the learner obeys certain natural constraints. Exactly learnable families are characterized for prudent learners with the following types of constraints: (0) conservative, (1) conservative and consistent, (2) conservative and responsive, and (3) conservative, consistent and responsive. The class of learnable families is shown to strictly increase going from (3) to (2) and from (2) to (1), while it stays the same going from (1) to (0). It is also shown that, when exactness is not required, prudence, consistency and responsiveness, even together, do not restrict the power of conservative learners.

1 Introduction

Gold [Gol67] presented a basic paradigm of inductive inference. While inference from complete data was shown by him to be powerful, interesting classes of language families were first shown to inferable from *positive data*, i.e., where only strings in the language to be learned are presented, in [Ang80a, Ang80b]. Angluin [Ang80b] showed that the families of languages learnable from positive data are exactly those for which there is a procedure that, for each language in the family, enumerates a finite set of which that language is a least upper bound.

Within language learning, Angluin [Ang80b] also initiated the study of *conservative* learners, which do not change a guess unless evidence inconsistent with it is encountered. It can easily be shown that a family can be learned by a conservative learner if and only if it can be learned by a *non overgeneralizing* learner, that is, one that never guesses a subset of a previous guess. Learning without overgeneralization is of interest to research in natural language acquisition since it has been

*This work was supported in part by the National Science Foundation grant MIP-86-02256.

†Institute for Research in Cognitive Science, University of Pennsylvania, 3401 Walnut Street–Suite C, Philadelphia PA 19104, USA

‡Department of Computer Science, Cornell University, Ithaca, New York 14853, USA, and Dipartimento di Elettronica ed Informatica, Universitã di Padova, 35131 Padova, Italy

hypothesized (for example, [Ber85]) that children learn in this fashion. (However, see [MW84] for evidence to the contrary.) Angluin [Ang80b] established that conservativeness actually restricts the class of learnable families, and derived a sufficient (but not necessary) condition for conservative learning.

In a framework similar to that in [Ang80a, Ang80b], defined in Section 2, we further investigate conservative learning. We consider the case when the language to be learned is *recursive* and where the learner has to identify the language by a *total Turing machine.* A variety of natural constraints on learners have been proposed in literature [OSW86], such as *prudence* (the learner must learn all the languages it ever guesses), *consistency* (at any stage of the learning process, the guessed language must contain all the evidence), and *responsiveness* (a guess must be made for any new piece of evidence). A family is said to be *exactly* learned if the learner learns it and nothing larger. In Section 3, we investigate the impact of all these constraints on conservative learning and demonstrate a nested relation between the classes of families exactly learnable by prudent learners of four types defined by selectively and together imposing responsiveness and consistency.

In Section 4, we first give a uniform characterization of prudent, exact conservative learning of the four types defined in Section 3, based on the concept of least upper bound. While a consistent guess of a conservative learner must always be a least upper bound of the evidence seen, simply guessing a least upper bound does not guarantee learning. We develop a learning algorithm based on the judicious use of a function that computes least upper bounds. One of the corollaries of our characterization is that prudent, exact conservative learnability of a given family of languages $\mathcal{F}$ is equivalent to the existence of a recursive enumeration of finite-set/total-machine pairs such that (i) in each pair, the machine accepts a language of $\mathcal{F}$ that is a least upper bound in $\mathcal{F}$ of the corresponding finite set, and (ii) there is at least one pair for each language in $\mathcal{F}$. Interestingly, if the condition on the enumeration is weakened so that the first component of the pair is a machine that recursively enumerates a finite set, then the existence of the enumeration characterizes general learning.

Finally, in Section 5, we investigate the impact of dropping the requirement of prudence on the learner and of exactness on the family learned. We show that every family that can be learned conservatively, can also be learned conservatively and prudently, although not necessarily exactly. In fact, the hierarchy obtained for prudent, exact conservative learning collapses when exactness is not stipulated, and prudence, consistency, and responsiveness, even together, fail to constrain the conservative learners.

2 Background

Let Σ be a finite alphabet and Σ^* be the set of all finite strings formed by concatenating elements of Σ. Let $M_1, M_2, M_3, \ldots$ be any standard enumeration of all Turing machines over Σ. Let Z_+ be the set of positive integers. For any *index* $I \in Z_+$, let W_I denote the *language* (subset of Σ^*) accepted by the machine M_I. Thus, the W_Is form a complete enumeration of all *recursively enumerable (r.e.)* languages. A collection of *non-empty, recursive* languages is called a *family*. An index I is *total* if the corresponding machine M_I is total and accepts a non-empty language. If $I_1, I_2, \ldots$ is a recursive enumeration of total indices, then $\mathcal{F} = W_{I_1}, W_{I_2}, \ldots$ is called an *indexed family of non-empty recursive languages* (hereafter, simply an *indexed family*). We denote by $\Delta_{\mathcal{F}}$ the class of all non-empty finite subsets of Σ^* that are contained in some language in family $\mathcal{F}$.

A *text* is an infinite sequence of strings from Σ^*; t ranges over texts, t_n is the nth string in text t, and $\bar{t}_n$ is the initial prefix of length n of the text t; t is *for* L if and only if the set of strings in t equals L. SEQ is the set of finite sequences of strings from Σ^*; *content*(σ) is the set of strings in $\sigma \in$ SEQ.

An *inductive inference machine (IIM)* M is an algorithmic procedure (say, a Turing machine) whose input is a text $t_1, t_2, \ldots$ and whose output is a sequence of nonnegative integers $M(\bar{t}_1), M(\bar{t}_2), \ldots$ constrained to be either 0 or total indices. The procedure works in stages, but it may never complete some stage. At the nth stage, t_n is input and $M(\bar{t}_n)$ is output. The intended interpretation is as follows: If $M(\bar{t}_n) = 0$, then the IIM makes no guess; otherwise, it guesses the language $W_{M(\bar{t}_n)}$.

For an IIM M, we define GUESS(M)=$\{W_{M(\sigma)} : \sigma \in \text{SEQ}\}$. We observe that GUESS($M$) is an indexed family. An IIM M is said to learn the language L if and only if, for each text t for L, there is a k such that $W_{M(\bar{t}_k)} = L$ and, for all $n > k$, $M(\bar{t}_n) \in \{M(\bar{t}_k), 0\}$. (Thus, if t is a text for a language that M learns, then $M(\bar{t}_n)$ is defined for any n.) Intuitively, from some point onward, the value of the guess stabilizes to a total index for the input language. (This is similar to the TxtEx-identification criterion [Gol67].) We denote by LEARN(M) the family of all languages learned by M, and say that M learns a family $\mathcal{F}$ if and only if $\mathcal{F} \subseteq$ LEARN(M).

An IIM M is *consistent* if it never guesses a language that does not contain all the data on which the guess is based. Formally, if M completes the nth stage and $M(\bar{t}_n) \neq 0$, then *content*$(\bar{t}_n) \subseteq W_{M(\bar{t}_n)}$. M is *conservative* if it does not change its output unless it is inconsistent with the data. Formally, if M completes the nth stage, $M(\bar{t}_n) \neq 0$, and *content*$(\bar{t}_{n+1}) \subseteq W_{M(\bar{t}_n)}$, then $M(\bar{t}_{n+1}) = M(\bar{t}_n)$. M is *responsive* if, for all $n > 0$, whenever *content*$(\bar{t}_n) \subseteq \Delta_{\text{GUESS}(M)}$, then M completes the nth stage and $M(\bar{t}_n) \neq 0$. M is *prudent* if LEARN(M) = GUESS(M). A family $\mathcal{F}$ is said to be *exactly* learned by M if $\mathcal{F}$ = LEARN(M).

A central role is played by the relationship that holds between a finite set T and a language L in a family $\mathcal{F}$ when T is a subset of L but T is not a subset of any language in $\mathcal{F}$ properly contained in L. This relation will be expressed by saying either that T is a *tell-tale* subset of L in $\mathcal{F}$, or that L is a *least upper bound (l.u.b.)* of T in $\mathcal{F}$. As we shall see, various forms of computable mappings from (indices for) languages to tell-tale subsets and from finite sets to (indices for) l.u.b. languages are intimately related to various forms of learning. The importance of tell-tale subsets was first indicated by the following result.

Theorem 1 *[Ang80b] There is an IIM M such that an indexed family $\mathcal{F}$ =* LEARN(M) *if and only if there is an effective procedure that, given as input* *a total index I such that $W_I \in \mathcal{F}$, recursively enumerates a tell-tale subset T_I of W_I. (The procedure need not signal when T_I has been entirely output.)*

3 Variants of Conservative Learning

Angluin [Ang80b] showed that, in the absence of the conservativeness constraint, if $\mathcal{F}$ is learnable it is also learnable by a consistent and responsive algorithm. In this section, we begin to analyze the relationships between the classes of families learnable by various types of conservative learners. It is convenient to introduce the following notation.

Definition 1 An IIM M is said to be of *type 0* if conservative, of *type 1* if conservative and consistent, of *type 2* if conservative and responsive, and of *type 3* if conservative, consistent, and responsive. For $X = 0, 1, 2, 3$, we denote by $\mathcal{L}_X$ the class of all families learnable by an IIM of type X, $\mathcal{L}_X^p$ the class of all families learnable by a prudent IIM of type X, and $\mathcal{L}_X^{ep}$ the class of all families learnable exactly by a prudent IIM of type X.

Table 1 shows the various classes defined above.

Table 1: Classes of conservatively learnable families under various constraints

		Consistent	Responsive	Consistent & Responsive
	$\mathcal{L}_0$	$\mathcal{L}_1$	$\mathcal{L}_2$	$\mathcal{L}_3$
Prudent	$\mathcal{L}_0^p$	$\mathcal{L}_1^p$	$\mathcal{L}_2^p$	$\mathcal{L}_3^p$
Prudent & Exact	$\mathcal{L}_0^{ep}$	$\mathcal{L}_1^{ep}$	$\mathcal{L}_2^{ep}$	$\mathcal{L}_3^{ep}$

Consistency and responsiveness constrain the power of exact, prudent learners as stated in the next theorem. Figure 1 displays the relationships established by the following theorem.

Theorem 2 $\mathcal{L}_0^{ep} = \mathcal{L}_1^{ep} \supset \mathcal{L}_2^{ep} \supset \mathcal{L}_3^{ep}$.

Proof: ($\mathcal{L}_0^{ep} = \mathcal{L}_1^{ep}$) Clearly, $\mathcal{L}_0^{ep} \supseteq \mathcal{L}_1^{ep}$. To show that $\mathcal{L}_0^{ep} \subseteq \mathcal{L}_1^{ep}$, we consider that an IIM of type 0 can be modified to refrain from making a guess whenever it would make an inconsistent guess. It is easy to see that the modified IIM learns the original family and in addition is of type 1.

($\mathcal{L}_1^{ep} \supset \mathcal{L}_2^{ep}$) Clearly, $\mathcal{L}_2^{ep} \subseteq \mathcal{L}_0^{ep} = \mathcal{L}_1^{ep}$. To see that $\mathcal{L}_2^{ep} \neq \mathcal{L}_1^{ep}$, consider the indexed family $\mathcal{F}$: for all $k \geq 1$, $W_{I_k} = \{0\} \cup \{k, k+1, \ldots\}$. A prudent IIM M of type 1 learns exactly $\mathcal{F}$ for which $M(\bar{t}_n) = I_{min(content(\bar{t}_n) - \{0\})}$, with $min(\emptyset)$ defined as 0. However, $\mathcal{F}$ can not be learned exactly by a prudent IIM of type 2, because on the input of a

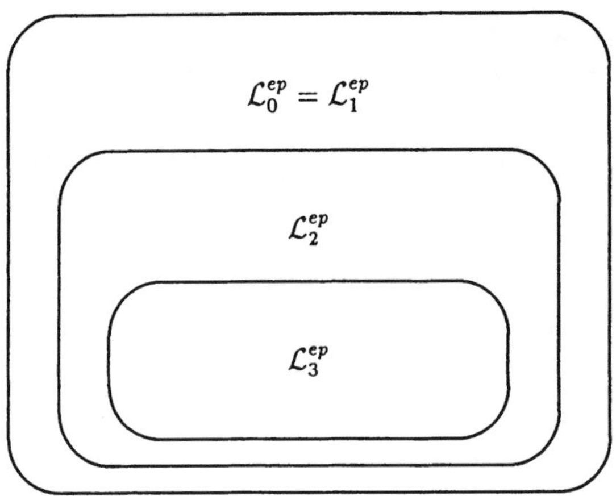

Figure 1: Relation between classes exactly learned by prudent learners of various types

text t for any language in $\mathcal{F}$ for which $t_1 = 0$, no index for a language in $\mathcal{F}$ can be guessed safely at the first stage.

($\mathcal{L}_2^{ep} \supset \mathcal{L}_3^{ep}$) Clearly, $\mathcal{L}_2^{ep} \supseteq \mathcal{L}_3^{ep}$. To see that $\mathcal{L}_3^{ep} \neq \mathcal{L}_2^{ep}$, consider the indexed family $\mathcal{F}$: for all $k > 1$, $W_{I_k} = \{0\} \cup \{k, k+1, \ldots\}$, and $W_{I_1} = \{5\}$. By the argument given in the preceding paragraph, there is no prudent IIM of type 3 that learns exactly $\mathcal{F}$. We do have a prudent IIM of type 2 that learns exactly $\mathcal{F}$, which behaves as follows: If $t_1 = t_2 = \cdots = t_n = 0$ (in a text t for any language in $\mathcal{F}$ excluding W_{I_1}), or $t_1 = t_2 = \cdots = t_n = 5$ (in a text t for W_{I_1}), then $M(\bar{t}_n) = I_1$, else $M(\bar{t}_n) = I_{min(content(\bar{t}_n) - \{0\})}$. ∎

4 Characterization of Prudent, Exact Conservative Learning

In the present section, we develop, in a unified way, a characterization of $\mathcal{L}_X^{ep}$, for $X = 0, 1, 2$, and 3, in terms of the computability of l.u.b. languages for certain finite sets. We then reformulate these characterizations for some specific cases. It is a simple observation that a consistent guess made by a conservative IIM M in response to a set of input strings S subset of some $L \in \text{LEARN}(M)$, must be a least upper bound for S in the family LEARN(M). (Berwick [Ber85] refers to this observation as the *subset principle*.) However, systematically guessing an l.u.b. of the available evidence does not guarantee learning, even when $\mathcal{F}$ is conservatively learnable [Kap91]. Thus, it is natural to investigate to which extent the ability of computing l.u.b. languages is necessary or sufficient for conservative learning.

We establish that a computable function that associates an l.u.b. to each finite subset of a language in $\mathcal{F}$ is sufficient to construct a conservative IIM, provided that each language in $\mathcal{F}$ is associated with at least one of its subsets. However, the existence of such an l.u.b. function is not necessary, except for the consistent and responsive case. Indeed, two issues have to be confronted when trying to construct an l.u.b. function from a conservative IIM. First, with some input set, a non-responsive IIM may associate no language, and a non-consistent IIM may associate a language that does not contain the set itself. Second, when the family being learned contains a finite language L, even at a stage when all the strings of L have appeared in the input, the IIM's guess may well differ from L. We deal with these issues by introducing, in the formal definition of l.u.b. functions given below, a parameter k that models the possible delay in the IIM's convergence to a correct guess.

Without loss of generality, for the prudent, exact learning case we can assume that the family $\mathcal{F}$ to be learned is an indexed family. We recall that $\Delta_{\mathcal{F}}$ denotes the class of all non-empty finite sets contained by some language in $\mathcal{F}$ and let $\Delta = \Delta_{\{\Sigma^*\}}$.

Definition 2 A partial recursive function $f : \Delta \times Z_+ \mapsto Z_+ \cup \{0\}$ is a *least upper bound (l.u.b.)* function of *type* X ($X = 0, 1, 2, 3$) for an indexed family $\mathcal{F}$ if, for $S \in \Delta_{\mathcal{F}}$ and $k \in Z_+$, $f(S,k)$ is defined, either as 0 or as a total index such that $W_{f(S,k)} \in \mathcal{F}$ and satisfies the following conditions:

type 0: if $S \subseteq W_{f(S,k)}$, then $W_{f(S,k)}$ is an l.u.b. of S;
type 1: f is of type 0 and ($f(S,k) = 0$ or $S \subseteq W_{f(S,k)}$);
type 2: f is of type 0 and $f(S,k) \neq 0$;
type 3: f is of type 0 and $S \subseteq W_{f(S,k)}$.

An l.u.b. function f is *full-range* if each L in $\mathcal{F}$ has a subset S such that, for some $k \in Z_+$, $W_{f(S,k)} = L$.

In terms of the concepts just introduced, we next state the characterization theorem.

Theorem 3 *Let* $\mathcal{F} = W_{I_1}, W_{I_2}, \ldots$ *where* $I_1, I_2, \ldots$ *is a recursive enumeration of total indices. For* $X = 0, 1, 2, 3$, $\mathcal{F} \in \mathcal{L}_X^{ep}$ *if and only if there exists a full-range l.u.b. function of type* X *for* $\mathcal{F}$.

Proof: The proof is organized in three parts. In part (a) we show how to obtain a full-range l.u.b. function of type X from a prudent IIM M of type X that learns exactly $\mathcal{F}$, for $X = 0, 1, 2, 3$. In part (b), we prove that any full-range l.u.b. function can effectively be converted into one with a certain special property. Finally, in part (c), we show how an l.u.b. function of type X of the special kind obtained in part (b) can be used to construct a prudent IIM M' of type X that learns exactly $\mathcal{F}$, thereby completing the proof.

(a) *From M to a full-range l.u.b. function.* The following procedure defines a function $f: \Delta \times Z_+ \mapsto Z_+ \cup \{0\}$. Let $S \in \Delta$ and $k \in Z_+$. Scan $\mathcal{F} = W_{I_1}, W_{I_2}, \ldots$ until an I_j is found such that $S \subseteq W_{I_j}$. (This scan terminates if and only if $S \in \Delta_{\mathcal{F}}$.) Then let $f(S,k)$ be the output of M when the input is the lexicographic enumeration $s_1, s_2, \ldots, s_n$ of S ($n = |S|$) followed by $k-1$ repetitions of s_n. Since $\mathcal{F}$ is learned, for any $S \in \Delta_{\mathcal{F}}$, this output must be defined. It is a simple exercise to show that if M is an IIM of type X, then f is an l.u.b. function of type X.

To establish the full-range property of f, we argue as follows. If $L \in F$ is infinite, let $t = t_1, t_2, t_3, \ldots$ be a text for L corresponding to its lexicographic presentation. Consider the behavior of M on t. At some stage n, M must output an index of L. Therefore $W_{f(content(\bar{t}_n),1)} = L$. If $L \in \mathcal{F}$ is finite, let $s_1, s_2, \ldots, s_n$ be the lexicographic enumeration of L, and let $s_1, s_2, \ldots, s_n, s_n, s_n, \ldots$ be a text for L presented to M. At some stage h, M must output an index for L. If $h \leq n$, then $W_{f(\{s_1,\ldots,s_h\},1)} = L$, else, $W_{f(\{s_1,\ldots,s_n\},h-n+1)} = L$.

(b) *From a full-range l.u.b. function to a special l.u.b. function.* Let f be an l.u.b. function for an indexed family $\mathcal{F}$. Let $L \in F$ and R be a finite subset of L. We say that R is *reserved* for L beyond m if, for every finite S such that $R \subseteq S \subseteq L$, and every $k \geq m$, $W_{f(S,k)} = L$. A *special* l.u.b. function is one that reserves a set for every L in $\mathcal{F}$. Thus, if f is special, then it is also full-range.

Given any full-range l.u.b. function f, we can construct another one, $\bar{f}$, which is special. To define $\bar{f}(S,k)$, consider a recursive enumeration $(D_1,h_1),(D_2,h_2),\ldots$ of $\Delta \times Z_+$ such that if $h < h'$ then (D,h) appears before (D,h'). First, $\mathcal{F}$ is scanned to determine whether $S \in \Delta_{\mathcal{F}}$. If so, consider the set $J(S,k) = \{j : j \leq k, D_j \subseteq S, \text{ and } S \subseteq W_{f(D_j,h_j)}\}$ (where, for convenience, we let $W_0 = \emptyset$). Observe that whenever $D_j \subseteq S \in \Delta_{\mathcal{F}}$, $f(D_j,h_j)$ must be defined. If $J(S,k)$ is empty, we let $\bar{f}(S,k) = f(S,k)$. Otherwise, we let $i = min\ J(S,k)$ and $\bar{f}(S,k) = f(D_i,h_i)$. In this case, since $D_i \subseteq S \subseteq W_{f(D_i,h_i)}$, we have that $W_{\bar{f}(S,k)} = W_{f(D_i,h_i)}$ is an l.u.b. of D_i and hence of S. Clearly, $\bar{f}$ is computable with the help of the algorithm that computes f and of the recursive enumeration of indices $I_1, I_2, \ldots$. It is also simple to see that $\bar{f}$ is an l.u.b. function of the same type as f. We now show that $\bar{f}$ is special.

Since f is full-range, for any given $L \in \mathcal{F}$, we can properly define the positive integer $m = min\ \{i : W_{f(D_i,h_i)} = L\}$. We now observe that, if $j < m$ and $D_j \subseteq L$, then $W_{f(D_j,h_j)} \not\supseteq L$. In fact, $W_{f(D_j,h_j)} = L$ would violate the definition of m, since $j < m$. Moreover, $W_{f(D_j,h_j)} \supset L$ would violate the basic l.u.b. property of f, since $W_{f(D_j,h_j)} \supset L \supseteq D_j$ implies that $W_{f(D_j,h_j)}$ is not an l.u.b. of D_j. Thus, there exists a finite set V containing a string $x_j \in (L \setminus W_{f(D_j,h_j)})$, for every $j < m$ such that $D_j \subseteq L$. We claim that the set $R = D_m \cup V$ is reserved by $\bar{f}$ for L, beyond m. In fact, let $S \in \Delta_{\mathcal{F}}$ be such that $R \subseteq S \subseteq L$ and $k \geq m$. Then, the set $J(S,k) = \{j : j \leq k, D_j \subseteq S, \text{ and } S \subseteq W_{f(D_j,h_j)}\}$ has minimum m. Indeed, it contains m (since $D_m \subseteq S \subseteq W_{f(D_m,h_m)} = L$) and no $j < m$ (since if $j < m$ and $D_j \subseteq S$ then $S \not\subseteq W_{f(D_j,h_j)}$, because $x_j \in S$ but $x_j \notin W_{f(D_j,h_j)}$). In conclusion, $\bar{f}(S,k) = min\ J(S,k) = m$ and R is reserved for L, as claimed.

(c) *From a special l.u.b. function to M'.* Given a special l.u.b. function $\bar{f}$ of type X for $\mathcal{F}$, consider the IIM M' that works as follows: At stage 1, first scan $\mathcal{F}$ to determine whether $\{t_1\} \in \Delta_{\mathcal{F}}$. If so, let $M'(\bar{t}_1) = \bar{f}(\{t_1\}, 1)$. At stage $n > 1$, maintain the previous guess if consistent with the previous input as well as with t_n. Otherwise, scan the family $\mathcal{F}$ to determine whether $content(\bar{t}_n) \in \Delta_{\mathcal{F}}$. If so, let $M'(\bar{t}_n) = \bar{f}(content(\bar{t}_n), n)$. It is easy to see that M' is an IIM of type X and that GUESS $(M') \subseteq \mathcal{F}$. We next establish that M' is prudent and learns exactly the indexed family $\mathcal{F}$.

Let $L \in F$ and let R be reserved beyond m for L. Given a text $t_1, t_2, \ldots$ for L, let h be the smallest integer such that $h \geq m$ and $R \subseteq content(\bar{t}_h)$. Then, either $W_{M'(\bar{t}_h)} = L$, in which case M' has already reached the correct guess at stage h, or there is an integer $r > h$ such that t_r is not in $W_{M'(\bar{t}_h)}$. Then $M'(\bar{t}_r) = \bar{f}(content(\bar{t}_r), r)$ is an index for L because R is reserved by $\bar{f}$ beyond m. In conclusion, L is learned by M. ∎

A simple adaptation of part (a) of the proof of Theorem 3 yields the following corollary.

Corollary 1 *If $\mathcal{F} \in \mathcal{L}_3^{ep}$ or if $\mathcal{F} \in \mathcal{L}_X^{ep}$ $(X = 0, 1, 2)$ and does not contain any finite language, then there is a full-range l.u.b. function for $\mathcal{F}$ that does not depend on k.*

For $\mathcal{L}_1^{ep}$, the characterizing condition of Theorem 3 admits the following useful reformulation.

Corollary 2 *An indexed family $\mathcal{F} \in \mathcal{L}_1^{ep}$ if and only if there is a recursive enumeration $(T_1, I_1), (T_2, I_2), \ldots$ of pairs such that, for each $h \in Z_+$, T_h is a finite set, I_h is a total index, W_{I_h} is an l.u.b. of T_h in $\mathcal{F}$, and $\mathcal{F} = \{W_{I_1}, W_{I_2}, \ldots\}$.*

It is an interesting exercise to restate the characterization of general learning in the form given below, which affords a direct comparison with Corollary 2.

Theorem 1 (restated) *There is an IIM M such that an indexed family $\mathcal{F} = \text{LEARN}(M)$ if and only if there is a recursive enumeration $(E_1, I_1), (E_2, I_2), \ldots$ of pairs such that, for each $h \in Z_+$, E_h is a procedure that recursively enumerates a finite set T_h, I_h is a total index, W_{I_h} is an l.u.b. of T_h in $\mathcal{F}$, and $\mathcal{F} = \{W_{I_1}, W_{I_2}, \ldots\}$.*

5 Prudence does not restrict Conservative Learners

In this section, we study conservative learnability when prudence or exactness are not required. In the case of general learning of r.e. languages, Fulk [Ful90] showed that prudence does not restrict the class of learnable families. Specifically, for every IIM M, there is a TxtEx-prudent IIM M' with LEARN$(M) \subseteq$ LEARN(M'). In the same spirit, although with different techniques, we have:

Theorem 4 $\mathcal{L}_1 = \mathcal{L}_1^p$.

Proof: Consider a family $\mathcal{F} \in \mathcal{L}_1$ and a conservative IIM of type 1 M that learns $\mathcal{F}$. For $D \in \Delta$ and $k \in Z_+$, let $< D, h >$ denote the lexicographic ordering of D followed by $h-1$ repetitions of the last string. Let $(S_1, J_1), (S_2, J_2), \ldots$ be the recursive enumeration, called the S-enumeration, obtained as follows. Scan a recursive enumeration $(D_1, h_1, k_1), (D_2, h_2, k_2), \ldots$ of $\Delta \times Z_+ \times Z_+$, and, if for some i, M halts on input $< D_i, h_i >$ in at most k_i steps with $M(< D_i, h_i >) \neq 0$, then output the pair $(D_i, M(< D_i, h_i >))$.

As M is not necessarily prudent, W_{J_i} need not be an l.u.b. of S_i. However, based on the S-enumeration, we will construct another enumeration of the type considered in Corollary 2. To this end, let us say that (S_r, J_r) is *certified* with *certificate* V_r if there is a finite set $V_r \subseteq W_{J_r}$ such that, for all $p < r$ for which $S_p \subseteq W_{J_r}$, $S_r \cup V_r \not\subseteq W_{J_p}$. It is straightforward to construct a certification procedure that, on input r, will output a certificate V_r if there is one, and run for ever otherwise. Using this procedure in a dove-tail fashion, we can produce a recursive enumeration $(T_1, I_1), (T_2, I_2), \ldots$, called the T-enumeration, containing the pair $(S_r \cup V_r, J_r)$ exactly when the pair (S_r, J_r) is certified with certificate V_r. We now state and prove two claims for the T-enumeration.

Claim 1 *For each $h \in Z_+$, W_{I_h} is an l.u.b. of T_h in $\{W_{I_1}, W_{I_2}, \ldots\}$.*

Proof: Let $(S_1', J_1'), (S_2', J_2'), \ldots$ be the subsequence of the S-enumeration containing the certified pairs, and let V_i' be the certificate produced for pair (S_i', J_i'). Then, the sequence $(S_1' \cup V_1', J_1'), (S_2' \cup V_2', J_2'), \ldots$ is a rearrangement of the T-enumeration. Therefore, the claim can be reformulated as follows assuming the T-enumeration to be infinite (the finite case can be argued similarly): For each $r \in Z_+$, and each k such that $1 \leq k \leq r$, $W_{J_k'}$ is an l.u.b. of $S_k' \cup V_k'$ in $\{W_{J_1'}, \ldots, W_{J_r'}\}$. For $r = 1$, the latter statement is trivially true. Let us inductively assume it holds for $r-1$ and establish it for r. If the statement fails for r, then for some $p < r$ either (a) $(S_p' \cup V_p') \subseteq W_{J_r'} \subset W_{J_p'}$ or (b) $(S_r' \cup V_r') \subseteq W_{J_p'} \subset W_{J_r'}$. In case (a), we have $S_p' \subseteq W_{J_r'}$ (by hypothesis) and $(S_r' \cup V_r') \subset W_{J_p'}$ (since $(S_r' \cup V_r') \subset W_{J_r'}$ and, by hypothesis, $W_{J_r'} \subset W_{J_p'}$). In case (b), we have $S_p' \subseteq W_{J_r'}$ (since $S_p' \subseteq W_{J_p'}$ and, by hypothesis, $W_{J_p'} \subset W_{J_r'}$) and $(S_r' \cup V_r') \subseteq W_{J_p'}$ (by hypothesis). In either case, we reach a contradiction with the fact that V_r' is a certificate for (S_r', J_r') and the claim remains established.

Claim 2 $\mathcal{F} \subseteq \{W_{I_1}, W_{I_2}, \ldots\}$.

Proof: For $L \in \mathcal{F}$, let (S_i, J_i) be the first pair in the S-enumeration such that $W_{J_i} = L$. The existence of such a pair can be established by arguments similar to those used in part (a) of the proof of Theorem 3. It suffices to show that (S_i, J_i) gets certified. If not, there is another pair (S_k, J_k), with $k < i$, such that $S_k \subseteq W_{J_i}$ and $W_{J_i} \subset W_{J_k}$. By the construction of the S-enumeration, this implies that for some $h \in Z_+$, $M(< S_k, h >) = J_k$. Thus, if t is a text for $W_{J_i} = L$ with $< S_k, h >$ as a prefix, for $n = |S_k| + h - 1$ $M(\bar{t}_n) = J_k$. Moreover, since $L \subset W_{J_k}$ and M is conservative, for $n > |S_k| + h - 1$, $M(\bar{t}_n) = J_k$. In conclusion, M does not learn $L \in \mathcal{F}$, against our assumptions.

By Claim 1 and Corollary 2, there is a prudent IIM M' such that LEARN $(M') = \{W_{I_1}, W_{I_2}, \ldots\}$. By Claim 2, $\mathcal{F} \subseteq$ LEARN(M') so that $\mathcal{F} \in \mathcal{L}_1^p$. Thus $\mathcal{L}_1 \subseteq \mathcal{L}_1^p$. As $\mathcal{L}_1^p \subseteq \mathcal{L}_1$, we conclude that $\mathcal{L}_1 = \mathcal{L}_1^p$. ∎

We next show that, when exactness is not required, responsiveness and consistency do not constrain the power of prudent learners.

Theorem 5 $\mathcal{L}_0^p = \mathcal{L}_1^p = \mathcal{L}_2^p = \mathcal{L}_3^p$.

Proof: That $\mathcal{L}_0^p = \mathcal{L}_1^p$ follows from the fact that a prudent IIM M of type 0 can be modified to refrain from making an inconsistent guess. The modified IIM is still prudent and learns the original family. To show that $\mathcal{L}_1^p = \mathcal{L}_2^p = \mathcal{L}_3^p$, we observe that, by definition, $\mathcal{L}_1^p \supseteq \mathcal{L}_2^p \supseteq \mathcal{L}_3^p$ and prove that $\mathcal{L}_1^p = \mathcal{L}_3^p$.

Let $\mathcal{F} \in \mathcal{L}_1^p$ and let M be a prudent IIM of type 1 that learns $\mathcal{F}$. We shall prove that $\mathcal{F} \in \mathcal{L}_3^p$ by constructing a full-range l.u.b. function of type 3, f, for a family $\mathcal{F}' \supseteq F$ and invoking Theorem 3. We will use the fact that, since M is prudent, $\Delta_{\text{LEARN}(M)}$ is recursively enumerable. For $S \in \Delta$, let $< S >$ denote the lexicographic ordering of S and, for a finite language L, let $index(L)$ denote a total index such that $W_{index(L)} = L$. With these preliminaries, for $S \in \Delta$, let $f(S)$ be defined (recursively) as follows.

If $S \notin \Delta_{\text{LEARN}(M)}$ then (a) let $f(S)$ diverge. Else ($S \in \Delta_{\text{LEARN}(M)}$) if $M(< S >) \neq 0$ then (b) let $f(S) = M(< S >)$. Else ($M(< S >) = 0$) if there is any non-empty Q such that $Q \subset S \subseteq W_{f(Q)}$, then (c) let $f(S) = f(Q)$; if more than one Q qualifies, choose a minimal one arbitrarily. Else (d) let $f(S) = index(S)$.

We observe that, for any $S \in \Delta_{\text{LEARN}(M)}$, $f(S)$ is a total index and consider the family $\mathcal{F}' = \{W_{f(S)} : S \in \Delta_{\text{LEARN}(M)}\}$. We show that LEARN$(M) \subseteq \mathcal{F}'$ and hence $\mathcal{F} \subseteq \mathcal{F}'$. Consider any $L \in$ LEARN(M).

If L is infinite, then for a suitable prefix σ of the lexicographic presentation of L, $W_{M(\sigma)} = L$. Since $content(\sigma) \in \Delta_{\text{LEARN}(M)}$ and $M(< content(\sigma) >) = M(\sigma) \neq 0$, f gets defined in clause (b) as $f(content(\sigma)) = M(\sigma)$, so that $L = W_{f(content(\sigma))} \in \mathcal{F}'$.

If L is finite, $f(L)$ can get defined in any of the clauses (b), (c), and (d). If $f(L)$ is defined in (b), then $M(< L >) \neq 0$ and $W_{M(<L>)} = L$, since M must guess an l.u.b. . If $f(L)$ is defined in (c), then $f(L) = f(Q)$ for some minimal Q such that $Q \subset L \subseteq W_{f(Q)}$. The minimality of Q implies that $f(Q)$ itself has been defined either in clause (b) or clause (d). The latter case is excluded because it would imply $W_{f(Q)} = Q$, whereas by assumption $Q \subset W_{f(Q)}$. Therefore, $W_{f(Q)} = W_{M(<Q>)}$ must be an l.u.b. of Q in $\mathcal{F}$. Since $L \in \mathcal{F}$ and $Q \subset L \subseteq W_{f(Q)}$, this implies $W_{f(Q)} = L$. Finally, if $f(L)$ is defined in (d), $W_{f(L)} = W_{index(L)} = L$. In any case, $L \in \mathcal{F}'$.

We now show that f is an l.u.b. function of type 3 for $\mathcal{F}'$. Otherwise, there must be a finite set S and a language $L \in \mathcal{F}'$ such that $S \subseteq L \subset W_{f(S)}$. Further, $f(S)$ must have been defined in clause (b) or clause (c), since in clause (d) $S = W_{f(S)}$.

If $f(S)$ is defined in clause (b) as $f(S) = M(< S >)$, then $W_{f(S)} \in \text{LEARN}(M)$, since M is prudent, and $L \notin \text{LEARN}(M)$, since M is conservative. Therefore, $L \in (\mathcal{F}' \setminus \text{LEARN}(M))$. By definition of f and $\mathcal{F}'$, $L \in (\mathcal{F}' \setminus \text{LEARN}(M))$ implies that $f(L)$ is defined in clause (d). However, this leads to a contradiction. In fact, either $L = S$ and $f(L)$ is defined in clause (b), by assumption, or $L \supset S$ and $f(L)$ is defined in clause (c), because the set $\{Q : Q \subset L \subseteq W_{f(Q)}\}$ contains at least S.

If $f(S)$ is defined in clause (c) as $f(S) = f(Q)$, we observe that, due to the minimality of Q, $f(Q)$ is defined in clause (b) and the argument made for S in the previous case can now be made for Q.

Having established that f is a type 3 l.u.b. function for $\mathcal{F}'$, we conclude by observing that, by the definition of $\mathcal{F}'$, f is also full-range. ∎

Considering that $\mathcal{L}_0 = \mathcal{L}_1$ (by an easy argument), $\mathcal{L}_1 \supseteq \mathcal{L}_2 \supseteq \mathcal{L}_3$ (by a simple consequence of the definition), $\mathcal{L}_3 \supseteq \mathcal{L}_3^p$ (again, by definition), $\mathcal{L}_3^p = \mathcal{L}_2^p = \mathcal{L}_1^p = \mathcal{L}_0^p$ (by Theorem 5), and $\mathcal{L}_1^p = \mathcal{L}_1$ (by Theorem 4), we obtain the following result.

Theorem 6 *For $X, Y \in \{0,1,2,3\}$, $\mathcal{L}_X = \mathcal{L}_Y^p$.*

In summary, when exactness is not required, any combination of prudence, consistency, and responsiveness does not restrict conservative learnability.

It is natural to wonder, for any $X \in \{0,1,2,3\}$, what is the relation between $\mathcal{L}_X^{ep}$ and $\mathcal{L}_X$. Clearly, $\mathcal{L}_X^{ep} \subset \mathcal{L}_X$ since $\mathcal{L}_X$ contains non-r.e. families whereas $\mathcal{L}_X^{ep}$ does not. However, the question remains whether any indexed family $\mathcal{F}$ that belongs to $\mathcal{L}_X$ also belongs to $\mathcal{L}_X^{ep}$. (An analogous result does hold for TxtEx-identification [OSW82].)

In conclusion, we would like to emphasize that the characterizations of Section 4 have been instrumental in establishing the results of the present section. We expect our characterization of conservative learning of various types to provide a means for exploring further the consequences of various constraints on learners in the Gold [Gol67] model of inductive inference.

References

[Ang80a] Dana Angluin. Finding patterns common to a set of strings. *Journal of Computer System Sciences*, 21:46–62, 1980.

[Ang80b] Dana Angluin. Inductive inference of formal languages from positive data. *Information and Control*, 48:117–135, 1980.

[Ber85] Robert Berwick. *The Acquisition of Syntactic Knowledge.* MIT press, Cambridge, MA, 1985.

[Ful90] Mark A. Fulk. Prudence and other conditions on formal language learning. *Information and Computation*, 85:1–11, 1990.

[Gol67] E. M. Gold. Language identification in the limit. *Information and Control*, 10:447–474, 1967.

[Kap91] Shyam Kapur. *Computational Learning of Languages.* Technical Report 91-1234, Cornell University, Ithaca, NY, September 1991. Ph.D. Thesis.

[MW84] Irene Mazurkewich and Lydia White. The acquisition of dative-alternation: unlearning overgeneralizations. *Cognition*, 16(3):261–283, 1984.

[OSW82] Daniel N. Osherson, Michael Stob, and Scott Weinstein. Learning strategies. *Information and Control*, 53:32–51, 1982.

[OSW86] Daniel N. Osherson, Michael Stob, and Scott Weinstein. *Systems that Learn: An Introduction to Learning.* MIT press, Cambridge, MA, 1986.

INVITED LECTURE

The Log-Star Revolution

TORBEN HAGERUP

Max-Planck-Institut für Informatik
W–6600 Saarbrücken, Germany

Abstract. The last approximately one year has witnessed a dramatic change in the way theoreticians think about computing on the randomized concurrent-read concurrent-write parallel random access machine (CRCW PRAM). Today we have superfast algorithms that were inconceivable a few years ago. Many of these having running times of the form $O((\log^* n)^c)$, for some small constant $c \in \mathbb{N}$, the name "log-star revolution" seems appropriate. This paper tries to put some of the most important results obtained next to each other and to explain their significance. In order to keep the exposition properly focussed, we restrict our attention to problems of a very fundamental nature that, in an ideal environment, would be handled by the operating system of a parallel machine rather than by each applications programmer: Processor allocation, memory allocation and the implementation of a particular conflict resolution rule for concurrent writing. The main contention of the paper is that the theoretical groundwork for providing such an ideal environment has been laid.

Our goal is to provide the reader with an appreciation of the log-star revolution and to enable him or her to carry on the torch of revolution by exploring the largely unchartered new territories. The emphasis is on ideas, not on rigor.

Warning: This paper deals exclusively with the randomized CRCW PRAM, a species of parallel machines that may never have any close relation to the realities of parallel computing. If you, dear reader, are very practically inclined, perhaps you should better stop reading here in order to not feel cheated later. The author confesses to a mathematical fascination with the subject rather than to a firm belief in any future practical impact (the possibility, definitely, is there). Having thus apologized once for playing a rather esoteric game, the author cordially invites the reader to accept the rules and to enjoy the game.

1 Preliminaries

A CRCW PRAM is a synchronous parallel machine with processors numbered $1, 2, \ldots$ and with a global memory that supports concurrent (i.e., simultaneous) access, even to a single cell, by arbitrary sets of processors. The meaning of concurrent reading is clear; for the time being, assume that if several processors attempt to write to the same cell in the same step, then one of them will succeed and write its value, but we cannot tell which processor this will be (with this convention the model is known as the ARBITRARY PRAM). In section 7 we will consider other ways of defining the semantics of concurrent writing. Throughout the paper, we will use

---

Supported in part by the Deutsche Forschungsgemeinschaft, SFB 124, TP B2, VLSI Entwurfsmethoden und Parallelität, and in part by the ESPRIT II Basic Research Actions Program of the EC under contract No. 3075 (project ALCOM).

as our model of computation the randomized CRCW PRAM. We assume that each processor is able to choose a random integer in the range $1..k$ in constant time, for arbitrary $k \in \mathbb{N}$ (whenever we speak of "choosing at random", we mean choosing from the uniform distribution and independently of other such choices).

A parallel algorithm that uses p processors and t time steps is said to have a *time-processor product* of pt or to execute pt *work*. It is a simple matter to simulate the parallel algorithm sequentially in time $O(pt)$. Therefore $pt = \Omega(T)$, where T is the sequential complexity of the problem under consideration, and the parallel algorithm is said to be (work-) *optimal* or to exhibit *optimal speedup* if $pt = O(T)$. An optimal parallel algorithm uses the minimum number of processors for the given running time, and vice versa (up to a constant factor). Optimal parallel algorithms are clearly very attractive, and a major goal in parallel computing is to discover such algorithms. In most cases considered in this paper, a parallel algorithm is optimal if and only if it executes $O(n)$ work.

Traditionally the performance of a parallel algorithm has been stated by giving the time and the number of processors used. We often find it more convenient to give the time and the work instead. Note that we can always replace resource bounds of p processors and time t by $\lceil p/k \rceil$ processors and $O(kt)$ time, for any $k \in \mathbb{N}$, simply by letting each physical processor simulate k virtual processors. In the cases of interest, this transformation is work-preserving, up to a constant factor, which is the main reason for considering the work measure to be fundamental. As a special case of the transformation, we can always go from $O(n)$ processors to n processors while increasing the running time by at most a constant factor.

A second measure applied to parallel algorithms, less well-defined than that of work and sometimes confused with it, is the number of *operations* executed. It indeed expresses the total number of (machine-level) operations carried out by the processors over the course of an execution of the algorithm, but counting only "useful" operations, which might loosely be characterized as those that would also be present in a sequential simulation of the algorithm at hand. We thus sum over the processors the length of time that each processor is active, but exclude idle time and time spent in processor allocation. By definition, the work executed by a parallel algorithm upper-bounds the number of operations. The ultimate goal of processor allocation is to bring the work within a constant factor of the number of operations without a significant penalty in the running time. For example, an attempt to design an optimal parallel algorithm for a problem of sequential complexity T might first attack the problem of finding a parallel algorithm that executes $O(T)$ operations, and only afterwards worry about any processor allocation problems encountered.

A basic fact of life in the CRCW PRAM world, and one exploited heavily in [21], is that with exponentially many processors, we can do essentially anything in constant time. For example, the following lemma can be proved using techniques of [21] or analogously to a related lemma of Ragde [22, p. 747].

Lemma 1.1: For every fixed $\delta > 0$ and for $p \geq n$, the prefix sums of n integers, each of at most $(\log p)^{1-\delta}$ bits, can be computed in $O(1 + \log n / \log\log p)$ time with p processors.

An equally basic fact in the world of randomized computing are the Chernoff bounds (see, e.g., [18]). The following form will be particularly useful to us.

Lemma 1.2: Let S be a binomially distributed random variable with $E(S) > 0$. Then for all β with $0 \leq \beta \leq 1$,

$$\Pr\left(\frac{|S - E(S)|}{E(S)} \geq \beta\right) \leq 2e^{-\beta^2 E(S)/3}.$$

We finally need to explain the name of the game. The function $\log^*$ (pronounced "log-star") maps $\mathbb{N}$ to $\mathbb{N}$ as follows: Let $g(1) = 2$, and for $k \in \mathbb{N}$, let $g(k+1) = 2^{g(k)}$ ($g(k)$ is a "tower" of k 2's). Then for all $n \in \mathbb{N}$, $\log^* n = \min\{k \in \mathbb{N} : g(k) \geq n\}$. The function $\log^*$ grows very slowly, so that at least theoretically, a parallel algorithm with a running time of $(\log^* n)^{O(1)}$ does not leave much to be desired in terms of speed.

Sections 2–5 of the present paper are devoted to the processor allocation problem; we come quite some way towards the ultimate goal mentioned above. Section 6 considers the related memory allocation problem, and Section 7 discusses what to do if the available hardware handles concurrent writing differently from what is required by the algorithm at hand.

2 Processor allocation — problem formulation

What is the processor allocation problem? In this section we attempt to formalize the issue in a way that models what one wants in a great many cases. Searching for a solution in a top-down manner, we are quickly led to consider two problems of independent interest, those of *approximate compaction* and *estimation*. These problems are treated in Sections 3 and 4, respectively, after which we return to the processor allocation problem in Section 5 and combine the results obtained in previous sections.

A parallel algorithm applied to a particular input conceptually uses a number of processors that bears no relation to the number of physical processors of the machine on which the algorithm is run; indeed, the algorithm may be run on machines of different sizes. A first level of abstraction therefore should allow the programmer to operate with the concept of *virtual processors*, each of which executes a particular *task*. In a static environment, this is very easy to handle at the operating-system level: If the number of virtual processors used in a particular run is smaller than the number of available physical processors, each virtual processor is simulated by a physical processor, and some physical processors remain idle. If the number of virtual processors exceeds the number of physical processors, the virtual processors are distributed about evenly among the physical processors, and each physical processor simulates the virtual processors assigned to it in a step-by-step fashion. Compared with the execution on a machine with sufficiently many physical processors, this slows down the computation by a factor proportional to the ratio between the numbers of virtual and physical processors, the best that one can hope for.

In a more dynamic environment a virtual processor may suddenly discover that the task assigned to it is larger than expected and cannot be executed sufficiently fast by the processor working on its own. At the same time other virtual processors may have nothing to do because their tasks turned out to be trivial. In such a situation we would clearly like to get the work to where the processing power is, i.e., to let processors with too much work recruit idle processors to do some of their work. We propose to model this by means of a *processor allocation operation* executed in unison by all currently existing virtual processors. Each virtual processor P contributes a single nonnegative integer x as an argument to the allocation operation. $x = 0$ means that the task assigned to P has been completed and that P can be removed. $x = 1$ means that P wants to carry on alone with whatever it is doing, and $x \geq 2$ is a request that $x - 1$ additional virtual processors be assigned to P's task (which is too much for P itself). In fact, we will not guarantee that a processor contributing an argument $x \geq 1$ continues to be assigned to the same task (hence all information must be kept in shared memory, accessible to a possible successor processor, as opposed to in local registers). Unifyingly, we can therefore say that each virtual processor that contributes an argument x to the processor allocation is to be replaced by a *team* of x virtual processors.

What form should the output of a processor allocation operation take? After its execution, each virtual processor clearly needs to know the task to which its team is assigned. In fact, it suffices to know who requested the team, since the processor number of the requesting virtual processor (call this the *number* of the team) allows each processor in the team to access all information available to the requesting processor, and therefore to continue the execution of its task. We must also require that the processors forming a team are consecutively numbered and that each processor in the team knows the smallest processor number of a member of the team (call this the *base* of the team), since a disorganized mob of processors seldom is able to do much good (actually, [13] provides an exception to this rule).

The desired output condition is therefore that each virtual processor knows the number and base of its team. Since this representation of the information is rather unhandy and not very precisely defined (what is a virtual processor? what does it mean that it knows something?), let us instead store the information in an array. We are led to consider the following *interval marking* problem:

Definition: The *interval marking* problem of size n is the following: Given n nonnegative integers $x_1, \ldots, x_n$, compute nonnegative integers $s, z_1, \ldots, z_s$ such that

(1) For all integers i, j, k with $1 \le i \le j \le k \le s$, if $z_i = z_k \neq 0$, then $z_j = z_i$;
(2) For $i = 1, \ldots, n$, $|\{j : 1 \le j \le s \text{ and } z_j = i\}| = x_i$;
(3) $s = O(\sum_{j=1}^{n} x_j)$.

We assume that the virtual processors executing a processor allocation operation are numbered $1, \ldots, n$. The jth such processor contributes the argument x_j, as described above. The operation must compute s, the new number of virtual processors, and $z_1, \ldots, z_s$, their team numbers. For $j = 1, \ldots, s$, $z_j = i \neq 0$ signifies that the jth new virtual processor is assigned to team number i. We allow $z_j = 0$, meaning that the jth processor is idle, but require (condition (3) above) that the number of idle processors is at most a constant factor larger than the number of assigned processors; this implies that the physical processors are able to simulate the active virtual processors optimally, except for a constant factor. Condition (1) states that the members of a team are indeed consecutively numbered, and condition (2) expresses that each team is of the size requested.

For the sake of simplicity, the solution to an instance of the interval marking problem does not explicitly inform the new virtual processors of the bases of their teams. For each team, however, this information can be deduced in constant time by the processor whose number equals the base of the team (call this processor the team *leader*), after which it can be displayed in a cell indexed by the number of the team and read by every member of the team.

For reasons of convenience, we straight away reduce the interval marking problem to the closely related *interval allocation* problem. The essential difference is that the interval allocation problem requires only team leaders to be informed of their team numbers. Here is the formal definition:

Definition: The *interval allocation* problem of size n is the following: Given n nonnegative integers $x_1, \ldots, x_n$, compute n nonnegative integers $y_1, \ldots, y_n$ such that

(1) For $j = 1, \ldots, n$, $x_j = 0 \Leftrightarrow y_j = 0$;
(2) For $1 \le i < j \le n$, if $0 \notin \{x_i, x_j\}$, then $\{y_i, \ldots, y_i + x_i - 1\} \cap \{y_j, \ldots, y_j + x_j - 1\} = \emptyset$;
(3) $\max\{y_j : 1 \le j \le n\} = O(\sum_{j=1}^{n} x_j)$.

Note that whereas $x_1, \ldots, x_n$ have the same meaning as in the definition of the interval marking problem, $y_1, \ldots, y_n$ do not correspond to $z_1, \ldots, z_s$. The intended interpretation now is that if $y_i \neq 0$, then the processors in team number i are those with processor numbers in the set $\{y_i, \dots, y_i + x_i - 1\}$. Shifting our point of view slightly, we consider the ith input number as a request for a *block* of x_i consecutively numbered "resource units", and we say that the block $\{y_i, \dots, y_i + x_i - 1\}$ is allocated to i. Condition (2) states that allocated blocks may not overlap, and condition (3), similar to condition (3) in the definition of the interval marking problem, requires the allocated blocks to be optimally packed, except for a constant factor.

Can we get from a solution to the interval allocation problem with a certain input $x_1, \dots, x_n$ to a solution to the interval marking problem with the same input $x_1, \dots, x_n$? First observe that for $i = 1, \dots, n$, we can easily compute the index j of the first output variable z_j to receive the value i (in fact, $j = y_i$). The problem is to copy the value i to all the following output variables $z_{j+1}, z_{j+2}, \dots$ with indices in the block allocated to i. Luckily, this problem fades away in the light of a result due to Berkman and Vishkin [6] and Ragde [22].

Definition: The *all nearest zero bit* problem of size n is, given a bit vector A of size n, to mark each position in A with the position of the nearest zero in A to its left, if any.

Lemma 2.1 [6]: All nearest zero bit problems of size n can be solved in $O(\log^* n)$ time using $O(n)$ work.

We leave the remaining details of the above reduction to the reader and now turn our attention to the interval allocation problem.

Let $x_1, \dots, x_n$ be an input to the interval allocation problem. We will assume that $x_j \leq n$, for $j = 1, \dots, n$. This is not a serious restriction; getting rid of it involves scaling the input numbers appropriately. For the moment, let us make a much more restrictive assumption: $x_j \in \{0, 1\}$, for $j = 1, \dots, n$. In this situation we can provide an array A of $c \sum_{j=1}^{n} x_j$ cells (each representing one resource unit), for a suitable constant $c \in \mathbb{N}$, and attempt to carry out the interval allocation via a random "dart throwing" process: For $j = 1, \dots, n$, if $x_j = 1$, then a processor associated with j chooses a random cell in A and attempts to "capture" it by writing to it. We let this succeed if and only if no other processor chooses the same cell of A, in which case the jth request has been satisfied. Unsuccessful processors try again in a second round, etc. We shall have occasion to study a dart throwing process of this kind in detail in Section 3. For the time being, however, recall that the assumption $x_j \in \{0, 1\}$ is not necessarily satisfied. If all nonzero requests were for the same number l of resource units, we could still use the same method, now with a smaller array A, each of whose cells represents a block of l (consecutively numbered) resource units. It therefore seems natural to group requests of the same size together and to attempt to satisfy the requests on a group-by-group basis, essentially independently for each group. Note, however, that we have to allocate an array A and the corresponding resource units to each size group. Since there are potentially n different nonempty size groups, this is an unfeasible task (indeed, it is the original interval allocation problem). On the other hand and by Lemma 1.1, we are able to perform the allocation if there are only $(\log n)^{O(1)}$ different size groups (this takes some rounding to reduce the number of bits in the input to the prefix summation). The final observation that makes everything fall into place is that we can easily arrange to have only $O(\log n)$ size groups: Simply initially round each nonzero request to the nearest larger power of 2. This at most doubles the sum of all requests, allowed by condition (3), and surely no request will mind being "over-satisfied".

Presented with an interval allocation problem with input $x_1, \dots, x_n$, we hence proceed as follows: For $i = 1, \dots, m = \lceil \log(n+1) \rceil$, let $B_i = \{j : 1 \leq j \leq n \text{ and } 2^{i-1} \leq x_j < 2^i\}$ and take $b_i = |B_i|$. A first *estimation phase*, described in Section 4, computes estimates $\hat{b}_1, \dots, \hat{b}_m$

of $b_1,\ldots,b_m$ that are correct up to a constant factor, i.e., we require that for $i = 1,\ldots,m$, $\hat{b}_i \geq b_i$, but $\hat{b}_i = O(b_i)$. Using Lemma 1.1 and for a suitable constant $c \in \mathbb{N}$, we then allocate $c\hat{b}_i$ blocks of 2^i (consecutively numbered) resource units each and an array A_i of $c\hat{b}_i$ cells to B_i, for $i = 1,\ldots,m$. A second *compaction phase*, described in Section 3, finally places the requests with indices in B_i in distinct cells of A_i, for $i = 1,\ldots,m$. If a block of 2^i resource units is associated with each cell of A_i and allocated to the request placed in the cell, if any, this satisfies all requests, and we are done.

The above reduction of interval allocation to estimation and compaction, due to Hagerup [14], was first published by Gil, Matias and Vishkin [9]. Certain other results in [9] relevant to the present paper will not be mentioned explicitly.

3 Approximate compaction

Although the problem encountered in Section 2 was to compact simultaneously into several arrays, in this section we limit ourselves to compaction into a single array. We are hence given an upper limit $d \leq n$ and an array A of size n containing at most d *objects* stored in distinct cells (we always assume that a cell contains at most one object), and the task is to move the at most d objects to (distinct cells in) a destination array B of size cd, for a suitable constant $c \in \mathbb{N}$. A formal definition can take the following form:

Definition: The *linear compaction* problem of size n is the following: Given an integer $d \leq n$ and n bits $x_1,\ldots,x_n$ with $\sum_{j=1}^{n} x_j \leq d$, compute n nonnegative integers $y_1,\ldots,y_n$ such that
(1) For $i = 1,\ldots,n$, $x_i = 0 \Leftrightarrow y_i = 0$;
(2) For $1 \leq i < j \leq n$, if $y_i \neq 0$, then $y_i \neq y_j$;
(3) $\max\{y_j : 1 \leq j \leq n\} = O(d)$.

Our approach to solving the linear compaction problem will be that of random dart throwing in several stages. All objects are initially *active*. When an active object in some stage succeeds in choosing a cell in B that was neither occupied in a previous stage nor chosen by some other object in the present stage, the object moves to that cell and becomes *inactive*. The goal is to deactivate all objects.

A first observation is that if $c \geq 2$, then a fixed object is successful already in the first stage with probability at least 1/2. This is because the total number of darts thrown is at most half the size of B. The first stage therefore reduces the number of active objects by about half. We could continue in this way, in each stage reducing the number of active objects by a constant factor, but this would require $\Theta(\log n)$ stages. When the number of active objects has decreased somewhat, however, the destination array B will not be congested even if each remaining active object throws several darts simultaneously, thereby increasing its chance of hitting a free cell somewhere. In fact, when the number of active objects has dropped below d/v, for some $v \in \mathbb{N}$, we can allow each active object to throw v darts simultaneously. The total number of darts thrown plus the number of cells in B that were occupied in previous stages is then at most $2d$. Hence provided only that $c \geq 4$, each dart will still be successful with probability at least 1/2. If we assume the outcomes (success or failure) of the v dart-throwing experiments carried out by a fixed active object to be independent events (they are not, but "almost"), then an active object remains active with probability at most 2^{-v}. Hence in one stage we can reduce the number of active objects from d/v to about $d/2^v$. Using this for $v = 1, 2, 2^2, 2^{2^2}, \ldots$, we easily see that the number of stages needed to deactivate all objects is $O(\log^* n)$.

The remaining problem is to provide each active object with the processing power needed to throw v darts in constant time. This would be easy if we had vn processors, but we prefer to get by with n processors. In order to solve the problem, we slightly modify the set-up, as described so far. Where above we assumed the number of active objects before the stage under consideration to be at most d/v, we will now assume this number to be at most d/v^5; this makes no essential difference to the argument that the number of stages is $O(\log^* n)$ (use some initial "halving" stages, as described above, in order to "get started").

Although it looks like reducing a problem to itself, we will allocate processors to active objects by means of dart throwing. Let the active objects throw one dart each into an array C of size $4d/v$, each cell of which represents v processors. Since the number of cells in C is so much larger than the total number of darts thrown, which is at most d/v^5, each fixed dart is successful with high probability. In fact, the probability that an active object fails to procure the necessary v processors (call such an object *unlucky*) is at most $\frac{d/v^5}{4d/v} = \frac{1}{4}v^{-4}$. What about the unlucky objects? Here we get unexpected help from a simple but powerful result due to Ragde:

Lemma 3.1 [22]: Let $n, v \in \mathbb{N}$ and let D be an array of size n containing at most v objects. Then n processors can move the objects in D to an array of size v^4 in constant time.

Suppose that we divide the array A into *subarrays* of size v^5 each. The number of available processors per subarray is v^5. On the other hand, the expected number of unlucky objects per subarray is at most $\frac{1}{4}v^{-4} \cdot v^5 = v/4$. Call a subarray *heavy* if it contains more than v unlucky objects. If a subarray is not heavy, then Lemma 3.1 can be used to move the unlucky objects in the subarray to an array of v^4 cells, each representing v processors, after which all processor allocation within the subarray has been successfully completed. Since this does not work for heavy subarrays, all active objects in heavy subarrays may survive to the next stage, but the probability that a fixed subarray becomes heavy is very small. In fact, if we assume than an object in a subarray becomes unlucky independently of what happens to other objects in the subarray (again, this is an approximation), then the Chernoff bound in Lemma 1.2 implies that a subarray becomes heavy with probability $2^{-\Omega(v)}$. Using this, one can show that the number of objects in heavy subarrays is comparable to the number of objects that remain active anyway, and that the necessary number of stages is still $O(\log^* n)$.

The above algorithm for linear compaction appeared implicitly in Raman's paper [23] and explicitly in that of Matias and Vishkin [19]. It marked the initial breakthrough that launched the log-star revolution. Hagerup [14, 16] and Goodrich [10] reduced the number of processors needed from n to $O(n/\log^* n)$, thereby obtaining an optimal algorithm. They also improved the bound on the failure probability of the algorithm from $d^{-\alpha}$, for arbitrary but fixed α, to what is shown in Theorem 3.2 below. The main problem for the analysis in [19] and [14, 16] was how to cope with events that are almost, but not quite independent (many of which were mentioned in the discussion above). Hagerup [13] observed that Azuma's inequality (see, e.g., [20]) from the theory of martingales is admirably well suited to deal with precisely such situations.

Theorem 3.2 [14]: There is a constant $\epsilon > 0$ such that linear compaction problems of size n can be solved in $O(\log^* n)$ time using $O(n)$ work with probability at least $1 - 2^{-n^{\epsilon}}$.

What is the value of ϵ in Theorem 3.2? Goodrich [10] states the theorem with $\epsilon = \frac{1}{25}$. Using different methods, the author has established that Theorem 3.2 holds for all $\epsilon < \frac{1}{4}$. All in all, however, the search for the optimal value of ϵ seems a tedious and not very rewarding task. The format with an unspecified constant ϵ will be kept throughout this paper.

4 Estimation

Recall that Section 2 left us with the problem of estimating $b_1, \ldots, b_m$, where $x_1, \ldots, x_n$ are integers in the range $1..m$ and $b_i = |\{j : 1 \leq j \leq n \text{ and } x_j = i\}|$, for $i = 1, \ldots, m$. Since the estimation of b_i can be carried out essentially independently of that of b_j for $i \neq j$, we will consider the problem of just estimating the number of 1's in a bit vector. Let us first fix a useful piece of terminology: Given nonnegative real numbers b, $\hat{b}$ and ρ, let us say that $\hat{b}$ *approximates* b with (relative) *error bound* ρ if $b \leq \hat{b} \leq (1+\rho)b$.

Consider now the following statement, parameterized by a real number $\rho \geq 0$:

(*) There is a constant $\epsilon > 0$ such that the number of 1's in a bit vector of length n can be estimated with error bound ρ in constant time using n processors with probability at least $1 - 2^{-n^\epsilon}$.

Hagerup [14] proved that (*) holds for some constant ρ that was left unspecified; this suffices for our application in Section 5. Goodrich [10] improved this by showing that (*) in fact holds for every constant $\rho > 0$. It is easy to extend the method of [14] to $\rho = (\log n)^{-k}$, for every fixed $k \in \mathbb{N}$. This improvement does not lie straight on our chosen path. However, since several colleagues expressed interest in the extension since it was first presented [15], we here describe the algorithm of [14] with the necessary very small modifications. It is interesting to note that the approach of Goodrich [10] apparently is not strong enough to show a result of this type. Another relevant remark is that results closely related to (*) but using boolean circuits rather than PRAMs as the model of computation have been known for a long time [24, 1]. In fact, disregarding the number of processors used as long as it is polynomial in n, (*) is directly implied by the old circuit results via the obvious simulation of circuits by PRAMs. This provides an alternative avenue for proving the results of this section (not ignoring processor counts).

Let A be a bit vector of length n, let b be the number of 1's in A and let $\rho = (\log n)^{-k}$, for fixed $k \in \mathbb{N}$. It is natural to attempt to solve the problem of determining the value of b by reducing it to the corresponding decision problem, i.e., by asking questions of the form "Is $b > t$?", where t is some real number. If we do this in parallel for $t = 1, 1+\rho, \ldots, (1+\rho)^i, \ldots, n$ and take as our estimate of b the first value of t for which the answer comes out negative, then we have clearly estimated b with error bound ρ. Note that the number of tests is polylogarithmic in n, i.e., small. Note also that accepting a relative error of $O(\rho)$ rather than ρ, we can allow each test an "uncertainty interval" of relative size ρ, i.e., if the relative difference between b and t is smaller than ρ, then the test may come out either way.

Consider now just a single test of the form "Is $b > t$?". In fact, we will assume that $t = n/2$; we are hence asking whether A is more than half full (of 1's).

A seemingly useless observation is that if we divide A into subarrays of size polylogarithmic in n, then we can use Lemma 1.1 to count the number of 1's in each subarray in constant time. This uses $O(n^\gamma)$ processors per subarray, for arbitrary fixed $\gamma > 0$, and hence $O(n^{1+\gamma})$ processors altogether. What makes this observation seem useless is the fact that the subarray counts are still far too many to add up. Suppose, however, that we view the local counts culled from the subarrays in a completely different way, as the outcome of a random sampling. At this point make the additional assumption that the 1's in A are distributed randomly, i.e., that all distributions of b 1's in A (in b distinct cells) are equally likely.

Very crudely, we expect any fixed subarray to be more than half full if and only if A is more than half full. While this statement about the number S of 1's in the subarray undoubtedly holds true for the expected value $E(S)$, it is not a priori clear that the behavior of S is sufficiently predictable for S to be of value as an indicator. However, since S is almost binomially distributed, it is is sharply concentrated around its mean. How sharply? Well, look at the Chernoff bound

$$\Pr\left(\frac{|S - E(S)|}{E(S)} \geq \beta\right) \leq 2e^{-\beta^2 E(S)/3}.$$

We can tolerate fluctuations of S around $E(S)$ of the same relative magnitude as the uncertainty allowed for our test, i.e., we must take $\beta \approx (\log n)^{-k}$. Note also that we are interested in the behavior of S when $E(S) \approx w/2$, where w is the size of a subarray, and recall that w is polylogarithmic in n, but otherwise arbitrary. It now follows from the above Chernoff bound that if we deem A to be more than half full if and only if a fixed subarray is more than half full, then except for an (allowed) relative uncertainty of $(\log n)^{-k}$, the test comes out wrong with probability at most $e^{-(\log n)^\alpha}$, where α can be chosen as an arbitrary constant. Incidentally, note here the bottleneck that prevents us from obtaining more accurate estimates.

In order to obtain a still more reliable test (with the same relative uncertainty, however), we begin with another seemingly useless observation, namely that a completely different procedure for making certain statements about b is provided by Lemma 3.1. If we try to move the 1's in A to an array of size l^4, for some $l \in \mathbb{N}$, then this will certainly fail if $b > l^4$, while Lemma 3.1 guarantees that it will succeed if $b \leq l$. We can hence distinguish between l and l^4, a far cry from estimating b with error bound ρ. But suppose now that we call a subarray *heavy* if it is more than half full and that we apply Lemma 3.1 not to the 1's in A, but to the heavy subarrays. If A is slightly less than half full, almost no subarrays will be heavy, and the compaction will easily succeed (for $l^4 = \sqrt{n}$, say). If A is slightly more than half full, almost all subarrays will be heavy, and the compaction will not succeed. Now the failure probability can be shown to be at most of the form 2^{-n^ϵ}, for fixed $\epsilon > 0$.

We made three assumptions above:
(1) $t = n/2$;
(2) $\Theta(n^{1+\gamma})$ processors are available;
(3) The 1's in A are randomly distributed.
A small modification to the algorithm allows us to justify all three assumptions: For a certain q with $0 < q \leq 1$, the following is done independently for each 1 in A: With probability q, the 1 is written to a randomly chosen cell in a new array B of n^ζ cells, for a suitably chosen constant ζ with $0 < \zeta \leq 1/5$ (ignore for the time being that several 1's may be written to the same cell of B). Then the test described above is applied to B instead of to A.

Let S be the number of 1's written to B and observe that $E(S) = qb$. Unless $t < n^\zeta$ (we deal with this case below), we can choose q to make $qt = n^\zeta/2$. Then in the cases of interest to us, S is very sharply concentrated around its mean; in fact, we may as well assume that $S = qb$ always. Recall that we wanted to test whether $b > t$. But $b > t$ if and only if $S = qb > qt = n^\zeta/2$, and the latter condition is precisely what we tested above.

We finally tie in two loose ends. If $b < n^\zeta$, we can begin by using Lemma 3.1 to move the 1's in A to an array of size $n^{4\zeta}$. Since $\zeta \leq 1/5$, it is then a trivial matter to associate n^ζ processors with each 1. Letting all of these carry out the above algorithm independently, we have in effect multiplied b by n^ζ, thereby reducing the case of small values of b to the case treated above.

Of course the 1's written to B will partly be written to the same cells, thereby hiding each other. This effect is very easy to analyze, however, and because of the large numbers involved the decrease in the number of 1's going from A to B is also very predictable. Therefore it is no problem to compensate for this effect.

So far we have used n processors for each test "Is $b > t$?". We execute some number h

of tests in parallel, where $h = (\log n)^{O(1)}$. Letting each bit in A "participate" in just one test chosen at random and multiplying the estimate for b found by h to obtain the final answer, we can get by with n processors altogether. The reader will easily be able to modify the procedure to estimate all of $b_1, \ldots, b_m$ in parallel, provided that $m = O(n^{1-\delta})$ for some fixed $\delta > 0$. Keeping track of the probabilities involved, one obtains

Theorem 4.1: For all fixed $\delta > 0$ and $k \in \mathbb{N}$, there is a constant $\epsilon > 0$ such that given n integers $x_1, \ldots, x_n$ in the range $1..m$, where $m = O(n^{1-\delta})$, the quantities $b_1, \ldots, b_m$, where $b_i = |\{j : 1 \le j \le n \text{ and } x_j = i\}|$, for $i = 1, \ldots, m$, can be estimated with error bound $(\log n)^{-k}$ in constant time using n processors with probability at least $1 - 2^{-n^{\epsilon}}$.

As a corollary to Theorem 4.1, we can improve a recent result of Goodrich [11] on *approximate selection*. Assume that we are given a set M of n distinct numbers and an integer l with $1 \le l \le n$. Our task is to compute an element in M whose rank in M is as close as possible to l. Goodrich shows that the resources stated in Theorem 4.1 are sufficient to compute an element whose rank approximates l with at most any fixed relative error. Essentially the same algorithm together with Theorem 4.1 yields Theorem 4.2 below. Note, in particular, that we can select the element of rank exactly l whenever l is polylogarithmic in n.

Theorem 4.2: For every fixed $k \in \mathbb{N}$, there is a constant $\epsilon > 0$ such that given a set M of n distinct numbers and an integer l with $1 \le l \le n$, an element in M whose rank r in M satisfies $l \le r \le l(1 + (\log n)^{-k})$ can be computed in constant time with n processors with probability at least $1 - 2^{-n^{\epsilon}}$.

5 Processor allocation — results

Consider again the algorithm for interval allocation described at the end of Section 2. The estimation phase can be carried out using Theorem 4.1, and it is easy to see that the algorithm of Theorem 3.2 with very slight changes solves the "multi-compaction" problem presented by the compaction phase. We hence have

Theorem 5.1 [14]: There is a constant $\epsilon > 0$ such that interval allocation problems of size n can be solved in $O(\log^* n)$ time using $O(n)$ work with probability at least $1 - 2^{-n^{\epsilon}}$.

We have already almost described how interval marking reduces to interval allocation. One additional detail to note is that a suitable value for the output variable s of the interval marking problem clearly is computed as part of the corresponding execution of the interval allocation algorithm (Goodrich [10] refers to this problem as "summation approximation"). Since interval marking was our formalization of the processor allocation problem, we are now able to derive results that relate work and number of operations. Let us first be slightly more explicit about the processor allocation operation by writing a call of it as *Allocate*$(x_1, \ldots, x_r)$, where x_j is the argument contributed by the jth virtual processor. A parallel algorithm will be called *standard* if its processor allocation can be done exclusively by means of calls of *Allocate*. The class of standard algorithms is very rich; in fact, the author is unable to offer an example of a nonstandard algorithm. For a standard algorithm, we can define the number of operations precisely. First define the *demand* of a call *Allocate*$(x_1, \ldots, x_r)$ as $\sum_{j=1}^{r} x_j$ and define the *cost* of a step of the algorithm as the demand of the last call of *Allocate* executed prior to the step under consideration, or as 1 if there was no such call (we assume that initially there is exactly one virtual processor). The number of operations executed by the algorithm is now simply the total cost of all its steps. In other words, we charge 1 for each virtual processor for each time step in which the processor exists.

Theorem 5.2: For every standard algorithm $\mathcal{A}$ that uses t time steps, t' calls of *Allocate* and $n = t^{1+\Omega(1)}$ operations, there is a constant $\epsilon > 0$ such that $\mathcal{A}$ can be executed in $O(t + t' \log^* n)$ time using $O(n)$ work with probability at least $1 - 2^{-n^\epsilon}$.

Proof: We implement the operation *Allocate* as described in previous sections and use $p = \Theta(n/(t + t' \log^* n))$ physical processors. For $i = 1, \ldots, t$, let n_i be the cost of the ith step of $\mathcal{A}$ and recall that $\sum_{i=1}^{t} n_i = n$. By requirement (3) in the definition of the interval marking problem, the ith step of $\mathcal{A}$ (not counting calls of *Allocate*) can be executed (with p processors) in time $O(\lceil n_i/p \rceil)$. Furthermore, a call of *Allocate* following the ith step of $\mathcal{A}$ can be executed in time $O(n_i/p + \log^* n)$. Over all steps, the time needed sums to $O(n/p + t + t' \log^* n) = O(t + t' \log^* n)$. Using the assumption $n = t^{1+\Omega(1)}$ (which is probably superfluous), one can show the probability that any (randomized) part of the algorithm fails to be at most 2^{-n^ϵ}, for some fixed $\epsilon > 0$.

The big message of Theorem 5.2 is that one can forget about all problems of processor allocation if one is willing to accept a slowdown of $\Theta(\log^* n)$, but no penalty in the total work executed. Processor allocation by means of Theorem 5.1 was first used in [14] and [5]. A result similar to Theorem 5.2 was proved by Goodrich [10] for the special case in which a call *Allocate*$(x_1, \ldots, x_r)$ is allowed only if $x_j > 2$ for at most one $j \in \{1, \ldots, r\}$. Matias and Vishkin [19] considered an even more restrictive special case of the problem and gave a solution that beats Theorem 5.2 whenever it is applicable. The special case is defined by the requirement that every call *Allocate*$(x_1, \ldots, x_r)$ must have $x_j \leq 1$, for $j = 1, \ldots, r$. In other words, the algorithm starts with a pool of active virtual processors, and these gradually become inactive. Once inactive, a processor does not return to the active state. Let us say that a standard algorithm with this property has *nonincreasing parallelism*.

Theorem 5.3 [19]: For every algorithm $\mathcal{A}$ with nonincreasing parallelism that uses t time steps and $n = t^{1+\Omega(1)}$ operations, there is a constant $\epsilon > 0$ such that $\mathcal{A}$ can be executed in $O(t + \log^* n \log\log^* n)$ time using $O(n)$ work with probability at least $1 - 2^{-n^\epsilon}$.

Proof: We use $p = \Theta(n/(t + \log^* n \log\log^* n))$ physical processors and ignore calls of *Allocate*, except that processors becoming inactive mark themselves as such. For $i = 1, \ldots, t$, let n_i be the number of processors active in the ith step of $\mathcal{A}$. After the ith step of $\mathcal{A}$, Theorem 4.1 is used to estimate n_{i+1}, which can surely be done in time $O(\lceil n_i/p \rceil)$ (in fact, it can be done in constant time). In some cases we also subsequently perform a *cleanup step*, i.e., we use Theorem 3.2 to move the active processors to an array of size $O(n_{i+1})$. Note that if we guarantee that a cleanup step is executed at least as soon as the number of active processors has decreased by some fixed factor since the last cleanup step, then the total time spent in actual simulation of steps of $\mathcal{A}$ is $O(n/p + t)$. A cleanup step executed after the ith step of $\mathcal{A}$ uses $O(n_i/p + \log^* n)$ time. Hence we can execute cleanup steps after all steps with $n_i > p \log^* n$ for a total cost of $O(n/p)$. On the other hand, as soon as n_i has dropped below p, there is a physical processor for each virtual processor, and the execution can be trivially finished in $O(t)$ time. All that remains is the part of the execution with $p \leq n_i \leq p \log^* n$. But during this part of the execution the number of active processors decreases by a constant factor only $O(\log\log^* n)$ times, giving a contribution of $O(\log^* n \log\log^* n)$ to the total running time. All other contributions sum to $O(n/p + t) = O(t + \log^* n \log\log^* n)$.

6 Memory allocation

We consider two very different memory allocation problems. The first problem is to allocate memory blocks of varying sizes to requesting tasks with at most a constant-factor waste due to gaps between blocks. This problem is quite similar to the processor allocation problem discussed in Section 2, and since a satisfactory solution is provided by Theorem 5.1, we need not concern ourselves any further with this problem.

The second memory allocation problem is to simulate a machine with a large but sparsely used memory on a machine with a smaller memory. By way of motivation, imagine the symbol table of a compiler. A trivial implementation of the symbol table might store the information associated with a particular variable in a record indexed directly by the character string that forms the name of the variable. This, however, would require a huge array of size equal to the number of potential variable names, clearly an unfeasible approach. A symbol table serves precisely to realize such a huge but sparsely used array within a reasonable amount of space.

An abstract formulation of the problem calls for the realization of a *dictionary* data structure that implements a set of records, each with an integer key, under the operations *Insert* (insert a new record with a given key), *Delete* (remove the record with a given key), and *Lookup* (inspect the record with a given key). At any time, the total space used by the dictionary must be proportional to the number of records currently stored in the dictionary. In the context of the memory simulation discussed above, keys correspond to addresses of the simulated memory, *Insert* stores a value in a given memory cell, *Delete* declares a memory cell to be no longer used (allowing the space associated with the cell to be reclaimed), and *Lookup* retrieves the value stored in a given cell. In the parallel setting, n *Insert*, *Delete* and *Lookup* operations are to be processed simultaneously, and the bounds mentioned below pertain to this problem.

Gil, Matias and Vishkin [9] sketched a parallel dictionary that works in $O(\log^* n)$ time using $O(n)$ work with high probability, and Bast, Dietzfelbinger and Hagerup [4] developed a *real-time* dictionary, which uses n processors and works in constant time with high probability (the latter dictionary uses slightly more than $\Theta(N)$ space to store N keys). As demonstrated by Dietzfelbinger and Meyer auf der Heide [7], the key problem in the construction of a (parallel) dictionary turns out to be the design of a *static* dictionary, which must support only *Lookup* operations. More precisely, the static dictionary problem is to provide a data structure for a given set of records that supports *Lookup* operations in constant sequential time, and the parameters of interest are the resources needed to construct a static dictionary for a given set of n records. We consider here only the static dictionary problem, and space considerations force us to be more superficial than in previous sections.

The best algorithm known for the static dictionary problem was described by Bast and Hagerup [5]. We sketch this algorithm and begin by restating the problem as follows: Given a multiset X of n integers, called *keys*, compute a *perfect hash function* for X, i.e., an injective function $h : X \to \{1, \ldots, s\}$, where $s = O(n)$. Note that the input X is a *multi*set rather than a simple set. In other words, a key may occur several times in the input, which is important if the dictionary is to be used as a component of a memory simulation for a CRCW PRAM (see Section 7), as well as in most other applications.

Most recent static dictionaries are based on the 2-level hashing scheme introduced by Fredman, Komlós and Szemerédi [8], which we briefly describe. A *primary hash function* f is used to partition X into n *buckets*, where each bucket is the collection of keys mapped by f to a particular value. Each bucket is provided with its own space, and for each bucket a *secondary hash function* maps the bucket injectively into its private space. In the following we abbreviate primary and secondary hash functions to *primaries* and *secondaries*, respectively. In order to hash a key x, first $f(x)$ is evaluated in order to determine the bucket containing x, and then the secondary for that bucket is applied to x. The scheme clearly realizes a hash function operating injectively on X.

Let $b_1, \ldots, b_n$ be the sizes of the buckets induced by a primary f. Fredman, Komlós and Szemerédi proved the existence of a constant $C > 0$ such that if f is chosen at random from a suitable class of hash functions and if X is a simple set, then $\sum_{i=1}^{n} b_i^2 \leq Cn$ with probability at least 1/2. Furthermore, if a bucket of size b_i is given $2b_i^2$ memory cells as its private space, then a random secondary for that bucket is injective with probability at least 1/2.

Following [5], we proceed as follows. First, since the scheme of Fredman, Komlós and Szemerédi works only for simple input sets, it is necessary to replace the input multiset X by the corresponding simple set, i.e., to remove all except exactly one occurrence of each input value from X. The next task is to find a *good* primary, which essentially is a primary whose induced bucket sizes $b_1, \ldots, b_n$ satisfy $\sum_{i=1}^{n} b_i^2 \leq Cn$. The approach of earlier papers on hashing was to simply choose a random primary, which would be good with probability at least 1/2. In order to achieve a better probability of success, we try out a large number of candidate primaries in parallel and test each of these using random sampling. For each candidate primary, this involves using Theorem 4.1 to estimate $b_1, \ldots, b_n$ by extrapolation from a small random sample of the keys, as well as testing the condition $\sum_{i=1}^{n} b_i^2 \leq Cn$ using Theorem 5.1. With high probability, at least one primary will be found to be good.

Given a good primary, we proceed to estimate its induced bucket sizes. Since there are n quantities to be estimated rather than $O(n^{1-\delta})$, we cannot use Theorem 4.1. Indeed, we are not able to guarantee that the estimates computed are correct up to a constant factor, and the final part of the algorithm, which constructs the injective secondaries, must be designed to cope with inaccurate estimates.

Three of the four main components of the above algorithm, the elimination of duplicate keys, the estimation of bucket sizes and the construction of injective secondaries, can be viewed as instances of an "generic log* algorithm" introduced by Bast and Hagerup [5]. The generic log* algorithm generalizes the approach taken in Section 3 and tries to realize its full potential. Allowing ourselves considerable inaccuracy in order to convey the basic intuition, we can describe the generic log* algorithm as follows. The input consists of $O(n)$ *active objects*, and the goal is to deactivate all of these. This is done by means of a constant-time procedure *Deactivate* that must have the following property, likely to look familiar in the light of our investigations in Section 3: For some constant $k \in \mathbb{N}$, if *Deactivate* is called when the number of active objects has already (through previous calls of *Deactivate*) decreased to $O(n/v^k)$, for some $v \in \mathbb{N}$, then a fixed active object is deactivated by the call with probability at least $1 - 2^{-v}$. In order to carry out its task, *Deactivate* is allowed to allocate resources in the form of consecutively numbered memory cells and/or processors to requesting active objects *at no cost*, provided only that the total resources used sum to $O(n/v^2)$.

In order to derive a concrete instance of the generic log* algorithm, we must define what an object is and what it means for it to be active, and we must realize the generic procedure *Deactivate* in flesh and blood. The good news is that if we succeed in this enterprise, the analysis of the generic algorithm guarantees that for some constant $\epsilon > 0$, instances of size n of the problem under consideration can be solved in $O(\log^* n)$ time using $O(n)$ work with probability at least $1 - 2^{-n^{\epsilon}}$ (note that in actuality certain other conditions must also be checked). As an illustration of this, let us cast the simplest part of the static dictionary algorithm, the construction of injective secondaries, in the mold provided by the abstract framework.

First it is necessary to require slightly more of a good primary. We replace the condition $\sum_{i=1}^{n} b_i^2 = O(n)$ by $\sum_{i=1}^{n} b_i^4 = O(n)$, which makes no essential difference. Also, in the interest of simplicity, we will assume the exact bucket sizes $b_1, \ldots, b_n$ to be known to the algorithm. Now define an object to be either a key or a bucket, and define a bucket as well as all keys in the bucket to be active as long as no injective secondary for that bucket has been found. Called when the number of active objects has decreased to $O(n/v^6)$ (i.e., we take $k = 6$), *Deactivate* first uses the free allocation mentioned above to allocate v processors to each active key and an array of $2vb_i^2$ cells to the ith bucket, for $i = 1, \ldots, n$, provided that the ith bucket is still active. For each active bucket, it is now easy to carry out the following in constant time: Choose v random secondaries, distribute $2b_i^2$ cells to each secondary and use these to test the secondary for being injective. Since a random secondary is injective with probability at least 1/2, the probability that this experiment finds no injective secondary is at most 2^{-v}. Hence the probability that an active bucket or an active key remains active after the call is at most 2^{-v}, as required. We still need to check that the free resources allocated by *Deactivate* sum to $O(n/v^2)$. Since we are dealing with altogether $O(n/v^6)$ active objects, this is obvious in the case of free processors. As concerns free memory cells, it suffices to prove that for every set $I = \{i_1, \ldots, i_r\} \subseteq \{1, \ldots, n\}$ with $r = O(n/v^6)$, we have $\sum_{i \in I} b_i^2 = O(n/v^3)$. But by the Cauchy-Schwarz inequality $|x \cdot y| \leq |x||y|$, applied to the vectors $x = (b_{i_1}^2, \ldots, b_{i_r}^2)$ and $y = (1, 1, \ldots, 1)$,

$$\sum_{i \in I} b_i^2 \leq \sqrt{\sum_{i \in I} b_i^4} \cdot \sqrt{r} = O\left(\sqrt{n} \cdot \sqrt{\frac{n}{v^6}}\right) = O\left(\frac{n}{v^3}\right).$$

Theorem 6.1 [5]: There is a constant $\epsilon > 0$ such that given a multiset X of n integers, a perfect hash function for X can be computed in $O(\log^* n)$ time using $O(n)$ work with probability at least $1 - 2^{-n^\epsilon}$.

The big message of this section is that accepting a slowdown of $O(\log^* n)$, one need not worry about *which* memory cells are used by a parallel algorithm, only about their number.

7 Simulation

Recall that up to this point, we have assumed that if several processors write to the same memory cell in the same step, then some (arbitrary) processor among them succeeds and writes its value (the ARBITRARY rule). This is clearly not the only meaningful convention concerning concurrent writing, and many *variants* of the CRCW PRAM, each with a different *write conflict resolution rule*, have been defined and used in concrete algorithms. At this point it is unclear which of these variants, if any, will eventually be realized in physical machines. Research into the relative power of different variants of the CRCW PRAM therefore serves two practical purposes: Firstly, it may guide decisions in the design of real PRAM-like machines relating to write conflict resolution ("How much additional computing power will I gain by providing the hardware necessary to support this fancy conflict resolution?"). Secondly, it may enable algorithms designed for one variant to be run on a machine realizing a different variant.

In addition to the ARBITRARY PRAM that was already introduced, this section studies the MINIMUM and TOLERANT PRAMs, whose associated write conflict resolution rules are as follows:

MINIMUM: The smallest value that some processor attempts to store in a given cell in a given step gets stored in the cell;

TOLERANT: If two or more processors attempt to write to a given cell in a given step, then the contents of that cell do not change.

An n-processor MINIMUM PRAM is clearly able to compute the minimum of n numbers in constant time, a fairly powerful operation. One algorithm that exploits the capabilities of the MINIMUM PRAM in a natural way is the minimum spanning forest (MSF) algorithm of Awerbuch and Shiloach [2] (their paper actually describes a slightly less natural implementation of the algorithm on a weaker variant). The algorithm computes an MSF by maintaining a collection of subtrees of an MSF and repeatedly letting subtrees "hook" themselves on to neighboring trees. If any edge may cause a hooking, the resulting graph is a spanning forest, but not necessarily an MSF. On the MINIMUM PRAM, however, it is easy to ensure that the edge that hooks a tree T to a neighboring tree will be the edge of smallest cost leaving T, in which case an MSF will result.

The TOLERANT PRAM makes much less stringent demands on the underlying hardware. Grolmusz and Ragde, who introduced this model [12], in fact characterized it as not being a true concurrent-write machine, the requirement simply being that the connection between processors and memory must "fail gracefully" in the case of a write conflict. The ARBITRARY PRAM is intermediate in power between the MINIMUM PRAM and the TOLERANT PRAM. We next give this statement a precise meaning.

Assume that A and B are (names of) variants of the CRCW PRAM such that an arbitrary step of an n-processor machine of type A can be simulated by a constant number of steps on an n-processor machine of type B, for arbitrary $n \in \mathbb{N}$. Then surely the variant B is no weaker than the variant A; we will express this relation as "$A \leq B$". "$\leq$" clearly is reflexive and transitive, i.e., it is a partial order. We show that

$$\text{TOLERANT} \leq \text{ARBITRARY} \leq \text{MINIMUM}.$$

The relation ARBITRARY $\leq$ MINIMUM is a triviality, since the MINIMUM conflict resolution rule is a valid specialization of the ARBITRARY conflict resolution rule. As for the other relation, one write step of a TOLERANT PRAM can be simulated by a constant number of steps on an ARBITRARY PRAM with the same number of processors as follows: Identify the processors of the two machines. Each processor has the task of attempting to write a particular value to a particular cell. It begins by reading and saving the old contents of the cell in question. Instead of writing the new value immediately, it then writes its processor number to the cell. This may involve concurrent writing, but on the simulating ARBITRARY PRAM exactly one processor will see its own processor number appear in the cell; call this processor the *leader*. The leader proceeds to carry out its task, i.e., to store its new value in the cell. If the leader was the only processor writing to the cell in question, the cell now contains the correct value. If not, at least one processor will have seen a foreign processor number (namely that of the leader) appear in the cell. The simulation is finished by letting each such processor restore the cell to its original state. This may involve concurrent writing (if there is more than one nonleader), but all values written will be identical.

In more precise terms, we require a simulation of a machine with m global memory cells to set aside a corresponding block of m memory cells. After each simulated step, each cell in the block of the simulating machine must contain the same value as the corresponding cell in the global memory of the simulated machine. We would prefer the simulation to use as little additional space as possible. The *slowdown* of a simulation is the maximum number of steps needed by the simulating machine to simulate one step of the simulated machine. The slowdown of a simulation of an n-processor machine on a p-processor machine clearly is $\Omega(\lceil n/p \rceil)$. We therefore call the simulation *optimal* if its slowdown is $O(n/p)$. It is easy to see that an optimal simulation is work-preserving. In the following we will consider *randomized* simulations, for which some of the above definitions need to be modified in a straightforward way.

It turns out that the majority of variants of the CRCW PRAM found in the literature are in fact *totally* ordered by the relation "$\leq$". These variants can therefore be arranged in a linear sequence going from the strongest to the weakest variants. The MINIMUM and TOLERANT variants are near opposite ends of this sequence. A fast and optimal simulation of MINIMUM on TOLERANT therefore satisfies the goal set out in the beginning of this section very well. On the one hand, if the simulation is judged to be efficient in practical terms, a machine builder may be well advised to design a machine in the simplest possible way, as long as it still has at least the capabilities of the TOLERANT PRAM (superfast or optimal simulations on variants even weaker than TOLERANT are not known). On the other hand, practically every published PRAM algorithm can be efficiently executed on such a machine. The remainder of this section describes an optimal simulation of a MINIMUM PRAM with n processors and m global memory cells on a TOLERANT PRAM with $O(n/\log^* n)$ processors and $O(n+m)$ global memory cells. The slowdown of the simulation is $O(\log^* n)$ with high probability. This improves the best previous simulation, due to Gil, Matias and Vishkin [9], who were unable to achieve $O(\log^* n)$ slowdown and $O(n+m)$ space simultaneously, and whose simulation is less reliable. The simulation presented here is a further development of one described in [17]. Concurrent writing can be limited to $O(n)$ cells.

The input to the computation is a sequence $(\xi_1, \theta_1), \ldots, (\xi_n, \theta_n)$ of n pairs called *requests*. The intended meaning of the request (ξ_j, θ_j) is that the jth processor of the simulated machine attempts to write the value θ_j to the global memory cell numbered ξ_j. Define a *color class* to be a maximal group of requests with the same first component, and call the common first component the *address* of the color class. The task of the simulation is to store the minimum second component within each color class in the cell indicated by the address of the color class, called the *target cell* of that address.

We begin by describing a nonoptimal simulation that uses $\Theta(n)$ processors, rather than $O(n/\log^* n)$, and later indicate how to achieve optimality. Our first goal is to "collect" each color class, i.e., to move its members to consecutive memory locations. This could be achieved by sorting the requests by their addresses, but this would be much too slow and also too inefficient. We therefore relax the conditions enforced by sorting in two directions: Firstly, we do not insist that the color classes are arranged in the order of increasing addresses, which is irrelevant for the simulation anyway, as long as all elements of a color class occur together. Secondly, we do not insist that the permuted requests are stored compactly in an array of size exactly n, but allow them to spread over an array of size $O(n)$, with unused locations containing a special value *nil*. In precise terms, we are asking for a padded representation of size $O(n)$ of a duplicate-grouping permutation for $\xi_1, \ldots, \xi_n$.

Definition: Given a sequence $\xi_1, \ldots, \xi_n$ of n elements, a permutation $\pi_1, \ldots, \pi_n$ of $1, \ldots, n$ is *duplicate-grouping* for $\xi_1, \ldots, \xi_n$ if for all $i, j, k \in \mathbb{N}$, if $1 \leq i \leq j \leq k \leq n$ and $\xi_{\pi_i} = \xi_{\pi_k}$, then $\xi_{\pi_j} = \xi_{\pi_i}$. A *padded representation* of size s of $\pi_1, \ldots, \pi_n$ is a vector of size s whose non-*nil* elements, taken in order, precisely form the sequence $\pi_1, \ldots, \pi_n$.

Even after these concessions, we are able to solve the problem at hand directly only if the integers $\xi_1, \ldots, \xi_n$ come from a range of size $O(n)$, i.e., if $m = O(n)$. At this point, we can put the static dictionary algorithm outlined in Section 6 to good and somewhat unexpected use. We are not quite able to implement that algorithm on the less powerful TOLERANT PRAM, but the following weaker statement suffices for our purpose.

Lemma 7.1: There is a constant $\epsilon > 0$ such that a perfect hash function for a sequence of n integers can be constructed on a TOLERANT PRAM using $O(\log^* n)$ time, n processors and $O(n)$ space with probability at least $1 - 2^{-n^{\epsilon}}$.

The algorithm of the above lemma yields an injective function $h : \{\xi_1, \ldots, \xi_n\} \to \{1, \ldots, s\}$, where $s = O(n)$. Replacing ξ_j by $h(\xi_j)$, for $j = 1, \ldots, n$, we can reduce the case of general values of m to the case $m = O(n)$ (the important property of h being that $\xi_i = \xi_j \Leftrightarrow h(\xi_i) = h(\xi_j)$, for $i, j = 1, \ldots, n$). We now appeal to the following result, whose proof is similar to that of Theorem 6.6 in [14].

Lemma 7.2 [3]: There is a constant $\epsilon > 0$ such that the following problem can be solved on a TOLERANT PRAM using $O(\log^* n)$ time, n processors and $O(n)$ space with probability at least $1 - 2^{-n^{\epsilon}}$: Given a sequence $x_1, \ldots, x_n$ of n integers in the range $1 \,.\,.\, n$, compute a padded representation of size $O(n)$ of a duplicate-grouping permutation for $x_1, \ldots, x_n$.

After the application of Lemma 7.2, we can assume the original requests $(\xi_1, \theta_1), \ldots, (\xi_n, \theta_n)$ to be stored in an array of size $O(n)$ that can be partioned into *segments* of consecutive cells such that the non-*nil* elements in a fixed segment are precisely the elements of one color class. Using Lemma 2.1, segments with this property can be explicitly computed, a necessary technicality. The remainder of the simulation can be viewed as an instance of the generic $\log^*$ algorithm described in Section 6. An object, in the sense of the generic algorithm, is a request, and all requests in a given color class are deactivated when the minimum second component occurring in the segment of that color class has been determined and written to the appropriate target cell. All that remains is to describe a procedure *Deactivate* to carry out this task. Recall that when the number of active objects has decreased to $O(n/v^k)$, for some $k \in \mathbb{N}$, then a call of *Deactivate* must deactivate a fixed color class with probability at least $1 - 2^{-v}$. Since we can execute v independent trials as in Section 6, however, it suffices to show that the minimum of n numbers can be computed by n processors in constant expected time. This is implied by Theorem 5.6 in [17].

This ends the description of the nonoptimal algorithm. We have shown that an n-processor MINIMUM PRAM can be simulated with slowdown $O(\log^* n)$ on a TOLERANT PRAM of the same size. Our remaining task is to reduce the number of processors of the simulating TOLERANT PRAM from $\Theta(n)$ to $O(n/\log^* n)$ in order to obtain an optimal simulation. This final part of the paper is quite technical and should be omitted by the casually interested reader.

Let ξ be the address of a color class C and let γ be the target cell of ξ. We will assume that γ can be initialized to the value ∞ prior to the write step under consideration and leave the removal of this assumption as an exercise for the reader. The simulation gradually decreases the value of γ by successively storing values $\theta_{i_1}, \theta_{i_2}, \ldots$ in γ, where $\theta_{i_1}, \theta_{i_2}, \ldots$ are second components of requests in C and $\theta_{i_1} > \theta_{i_2} > \cdots$. Call a request (ξ, θ) *active* as long as the value stored in γ is larger than θ. Our goal is to describe a preprocessing phase that reduces the number of active requests from n to $O(n/\log^* n)$, at which point we can apply Theorem 3.2 to compact the remaining active requests and finish the simulation by means of the nonoptimal algorithm described above.

We first draw a random sample from the set of all requests by including each request in the sample independently of all other requests and with probability $1/\log^* n$. With high probability, the sample is of size $O(n/\log^* n)$, so that it can be given as input to the nonoptimal algorithm. This deactivates all requests in the sample, as well as many other requests. In fact, since one element in $\log^* n$ is picked for the sample, we expect about $\log^* n$ requests to remain active in each color class of size $\geq \log^* n$. More precisely, if we define a *bad* request to be an active request in a color class with more than $(\log^* n)^{5/4}$ active requests, then with high probability the total number of bad requests is $O(n/\log^* n)$. Although the bad requests cannot be identified and removed, we shall ignore them in the following. After removing inactive requests by means of Theorem 3.2, we hence assume the size of each color class to be at most $(\log^* n)^{5/4}$.

The next major step is an attempt to identify and split the color classes of size $\geq (\log^* n)^{3/4}$, which will be called *large*. Draw a random sample as above, now including each request with probability $1/(\log^* n)^{1/2}$. Each large color class is likely to have at least one representative in the sample; we will try to elect a unique *leader* in each such color class. Store the sample in a rectangular *tableau* of $r = \Theta(n/(\log^* n)^{3/2})$ rows and $s = \log^* n$ columns (an $r \times s$ tableau) and permute the entries in the tableau randomly using Theorem 3.8 in [16]. Then process the tableau one column at a time. To process a column is to process each of its entries. To process a request (ξ_j, θ_j) is to attempt to store the value j (calling for the request (ξ_j, θ_j) to be the leader) in an auxiliary cell associated with the address ξ_j. Observe what happens: If some column in the tableau contains exactly one element of some color class, then a leader is certain to be elected in that color class. But since the expected number of elements of a fixed color class in the tableau is at most $(\log^* n)^{3/4}$, while the number of columns is $s = \log^* n$, it is unlikely that no element of a fixed color class represented in the tableau will be alone in its column (see Lemma 4.5 in [17]). It follows that leaders are elected in most large color classes; as above, we will ignore color classes for which this is not the case.

Use Theorem 3.2 to move the leaders to an array of size $O(n/(\log^* n)^{1/2})$ and allocate a block of $2(\log^* n)^{1/2}$ cells to each leader. Each element of a color class for which a leader was elected now assigns itself at random to one of the cells allocated to the leader of its color class. This in effect replaces each such color class by a number of smaller color classes. Since by assumption no color class was of size $> (\log^* n)^{5/4}$ before this step, we can ignore all color classes of size $> (\log^* n)^{3/4}$ after the step.

The preprocessing phase is finished by $T = \log\log^* n$ *stages*, each of which with high probability reduces the number of remaining active requests by half. After Stage T, the total number of active requests (including those ignored above) will be $O(n/\log^* n)$, as desired.

At the beginning of Stage t, for $t = 1, \ldots, T$, the active requests are stored in an $r \times s_t$ tableau A_t, where $r = n/\log^* n$ and $s_t = 64a^t \log^* n$. Here a is a constant with $\frac{1}{2} \leq a < 1$, chosen to make $a^T \geq (\log^* n)^{-1/4}$. The stage proceeds much as the leader election above, except that processing a request (ξ_j, θ_j) now means writing θ_j to the target cell γ of ξ_j, provided that γ does not already contain a value smaller than θ_j. If any value is written to γ at all in the stage, the value of γ at the end of the stage is no larger than a second component chosen at random among those of active requests in the color class under consideration, so that about half of the color class is likely to be deactivated. On the other hand, repeating an argument used in the discussion of the leader election and observing that the size of every color class is bounded by s_t, one easily shows that it is unlikely that no value is written to γ. Processing the tableau in this manner a constant number of times, we can ensure that with high probability the stage reduces the number of active requests by half.

Stage t is completed by moving the at most $a^t \log^* n$ remaining active requests in A_t to the smaller tableau A_{t+1}. If rows were deactivated at the same pace, this would be easy and could be done independently for each row. That, however, is too much to hope for, and we resort to a technique of [5]. First attempt to move each active request in A_t to a random position in the left half of A_{t+1} (dart throwing). Since the number of cells in A_{t+1} is at least 32 times larger than the number of active requests, the probability that a fixed request fails to find a free position (call such a request *unlucky*) is at most 1/16. In other words, the expected number of unlucky requests in a fixed row of A_t is at most $s_t/16$. Call a row of A_t *heavy* if it contains more than $s_t/4$ unlucky requests. The unlucky requests in a nonheavy row can simply be copied to the corresponding row in the right half of A_{t+1}, which contains at least $s_t/4$ cells and has not been used so far. On the other hand, heavy rows are so unlikely that unlucky requests in heavy rows can be ignored; with high probability, their number sums to $O(n/\log^* n)$ over all stages.

Each of the T stages needs a random permutation, whose computation takes $\Theta(\log^* n)$ time. These permutations, however, can be precomputed before the first stage, so that this is not a problem. The total time needed by the preprocessing is $O(\sum_{t=1}^{T} s_t) = O(\log^* n)$. This ends the description of the preprocessing phase. We have proved:

Theorem 7.3: There is a constant $\epsilon > 0$ such that one step of a (randomized) n-processor MINIMUM PRAM with m global memory cells can be simulated by $O(\log^* n)$ steps of a randomized TOLERANT PRAM with $O(n/\log^* n)$ processors and $O(n+m)$ global memory cells with probability at least $1 - 2^{-n^{\epsilon}}$.

As a final remark, the slowdown of the above simulation can be reduced to a constant if we allow the simulating TOLERANT PRAM to have $\Theta(n \log^{(k)} n)$ processors and $\Theta(n \log^{(k)} n + m)$ space, for some constant $k \in \mathbb{N}$. Here $\log^{(k)}$ denotes k-fold repeated application of the function log, i.e., $\log^{(1)} n = \log n$, and $\log^{(k)} n = \log \log^{(k-1)} n$, for $k \geq 2$.

References

[1]: M. AJTAI AND M. BEN-OR, A Theorem on Probabilistic Constant Depth Computations, *in* Proc. 16th Annual ACM Symposium on Theory of Computing (1984), pp. 471–474.

[2]: B. AWERBUCH AND Y. SHILOACH, New Connectivity and MSF Algorithms for Ultracomputer and PRAM, *in* Proc. International Conference on Parallel Processing, 1983, pp. 175–179.

[3]: H. BAST, personal communication.

[4]: H. BAST, M. DIETZFELBINGER, AND T. HAGERUP, A Parallel Real-Time Dictionary, in preparation.

[5]: H. BAST AND T. HAGERUP, Fast and Reliable Parallel Hashing, manuscript. A preliminary version appears in Proc. 3rd Annual ACM Symposium on Parallel Algorithms and Architectures (1991), pp. 50–61.

[6]: O. BERKMAN AND U. VISHKIN, Recursive *-Tree Parallel Data-Structure, *in* Proc. 30th Annual Symposium on Foundations of Computer Science (1989), pp. 196–202.

[7]: M. DIETZFELBINGER AND F. MEYER AUF DER HEIDE, A New Universal Class of Hash Functions and Dynamic Hashing in Real Time, *in* Proc. 17th International Colloquium on Automata, Languages and Programming (1990), Springer Lecture Notes in Computer Science, Vol. 443, pp. 6–19.

[8]: M. L. FREDMAN, J. KOMLÓS, AND E. SZEMERÉDI, Storing a Sparse Table with $O(1)$ Worst Case Access Time, *J. ACM* **31** (1984), pp. 538–544.

[9]: J. GIL, Y. MATIAS, AND U. VISHKIN, Towards a Theory of Nearly Constant Time Parallel Algorithms, *in* Proc. 32nd Annual Symposium on Foundations of Computer Science (1991), pp. 698–710.

[10]: M. T. GOODRICH, Using Approximation Algorithms to Design Parallel Algorithms that May Ignore Processor Allocation, *in* Proc. 32nd Annual Symposium on Foundations of Computer Science (1991), pp. 711–722.

[11]: M. T. GOODRICH, Fast Parallel Approximation Algorithms for Problems with Near-Logarithmic Lower Bounds, Tech. Rep. no. JHU–91/13 (1991), Johns Hopkins University, Baltimore, MD.

[12]: V. GROLMUSZ AND P. RAGDE, Incomparability in Parallel Computation, *in* Proc. 28th Annual Symposium on Foundations of Computer Science (1987), pp. 89–98.

[13]: T. HAGERUP, Constant-Time Parallel Integer Sorting, *in* Proc. 23rd Annual ACM Symposium on Theory of Computing (1991), pp. 299–306.

[14]: T. HAGERUP, Fast Parallel Space Allocation, Estimation and Integer Sorting, Tech. Rep. no. MPI–I–91–106 (1991), Max-Planck-Institut für Informatik, Saarbrücken.

[15]: T. HAGERUP, On Parallel Counting, presented at the Dagstuhl Workshop on Randomized Algorithms, Schloß Dagstuhl, June, 1991.

[16]: T. HAGERUP, Fast Parallel Generation of Random Permutations, *in* Proc. 18th International Colloquium on Automata, Languages and Programming (1991), Springer Lecture Notes in Computer Science, Vol. 510, pp. 405–416.

[17]: T. HAGERUP, Fast and Optimal Simulations between CRCW PRAMs, these proceedings.

[18]: T. HAGERUP AND C. RÜB, A Guided Tour of Chernoff Bounds, *Inform. Proc. Lett.* **33** (1990), pp. 305–308.

[19]: Y. MATIAS AND U. VISHKIN, Converting High Probability into Nearly-Constant Time — with Applications to Parallel Hashing, *in* Proc. 23rd Annual ACM Symposium on Theory of Computing (1991), pp. 307–316.

[20]: C. MCDIARMID, On the Method of Bounded Differences, *in* Surveys in Combinatorics, 1989, ed. J. Siemons, London Math. Soc. Lecture Note Series 141, Cambridge University Press, pp. 148–188.

[21]: I. PARBERRY, Parallel Complexity Theory, Pitman, London, 1987.

[22]: P. RAGDE, The Parallel Simplicity of Compaction and Chaining, *in* Proc. 17th International Colloquium on Automata, Languages and Programming (1990), Springer Lecture Notes in Computer Science, Vol. 443, pp. 744–751.

[23]: R. RAMAN, The Power of Collision: Randomized Parallel Algorithms for Chaining and Integer Sorting, *in* Proc. 10th Conference on Foundations of Software Technology and Theoretical Computer Science (1990), Springer Lecture Notes in Computer Science, Vol. 472, pp. 161–175.

[24]: L. STOCKMEYER, The Complexity of Approximate Counting, *in* Proc. 15th Annual ACM Symposium on Theory of Computing (1983), pp. 118–126.

COMPLEXITY & COMMUNICATION

Separating Counting Communication Complexity Classes

Carsten Damm
Fachbereich Informatik
Postfach 1297
Humboldt-Universität zu Berlin
O-1086 Berlin, Germany

Matthias Krause
Lehrstuhl Informatik II
Universität Dortmund
Postfach 500 500
W-4600 Dortmund 50, Germany

Christoph Meinel
Fachbereich Informatik
Postfach 1297
Humboldt-Universität zu Berlin
O-1086 Berlin, Germany

Stephan Waack
Karl-Weierstraß-Institut
für Mathematik
Mohrenstr. 39
O-1086 Berlin, Germany

November 4, 1991

Abstract

We develope new lower bound arguments on communication complexity and establish a number of separation results for Counting Communication Classes. In particular, it will be shown that for Communication Complexity MOD_p-P and MOD_q-P are uncomparable via inclusion for all pairs of distinct primes p, q. Further we prove that the same is true for PP and MOD_p-P for any prime number p. Our results are due to mathematical characterization of modular and probabilistic communication complexity by the minimum rank of matrices belonging to certain equivalence classes. We use arguments from algebra and analytic geometry.

Keywords

complexity of Boolean function, communication complexity and distributed computing, probabilism, lower bound arguments, separation of complexity classes

1 Introduction

Communication complexity theory has become an autonomous and important discipline of theoretical computer science. This is essentially due to to the fact that a lot of interesting structural questions concerning the complexity of Turing machines, PRAMs, VLSI- and combinational circuits could be answered by reducing them to analyzing the computational power of several kinds of communication games. Among others this regards time–area tradeoffs for VLSI–computations [AUY 83], time–space tradeoffs for Turing machines , and width–length tradeoffs for oblivious and usual branching programs and Ω–branching programs [AM 86][KMW 89][KW 89][K 90]. Further in this line we have to mention characterizations of circuit depth and lower bounds on the depth of monotone circuits [RW 90], lower bounds on depth–two circuits having symmetric gates [S 90][KW 91], and structural results in the field of designing pseudorandom sequences [BNS 91].

Besides the central problem of establishing mathematical arguments for estimating the computational power of certain communication games (decidable papers in this line are, e.g.,[Y 83][MS 82][BNS 87]) a lot of papers are dealing with comparing the power of deterministic, nondeterministic, probabilistic and alternating communication protocols. Effort in this direction has reached a new stage by using the framework of communication complexity classes and rectangular reducibility introduced and developed in [BNS 86] and [HR 88]. This framework is based on the observation that the whole richness of complexity classes and hierarchies of complexity classes as we know from Turing machines can be defined for communication complexity in a very natural way.

In this paper we are dealing with the mathematical characterization of counting communication classes. We develope new lower bound arguments and present a number of new separation results for probabilistic and modular computations. Among others one motivation for investigating counting communication complexity is due to the open problem of separating ACC. In [Y 90] it has been shown that all problems computable by constant depth, polynomial size circuits with MOD_m-gates for arbitrary integers m, are contained in certain counting communication complexity classes.

In this paper we suppose that decision functions are given in a distributed form $f : S \times T \longrightarrow \{0,1\}$, where S, T are finite sets.

The communication complexity of f is the minimal number of information bits two processors have to exchange for cooperatively computing f under the restriction that one processor can access only to S and the other only to T.

In particular, a communication protocol P of length m over $S \times T$ consists of two processors P_l and P_r with unbounded computational power and a communication tape of length m. A computation of the protocol P on an input $(s,t) \in S \times T$ is going in m rounds. Starting with P_l the processors write alternatingly bits to the communication tape, in each round exactly one bit. The m–th bit communicated is the output produced.

In our framework we suppose protocols to work nondeterministically. Let us denote by

$Acc_P(s,t)$ and $Rej_P(s,t)$ the sets of accepting and rejecting P–computations on input (s,t). Clearly, $Acc_P(s,t)$ and $Rej_P(s,t)$ can be thought of as to be subsets of $\{0,1\}^m$.

As it is common use we call an acceptation mode an *counting acceptation mode* if the decision wether P computes 1 or 0 on input (s,t) only depends on the values $acc_P(s,t) = \#Acc_P(s,t)$ and $rej_P(s,t) = \#Rej_P(s,t)$.

In this paper our special interest is devoted to the following types of counting acceptation modes.

Definition 1.1 *A protocol P is called Modulo-p protocol, p prime number, if it is appointed that P computes 1 on inputs (s,t) if and only if $acc_P(s,t)$ is not divisible by p. P is called Majority protocol, if it is appointed that P computes 1 if and only if $acc_P(s,t) > rej_P(s,t)$.*

By increasing the length of P by at most two it can be always achieved that for all inputs (s,t) $acc_P(s,t) \neq rej_P(s,t)$.

The *communication complexity* of a given decision function $f : S \times T \longrightarrow \{0,1\}$ with respect to a predefined acceptation mode AM is defined to be the minimal length an AM–protocol over $S \times T$ must have for computing f. Clearly, the communication complexity of f is not greater than $2\min\{\log_2(\#S), \log_2(\#T)\}$. Thus, for all sequences of Boolean functions $F = (f_n : \{0,1\}^n \times \{0,1\}^n \longrightarrow \{0,1\})_{n \in \mathbb{N}}$ the communication complexity is $O(n)$. We say that F is feasible computable by communication protocols if the communication complexity of F is $O(\log^k(n))$ for some $k \in \mathbb{N}$. We consider the following counting communication complexity classes.

$C\text{-}P = \{F, F$is feasible computable by deterministic protocols$\}$,

$C\text{-}NP = \{F, F$is feasible computable by nondeterministic protocols$\}$,

$C\text{-}co\text{-}NP = \{F, F$is feasible computable by co-nondeterministic protocols$\}$,

$C\text{-}\oplus P = \{F, F$is feasible computable by Parity- protocols$\}$,

$C\text{-}MOD_p\text{-}P = \{F, F$is feasible computable by Modulo-p protocols (p prime)$\}$,

$C\text{-}PP = \{F, F$is feasible computable by Majority protocols$\}$.

Let us mention some previously obtained separation results for these classes. The first is that $C\text{-}NP$, $C\text{-}co\text{-}NP$, and $C\text{-}\oplus P$ are pairwise uncomparable via inclusion [BFS 86][HR 88]. It is not hard to see that $C\text{-}NP$, as well as $C\text{-}co\text{-}NP$ are subsets of $C\text{-}PP$. It holds that the union of $C\text{-}NP$ and $C\text{-}co\text{-}NP$ is properly contained in $C\text{-}PP$. This follows from the fact that the COMPARISIÔN Problem $COMP = (comp_n : \{0,1\}^n \times \{0,1\}^n \longrightarrow \{0,1\})_{n \in \mathbb{N}}$ is contained in $C\text{-}PP$ [Y 83]. (For all $n \in \mathbb{N}$, $x, y \in \{0,1\}^n$ let $comp_n(x,y) = 1$ if and only if x is lexicographically less than y.) It is quite straightforward to show that COMP belongs neither to $C\text{-}NP$ nor to $C\text{-}co\text{-}NP$ (see chapter 2).

A further important result has been shown in [HR 88] saying that $C\text{-}\oplus P$ is not contained in $C\text{-}PP$. It is obtained by a probabilistic lower bound argument for Majority communication complexity which works only for the case if the $\{+,-\}$-communication matrix of the decision function is orthogonal, i.e., an Hadamard–matrix.

In the next section we present another lower bound argument for Majority- (or probabilistic) communication complexity which is based on analytic geometry in Euclidian vector spaces and which allows to establish maximal lower bounds for a much wider range of problems. By help of this new method together with a lower bound bound argument for Modulo-p communication complexity we obtain a number of new separation results for counting communication complexity classes. In particular we will show that for all pairs of distinct prime numbers p and q the classes $C\text{-}MOD_p\text{-}P$ and $C\text{-}MOD_q\text{-}P$ are uncomparable via inclusion (Theorem 1) and that for all prime numbers p the classes $C\text{-}PP$ and $C\text{-}MOD_p\text{-}P$ have the same property (Theorem 2).

The paper is organized as follows. In the next section we provide the theoretical background for proving our lower bounds. In sections 3 and 4 we prove Theorem 1 and 2, respectively. At the end some related open problems are stated.

2 Theoretical Background

A well-known method of representing a decision functions $f : S \times T \longrightarrow \{0,1\}$ by "more algebraic" objects is to consider the *communication matrix*, $M(f)$, defined as

$$M(f)_{s,t} = f(s,t) \quad \text{for all} \quad s \in S, \quad t \in T.$$

We shall use the terminus *matrix rank* in the following way.

Definition 2.1 *Given a semiring H and a $N \times M$–matrix A we denote by $rank_H(A)$ the minimum number K so that there is a $N \times K$–matrix B and a $K \times M$–matrix C fulfilling $A = BC$.*

For example, the complexity of any decision function f with respect to Parity communication protocols is given by $\Theta(\log_2(rank_{\mathbb{F}_2}(M(f))))$. Similar characterizations can be derived for nondeterministic and co– nondeterministic communication complexity by using the Boolean semiring. (Looking at the corresponding communication matrices it is now easy to see that the COMPARISION problem belongs neither to $C\text{-}\oplus P$, nor to $C\text{-}NP$, nor to $C\text{-}co\text{-}NP$). For "higher level" acceptation modes such as Majority– or Modulo-p acceptation characterizations of this easy type do not exist. Up to now, whenever we say that given numbers, vectors, matrices, or functions are *equivalent* we do this with respect to the following "universal" equivalence relation.

Definition 2.2 *Let N and M be natural numbers. Given two real $N \times M$–matrices $A = (A_{i,j})$ and $B = (B_{i,j})$ we say that A is equivalent to B (for short, $A \sim B$) if for all i, $1 \leq i \leq N$, and j, $1 \leq j \leq M$,*

$$A_{i,j}B_{i,j} \geq 0 \qquad \textit{and} \qquad A_{i,j}B_{i,j} = 0 \;\; \textit{iff} \;\; A_{i,j} = B_{i,j} = 0.$$

Given a set S and two functions $f, g : S \longrightarrow \mathbb{R}$ we write $f \sim g$ if for all $s \in S$ it holds $f(s) \sim g(s)$.

Now observe the following characterization of the complexity classes $C\text{-}MOD_p\text{-}P$, p prime number.

Proposition 2.1 *Let Σ be a finite alphabet and $F = (f_n : \Sigma^n \times \Sigma^n \longrightarrow \{0,1\})_{n \in \mathbb{N}}$ be a decision problem. Then F belongs to $C\text{-}MOD_p\text{-}P$ if and only if for all $n \in \mathbb{N}$ there is a function $g_n : \Sigma^n \times \Sigma^n \longrightarrow \mathbb{F}_p^n$ fulfilling (a) $f_n \sim g_n$,*

(b) The $\mathbb{F}_p^n$-rank of the communication matrix of g_n is quasipolynomially bounded.

As we want to apply our universal equivalence relation also to $\mathbb{F}_p$ we identify the elements of $\mathbb{F}_p$ with the integers $0, \ldots, p-1$. For reading the proof we refer to the detailed version of this paper. The idea is to exhibit the fact that for all protocols P of length m the functions $Acc_P, Rej_P : S \times T \longrightarrow \mathbb{N}$ have communication matrices of $\mathbb{Z}$-rank at most 2^m.□.

Thus, for showing that a decision problem $F = (f_n : \{0,1\}^n \times \{0,1\}^n \longrightarrow \{0,1\})_{n \in \mathbb{N}}$ does not belong to $C\text{-}MOD_p\text{-}P$ we have to verify that all $\mathbb{F}_p$–matrices equivalent to $M(f_n)$ have at least subexponential rank. This task becomes much easier if we consider the following lemma which can be easily proved using FERMAT's Theorem.

Lemma 2.1 *Let S, T be finite sets, $f : S \times T \longrightarrow \{0,1\}$ be a decision function, and P be a communication protocol of length m which computes f via MOD_p- acceptation. Then there is a protocol Q of length $(p-1)m$ computing f via MOD_p- acceptation which has the additional property that the communication matrix of f and the acceptation matrix of Q are congruent modulo p.* □

For demonstrating the limitation of the computational power of Majority protocols we have to investigate the *difference matrix* of communication protocols.

Definition 2.3 *Let S, T be finite sets and P be any communication protocol over $S \times T$. The $S \times T$–matrix $(acc_P(s,t) - rej_P(s,t))_{s \in S, t \in T}$ is called the difference matrix of P.*

Definition 2.4 *Let $a \in \mathbb{N}$ be a natural number. Any real matrix A is said to be a-bounded if the norm of each coefficient of A is bounded by $\frac{1}{a}$ from below and by a from above.*

Difference matrices of communication protocols have the following important property.

Lemma 2.2 *Let P be a protocol over $S \times T$ of length m and D be the difference matrix of P. Then D is 2^m-bounded, and it holds $\text{rank}_{\mathbb{Z}}(D) \leq 2^m$.*□

Now we need a somewhat another type of communication matrices.

Definition 2.5 *For all decision functions $f : S \times T \longrightarrow \{0,1\}$ we denote by $M^{\pm}(f)$ the $\{+,-\}$-matrix of f which is defined as*

$$M^{\pm}(f)_{s,t} = 1 \quad if \quad f(s,t) = 1 \qquad and \qquad M^{\pm}(f)_{s,t} = -1 \quad if \quad f(s,t) = 0.$$

Observe that a Majority protocol P computes a given decision function $f : S \times T \longrightarrow \{0,1\}$ if and only if the difference matrix of P is equivalent to $M^{\pm}(f)$. (Hereby we have to obey our convention $acc_P(s,t) \neq rej_P(s,t)$ for all $s \in S, t \in T$.)

Proposition 2.2 *If a decision function $f : S \times T \longrightarrow \{0,1\}$ can be computed by a Majority protocol P of length m then there is an 2^m-bounded $S \times T$-matrix D fulfilling*

$$rank_{\mathbb{R}}(D) \leq 2^m \qquad and \qquad D \sim M^{\pm}(f). \ \square$$

Thus, if we want to show $F = (f_n : \{0,1\}^n \times \{0,1\}^n \longrightarrow \{0,1\})_{n \in \mathbb{N}} \notin C\text{-}PP$ we have to prove that all quasipolynomially bounded matrices equivalent to $M^{\pm}(f_n)$ have at least subexponential rank. This can be achieved by using the following technical result which generalizes the Main Lemma in [K 91].

Proposition 2.3 *Let $N, a, b \in \mathbb{N}$ and A be a real orthogonal a-bounded $N \times N$-matrix Then for any b-bounded integer matrix D fulfilling $A \sim D$ holds $rank_{\mathbb{R}}(D) \geq \frac{N}{a^4 b^4}$.*

Sketch of the Proof Let us denote by $E^N, (.,.)$ the N-dimensional Euclidian vector space. For all vectors $v, w \in E^N$ we denote by $\beta(v,w) \in [0,\pi]$ the angle formed by v, w, and by $\cos(v,w)$ the cosinus of $\beta(v,w)$. Remember that

$$\cos(v,w) = \frac{(v,w)}{\|v\| \, \|w\|}$$

One main technical result we need is that for all equivalent a-bounded $v, w \in E^N$ holds

Lemma 2.3 $\cos(v,w) \in [c_a, 1]$, *where* $c_a = \frac{2a}{a^2+1}$.

Due to lack of space the proof of this fact has to be omitted.

For each linear subspace U of E^N and each $v \in E^N$ we denote by $pr_U(v) \in E^N$ the orthogonal projection of v to U and by $\beta(v,U)$ the infimum taking over all angles occuring between v and vectors from U. Remember that $\beta(v,U) = \beta(v, pr_U(v))$.

For all real $N \times N$-matrices A consider the norm $\|A\| = max\{\|Av\| \ , \quad v \in E^N, \ \|v\| = 1\}$.

Lemma 2.4 *Let A, Dbe equivalent real $N \times N$-matrices for which the norm of each coefficient is between $\frac{1}{a}$ and b. Then $rank_{\mathbb{R}}(D) \geq \frac{N^2}{\|A\|^2 a^3 b}$.*

Proof: Let m be the rank of D and U the m–dimensional subspace generated by the rows of D. We denote by $a_1, \ldots, a_N$ the rows of the matrix A. By Lemma 2.3 we know that the cosinus of the angle between two equivalent vectors with coefficients norms in $[\frac{1}{a}, b]$ cannot be smaller than $c_{ab} = \frac{2\sqrt{ab}}{ab+1}$.
Consequently, for all $j = 1, \ldots, N$ we have the relation

$$c_{ab} \leq \cos(\beta(a_j, U)) = \cos(\beta(a_j, pr_U(a_j))).$$

Now let us fix an orthonormal basis $u_1, \ldots, u_N$ in such a way, that $u_1, \ldots, u_m$ generate the subspace U. Then for all $v \in E^N$ we have

$$pr_U(v) = \sum_{i=1}^{m}(v, u_i)u_i\ , \quad \text{and} \quad \|pr_U(v)\| = \sqrt{\sum_{i=1}^{m}(v, u_i)^2}\ .$$

Consequently, for all $j = 1, \ldots, N$ we can write

$$\cos(\beta(a_j, pr_U(a_j))) = \frac{(a_j, pr_U(a_j))}{\|a_j\|\,\|pr_U(a_j)\|} = \frac{\sum_{i=1}^{m}(a_j, u_i)^2}{\|a_j\|\,\|pr_U(a_j)\|} = \frac{\|pr_U(a_j)\|^2}{\|a_j\|\,\|pr_U(a_j)\|} = \frac{\|pr_U(a_j)\|}{\|a_j\|}.$$

Using the fact that for all $j = 1, \ldots, N$ $\|a_j\| \geq \frac{\sqrt{N}}{a}$ we obtain

$$c_{ab}^2 \leq \cos^2(\beta(a_j, U)) \leq \frac{a^2\,\|pr_U(a_j)\|^2}{N}.$$

Taking the sum over all $j = 1, \ldots, N$ we get

$$\frac{N^2 c_{ab}^2}{a^2} \leq \sum_{j=1}^{N}\sum_{i=1}^{m}(a_j, u_i)^2 = \sum_{i=1}^{m}\sum_{j=1}^{N}(a_j, u_i)^2 = \sum_{i=1}^{m}\|Au_i\|^2 \leq m\,\|A\|^2.$$

From $c_{ab}^2 = \frac{4a^2b^2}{(a^2b^2+1)^2} \geq \frac{1}{a^2b^2}$. we obtain $m \geq \frac{N^2}{\|A\|^2 a^4 b^2}$.□
It is not hard to show that for all real b–bounded $N \times N$ matrices A holds $\|A\| \leq bN$. However, if A is orthogonal then $\|A\|$ is sufficiently small.

Lemma 2.5 *Let A be a real orthogonal b-bounded $N \times N$–matrix. Then $\|A\| \leq b\sqrt{N}$.*□

The proof of Proposition 2.3 follows straightforwardly.□
As a conclusion we obtain the following lower bound argument.

Corollary 2.1 *For a given $f : \{0,1\}^n \times \{0,1\}^n \longrightarrow \{0,1\}$ suppose that we can construct a real a-bounded $2^n \times 2^n$– matrix $A \sim M_-^+(f)$ which has an orthogonal submatrix of size $T \times T$. Then each Majority protocol computing f has length at least $\frac{1}{5}(\log_2(T) - 4\log_2(a))$.*

Proof: Suppose that D is the difference matrix of a Majority protocol of length m which computes f. By Proposition 4.2 and Proposition 4.3 we have $2^m \geq \frac{T}{a^4 2^{4m}}$, thus $5m \geq \log_2(T) - 4\log_2(a)$. □

3 Separating Modulo-p Classes

Theorem 1 *For all pairs of distinct prime numbers p, q and the corresponding communication complexity classes $C\text{-}MOD_p\text{-}P$ and $C\text{-}MOD_q\text{-}P$ it holds $C\text{-}MOD_p\text{-}P \not\subseteq C\text{-}MOD_q\text{-}P$.* ■

Consider the following decision problem ORT^p, p prime.

Definition 3.1 *$ORT^p = (ORT^p_n : \mathbb{F}_p^n \times \mathbb{F}_p^n \longrightarrow \{0,1\})_{n \in \mathbb{N}}$ is defined as follows. For all vectors $x, y \in \mathbb{F}_p^n$ let*

$$ORT^p_n(x,y) = 1 \iff (x,y)_p = 0$$

Lemma 3.1 *For all primes p holds $ORT^p \in C\text{-}MOD_p\text{-}P$.*

Proof: Observe that the scalar product in $\mathbb{F}_p^n$ is equivalent to $\neg ORT^p_n(v,w)$ and induces a communication matrix of $\mathbb{F}_p$-rank n. By Proposition 2.1 the Modulo p communication complexity of the complement of ORT^p is $O(\log(n))$. As $C\text{-}MOD_p\text{-}P$ is closed under complementation we are done. □.

Lemma 3.2 *Let p, q be distinct prime numbers, and for each $n \in \mathbb{N}$ let Q_n be a MOD_q-protocol computing ORT^p_n. Then $length(Q_n) = \Omega(n)$.*

Proof: For all $n \in \mathbb{N}$ we denote by M^n the communication matrix of ORT^p_n. Due to Lemma 2.2 we can suppose that for all $n \in \mathbb{N}$ the acceptation matrix of Q_n is congruent to M^n modulo q. Thus, it is sufficient to show $rank_{\mathbb{F}_q}(M^n) = exp(\Omega(n))$. Represent each onedimensional subspace of $\mathbb{F}_p^n$ by a nontrivial vector and denote the set of these vectors by U. Let D be the submatrix of M^n formed by U. Obviously, D is a $d \times d$-matrix over $\{0,1\}$, where $d = \frac{p^n-1}{p-1}$

Observe that for all $y, z \in U$ the set of all vectors x fulfilling $ORT^p_n(y,x) = 1$ and $ORT^p_n(z,x) = 1$ is a linear subspace of $\mathbb{F}_p^n$ whose dimension is $n-1$ if $y = z$ and $n-2$ otherwise.

Hence, $(D^2)_{y,y} = \frac{p^{n-1}-1}{p-1} =: a$ and $(D^2)_{y,z} = \frac{p^{n-2}-1}{p-1} =: b$, for $y \neq z$.

Let us call a quadratic matrix to be a (s,t)-matrix if all coefficients on the main diagonal are equal to s and all remaining coefficients are equal to t. Thus, D^2 is an (a,b)-matrix. It can easily be verified that the inverse of a (s,t)-matrix is, if exists, again a matrix of this form. It is not hard to show that D^2 mod $\mathbb{F}_q$ is invertible (over $\mathbb{F}_q$) if and only if $p^{n-2} + db \not\equiv 0 \bmod q$.

Now suppose that $n-2$ is divisible by $q-1$, i.e., $n-2 = (q-1)\lambda, \lambda \in \mathbb{Z}$. Then, by FERMAT's Theorem $p^{n-2} = p^{\lambda(q-1)} \equiv 1 \bmod q$ and therefore $b \equiv 0 \bmod q$ and $p^{n-2} + db \equiv 1 \bmod q$. Hence, for those n the matrix D is invertible. □

4 Probabilism versus Modulo-p

Theorem 2 *For all prime numbers p the following is true.*

$$C\text{-}MOD_p\text{-}P \not\subseteq C\text{-}PP \quad \text{and} \quad C\text{-}PP \not\subseteq C\text{-}MOD_p\text{-}P.$$

At first observe that the following decision problem SEQUENCE EQUALITY TEST (SEQ) separates $C\text{-}MOD_p\text{-}P$ from $C\text{-}PP$ for all primes p.

$SEQ = (SEQ_{2n} : \{0,1\}^n \times \{0,1\}^n \longrightarrow \{0,1\})_{n \in \mathbb{N}}$, where for all $n \in \mathbb{N}$ and $v, w \in \{0,1\}^n$ let $SEQ_{2n}(v,w) = 1$ iff $v = w$. It can easily be shown that SEQ_{2n} can be computed by co-nondeterministic protocols of length $\log(n)+3$, and, thus $SEQ \in C\text{-}PP$. On the other hand observe that the communication matrix of SEQ_n is always the unique matrix. Exhibiting Proposition 2.1 we obtain $SEQ \notin C\text{-}MOD_p\text{-}P$.

We show now that for all primes p and natural numbers n Majority protocols computing ORT_n^p have length $\Omega(n)$. Due to Corollary 2.1 it is sufficient to show

Lemma 4.1 *Let p be an arbitrary prime and n an even natural number, $n = 2k$. Further let B denote the following real $\mathbb{F}_p^n \times \mathbb{F}_p^n$ - matrix. For all vectors $v, w \in \mathbb{F}_p^n$ let*

$$B_{v,w} = p \quad \textit{iff} \quad (v,w)_p = 0 \quad \textit{and} \quad B_{v,w} = -\frac{p^k - p}{p^k - p^{k-1}} \quad \textit{iff} \quad (v,w)_p \neq 0$$

Then B is p–bounded and equivalent to the $\{+,-\}$ - matrix of ORT_n^p, and it has an orthogonal submatrix of size $\frac{p^n-1}{p-1} \times \frac{p^n-1}{p-1}$.

Proof: The equivalence follows immediately from the definition of ORT_n^p. For constructing an orthogonal submatrix fix the same subset U of $\mathbb{F}_p^n$ as in the proof of Lemma 3.2.

Denote by C the submatrix of B induced by U. Clearly, C is symmetric as B is. We show that C is orthogonal, i.e., that C^2 is a regular diagonal matrix. As C is symmetric it holds $C^2_{v,v} > 0$ for all $v \in U$.

Let $v \neq w \in U$ be arbitrarily fixed. We have to prove $C^2_{v,w} = 0$. Observe that

$$C^2_{v,w} = a^2 X + p^2 Y - apZ,$$

where, for short, $a = \frac{p^k - p}{p^k - p^{k-1}}$ and X, Y, Z denote the number of $u \in U$ which are orthogonal neither to v nor to w, to both v and w, to exactly one of the vectors v and w, respectively. It can easily be derived that

$$X = \frac{p^n - 1}{p-1} - 2\frac{p^{n-1} - 1}{p-1} + \frac{p^{n-2} - 1}{p-1},$$

$$Y = \frac{p^{n-2} - 1}{p-1}, \quad \text{and} \quad Z = 2\left(\frac{p^{n-1} - 1}{p-1} - \frac{p^{n-2} - 1}{p-1}\right).$$

Consequently, we have to show

$$a^2(p^n - 2p^{n-1} + p^{n-2}) - 2a(p^n - p^{n-1}) + p^n - p^2 = 0$$

Using $p^n - 2p^{n-1} + p^{n-2} = (p^k - p^{k-1})^2$ we obtain that this is equivalent to

$$a^2 - 2\frac{p^k}{p^k - p^{k-1}}a + \frac{p^{2k} - p^2}{(p^k - p^{k-1})^2} = 0.$$

Applying the standard method for solving quadratic equations we have to show that

$$a \in \{\frac{p^k + p}{p^k - p^{k-1}}, \frac{p^k - p}{p^k - p^{k-1}}\}.$$

But this is true by definition of a.□

5 Open Problems

Concerning communication complexity classes from our point of view there are three central open separation problems. The first is to separate $C\text{-}PP$ from the class $C\text{-}UPP$ of decision problems feasible computable by randomized communication protocols where the error is allowed to be arbitrarily small.

The second problem is to separate the communication polynomial time hierarchy. If applying the well-known Valiant-Vazirani Result [VV 86] to communication complexity one can prove $C\text{-}PH \subseteq BP(C\text{-}MOD_p\text{-}P)$ [H 90]. Characterizations of this kind are the basis of the proof that certain decision problems cannot be computed by constant depth (quasi)polynomial size circuits with MOD_p- , AND-, and OR-gates (p prime) [R 86][S 87]. In particular, these results are due to arguments providing that certain decision functions cannot be approximated sufficiently good by small degree polynomials. Representing Boolean functions as small degree polynomial implies small rank of related communication matrices. Unfortunately, up to now we dont know arguments providing that certain matrices cannot be approximated sufficiently good by small rank matrices.

The third open problem is to separate the union of all counting communication complexity classes. This would imply the separation of $pureACC$ [Y 90], and, thus, solve a central open problem in complexity theory.

6 References

[AUY 83] A.V.Aho, J.D.Ullman, M.Yannakakis, *On Notions of Information Transfer in VLSI Circuits*, Proc. 15-th ACM STOC, 133–139, 1983.

[AM 86] N.Alon, W.Maass, *Meanders, Ramsey Theory, and Lower Bounds for Branching Programs,* Proc. 27th IEEE FOCS, 30–39, 1986.

[BFS 86] L.Babai, P.Frankl, J.Simon, *Complexity Classes in Communication Complexity Theory,* Proc. 27-th IEEE FOCS, 337–347, 1986.

[BNS 89] L.Babai, N.Nisan, M.Szegedy, *Multiparty Protocols and Logspace-hard Pseudo-random Sequences*, Proc. 21-th ACM STOC, 1–11, 1989.

[HR 88] B.Halstenberg, R.Reischuk, *Relations between Communication Complexity Classes*, Proc. 3-th IEEE Structure in Complexity Theory Conference, 19–28, 1988.

[H 90] B.Halstenberg, *The Polynomial Communication Hierarchy and Protocols with Moderately Bounded Error,* Technical Report TI 1/90 of TH Darmstadt, 1990.

[K 90] M.Krause, *Separating* $\oplus L$ *from L, NL, co-NL, and* $AL=P$ *for Oblivious Turing Machines of Linear Access Time*, Proc. 15-th Symp. MFCS, 385–392, 1990.

[K 91] M.Krause, *Geometric Arguments yield better Bounds for Threshold Circuits and Distributed Computing,* Proc. 6-th IEEE Structure in Complexity Conference, 314–321, 1991.

[KMW 89] M.Krause, Ch.Meinel, S.Waack, *Separating Complexity Classes related to certain Input-Oblivious, Logarithmic Space Bounded Turing Machines,* Proc. 4-th IEEE Structure in Complexity Theory Conference, 1989.

[KW 89] M.Krause, S.Waack, *On oblivious Branching Programs of Linear Length,* Proc. 7-th Conference FCT, LNCS 380, 287–297, 1989.

[KW 91] M.Krause, S.Waack, *Variation ranks of Communication Matrices and Lower Bounds for Depth Two Circuits having Symmetric Gates with unbounded Fan-in*, Proc. of 32-th IEEE FOCS, 1991, 777–782.

[MS 82] K.Mehlhorn, E.M.Schmidt, *Las vegas is Better than Determinism in VLSI and Distributed Computing,* Proc.14-th ACM STOC, 330–337, 1982.

[PS 84] R.Paturi, J.Simon, *Probabilistic Communication Complexity*, Proc. 25–th IEEE FOCS, 118–126, 1984.

[R 86] A.A.Razborov, *Lower Bounds for the Size of Circuits of bounded Depth with Basis* $\{\oplus, \wedge\}$, Math. Zametki 41(4), 1987, 598-607 in Russian, Engl.transl. in Mathematical Notes 41(4) 1987, 333-338.

[RW 90] R.Raz,A.Widgerson, *Monotone Circuits for Matching require linear Depth,* Proc. 20-th ACM STOC, 287–292, 1990.

[S 87] R.Smolensky, *Algebraic Methods in the Theory of Lower Bounds for Boolean Circuit Complexity,* Proc. 19th ACM STOC, 77–82, 1987.

[S 90] R.Smolensky, *On Interpolation by Analytic Functions with special Properties and some weak Lower Bounds on the Size of Circuits with Symmetric Gates,* Proc. 31-th IEEE FOCS, 628–631, 1990.

[VV 86] L.G.Valiant, V.V.Vazirani, *NP is as easy as detecting unique solutions,* Theoretical Computer Science 47 (1986), 85–93, North Holland.

[Y 83] A.C.C.Yao, *Lower Bounds by Probabilistic Arguments*, Proc. 27th IEEE FOCS, 420–428, 1983.

[Y 90] A.C.C.Yao, *On ACC and Threshold Circuits,* Proc. 31-th IEEE FOCS, 619–627, 1990.

A NONLINEAR LOWER BOUND ON THE PRACTICAL COMBINATIONAL COMPLEXITY*

Xaver Gubáš+, Juraj Hromkovič* Juraj Waczulík**

* Departement of Mathematics and Computer Science, University of Paderborn Postfach 1621, 4790 Paderborn, Germany

+ Department of Computer Science, Comenius University, 842 15 Bratislava, ČSFR.

** Computer Science Institute of the Comenius University, 842 15 Bratislava, ČSFR.

Abstract

An infinite sequence $F = \{f_n\}_{n=1}^{\infty}$ of one-output Boolean functions with the following three properties is constructed:

(1) f_n can be computed by a Boolean circuit with $O(n)$ gates.

(2) For any positive, nondecreasing, and unbounded function $h : N \to R$, each Boolean circuit having an $n/h(n)$ separator requires nonlinear number of gates to compute f_n.

(3) Each planar Boolean circuit requires $\Omega(n^2)$ gates to compute f_n.

Thus, one can say that f_n has linear combinational complexity and a nonlinear practical combinational complexity because the constant-degree parallel architectures used in practice have their separators in $0(n/\log_2 n)$.

1 Introduction

One of the most challenging problems in complexity theory is to prove a nonlinear lower bound on the combinational complexity (the number of gates in Boolean circuits) of a specific Boolean function. The highest lower bounds are only linear ones (for the base of all Boolean functions of two variables in [Bi84, HS73, HHS75, KM65, Pa77, Sch80, St77], for some special complete bases in [Re81, Sch74, So65]) despite of the well-known fact that almost all Boolean functions of n variables require $\Omega(2^n/n)$ combinational complexity (see Shannon [Sh49] and Lupanov [Lu58]).

One attempt to attack this challenging problem was made in [Hr90], where it has been shown that each unbounded fan-in, fan-out Boolean circuit with $0(n^a)$ vertex-separator for an $a < 1$ must have $\Omega(n^{1/a})$ processors (gates) to compute some specific one-output Boolean functions. The above mentioned result has led to the formulation of the following two research problems.

1° To improve the result proved in [Hr90] by proving a nonlinear lower bound on the number of gates of Boolean circuits with $O(n/f(n))$ separators for some functions $f(n)$ increasing slower than n^a for any $a > 0$.

*Supported in part DFG-Grant DI 412/2-1.

2° To decide whether there exists an one-output Boolean function with a linear combinational complexity and a nonlinear "practical" combinational complexity, where the practical combinational complexity of a Boolean function f is the minimum over the combinational complexities of all Boolean circuits with sublinear separators computing f.

The importance of problem 2° consists in the fact that the nonexistence of such a Boolean function together with the solution of the problem 1° would imply a nonlinear lower bound on the combinational complexity of a specific Boolean function, and the existence of such a Boolean function implies that the strongly connected topology (for example, magnifiers) brings an additional computing power to the circuits (parallel architectures). On the other hand the existence of such a Boolean function means also that there are functions which are theoretically easy to compute (with a linear number of gates) but which are not easy (a nonlinear number of gates is required) for any parallel architecture currently used, because the circuits and the architectures used in practice have the separators in $O(n/\log_2 n)$.

The main result of this paper gives the solution of both problems 1° and 2° by constructing an infinite sequence $F = \{f_n\}_{n=1}^{\infty}$ of one-output Boolean functions with the following two properties:

(i) f_n can be computed by a Boolean circuit with $0(n)$ gates, i.e. $\{f_n\}_{n=1}^{\infty}$ has a linear combinational complexity

(ii) for any positive, nondecreasing and unbounded function: $h : N \to R$ there exists a nondecreasing unbounded function $g : N \to R$ such that each Boolean circuit having an $n/h(n)$ separator requires a nonlinear $\Omega(n \cdot g(n))$ number of gates to compute f_n. Moreover, if $h(n) = n^c$ for some $0 < c < 1$ then $g(n) = n^{c/(1-c)}$, and if $h(n)$ is polylogarithmic then $g(n) \in \Omega(h(n))$.

An interesting consequence of this result is that we construct a function with a linear combinational complexity and a quadratic planar complexity (see [McC85, McCP84, We87, page 344] for the definition of planar Boolean circuits and [LT79] for the fact that planar graphs have $O(\sqrt{n})$ separators). This paper is organized as follows. In the next section we give the construction of a special sequence of Boolean functions F with linear communication complexity. In section 3 it is shown that F has linear combinational complexity (i.e., that F has the property (i)). The lower bound (ii) for computing $F = \{f_n\}_{n=1}^{\infty}$ is proved in Section 4 by showing that each Boolean function with linear communication complexity requires a nonlinear number of gates to be computed by Boolean circuits with sublinear separators. The last Section 5 discusses some further consequences of the lower bound proof technique presented in Section 4.

2 CONSTRUCTION OF F

To construct $F = \{f_n\}_{n=1}^{\infty}$ we use a result of several authors (see, for example [GG81]) providing a constructive proof of the magnifiers among 3-regular graphs. Thus, we can assume that there are a constant c and an algorithm which for a given even positive integer $n \geq 6$ constructs a graph $G_n = (V_n, E_n)$ with the following three properties:

(i) G_n is a 3-regular graph

(ii) $|V_n| = n$

(iii) For each $X \subseteq V_n, |X| \leq \lfloor n/2 \rfloor + 1$ there are at least $c \cdot |X|$ edges leading among the vertices in X and the vertices in $V_n - X$.

Now, we use a construction similar to the construction of Lipton and Sedgewick [LS81] in order to construct $f_n : \{0,1\}^n \to \{0,1\}$ from G_n. We note that each 3-regular graph is 3-colourable (Brook's Theorem [Br41]) which is used for the following construction.

The construction of f_n from $G_n = (V_n, E_n)$ is done in the following 4 steps.

1. Denote the n vertices of G_n by n variables $x_1, x_2, \ldots x_n$ in an arbitrary way.
2. Colour the vertices of G_n using 3 colours $\bar{1}, \bar{2}, \bar{3}$ by giving a function $h : V_n \to \{\bar{1}, \bar{2}, \bar{3}\}$ with the property $h(r) \neq h(s)$ for each $(r, s) \in E_n$.
3. For all $i, j \in \{1, 2, 3\}, i < j$, define
   $f_n(i,j) = \bigwedge_{(v,u) \in E_{i,j}} (v \vee u)$, where
   $E_{i,j} = \{(v,u) | (v,u) \in E_n \wedge h(v) = \bar{i} \wedge h(u) = \bar{j}\}$.
4. Define: $f_n(x_1, \ldots, x_n) = f_n(1,2)(x_1, \ldots, x_n) \vee f_n(1,3)(x_1, \ldots, x_n) \vee f_n(2,3)(x_1, \ldots, x_n)$

Note, that f_n is a monotone function. We have defined f_n only for even n, but one can extend the definiton for odd n's in several distinct ways. For instance, $f_{n+1}(x_1, \ldots, x_{n+1}) = f_n(x_1, \ldots, x_n) \vee x_{n+1}$.

Now, we shall show that the function f_n constructed in the above described way has linear information content. We shall use this fact later to show that a nonlinear number of gates is required to compute F on circuits with sublinear separators.

Let us first say what does the fact "f_n has the information content $I_f(n)$" mean?
Let $X = \{x_1, \ldots x_n\}$, and let $P_X = \{\Pi_X = (X_L, X_R) | X_L \cap X_R = \emptyset, X = X_L \cup X_R, \lfloor n/3 \rfloor \leq |X_L| \leq |X_R|\}$ be the set of all partitions Π_X of X. Let, for each word $a = a_1 a_2 \ldots a_n \in \{0,1\}^n$ and each $\Pi_X = (\{x_{i_1}, x_{i_2}, \ldots x_{i_r}\}, \{x_{j_1}, x_{j_2}, \ldots, x_{j_s}\})$ (note that $r + s = n$), $\Pi_X(a, L) = a_{i_1} a_{i_2} \ldots a_{i_r}$, $\Pi_X(a, R) = a_{j_1} a_{j_2} \ldots a_{j_s}$, and $\Pi_X(a, L) \cdot \Pi_X(a, R) = a$. Now, to show that f_n has the information content at least $I(n)$ means that for each $\Pi_X \in P_X$ we have to find a set $S(\Pi_X, f_n) \subseteq \{0,1\}^n$ (called the fooling set for Π_X and f_n) with the following properties:

1° $|S(\Pi_X, f_n)| \geq 2^{I(n)}$

2° For every two words $a, b \in S(\Pi_X, f_n)$:
   $f_n(a) \neq f_n(\Pi_X(a, L) \cdot \Pi_x(b, R))$ or
   $f_n(a) \neq f_n(\Pi_X(b, L) \cdot \Pi_x(a, R))$.

The information content of languages (sequences of Boolean functions) was successfully used to prove lower bounds on the area complexity and on the area-time squared complexity of VLSI circuits. We note that the detailed description of information content as a complexity measure and its relation to communication complexity and to other complexity measures for VLSI computations can be found in [Hr88b, Hr89].

Now, let us show that the information content $I_f(n)$ of f_n constructed from the graph $G_n = (V_n, E_n)$ is in $\Omega(n)$. Let $\Pi_X = (X_L, X_R)$ be a partition of $X = \{x_1, \ldots x_n\}$, where $X_L = \{x_{i_1}, x_{i_2}, \ldots, x_{i_s}\}$, and $X_R = \{x_{j_1}, x_{j_2}, \ldots, x_{j_r}\}$.

Let $E_n(\Pi_X) = E_n \cap (X_L \times X_R)$. Obviously, the property (iii) of G_n implies that $E_n(\Pi_X) \geq c \cdot n/3$. Now, we can assume there are $i, j \in \{1,2,3\}, i \neq j$, such that the number $d(n)$ of edges $(x_{r_1}, x_{s_1}), \ldots (x_{r_{d(n)}}, x_{s_{d(n)}})$ leading between X_L and X_R and coloured by the colours $\bar{i}$ and $\bar{j}$ (either x_{r_m} is coloured by $\bar{i}$ and x_{s_m} by $\bar{j}$ for each $m \in \{1, \ldots, d(n)\}$ or x_{r_m} is coloured by $\bar{j}$ and x_{s_m} is coloured by $\bar{i}$ for each $m \in \{1, \ldots, d(n)\}$) is at least $c \cdot n/9$. Thus, we can write $f_n(i,j) = \bigwedge_{k=1}^{d(n)} (x_{r_1} \vee x_{s_1}) \wedge f_n'(i,j)$, where $A_L = (x_{r_1}, \ldots, x_{r_{d(n)}}) \in (X_L)^{d(n)}$ and $A_R = (x_{s_1}, \ldots, x_{s_{d(n)}}) \in (X_R)^{d(n)}$. Let $\overline{X}_L \subseteq X_L (\overline{X}_R \subseteq X_R)$ be the set of all distinct variables in the vector A_L (A_R). Let $X_L' \subseteq \overline{X}_L$ and $X_R' \subseteq \overline{X}_R$ be such subsets of the set of input variables that for $\forall x, z \in X_L', \forall y, w \in X_R' : ((x,y) \in E_n(\Pi_X)$ and $(z,w) \in E_n(\Pi_X))$ implies that $(x,w) \notin E_n(\Pi_X)$ and $(z,y) \notin E_n(\Pi_X)$. Obviously, taking X_L' and X_R' as large as possible we have $b(n) = |X_L'| = |X_R'| \geq d(n)/13 \geq c \cdot n/117$. Thus, we can write $f_n(i,j) = \bigwedge_{p=1}^{b(n)} (x_{u_1} \vee x_{v_1}) \bigwedge \overline{f}_n(i,j) \wedge f_n'(i,j)$, where $X_L' = \{x_{u_1}, \ldots, u_{u_{b(n)}}\}$ and $X_R' = \{x_{u_1}, \ldots, x_{v_{b(n)}}\}$. Now, let us describe the construction of the fooling set $S(\Pi_X, f_n)$ for Π_x and f_n as a subset of $\{0,1\}^n$ with the following properties:

(1) Choose such 4 variables y_1, y_2, y_3, y_4 from X that $y_1 = y_2 = y_3 = y_4 = 0$ implies $f_n(k,l) = 0$ for each $(k,l) \neq (i,j), k, l \in \{1,2,3\}, k < l$ and $y_1 = \ldots = y_4 = 0$ does not imply $f_n(i,j) = 0$. Fix the zero values of y_1, y_2, y_3, y_4 in all words in $S(\Pi_X, f_n)$.

(2) Fix the value 1 for all variables in $X' = X - (X_L' \cup X_R' \cup \{y_1, \ldots, y_4\})$.

(3) The variables in X' may have both values 0 and 1 with the following restrictions: For all $k \in \{1, \ldots b(n)\}$ if $x_{u_k} = 1(0)$ then $x_{v_k} = 0(1)$.

Now, let us show that $S(\Pi_X, f_n)$ is a fooling set for Π_X and f_n. The property (1) secures that $f_n(a) = f_n(i,j)(a)$ for all $a \in S(\Pi_X, f_n)$. The property (2) secures that $f_n'(i,j)(a) = 1$ and $\overline{f}_n(i,j) = 1$ for all $a \in S(\Pi_X, f_n)$ which implies $f_n(a) = \bigwedge_{k=1}^{b(n)} (a_{u_k} \vee a_{v_k})$ for each $a = a_1 \ldots a_n \in S(\Pi_X, f_n)$. Following this fact and the property (3) we have that for all $a, b \in S(\Pi_X, f_n)$ either $1 = f_n(a) \neq f_n(\Pi_X(a,L) \cdot \Pi_X(b,R)) = \bigwedge_{k=1}^{b(n)} (a_{u_1} \vee b_{v_1}) = 0$ or $1 = f_n(a) \neq f_n(\Pi_X(b,L) \cdot \Pi_X(a,R)) = \bigwedge_{k=1}^{b(n)} (b_{u_1} \vee a_{v_1}) = 0$, i.e. $S(\Pi_X, f_n)$ is a fooling set.

Now, let us estimate the cardinality of $S(\Pi_X, f_n)$.
Since $b(n) \geq cn/117$ we have $|S(\pi_x, f_n)| \geq 2^{cn/117}$.

Thus we have constructed a sequence $F = \{f_n\}_{n=1}^{\infty}$ of Boolean functions with the information content $I_f(n) \in \Omega(n)$.

3 UPPER BOUND

In this section we show that the combinational complexity of the function sequence F constructed in the previous section is linear.

Boolean circuits are considered as the usual Boolean circuits model [We87] whose gates have fan-in bounded by 2 and an unbounded fan-out. The gates can realize any Boolean function of two variables. The combinational complexity C of Boolean circuits is considered as a number of gates and input vertices here. Usually the input vertices are not included but this $+n$ factor is not essential for our asymptotical considerations. For any Boolean circuit B we shall denote by $C(B)$ the combinational complexity of B.

Since the number of vertices in G_n is n and the degree of G_n is bounded by 3 we have that the number of edges in G_n is at most $3n/2$. Thus, the function f_n expressed as the formula $f_n(1,2) \vee f_n(1,3) \vee f_n(2,3)$ has a linear size (i.e., it contains at most a linear number of symbols). Obviously, there is a Boolean circuit with a linear number of gates realizing f_n. We note that this circuit has bounded fan-in and fan-out of all gates by 2, and that the depth of this circuit is logarithmic (note, that if we use the unbounded fan-in circuits then the depth is only three). The result of the next section also claims, that this circuit does not have any sublinear separator.

4 LOWER BOUND

To formulate our lower bound for F we need first to make precise the meaning of the notion "separator" used in this paper.

Definition 4.1 Let G be a graph having n vertices, and let $f : N \to N$ be a function. We say that G has an f-separator if there are such $f(n)$ vertices in G that their deletion divides G into two components G_1 and G_2 with the following two properties:

(a) for $i = 1, 2$, the number of vertices in G_i is at most $n/2$

(b) for $i = 1, 2$, G_i has an f-separator.

If the graph G corresponds to a Boolean circuit B we say also that the circuit B has an f-separator.

In this section we shall prove the following result.

Theorem 4.2 Let $h : N \to R$ be a positive, nondecreasing and unbounded function.
Let $\{B_n\}_{n=1}^{\infty}$ be such a sequence of Boolean circuits that for each n B_n computes f_n and B_n has an $C(B_n)/h(C(B(n))$-separator. Then

$$C(B_n) \in \Omega(ng(n)),$$

where $g : N \to R$ is such a function that $\lim_{n \to \infty} g(n) = \infty$.

To prove Theorem 4.2 we shall use the complexity measure "information transfer" of Boolean functions defined in [Hr90] as follows.

Definition 4.3 Let S be a Boolean circuit computing a Boolean function $f : \{0,1\}^n \to \{0,1\}$ with the set of input variables X. A group of edges of S is any nonempty subset of a set of edges leading from one vertex (gate) of S. Let, for any partition $\Pi_X = (X_L, X_R) \in P_X, T(S, \Pi_X)$ be the minimal number of groups of edges that must be removed from S in order to divide S into two such components that one (the left) component contains exactly the input variables corresponding to the input vertices in X_L and the other (right) component contains the input vertices corresponding to the input variables in X_R. We define the transfer complexity of S as $R(S) = \min\{T(S, \Pi_X) | \Pi_X \in P_X\}$. Finally, the transfer complexity of f is $T(f) = \min\{T(S) | S \text{ computes } f\}$.

In [Hr90] it is shown that the communication complexity defined in [PS84] provides a lower bound on $T(f)$, and Aho et. al. [AUY83] (see also [Hr89]) have shown that the information content provides a lower bound on communication complexity. So, we have the following

Lemma 4.4 For any Boolean function $h_n : \{0,1\}^n \to \{0,1\}$ $T(h_n) \geq I_h(n)$.

If one wants to construct a direct proof of Lemma 4.4 then one can do it very simply by realizing the following two facts:

1° In any Boolean circuit each edge transfer exactly one Boolean value during the whole computation on an input.

2° Each group of edges of any Boolean circuit S transfers the same value during the computation of S on an input.

Now, we are able already to give the proof of Theorem 4.2. We note that the proof differs from the lower bound proof in [Hr90] in the following sense. In [Hr90] such a cut (a set of groups of edges whose removal divides the circuit into two components) Q of a circuit B is found that Q corresponds to a partition of input variables but the number of group of edges in this cut is potentially greater than linear (according to the number of processors in B). To overcome this problem a new circuit B' computing the same function as B (potentially with a few more gates than B) is constructed. This new circuit B' has a cut Q' having a sublinear number of edges which implies directly that the function computed by B' (and also by B) has communication complexity (transfer complexity) bounded by the number of groups of edges in Q'. Thus, the circuits with linear number of gates are not able to compute functions with linear communication complexity. Here, we shall find directly a sublinear cut for each circuit with a sublinear separator.

The proof of Theorem 4.2 Let, for every $n \in N$, B_n be a circuit compuing f_n, and let G_n be the graph corresponding to B_n. Let $G_n = (V_n, E_n)$ have m vertices and an $m/h(m)$-separator. Now, we shall divide G_n into two components $G^1 = (V^1, E^1)$ and $G^2 = (V^2, E^2)$ such that G^i contains at least $n/3$ input vertices of B_n for each $i \in \{1,2\}$. (Obviously, the division of G_n into G^1 and G^2 will correspond to a partition Π_X of the set X of input variables of f_n). We shall try to do this division by removing the smallest number of groups of edges of B_n as possible.

Let us start by dividing G_n in two components, each of at most $m/2$ vertices, by removing edges adjacent to at most $m/h(m)$ vertices. If each of the components contains at least $n/3$ input vertices of B_n then the division process is completed. Let H be a component containing more than $2n/3$ input vertices. Then we divide H by removing edges adjacent to at most $m/2h(m/2)$ vertices into two components, each having at most $m/4$ vertices. We continue to divide the component with at least $2n/3$ input vertiçes. This process will stop after at most $\log_2 m$ divisions because each component containing $2n/3$ input vertices has to have at least $2n/3 \geq m/2^{\log m} = 1$ vertices.

Clearly, if we divide H with at least $2n/3$ input vertices into two components H_1 and H_2 both with smaller than $2n/3$ number of input vertices then at least one of these two components, say H_1, has at least $n/3$ input vertices. Considering $G^1 = H_1$ as one component and the union of all other components arising during the deletion process as the second component G_2 we have obtained the required division of G corresponding to a partition Π_X of X.

Now, let us sum up the number of removed nodes. Since the division process runs at most $\log_2 m$ times the number of removed nodes is at most

$$z(m) \leq \sum_{i=0}^{\log_2 m} m/2^i h(m/2^i) = m \cdot \sum_{i=0}^{\log_2 m} (2^i h(m/2^i))^{-1}.$$

Since the fan-in of the vertices in B_n is bounded by 2 it is sufficient to remove at most $3 \cdot z(m)$ groups of edges of B_n in order to divide G_n into G^1 and G^2.

In Section 2 we have proved that $I_f(n) \geq bn$ for some constant b and all $n \geq n_0$ for some constant n_0. Following Lemma 4.4 we obtain $T(h_n) \geq bn$ for all $n \geq n_0$, and following Definition 4.3 we have $T(h_n) \leq 3z(m)$. Thus, we obtain $3z(m) \geq b \cdot n$ for all $n \geq n_0$.

Now, let us assume that m is a function of n of the form $m(n) = n \cdot g(n)$ (note that obviously $m(n) \geq 2n - 1$).

Thus, we have

$$3 \cdot n \cdot g(n) \sum_{i=0}^{\log_2(n \cdot g(n))} (2^i h(n \cdot g(n)/2^i))^{-1} \geq bn \quad \text{for all } n \geq n_0,$$

i.e., $g(n) \cdot H(n) \geq b/3$ for $H(n) = \sum_{i=0}^{\log_2(n \cdot g(n))} (2^i h(n \cdot g(n)/2^i))^{-1}$

Now, let us show that $\lim_{n \to \infty} H(n) = 0$ which completes the proof of Theorem 4.2.

By the substitution $r = \log_2(n \cdot g(n))$ we obtain

$$H(n) = \sum_{i=0}^{r} \frac{1}{2^i h(2^{r-i})} = \frac{1}{2^r} \sum_{j=0}^{r} \frac{2^j}{h(2^j)} = \frac{1}{2^r} \Delta$$

Considering the lower integral sum for Δ we have

$$\frac{\Delta}{2} = \frac{1}{2} \cdot \frac{1}{h(1)} + 1 \cdot \frac{1}{h(2)} + 2 \cdot \frac{1}{h(2^2)} + \ldots + (2^m - 2^{m-1}) \cdot \frac{1}{h(2^m)}$$

i.e., $H(n) \leq \frac{1}{2^r} \int_{1/2}^{2^r} \frac{dx}{h(x)}$

Now, $\lim_{n \to \infty} H(n) \leq \lim_{t \to \infty} \frac{1}{t} \int_{1/2}^{t} \frac{dx}{h(x)} = \lim_{t \to \infty} \frac{(\int_{1/2}^{t} \frac{dx}{h(x)})'}{(t)'} = \lim_{t \to \infty} \frac{1}{h(t)} = 0.$ □

Now, the question arises how the function $g(n)$ in the lower bound $\Omega(ng(n))$ is connected with the function $h(n)$ from the $n/h(n)$-separator. Following the result from [Hr90] we see that if $h(n) = n^b$ for a $b < 1$ then $g(n) = n^{b/(1-b)}$. Thus, for example, if B_n is planar then $h(n) \in \Omega(n^{1/2})$ which implies $g(n) \in \Omega(n)$. Now, we present a result showing that if $h(n)$ is bounded by a polylogarithmic function, then $g(n) \in \Omega(h(n))$.

Lemma 4.5
Let $h : N \to R$ be a positive, nondecreasing, and unbounded function which is bounded by a polylogarithmic function. Let $\{B_n\}_{n=1}^{\infty}$ be such a sequence of Boolean circuits that for each n B_n computes f_n

and B_n has an $n/h(n)$-separator. Then

$$C(B_n) \in \Omega(nh(n)).$$

Proof. To prove Lemma 4.3 it is sufficient to show that the sum $z(m)$ can be bounded as follows

$$z(m) \leq m \cdot \sum_{i=0}^{(\log_2 m)/2} (2^i h(m/2^i))^{-1} + m \cdot \sum_{i=(\log_2 m)/2}^{\log_2 m} (2^i h(m/2^i))^{-1} \leq$$
$$\leq m \cdot (1 \cdot h(m/2^{(\log_2 m)/2}))^{-1} + m \cdot (2^{(\log_2 m)/2} \cdot h(1))^{-1} \leq$$
$$\leq m/h(\sqrt{m}) + h(1) \cdot \sqrt{m} \in 0(m/h(\sqrt{m}))$$

for each h bounded by a polylogarithmic function.

Since $3 \cdot z(m) \geq bn$, and $m \geq n$ we obtain $m \in \Omega(n \cdot h(n))$. □

5 CONCLUSION

In this paper we have found the function F that has linear combinational complexity but that cannot be computed with linear number of gates by any circuit with a sublinear separator.

We note that the lower bound proof in Section 4 gives much more than the lower bound for F because the only fact about F used in the proof is that F has linear information content. So, the proof of Theorem 4.2 works also for the following assertion.

Theorem 5.1 Let $P = \{p_n\}_{n=1}^{\infty}$ be a sequence of Boolean functions with linear information content, and let $h : N \to R$ be a positive, nondecreasing, and unbounded function. Let $\{B_n\}_{n=1}^{\infty}$ be such a sequence of Boolean circuits that B_n computes p_n for any n, and B_n has an $n/h(n)$-separator. Then

$$C(B_n) \in \Omega(ng(n)),$$

where $g : N \to R$ is a function with the property $\lim_{n\to\infty} g(n) = \infty$.

We note that we know already many Boolean functions with linear information content (see, for instance, [AUY83,Hr88a, Hr89, KJ89, LS81, PS84, Ul84]). We let the question, whether one of these Boolean function has a linear combinational complexity, open here. We conjecture that some of these functions are good candidates for a nonlinear lower bound on the combinational complexity.

Another consequence of our results is the largest difference between the planar combinational complexity and the combinational complexity of a specific Boolean function. F has linear combinational complexity and quadratic planar complexity (in [Hr90] it is proved that each Boolean function with linear information content has quadratic planar complexity). The largest gap known till now was $n/(\log_2 n)^3$ ($\Omega((n/\log_2)^2)$ planar complexity and $O = (n \log_2 n)$ combinational complexity) proved in [LRSHS88].

Acknowledgement

We would like to thank to Burkhard Monien for a motivating discussion and Martin Dietzfelbinger and Friedhelm Meyer auf der Heide for a helpful discussion resulting in Lemma 4.5.

REFERENCES

AUY83 Aho, A.V. - Ullman, J.D. - Yannakakis, M.: On notions of information transfer in VLSI circuits. In: Proc. 15th ACM STOC, ACM Press 1983, pp. 133-139.

Bl84 Blum, N.: A Boolean function requiring $3n$ networksize. Theoretical Computer Science 28 (1984), pp. 337-345.

Br41 Brooks, R.L.: On coloring the nodes of network. In: Proc. Cambridge Philos. Soc. 37 (1941), 194-197.

GG81 Gaber - Galil, Z.: Explicit constructions of linear-sized superconcentrators. J. Comp. Syst. Sci. 22 (1981), 407-420.

HS73 Harper, L.H. - Savage, J.E.: Complexity made Simple. In: Proc. of the Intern. Symp. on Combinatorial Theory, Rom, Sept. 1973, pp. 2-15.

HHS75 Harper, L.H. - Hsiek, W.N. - Savage, J.E.: A class of Boolean functions with linear combitional complexity. Theoretical Computer Science 1, No. 2 (1975), pp. 161-183.

Hr88a Hromkovič, J.: The advantages of a new approach to defining the communication complexity of VLSI. Theor. Comp. Science 57 (1988), 97-111.

Hr88b Hromkovič, J.: Some complexity aspects of VLSI computations. Part 1. A framework for the study of information transfer in VLSI circuits. Computers and Artificial Intelligence 7 (1988), No. 3, pp. 229-252.

Hr89 Hromkovič, J.: Some complexity aspects of VLSI computations. Part 6. Communication complexity. Computers and Artificial Intelligence 8 (1989), No. 3, pp. 209-225.

Hr90 Hromkovič, J.: Nonlinear lower bounds on the number of processors of circuits with sublinear separators. Information and Computation, to appear.

KJ90 Kumičáková-Jirásková, G.: Chomsky hierarchy and communication complexity. J. Inf. Process. Cybern. EIK 25 (1989), No. 4, 157-164.

KM65 Kloss, B.M. - Malyshev, V.A.: Bounds on complexity of some classes of functions. Vest. Mosk. Univ., Seria matem., mech., 1965, No. 4, pp. 44-51 (in Russian).

LS81 Lipton, R.J. - Sedgewick, R.: Lower bounds for VLSI. In: Proc. ACM STOC'81, ACM 1981, pp. 300-307.

LT79 Lipton, R.J. - Tarjan, R.E.: A separator theorem for planar graphs. SIAM J. Appl. Math. 36 (1979), No. 2, pp. 177-189.

LRSH88 Lozkin, C.A. - Rybko, A.N. - Sapoženko, A.A. - Hromkovič, J. - Škalikova, N.A.: On a approach to a bound on the area complexity of combinational circuits. In: Mathematical problems in computation theory. Proc. of the Banach Center Publications. Warsaw 1988, pp. 501-510 (in Russian), the full version of this paper is accepted for publication in Theor. Comp. Sci.

Lu58 Lupanov, O.B.: Ob odnom metode siteza skhem (zv. VUZ radiofizika) 1 (1958), pp. 120-140 (in Russian).

McC85 McColl: Planar circuits have short specifications. In: Proc. 2nd STACS'85, Lecture Notes in Computer Science 18, Springer-Verlag 1985, pp. 231-242.

McCP84 McColl – Paterson, M.P.: The planar realization of Boolean functions. Technical Report, University of Warwick 1984.

Pa77 Paul, W.J.: A $2,5n$ - lower bound on the combinational complexity of Boolean functions. SIAM J. Comp. 6 (1977), pp. 427-443.

PS84 Papadimitriou, Ch. – Sipser, M.: Communication complexity. J. Computer System Sci. 28 (1984), pp. 260-269.

Re81 Redkin, N.P.: Proof of minimality of circuits consisting of functional elements. Problemy kibernetiki 38 (1981), pp. 181-216 (in Russian).

Sch74 Schnorr, G.P.: Zwei lineare untere Schranken für die Komplexität Boolescher Funktionen. Computing 13 (1974), pp. 155-171.

Sch80 Schnorr, G.P.: A $3 \cdot n$ lower bound on the network complexity of Boolean functions. Theor. Comp. Science 10 (1980), pp. 83-92.

Sh49 Shannon, C.E.: The synthesis of two-terminal switching circuits. Bell System Techn. J. 28 (1949), pp. 59-98.

So65 Sopranenko, E.P.: Minimal realizations of functions by circuits using functional elements. Probl. Kibernetiki 15 (1965), pp. 117-134 (in Russian).

Tu89 Turán, G.: Lower bounds for synchronous circuits and planar circuits. Infor. Proc. Lett. 30 (1989), pp. 37-40.

Ul84 Ullman, J.: Computational Aspects of VLSI. Principles of Computer Science Series. Computer Science Press 1984.

We87 Wegener, I.: The Complexity of Boolean Functions. Wiley-Teubner Series in Computer Science, John Wiley and Sons Ltd., and B.G. Teubner, Stuttgart 1987.

Ya81 Yao, A.C.: The entropic limitation of VLSI computations. In: Proc. 13th Annual ACM STOC, ACM 1981, pp. 308-311.

STRUCTURAL COMPLEXITY 2

Characterizations of some complexity classes between Θ_2^p and Δ_2^p.

*Jorge Castro**
Dept. Llenguatges i Sistemes Informàtics
Universitat Politècnica de Catalunya
Pau Gargallo 5
08028 Barcelona, Spain
E-mail: castro@lsi.upc.es

Carlos Seara
Dept. Matemàtica Aplicada II
Universitat Politècnica de Catalunya
Pau Gargallo 5
08028 Barcelona, Spain
E-mail: seara@ma2.upc.es

Abstract

We give some characterizations of the classes $P^{NP}[O(\log^k n)]$. First, we show that these classes are equal to classes $AC^{k-1}(NP)$. Second, we prove that they are also equivalent to some classes defined in the Extended Boolean hierarchy. As a last characterization, we show that there exists a strong connection between classes defined by polynomial time Turing machines with few queries to an NP oracle and classes defined by small size circuits with NP oracle gates. With these results we solve open questions that arose in [Wa-90] and [AW-90]. Finally, we give an oracle relative to which classes $P^{NP}[O(\log^k n)]$ and $P^{NP}[O(\log^{k+1} n)]$ are different.

1. Introduction

Wagner gave in his paper [Wa-90] some characterizations of the class Θ_2^p. One of them characterizes Θ_2^p as the class of languages recognized by polynomial time Turing machines which make at most $O(\log n)$ queries to an NP oracle on inputs of length n. Another one characterizes Θ_2^p as the class of languages obtained by placing polynomial bounds on the levels of the Extended Boolean hierarchy.

Wilson introduced the notion of relativized circuits by allowing oracle gates in the circuits, and he has studied properties of relativized versions of NC and AC. He defined an oracle gate as a k-input, one-output gate which, on an input x of length k, will produce the value 1 on its output edge if and only if x is in the specified oracle set. For NC classes each oracle gate counts as a depth logarithmic in the number of its input wires, while for AC classes these gates count 1. The motivations of these definitions are argued in [Wi-87] and [Wi-90] (compare also with the notion of NC^1-reducibility introduced in [Co-85]). In this paper we only consider relativizations to sets in NP; note that with this restriction we can only obtain classes of languages which are included in Δ_2^p. The first motivation of our work was to find relationships between these classes and those mentioned in [Wa-90].

Below we show connections between classes of languages recognized by polynomial time Turing machines which make polylog queries to an NP oracle and other classes defined on different ways. More exactly, in section 3 we show that the class of languages defined by P^{NP} Turing machines with at most $O(\log^k n)$ queries is the same as the class defined by AC^{k-1} circuits with oracle gates in NP. In section 4 we characterize the Extended Boolean hierarchy for some superpolynomial bounding functions such as the classes just mentioned. In section 5 we show a last characterization as the class of languages which are reducible to languages in

* Work partially supported by the ESPRIT II Basic Research Actions Program of the EC under contract no. 3075, project ALCOM.

NP via circuits of polylog size. With these results we solve questions asked by Wagner and by Allender and Wilson in their papers [Wa-90] and [AW-90].

In the final section we prove that there is an oracle relative to which classes $P^{NP}[O(\log^k n)]$ and $P^{NP}[O(\log^{k+1} n)]$ are different for any integer $k \geq 1$. Taking account of characterizations showed in previous sections we prove that, relative to the same oracle, $AC^{k-1}(NP) \subset AC^k(NP)$, $NC^k(NP) \subset AC^k(NP)$ and $NC^k(NP) \subset NC^{k+1}(NP)$; where "$\subset$" denotes proper containment. With these results we improve the relativized separations for Θ_2^p and Δ_2^p given by Buss and Hay and by Lozano and Toran in their papers [BH-91] and [LT-91].

2. Definitions and notation

We fix the notation that we shall use in the following sections.

Definition For any class C, P^C denotes the class of languages accepted by deterministic polynomial time machines using a set in C as oracle.

$$L \in P^C \iff \exists B \in C : L \leq_T^P B.$$

Definition For all $k \geq 1$, $P^C[O(\log^k n)]$ is the class of languages in P^C that are accepted by machines that make at most $O(\log^k n)$ queries on inputs of length n.

$$L \in P^C[O(\log^k n)] \iff \exists B \in C : L \leq_{T[O(\log^k n)]}^P B.$$

Since queries to any oracle of NP can be translated into queries for SAT, $P^{NP} = P^{SAT}$. We use these terms as synonyms and we also use the terms $P^{NP}[O(\log^k n)]$ and $P^{SAT}[O(\log^k n)]$ as equivalent. For any string x, we write $|x|$ for the length of x.

Definition $P_{\|}^{NP}$ is the class of languages accepted by deterministic polynomial time Turing machines which make one round of parallel queries to a set $B \in NP$. By a round of parallel queries to B we mean that the Turing machine writes a set of strings separated by delimiters on a query tape and then invokes an oracle for B; the oracle returns a string of YES/NO answers on an answer tape which specify membership of each query string to B. Note that within one round of parallel queries, all of them must be formulated before any answer are known. In general, we denote by $P_{\|O(f(n))}$ the class of sets recognized by deterministic polynomial time Turing machines which make $O(f(n))$ adaptive rounds of parallel queries on inputs of length n.

The classes $\Theta_2^p = P^{NP}[O(\log n)]$ and $\Delta_2^p = P^{NP}$ are well known, and the first one has a lot of different characterizations as it is shown in [Wa-90]. The Boolean hierarchy has been studied independently by different authors who gave equivalent definitions for it. At the beginning of the Section 4 we outline some interesting properties of this hierarchy. In the Section 6 of [Wa-90] it is introduced the Extended Boolean hierarchy as the hierarchy of classes $NP(r)$ (r is a function) defined as follows:

$$A \in NP(r) \iff \exists B \in NP \text{ such that } c_B(x, i+1) \leq c_B(x,i) \text{ for all } i,\ i \leq r(|x|) \text{ and }$$
$$c_A(x) \equiv \sum_{i=1}^{r(|x|)} c_B(x,i) \bmod 2;$$

where c_A denotes the characteristic function of the set A. It was known that $NP(n^{O(1)}) = \Theta_2^P$ and $NP\left(2^{n^{O(1)}}\right) = \Delta_2^P$, but similar results for other superpolynomial bounding functions were not known.

For definitions and notation about circuits we follow Wilson [Wi-87] [Wi-90]. A Boolean circuit with bounded fan-in is an acyclic directed graph whose nodes are labelled with an operator. Nodes of indegree zero are the input and constant ones, nodes of indegree one are labelled by negations and nodes of indegree two are labelled by *and* and *or* operators. Circuits with unbounded fan-in have no restriction on the indegree of nodes labelled *and* or *or*. Since we are interested in deciding set membership, the circuits have only a single output gate; the circuit accepts an input string if the length of the string in binary is the same as the number of input gates of the circuit and the circuit outputs 1 when the string is given on its input gates.

The *size* of the circuit is the number of nodes of the circuit and its *depth* is the length of the longest directed path from the input to the output in the graph. The size represents a measure of the hardware resource and the depth is a measure of the parallel time. A circuit family $\{C_n\}$, $n \geq 1$ accepts a set L if, for all n, C_n has n input nodes and accepts only those strings in L of length n. Note that, if the nth circuit could be independent of the $(n-1)^{th}$ circuit then there would exist a family of circuits which would accept a nonrecursive set. Therefore, it is necessary to require some kind of uniformity in the description of the circuits $\{C_n\}$, $n \geq 1$. We shall use in this paper one of the most frequently used concepts of uniformity:

Definition A circuit family $\{C_n\}$ is $O(\log n)$-uniform if there is a logspace deterministic Turing machine which on any input of length n outputs an encoding of C_n.

Classes of languages $NC = \bigcup_{k \geq 0} NC^k$ and $AC = \bigcup_{k \geq 0} AC^k$ are defined as follows:

$$L \in NC^k(AC^k) \iff \exists \{C_n\}, n \geq 1, \text{ bounded fan-in (unbounded fan-in) circuits } O(\log n)\text{-uniform with size } O(n^{O(1)}) \text{ and depth } O(\log^k n) \text{ which recognizes } L.$$

In this paper we work with relativized circuits. In these circuits oracle gates are allowed which can determine the membership of a string in an oracle set. In an NC circuit an oracle gate which has k input bits is defined to have size 1 and depth $\lceil \log k \rceil$. In an AC circuit (unbounded fan-in) the size and depth of these gates is 1.

We shall use the notation $AC^k(A)$ for the class of languages recognized by a family of unbounded fan-in circuits with polynomial size and $O(\log^k n)$ depth using A as oracle; and we define $AC^k_{[i]}(A)$ as the class obtained if we restrict the circuits to at most i oracle gates on each path from an input node to the output node.

3. Relating $P^{NP}[O(\log^k n)]$ with $AC^{k-1}(NP)$

Let us begin characterizing Θ_2^p as the class of languages recognized by families of circuits of polynomial size, constant depth and with at most one oracle gate on each path from an input node to the output node.

Proposition 1 $P^{NP}[O(\log n)] = AC^0_{[1]}(NP)$.

Proof Let M^{SAT} be a fixed polynomial time Turing machine that makes at most $c \log n$ queries to SAT on inputs of length n. We define the language:

$$L = \{\langle x, y\rangle \mid \text{the answer to the query number } |y|+1 \text{ of } M^{SAT} \text{ running on } x \text{ is "YES" supposing that the string } y \text{ records the answers to the first } |y| \text{ queries}\}.$$

We consider also the language:

$$L' = \{\langle x, a\rangle \mid M \text{ accepts } x \text{ using } a \text{ as oracle string (the answer to the } k^{th} \text{ query is interpreted as the } k^{th} \text{ bit of } a)\}.$$

It is easy to see that the language L is in NP and the language L' is in P. We shall use these languages as oracles to simulate the machine M^{SAT} by a family of circuits. The Figure 1 shows the member of this family which operates on inputs of length n.

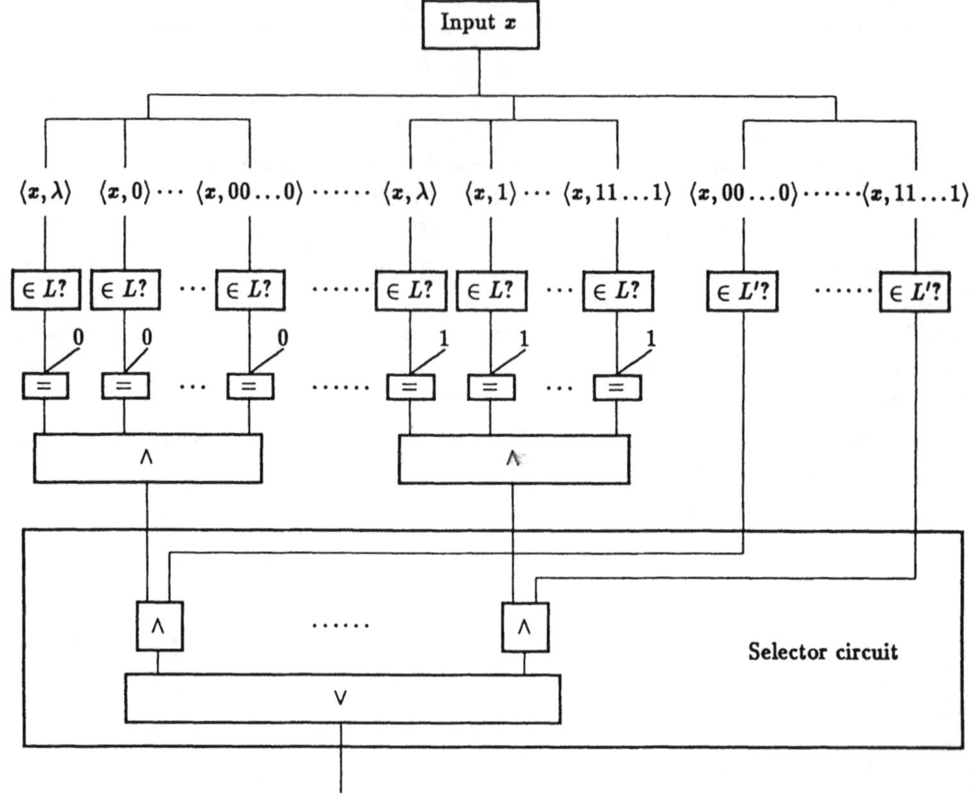

Figure 1.

On input x, this circuit has a first level of queries $\langle x, y\rangle \in L$ for all y, $|y| \leq c \log n$; with the answers it can determine the right answers to the queries of M^{SAT} on x. Moreover, in the same level, the circuit asks if $\langle x, a\rangle$ is in L' for all a, $|a| \leq c \log n$. Note that in this level there are only a polynomial number of queries because the length of y and a are logarithmic. When the circuit knows what is the right string y of answers of M^{SAT} on x, it chooses the suitable a with the help of a selector circuit.

The other inclusion is trivial considering that $AC^0_{[1]}(NP)$ is included in $P^{NP}_{||}$ and this one is equal to $P^{NP}[O(\log n)]$ (Th. 3.4 [Wa-90]). □

Buss and Hay [BH-91] showed that, for polynomial time Turing machines, a constant number of rounds of parallel queries to SAT can be reduced to one round of parallel queries. Using this fact it is not difficult to prove that $AC^0(NP)$ is equal to $P^{NP}[O(\log n)]$: observe that every language belonging to the first class is recognized by a polynomial time Turing machine which makes a constant number of rounds of parallel queries to SAT and then, by [BH-91] it can be recognized by another one which makes only one round of parallel queries. Finally, considering the equality $P^{NP}_{||} = P^{NP}[O(\log n)]$ mentioned before, we have:

Proposition 2 $P^{NP}[O(\log n)] = AC^0(NP)$. □

Now, we shall generalize this last result by allowing more queries to SAT, more exactly, we ask about $O(\log^k n)$ queries to SAT. First we prove the following proposition.

Proposition 3 For all $k \geq 1$, $P^{NP}[O(\log^k n)] \subseteq AC^{k-1}(NP)$.

Proof We proceed by induction on k:

$k = 1$, it is part of proposition 2.

We suppose that the result is true up to $k-1$ by induction hypothesis. Fix a polynomial time Turing machine M with at most $c \log^k n$ queries to SAT on inputs of length n, and let us consider the following languages:

$$L_{1,n} = \{y \mid M^{SAT} \text{ starting at the configuration } y \text{ reaches after } c\log^{k-1} n \text{ queries a configuration which has a 1 as the first bit}\}$$

$$L_{2,n} = \{y \mid M^{SAT} \text{ starting at the configuration } y \text{ reaches after } c\log^{k-1} n \text{ queries a configuration which has a 1 as the second bit}\}$$

...

$$L_{p(n),n} = \{y \mid M^{SAT} \text{ starting at the configuration } y \text{ reaches after } c\log^{k-1} n \text{ queries a configuration which has a 1 as the } p(n)^{th} \text{ bit}\}$$

where $p(n)$ is a polynomial bounding the length of configurations corresponding to inputs of length n.

Note that in these languages there may exist configurations y such that M^{SAT} does not reach them on any input. It is clear that all these languages are in $P^{NP}[O(\log^{k-1} n)]$ and then, by induction hypothesis, all of them have $AC^{k-2}(NP)$ circuits. Now, let C_n be the circuit on inputs of length n described by Figure 2.

This circuit on input x computes in $\log n$ levels the configuration f of M^{SAT} on x after all the queries have been done. All the levels are equal and they are composed by $AC^{k-2}(NP)$ circuits for $L_{1,n}, L_{2,n}, \ldots, L_{p(n),n}$ placed in parallel. The circuit C_n ends with a small circuit $O \in AC^0(P)$ which computes whether $x \in L\left(M^{SAT}\right)$ knowing what is the configuration f.

Clearly, the language accepted by the family of circuits $\{C_n\}$, $n \geq 1$ is $L\left(M^{SAT}\right)$, this family is logspace uniform, have polynomial size and $O(\log^{k-1} n)$ depth. □

Now, we shall see that the inclusions in proposition 3 are actually equations.

Theorem 4 For all $k \geq 1$, $P^{NP}[O(\log^k n)] = AC^{k-1}(NP)$.

Proof We only have to prove that $AC^{k-1}(NP) \subseteq P^{NP}[O(\log^k n)]$. Given a circuit with oracle gates we define the level of a query as follows: queries at the first level are the queries that depend on no other queries, queries at the second level all depend on some query at the first level, and so on.

Let $\{C_n\}$ be a family of circuits in $AC^{k-1}(NP)$ with depth $d(n) \in O(\log^{k-1} n)$ and $p(n)$ a polynomial bound on the number of queries in each level. We define X as the set formed by sequences $\langle x, i_1, i_2, \ldots, i_{d(n)}, o\rangle$ belonging to $\{0,1\}^* \times \{0,1,\ldots,p(n)\}^{d(n)} \times \{0,1\}$ (where $n = |x|$) such that there exist strings $A_1, A_2, \ldots, A_{d(n)}$ in $\{0,1\}^{\leq p(n)}$, each of them representing possible answers to the queries at levels $1, 2, \ldots, d(n)$ respectively and verifying the following:

$A_{d(n)}$ has $j_{d(n)}$ "YES" answers and these answers are correct if the inputs to queries of level $d(n)$ are computed from $x, A_1, A_2, \ldots, A_{d(n)-1}$.

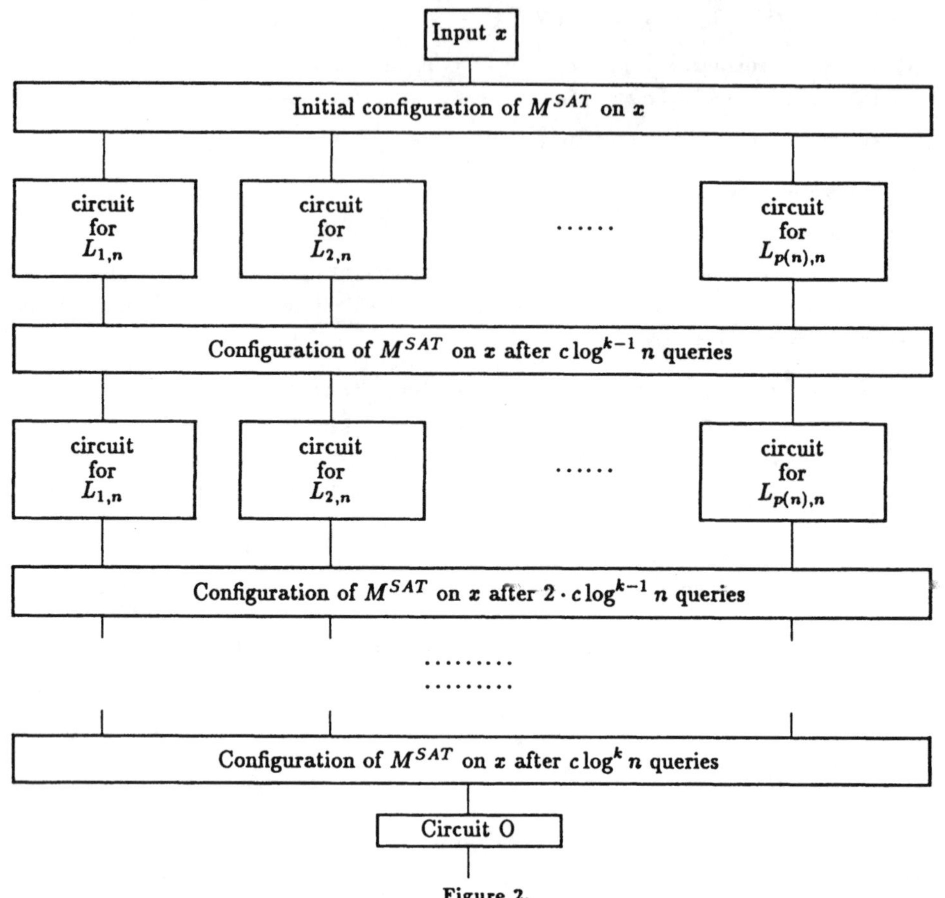

Figure 2.

$A_{d(n)-1}$ has $j_{d(n)-1}$ "YES" answers and these answers are correct if the inputs to queries of level $d(n)-1$ are computed from $x, A_1, A_2, \ldots, A_{d(n)-2}$.

...

...

A_1 has j_1 "YES" answers and these answers are correct if the input to the circuit is x. Finally, $\langle j_1, j_2, \ldots, j_{d(n)}, o' \rangle \geq \langle i_1, i_2, \ldots, i_{d(n)}, o \rangle$, where $\geq$ denotes the lexicographical order and $o' \in \{0,1\}$ is the output of the circuit C_n on input x taking $A_1, A_2, \ldots, A_{d(n)}$ as the answers to the queries.

Facts:

1. $X \in NP$.
2. X is closed by lexicographical $\leq$ order: if $\langle a_1, a_2, \ldots, a_{d(n)}, o_a \rangle \leq \langle b_1, b_2, \ldots, b_{d(n)}, o_b \rangle$ and $\langle x, b_1, b_2, \ldots, b_{d(n)}, o_b \rangle \in X$, then $\langle x, a_1, a_2, \ldots, a_{d(n)}, o_a \rangle$ is also in X.
3. Fixed x of length n, we define

$$\langle m_1, m_2, \ldots, m_{d(n)}, o_m \rangle = \max \left\{ \langle i_1, i_2, \ldots, i_{d(n)}, o_i \rangle \;\middle|\; \langle x, i_1, i_2, \ldots, i_{d(n)}, o_i \rangle \in X \right\}.$$

This maximum verifies:

m_1 is the number of "YES" answers of C_n on input x at the first level of queries.

m_2 is the number of "YES" answers of C_n on input x at the second level of queries.

......

$m_{d(n)}$ is the number of "YES" answers of C_n on input x at the $d(n)^{th}$ level of queries.

o_m is the value computed by the circuit.

Fact 1 is easy to prove if we note that the circuits have polynomial size and, in each level of queries, we only have to check the "YES" answers (of queries to SAT) supposing that it is known what were the answers of queries in previous levels.

Fact 2 is an easy consequence of the definition of X.

To show fact 3 just observe that the tuple $\langle k_1, k_2, \ldots, k_{d(n)}, o_k\rangle$ where k_1 is the number of "YES" answers of C_n on input x at the first level of queries, k_2 is the number of "YES" answers of C_n on input x at the second level of queries... and o_k is the value computed by C_n on input x, satisfies that $\langle x, k_1, k_2, \ldots, k_{d(n)}, o_k\rangle \in X$. Moreover, if there was $\langle x, i_1, i_2, \ldots, i_{d(n)}, o_i\rangle \in X$ such that $\langle x, i_1, i_2, \ldots, i_{d(n)}, o_i\rangle > \langle x, k_1, k_2, \ldots, k_{d(n)}, o_k\rangle$ it would imply that either

$$\exists j, 1 \leq j \leq d(n) : i_1 = k_1, \ldots, i_{j-1} = k_{j-1}, i_j > k_j$$

or otherwise $o_i > o_k$. In the first case, $i_1 = k_1, \ldots, i_{j-1} = k_{j-1}$ implies that the queries answered "YES" in the first $j-1$ levels of queries must be exactly the same (remember the meaning of k_l's). Therefore, queries to oracle gates of level j are the same in both cases and $i_j > k_j$ is an obvious contradiction. On the other hand, in the second case, if $o_i > o_k$ and for all j, $1 \leq j \leq d(n)$ is $i_j = k_j$, using a similar argument we also get a contradiction.

Now, using facts 1, 2 and 3, it is easy to prove that $AC^{k-1}(NP) \subseteq P^{NP}[O(\log^k n)]$. Given an input x of length n to the circuit C_n we can determine if C_n accepts x with a Turing machine which proceeds as follows:

Using binary search, it determines

$$\langle m_1, m_2, \ldots, m_{d(n)}, o_m\rangle = \max\left\{\langle i_1, i_2, \ldots, i_{d(n)}, o_i\rangle \mid \langle x, i_1, i_2, \ldots, i_{d(n)}, o_i\rangle \in X\right\};$$

this can be done with at most $d(n)\log p(n) + 1$ queries to X (observe that $d(n)\log p(n) \in O(\log^k n)$). Then it accepts x iff $o_m = 1$. This Turing machine recognizes the language accepted by the circuits.

Finally, knowing that $X \in NP$ by fact 1, we may conclude that

$$L(\{C_n\}, n \geq 1) \in P^{NP}[O(\log^k n)].$$

□

Similar sets, but simpler, were defined by Buss and Hay to show that, for reducibility to the NP-complete problem SAT, polynomial time truth-table reducibility via Boolean circuits is equivalent to logspace truth-table reducibility via Boolean formulas (Th. 2 [BH-91]).

Corollary 5 For any integer $k \geq 1$, $P^{NP}[O(\log^k n)] = P^{NP}_{\|O(\log^{k-1} n)}$.

Proof ($\subseteq$) Observe that $AC^{k-1}(NP) \subseteq P^{NP}_{\|O(\log^{k-1} n)}$, therefore this direction follows directly from theorem 4.

($\supseteq$) (sketch) Given a polynomial time Turing machine M that makes at most $O(\log^{k-1} n)$ rounds of parallel queries on inputs of length n, we can define (doing a few evident changes) the set X as in the proof of theorem 4 and verifying the same properties. Now, this direction is also clear. □

4. The Extended Boolean hierarchy with superpolynomial bounding functions

Cai and Hemachandra introduced the Boolean hierarchy in [CH-86]. They considered the boolean functions:

$$\begin{aligned} h_1(x_1) &= x_1 \\ h_{2k}(x_1, x_2, \ldots, x_{2k}) &= h_{2k-1}(x_1, x_2, \ldots, x_{2k-1}) \wedge \neg x_{2k} \\ h_{2k+1}(x_1, x_2, \ldots, x_{2k+1}) &= h_{2k}(x_1, x_2, \ldots, x_{2k}) \vee x_{2k+1} \end{aligned}$$

for all k, $k \geq 1$. With these functions they define the classes of languages $NP(k)$:

$$A \in NP(k) \iff \exists B_1, \ldots, B_k \in NP \text{ such that } c_A(x) = h_k(c_{B_1}(x), \ldots, c_{B_k}(x)) \text{ for all } x.$$

Finally, they defined the Boolean hierarchy

$$BH = \bigcup_{k \geq 1} NP(k).$$

Köbler, Schöning, Wagner and Wechsung gave in their papers [KSW-87] and [WW-85] different definitions of the Boolean hierarchy. With the results presented in [WW-85] it could be proved that all three definitions are equivalent; they gave the following characterization for $NP(k)$:

$$A \in NP(k) \iff \exists B \in NP \text{ such that } c_B(x, i+1) \leq c_B(x, i) \text{ for all } i, i \leq k \text{ and } c_A(x) \equiv \textstyle\sum_{i=1}^{k} c_B(x, i) \bmod 2.$$

From these results and those presented in [Be-87], it could be proved that the Boolean hierarchy coincides with the constant query classes:

$$BH = P_{\|}^{NP}[O(1)] = P^{NP}[O(1)].$$

The Extended Boolean hierarchy considers classes $NP(r)$ for non-constant bounding functions r. Naturally, classes $NP(r)$ are defined as

$$A \in NP(r) \iff \exists B \in NP \text{ such that } c_B(x, i+1) \leq c_B(x, i) \text{ for all } i, i \leq r(|x|) \text{ and } c_A(x) \equiv \textstyle\sum_{i=1}^{r(|x|)} c_B(x, i) \bmod 2.$$

It is shown in [BH-91] and [Wa-90] that $NP\left(n^{O(1)}\right) = \Theta_2^p$. Also, in the second paper it is proved that

$$NP\left(2^{n^{O(1)}}\right) = \Delta_2^p.$$

Wagner asked in his paper [Wa-90] what could be said about other superpolynomial bounding functions. We answer this question below, using ideas from the proof of theorem 4.

Theorem 6 For all $k \geq 1$,

$$P^{NP}[O(\log^k n)] = NP\left(n^{O(\log^{k-1} n)}\right).$$

Proof ($\supseteq$) Given a set $A \in NP(b(n))$, where $b(n) \in n^{O(\log^{k-1} n)}$, we know by definition that there exists a set $B \in NP$ such that

$$c_B(x, i+1) \leq c_B(x, i) \quad \text{for all } i \leq b(|x|) \quad \text{and}$$
$$c_A(x) \equiv \sum_{i=1}^{b(|x|)} c_B(x, i) \bmod 2.$$

We consider a Turing machine M with oracle B which on input x works as follows:

First, it determines the maximum i such that $\langle x, i\rangle \in B$. By the first property of B this can be done with a binary search with only $\log b(|x|)$ queries to B.

Second, using the second property of B, M accepts x iff i is odd.

Clearly, $L(M) = A$, M works in polynomial time and makes at most $O(\log^k n)$ queries on inputs of length n.

($\subseteq$) By the theorem 4, it is sufficient to prove that $AC^{k-1}(NP) \subseteq NP\left(n^{O(\log^{k-1} n)}\right)$.

Given $\{C_n\}$, $n \geq 1$ a family of circuits in $AC^{k-1}(NP)$ with depth $d(n) \in O(\log^{k-1} n)$ and $p(n)$ a polynomial bound on the number of queries in each level, we define the set X as in the proof of theorem 4. Remember facts 1, 2 and 3 that this set verifies. From fact 3, it is easy to see that

$$x \in L(C_n) \iff \left(m = \max\left\{\langle i_1, i_2, \ldots, i_{d(n)}, o_i\rangle \mid \langle x, i_1, i_2, \ldots, i_{d(n)}, o_i\rangle \in X\right\}\right) \text{ is odd.}$$

Obviously,

$$m \leq 2p(n)^{d(n)} \in n^{O(\log^{k-1} n)}.$$

Now, considering $B = \left\{\langle x, i\rangle \mid \langle x, i\rangle = \langle x, \langle i_1, i_2, \ldots, i_{d(n)}, o_i\rangle\rangle \in X\right\}$ and using facts 1 and 2, we may conclude that

$$L(\{C_n\}, n \geq 1) \in NP\left(n^{O(\log^{k-1} n)}\right).$$

□

5. $P^{NP}[O(\log^k n)]$ and circuits of small size

Allender and Wilson showed in their paper [AW-90] that for functions $b(n)$ bounded by a polynomial in n, $NP(O(b(n)))$ is equal to the class of languages which are reducible to languages in NP via reductions of size $\log b(n) + O(1)$. Thus, as a consequence, they obtained circuit characterizations of $P^{NP}[O(\log n)]$. However, it was not know whether similar results hold for other kinds of bounding functions. Now, we shall prove that the same results remain true for bounding functions $b(n) \leq 2^{n^{O(1)}}$.

Proposition 7 For any bounding function $b(n) \leq 2^{n^{O(1)}}$, $NP(O(b(n)))$ is equal to the class of languages which are reducible to languages in NP via reductions of size $\log b(n) + O(1)$.

Proof ($\subseteq$) This direction can be done following exactly the (left to right) proof of theorem 4 of [AW-90]. Given $A \in NP(b(n))$ let B be the set in NP such that

1. $x \in A \iff \max\{i \leq b(n) \mid \langle x, i\rangle \in B\}$ is odd.
2. If $\langle x, i\rangle \in B$, then $\langle x, i-1\rangle \in B$.

First, query if $\langle x, 10\ldots 0\rangle$ is in B; call the answer to this query b_1. Next query if $\langle x, b_1 1\ldots 0\rangle$ is in B. It is clear how to proceed. Note that exactly $\lceil \log b(n)\rceil$ gates are necessary.

($\supseteq$) Let A be reducible to SAT via circuits $\{C_n\}$, $n \geq 1$ of size $\log b(n) + O(1)$. Let $d(n)$ be the depth of circuit C_n and let $p_1(n), p_2(n), \ldots, p_{d(n)}(n)$ be the number of queries of that circuit on levels $1, 2, \ldots, d(n)$ respectively. Note that these functions can not be superpolynomial and:

$$p_1(n) + p_2(n) + \cdots + p_{d(n)}(n) \leq \log b(n) + a,$$

where a is a constant. We define the set X as in the proof of theorem 4 (but with the new bounds $p_1(n), \ldots, p_{d(n)}(n)$ on the levels of queries). Clearly, facts 1, 2 and 3 remain true. Now, as in the proof of theorem 6, we have

$$x \in L(C_n) \iff \left(m = \max\{\langle i_1, i_2, \ldots, i_{d(n)}, o_i\rangle \mid \langle x, i_1, i_2, \ldots, i_{d(n)}, o_i\rangle \in X\}\right) \text{ is odd;}$$

where $i_j \leq p_j(n)$. Thus

$$m \leq 2p_1(n)p_2(n)\cdots p_{d(n)}(n) \in O(b(n)).$$

□

Now, from theorem 6 and proposition 7 we have a new characterization for $P^{NP}[O(\log^k n)]$.

Corollary 8 For all $k \geq 1$, $P^{NP}[O(\log^k n)]$ is equal to the class of languages which are reducible to languages in NP via circuits of size $O(\log^k n)$. □

This corollary gives a strong connection between classes defined by polynomial Turing machines with few queries to an NP oracle and classes defined by small size circuits with NP oracle gates.

6. Separation with oracles

In [LT-91] there is exhibited a set A such that the classes Θ_2^p and Δ_2^p are different relativized to A (this is also shown in [BH-91]). Here, applying similar methods, we extend that result by showing that there exists a relativized world where $P^{NP}[O(\log^k n)]$ is different from $P^{NP}[O(\log^{k+1} n)]$ for any integer $k \geq 1$. As a consequence, we can obtain relativized separations between some circuit classes.

Theorem 9 There exists a recursive set A so that for any integer $k \geq 1$:

$$P^{NP(A)}[O(\log^{k+1} n)] - P^{NP(A)}[O(\log^k n)] \neq \emptyset.$$

Proof We will introduce a language $L_{k+1}(A)$ with the property that for any A, $L_{k+1}(A) \in P^{NP(A)}[O(\log^{k+1} n)]$. Afterwards, a specific oracle A will be constructed so that for all $k \geq 1$, $L_{k+1}(A) \notin P^{NP(A)}[O(\log^{k} n)]$. Let $L_{k+1}(A)$ be the set defined as follows:

$$L_{k+1}(A) = \left\{0^n \mid A^{=\log^{k+1} n} = \emptyset \text{ or the minimum word of length } \log^{k+1} n \text{ in } A \text{ is even}\right\}.$$

First, we can see that for any A, $L_{k+1}(A) \in P^{NP(A)}[O(\log^{k+1} n)]$: Note that the language $S = \{y \mid \exists x \in A, |x| = |y|, x \leq y\}$ belongs to $NP(A)$, and for any integer $m \geq 0$, it verifies:

i) $\min S^{=m} = \min A^{=m}$

ii) if z and y are words of length m such that $z \geq y$ and $y \in S$, then z belongs to S.

Now it is clear how to proceed, doing a binary search over the words of length $\log^{k+1} n$ we only have to do $\log^{k+1} n$ queries to S to find the minimum.

Second, we construct a specific oracle A such that $L_{k+1}(A) \notin P^{NP(A)}_{\|O(\log^{k-1} n)}$. Note that this class is the same as $P^{NP(A)}[O(\log^{k} n)]$, as it is shown in corollary 5. Let us consider $M_1, M_2, \ldots$ and $N_1, N_2, \ldots$ enumerations of the deterministic and nondeterministic Turing machines respectively, where a machine i has running time bounded by a polynomial p_i. It is well known that, for every set B, the set:

$$K(B) = \{\langle j, y, 0^t \rangle \mid N_j \text{ with oracle } B \text{ accepts } y \text{ in at most } t \text{ steps}\}$$

is complete for the class NP relative to B. So, we can suppose that deterministic machines always make queries of type $\langle j, y, 0^t \rangle$ to the oracle $K(A)$.

A is constructed in stages. At stage $s = \langle i, k, c \rangle$, if the machine M_i on input 0^{n_s} makes at most $c \log^{k-1} n_s$ rounds of parallel queries, we will add words of length $\log^{k+1} n_s$ to A, diagonalizing away from the language recognized by M_i. We will also keep a list A' of words that will never be included in A: for every query $q = \langle j, y, 0^t \rangle$ of M_i to $K(A)$ answered positively, we will include in A' the list of words that are not in A and are queried in a certain accepting path of N_j on input y.

Let us suppose that M_i makes at most $c \log^{k-1} n_s$ rounds of parallel queries (otherwise, we finish stage s). We say that machine M_i behaves correctly with input 0^{n_s} and oracle $K(A)$ if it accepts (rejects), and the minimum word of length $\log^{k+1} n_s$ in A is even (odd). If so, we will add a new odd (even) word w to A of length $\log^{k+1} n_s$ and smaller than the existing ones, in such a way that it does not affect the positive answers of the queries to $K(A)$. Machine M_i, with the new oracle, either behaves incorrectly or has a round of queries r, $(1 \leq r \leq c \log^{k-1} n_s)$ such that, at any previous round r', $1 \leq r' < r$, the answers to the queries remains the same, but at round r there is, at least, one more query answered positively. In other words, if we denote by $\langle a_1, a_2, \ldots, a_{d(n_s)} \rangle$, $d(n_s) = c \log^{k-1} n_s$, the sequence of numbers of queries answered positively in each round by the machine M_i with oracle $K(A)$, the new sequence $\langle b_1, b_2, \ldots, b_{d(n_s)} \rangle$, corresponding to M_i with the new oracle $K(A \cup \{w\})$, will be greater (in lexicographical order) than $\langle a_1, a_2, \ldots, a_{d(n_s)} \rangle$. Since the number of different sequences is $p_i(n_s)^{d(n_s)} \leq 2^{b \log^k n_s}$, repeating the above procedure $2^{b \log^k n_s}$ many times we can diagonalize away from M_i.

stage 0

$A(0) := \emptyset$

$n_0 := 0$

endstage
stage $s = \langle i, k, c\rangle$
Let n_s be the smallest integer such that
(1) $\log^{k+1} n_s$ is larger than anything queried or added to A at any previous stage
(2) $2 \cdot (2^{b\log^k n_s} \cdot p_i^2(n_s) + 2^{b\log^k n_s}) < 2^{\log^{k+1} n_s}$
Check that $M_i(0^{n_s})$ has at most $c\log^{k-1} n_s$ rounds. If not, skip the stage
$A' := \emptyset$
$A(s) := A(s-1)$
$d := 1$; $u_d := 1^{\log^{k+1} n_s}$
repeat
$A(s) := A(s) \cup \{u_d\}$
compute all the queries made by M_i on input 0^{n_s} and oracle $A(s)$
for all the queries $q_l = \langle j_l, y_{j_l}, 0^{t_{j_l}}\rangle$ **do**
if N_{j_l} with oracle $A(s)$ accepts y_{j_l} in less than t_{j_l} steps **then**
$A' := A' \cup \{$words queried in the minimum accepting path for $N_{j_l}^{A(s)}$ that are not in $A(s)\}$
endif
endfor
$d := d + 1$
$u_d := \max\{x : |x| = \log^{k+1} n_s$ and (x is even $\iff$ u_{d-1} is odd) and $x \notin A'$ and $x < u_{d-1}\}$
until $0^{n_s} \in L(M_i, K(A(s))) \iff 0^{n_s} \in L(M_i, K(A(s) \cup \{u_d\}))$
if $M_i(K(A(s)))$ behaves correctly **then**
$A(s) := A(s) \cup \{u_d\}$
endif
endstage.

Note that the selected word u_d included in $A(s)$ in each iteration always exists since in A' there are at most $2^{b\log^k n_s} \cdot p_i^2(n_s)$ reserved words and before including u_d we could have included at most $2^{b\log^k n_s}$ words of length $\log^{k+1} n_s$ in A. □

Corollary 10 There exists an oracle A such that:
1. $AC^0(NP(A)) \subset AC^1(NP(A)) \subset \cdots \subset P^{NP(A)}$
2. For any integer $k \geq 1$: $NC^k(NP(A)) \subset AC^k(NP(A))$
3. $NC^1(NP(A)) \subset NC^2(NP(A)) \subset \cdots \subset P^{NP(A)}$

where "$\subset$" denotes proper containment.

Proof The first statement follows directly from theorem 4 and theorem 9. For the second point we only give an outline of the proof: it is not difficult to prove that $NC^k(NP) \subseteq P^{NP}_{\|O(\log^k n/\log\log n)}$. Now, note that the set A constructed in the proof of theorem 9 gives, in fact, a relativized separation between $P^{NP}_{\|O(\log^k n/g(n))}$ and $P^{NP}[O(\log^{k+1} n)]$ for any function $g(n) \in \omega(1)$: the condition (2) of the algorithm over n_s will be also satisfiable if on the left hand of the inequality we change $\log^k n_s$ by $\log^{k+1} n_s/g(n_s)$. Finally, the third statement is an obvious consequence of the previous ones. □

Conclusions

We have shown different characterizations for some complexity classes between Θ_2^p and Δ_2^p. In particular, we have determined what the classes of languages AC^k reducible to NP are and we have obtained circuit characterizations for the Extended Boolean hierarchy with superpolynomial bounding functions. However, all the circuit characterizations are based on AC^k circuits or on circuits of small size; we have not obtained characterizations based on NC^k circuits, but will continue our work in this direction.

It seems an interesting problem to study the NC^k complexity classes with oracle gates in NP and to show relationships between these classes and those present in this paper. Concretely, it would be nice to show that $NC^1(NP) = \Theta_2^p$: in this case the well known inclusions $AC^0 \subset NC^1 \subseteq L$ would collapse when the relativized versions to NP were considered, in contrast with the general relativized case (see [Wi-87]). On the other hand, if the equality $NC^1(NP) = \Theta_2^p$ fails we would find an irregular situation: while it holds that $AC^0 \subset NC^1 \subseteq L$ unrelativized, the first and the last relativized to NP counterparts would coincide, but the middle one would not.

Acknowledgements

We are very grateful to José L. Balcázar for reading earlier versions of this paper and for many interesting suggestions. Fred Green deserve our thanks for suggesting us to study the subject of the last section. We would like to thank Eric Allender, Birgit Jenner and Chris Wilson for helpful discussions which took place in Barcelona, during the Structure Conference 1990 and specially, we are indebted to Eric Allender for pointing out an error in a preliminary version.

References

V-90] E. Allender, C. B. Wilson, *Width-Bounded Reducibility and Binary Search over Complexity Classes,* Proc. 5th IEEE Conf. on Structure in Complexity Theory, (1990), pp. 122-129.

e-87] R. J. Beigel, *Bounded queries to SAT and the Boolean hierarchy,* To appear in TCS.

H-91] S. Buss, L. Hay, *On truth-table reducibility to SAT,* Information and Computation, vol. 91 (1991), pp. 86-102.

H-86] J. Cai, L. A. Hemachandra, *The Boolean hierarchy: hardware over NP,* Proc. 1st IEEE Conf. on Structure in Complexity Theory, (1986), pp. 105-124.

o-85] S. Cook, *A taxonomy of problems with fast parallel algorithms,* Information and Control, vol. 64 (1985), pp. 2-22.

N-87] J. Köbler, U. Schöning, K. W. Wagner, *The difference and the truth-table hierarchies for NP,* R.A.I.R.O. 21 (1987), pp. 419-435.

T-91] A. Lozano, J. Torán, *Self-reducible sets of small density,* Math. Systems Theory 24 (1991), pp. 83-100

a-90] K. W. Wagner, *Bounded query classes,* SIAM J. Comput., vol. 19, 5 (1990), pp. 833-846.

N-85] G. Wechsung, K. W. Wagner, *On the Boolean closure of NP,* manuscript 1985 (extended abstract as: Wechsung G., On the Boolean closure of NP, Proc. Conf. Fundam. Comp. Theory, Cottbus 1985, LNCS 199 (1985), pp. 485-493).

Vi-87] C. B. Wilson, *Relativized NC,* Math. Systems Theory, vol. 20 (1987), pp. 13-29.

Vi-90] C. B. Wilson, *Decomposing NC and AC,* SIAM J. Comput., vol. 19, 2 (1990), pp. 384-396.

On Complexity Classes and Algorithmically Random Languages

– Extended Abstract –*

Ronald V. Book †

Department of Mathematics, University of California

Santa Barbara, CA 93106, USA

Jack H. Lutz ‡

Department of Computer Science, Iowa State University

Ames, Iowa 50011, USA

Klaus W. Wagner

Institut für Informatik, Universität Würzburg

W-8700 Würzburg, Germany

Abstract

Every class **C** of languages satisfying a simple topological condition is shown to have probability one if and only if it contains some language that is algorithmically random in the sense of Martin-Löf. This result is used to derive separation properties of algorithmically random oracles and to give characterizations of the complexity classes **P**, **BPP**, **AM**, and **PH** in terms of reducibility to such oracles. These characterizations lead to the following result:

(i) **P** = **NP** if and only if there exists an algorithmically random set that is $\leq^{\mathrm{P}}_{btt}$-hard for **NP**.

(ii) **P** = **PSPACE** if and only if there exists an algorithmically random set that is $\leq^{\mathrm{P}}_{btt}$-hard for **PSPACE**.

(iii) The polynomial-time hierarchy collapses if and only if there exists $k > 0$ such that some algorithmically random set is Σ^{P}_{k}-hard for **PH**.

(iv) **PH** = **PSPACE** if and only if there exists a algorithmically random set that is **PH**-hard for **PSPACE**.

---

*Full version of this paper as Technical Report No. 32/1991, Bayerische Julius-Maximilians-Universität Würzburg, Institut für Informatik

†The work of the first author was supported in part by the Alexander-von-Humboldt-Stiftung and by the National Science Foundation under Grant CCR-8913584 while he visited the Lehrstuhl für Theoretische Informatik, Institut für Informatik, Universität Würzburg, Germany.

‡The work of the second author was supported in part by the National Science Foundation under Grant CCR-8809238 and in part by DIMACS, where he was a visitor while a portion of his work was done.

1 Introduction

Many results in complexity theory involve conditions which are satisfied by "almost every" oracle. Two of the best-known examples are the following:

(i) For almost every oracle A, $\mathbf{P}(A) \neq \mathbf{NP}(A) \neq$ co$-\mathbf{NP}(A)$ [BG81].

(ii) For every language B, $B \in \mathbf{BPP}$ if and only if for almost every oracle A, $B \in \mathbf{P}(A)$ [BG81, Amb86].

In such results, the assertion that "almost every oracle A has property θ" means that $\theta(A)$ is true with probability one when the oracle $A \subseteq \{0,1\}^*$ is selected probabilistically by using an independent toss of a fair coin to decide membership of each string in A.

The class **RAND** of *algorithmically random languages*, defined by Martin-Löf [Mar66] (and in Section 3 below) contains almost every oracle. Thus, for every property θ which is satisfied by almost every oracle, there exists an oracle $A \in \mathbf{RAND}$ satisfying $\theta(A)$.

In this paper we prove that the converse holds for a wide variety of properties θ. Specifically in Section 3 below, we prove the following. Assume that the class of all oracles A satisfying $\theta(A)$ is a union of recursively closed sets (in the Cantor topology on the set of all languages) and is closed under finite variation. Then $\theta(A)$ holds for *some* $A \in \mathbf{RAND}$ if and only if $\theta(A)$ holds for *almost every* oracle A.

To date, most complexity theory results concerning almost every oracle are either oracle separation results, like (i) above, or characterizations of complexity classes, like (ii) above. In Section 4 we illustrate the Main Theorem in both of these contexts. We show how, in many cases, separations for relativized complexity classes for almost every oracle immediately imply separations for *every* algorithmically random oracle. In addition, we show how characterizations of reducibility to *some* algorithmically random oracle yield characterizations of complexity classes in terms of reducibility to almost every oracle. These characterizations yield the following result (Theorem 12):

(i) $\mathbf{P} = \mathbf{NP}$ if and only if there exists an algorithmically random set that is $\leq^{\mathrm{P}}_{btt}$-hard for **NP** if and only if every algorithmically random set is $\leq^{\mathrm{P}}_{btt}$-hard for **NP**.

(ii) $\mathbf{P} = \mathbf{PSPACE}$ if and only if there exists an algorithmically random set that is $\leq^{\mathrm{P}}_{btt}$-hard for **PSPACE** if and only if every algorithmically random set is $\leq^{\mathrm{P}}_{btt}$-hard for **PSPACE**.

(iii) The polynomial-time hierarchy collapses if and only if there exists $k > 0$ such that some algorithmically random set is Σ^{P}_k-hard for **PH** if and only if there exists $k > 0$ such that every algorithmically random set is Σ^{P}_k-hard for **PH**.

(iv) $\mathbf{PH} = \mathbf{PSPACE}$ if and only if there exists a algorithmically random set that is **PH**-hard for **PSPACE** if and only if every algorithmically random set is **PH**-hard for **PSPACE**.

2 Preliminaries

For the most part our notation is standard, following that used by Balcázar, Díaz, and Gabarró [BDG88, BDG90]. We assume that the reader is familiar with the standard recursive reducibilities and the variants obtained by imposing resource bounds such as time or space of the algorithms that computed these reducibilities.

A *word* (string) is an element of $\{0,1\}^*$. The length of a word $w \in \{0,1\}^*$ is denoted $|w|$.

We assume a fixed pairing function, $\langle\cdot,\cdot\rangle : \{0,1\}^* \times \{0,1\}^* \longrightarrow \{0,1\}^*$, that is computable in polynomial time and whose projections (inverses) are computable in polynomial time.

The power set of a set A is denoted by $\mathcal{P}(A)$.

The *characteristic sequence* of a language A is a (one-way) infinite sequence ξ_A on $\{0,1\}$. We freely identify a language with its characteristic sequence and the class of all languages on the fixed finite alphabet $\{0,1\}$ with the set $\{0,1\}^\omega$ of all such infinite sequences; the usage is based on context so that there should be no ambiguity on the part of the reader.

For each string w, $C_w = \{w\} \cdot \{0,1\}^\omega$ is the *cylinder* defined by w. Note that C_w is the set of all sequences ξ such that w is a *prefix* of ξ.

An *open set* is a (finite or infinite) union of cylinders. (This definition gives the usual product topology, also known as the Cantor topology, on $\{0,1\}^\omega$.) A *closed set* is the complement of an open set.

If X is a set of strings (i. e., a language) and C is a set of sequences (i. e., a class of languages), then $X \cdot \mathrm{C}$ denotes the set $\{w\xi \mid w \in X,\ \xi \in \mathrm{C}\}$. (Note that $X \cdot \mathrm{C}$ is itself a class of languages under our identification of languages with sequences.)

We are frequently concerned with classes of the form $X \cdot \{0,1\}^\omega$ for some X. Notice that $X \cdot \{0,1\}^\omega = \bigcup_{w \in X} C_w$ so that $X \cdot \{0,1\}^\omega$ is an open set.

We assume an effective enumeration of the recursively enumerable languages as W_1, $W_2, \ldots$, and of the finite languages as D_1, $D_2, \ldots$.

A set C of languages is *recursively open* if it is of the form $\mathrm{C} = W_j \cdot \{0,1\}^\omega$ for some recursively enumerable set W_j. A set C of languages is *recursively closed* if it is the complement of some recursively open set.

For a class C of languages we write Prob[C] for the probability that $A \in \mathrm{C}$ when A is chosen by a random experiment in which an independent toss of a fair coin is used to decide whether a string is in A. This probability is defined whenever C is measurable in the usual product topology of $\{0,1\}^\omega$. In particular, if C is a countable union or intersection of (recursively) open or closed sets, then C is Borel, hence measurable, so Prob[C] is defined. Note that there are only countably many recursively open sets, so every intersection of recursively open sets can be expressed as a countably intersection of such sets, and hence is Borel; similarly every union of recursively closed sets is Borel.

A class C is *closed under finite variation* if $A \in \mathrm{C}$ holds whenever $B \in \mathrm{C}$ and A an B have finite symmetric difference.

3 Main Results

It will be convenient to use the following normal form lemma for recursively closed languages.

Lemma 1 *For every recursively closed class* **F** *of languages, there exists a total recursive function* $f : \mathsf{N} \longrightarrow \mathsf{N}$ *with the following properties:*

(i) $\mathbf{F} = \bigcap_{k \geq 0} D_{f(k)} \cdot \{0,1\}^{\omega}$.

(ii) $(\forall\, k \in \mathsf{N}),\ D_{f(k+1)} \cdot \{0,1\}^{\omega} \subseteq D_{f(k)} \cdot \{0,1\}^{\omega}$.

Now we recall (in our notation) Martin-Löf's definition of algorithmic randomness.

Definition [Mar66]. *A* constructive null cover *of a class* C *of languages is a sequence of recursively open sets* $W_{g(1)} \cdot \{0,1\}^{\omega}, W_{g(2)} \cdot \{0,1\}^{\omega}, \ldots$ *specified by a total recursive function* g *with the properties that for every* k,

(i) $\mathbf{C} \subseteq W_{g(k)} \cdot \{0,1\}^{\omega}$, *and*

(ii) $\mathrm{Prob}[W_{g(k)} \cdot \{0,1\}^{\omega}] \leq 2^{-k}$.

If a class C *has a constructive null cover, then* C *is a* constructive null set. *A language* B *is* algorithmically random *if there does not exist a constructive null set containing* B. *Let* **RAND** *denote the class of languages that are algorithmically random.*

Martin-Löf [Mar66] showed that $\mathrm{Prob}[\mathbf{RAND}] = 1$ and that no recursively enumerable language is in **RAND**.

In what follows we will use a necessary condition for algorithmic randomness:

Definition *A* strong constructive null cover *of a class* C *of languages is a sequence of open sets* $D_{g(1)} \cdot \{0,1\}^{\omega},\ D_{g(2)} \cdot \{0,1\}^{\omega},\ \ldots$ *specified by a total recursive function* g *with the properties that for every* k,

(i) $\mathbf{C} \subseteq D_{g(k)} \cdot \{0,1\}^{\omega}$, *and*

(ii) $\mathrm{Prob}[D_{g(k)} \cdot \{0,1\}^{\omega}] \leq 2^{-k}$.

Since every strong constructive null cover is easily transformed into a constructive null cover, we have the following fact.

Lemma 2 *If* **F** *is a recursively closed set of languages with* $\mathrm{Prob}[\mathbf{F}] = 0$, *then* **F** *is a (strong) constructive null set.*

Corollary 3 *If* C *is a union of recursively closed sets of languages and* $\mathrm{Prob}[\mathbf{C}] = 0$, *then* C *is a constructive null set.*

Now we come to the main result.

Theorem 4 *Let* C *be a class of languages with the following two properties.*

(i) C *is a union of recursively closed sets.*

(ii) C *is closed under finite variation.*

Then the following are equivalent:

(a) Prob[C] = 1.

(b) C ∩ **RAND** ≠ ∅.

The following dual of Theorem 4 is also useful.

Corollary 5 *Let* C *be a class of languages with the following two properties.*

(i) C *is an intersection of recursively open sets.*

(ii) C *is closed under finite variation.*

Then the following conditions are equivalent:

(a) Prob[C] = 1.

(b) **RAND** ⊆ C.

4 Applications

We illustrate the power of the Main Theorem with applications of two types; namely, oracle separations and characterizations of complexity classes.

Since we are concerned with the use of oracles, we consider complexity classes that can be specified so as to "relativize," that is, that allow for Turing reducibilities. But we want to do this in a general setting and so we introduce a few definitions.

We assume a fixed enumeration M_0, M_1, M_2, ... of nondeterministic oracle Turing machines.

A *relativized class* is a function $\mathrm{C} : \mathcal{P}(\{0,1\}^*) \longrightarrow \mathcal{P}(\mathcal{P}(\{0,1\}^*))$. A *recursive presentation* of a relativized class C of languages is a total recursive function $f : \mathsf{N} \longrightarrow \mathsf{N}$ such that for every language A, $\mathrm{C}(A) = \{L(M^A_{f(i)}) \mid i \in \mathsf{N}\}$. A relativized class is *recursively presentable* if it has a recursive presentation.

A *reducibility* is a relativized class. A *bounded reducibility* is a relativized class which is recursively presentable.

If $\mathbf{R}$ is a reducibility, then we use the notation $A \leq^{\mathrm{R}} B$ to indicate that $A \in \mathbf{R}(B)$. In addition we write $\mathbf{R}^{-1}(A)$ for $\{B \mid A \leq^{\mathrm{R}} (B)\}$.

Typical bounded reducibilities include $\leq_m^{\mathrm{P}}$, $\leq_{btt}^{\mathrm{P}}$, $\leq_T^{\mathrm{P}}$, $\leq_T^{\mathrm{NP}}$, $\leq_T^{\mathrm{SN}}$, $\leq_m^{\mathrm{logspace}}$, etc. The relations $\leq_m$ and $\leq_T$ are reducibilities which are not bounded.

In many contexts it is useful to restrict attention to reducibilities that are reflexive and transitive, but we do not need such restrictions here.

If $\mathbf{R}$ is a reducibility and $\mathbf{C}$ is a set of languages, then a language A is $\leq^{\mathrm{R}}$*-complete* for $\mathbf{C}$ if $A \in \mathbf{C} \subseteq \mathbf{R}(A)$. A relativized class $\mathbf{C}$ is *recursively presentable with a* $\leq^{\mathrm{R}}$*-complete language* if there exist a recursive presentation f of $\mathbf{C}$ and a constant $c \in \mathsf{N}$ such that for every language A, $L(M_{f(c)}^A)$ is $\leq^{\mathrm{R}}$-complete for $\mathbf{C}(A)$.

If $\mathbf{R}$ is a reducibility and $\mathbf{C}$ is a set of languages, write $\mathbf{R}(\mathbf{C})$ for $\bigcup_{A \in \mathbf{C}} \mathbf{R}(A)$. A relativized class $\mathbf{C}$ is *closed* under a reducibility $\mathbf{R}$ if $\mathbf{R}(\mathbf{C}(A)) \subseteq \mathbf{C}(A)$ for every language A.

While the next result is quite general, it does apply to a number of specific situations that are of interest in complexity theory.

Theorem 6 *Let* $\mathbf{C}$ *and* $\mathbf{D}$ *be relativized complexity classes and let* $\mathbf{R}$ *be a reducibility. Suppose that each of the following holds:*

(i) $\mathbf{C}$ *is recursively presentable with an* $\leq^{\mathrm{R}}$*-complete language.*

(ii) $\mathbf{D}$ *is closed under* $\mathbf{R}$ *and is recursively presentable.*

(iii) $\mathbf{C}$ *and* $\mathbf{D}$ *are invariant under finite variations of the oracle.*

Then the following conditions are equivalent:

(a) $\mathbf{C}(A) \not\subseteq \mathbf{D}(A)$ *for almost every* A.

(b) $\mathbf{C}(A) \not\subseteq \mathbf{D}(A)$ *for every* $A \in \mathbf{RAND}$.

From Theorem 6 and known probability one oracle separations, it follows immediately that *every* algorithmically random set A satisfies

(i) $\mathbf{P}(A) \neq \mathbf{NP}(A) \neq \text{co}-\mathbf{NP}(A)$ [BG81],

(ii) complete separation of $\mathbf{BH}(A)$ into infinitely many levels [Cai87],

(iii) $\mathbf{PH}(A) \neq \mathbf{PSPACE}(A)$ [Cai89],

etc. Similarly, if with probability one, the relativized polynomial-time hierarchy is separated into infinitely many levels, then this separation is achieved relative to *every* algorithmically random set.

We wish to develop characterizations of complexity classes. Let $\mathbf{REC}$ denote the class of recursive languages.

Lemma 7 *If* $A \in \mathbf{REC}$ *and* $\mathbf{R}$ *is a bounded reducibility which is invariant under finite variations of the oracle, then* $\mathbf{R}^{-1}(A)$ *is a union of recursively closed sets.*

For each relativized class **C**, let almost$-$**C** $= \{A \mid \mathrm{Prob}[\mathbf{C}^{-1}(A)] = 1\}$.

Theorem 8 *If* **R** *is a bounded reducibility which is invariant under finite variations of the oracle, then* almost$-$**R** $=$ **R(RAND)** $\cap$ **REC**.

Now we turn to characterizations of complexity classes. For the sake of brevity, we give just four applications, characterizing the classes **P**, **BPP**, **AM**, and **PH** in terms of reducibilities to algorithmically random languages.

Theorem 9 (a) $\mathbf{P} = \mathbf{P}_m(\mathbf{RAND}) \cap \mathbf{REC} = \mathbf{P}_{btt}(\mathbf{RAND}) \cap \mathbf{REC}$
$= \mathbf{P}_{\log n-T}(\mathbf{RAND}) \cap \mathbf{REC}$.

(b) $\mathbf{BPP} = \mathbf{P}_{tt}(\mathbf{RAND}) \cap \mathbf{REC} = \mathbf{P}_T(\mathbf{RAND}) \cap \mathbf{REC}$.

(c) $\mathbf{AM} = \mathbf{NP}_T(\mathbf{RAND}) \cap \mathbf{REC}$.

(d) $\mathbf{PH} = \mathbf{PH}(\mathbf{RAND}) \cap \mathbf{REC}$.

Note that $\mathbf{BPP} = \mathbf{P}_T(\mathbf{RAND}) \cap \mathbf{REC}$ has already been proved in [Ben88].

The class **RAND** is considered to be the class of those languages having the greatest possible information content. It is well known that there is a constant c such that for all languages A and all n, the Kolmogorov complexity of the finite language $A_{\leq n} = \{x \in A \mid |x| \leq n\}$ is not greater than $2^{n+1} + c$. Martin-Löf [Mar71] proved that every language A in **RAND** has nearly maximal information content in the sense that the Kolmogorov complexity of $A_{\leq n}$ is strictly greater than $2^n - 2n$ for all but finitely many n. However, while the amount of such a language is very great, one can interpret the results presented here as indicating that this information is encoded in such a way that little of it is computationally useful (from the standpoint of structural complexity theory).This interpretation is reinforced by the following results.

Recall that a set S is *sparse* if there exists a polynomial q such that for all n, $|\{x \in S \mid |x| \leq n\}| \leq q(n)$. Sparse sets are considered to be sets with small information content since their census function is bounded above by a polynomial. The following is a summary of certain known results about reducibilities and sparse sets:

Proposition 10 (i) [OW91] **P** $=$ **NP** *if and only if there exists a sparse set that is* $\leq^{\mathrm{P}}_{btt}$*-hard for* **NP**.

(ii) [OL91] **P** $=$ **PSPACE** *if and only if there exists a sparse set that is* $\leq^{\mathrm{P}}_{btt}$*-hard for* **PSPACE**.

(iii) [KL82] *If there exists a sparse set that is* $\leq^{\mathrm{P}}_{T}$*-hard for* **NP**, *then the polynomial-time hierarchy collapses.*

(iv) [KL82] *If there exists a sparse set that is* $\leq^{\mathrm{P}}_{T}$*-hard for* **PSPACE**, *then* **PH** $=$ **PSPACE**.

It is easy to see that parts (i) and (ii) of Proposition 10 can be extended in the following way:

(i') **P** = **NP** if and only if there exists a sparse set that is $\leq^{\mathrm{P}}_{btt}$-hard for **NP** if and only if every sparse set is $\leq^{\mathrm{P}}_{btt}$-hard for **NP**.

(ii') **P** = **PSPACE** if and only if there exists a sparse set that is $\leq^{\mathrm{P}}_{btt}$-hard for **PSPACE** if and only if every sparse set is $\leq^{\mathrm{P}}_{btt}$-hard for **PSPACE**.

Balcázar, Book, and Schöning [BBS86] and Long and Selman [LS86] extended the results of Karp and Lipton to obtain the following facts.

Proposition 11 (i) *The polynomial-time hierarchy collapses if and only if there exists a sparse set S such that the polynomial-time hierarchy relative to S collapses if and only if for every sparse set S, the polynomial-time hierarchy relative to S collapses.*

(ii) **PH** = **PSPACE** *if and only if there exists a sparse set S such that* **PH**(S) = **PSPACE**(S) *if and only if for every sparse set S* **PH**(S) = **PSPACE**(S).

From Proposition 11, it is clear that the following hold:

(i') The polynomial-time hierarchy collapses if and only if there exists $k > 0$ such that some sparse set is Σ^{P}_k-hard for **PH** (i. e., for some sparse S, **PH** $\subseteq \Sigma^{\mathrm{P}}_k(S)$) if and only if there exists $k > 0$ such that every sparse set is Σ^{P}_k-hard for **PH**.

(ii') **PH** = **PSPACE** if and only if there exists a sparse set that is **PH**-hard for **PSPACE** if and only if every sparse set is **PH**-hard for **PSPACE**.

From Theorem 9, we obtain a result that is parallel to the above variants on Propositions 10 and 11.

Theorem 12 (i) **P** = **NP** *if and only if there exists an algorithmically random set that is $\leq^{\mathrm{P}}_{btt}$-hard for* **NP** *if and only if every algorithmically random set is $\leq^{\mathrm{P}}_{btt}$-hard for* **NP**.

(ii) **P** = **PSPACE** *if and only if there exists an algorithmically random set that is $\leq^{\mathrm{P}}_{btt}$-hard for* **PSPACE** *if and only if every algorithmically random set is $\leq^{\mathrm{P}}_{btt}$-hard for* **PSPACE**.

(iii) *The polynomial-time hierarchy collapses if and only if there exists $k > 0$ such that some algorithmically random set is Σ^{P}_k-hard for* **PH** *if and only if there exists $k > 0$ such that every algorithmically random set is Σ^{P}_k-hard for* **PH**.

(iv) **PH** = **PSPACE** *if and only if there exists a algorithmically random set that is* **PH**-*hard for* **PSPACE** *if and only if every algorithmically random set is* **PH**-*hard for* **PSPACE**.

The similarity between Theorem 12 and the variants of Propositions 10 and 11 is striking. In Theorem 12, the sets having the greatest possible information content, algorithmically random sets, serve as the oracle sets. In Propositions 10 and 11, the sparse sets, sets having very small information content, serve as oracle sets. But the conclusions are the same. One can interpret the results presented in Theorem 12 as indicating that the information in algorithmically random sets is encoded in such a way that little of it is computationally useful from the standpoint of structural complexity theory, since one may as well use a sparse set. This suggests that a theory that relates the information content of oracle sets to the computational power of reducibilities needs to be developed; the results presented here should be viewed as only "first steps."

There are some open questions.

1) If **C** is a relativizable class of languages, under what conditions is it the case that $\mathbf{C}(\mathbf{RAND}) \cap \mathbf{REC} = \mathbf{B}\cdot\mathbf{C}$? This equation is known to be true for $\mathbf{C} = \mathbf{P}$, $\mathbf{C} = \mathbf{NP}$, and $\mathbf{C} = \mathbf{PH}$ by the results stated above. If **C** is a relativizable class of languages, under what conditions is it the case that $\mathbf{BP}\cdot\mathbf{C} = \mathbf{C}$? It is known to be true for $\mathbf{C} = \mathbf{PH}$. It is clear that $\mathbf{BP}\cdot\mathbf{PSPACE} = \mathbf{PSPACE}$. Is $\mathbf{PSPACE}(\mathbf{RAND}) \cap \mathbf{REC}$ equal to **PSPACE** ?

2) While we know that $\mathbf{P}_{\log n-T}(\mathbf{RAND}) \cap \mathbf{REC} = \mathbf{P}$, what can be said about $\mathbf{P}_{\mathrm{poly}(\log n)-T}(\mathbf{RAND}) \cap \mathbf{REC}$ except that it is included in **BPP** ?

References

[Amb86] K. Ambos-Spies. Randomness, relativations, and polynomial reducibilities. In *Lecture Notes in Computer Sci. 223*, pages 23–34. Proc. 1st Conf. Stucture in Complexity Theory, Springer-Verlag, 1986.

[BBS86] J. Balcázar, R. Book, and U. Schöning. The polynomial-time hierarchy and sparse oracles. *J. Assoc. Comput. Mach.*, 33:603–617, 1986.

[BDG88] J. Balcázar, J. Díaz, and J. Gabarró. *Structural Complexity I.* Springer-Verlag, 1988.

[BDG90] J. Balcázar, J. Díaz, and J. Gabarró. *Structural Complexity II.* Springer-Verlag, 1990.

[Ben88] C. Bennett. Logical depth and physical complexity. In R. Herken (ed.), *The Universal Turing Machine: A Half-Century Survey*, pages 227–257. Oxford University Press, 1988.

[BG81] C. Bennett and J. Gill. Relative to a random oracle $P^A \neq NP^A \neq co-NP^A$ with probability 1. *SIAM J. Computing*, 10:96–113, 1981.

[Cai87] J.-Y. Cai. Probability one separation of the boolean hierarchy. In *Lecture Notes in Computer Sci. 38*, pages 148–158. STACS 87, Springer Verlag, 1987.

[Cai89] J.-Y. Cai. With probability one, a random oracle separates PSPACE from the polynomial-time hierarchy. *J. Comput. Systems Sci.*, 38:68–85, 1989.

[KL82] R. Karp and R. Lipton. Turing machines, that take advice. *L'Enseignement Mathématique*, 28 2nd series:191–209, 1982.

[LS86] T. Long and A. Selman. Relativizing complexity classes with sparse oracles. *J. Assoc. Comput. Mach.*, 33:618–627, 1986.

[Mar66] P. Martin-Löf. On the definition of random sequences. *Info. and Control*, 9:602–619, 1966.

[Mar71] P. Martin-Löf. Complexity oscillations in infinite binary sequences. *Zeitschrift für Wahrscheinlichkeitstheorie und Verwandte Gebiete*, 19:225–230, 1971.

[OL91] M. Ogiwara and A. Lozano. On one query self reducible sets. In *Proc. 6th IEEE Conference on Structure in Complexity Theory*, pages 139–151, 1991.

[OW91] M. Ogiwara and O. Watanabe. On polynomial bounded truth table reducibility of NP sets to sparse sets. *SIAM J. Computing*, 20:471–483, 1991.

New Time Hierarchy Results for Deterministic TMs

- Extended Abstract -

Krzysztof Loryś *

Institute of Computer Science, University of Wrocław

ul. Przesmyckiego 20, 51-151 Wrocław, Poland

Abstract

We show a method of maintaining a distributed counter by 2-dimensional deterministic Turing machines with at least two tapes. Our method yields a tight time hierarchy for these machines solving an open problem posed by M. Fürer ([Fü1],[Fü2]). Moreover, we improve the best known time hierarchy theorem for one-tape off-line deterministic TMs; among others we show that this hierarchy is tight for functions bounded by polynomials.

1 Introduction

The problem of time hierarchies for Turing machines (TMs) is of fundamental significance in computational complexity theory. Despite many efforts made in this area, many basic questions, including the existence of tight hierarchy for multitape Turing machines, remain open for years. In this paper we consider this problem for deterministic TMs.

The table below presents the best known results for different types of DTMs with all tapes one-dimensional. More precisely, the table shows the conditions on functions T and T_1 which suffice to separate the class of languages recognized in time T_1 from the class of languages recognized in time T. We omit here the condition of time constructibility of function T_1, which is common for each case.

type of DTM	condition	ref.
single-tape	$T \log T \in o(T_1)$	[Ha]
off-line one-tape	$T \log T \in o(T_1)$	[Ha]
k-tape ($k \geq 2$)	$T \in o(T_1)$	[Fü]
multitape	$T \log T \in o(T_1)$	[HeSt]

By a single-tape machine we mean a machine with one read-write tape which initially contains an input word. An off-line machine have a separate read-only input tape and one read-write work tape. All tapes of k-tape and multitape machines, including an input tape, are read-write.

We say that a hierarchy is *tight* when it separates time T_1 from T for $T \in o(T_1)$. So the tight hierarchy is known only for DTMs with a fixed number (at least 2) of tapes.

*This work was supported by the Alexander-von-Humboldt-Stiftung while the author was visited the Lehrstuhl für Theoretische Informatik, Institut für Informatik, Universität Würzburg, 8700 Würzburg, Germany

The main obstacle in obtaining it for multitape DTMs is that we cannot construct a universal machine efficiently simulating DTMs of any number of tapes. The best simulation method of Hennie and Stearns [HeSt] needs $\log T$ steps per each of T simulated steps. This obstacle does not exist in the case of DTMs with fixed number of tapes. But then there is another problem: a universal machine should be able to count efficiently simulated steps, having no separate tape for a counter. The simple attempt to do this is to drag the counter according to the head moves. But this again costs $\log T$ step per each simulated step.

A significant improvement in this method was made by Paul [Pa]. He has noticed that it is thriftless to move the whole counter after each simulated step, because the ith digit of the counter is being changed only after each 2^i steps. He showed that dividing the counter into smaller subcounters of appropriate length and moving them only when it is necessary it is possible to reduce the cost of maintaining the counter by (at least) two-tape DTMs to $\log^* T$ per each simulated step.

But the first who has obtained the tight hierarchy was Martin Fürer, who has invented a very clever distributed counter, causing only a linear delay in simulation. Fürer's construction needs at least two tapes, one of them has to be one-dimensional, and does not work when TMs have only tapes of higher dimensions. For such machines one can use Paul's method (see [Fü1]) which gives time hierarchy with gap's $\log^* T$. In [Fü1] Fürer expressed his doubts whether in this case there exists the tight hierarchy, and in [Fü2] he posed an open question whether his technique of distributed counting is sufficient to obtain such a hierarchy. In this paper we give an affirmative answer to this question.

Theorem 1 Let $\mathrm{DTIME}_k^2(T)$ denote the class of languages recognized by k tape 2-dimensional DTMs working in time T. If $k \geq 2$ and T_1 is fully time-constructible then for any $T \in o(T_1)$

$$\mathrm{DTIME}_k^2(T) \subsetneq \mathrm{DTIME}_k^2(T_1).$$

In fact we show a bit stronger theorem. Namely, the tight hierarchy holds for any type of DTMs with $k \geq 2$ tapes, if at least one of them is 2-dimensional.

The second result of this paper concerns one-tape off-line DTMs. In this case (as well as in the case of single-tape and multitape DTMs) the gap $\log T$ in time hierarchy can be, by using padding method of Ruby and Fischer [RuFi], reduced to $(\log T)^e$ in range below 2^n, where e is any positive constant. But the tight hierarchy was not known for any range. Note that for such machines we cannot use the method which for single-tape machines yields the tight hierarchy below n^2 (see[He]). This method consists in showing for each well-behaved T_1 below n^2 a natural language recognizable by a single-tape DTM in time T_1 but not recognizable by any single-tape DTM working in time $T \in o(T_1)$. For off-line DTMs the only nonlinear lower bound for a natural language was obtained recently by Maass, Schnitger and Szemeredi [MaScSz] but it does not match the best known upper bound and therefore it cannot be applied directly to the time hierarchy problem.

We show that the condition $(\log T/\log n) \cdot T \in o(T_1)$ suffices to separate the corresponding classes of languages.

Theorem 2 If T_1 is fully time-constructible and $(\log T/\log n) \cdot T \in o(T_1)$ then one-tape off-line DTMs working in time T_1 recognize more languages than working in time T.

This theorem yields the tight hierarchy in the area of functions bounded by polynomials and tighten also the previously known time hierarchy for functions which grow a bit faster than polynomials, e.g. $n^{\log\log n}$.

We are still not able to tighten time hierarchy for single-tape DTMs. The distributed counter, we can construct, can be cheaply maintained by these machines only when they have space bounded by $T/((\log T)^2 \log\log T)$. Remind that the single-tape machines can not use more $T/\log T$ cells making T steps, unless the computation is nonaccepting (see [PaPrRe]). As a corollary we obtain a closure under complement of some time-space bounded classes.

Theorem 3 Let $\mathrm{DTISP}_1(T,S)$ denote the class of languages recognized by single-tape DTMs working in time T and space S. If $S \leq T/((\log T)^2 \log\log T)$ and T is fully time constructible in space S by a single-tape DTM then the class $\mathrm{DTISP}_1(T,S)$ is closed under complement.

The rest of paper is organized as follows. In the next section we recall the main properties of Fürer's construction and in two subsequent sections we discuss the key points in maintaining distributed counters by 2-dimensional and by one-tape off-line DTMs. We omit the standard diagonalization part of proofs of hierarchy theorems. We also omit the construction of a distributed counter for single-tape DTM, which is quite involved (it will be included in the final version of the paper).

2 Fürer's counter

In this section we outline shortly the construction of Fürer's counter (FC, for short) and recall its basic properties.

The counter has a form of full binary tree, in which each node contains a B-ary digit (for some fixed B). The number B is called *base* of the counter. Let for each node v, $h(v)$ denote the height of v (i.e. the distance from v to the leaves), and let $d(v)$ denote the digit stored in v. The *content* of the counter is the number $\sum_{v \in Tree} d(v) \cdot B^{h(v)}$.

If we want to increase the counter by one , we may add one to any of the leaves. If we have chosen a leaf v already containing $B-1$ then we must find the nearest v's ancestor u such that $d(u) < B-1$, add one to it and put zero to all nodes lying on the path from u to v.

The counter's nodes are stored in inorder on a tape of a TM. On a separate track of the tape there are stored information facilitating traversing the tree by TM, e.g. below each node can be stored letter L or R, depending on whether the node is a left or right son of its father (see figure below).

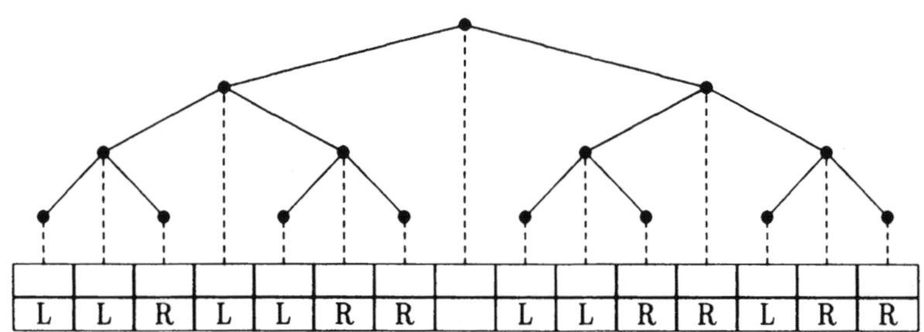

The length of the counter is the number of nodes in the tree. Thus only the numbers $2^k - 1$ (for $k \in N$) can be lengths of the counters. We say that the counter is *overfulled* if TM attempts to add one to the root which already contains $B-1$.

Lemma 4 Let B $\geq$ 3.

(a) Full counter (i.e. with all nodes containing $B-1$) of length n contains the number $O(n^{log_2 B})$.

(b) Let us suppose that we start with the empty counter (i.e. with the content= 0) of length n and after m increments the counter is overfulled. Then $m = \Omega(n^{log_2 B})$.

(c) m successive increments of the initially empty counter of length n can be performed even by single-tape DTM in $O(m)$ steps.

(d) The counter of length n can be constructed by a two-tape DTM in time $O(n)$.

(e) The content of the counter of length n can be computed by a two-tape DTM in time $O(n)$.

3 Distributed counter for 2-dimensional TMs

In this section we sketch the construction of a distributed counter for 2-dimensional DTMs with at least two tapes which needs in average O(1) cost per each simulated step. We start with an answer to the following question:

(A) Why Fürer's counters (FCs) cannot be directly used by 2-dimensional TMs?

Fürer's construction uses essentially the fact that TM needs at least $n/2$ steps to leave the area occupied by an FC of length n when it starts with the head placed over the middle cell of this area. Due to this fact the FC is being moved at most $O(T/n)$ times while simulating T steps. Moreover, moving such a counter needs $O(n)$ steps.

Quite a different situation we have in the case of two dimensions. If we want to ensure that a counter of similar kind as the Fürer's one will not have to be moved in next $n/2$ steps, it should occupy a square area of the side length equal to n. But construction of such counter takes $\Omega(n^2)$ steps.

(B) New Strategy.

TM uses two types of counters:
- binary counter,
- FCs.

There is only one global binary counter (GC), which initially contains T - the number of states to be simulated. From time to time TM decreases GC and halts when it becomes non-positive.

There are many FCs (in fact infinitely many) of different size. At the beginning each cell of coordinates (x,y) s.t. $x + y \equiv 0 \bmod 2$ contains FC of size 1; initially these counters contain 0. The base of FCs is equal to 8.

After each simulated step TM adds 1 to any of the counters (i.e. to any of FCs), so it has to go at most one cell in any direction to find a counter.

If necessary, TM during its work creates larger FCs. Each such FC will occupy a segment of a row of the tape; e.g. one row can look as below:

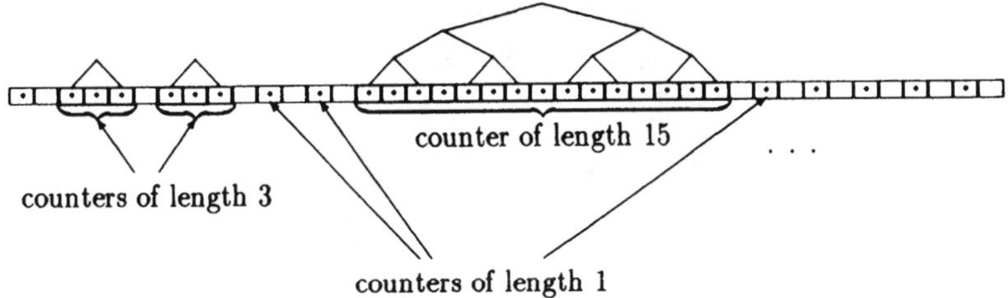

The leaves of any FC occupy only these cells, which initially contained FCs of size 1.

Of course, TM has to control the growth of FCs and from time to time it has to subtract the contents of FCs from GC. To this end it divides the whole tape into squares of side length equal to log T. At the beginning there is only one square marked (it contains the cell seen by the head), and TM marks a new square whenever the head enters the area lying outside the marked squares.

The marking of one square costs O(log T) and TM is not able to mark more than 4 new squares while simulating less than log T steps. So the total cost of marking squares is linearly bounded by the number of simulated steps.

The Global Counter (GC) is kept by TM close to the active square (i.e. the square inside which there is the head). Note that GC has the length $\approx$ log T, so it cannot be moved too often. It will be moved when TM's head crosses a boundary of the area consisting of the active square and eight squares adjacent to it. This ensures that the shift will be done no more often than every log T steps. Then TM will also clear all FCs in this area simultaneously computing their total contents (i.e. the number of steps simulated in this area), and then subtract this number from GC. Note that during the simulation at most nine squares can contain nonempty FCs at the same time. Since the total contents of all FCs in one square is bounded by $c{\cdot}(\log T)^4$, the number of simulated steps is O(T).

(C) More Details.

Now we describe how TM adds 1 to FCs after each simulated step.
TM remembers in its states whether the head is over a cell of coordinates (x,y) fulfilling the condition $(x+y) \equiv 0 \bmod 2$. If so (this means that this cell corresponds to a leaf of certain FC), then TM adds 1 to this leaf. If not, then TM adds 1 to any of adjacent cells. If this leaf has just contained 7, then TM stores there 0 and adds 1 to the father of the leaf.
Let us suppose that it was a leaf of an FC of height i (so the length of the FC is equal to $2^i - 1$) and let us denote this FC by R. If it turns out that R is overfulled then TM looks 2^i cells lying on the right- and 2^i cells lying on the left-hand side of R.

(1) If in this area begins an FC of height $\geq i+1$, then TM finds the first node of height $i+1$ (it lies no further than $2^{i+1}-1$ cells from R) and adds 1 to it. Of course, if it just contains 7 then TM looks for its father, etc...

(2) If in this area lie only small FCs (i.e. of height $\leq i$, then TM tries to enlarge R. To this end it chooses a segment of $2^{i+1}-1$ cells in this area and computes the total contents X of all FCs lying in this segment. (Note that the last FC can lie partly outside the segment). Then TM destroys all these counters, builds an FC of height $i+1$ and puts X into it. This can be easily done if X is computed in 8-ary system. It is important that X is $\leq 4 \cdot 8^{i+1}$, so initially the root of the new FC will contain a digit no greater than 4. Therefore this counter can be overfulled only after next $4 \cdot 8^{i+1}$ steps.

(3) If R cannot be enlarged (because it would enter the area beyond the active square), then TM computes the total contents of all counters lying in the active square (clearing them) and subtracts this value from GC.

We will not present here a detailed cost analysis, but rather point out only crucial elements of it. Let K be a number of simulated steps. Then

(i) K is $O(T)$,

(ii) the cost of carry propagation is in average bounded by a constant,

(iii) the point (2) above is performed for FCs of height i at most $K/(4 \cdot 8^{i+1})$ times and it costs $c \cdot 2^{i+1}$, where c is a constant,

(iv) the point (3) above is performed at most $K/(c \cdot (\log T)^3)$ times and it costs $O((\log T)^2)$.

At the end we describe shortly the way in which TM clears FCs and computes their total contents when the head leaves the area of nine squares with GC placed in the middle square. The cells which TM has to visit are of two kinds:

(i) cells visited by the simulated machine,

(ii) cells visited by the simulating machine only when it has been using FCs.

It is easy to see that the number of cells of both types is bounded by the number of steps simulated in this area since its last clear out. Moreover these cells create one connected area. It is not difficult to construct an algorithm which visits all cells in such an area in time proportional to its size and leaves the head in the "starting" cell. Thus TM can perform this algorithm and switch into a procedure computing a contents of an FC whenever it come across a cell in which an FC begins.

4 One-tape off-line DTMs

In this section we describe how one-tape off-line DTM can maintain efficiently a distributed Fürer's counter (FC). In fact the the method works only when TM has inputs of special form, but this is sufficient to obtain the mentioned time hierarchy results.

The general outline is similar as in the case of two-tape machines. TM places an FC such that the head reads its middle cell and then simulates another machine until the counter becomes overfulled or the head goes outside the part of the tape occupied by the counter. In both cases TM do the following:

(1) computes the content of FC and subtracts this value from GC (the global binary counter),

(2) moves appropriately FC,

(3) moves appropriately GC.

But now (1)-(3) cannot be performed in such easy way as in the case of two-tape machines. At the beginning TM stores the input head position on the work tape and places the input head at left end of input tape, because then it uses the input tape as a unary counter.

Ad.1. TM goes along FC looking for a nodes with non-zero contents. When it finds such a node x that:

- if x is a leaf of FC (TM knows it, because every second cell corresponds to a leaf of FC), then it put 0 into it and moves input head an appropriate number of cells to the right,
- if x is not a leaf then TM computes the heigth of x (or even better its weight). This can be done by traversing the path from one of the leaves, which are descendants of x to the node x (note that such two leaves are adjacent to x on the tape). This can be done in the same way as while moving carries and it costs O(weight of x). But this means that in average it costs O(1) per each simulated step. Having the weight written in binary on the work tape TM can easily move the input head an appropriate number of cells to the right.

Whenever during this procedure input head reaches the right end of the input word, then TM moves the work head over GC and subtracts n from it.

Ad.2. TM uses FC of length n (this is as well the length of input word). Since it is possible that such a counter would be moved every $n/2$ steps and since such moving is expensive we assume that input words have a special form. Namely, they can be treated as words written down on two tracks of the tape and on the second track there is written the empty FC. For simplicity we assume that input words have length $2^k - 1$ ($k \in N$), so it corresponds to the lengths of FCs. Of course, at the beginning TM has to check whether given input word has a proper form; if not then TM rejects. This can be performed in $O(n)$ steps. Then, whenever TM has to move FC, it simply copies it from the second track.

Ad.3. The length of GC is bounded by $\log T$. The area occupied by GC is divided into parts of length no greater than $\log n$, so the contents of each part can be unary represented by a position of input head. Thus TM moves each part separately using input tape. Since every time TM moves GC no more than O(n) cells, the cost of this is bounded by $(\log T/ \log n) \cdot n$. Finally let us note yet that this procedure is performed at most O(T/n) times.

Let us note that using FC of length equal to length of input word we cannot obtain the hierarchy below n^2, because if the head does not leave the area of FC then TM can simulate even $n^{log_2 B}$ steps before FC becomes overfulled. To avoid this drawback we can assume that input words are partitioned into parts of length $\sqrt{n}$, (each with an empty FC) and fix the base to be equal to 4.

An analogous method yields the tight hierarchy for single-tape DTMs with auxiliary pushdown. In this case the points (1)-(3) can be even easier realized and we obtain the tight hierarchy with no restriction on the growth of functions.

Theorem 5 Let PD-T-DTIME(T) denote the class of languages recognized by single-tape DTMs with auxiliary pushdown working in time T. If T_1 is fully-time constructible and $T \in o(T_1)$ then

$$\text{PD-T-DTIME}(T) \subsetneq \text{PD-T-DTIME}(T_1).$$

Acknowledgements

This work was done while I have been visiting the Lehrstuhl für Theoretische Informatik, Universität Würzburg, where I have enjoyed working in a stimulating and hospitable atmosphere. I am much indebted to all my colleagues who created it, and especially to prof. Klaus Wagner.

I have benefited from comments of Ronald Book, Gerhard Buntrock, Hans-Joerg Burtschick, Ulrich Hertrampf, Mirosław Kowaluk and Heribert Vollmer. Thanks to all of them.

References

[Fü1] M. Fürer. The tight deterministic time hierarchy. 14th STOC (1982), 8–16.

[Fü2] M. Fürer. Data structures for distributed counting. JCSS 28 (1984), 231–243.

[Ha] J. Hartmanis. Computational complexity of one-tape Turing machine computations. JACM 15 (1968), 325–339.

[He] F.C. Hennie. One-tape off-line Turing machine computations. Inform. and Contr. 8 (1965), 553–578.

[HeSt] F.C. Hennie and R.E. Stearns. Two-tape simulation of multitape Turing machines JACM 13 (1966), 533–546.

[MaScSz] W. Maass, G. Schnitger and E. Szemeredi. Two tapes are better than one for off-line Turing machines. 19th STOC (1987), 94–100.

[Pa] W.J. Paul. On time hierarchies. 9th STOC (1977), 218–222.

[PaPrRe] W.J. Paul, E.J. Preuß and R. Reischuk. On alternation. Acta Informatica 14 (1980), 243-255.

[RuFi] S.S. Ruby and P.C. Fischer. Translational methods and computational complexity. 6th Ann. Symp. on Circiut Theory and Log. Design (1965), 173–178.

DISTRIBUTED SYSTEMS

Unconditional Byzantine Agreement for any Number of Faulty Processors

— Extended Abstract —

Birgit Pfitzmann
Institut für Informatik
Universität Hildesheim
Samelsonplatz 1
W-3200 Hildesheim, Germany
pfitzb@infhil.uucp

Michael Waidner
Institut für Rechnerentwurf
und Fehlertoleranz
Universität Karlsruhe
Postfach 6980
W-7500 Karlsruhe, Germany
waidner@ira.uka.de

Abstract. We present the first Byzantine agreement protocol which tolerates *any* number of maliciously faulty processors *without* relying on computational assumptions (such as the unforgeability of digital signatures).

Our protocol needs reliable broadcast and secret channels in a precomputation phase. For a security parameter σ, it achieves Byzantine agreement with an error probability of at most $2^{-\sigma}$, whereas all computations are polynomial in σ and the number of processors.

The protocol is based on an unconditionally secure authentication mechanism, called pseudosignatures. Pseudosignatures are a generalization of a mechanism by CHAUM and ROIJAKKERS and might be useful in other protocols, too.

1 Introduction

Byzantine agreement protocols (BAPs) are an important primitive for distributed computations. They are intended to achieve reliable broadcast where this is not available physically and where some processors may be faulty. Correct agreement on the value $v \in D$ (where $|D| \geq 2$) of a transmitter means [PeSL_80]:

Consistency All good processors agree on the same value $v \in D$.

Correctness If the transmitter is good, v is the value it meant to send.

Let n be the number of processors and t an upper bound on the number of faulty processors. Faulty processors are assumed to be malicious. Except for preventing good processors from communicating, they can do whatever they like. In particular, it is *not* presupposed that faulty processors are limited to polynomial-time computations.

On the presupposition of an **authentication mechanism,** any $t < n$ can be tolerated [PeSL_80, DoSt_83] (**authenticated BAP**): Such a mechanism enables each good processor P to *authenticate* its messages so that any recipient R can locally verify whether a received message really comes from P. This must be possible even if R receives the message indirectly, via an arbitrary number of intermediate processors. Of course, maliciously faulty processors will try to forge such authentications.

Till now, there has been no *completely* verified authentication mechanism, thus no *completely* verified authenticated BAP: The most secure substitute were **digital signature** schemes [DiHe_76, GoMR_88], and all known schemes are based on *unproven* computational assumptions (e.g., "factoring is hard" [GoMR_88] or "one-way functions exist" [Romp_90], with respect to polynomially bounded faulty processors). But even if one of these computational assumptions could be proved (which would imply $\mathcal{P} \neq \mathcal{NP}$), digital signatures would still require that faulty processors are *polynomially* bounded.

A BAP which tolerates computationally *unbounded* faulty processors is called **unconditional.**

There was a folklore believe that unconditional BA can be achieved iff $t < n/3$. This was caused by [PeSL_80], which proved this impossibility result in a deterministic (i.e., errorfree) sense.

The first unconditional BAPs for numbers $t \geq n/3$ were presented in [BaPW_91, Waid_91: Ch. 2]: They tolerate any $t < n/2$ with an exponentially small error probability, i.e., $\leq 2^{-\sigma}$, in $O(\sigma)$ rounds [BaPW_91], or even in an expected constant number of rounds [Waid_91: Ch. 2.2.3]. All computations are polynomial in σ and n. It is presupposed that secret channels and reliable broadcast are available in a precomputation phase.

In the same model, we present the first unconditional BAP which tolerates *any* number $t < n$ (in $O(t)$ rounds).

Our BAP has an exponentially small error probability, too. One should not confuse this error probability with the computational assumptions of digital signatures: The former is a property which is rigorously proved, whereas the latter might be completely wrong.

Because of secret channels, faulty processors receive no information about the messages good processors exchange. For $t \geq n/3$ and an arbitrarily small error probability, secret channels are necessary. This can be derived from [GrYa_89].

Reliable broadcast in a precomputation phase is a presupposition for authenticated BAPs implemented with digital signatures, too: For such BAPs, all good processors must agree on the test keys of all processors. Thus, these test keys must be broadcast reliably before the agreement protocol can start. (Technically, the operators of the processors can perform the precomputation phase in a highly secure environment, e.g., with some other processors which are connected via a physically reliable broadcast network.)

The basis of our BAP is an unconditionally secure authentication mechanism, called **pseudosignatures** (Section 2), which is a generalization of a mechanism from [ChRo_90]. In Section 3 we obtain our BAP by fitting pseudosignatures into an authenticated BAP (that of [DoSt_83]).

Our constructions are described and proved in full detail in [Waid_91: Ch. 3].

Notations and assumptions: Let $P = \{P_1, \ldots, P_n\}$ be the set of processors, and $\sigma \in \mathbb{N}$ a security parameter so that an error probability of $2^{-\sigma}$ is acceptable. All our protocols are polynomial in σ and n.

If a statement is true with probability exponentially close to 1, we say that it holds **almost certainly**.

As usual with BAPs, we assume that each pair of good processors can communicate securely. We also assume that the network is synchronous. Additionally, we assume that in a *precomputation phase*, i.e., *before* the value which is to be distributed by a BAP is chosen,

- each processor can reliably broadcast some information, and
- each pair of processors has a secret and secure channel.

With these secret and secure channels, the assumption that each pair of good processors can communicate securely during the agreement protocol can be reduced to the assumption that faulty processors cannot prevent good processors from communicating [GiMS_74].

2 Pseudosignatures

Informally, pseudosignatures satisfy the following three properties:

Correctness If a good processor S pseudosigns a message, each good processor R would *accept* this message from S, almost certainly.

Unforgeability If a good processor R accepts a message from a good processor S, then S has pseudosigned this message, almost certainly.

Transferability For a given parameter $\lambda \in \mathbb{N}$, a message pseudosigned by a good S can be transferred λ times, e.g., via

$$S = P_{i_0} \rightarrow P_{i_1} \rightarrow \ldots \rightarrow P_{i_\lambda} ,$$

so that for each $j < \lambda$, if P_{i_j} accepts the message, then it knows that $P_{i_{j+1}}$ will accept the message, too, almost certainly.

If P_{i_λ} forwards the message, it is not guaranteed that the (λ+1)-st recipient accepts the message, too. Thus λ must be chosen carefully, depending on the specific protocol (e.g., if the protocol terminates deterministically in at most R rounds, $\lambda = R$ suffices).

For comparison, conventional digital signatures should satisfy these requirements, too, and should even be transferable an arbitrary number of times. But their unforgeability relies on unproven computational assumptions and they cannot withstand computationally unbounded adversaries (cf. Sect. 1).

In the following, we present unconditionally secure pseudosignatures. Our scheme is a generalization of the scheme from [ChRo_90], which dealt with the special case $\lambda = 2$.[1]

2.1 General structure

Conventional digital signatures cannot be unconditionally secure [DiHe_76]: The test keys must be public, i.e., a forger knows how a good processor P verifies the authenticity of a signed message. Thus, signatures can always be forged by brute force. (If testing a signature is in $\mathcal{P}$, then forging signatures is in $\mathcal{NP}$.)

For pseudosignatures, each processor P_i therefore needs a *different* test key, which is known only to P_i and to the pseudosigner, S [ChRo_90]. Thus, a *basic* pseudosignature ψ consists of n parts, called minisignatures; one for the test-key T_i of each P_i.

Forging a minisignature must be almost certainly impossible. This is achieved by using **authentication codes** [GiMS_74]: We use an **authentication key** as the test key T_i, and the encoding of the message v with T_i, $T_i(v)$, as the minisignature on v. There are efficiently computable authentication codes with a probability of successful forgery[2] of at most $2^{-\tau}$, where the key length is logarithmic in the length of v and linear in τ, and the length of each minisignature is τ+1 bit [WeCa_81]. For reasons of security, an authentication key can be used at most once.

Obviously, if S is faulty and knows which test keys belong to which processors, it can construct a basic pseudosignature which is acceptable for P_i, but not for P_j.

---

1 In Section 5 of [ChRo_90], a scheme is sketched which should work for the general case $\lambda > 2$. Implicitly, this scheme assumes that the original sender of the message is good; thus this extension is not applicable here. (For readers familiar with [ChRo_90: Sect. 5]: The hash functions used there are useful only if they are *unknown* to the faulty processors.)

2 The probability of successful forgery is the probability that an adversary finds the correct value $T_i(v)$, provided it knows the value $T_i(v')$ for at most one $v' \neq v$.

Therefore, the relation between test keys and processors is hidden: During an **initialization phase**, each processor chooses a test key randomly and sends it to S [ChRo_90]. This is done via a network which offers unconditional sender untraceability and perfect privacy. Thus, the pseudosigner only receives the *set* $\{T_1, \ldots, T_n\}$, and all other processors do not receive anything (Section 2.2).

Of course, this initialization does not prevent a faulty S from disturbing some minisignatures at random, i.e., from generating a basic pseudosignature which is acceptable for some processors, but not for all.

To prevent this attack, for each *complete* pseudosignature, m sets $\{T_{h,1}, \ldots, T_{h,n}\}$, $h = 1, \ldots, m$, are initialized independently, and the pseudosignature ψ on the message v consists of the m basic pseudosignatures $\psi_h := \{T_{h,1}(v), \ldots, T_{h,n}(v)\}$ [ChRo_90].

Now assume that a good processor P_i receives a pseudosignature $\psi = (\psi_1, \ldots, \psi_m)$ on a message v. Let a_i denote the number of different indices $h \in \{1, \ldots, m\}$ so that ψ_h contains a minisignature which passes the test with $T_{h,i}$.

If P_i receives (ψ, v) from S directly, $a_i = m$ must hold. Otherwise, S is faulty, and there is no need to accept anything from a faulty processor. Therefore, if $a_i = m$, P_i **1-accepts** (ψ, v).

If P_i receives (ψ, v) from another processor $P_j \neq S$ which has 1-accepted (ψ, v), we cannot necessarily expect $a_i = m$. If, e.g., P_i and P_j are the only good processors and S disturbs exactly one minisignature of a good processor, then either P_i or P_j 1-accept (ψ, v), but not both. Therefore, the acceptance rule is weakened for the second recipient. In [ChRo_90], P_i **2-accepts** (ψ, v) if $a_i \geq m/2$. Similarly, one can require $a_i \geq 1$: Then the probability that P_j 1-accepts (ψ, v) from S, and P_i does not even 2-accept (ψ, v), is at most 2^{-m}.

The scheme described so far is equivalent to the one proposed in [ChRo_90] (except that they only considered binary messages).

We extend this scheme from $\lambda = 2$ to arbitrary (but fixed) values of λ in a natural way: For a parameter $\Delta \in \mathbb{N}$ (which determines the error probability of transferability), we choose $m := (\lambda-1)\Delta+1$. If $a_i \geq m-(h-1)\Delta$, then P_i ***h*-accepts** (ψ, v), $h = 1, \ldots, \lambda$. For $h = 1$, this means $a_i = m$, for $h = \lambda$ this means $a_i \geq 1$.

In the following, we discuss the extended scheme in greater detail. We add some parts which were not completely described or proved in [ChRo_90], in particular, a provably correct initialization phase (Sect. 2.2). For all the three properties of pseudosignatures we prove upper bounds on the error probabilities, even against active attacks. This is especially important for transferability, where active attacks on recipients of pseudosignatures must be considered. Details omitted in this extended abstract can be found in [Waid_91: Ch. 3].

2.2 Initialization phase

During the initialization phase, the intended pseudosigner S is to receive the m sets $\{T_{h,1}, \ldots, T_{h,n}\}$, $h = 1, \ldots, m$. This is achieved by performing a **basic initialization protocol** m times in parallel. A similar initialization protocol is described in [ChRo_90], but without all necessary details, and without any proof of correctness.

The initialization phase needs secure and secret channels between each pair of processors and a reliable broadcast network. We say that a processor **publishes** a message M if M is sent via this reliable broadcast network.

Basis of the initialization protocol is the **DC-protocol** [Chau_88]:

Let $u \in \mathbb{N}$, and $F = GF(2^u)$, the finite field with 2^u elements. For each execution of the DC-protocol, each pair of processors $\{P_i, P_j\}$ needs a common secret DC-key, $K_{i,j} \in F$, and each processor P_i has a local input $x_i \in F$. In the following, x_i will be the encoding of an authentication key, thus u must be chosen sufficiently large (at least τ). Each processor P_i computes and publishes its local sum $O_i := x_i + K_{i,1} + \ldots + K_{i,n}$ and computes the global sum $\Sigma := O_1 + \ldots + O_n$ from all published local sums.

Obviously, Σ is equal to $x_1 + \ldots + x_n$, and one can prove that the ensemble of the local sums does not contain *any* information about the x_i's except Σ [Chau_88]. If S is not faulty and does *not* publish its local sum, even Σ is hidden from the faulty processors.

Thus, the DC-protocol enables S to receive the sum $\Sigma = x_1 + \ldots + x_n$, while no processor except S obtains any information about Σ, and not even S obtains any information about which x_i was chosen by which processor.

Finally, S should not receive the *sum*, but the *set* of all x_i's. To achieve this, we use a multiple-access protocol from [BoBo_90]: In n consecutive iterations of the DC-protocol, S is to receive the first n odd power sums of the x_i's, i.e., $x_1 + \ldots + x_n$, $x_1^3 + \ldots + x_n^3$, ..., $x_1^{2n-1} + \ldots + x_n^{2n-1}$. If all processors behave well, then from these n power sums, S can compute the set of all x_i's which were chosen by an odd number of processors, almost certainly in polynomial time [Burt_71, Rabi_80: Th. 6].

If the faulty processors did not disturb the initialization, the basic initialization protocol would be trivial now: Each processor P_i would choose x_i as an encoding of a randomly chosen authentication key. Almost certainly, all good processors choose different keys. Thus S would obtain a set which contains at least the authentication keys of all good processors, almost certainly. However, faulty processors can easily disturb the initialization, e.g., by publishing wrong O_i's. Then, S receives garbage.

To handle this denial-of-service attack, S must be able to *detect* such disturbances, and it must be possible to *localize* disturbers:

To be able to *detect* disturbances, S is to receive the first $2n$ odd power sums, instead of the first n only. S uses the first n ones as described above and tests whether the second n ones are compatible with the set of the x_i's just computed. It can be seen that if the received sums pass this test, then almost certainly no disturbance has occured.

Disturbers are *localized* as described in [Chau_88]: After a detected disturbance, all data (i.e., messages published, DC-keys, local inputs x_i) are verified publicly. (N.B.: We presuppose reliable broadcast during this precomputation phase.) Each processor which behaves obviously badly is *eliminated*. If two processors P_i, P_j do not agree on their common DC-key $K_{i,j}$, it cannot be decided which of them was faulty. Thus only this disputed DC-key is *eliminated* once and for all (i.e., P_i and P_j never use a common DC-key again). If there are two groups of processors with no keys left between them, we define that these two groups perform the DC-protocol *independently*. After less than $n(n-1)/2$ detected disturbances, there are no keys left between good and faulty processors. Thus the good processors perform the DC-protocol by themselves and can never be disturbed again.

To initialize a complete pseudosignature, m basic initializations are performed. The result of this initialization is described by the following Theorem 1.

Theorem 1: (Correctness of pseudosignatures)

After the initialization phase, the following holds with probability $\geq 1 - mn^4\, 2^{-\tau+1}$: If the pseudosigner S is good and forms the pseudosignature ψ on message v, then each good processor will 1-accept the pair (ψ, v). ♦

In the worst case, the initialization phase needs $2n^2$ rounds and messages of polynomial length. If w pseudosignatures are initialized sequentially, and if eliminations of DC-keys or of faulty processors are valid for all initializations, all w initializations together need at most $O(w+n^2)$ rounds.

2.3 Pseudosignatures with λ levels of acceptance

Pseudosignatures with λ levels of acceptance have already been defined in Section 2.1. It only remains to choose a value for Δ: To achieve an error probability for the transferability of at most $2^{-\kappa}$, we choose $\Delta := \lceil 2\,\lambda\kappa \ln 2 \rceil$.

For this parameter, we prove the following Theorem 2, which essentially says that pseudosignatures are unconditionally unforgeable and transferable. Theorem 2 considers passive attacks only (i.e., there is no interaction between good and faulty processors). Active attacks are considered in the second half of this section in general, and in Section 3 for a specific BAP.

Our definition of Δ implies $m \le 2\lambda^2\kappa$, thus a pseudosignature ψ is of length $\le 2n\lambda^2\kappa(\tau+1)$ bit.

Theorem 2: (Unforgeability and transferability in spite of passive attacks)

a) Assume S is good, and consider a good processor P and an arbitrary pair (ψ, v). Then P λ-accepts ψ as a pseudosignature on v with probability $\le mn\ 2^{-\tau}$. This is true even if a correct pseudosignature (ψ', v') on a message $v' \ne v$ is known.

b) Let P and Q be two good processors, and $k < \lambda$. Then for each pair (ψ, v), the probability that P k-accepts, while Q not even $(k+1)$-accepts, is at most $2^{-\kappa}$. ♦

Sketch of proof

a) Follows from the unforgeability of authentication codes (Sect. 2.1).

b) We consider the worst case only: P and Q are the only good processors. Thus S is faulty, and for each of the m parts of the pseudosignature, the faulty processors know which two authentication keys belong to P and Q. But they do not know which key was chosen by P and which by Q. For each of the m parts, the faulty processors can decide to change none, one, or two of the two correct minisignatures. For reasons of symmetry, we can describe this decision by just two parameters:

g denotes the number of parts where both minisignatures are changed.

h denotes the number of parts where exactly one minisignature is changed.

Let $x := (k-1)\Delta$; this is the maximum number of minisignatures for P which may be changed so that P still k-accepts ψ. Let B_h denote the binomial distribution with parameter h and probability $1/2$, i.e.,

$$B_h(k) = \sum_{z \le k} \binom{h}{z} 2^{-h}.$$

With probability $\binom{h}{y} 2^{-h}$, within the h parts where one minisignature was changed, exactly y changed minisignatures belong to P. The faulty processors are successful if they have changed at most x minisignatures for P and more than $x+\Delta$ minisignatures for Q, altogether, i.e., $y \le \min\{x-g, g+h-x-\Delta-1\}$.

Therefore, the faulty processors are successful with probability

$$\mathrm{P}(g,h) := B_h(\min\{x-g, g+h-x-\Delta-1\}) \le B_h\left(\frac{(x-g)+(g+h-x-\Delta-1)}{2}\right) = B_h\left(\frac{h}{2} - \frac{\Delta+1}{2}\right). \qquad (*)$$

For each $\gamma \ge 0$, it follows from [ErSp_74: (3.8)] that

$$B_h\left(\frac{h}{2} - \gamma - \frac{1}{2}\right) \le \sum_{z < h/2-\gamma} \binom{h}{z} 2^{-h} < \exp\left(-\frac{2\gamma^2}{h}\right).$$

We apply this to (*). We want to show that $\mathrm{P}(g,h) \le 2^{-\kappa}$.

$$\mathrm{P}(g,h) \le B_h\left(\frac{h}{2} - \frac{\Delta}{2} - \frac{1}{2}\right) < \exp\left(-\frac{\Delta^2}{2h}\right).$$

Thus, $P(g,h) \le 2^{-\kappa} \Leftarrow \Delta^2 \ge 2h\,\kappa \ln 2$. With $m \ge h$, we obtain

$$P(g,h) \le 2^{-\kappa} \Leftarrow \Delta^2 \ge 2m\,\kappa \ln 2 = 2\,((\lambda-1)\Delta + 1)\,\kappa \ln 2,$$

and one can easily verify that the last inequality is satisfied if $\Delta \ge 2\,\lambda\kappa \ln 2$. ❐

Theorem 2 deals with passive, i.e., non-interactive, attacks only. In general, a protocol enables interactive attacks: After an attack, the faulty processors receive some information about how successful their attack was. Depending on this information, they can repeat their attack.

As an example, consider the following attack on a recipient of a pseudosignature: Let ψ be a correct pseudosignature, and let P be a good processor. For each interaction, the faulty processors change exactly one minisignature in ψ and send the result to P, which answers with its level of acceptance. Processor P 1-accepts iff none of its minisignatures was changed. Therefore, after a certain number of interactions, the faulty processors know exactly which minisignatures belong to P. Hence, they can change ψ so that P will 1-accept the changed ψ, whereas all other good processors will not even λ-accept it.

In order to limit the success probability of an active attack, the lenght or kind of the attack must be restricted. We do this by assuming that pseudosignatures are used in a *deterministically polynomial* protocol (polynomial in n and a security parameter ω). Thus, the faulty processors can repeat their attack a polynomial number of times, u. We also assume that the security parameters τ, κ can be chosen arbitrarily. ([Waid_91: Sect. 3.1.6] contains a more formal description of this.) In this case, we can estimate the probability of success of the faulty processors by brute force:

First, consider unforgeability: In up to u iterations, the faulty processors select a pair (ψ, v), send it to P, and receive either "λ-accepted" or "rejected". They are successful if P λ-accepts at least one of these pairs. Applying Theorem 2a to all the u pairs yields a success probability of at most $umn\,2^{-\tau}$. If we choose $\tau := \lceil \omega + \log_2(umn) \rceil$, this is at most $2^{-\omega}$.

Secondly, consider transferability: Here, the faulty processors send at most u pairs to P and at most u to Q. Each time, they receive one out of at most $\lambda+1$ different reactions ("k-accepted" for $k = 1, \ldots, \lambda$, or "rejected"). Without loss of generality, their strategy is deterministic[3] and can be described by a *strategy tree*: Each node contains a pair (ψ, v) and the processor $X \in \{P, Q\}$ which receives the pair. Each child of a node belongs to exactly one of the $\lambda+1$ possible reactions of X. The depth of the tree is at most $2u-1$, and therefore it contains at most $\lambda^{-1}(\lambda+1)^{2u+1}$ nodes, i.e., pseudosignatures. Applying Theorem 2b to each of these pseudosignatures yields an error probability of at most $\lambda^{-1}(\lambda+1)^{2u+1}2^{-\kappa}$. (Note that whether a pseudosignature is transferable from P to Q is statically fixed in each event; the whole information from the active attack lies in the decisions whether this node is reached at all in an actual protocol run.) Of course, the strategy tree contains an exponential number of nodes, but this influences the security parameter κ only logarithmically: If we choose $\kappa := \lceil \omega + (2u+1)\log_2(\lambda+1) - \log_2(\lambda) \rceil$, the success probability is at most $2^{-\omega}$.

In Section 3, we consider a specific BAP. In order to obtain more precise upper bounds on the error probabilities, we define the strategy tree specifically for this protocol and count the nodes more precisely.

---

3 If they use an indeterministic strategy, then for each indeterministic decision there is one with maximum probability of success. If they always choose this one, their strategy remains at least equally successful.

3 Byzantine Agreement

In the following, we modify a well-known authenticated BAP [DoSt_83: Th. 3] so that it can be implemented with pseudosignatures instead of conventional digital signatures (Sect. 3.1), and we prove that it achieves BA almost certainly (Sect. 3.2). At the end, we sketch how we can use a fixed number of initialized pseudosignatures to perform an arbitrary number of BAs (Sect. 3.3).

3.1 Byzantine agreement protocol with pseudosignatures

The following BAP is based on the authenticated BAP from [DoSt_83: Th. 3]; for each "signature", we have determined the level of acceptance, and we have kept the strategy tree described in Section 2.3 as small as possible.

Let σ be the security parameter of the protocol, i.e., we accept an error probability of at most $2^{-\sigma}$, and let $t < n$ be an upper bound on the number of faulty processors.

The protocol uses pseudosignatures with $\lambda = t+1$ levels of acceptance with parameters $\kappa := \lceil \sigma + 18 \log_2(n) + \log_2(72) \rceil$ and $\tau := \lceil \sigma + \log_2(\kappa) + 7 \log_2(n) + \log_2(24) \rceil$. The length of the messages which must be pseudosigned is equal to the length of the transmitter's message.

Each execution of the BAP needs its own precomputation phase, where $2n-1$ pseudosignatures are initialized[4]. The transmitter T acts as the pseudosigner for one of them, each other processor for two of them. We call the two pseudosignatures of $P \neq T$ its **A-signature** and **B-signature**, and the pseudosignature of T its **A-signature**. Pseudosignatures will only be used in triples (i, α, ψ), where i denotes the index of the supposed pseudosigner P_i, $\alpha \in \{A, B\}$ the "type", and ψ the pseudosignature itself. A processor P **k-accepts** a triple (i, α, ψ) as a pseudosignature on v if it k-accepts (ψ, v) with respect to the initialization denoted by (i, α).

Protocol messages always contain a value v and a set of up to $t+1$ triples (i, α, ψ). For each level of acceptance $k \in \{1, \ldots, t+1\}$, we call a protocol message M **k-consistent**[5] for P_i iff M is of the form $(v, \{(i_1, \alpha_1, \psi_1), \ldots, (i_k, \alpha_k, \psi_k)\})$ and

- according to the parameters i_j, M contains k pseudosignatures of k different processors,
- according to the parameters i_j and α_j, M contains the A-signature of the transmitter,
- P_i k-accepts all the triples (i_j, α_j, ψ_j) as pseudosignatures on v.

P_i **relays** a protocol message M containing v in the following way:

- If M is the first message P_i relays, P_i pseudosigns v using its A-signature, otherwise using its B-signature. P_i adds the resulting triple (i, α, ψ) to M.
- P_i forwards M (including its own triple (i, α, ψ) now) to each processor of whom no pseudosignature is contained in M yet.

During the protocol, each processor P_i has to relay at most two protocol messages, which are k-consistent for him.

The set ACC_i contains all the values which P_i has accepted from the transmitter. At the end, if ACC_i contains exactly one value v, P_i accepts this value as its **final value**. Otherwise, P_i

---

[4] In Section 3.3, we sketch how the BAP can be executed u times with a number of initialized pseudosignatures independent of u.

[5] In [DoSt_83], k-consistent protocol messages are defined in a slightly different way: There, each processor signs not just v, but the whole protocol message. Our definition results in smaller messages to be signed and does not influence the fault tolerance of the protocol.

knows that the transmitter was faulty. The set OLD_i ensures that a good processor P_i reacts to at most two protocol messages from each processor P_x. This keeps the strategy tree relatively small (see Sect. 3.2).

A good transmitter T starts with a value v, which is to be distributed.

Byzantine Agreement Protocol:

[0] Transmitter T relays v, i.e., T pseudosigns v with its A-signature and forwards it to all the other processors. Each $P_i \neq T$ initializes two empty sets OLD_i and ACC_i.

For $k = 1, \ldots, t{+}1$ and each processor $P_i \neq T$:

[k] a. P_i forms the set $N_{i,k}$ of *new* protocol messages:
For each protocol message M that P_i received in round $[k-1]$: Let P_x be the sender of M. If M fulfills the first and second condition of k-consistency and contains a triple (x, α, ψ) and $(x, \alpha) \notin OLD_i$, then P_i adds M to $N_{i,k}$ and (x, α) to OLD_i. Otherwise P_i ignores M. (In this case, P_i knows that P_x is faulty, and there is no need to consider messages from faulty processors.)

b. P_i forms the set $V_{i,k}$ of all k-consistent messages from $N_{i,k}$. Let $V_{i,k}$ be lexicographically ordered. P_i adds all the values which are included in a message in $V_{i,k}$ to ACC_i.

c. If $k \leq t$ and $V_{i,k} \neq \emptyset$, and if P_i has not relayed any message before, then P_i *relays* the first two messages in $V_{i,k}$ which contain different values, or just the first one, if all messages contain the same value.

d. If $k \leq t$ and $V_{i,k} \neq \emptyset$, and if P_i relayed exactly one message, which contained a value v', then P_i *relays* the first message in $V_{i,k}$ which contains a value $v'' \neq v'$, if there is one.

[end] The **final value** of T is its own local input v. If ACC_i contains exactly one value v, processor P_i takes this v as its final value; otherwise, P_i decides on "faulty transmitter". ♦

3.2 Proof of the Byzantine agreement protocol

The correctness of the protocol is proved in two steps:

Lemma 3 states that if pseudosignatures were ideally secure, then our BAP would achieve errorless BA. Theorem 4 states that the assumption made in Theorem 3 holds with probability $\geq 1-2^{-\sigma}$, i.e., the error probability of our BAP is at most $2^{-\sigma}$.

Lemma 3: (Deterministic version of Theorem 4)

For any $t < n$, the protocol described above achieves Byzantine agreement if for all pseudosignatures ψ of the protocol run, all $\alpha \in \{A, B\}$, all good processors P and Q, all values v, and all rounds $[k]$ the following is true:

A_1 If ψ is the α-signature of P on v, then Q 1-accepts ψ.

A_2 If in round $[k]$, P accepts ψ as an α-signature of Q on v, then Q has relayed a protocol message containing the α-signature of Q on v not later than in round $[k-1]$.

A_3 If P k-accepts ψ as an α-signature of some processor R, then Q would $(k{+}1)$-accept ψ. ♦

Lemma 3 and its proof are very similar to [DoSt_83: Th. 3]; thus we omit the proof.

Theorem 4: (Byzantine agreement)

For any $t < n$, the protocol described above achieves Byzantine agreement with an error probability of at most $2^{-\sigma}$. ♦

Sketch of proof

Let W_i be the probability that assumption A_i of Lemma 3 does not hold. Then, the error probability of the protocol is at most $W_1 + W_2 + W_3$.

Estimation of W_1: From Theorem 1, we know the probability corresponding to W_1 for each of the $\leq 2n$ pseudosignatures, and it is easy to see that $m \leq 2\lambda^2\kappa$. Thus we multiply the probability of Theorem 1 by $2n$, apply $\lambda \leq n$, and obtain $W_1 \leq \kappa n^7 2^{-\tau+3}$.

Estimation of W_2: In a similar way, we obtain $W_2 \leq \kappa n^6 2^{-\tau+3}$.

Estimation of W_3: Let W_3^* be the probability corresponding to W_3 for *one* of the $\leq n^2/2$ pairs $\{P, Q\}$ and *one* of the $\leq 2n$ pseudosignatures (i, α) initialized. Thus we have $W_3 \leq n^3 W_3^*$. (We need not consider the λ different levels of acceptance, since in each round $[k]$, P or Q either k-accept a pseudosignature or reject it.)

As in Section 2.3, we estimate W_3^* by considering an overly pessimistic case: P and Q are the only good processors, and the faulty processors know everything, except which of the two authentication keys of P and Q belong to P and which to Q, for the m parts of the pseudosignature considered. Again, we count pseudosignatures in a strategy tree B.

For our specific BAP, we define B slightly differently: Each node of B represents one round and is described by the protocol messages that P and Q received in that round. Since P and Q do not react to old messages (Step $[k.a]$), it is sufficient to consider *new* messages only.

The depth of B is t ($\leq n$), corresponding to the rounds $[0]$ ($\approx$ root), ..., $[t]$ ($\approx$ leaves) of the BAP. For each node of depth $k < t$, P and Q can react in round $[k+1]$ by relaying up to two of the new messages just received. Each child corresponds to one reaction and can be described unambiguously by specifying which messages P and Q relay. (The pseudosignature that P or Q adds to a message before relaying it is already known.) Call the child corresponding to the case that neither P nor Q relays a message the *leftmost* child.

During the protocol, each of P and Q considers at most two protocol messages from each of the $< n$ faulty processors, and each protocol message contains the pseudosignature we are interested in at most once. Therefore, the messages in each path in B contains $< 4n$ pseudosignatures. If w denotes the number of different paths in B, then, according to Theorem 2b, $W_3^* \leq w\, 4n\, 2^{-\kappa}$.

Altogether, P and Q relay at most two messages each. Thus on each path, there are at most 4 branches to the right (i.e., not to the leftmost child). In each round, P and Q accept at most two messages from each other processor, thus together at most $2n-2$ each. At the beginning, P and Q can relay up to two messages each. These are

$$\binom{2n-2}{0} + \binom{2n-2}{1} + \binom{2n-2}{2} \leq 2n^2$$

different possibilities for P and Q each. Therefore, until the first branch to the right, a node has $\leq x_0 := 4n^4$ children. Each branch to the right reduces the number of possible reactions. It is easy to see that after the first branch to the right, there are $\leq x_1 := 4n^3$ possibilities left, after the second, $\leq x_2 := 4n^2$, after the third, $\leq x_3 := 2n$, and of course, after the fourth, $x_4 = 1$. In this way, we obtain an upper bound on w: Each path can be uniquely described by specifying in which of the $n-1$ possible depths the $L \leq 4$ branches to the right occur, and for the i-th branch ($i = 1, \ldots, L$), which of the $\leq x_{i-1} - 1$ branches to the right is chosen. Thus, there are

$$w_L \leq \binom{n}{L} x_0 \cdots x_{L-1}, \; L = 0, 1, 2, 3, 4,$$

paths with exactly L branches to the right. After some computation, this yields $w \leq 9n^{14}$.

Thus, we obtain $W_3^* \leq w\, 4n\, 2^{-\kappa} \leq 36\, n^{15}\, 2^{-\kappa}$ and $W_3 \leq n^3\, W_3^* \leq 36\, n^{18}\, 2^{-\kappa}$.

Finally, this implies

$$W \le W_1 + W_2 + W_3 \le \kappa(8n^7+8n^6)\,2^{-\tau} + 36\,n^{18}\,2^{-\kappa},$$

and one can easily verify that both summands are at most $2^{-\sigma-1}$. ❒

3.3 Fixed use of the reliable broadcast network

The BAP of Section 3.1 needs $2n-1$ pseudosignatures for each execution. This suggests that the number of pseudosignatures initialized during the precomputation phase, i.e., using a physically reliable broadcast network, must be proportional to the intended number of protocol executions.

However, using a bootstrapping mechanism, we need to prepare just $O(n^2)$ executions in order to execute the protocol u times, for arbitrary u.

The parameter u influences the protocol performance via the security parameter σ only: σ must be logarithmic in u. Obviously, this is necessary, because the error probability of u executions is about u times the error probability of one execution. Therefore, we need $\sigma \ge \omega + \log_2(u)$ to achieve an error probability of at most $2^{-\omega}$.

Thus we have a similar situation as with conventional digital signatures.

The idea is the following:

Assume that in the precomputation phase, we have initialized a certain number of pseudosignatures. If the BAP must be executed, we first initialize the necessary $2n-1$ pseudosignatures. Unfortunately, at this time we cannot access the physically reliable broadcast network any longer. (Otherwise, Byzantine agreement would be trivial.) Thus, we use our BAP to implement reliable broadcast, i.e., we consume some of our already initialized pseudosignatures. To be able to repeat this procedure, we must initialize some additional pseudosignatures to replace the consumed ones. This is possible because with a fixed number of reliably broadcast messages, we can initialize as many pseudosignatures as we like.

4 Summary

In Section 2, we showed how pseudosignatures with an arbitrary number of levels of acceptance can be implemented. We also demonstrated that, even though active attacks must be considered, these pseudosignatures can be an unconditionally secure substitute for conventional digital signatures in a large class of deterministically polynomial protocols.

In Section 3, we presented a modification of a well-known authenticated Byzantine agreement protocol (BAP) and proved that it can be implemented securely with pseudosignatures. This results in the first BAP tolerating *any* number $t < n$ of faulty processors without relying on unproven computational assumptions.

We consider our results as constructive proofs of existence rather than as practical constructions. It is an interesting open question whether there are more efficient constructions, in particular for pseudosignatures.

Acknowledgements: We are pleased to thank *Birgit Baum-Waidner, Manfred Böttger, David Chaum, Klaus Echtle, Maarten van der Ham, Andreas Pfitzmann, Rüdiger Reischuk,* and *Sandra Roijakkers* for helpful comments and discussions.

References

BaPW_91 Birgit Baum-Waidner, Birgit Pfitzmann, Michael Waidner: Unconditional Byzantine Agreement with Good Majority; STACS '91, LNCS 480, Springer-Verlag, Heidelberg 1991, 285-295.

BoBo_90 Jurjen Bos, Bert den Boer: Detection of disrupters in the DC protocol; Eurocrypt '89, LNCS 434, Springer-Verlag, Berlin 1990, 320-327.

Burt_71 Herbert O. Burton: Inversionless Decoding of Binary BCH Codes; IEEE Transactions on Information Theory 17/4 (1971) 464-466.

Chau_88 David Chaum: The Dining Cryptographers Problem: Unconditional Sender and Recipient Untraceability; Journal of Cryptology 1/1 (1988) 65-75.

ChRo_90 David Chaum, Sandra Roijakkers: Unconditionally Secure Digital Signatures; Crypto '90, Santa Barbara, 11-15 August 1990, Abstracts, 209-217.

DiHe_76 Whitfield Diffie, Martin E. Hellman: New Directions in Cryptography; IEEE Transactions on Information Theory 22/6 (1976) 644-654.

DoSt_83 Danny Dolev, H. Raymond Strong: Authenticated Algorithms for Byzantine Agreement; SIAM J. Comput. 12/4 (1983) 656-666.

ErSp_74 Paul Erdös, Joel Spencer: Probabilistic Methods in Combinatorics; Probability and Mathematical Statistics 17, Academic Press, New York 1974.

GiMS_74 E. N. Gilbert, F. J. Mac Williams, N. J. A. Sloane: Codes which detect deception; The Bell System Technical Journal 53/3 (1974) 405-424.

GoMR_88 Shafi Goldwasser, Silvio Micali, Ronald L. Rivest: A Digital Signature Scheme Secure Against Adaptive Chosen-Message Attacks; SIAM J. Comput. 17/2 (1988) 281-308.

GrYa_89 Ronald L. Graham, Andrew C. Yao: On the Improbability of Reaching Byzantine Agreement; 21st STOC 1989, ACM Press, New York 1989, 467-478.

PeSL_80 Marshall Pease, Robert Shostak, Leslie Lamport: Reaching Agreement in the Presence of Faults; Journal of the ACM 27/2 (1980) 228-234.

Rabi_80 Michael O. Rabin: Probabilistic Algorithms in Finite Fields; SIAM J. Comput. 9/2 (1980) 273-280.

Romp_90 John Rompel: One-Way Functions are Necessary and Sufficient for Secure Signatures; 22nd STOC 1990, ACM Press, New York 1990, 387-394.

Waid_91 Michael Waidner: Byzantinische Verteilung ohne kryptographische Annahmen trotz beliebig vieler Fehler; Universität Karlsruhe, Fakultät für Informatik, Dissertation, October 1991; to appear.

WeCa_81 Mark N. Wegman, J. Lawrence Carter: New Hash Functions and Their Use in Authentication and Set Equality; Journal of Computer and System Sciences 22 (1981) 265-279.

Broadcasting in Butterfly and DeBruijn Networks

(Extended Abstract)

R. Klasing*, B. Monien*, R. Peine

Universität-GH Paderborn, FB 17,

Warburger Str. 100, W-4790 Paderborn, Germany

e-mail : Ralf.Klasing@uni-paderborn.de, Burkhard.Monien@uni-paderborn.de,

Regine.Peine@uni-paderborn.de

E. Stöhr

Karl-Weierstraß-Institut für Mathematik,

Mohrenstr. 39, O-1086 Berlin, Germany

e-mail: stoehr@dboadw11.bitnet

Abstract: Broadcasting is the process of message dissemination in a communication network in which a message originated by one processor is transmitted to all processors of the network. In this paper, we present a new lower bound of $1.7417m$ for broadcasting in the butterfly network of dimension m. This improves the best known lower bound of $1.5621m$. We also describe an algorithm which improves the upper bound from $2m$ to $2m-1$. This is shown to be optimal for small dimensions m. In addition, the presented lower bound technique is used to derive non-trivial lower bounds for broadcasting in the deBruijn network of dimension m. An upper bound of $1.5m+1.5$ is well-known for this network. Here, we are able to improve the lower bound from $1.1374m$ to $1.3171m$.

Classification: Theory of parallel and distributed computation.

1 Introduction

Broadcasting is the process of message dissemination in a communication network in which a message originated by one processor is transmitted to all processors of the network (for a survey, cf. [HHL88] and [FL91]). For this purpose, undirected (and directed) graphs are considered as models for communication networks. Thus, a given vertex in a graph has a message, which it wishes to disseminate to all other vertices; each vertex can transmit a message to exactly one vertex to which it is adjacent during one unit of time, and each vertex can either transmit or receive the message per unit of time.

Assume the message originates at vertex u of the connected graph G. The *broadcast time of the vertex u, b(u)*, is the minimum number of time units required to complete broadcasting from

*The work of these authors was supported by grant Mo 285/4-1 from the German Research Association (DFG).

vertex u. The *broadcast time of a graph* G, $b(G)$, is the maximum broadcast time over all vertices u in G.

Broadcasting in graphs of bounded degree has been studied in [BHLP88], [BP88], [CGV89], [HJM90], [HOS91], [LP88], [St91a] and [St91b]. The maximum degree is an important parameter in the design of interconnection networks. This is one of the motivations to investigate broadcasting in graphs with fixed maximum degree, in particular in bounded-degree "approximations" of the hypercube such as cube-connected cycles, butterfly, shuffle-exchange and deBruijn networks. Lower and upper bounds on the time required to broadcast in graphs with maximum degrees 3 and 4 were given by Liestman and Peters [LP88]. Their results were improved in [BHLP88] and [CGV89] where general lower bounds were obtained.

A trivial lower bound on the broadcast time is the diameter of a graph. For cube-connected cycles and shuffle-exchange networks the broadcast time and the diameter differ at most by a constant [HJM90], [LP88]. In [St91b], a non-trivial lower bound on the broadcast time of the Butterfly graph $BF(m)$ was presented. It was shown that $b(BF(m)) > 1.5621m$ for all sufficiently large m. Note that the diameter of $BF(m)$ is $D = \lfloor 3m/2 \rfloor$. Thus, $b(BF(m)) > cD$ for some constant $c > 1$. In Section 3, we improve the result of [St91b] and show that $b(BF(m)) > 1.7417m$ for all sufficiently large m.

In order to apply the lower bound technique to other networks, we point out the used properties of the graph:

a) degree 4,

b) there is a node from which *a lot* of vertices have a *large* distance (*large* $\approx$ diameter),

c) each edge is contained in a cycle of length at most 4.

Now, the lower bound argument is an extension of the one in [LP88], [BHLP88] and [CGV89]. It consists of finding an upper bound on the maximum number of nodes which can be informed in t time steps. But unlike the papers before, we are not only using property a), but we also exploit properties b) and c).

An upper bound on the broadcast time of the Butterfly graph was given in [St91a]. It was shown that $b(BF(m)) \leq 2m$. In Section 4, we present a different algorithm which only needs $2m - 1$ rounds.

In Section 5, we give an overview of the obtained lower and upper bounds for broadcasting in small butterfly networks. It points out the fact that the two bounds are not very far apart for small dimensions m. Actually, for $m \leq 4$, i.e. up to 64 nodes, our broadcasting algorithm turns out to be optimal.

DeBruijn networks have been proposed as a possible alternative for designing large interconnection networks [BP89], [SP89]. Broadcasting in these graphs was considered in [BP88] and [HOS91]. In [BP88], an upper bound of $1.5m+1.5$ for the (binary) deBruijn graph of dimension m, $DB(m)$, was stated. A lower bounds of $b(DB(m)) > 1.1374m$ can be derived from [LP88]. Applying the same techniques as for the butterfly network, we are able to show in Section 6 that $b(DB(m)) > 1.3171m$ for all sufficiently large m.

2 Definitions

(cf. [DB46, KLM90, MS90, P68]) The *butterfly network* $BF(m)$ *of dimension* m has vertex-set $V_m = \{0, 1, ..., m-1\} \times \{0,1\}^m$, where $\{0,1\}^m$ denotes the set of length-m binary strings. For each vertex $v = \langle i, \alpha \rangle \in V_m$, $i \in \{0, 1, ..., m-1\}$, $\alpha \in \{0,1\}^m$, we call i the level and α the position-within-level (PWL) string of v. The edges of $BF(m)$ are of two types: For each $i \in \{0, 1, ..., m-1\}$ and each $\alpha = a_0 a_1 ... a_{m-1} \in \{0,1\}^m$, the vertex $\langle i, \alpha \rangle$ on level i of $BF(m)$ is connected

- by a straight-edge with vertex $\langle i + 1(\bmod m), \alpha \rangle$ and
- by a cross-edge with vertex $\langle i + 1(\bmod m), \alpha(i) \rangle$

on level $i + 1(\bmod m)$. Here, $\alpha(i) = a_0 \ldots a_{i-1} \bar{a}_i a_{i+1} \ldots a_{m-1}$, where $\bar{a}$ denotes the binary complement of a. $BF(m)$ has $m2^m$ nodes, diameter $\lfloor 3m/2 \rfloor$ and maximum node degree 4.

The *(binary) deBruijn network* $DB(m)$ *of dimension* m is the graph whose nodes are all binary strings of length m and whose edges connect each string $a\alpha$, where α is a binary string of length $m-1$ and a is in $\{0,1\}$, with the strings αb, where b is a symbol in $\{0,1\}$. (An edge connecting $a\alpha$ with αb, $a \neq b$, is called a *shuffle-exchange* and an edge connecting $a\alpha$ with αa is called a *shuffle* edge.) $DB(m)$ has 2^m nodes, diameter m and maximum degree 4.

3 A New Lower Bound for the Butterfly Network

In this section, we will prove a lower bound of $1.7417m$ for broadcasting in the butterfly network of dimension m.

3.1 The Distance Argument

In the first proof, we will only use properties a) and b) of the butterfly network:

a) degree 4,

b) there is a node from which *a lot* of vertices have a *large* distance (*large* $\approx$ diameter).

The arguments of the proof in this section are the same as in [St91b]. But we exploit the method up to its full potential by making all the occuring estimations as sharp as possible. First, we will state property b) more exactly:

__Lemma 1:__ (Distances in the butterfly network)

Let $BF(m)$ be the butterfly network of dimension m. Let $v_0 = \langle 0, 00...0 \rangle$. Let $\varepsilon > 0$ be any positive constant. Then there exist $2^m - o(2^m)$ nodes which have at least a distance of $\lfloor 3m/2 - \varepsilon m \rfloor$ from v_0.

Proof: Let

$$L = \{ \langle \lfloor m/2 \rfloor, \delta \rangle \mid \delta \neq \alpha 0^k \beta \text{ for some } k \geq \varepsilon m/2,\ \alpha 0^k \beta \in \{0,1\}^m \}$$

be the subset of the level-$\lfloor m/2 \rfloor$ vertices of $BF(m)$. Then $|L| \geq 2^m - m2^{m - \varepsilon m/2}$. It is not very difficult to show that the distance between any vertex v from L and v_0 is at least $\lfloor 3m/2 - \varepsilon m \rfloor$. □

Now, we are able to show the first improved lower bound:

Theorem 1: (First lower bound for broadcasting in the butterfly network)

$b(BF(m)) > 1.7396m$ for all sufficiently large m.

Proof: To obtain a contradiction suppose that broadcasting can be completed on $BF(m)$ in time $3m/2 + tm$, $0 \leq t < 1/2$.

Since the butterfly graph is a Cayley graph [ABR90], and every Cayley graph is vertex symmetric [AK89], we can assume that the message originates at vertex $v_0 = \langle 0, 00...0 \rangle$, and the originator learns the message at time 0.

Let $A(i,t)$ denote the maximum number of nodes which can be reached in round t on a path of length i. Since $BF(m)$ has maximum degree 4, once a node has received the message it can only inform 3 additional neighbours in the next three rounds. Therefore, $A(i,t)$ is recursively defined as follows:

$$\begin{aligned}
&A(0,0) = 1,\\
&A(1,1) = 1,\\
&A(1,2) = 1,\ A(2,2) = 1,\\
&A(1,3) = 1,\ A(2,3) = 2,\ A(3,3) = 1,\\
&A(1,4) = 1,\ A(2,4) = 3,\ A(3,4) = 3,\ A(4,4) = 1,\\
&A(i,t) = A(i-1,t-1) + A(i-1,t-2) + A(i-1,t-3) \quad \text{for } t \geq 5.
\end{aligned}$$

It can easily be shown by induction (cf. [BP85]) that

$$A(n, n+l) \leq 2 \cdot \sum_{\substack{p+2q=l,\\ 0 \leq p,q \leq n}} \binom{n}{p+q} \cdot \binom{p+q}{q}.$$

In the following, we omit $\lfloor\ \rfloor$ and $\lceil\ \rceil$ for the sake of simplification. It should be added everywhere according to the context. Let $\varepsilon > 0$ be any positive constant. From Lemma 1, we know that for any broadcasting scheme

$$\sum_{n=3m/2-\varepsilon m}^{3m/2+tm} \ \sum_{l=0}^{3m/2+tm-n} A(n, n+l) \geq 2^m - o(2^m).$$

For ε tending towards 0, we have

$$\begin{aligned}
2^m - o(2^m) &\leq \sum_{n=3m/2}^{3m/2+tm} \ \sum_{l=0}^{3m/2+tm-n} A(n,n+l)\\
&\leq \sum_{n=3m/2}^{3m/2+tm} \ \sum_{l=0}^{3m/2+tm-n} 2 \cdot \sum_{\substack{p+2q=l,\\ 0 \leq p,q \leq n}} \binom{n}{p+q} \cdot \binom{p+q}{q}\\
&\leq 2 \cdot \sum_{n=3m/2}^{3m/2+tm} \ \sum_{0 \leq p+2q \leq 3m/2+tm-n} \binom{n}{p+q} \cdot \binom{p+q}{q}\\
&\leq cm^3 \cdot \max_{\substack{3m/2 \leq n \leq 3m/2+tm,\\ 0 \leq p+2q \leq 3m/2+tm-n}} \binom{n}{p+q} \cdot \binom{p+q}{q}
\end{aligned}$$

for some constant c. It can easily be verified that the above maximum is obtained for $n = \lfloor 3m/2 \rfloor$, $p + 2q = \lfloor tm \rfloor$ when $t < 1/2$. Therefore,

$$\max_{\substack{3m/2 \leq n \leq 3m/2+tm,\\ 0 \leq p+2q \leq 3m/2+tm-n}} \binom{n}{p+q} \cdot \binom{p+q}{q} = \max_{0 \leq i \leq tm/2} \binom{3m/2}{tm-i} \cdot \binom{tm-i}{i}.$$

The latter term is maximized for $i = i_0 m$ where

$$i_0 = \frac{1}{4} + \frac{t}{2} - \sqrt{\left(\frac{1}{4} + \frac{t}{2}\right)^2 - \frac{t^2}{3}}$$

For large m, an approximate expression for the factorial is given by Stirling's formula $m! \approx m^m e^{-m} \sqrt{2\pi m}$. Using Stirling's formula we obtain

$$\binom{3m/2}{tm - i_0 m} \cdot \binom{tm - i_0 m}{i_0 m} \approx \frac{(3/2)^{3m/2}}{(3/2 - t + i_0)^{3m/2 - tm + i_0 m} (i_0)^{i_0 m} (t - 2i_0)^{tm - 2i_0 m}}.$$

Thus,

$$cm^3 \cdot \frac{(3/2)^{3m/2}}{(3/2 - t + i_0)^{3m/2 - tm + i_0 m} (i_0)^{i_0 m} (t - 2i_0)^{tm - 2i_0 m}} \geq 2^m - o(2^m).$$

Taking the m-th root on both sides, we have for large m

$$\frac{(3/2)^{3/2}}{(3/2 - t + i_0)^{3/2 - t + i_0} (i_0)^{i_0} (t - 2i_0)^{t - 2i_0}} \geq 2.$$

The latter inequality is not true for $t \leq 0.2396$. This contradiction establishes the Theorem. □

3.2 The Small Cycles Argument

Now, we will exploit all three properties of the butterfly network:

a) degree 4,

b) there is a node from which *a lot* of vertices have a *large* distance (*large* $\approx$ diameter),

c) each edge is contained in a cycle of length at most 4.

First, we have to show that property c) holds for $BF(m)$:

<u>**Lemma 2:**</u> (Small cycles in the butterfly network)

Let $BF(m)$ be the butterfly network of dimension m. Then each node is contained in 2 edge-disjoint cycles of length 4.

Proof: Let $v = \langle i, \alpha \rangle = \langle i, a_0 a_1 \ldots a_{m-1} \rangle$ be a node in $BF(m)$. Then v is contained in the two edge-disjoint cycles

$$\langle i, \alpha \rangle - \langle i \oplus 1, \alpha \rangle - \langle i, \alpha(i) \rangle - \langle i \oplus 1, \alpha(i) \rangle - \langle i, \alpha \rangle$$

and

$$\langle i, \alpha \rangle - \langle i \ominus 1, \alpha \rangle - \langle i, \alpha(i \ominus 1) \rangle - \langle i \ominus 1, \alpha(i \ominus 1) \rangle - \langle i, \alpha \rangle$$

of length 4, where $\oplus$, $\ominus$ denote addition and subtraction modulo m. □

Now, we are able to show the second improved lower bound:

Theorem 2: (Second lower bound for broadcasting in the butterfly network)
$b(BF(m)) > 1.7417m$ for all sufficiently large m.

Proof: The proof is similar to that of Theorem 1. Only the recursive definition of $A(i,t)$ has to be changed slightly. The intial values of $A(i,t)$, for $0 \leq t \leq 5$, stay the same, but the recurrence relation is different. Since all edges incident to a node are covered by two cycles of length (at most) 4, once a node v has received the message on the first cycle C_1, "the best way" to spread the information is to send it to the two sons of v on the second cycle C_2 before sending it to the son of v on C_1. (It is possible to formalize this argument, but because of lack of space we omit this.) That way, when v gets the message in round r, the two sons of v on C_2 will receive the information in rounds $r+1$ and $r+2$, and the last node on C_2 will receive the information in round $t+4$. Therefore, we obtain the following recursion:

$$A(i,t) = A(i-1,t-1) + A(i-1,t-2) + A(i-2,t-4) \quad \text{for } t \geq 6.$$

Again, it can easily be shown by induction that

$$A(n,n+l) \leq 2 \cdot \sum_{\substack{p+2q=l,\\ 0\leq p,q\leq n}} \binom{n-q+1}{q} \cdot \binom{n-2q+1}{p}.$$

Similar estimations as in the proof of Theorem 1 show that the condition

$$\sum_{n=3m/2}^{3m/2+tm} \sum_{l=0}^{3m/2+tm-n} A(n,n+l) \geq 2^m - o(2^m)$$

from Lemma 1 can only be true if $t > 0.2417$. □

4 A New Upper Bound for the Butterfly Network

In the following we present the algorithm for broadcasting in the Butterfly network $BF(m)$ in time $2m-1$.

Theorem 3: (Upper bound for broadcasting in the Butterfly network)
The broadcast time of the Butterfly network $BF(m)$ is at most $2m-1$.

Proof: First, note that $BF(m)$ contains two isomorphic subgraphs F_0 and F_1. The subgraph F_0 has vertex set $\{\langle\, l\,,\, \alpha 0\,\rangle \mid 0 \leq l \leq m-1,\ \alpha \in \{0,1\}^{m-1}\}$, and the subgraph F_1 has vertex set $\{\langle\, l\,,\, \alpha 1\,\rangle \mid 0 \leq l \leq m-1,\ \alpha \in \{0,1\}^{m-1}\}$. Obviously, $F_0 \cap F_1 = \emptyset$.

Again, we assume that the message originates at vertex $v_0 = \langle 0, 0 \ldots 0 \rangle$, and the originator learns the message at time 0.
In the first step, vertex v_0 informs its neighbour $v_1 = \langle\, m-1\,,\, 0\ldots01\,\rangle$. Now, in F_0 as well as in F_1 one vertex has the message. Then broadcasting in F_0 and F_1 will be done as follows:

Broadcasting in F_0:

Phase 1 : Inform all vertices at level $m-1$ in the following way:
From any vertex $\langle\, l\,,\, \alpha 0\,\rangle$, $0 \leq l \leq m-2$, $\alpha = \alpha_0 \ldots \alpha_{m-2} \in \{0,1\}^{m-1}$, of F_0 that receives the message at time t, the vertex $\langle\, l+1\,,\, \alpha 0\,\rangle$ receives the message at time $t+1$ and the vertex $\langle\, l+1\,,\, \alpha_0\alpha_1 \ldots \bar{\alpha}_l \ldots \alpha_{m-2}0\,\rangle$ receives the message at time $t+2$.

As soon as the vertex $\langle\, m-1\,,\, \alpha 0\,\rangle$ with $\alpha \in \{0,1\}^{m-1}$ receives the information it informs its neighbours as described in Phase 2:

Phase 2 : Consider the following path P_α in F_0 of length m from $\langle m-1\,,\,\alpha 0\,\rangle$ to $\langle m-1\,,\,\bar{\alpha}0\rangle$, $\alpha = \alpha_0 \dots \alpha_{m-2} \in \{0,1\}^{m-1}$:

$$P_\alpha = (\langle\, m-1\,,\,\alpha 0\,\rangle, \langle\, m-2\,,\,\alpha_0\alpha_1\dots\alpha_{m-3}\bar{\alpha}_{m-2}\,0\,\rangle, \langle\, m-3\,,\,\alpha_0\alpha_1\dots\alpha_{m-4}\bar{\alpha}_{m-3}\,\bar{\alpha}_{m-2}\,0\,\rangle, \dots, \dots, \langle\,0\,,\,\bar{\alpha}_0\,\bar{\alpha}_1\,\dots\bar{\alpha}_{m-2}\,0\,\rangle, \langle\, m-1\,,\,\bar{\alpha}_0\,\bar{\alpha}_1\,\dots\bar{\alpha}_{m-2}\,0\,\rangle)$$

This path traverses $m-1$ cross-edges and one straight-edge.

Along this path all vertices are informed by sending the message from both endpoints as follows:

Case 1: $\alpha = \tilde{\alpha}0,\ \tilde{\alpha} \in \{0,1\}^{m-2}$

From the vertex $\langle m-1\,,\,\alpha 0\rangle$ that receives the message at time t, $t \le 2m-2$, in Phase 1, the vertex $\langle\, m-2\,,\,\alpha_0\alpha_1\dots\alpha_{m-3}\bar{\alpha}_{m-2}\,0\,\rangle$ receives the message at time $t+1$. From the vertex $\langle\, m-1\,,\,\bar{\alpha}0\,\rangle$ that receives the message at time $\tilde{t}$, $\tilde{t} \le 2m-2$, in Phase 1, the vertex $\langle\,0\,,\,\bar{\alpha}_0\,\bar{\alpha}_1\,\dots\bar{\alpha}_{m-2}\,0\,\rangle$ receives the message at time $\tilde{t}+1$.

Every other vertex of P_α receiving the message at time t with $t \le 2m-2$ in Phase 2 from one of its neighbours of P_α sends the message to its other neighbour of P_α at time $t+1$.

Case 2: $\alpha = \tilde{\alpha}1,\ \tilde{\alpha} \in \{0,1\}^{m-2}$

The vertex $\langle\, m-1\,,\,\alpha 0\,\rangle$ receives the message at time t in Phase 1. So the neighbour $\langle\, m-2\,,\,\alpha_0\alpha_1\dots\alpha_{m-3}\bar{\alpha}_{m-2}\,0\,\rangle$ receives the message at time $t-2$ in Phase 1.
This vertex informs the vertex $\langle\, m-3\,,\,\alpha_0\alpha_1\dots\alpha_{m-4}\bar{\alpha}_{m-3}\,\bar{\alpha}_{m-2}\,0\,\rangle$ at time $t+1$, if $t \le 2m-2$.

From the vertex $\langle m-1\,,\,\bar{\alpha}0\rangle$ that receives the message at time $\tilde{t}$, $\tilde{t} \le 2m-3$, in Phase 1, the vertex $\langle\,0\,,\,\bar{\alpha}_0\,\bar{\alpha}_1\,\dots\bar{\alpha}_{m-2}\,0\,\rangle$ receives the message at time $\tilde{t}+2$.

Every other vertex of P_α receiving the message at time t with $t \le 2m-2$ in Phase 2 from one of its neighbours of P_α sends the message to its other neighbour of P_α at time $t+1$.

Broadcasting in F_1:

The broadcasting scheme in F_1 is nearly the same as the one in F_0. The difference is that in F_1 in Phase 1, all vertices at level 0 are informed. So the vertex $\langle\, l\,,\,\alpha 1\,\rangle$, $1 \le l \le m-1$, $\alpha = \alpha_0\dots\alpha_{m-2} \in \{0,1\}^{m-1}$, of F_1 informs at first its neighbour $\langle\, l-1\,,\,\alpha 1\,\rangle$ and then its neighbour $\langle\, l-1\,,\,\alpha_0\alpha_1\dots\bar{\alpha}_l\dots\alpha_{m-2}1\,\rangle$.
In Phase 2, we consider the path R_α in F_1 of length m from $\langle\,0\,,\,\alpha 1\,\rangle$ to $\langle\,0\,,\,\bar{\alpha}1\,\rangle$, $\alpha \in \{0,1\}^{m-1}$. This path is defined as follows:

$$R_\alpha = (\langle\,0\,,\,\alpha 1\,\rangle, \langle\,1\,,\,\bar{\alpha}_0\,\alpha_1\,\dots\alpha_{m-2}\,1\,\rangle, \langle\,2\,,\,\bar{\alpha}_0\,\bar{\alpha}_1\,\alpha_2\,\dots\alpha_{m-2}\,1\,\rangle, \dots, \dots, \langle\, m-1\,,\,\bar{\alpha}_0\,\bar{\alpha}_1\,\dots\bar{\alpha}_{m-2}\,1\,\rangle, \langle\,0\,,\,\bar{\alpha}_0\,\bar{\alpha}_1\,\dots\bar{\alpha}_{m-2}\,1\,\rangle)$$

As in Phase 2 for broadcasting in F_0, the vertices of the path are informed along this path. □

Now let us look at the analysis of this algorithm. We will show that every node $\langle\, i\,,\,\alpha 0\,\rangle$ in F_0 with $0 \le i \le m-1$, $\alpha \in \{0,1\}^{m-1}$, receives the message in at most $2m-1$ rounds.

Let $\alpha \in \{0,1\}^{m-1}$ be any string of length $m-1$. By $\#_1(\alpha)$ we denote the number of 1's in α and by $\#_0(\alpha)$ we denote the number of 0's in α. So from the definition we have
$\#_1(\alpha) + \#_0(\alpha) = m-1$.
First we consider Phase 1. In the subgraph F_0 the vertex $w_0 = \langle\, m-1\,,\,\alpha 0\,\rangle$, $\alpha \in \{0,1\}^{m-1}$, is informed from v_0 along a path of length $m-1$. This path traverses $\#_1(\alpha)$ cross-edges and $\#_0(\alpha)$

straight-edges. Thus, the vertex w_0 is informed at time $1 + 2 \cdot \#_1(\alpha) + \#_0(\alpha) = m + \#_1(\alpha) \leq 2m - 1$.

In Phase 2, the vertices which did not get the message in Phase 1 are informed along the paths P_α with $\alpha \in \{0,1\}^{m-1}$. First we show that for all $\alpha, \beta \in \{0,1\}^{m-1}$, $\alpha \notin \{\beta, \bar{\beta}\}$, the paths P_α and P_β are nodedisjoint.

Assume P_α and P_β, $\alpha, \beta \in \{0,1\}^{m-1}$, $\alpha \notin \{\beta, \bar{\beta}\}$, have at least one common vertex $\langle l, \delta 0 \rangle$, $\delta \in \{0,1\}^{m-1}$, $l \in \{0, \ldots, m-1\}$.

Case 1: $l = m - 1$

Then on the path P_α you have $\langle l, \delta 0 \rangle = \langle m-1, \alpha 0 \rangle$ or $\langle l, \delta 0 \rangle = \langle m-1, \bar{\alpha} 0 \rangle$.
On the path P_β you have $\langle l, \delta 0 \rangle = \langle m-1, \beta 0 \rangle$ or $\langle l, \delta 0 \rangle = \langle m-1, \bar{\beta} 0 \rangle$.
Since $\alpha \notin \{\beta, \bar{\beta}\}$, P_α and P_β have no common vertex at level $m - 1$.

Case 2: $l \in \{0, \ldots, m-2\}$

Then you have $\langle l, \delta 0 \rangle = \langle l, \alpha_0 \ldots \alpha_{l-1} \bar{\alpha}_l \ldots \bar{\alpha}_{m-2} 0 \rangle$ (in P_α)
$\langle l, \beta_0 \ldots \beta_{l-1} \bar{\beta}_l \ldots \bar{\beta}_{m-2} 0 \rangle$ (in P_β)

So you have

$$\alpha_0 \ldots \alpha_{l-1} \bar{\alpha}_l \ldots \bar{\alpha}_{m-2} = \beta_0 \ldots \beta_{l-1} \bar{\beta}_l \ldots \bar{\beta}_{m-2}$$
$$\Leftrightarrow \alpha_0 \ldots \alpha_{l-1} \alpha_l \ldots \alpha_{m-2} = \beta_0 \ldots \beta_{l-1} \beta_l \ldots \beta_{m-2},$$

in contradiction to $\alpha \notin \{\beta, \bar{\beta}\}$.

With the same argumentation we can show that for all $\alpha \in \{0,1\}^{m-1}$, the paths P_α and $P_{\bar{\alpha}}$ have exactly two common nodes, namely $\langle m-1, \alpha 0 \rangle$ and $\langle m-1, \bar{\alpha} 0 \rangle$.

Now we have to consider the following two cases:

Case 1: $\alpha = \tilde{\alpha} 0$, $\tilde{\alpha} \in \{0,1\}^{m-2}$

From the vertex $\langle m-1, \alpha 0 \rangle$ the vertices $\langle m-i-1, \alpha_0 \ldots \alpha_{m-i-2} \bar{\alpha}_{m-i-1} \ldots \bar{\alpha}_{m-2} 0 \rangle$ of P_α with $1 \leq i \leq m - \#_1(\alpha) - 1$ are informed. Since $\langle m-1, \alpha 0 \rangle$ is informed at time $m + \#_1(\alpha)$ in Phase 1 the vertex $\langle m-i-1, \alpha_0 \ldots \alpha_{m-i-2} \bar{\alpha}_{m-i-1} \ldots \bar{\alpha}_{m-2} 0 \rangle$ with $i \in \{1, \ldots, m - \#_1(\alpha) - 1\}$ receives the message at time $m + \#_1(\alpha) + i \leq 2m - 1$ in Phase 2.
From the vertex $\langle m-1, \bar{\alpha} 0 \rangle$ the vertices $\langle j, \alpha_0 \ldots \alpha_{j-1} \bar{\alpha}_j \ldots \bar{\alpha}_{m-2} 0 \rangle$ of P_α with $0 \leq j \leq \#_1(\alpha) - 1$ are informed. The vertex $\langle m-1, \bar{\alpha} 0 \rangle$ is informed at time $m + \#_1(\bar{\alpha}) = 2m - \#_1(\alpha) - 1$ in Phase 1. So the vertex $\langle j, \alpha_0 \ldots \alpha_{j-1} \bar{\alpha}_j \ldots \bar{\alpha}_{m-2} 0 \rangle$ with $j \in \{0, \ldots, \#_1(\alpha) - 1\}$ receives the message at time $2m - \#_1(\alpha) + j \leq 2m - 1$ in Phase 2.
Since $\{\langle m-i-1, \alpha_0 \ldots \alpha_{m-i-2} \bar{\alpha}_{m-i-1} \ldots \bar{\alpha}_{m-2} 0 \rangle ; 1 \leq i \leq m - \#_1(\alpha) - 1\}$
$= \{\langle i, \alpha_0 \ldots \alpha_{i-1} \bar{\alpha}_i \ldots \bar{\alpha}_{m-2} 0 \rangle ; \#_1(\alpha) \leq i \leq m-2\}$,
from the vertex $\langle m-1, \alpha 0 \rangle$ you inform the vertices $\langle i, \alpha_0 \ldots \alpha_{i-1} \bar{\alpha}_i \ldots \bar{\alpha}_{m-2} 0 \rangle$ with $\#_1(\alpha) \leq i \leq m-2$. That means that all vertices of P_α are informed in at most $2m - 1$ rounds.

Case 2: $\alpha = \tilde{\alpha} 1$, $\tilde{\alpha} \in \{0,1\}^{m-2}$

In Phase 1, the vertices $\langle m-1, \alpha 0 \rangle$ and $\langle m-2, \alpha_0 \ldots \alpha_{m-3} \bar{\alpha}_{m-2} 0 \rangle$ of P_α have been informed. So from the vertex $\langle m-2, \alpha_0 \ldots \alpha_{m-3} \bar{\alpha}_{m-2} 0 \rangle$, the vertices

$\langle\, m-i-2\,,\, \alpha_0 \ldots \alpha_{m-i-3}\, \bar{\alpha}_{m-i-2} \,\ldots\, \bar{\alpha}_{m-2}\; 0 \,\rangle$ of P_α with $1 \leq i \leq m - \#_1(\alpha) - 1$ are informed. The vertex $\langle\, m-2\,,\, \alpha_0 \ldots \alpha_{m-3}\, \bar{\alpha}_{m-2}\; 0 \,\rangle$ received the message at time $m + \#_1(\alpha) - 2$. So the vertex $\langle\, m-i-2\,,\, \alpha_0 \ldots \alpha_{m-i-3}\, \bar{\alpha}_{m-i-2} \,\ldots\, \bar{\alpha}_{m-2}\; 0 \,\rangle$ with $i \in \{1, \ldots, m - \#_1(\alpha) - 1\}$ is informed at time $m + \#_1(\alpha) + i \leq 2m - 1$ in Phase 2.
From the vertex $\langle\, m-1\,,\, \bar{\alpha}0 \,\rangle$, the vertices $\langle\, j\,,\, \alpha_0 \ldots \alpha_{j-1}\, \bar{\alpha}_j \,\ldots\, \bar{\alpha}_{m-2}\; 0 \,\rangle$ of P_α with $0 \leq j \leq \#_1(\alpha) - 2$ are informed. The vertex $\langle\, m-1\,,\, \bar{\alpha}0 \,\rangle$ receives the message at time $2m - \#_1(\alpha) - 1$ in Phase 1. So the vertex $\langle\, j\,,\, \alpha_0 \ldots \alpha_{j-1}\, \bar{\alpha}_j \,\ldots\, \bar{\alpha}_{m-2}\; 0 \,\rangle$ with $j \in \{0, \ldots, \#_1(\alpha) - 2\}$ is informed at time $2m - \#_1(\alpha) - 1 + j + 2 = 2m - \#_1(\alpha) + j + 1 \leq 2m - 1$ in Phase 2.
Since $\{\langle\, m-i-2\,,\, \alpha_0 \ldots \alpha_{m-i-3}\, \bar{\alpha}_{m-i-2} \,\ldots\, \bar{\alpha}_{m-2}\, 0 \,\rangle\,;\, 1 \leq i \leq m - \#_1(\alpha) - 1\}$
$= \{\langle\, i\,,\, \alpha_0 \ldots \alpha_{i-1}\, \bar{\alpha}_i \,\ldots\, \bar{\alpha}_{m-2}\, 0 \,\rangle\,;\, \#_1(\alpha) - 1 \leq i \leq m - 3\}$,
from the vertex $\langle\, m-1\,,\, \alpha 0 \,\rangle$ you inform the vertices $\langle\, i\,,\, \alpha_0 \ldots \alpha_{i-1}\, \bar{\alpha}_i \,\ldots\, \bar{\alpha}_{m-2}\; 0 \,\rangle$ with $\#_1(\alpha) - 1 \leq i \leq m - 3$. That means that all vertices of P_α are informed in at most $2m - 1$ rounds.

With the same argumentation we can show that every node $\langle\, m-1\,,\, \alpha 0 \,\rangle$ in F_1 with $0 \leq i \leq m-1$, $\alpha \in \{0,1\}^{m-1}$ receives the message in at most $2m - 1$ rounds. So we have that the broadcast algorithm needs at most $2m - 1$ rounds. □

5 Overview for Small Butterfly Networks

To give an idea of how far the upper and lower bound for broadcasting in the butterfly network are still apart, we present an overview for small dimensions m in Table 1.

m	lower bound	upper bound	no. processors
2	3	3	8
3	5	5	24
4	7	7	64
5	8	9	160
6	10	11	384
7	11	13	896
8	13	15	2048
9	15	17	4608
10	16	19	10240
11	18	21	22528
12	19	23	49152
13	21	25	106496
14	23	27	229376
15	24	29	491520
16	26	31	1048576
17	27	33	2228224
18	29	35	4718592

Table 1: Broadcast time for small butterfly networks

The upper bound comes from the algorithm in Section 4. For the lower bound, we computed the $A(i,t)$'s from the proof of Theorem 2, and the number of nodes in the butterfly network at a certain distance from the originating node.

The overall picture is that the upper and lower bound are not very far apart for small dimensions m. Actually, for $m \leq 4$, i.e. up to 64 nodes, our broadcasting algorithm is optimal.

6 A New Lower Bound for the DeBruijn Network

In this section, we will show how to apply our lower bound techniques from Section 3 to the deBruijn network. Again, we will use the following properties of the network:

a) degree 4,

b) there is a node from which *a lot* of vertices have a *large* distance (*large* $\approx$ diameter),

c) each edge is contained in a cycle of length at most 4.

First, we will give evidence that properties b) and c) are valid for the deBruijn network:

Lemma 3: (Distances in the deBruijn network)

Let $DB(m)$ be the deBruijn network of dimension m. Let $v_0 = 00...0$. Let $\varepsilon > 0$ be any positive constant. Then there exist $2^m - o(2^m)$ nodes which have at least a distance of $\lfloor m - \varepsilon m \rfloor$ from v_0.

Proof: Let

$$L = \{\, v \mid v \neq \alpha 0^k \beta \ \text{ for some } \ k \geq \varepsilon m,\ \alpha 0^k \beta \in \{0,1\}^m \}$$

be the subset of vertices. Then $|L| \geq 2^m - m2^{m-\varepsilon m}$. Let $v \in L$. As the longest sequence α of consecutive 0's in v has at most length $\lfloor \varepsilon m \rfloor$, the bit string $v_0 = 00...0$ has to be rotated at least $\lfloor m - \varepsilon m \rfloor$ times to change the 1's left and right of α. Therefore, the distance between any vertex v from L and v_0 is at least $\lfloor m - \varepsilon m \rfloor$. □

Lemma 4: (Small cycles in the deBruijn network)

Let $DB(m)$ be the deBruijn network of dimension m. Then each edge is contained in a cycle of length at most 4.

Proof: Let $v = a\alpha b$ be a node in $DB(m)$. Then all edges incident to v are covered by the two cycles

$$a\alpha b \;—\; \alpha b\bar{a} \;—\; \bar{a}\alpha b \;—\; \alpha b a \;—\; a\alpha b$$

and

$$a\alpha b \;—\; \bar{b}a\alpha \;—\; a\alpha\bar{b} \;—\; ba\alpha \;—\; a\alpha b$$

of length at most 4. □

Now, we are able to state the new lower bound:

Theorem 4: (Lower bound for broadcasting in the deBruijn network)

$b(DB(m)) > 1.3171m$ for all sufficiently large m.

Proof: The proof is similar to that of Theorem 2. We suppose that broadcasting can be completed on $DB(m)$ in time $m + tm$, $0 \le t < 1/3$. The node $v_0 = 00\ldots0$ is taken as the originator of the message. The recursion formula for $A(i,t)$ is exactly the same as in the proof of Theorem 2. This time, the condition obtained from Lemma 3 is

$$\sum_{n=m}^{m+tm} \sum_{l=0}^{m+tm-n} A(n, n+l) \ge 2^m - o(2^m).$$

Similar estimations as before show that this cannot be true for $t \le 0.3171$. □

7 Conclusions

In this paper, we have proved new lower and upper bounds for broadcasting in the butterfly network. In order to apply the lower bound technique to other networks as well, we have pointed out the needed properties of the graph:

a) fixed degree d,

b) there is a node from which *a lot* of vertices have a *large* distance (*large* $\approx$ diameter),

c) each edge is contained in a cycle of length at most ℓ

(in our case, $d = 4$ and $\ell = 4$). That way, we were able to derive new lower bounds for broadcasting in the (binary) deBruijn network as well. In both cases, we were able to reduce the gap between the upper and lower bound considerably. But still a lot of work needs to be done to close it.

As far as improving the lower bounds is concerned, we think that there must be some other property of the graphs which makes broadcasting very difficult. One such property seems to be the many cycles of length m, but we could not make a proof out of it so far.

As for the upper bound in the butterfly network, we have seen that our algorithm performs very well for small networks. Actually, for dimension at most 4, it has turned out to be optimal. In the meantime, we have found a new algorithm which seems to take less than $2m - 1$ rounds for large m. But we are still working on the analysis.

Finally, with some more extensions, it is possible to apply our lower bound technique to networks of higher node degree, e.g. general butterfly networks as considered in [ABR90], or general deBruijn and Kautz networks as investigated in [BP88] and [HOS91]. So far, we have been able to come up with the general analysis, which is somewhat more complicated. But we still have to apply it to the networks of interest.

References

[AK89] S.B. Akers, B. Krishnamurthy, "A group-theoretic model for symmetric interconnection networks", *IEEE Transactions on Computers*, vol. 38, no. 4, pp. 555-566, 1989.

[ABR90] F. Annexstein, M. Baumslag, A.L. Rosenberg, "Group action graphs and parallel architectures", *SIAM J. Comput.*, vol. 19, no. 3, pp. 544-569, 1990.

[BHLP88] J.-C. Bermond, P. Hell, A.L. Liestman, J.G. Peters, "Broadcasting in bounded degree graphs", Technical report, Simon Fraser University, 1988, to appear in *SIAM Journal on Discrete Maths.*

[BP85] C.A. Brown, P.W. Purdom, *The Analysis of Algorithms*, Holt, Rinehart and Winston, New York, 1985, § 5.3.

[BP88] J.-C. Bermond, C. Peyrat, "Broadcasting in deBruijn networks", Proc. 19th Southeastern Conference on Combinatorics, Graph Theory and Computing, *Congressus Numerantium* 66, pp. 283-292, 1988.

[BP89] J.C. Bermond, C. Peyrat, "De Bruijn and Kautz networks: a competitor for the hypercube?", In F. Andre, J.P. Verjus, editors, *Hypercube and Distributed Computers*, pp. 279-294, North-Holland, 1989.

[CGV89] R.M. Capocelli, L. Gargano, U. Vaccaro, "Time Bounds for Broadcasting in Bounded Degree Graphs", *15th Int. Workshop on Graph-Theoretic Concepts in Computer Science* (WG 89), LNCS 411, pp. 19-33, 1989.

[DB46] N.G. De Bruijn, "A combinatorial problem", *Koninklijke Nederlandsche Akademie van Wetenschappen Proc.*, vol. 49, pp. 758-764, 1946.

[FL91] P. Fraigniaud, E. Lazard, "Methods and problems of communication in usual networks", *manuscript*, 1991.

[HHL88] S.M. Hedetniemi, S.T. Hedetniemi, A.L. Liestman, "A survey of gossiping and broadcasting in communication networks", *Networks*, vol. 18, pp. 319-349, 1988.

[HJM90] J. Hromkovic, C.D. Jeschke, B. Monien, "Optimal algorithms for dissemination of information in some interconnection networks", *Proc. MFCS 90*, LNCS 452, pp. 337-346, 1990.

[HOS91] M.C. Heydemann, J. Opatrny, D. Sotteau, "Broadcasting and spanning trees in de Bruijn and Kautz networks", 1991, to appear in *Discrete Applied Mathematics, special issue on Interconnection Networks.*

[KLM90] R. Klasing, R. Lüling, B. Monien, "Compressing cube-connected cycles and butterfly networks", *Proc. 2nd IEEE Symposium on Parallel and Distributed Processing*, pp. 858-865, 1990.

[LP88] A.L. Liestman, J.G. Peters, "Broadcast networks of bounded degree", *SIAM Journal on Discrete Maths*, vol. 1, no. 4, pp. 531-540, 1988.

[MS90] B. Monien, I.H. Sudborough, "Embedding one Interconnection Network in Another", *Computing Suppl.* 7, pp. 257-282, 1990.

[P68] M.G. Pease, "An adaption of the Fast Fourier transform for parallel processing", *Journal ACM*, vol 15, pp. 252-264, April 1968.

[SP89] M.R. Samatham, D.K. Pradhan, "The de Bruijn multiprocessor network: a versatile parallel procesing and sorting network for VLSI", *IEEE Transactions on Computers*, vol. 38, no. 4, pp. 567-581, 1989.

[St91a] E.A. Stöhr, "Broadcasting in the butterfly network", *IPL* 39, pp. 41-43, 1991.

[St91b] E.A. Stöhr, "On the broadcast time of the butterfly network", *Proc. 17th Int. Workshop on Graph-Theoretic Concepts in Computer Science* (WG 91), 1991, to be published in LNCS.

Interval Approximations of Message Causality in Distributed Executions

Claire DIEHL and Claude JARD

IRISA,
Campus de Beaulieu,
F-35042 Rennes, FRANCE.
jard@irisa.fr

Abstract

In this paper we study timestamping for the dynamic analysis of events ordering in message-passing systems. We use the partial order theory to shed new light on classical timestamping algorithms. Consequently we present a new stamp technique based on a special class of order: interval orders. This technique gives better results than the Lamport's logical clocks for the same cost. Lastly we characterize executions for which our timestamping is optimum.

1 The nature of distributed executions

1.1 General motivation

Interest for distributed computing is growing but mastering all its aspects is still a challenge. We need theories and algorithms to handle distributed executions.

In a distributed execution, the ordering of events must be precisely defined. The set of events occurring on a single processor is totally ordered, but the relations between events from distinct processors depends on communication. The observation of this ordering without disturbing the program behavior is still a difficult problem. But this is a necessary step for many applications: diagnosis, performance evaluation, monitoring, replay, animation... Anyway this can help us to have a better understanding of an execution.

As far as possible we will try to stay in a general framework: we consider a distributed system as a collection of n processors which communicate each other by exchanging messages. The number of processors is constant and known. Therefore we confuse the notion of process and the notion of processor. We do not make any assumption about how the computers are connected, provided that the network is connected.

1.2 Events and causality

When a program runs on a group of processors, it defines a set of *actions*. When an action is executed, it is called an *event*. Such an action can be an internal action on a processor, a sending or a receipt of a message to or from another processor. The set of events is ordered by the message causality relation. This order was first presented by Lamport in [6] and is usually called *happened before*. It can be defined as the transitive closure of these two rules: the events happening on a single processor are totally ordered, and the sending of a message occurs before its receipt.

In practice, only particular events, defined as observable, are traced. But all the communicating events are needed to compute the causality relation.

We will use the following notations: θ is the causality relation:

$$\theta(x,y) \iff x \textit{ causally precedes } y \iff$$

$$\textit{there exists a path from } x \textit{ to } y \textit{ in the time diagram of the execution}$$

E is the set of events used for the timestamping and $\mathcal{E} = (E,\theta)$ the associated order. O is the subset of the events which are effectively observed and $\mathcal{O} = (O,\theta)$ the associated suborder.

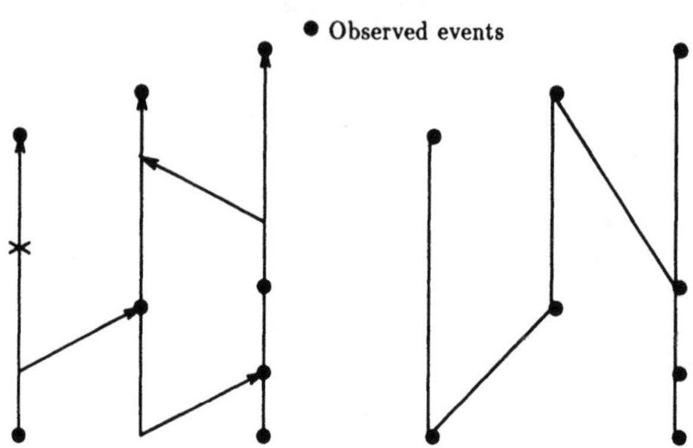

Figure 1: A distributed execution and its associated diagram

1.3 Preliminaries on partial orders

We will represent distributed computations using **diagrams**: the transitive reduction of the graph of the observed events ordering where the edge direction is implicit from bottom to top (see figure 1).

The **size** $|\mathcal{E}|$ of an order $\mathcal{E} = (E,\theta)$ is the total number of relations in θ. This is also the transitive closure edge number of the diagram representing $\mathcal{E}$.

A set of pairwise comparable elements is called a **chain** and a set of pairwise incomparable elements is an **antichain**.

The **width** of an order is the length of one of its largest antichains.

y **covers** x, or y is an **immediate successor** of x, and we note $\theta_{im}(x,y)$ when:

$$\theta_{im}(x,y) \iff \theta(x,y) \text{ and } \nexists z \in E,\ \ \theta(x,z) \text{ and } \theta(z,y).$$

We define the **predecessor set** for an element $x \in E$ as:

$$Pred_{\mathcal{E}}(x) = \{y \in E/\theta(y,x)\} \text{ and } Pred_{im\mathcal{E}}(x) = \{y \in E/\theta_{im}(y,x)\}$$

In order to describe the stamping mechanisms, we will use the following notations. If e is an event, we denote e^- (resp. e^+) the immediately preceding (resp. following) event on the same processor. If e is the sending of a message, e^r is the receipt of the same message. And if e is a receipt, e^e is the sending of the message (see figure 2).

Interval orders model the sequential structure of intervals of the real line. As there is a large range of applications using the time concept, they have been intensively studied in recent years [5, 8, 10].

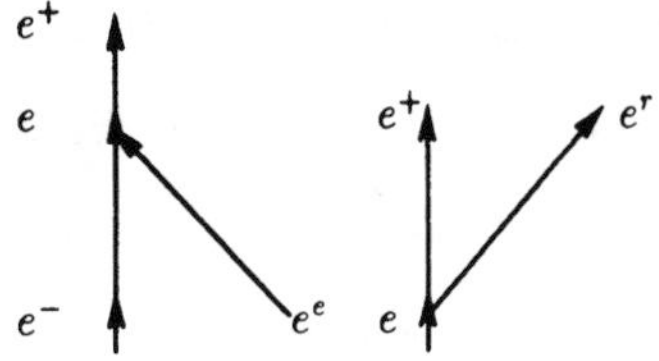

Figure 2: Notations

Definition 1 *(E,θ) is an* **interval order** iff *there is a set of intervals $(I_x)_{x\in E}$ on the real line such that:*

$$\theta(x,y) \iff I_x \text{ is on the left of } I_y$$

This definition gives a nice representation of such an order: we can associate a pair of real numbers to each element. There exist other characterizations of interval orders which can be easily obtained (see for instance [8]):

Theorem 1 *For an order $\mathcal{E} = (E,\theta)$ the following statements are equivalent:*

(i) $\mathcal{E}$ is an interval order.

(ii) $\mathcal{E}$ does not contain any suborder isomorphic to the order "$2 \oplus 2$" (see figure 3).

(iii) The maximal antichains of $\mathcal{E}$ can be linearly ordered such that, for every element x the maximal antichains containing x occur consecutively.

(iv) The sets of predecessors $Pred_{\mathcal{E}}(x)$ are linearly ordered by inclusion.

Figure 3: The order "$2 \oplus 2$"

An **extension** of $\mathcal{E} = (E,\theta)$ is an order including θ. A **linear extension** is an extension which totally orders E. An **interval extension** is an extension which is an interval order (see the definition just above).

2 Timestamping and partial orders

2.1 General properties

A timestamping technique associates a date $\delta(x)$ to each event x. If there exists an ordering of the stamps, there also exists a relation between the events.

Coding exactly the causality order of a computation is very expensive in general and can alter it: additional computations, larger messages slow down the execution speed and possibly alter the algorithm behavior. Then we are interested in just observing approximations of the causality order. A timestamping algorithm should have the following properties:

- The stamps ordering extends the causality order.
- The algorithm is incremental: the stamp of an event depends on its preceding events. If necessary it should depend on some events in its future, but this future should be close: in this case the algorithm is said to be *pseudo-incremental.*
- The computation of a stamp on a processor depends on local informations: the algorithm does not have to add any message.

In order to compare two different timestamping algorithms, we will compute the size of the order [2] produced by both algorithms on the same execution.

2.2 Vector stamping

Timestamping by vectors of integers is actually the illustration of a general technique for coding partial orders. **Dilworth theorem** [3] says that an order of width k can be decomposed into k chains. And with such a partition, we can build an order preserving mapping from E to $\mathbb{N}^k$. The following theorem is probably a special case of a general property but the proof helps us to understand the structure of the coding.

Theorem 2 *For* $\mathcal{E} = (E, \theta)$ *a partially ordered set of width* k *and* $(L_i)_{1\leq i\leq k}$*, a partition into chains of* E *the map:*

$$\begin{aligned} \phi\ :\ & E \longrightarrow \mathbb{N}^k \\ & x \longmapsto (|(Pred_{\mathcal{E}}(x) \cup \{x\}) \cap L_i|)_{1\leq i\leq k} \end{aligned}$$

satisfies[1]*:*

$$\forall x, y \in E,\ \ \theta(x,y) \Longleftrightarrow \phi(x) <_{\mathbb{N}^n} \phi(y)$$

Proof: let x and y be such that
$\theta(x,y)$, *i.e.* $x \in Pred_{\mathcal{E}}(y)$.
So $Pred_{\mathcal{E}}(x) \cup \{x\} \subset Pred_{\mathcal{E}}(y) \cup \{y\}$ and clearly
$\forall i \in \{1..k\},\ \phi(x)_i \leq \phi(y)_i$ and $\exists j \in \{1..k\},\ y \in L_j$ and $\phi(x)_j < \phi(y)_j$.
Thus we have shown $\Longrightarrow$.
Let x and y be such that
$\forall i \in \{1..k\},\ |(Pred_{\mathcal{E}}(x) \cup \{x\}) \cap L_i| \leq |(Pred_{\mathcal{E}}(y) \cup \{y\}) \cap L_i|$. Each L_i is a chain, hence $(Pred_{\mathcal{E}}(x) \cup \{x\}) \cap L_i \subseteq (Pred_{\mathcal{E}}(y) \cup \{y\}) \cap L_i$ and consequently
$Pred_{\mathcal{E}}(x) \cup \{x\} \subset Pred_{\mathcal{E}}(y) \cup \{y\}\ :\ \theta(x,y)$. □

When observing a distributed computation, there is a trivial partition of the set of events: each chain L_i contains the events occurring on the ith processor. If there are n processors in the system:

$$E = \biguplus_{1\leq i\leq n} L_i,\ L_i \text{ contains the events of the } i^{th} \text{ processor.}$$

We denote δ_i the vector whose all components are null except the ith which is equal to 1. The mechanism which constructs the preceding coding of the causality order is the following. Each processor has a clock vector of size n: $\delta(x) \in \mathbb{N}^n$.

- When the first event x of the ith processor occurs, $Pred_{\mathcal{E}}(x) = \emptyset$, its clock is initialized to δ_i.

[1] $<_{\mathbb{N}^k}$ is the canonical order on $\mathbb{N}^k$: $\forall x, y \in \mathbb{N}^k,\ x <_{\mathbb{N}^k} y \Longleftrightarrow \forall i \in \{1..k\},\ x_i \leq y_i$ and $\exists j \in \{1..k\},\ x_j < y_j$

- If the current event $x \in L_i$ is an internal event or the sending of a message, $Pred_{\mathcal{E}}(x) = Pred_{\mathcal{E}}(x^-) \cup \{x^-\}$ and therefore $\delta(x) = \delta(x^-) + \delta_i$.
- If the current event $x \in L_i$ is the receipt of a message, $Pred_{\mathcal{E}}(x) = Pred_{\mathcal{E}}(x^-) \cup Pred_{\mathcal{E}}(x^e) \cup \{x^-, x^e\}$.
  The sets L_j are totally ordered, hence
  $\forall j \in \{1..n\}, (Pred_{\mathcal{E}}(x^e) \cup \{x^e\}) \cap L_j \subseteq (\text{or} \supseteq) (Pred_{\mathcal{E}}(x^-) \cup \{x^-\}) \cap L_j$
  and hence
  $|Pred_{\mathcal{E}}(x) \cap L_j| = max(|(Pred_{\mathcal{E}}(x^e) \cup \{x^e\}) \cap L_j|, |(Pred_{\mathcal{E}}(x^-) \cup \{x^-\}) \cap L_j|)$:
  $\delta(x) = max(\delta(x^-), \delta(x^e)) + \delta_i$.

We recognize the vector clock mechanism proposed in 1988 by Fidge and Mattern [4, 7]. This timestamping mechanism codes the causality order: the ith component of an event stamp is the total number of events preceding it on the ith processor and characterizes the predecessors set (see figure 4). To compute this clock, each message has to hold a stamp of length n: this is the main drawback of this algorithm and this becomes unrealistic in large parallel systems. But, as said in [1], the size n is necessary if we look for embeddings of the causality order in $(\mathbb{N}, <_{\mathbb{N}^n})$. This is why we are going to study constant size stamps which will approximate the causality.

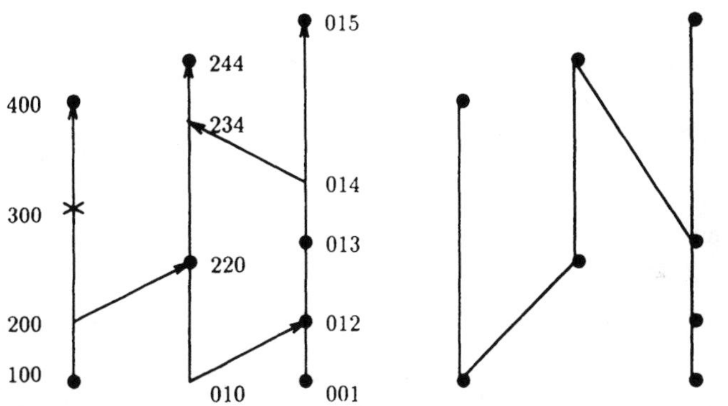

Figure 4: Mattern's vector stamping: $|\mathcal{O}| = 13$

2.3 Scalar stamping

Definition 2 *Let $\mathcal{E} = (E, \theta)$ be a partial order. For $x \in E$ the rank of x, $\rho(x)$ is the maximal length of a chain ending at x.*

The rank concept allows us to partition ordered sets:

Definition 3 *Let $\mathcal{E} = (E, \theta)$ be a partial order over a finite set. The partition:*

$$E = \biguplus_{0 \le i \le h} R_i \text{ where } \forall i \geq 0,\ R_i = \{x \in E / \rho(x) = i\}$$

is called the rank decomposition of $\mathcal{E}$.

We prefer the following constructive characterization. It corresponds to a classical breadth–first graph traversal.

Theorem 3 *Let $\mathcal{E} = (E, \theta)$ be a partial order over a finite set. The partition:*

$$E = \biguplus_{0 \leq i \leq h} A_i \ \textit{where} \ \ A_0 = Min(E, \theta) = \{x \in E / \ \nexists y \in E, \ \theta(y, x)\}$$
$$\textit{and}\ \forall i > 0, \ A_i = Min(E - \bigcup_{j<i} A_j, \theta)$$

is the rank decomposition (i.e. $\forall i, \ R_i = A_i$).

Proof: let $i \geq 0$ and $x \in A_i$. In view of the definition of the sets A_j we can construct a chain ending at x which contains at least one element in each set $A_j, j < i$. Since it is a chain, it cannot contains two elements from the same A_j (each A_j is an antichain). Thus the length of a maximal chain ending at x is i: $\rho(x) = i$. □

Thus the rank decomposition of a partially ordered set is a partition into antichains. Each antichain is not maximal but the total number of antichains is minimal. This theorem gives us an algorithm which easily computes the rank function. Let $x \in A_i, (i > 0)$, by definition of the sets A_j, at least one immediate predecessor of x is in A_{i-1}. Therefore:

$$\rho(x) = max\{\rho(y), \ \theta_{im}(y, x)\} + 1$$

In particular, when $\mathcal{E} = (E, \theta)$ is the causality order of a distributed execution, the rank is computed by the following rules:

- If e is the first event occurring on a site, except the receipt of a message then e is minimal *i.e.* $e \in A_0$ and $\rho(e) = 0$.
- If e is an internal event or the sending of a message, then $Pred_{im\mathcal{E}}(e) = \{e^-\}$ and $\rho(e) = \rho(e^-) + 1$.
- If e is the receipt of a message then $Pred_{im\mathcal{E}}(e) \subseteq \{e^-, e^e\}$ and $\rho(e) = max(\rho(e^-), \rho(e^e)) + 1$.

This is the same mechanism as the logical clock proposed by Lamport in [6]. Fidge and Mattern's vector clock is based on a chain partition of the event set, the scalar stamp of Lamport corresponds to an antichain decomposition. Figure 5 makes visible the structure of the extension produced by the Lamport's stamp algorithm. This is the structure of a **weak order** : a weak order is a totally ordered set where antichains have been put in place of some elements.

Proposition 1 *The Lamport's timestamping computes the rank of each event for the causality order and the resulting order is a weak order.*

3 Interval approximations

3.1 Improvement of Lamport's algorithm

The events ordering due to Lamport's clock can be viewed as an interval order. Actually, to each event e, we can associate an interval of the real line: $[\rho(e), \rho(e) + 1[$. The ordering of the intervals is the same as the ordering of the events. Moreover all these intervals have the same length, thus this ordering belongs to a subclass of interval orders called semiorder (a semiorder is an interval order for which there exists a representation with intervals of the same length). The scheme of the figure 6 shows the interval structure of this order.

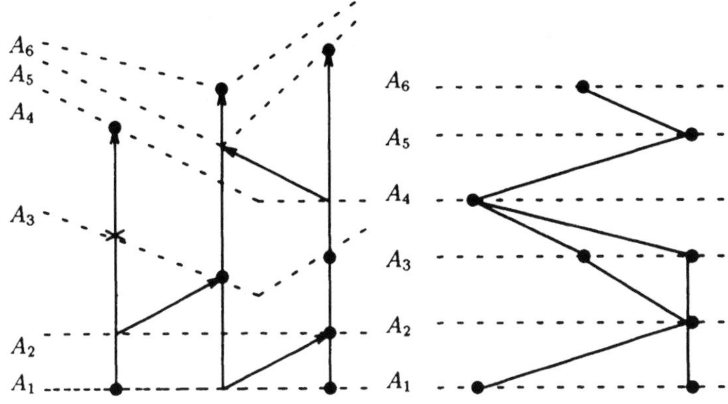

Figure 5: Lamport's scalar stamping:$|\mathcal{O}| = 26$

This remark gives us an idea to improve the Lamport's interval approximation. We associate to each event e the interval $[\delta(e)^-, \delta(e)^+[$ where $\delta(e)^- = \rho(e)$, the Lamport's stamp, and $\delta(e)^+ = min\{\rho(f)/\theta_{im}(e, f)\}$. The value of $\delta(e)^+$ preserves θ because it preserves the ordering of events on a same processor and the causality between a sending and its corresponding receipt. We increase the size of some intervals. Thus it is clear that it will reduce the size of the observed extension.

Proposition 2 *The set* $([\delta(e)^-, \delta(e)^+[)_{e \in E}$ *where:*

$$\delta(e)^- = \rho(e), \text{ Lamport's stamp}$$
$$\delta(e)^+ = min\{\delta(f)^-/\theta_{im}(e, f)\}$$

gives an interval extension (or an interval approximation) of the causality order.

An illustration of this stamp technique appears in figure 6. In terms of order inclusion this interval approximation is between the message causality and the Lamport's logical clocks. However this stamp mechanism is no longer incremental: the right interval endpoint does not depend on the past of the events but on a close future. To know the complete stamp of an event, we must wait for all its immediate successors to occur.

- The internal events and the receipts have only one immediate successor: the next event occurring on the same processor. When it occurs we can trace the stamp of the preceding event.

- A message sending e should have two immediate successors: the next event on the same site e^+ and the receipt event e^r. To compute $\delta(e)^+ = min(\delta(e^+)^-, \delta(e^r)^-)$, we trace its two possible values and we take the lowest.

Consequently we have to process the observed trace, but this can be done in parallel with the execution: we do not have to wait for the end of the computation. As we will see in the following sections, there is a better solution to this problem, when all the sendings are followed by a blocking acknowledgment. The disturbance due to the observation is the same than for the Lamport's logical clock: the stamp of each message has the same size (one integer).

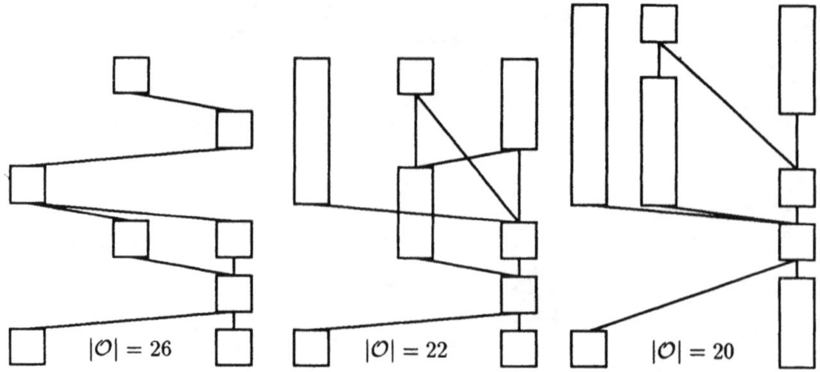

Figure 6: Interval representation of Lamport's stamping, its refinement and the "max+2" stamping

3.2 Computations coded by intervals

In the last section, we have found two interval extensions of the causality order: Lamport's stamping and its refinement. Are there timestamping algorithms which code exactly computations whose causality order is an interval order?

3.2.1 General case

Interval characterization

Papadimitriou and Yannakakis have developed in [9] an algorithm for recognizing interval orders. It constructs an interval representation which is founded on the following theorem.

Theorem 4 *Let $\mathcal{E} = (\mathcal{E}, \theta)$ be an interval order. The set of intervals $([\delta(x)^-, \delta(x)^+[)_{x \in E}$ where*[2]*:*

$$\delta(x)^- = max\{\delta(y)^- / Pred_{\mathcal{E}}(y) \subset Pred_{\mathcal{E}}(x)\} + 1$$
$$\delta(x)^+ = min\{\delta(y)^- / \theta(x, y)\}$$

is an interval representation of $\mathcal{E}$.

It corresponds to an ordering of the maximal antichains of $\mathcal{E}$ according to the property (iii) of theorem 1: $\delta(x)^-$ (resp. $\delta(x)^+$) is the number of the first (resp. the last) antichain containing x. But it seems to be difficult to compute this representation "on the fly" during a distributed execution. More generally there is the following result:

Theorem 5 *Let $\mathcal{E} = (\mathcal{E}, \theta)$ be an interval order. The set $([\delta(x)^-, \delta(x)^+[)_{x \in E}$ where:*

$$\delta^- \textit{ satisfies } \forall x, y \in E,\ \ \delta(x)^- \leq \delta(y)^- \Longrightarrow Pred_{\mathcal{E}}(x) \subseteq Pred_{\mathcal{E}}(y)$$
$$\textit{and } \delta(x)^+ = min\{\delta(y)^- / \theta_{im}(x, y)\}$$

is an interval representation of $\mathcal{E}$.

---

[2] By convention $max(\emptyset) = 0$

The proof is based on the proof of theorem 4 given in [9].

Proof: We have to prove that: $\forall x, y \in E,\ \theta(x,y) \Longleftrightarrow \delta(x)^+ \leq \delta(y)^-$.
$\Longrightarrow$ is obvious because of the definition of $\delta(x)^+$:
$min\{\delta(y)^-/\theta_{im}(x,y)\} = min\{\delta(y)^-/\theta(x,y)\}$.
Conversely let x and y be such that $\delta(x)^+ \leq \delta(y)^-$.
There exists z such that $\theta(x,z)$ and $\delta(x)^+ = \delta(z)^-$.
Therefore $\delta(z)^- \leq \delta(y)^-$ and $x \in Pred_{\mathcal{E}}(z) \subseteq Pred_{\mathcal{E}}(y) : \theta(x,y)$. □

For instance the function:

$$\delta(x)^- = |Pred_{\mathcal{E}}(x)|$$

satisfies the assumption of theorem 5. In fact let x and y be such that $|Pred_{\mathcal{E}}(x)| \leq |Pred_{\mathcal{E}}(y)|$. The predecessor sets of $\mathcal{E}$ are totally ordered by inclusion (see theorem 1) and therefore we necessarily have $Pred_{\mathcal{E}}(x) \subseteq Pred_{\mathcal{E}}(y)$.

Timestamping mechanism

We now assume that the causality order of the distributed computation is an interval order. This assumption is of course unrealistic, the order "$2 \bigoplus 2$" (theorem 1) is unavoidable during a distributed computation. But with this assumption, for each event e, the following mechanism computes the stamp $[|Pred_{\mathcal{E}}(e)|, min\{\delta(f)^-/\theta_{im}(e,f)\}[$ which codes exactly the causality order.

- If e is an internal event:
  $Pred_{\mathcal{E}}(e) = Pred_{\mathcal{E}}(e^-) \cup \{e^-\}$ and therefore
  $\delta(e)^- = \delta(e^-)^- + 1$ and $\delta(e)^+ = \delta(e^+)^-$ ($\delta(e)^+$ is known when e^+ occurs).
- If e is the sending of a message, as previously
  $\delta(e)^- = \delta(e^-)^- + 1$, but
  $\delta(e)^+ = min(\delta(e^+)^-, \delta(e^r)^-)$. We have to trace both values.
- If e is a receipt $\delta(e)^+ = \delta(e^+)^-$. To compute $\delta(e)^-$ we can have either e^- or e^e or the two immediate predecessors of e. There are two cases:
  - If $\theta(e^-, e^e)$ or $\theta(e^e, e^-)$ then $\delta(e)^- = max(\delta(e^e)^-, \delta(e^-)^-) + 1$
  - If e^- et e^e are concurrent then $\delta(e)^- = max(\delta(e^e)^-, \delta(e^-)^-) + 2$

Determining *on line* the relation existing between two events seems to be insolvable without the vector stamps. In the general case we cannot say whether we must perform
$\delta(e)^- = max(\delta(e^e)^-, \delta(e^-)^-) + 1$ or $+2$.

If the causality order of the execution is not an interval order it is clear that performing $+1$ or $+2$ preserves the message causality ordering: it orders the events on a same processor and it preserves the ordering of a sending and its receipt. By this way we produce interval extensions of the causality order. If we always perform $+1$, we compute the refinement of Lamport's stamping. If we always perform $+2$ then we observe another interval extension which is generally neither better nor worse than the previous one: it can have more or less edges (see figure 6). However there is a particular class of execution for which we are able to characterize the interval orders *on line*.

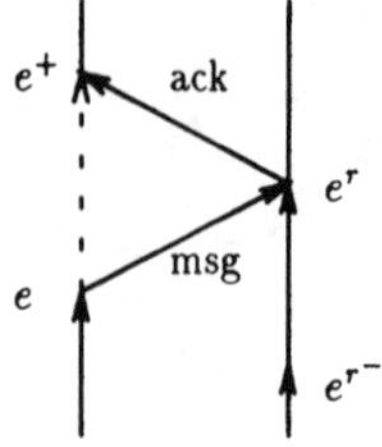

Figure 7: A synchronous execution

3.2.2 The RPC case

We are now considering distributed systems where the message-passing works like a Remote Procedure Call: the sender of a message is blocked until a return message is available. Each exchange is done like according to the figure 7: we merge the receipt of a message and the acknowledgment sending event. Such systems present two advantages for us. First we can really compute a stamp $\delta(e)^-$ which characterizes the interval orders and then the second stamp $\delta(e)^+$ is known on the processor where e occurs, possibly with some delay.

Actually we have seen in the preceding paragraph that if e is a receipt then we have to know whether e^- and e^e are concurrent to compute $|Pred_{\mathcal{E}}|$. But in this case there is the following result:

Proposition 3 *Let $\mathcal{E} = (E, \theta)$ be the causality order of an RPC distributed execution. If $\mathcal{E}$ is an interval order then the set of intervals $([\delta(e)^-, \delta(e)^+[)_{e \in E}$ where:*

$$\delta(e)^- = \begin{cases} \delta(e^e)^- + 1 & \text{if } e \text{ is the receipt of an acknowledgment} \\ max(\delta(e^-)^-, \delta(e^e)^-) + 2 & \text{if } e \text{ is the receipt of a message} \\ \delta(e^-) + 1 & \text{otherwise} \end{cases}$$
$$\delta(e)^+ = min\{\delta(f)^- / \theta_{im}(e, f)\}$$

is an interval representation of $\mathcal{E}$.

Proof: As before, if e is an internal event or a sending, then $|Pred_{\mathcal{E}}(e)| = |Pred_{\mathcal{E}}(e^-)| + 1$.

If e is the receipt of an acknowledgment then $\theta(e^-, e^e)$ and $Pred_{\mathcal{E}}(e) = Pred_{\mathcal{E}}(e^e) \cup \{e^e\}$. Thus $|Pred_{\mathcal{E}}(e)| = |Pred_{\mathcal{E}}(e^e)| + 1$.

If e the receipt of a message, then $|Pred_{\mathcal{E}}(e)| = max(|Pred_{\mathcal{E}}(e^-)|, |Pred_{\mathcal{E}}(e^e)|) + (1 \text{ or } 2)$ (because $\mathcal{E}$ is an interval order and the predecessors sets are included).

Thus performing
$\delta(e)^- = max(\delta(e^-)^-, \delta(e^e)^-) + 2$ may count an event twice. Therefore
$\delta(e)^- = |Pred_{\mathcal{E}}(e)| + K(e)$ where $K(e)$ corresponds to the number of sending events preceding e which have been counted twice.

Let e be the sending of a message, if e is counted once (resp. twice) when we compute $\delta(e^r)^-$ then, because of the return message, it is also counted once (resp. twice) when we compute $\delta(e^+)^-$. Therefore all the stamps of the successors of e count e either always once or always twice. This is the main difference from asynchronous systems. So we can say that $K(e)$ is an increasing function of $Pred_{\mathcal{E}}(e)$. In this way:

$$\begin{array}{lll} \forall e, f \in E, & Pred_{\mathcal{E}}(f) \subset Pred_{\mathcal{E}}(e) & \Longrightarrow K(f) \leq K(e) \\ & & \Longrightarrow \delta(f)^- < \delta(e)^- \\ i.e. & \delta(e)^- \leq \delta(f)^- & \Longrightarrow Pred_{\mathcal{E}}(e) \subseteq Pred_{\mathcal{E}}(f) \end{array}$$

because all the predecessors sets are ordered by inclusion.
Therefore δ^- satisfies the assumption of the theorem 5. □

The computation of the second stamp $\delta(e)^+$ is easy:

- If e is an internal event or the receipt of an acknowledgment then $\delta(e)^+ = \delta(e^+)^-$ and it is known with a delay.
- If e is a sending then $\theta(e^r, e^+)$ and $\delta(e)^+ = \delta(e^r)^-$. These values are known on the same site when the return message arrives: $\delta(e)^+ = \delta(e^+)^- - 1$.
- If e is the sending of an acknowledgment then $\delta(e^r)^- = \delta(e)^- + 1$ and necessarily $\delta(e)^+ = \delta(e)^- + 1$, this is known when e occurs.

The delay necessary to stamp most events is not a problem: in practice we only observe a few events and thus this delay becomes invisible. We write the corresponding algorithm:

The algorithm

```
h := 0;          (* logical clock *)
t := 0;          (* preceding value of h *)
e⁻ := "";        (* name of the preceding event *)
s := false;      (* true if the preceding event is an acknowledgment *)
At each event e ∈ E do begin
    t := h;
    case
        e is an internal event do
            h := h+1;
            if ¬s and e⁻ ∈ O then trace(e⁻, t, h);
            s := false;
        e is the sending of a message m do
            h := h+1;
            if ¬s and e⁻ ∈ O then trace(e⁻, t, h);
            s := false;
            !m(h); (* sending of m stamped by h *)
        e is the receipt of the message m stamped by h' do
            h := max(t, h')+2;
            if ¬s and e⁻ ∈ O then trace(e⁻, t, h);
            if e ∈ O then trace(e, h, h+1)
            s := true;
            !ack(h); (* sending of the acknowledgment stamped by h *)
        e is the receipt of an acknowledgment stamped by h' do
            h := h'+1;
            if e⁻ ∈ O then trace(e⁻, t, h')
            s := false;
    end
    e := e⁻;
end
```

If the message causal ordering of a distributed execution is an interval order then this algorithm allows us to observe it exactly. In the general case it produces an interval approximation which in that sense is optimum.

4 Conclusion and prospects

In this paper we have shown that the partial order theory can contribute to the elaboration of observation tools for distributed executions. Presently there is a gap between the two main stamp algorithms. Lamport's one (1978) is efficient but imprecise, Fidge and Mattern's one (1988) has a perfect precision but induces too much overhead.

We have used the class of interval orders to exhibit a new stamp technique. We have shown that it seems particularly suitable for distributed systems where message communications are performed by RPC.

In this way it seems interesting to look for other classes which can be coded by bounded size vectors. It also interesting to study the properties of the ordering induced by particular protocols. This research may lead to new stamping algorithms allowing to approach the causality order with a realistic overhead.

Acknowledgments

Special thanks are due to M. Habib, P. Baldy and C. Fioro, members of the CRIM laboratory in Montpellier, France. By long stimulating discussions, they provided most of the theoretical background on partial orders needed for this paper. We also thank B. Caillaud and JX. Rampon at IRISA for their advices and the careful reading of the paper. This work was partly funded by the French national project C^3 on concurrency.

References

[1] B. Charron-Bost. Concerning the size of logical clocks in distributed systems. *Information Processing Letters*, 39(1):11–16, july 1991.

[2] W.A. Cook, M. Kress, and M. Seiford. An axiomatic approach to distance on partial orderings. *R.A.I.R.O. Recherche opérationnelle/Operations Research*, 20(2):115–122, Mai 1986.

[3] R.P Dilworth. A decomposition theorem for partially ordered sets. *Annals of Math.*, 51:161–165, 1950.

[4] J. Fidge. Timestamps in message passing systems that preserve the partial ordering. In *Proc.* 11th *Australian Computer Science Conference*, pages 55–66, 1988.

[5] P.C Fishburn. *Interval orders and interval graphs.* Wiley, New York, 1985.

[6] L. Lamport. Time, clocks and the ordering of events in a distributed system. *Communications of the ACM*, 21(7):558–565, July 1978.

[7] F. Mattern. Virtual time and global states of distributed systems. In Cosnard, Quinton, Raynal, and Robert, editors, *Proc. Int. Workshop on Parallel and Distributed Algorithms Bonas, France, Oct. 1988*, North Holland, 1989.

[8] R.H. Möhring. Computationally tractable classes of ordered sets. In *Algorithms and Order*, pages 105–193, Kluwer Academic Publishers, 1989.

[9] C. H. Papadimitriou and M. Yannakakis. Scheduling interval-ordered task. *Siam J. Comput.*, 8(3):405–409, August 1979.

[10] J.X. Rampon. Mesures de concurrence et extensions d'intervalles. Ph. D Thesis, University of Montpellier II, February 1991.

COMPLEXITY

On the Approximability of the Maximum Common Subgraph Problem

Viggo Kann
Department of Numerical Analysis and Computing Science
Royal Institute of Technology
S-100 44 Stockholm
Sweden
viggo@nada.kth.se

Abstract

Some versions of the maximum common subgraph problem are studied and approximation algorithms are given. The maximum bounded common induced subgraph problem is shown to be MAX SNP-hard and the maximum unbounded common induced subgraph problem is shown to be as hard to approximate as the maximum independent set problem. The maximum common induced connected subgraph problem is still harder to approximate and is shown to be NPO PB-complete, i.e. complete in the class of optimization problems with optimal value bounded by a polynomial.
Key words: Approximation, graph problems, computational complexity.

1 Introduction

The SUBGRAPH ISOMORPHISM problem is a famous NP-complete problem. It is one of the first problems mentioned in *Computers and Intractability* by Garey and Johnson [11]. Given two graphs the problem is to decide whether the second graph is isomorphic to any subgraph of the first graph. The problem is shown to be NP-complete by the following simple reduction from the CLIQUE problem. Let the first input graph to SUBGRAPH ISOMORPHISM be the input graph to CLIQUE and let the second input graph be a K-clique where K is the bound in the CLIQUE problem. Now the K-clique is isomorphic to a subgraph of the first graph if and only if there is a clique of size K or more in the graph.

A related optimization problem is called MAXIMUM COMMON SUBGRAPH. In this problem we are given two graphs and we want to find the largest subgraphs which are isomorphic. The corresponding decision problem was shown to be NP-complete by Garey and Johnson using the same reduction as above [11]. The approximation properties of various versions of this problem are studied in this paper.

NP problems, like MAXIMUM COMMON SUBGRAPH, which are in fact optimization problems are called NPO problems (NP optimization problems). Provided that NP $\neq$ P there is no algorithm which finds the optimal solution to an NP-complete optimization problem in polynomial time. Still there can exist *polynomial time approximation algorithms* for the problem. However the approximability of different NPO problems differs enormously.

For example the TSP (Travelling Salesperson Problem) with triangular inequality can be solved approximately within a factor 3/2, i.e. one can in polynomial time find a trip of length at most 3/2 times the shortest trip possible [7]. The general TSP cannot be approximated within any constant factor if NP $\neq$ P [11]. Another example is the knapsack problem, which is NP-complete but can be approximated within every constant in polynomial time [12]. Such a scheme for approximating within every constant is called a PTAS (Polynomial Time Approximation Scheme).

In 1988 Papadimitriou and Yannakakis defined, using Fagin's logical classification of NP [9], the classes MAX NP and MAX SNP together with a concept of reduction, called *L-reduction*, which preserves approximability within constants [22]. All problems in MAX NP and MAX SNP can be approximated within a constant in polynomial time. Several maximization problems were shown to be complete in MAX SNP under L-reductions, for example maximum 3-satisfiability, maximum cut and maximum 3-set packing in a graph with bounded degree [3, 4, 14, 15, 16, 22, 23]. Minimization problems can also be placed in MAX SNP through L-reductions to maximization problems.

All attempts to construct a PTAS for a MAX SNP-complete problem have failed and hence it seems reasonable to conjecture that no such scheme exists, in particular since if one problem had a PTAS then every problem in MAX SNP would admit a PTAS.

Recently more work has been done on logical definability of NPO problems [17, 18, 21]. New classes over MAX SNP and MAX NP and analogous minimization classes have been defined. Not many of these classes seem to capture approximation properties, however. The value of classifying problems using logical definability can be discussed because the same problem may or may not be included in a class depending on how it is formulated [16].

Krentel has defined a class of optimization problems called OPTP[$\log n$], which consists of all NPO problems which are polynomially bounded, that is all problems satisfying $opt(I) \leq p(|I|)$ for all problem instances I, where p is a polynomial [19]. In this paper we shall call this class NPO PB. Some problems, for example the LONGEST INDUCED PATH problem, are NPO PB-complete under L-reductions [2]. An NPO PB-complete problem cannot be approximated within $O(n^{\varepsilon})$ for any $\varepsilon > 0$, unless NP = P. Note that this is a different reduction from the one Krentel used when he defined OPTP[$\log n$]-complete problems.

Several other attempts have been made to find a theory explaining why a problem enjoys particular approximation properties. See [6] for a survey of the field.

2 Definitions

Definition 1 [8] An NPO problem (over an alphabet Σ) is a tuple $F = (\mathcal{I}_F, S_F, m_F, opt_F)$ where

- $\mathcal{I}_F \subseteq \Sigma^*$ is the space of *input instances*. It is recognizable in polynomial time.
- $S_F(x) \subseteq \Sigma^*$ is the space of *feasible solutions* on input $x \in \mathcal{I}_F$. The only requirement on S_F is that there exist a polynomial q and a polynomial time computable predicate π such that for all x in $\mathcal{I}_F$ S_F can be expressed as $S_F(x) = \{y : |y| \leq q(|x|) \wedge \pi(x, y)\}$ where q and π only depend on F.
- $m_F : \mathcal{I}_F \times \Sigma^* \to \mathbb{N}$, the *objective function*, is a polynomial time computable function. $m_F(x, y)$ is defined only when $y \in S_F(x)$.
- $opt_F \in \{\max, \min\}$ tells if F is a *maximization* or a *minimization* problem.

Solving an optimization problem F given the input $x \in \mathcal{I}_F$ means finding a $y \in S_F(x)$ such that $m_F(x, y)$ is optimum, that is as big as possible if $opt_F = \max$ and as small as possible if $opt_F = \min$. Let $opt_F(x)$ be this optimal value of f. Approximating an optimization problem F given the input $x \in \mathcal{I}_F$ means finding any $y' \in S_F(x)$. How good the approximation is depends on the relation between $m_F(x, y')$ and $opt_F(x)$.

Definition 2 The *relative error* of a feasible solution with respect to the optimum of an NPO problem F is defined as

$$\mathcal{E}\left(opt_F(x), y\right) = \frac{|opt_F(x) - m_F(x, y)|}{opt_F(x)}$$

where $y \in S_F(x)$.

Definition 3 We say that a maximization problem F *can be approximated within* p if there exists a polynomial time algorithm A such that for all instances $I \in \mathcal{I}_F$, $A(I) \in S_F(I) \wedge opt_F(I)/m_F(A(I)) \leq p$.

Definition 4 [22] Given two NPO problems F and G and a polynomial time transformation $f : \mathcal{I}_F \to \mathcal{I}_G$. f is an *L-reduction* from F to G if there are positive constants α and β such that for every instance $I \in \mathcal{I}_F$

i) $opt_G(f(I)) \leq \alpha \cdot opt_F(I)$,

ii) for every solution of $f(I)$ with measure c_2 we can in polynomial time find a solution of I with measure c_1 such that $|opt_F(I) - c_1| \leq \beta\, |opt_G(f(I)) - c_2|$.

Papadimitriou and Yannakakis have shown that the composition of L-reductions is an L-reduction and that if F L-reduces to G with constants α and β and there is a polynomial time approximation algorithm for G with worst-case relative error ε, then there is a polynomial time approximation algorithm for F with worst-case relative error $\alpha\beta\varepsilon$ [22]. Thus the L-reduction preserves approximability within constants.

Definition 5 Definition of the problems mentioned in the text.

MAX CIS — *Maximum common induced subgraph.*
$\mathcal{I} = \{G_1 = <V_1, E_1>, G_2 = <V_2, E_2> \text{ graphs}\}$
$S(G_1, G_2) = \{V_1' \subseteq V_1, V_2' \subseteq V_2 : G_1|_{V_1'} \text{ and } G_2|_{V_2'} \text{ are isomorphic}\}$
$m(V_1', V_2') = |V_1'| = |V_2'|$
$opt = \max$
The isomorphism between $G_1|_{V_1'}$ and $G_2|_{V_2'}$ is expressed as a bijective function $f : V_1' \to V_2'$ such that for all $v_1, v_2 \in V_1$ we have $(v_1, v_2) \in E_1 \Leftrightarrow (f(v_1), f(v_2)) \in E_2$. We say that v *is matched with* $f(v)$ and that $f(v)$ *is matched with* v.

MAX CES — *Maximum common edge subgraph.*
$\mathcal{I} = \{G_1 = <V_1, E_1>, G_2 = <V_2, E_2> \text{ graphs}\}$
$S(G_1, G_2) = \{E_1' \subseteq E_1, E_2' \subseteq E_2 : G_1|_{E_1'} \text{ and } G_2|_{E_2'} \text{ are isomorphic}\}$
$m(E_1', E_2') = |E_1'| = |E_2'|$
$opt = \max$

MAX CIS $-B$ — *Maximum bounded common induced subgraph.* This is the same problem as MAX CIS but the degree of the graphs G_1 and G_2 is bounded by the constant B.

MAX CES $-B$ — *Maximum bounded common edge subgraph.* This is the same problem as MAX CES but the degree of the graphs G_1 and G_2 is bounded by the constant B.

MAX CICS — *Maximum common induced connected subgraph.* This is the same problem as MAX CIS but the only valid solutions are connected subgraphs.

MAX 3SAT $-B$ — *Maximum bounded 3-satisfiability.*
$\mathcal{I} = \{U$ set of variables, C set of disjunctive clauses, each involving at most three literals (a variable or a negated variable) and such that the total number of occurrences of each variable is bounded by the constant $B\}$
$S(U, C) = \{C' \subseteq C :$ there is a truth assignment for U such that every clause in C' is satisfied$\}$
$m(C') = |C'|$
$opt = \max$

MAX CLIQUE — *Maximum clique in a graph.*
$\mathcal{I} = \{G = <V, E>: G \text{ is a graph}\}$
$S(<V, E>) = \{V' \subseteq V : v_1, v_2 \in V' \wedge v_1 \neq v_2 \Rightarrow (v_1, v_2) \in E\}$
$m(V') = |V'|$
$opt = \max$

LIP — *Longest induced path in a graph.*
$\mathcal{I} = \{G = <V, E>: G \text{ is a graph}\}$
$S(<V, E>) = \{V' \subseteq V : G|_{V'} \text{ is a simple path}\}$
$m(V') = |V'|$
$opt = \max$

3 Approximation algorithms

Theorem 1 *Maximum bounded common induced subgraph* (MAX CIS $-B$) *can be approximated within* $B + 1$.

PROOF. We use that independent sets of the same size are always isomorphic. The following trivial algorithm finds an independent set V_1' in the graph $G_1 = <V_1, E_1>$.

- $V_1' \leftarrow \emptyset$
- Pick nodes from V_1 in any order and add each node to V_1' if none of its neighbours are already added to V_1'.

This algorithm will create a set of size $|V_1'| \geq |V_1|/(B+1)$ because for each node in V_1 either the node itself or one of its at most B neighbour nodes must be in V_1'.

Applying the algorithm to G_1 and G_2 gives us two independent sets V_1' and V_2'. If they are of different size, remove nodes from the largest set until they have got the same size. These sets are a legal solution to the problem of size at least $\min(|V_1|,|V_2|)/(B+1)$. Since the optimal solution has size at most $\min(|V_1|,|V_2|)$ the algorithm approximates the problem within the constant $B+1$. □

Lemma 1 *A maximum matching of a graph $G = <V,E>$ with degree at most B contains at least $|E|/(B+1)$ edges.*

PROOF. Let ν be the number of edges in the maximum matching and p be the number of nodes in the graph. If there exists a perfect matching, then $p = 2\nu$ and

$$\frac{|E|}{B+1} \leq \frac{p \cdot B/2}{B+1} < \frac{p}{2} = \nu.$$

If $p \geq 2\nu + 1$ the inequality $|E| \leq (B+1)\nu$ follows from [20, theorem 3.4.6]. □

Theorem 2 *Maximum bounded common edge subgraph* (MAX CES $-B$) *can be approximated within* $B+1$.

PROOF. We use the same idea as in the proof of Theorem 1, but create an independent set of *edges* instead. In polynomial time we can find a maximum matching of the graphs. The size of the smallest maximum matching is by Lemma 1 at least $\min(|E_1|,|E_2|)/(B+1)$ and the size of the optimal solution is at most $\min(|E_1|,|E_2|)$, so we can approximate this problem too within the constant $B+1$. □

Theorem 3 *An approximation algorithm for* MAX CLIQUE *gives us an equally good approximation algorithm for maximum common induced subgraph* (MAX CIS).

PROOF. From the input graphs G_1 and G_2 we form a *derived graph* $G = <V, E>$ in the following way (due to Barrow and Burstall [1]).

Let $V = V_1 \times V_2$ and call V a set of *pairs*. Call two pairs $<m_1, m_2>$ and $<n_1, n_2>$ *compatible* if $m_1 \neq n_1$ and $m_2 \neq n_2$ and if they preserve the edge relation, that is there is an edge between m_1 and n_1 if and only if there is an edge between m_2 and n_2. Let E be the set of compatible pairs.

A k-clique in the derived graph G can be interpreted as a matching between two induced k-node subgraphs. The subgraphs are isomorphic since the compatible pairs preserve the edge relation.

Thus, if we have a polynomial time approximation algorithm for the MAX CLIQUE problem which finds a solution within p we can apply it to the derived graph and use the answer to get an approximative solution for the MAX CIS problem of exactly the same size. Since the maximum clique corresponds to the maximum common induced subgraph this yields a solution within p for the MAX CIS problem.

For example we can use the MAX CLIQUE approximation algorithm by Boppana and Halldórsson [5]. This algorithm will, for a graph of size n, find a clique of size at least $O((\log n)^2/n)$ times the size of the maximum clique. □

Theorem 4 *An approximation algorithm for* MAX CLIQUE *gives us an equally good approximation algorithm for maximum common edge subgraph* (MAX CES).

PROOF. We use the same idea as in the preceding proof but the pairs are now pairs of directed edges instead of pairs of nodes. Let $V = A_1 \times A_2$, where A_i consists of two directed edges, $\overleftarrow{e}$ and $\overrightarrow{e}$, for each edge $e \in E_i$. We say that two pairs $<\overrightarrow{m_1}, \overrightarrow{m_2}>$ and $<\overrightarrow{n_1}, \overrightarrow{n_2}>$ are compatible if $\overrightarrow{m_1} \neq \overrightarrow{n_1}$, $\overrightarrow{m_1} \neq \overleftarrow{n_1}$, $\overrightarrow{m_2} \neq \overrightarrow{n_2}$, $\overrightarrow{m_2} \neq \overleftarrow{n_2}$ and they preserve the node relation, that is $\overrightarrow{m_1}$ and $\overrightarrow{n_1}$ are incident to the same node if and only if $\overrightarrow{m_2}$ and $\overrightarrow{n_2}$ are incident to the same node *in the same way*. For example, in Figure 1

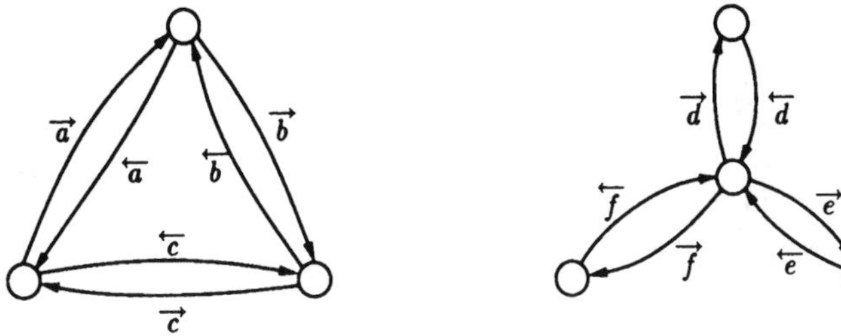

Figure 1: The digraphs resulting from a triangle and a 3-star in Theorem 4.

$<\overrightarrow{a}, \overleftarrow{d}>$ is compatible with $<\overrightarrow{b}, \overrightarrow{e}>$, $<\overrightarrow{b}, \overrightarrow{f}>$, $<\overleftarrow{b}, \overleftarrow{e}>$ and $<\overleftarrow{b}, \overleftarrow{f}>$ but not with e.g. $<\overleftarrow{b}, \overrightarrow{e}>$ or $<\overrightarrow{b}, \overleftarrow{e}>$.

A k-clique in the derived graph can be interpreted as a matching between two edge subgraphs with k edges in each subgraph. The subgraphs are isomorphic since the compatible pairs preserve the node relation.

In the same way as above we can use a MAX CLIQUE approximation algorithm to get an approximative solution for the MAX CES problem of the same size. □

4 Reductions between the problems

Theorem 5 MAX CIS $-B$ *is* MAX SNP-*hard when* $B \geq 25$.

PROOF. The problem MAX 3SAT -6, where each variable is in at most six clauses is known to be MAX SNP-complete [22]. We assume that each variable x_i occurs both as x_i in some clause and as $\overline{x}_i$ in some other clause, that no clause is trivially satisfied (e.g. $x_i \vee \overline{x}_i$) and that there are more clauses than variables. The problem is still MAX SNP-complete under these assumptions. We will show that there is an L-reduction f_1 from this problem to MAX CIS $-B$. Let $U = \{x_1, x_2, \ldots, x_n\}$ be the variables and $C = \{c_1, c_2, \ldots, c_m\}$ be the clauses of the input instance.

f_1 takes the sets U and C and constructs a MAX CIS $-B$ instance (the graphs G_1 and G_2) in the following way. G_1 and G_2 are similar and consist of $6n$ *literal nodes* (six for each variable), $18m$ *clique nodes* (18 for each clause) and a number of *clause nodes*. G_1 has $7m$ clause nodes (seven for each clause) and G_2 has m clause nodes (one for each clause). The clique nodes are connected in 18-cliques (m in each graph). In both graphs the six literal nodes for a variable x_i are connected in two 3-cliques — one 3-clique we call the x_i *clique* and the other 3-clique we call the $\overline{x}_i$ *clique*.

In G_2 each clause node is connected with one of the clique nodes in the corresponding clique and with all the literal nodes corresponding to the at most three literals which are contained in the corresponding clause in the MAX 3SAT -6 problem. This completes the description of graph G_2. G_1 has edges between each pair of literal nodes which corresponds to the same variable (i.e. building a 6-clique). Finally there are some edges from the clause nodes to the clique nodes and literal nodes in G_1. Number the seven clause nodes of clause c_j from 1 to 7 and the 18 clique nodes in the corresponding clique from 1 to 18. Now add edges between clause node i and clique node i for i from 1 to 7. Call the three literal 3-cliques corresponding to the three literals in c_i A, B and C. Add edges between clause node 1 and each node in A, 2 and B, 3 and A, 3 and B, 4 and C, 5 and A, 5 and C, 6 and B, 6 and C, 7 and A, 7 and B, 7 and C. If c_i only has two literals just add the edges from clause nodes 1, 2 and 3. If c_i only has one literal just add three edges, between node 1 and the literal 3-clique. See Figure 2 for an example.

The idea is that a truth assignment shall be encoded in the subgraph problem by including the corresponding literal 3-cliques in the subgraphs. For example, if x_4 is true then the literal 3-clique x_4 is in the subgraphs, and if x_4 is false then the literal 3-clique $\overline{x}_4$ is in the subgraphs. The included literal 3-cliques of G_1 and G_2 are matched with each other. A clause node in graph G_2 is included in the subgraph iff it is satisfied by the truth assignment. If a clause node in G_2 is included then it is matched

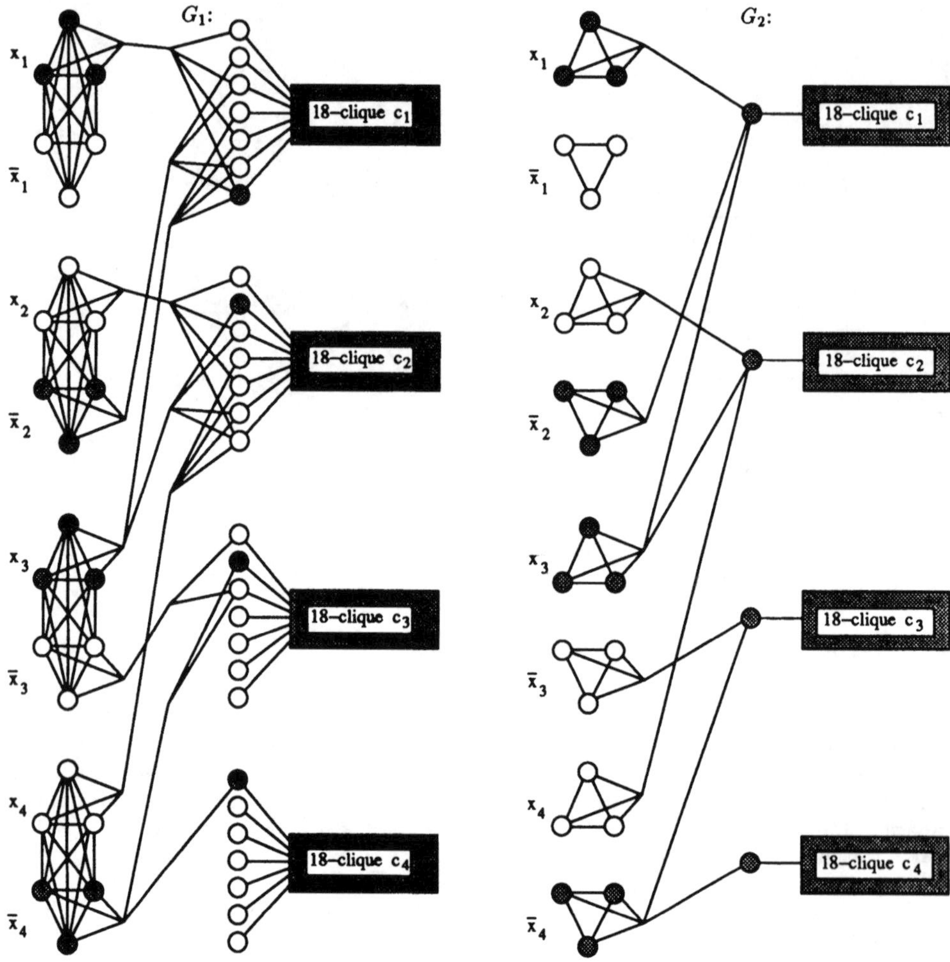

Figure 2: The constructed instance of the MAX CIS $-B$ problem for the MAX 3SAT -6 input $U = \{x_1, x_2, x_3, x_4\}, C = \{(x_1 \vee \overline{x}_2 \vee x_3), (x_2 \vee x_3 \vee x_4), (\overline{x}_3 \vee \overline{x}_4), (\overline{x}_4)\}$. One of the possible maximum common subgraphs is formed by including the shaded nodes.

with one of the corresponding seven clause nodes in G_2, namely with the node which is connected with exactly those literals in the clause which are true in the truth assignment. All the clique nodes are included in the subgraphs and are matched with each other clause-wise.

A solution of the MAX 3SAT -6 problem with k satisfied clauses will be encoded as two subgraphs (of G_1 and G_2), each with k clause nodes, $3n$ literal nodes and $18m$ clique nodes. Since a literal can be contained in at most five clauses at the same time, a literal node in the graph G_1 has degree at most $5 \cdot 4 + 5 = 25$, a clause node has degree at most 4 and a clique node has degree 17 or 18. In G_2 a literal node has degree at most 7, a clause node at most 4 and a clique node at most 18. Thus $B \geq 25$.

We will now prove that the maximum solution of the MAX 3SAT -6 problem will be encoded as a maximum solution of the MAX CIS $-B$ problem, and that given a solution of the MAX CIS $-B$ problem we can in polynomial time find an at least as good solution which is a legal encoding of a MAX 3SAT -6 solution.

Suppose we have any solution of the MAX CIS $-B$ problem, that is an induced subgraph of G_1, an induced subgraph of G_2 and a matching between each node in the G_1 subgraph and the corresponding node in the isomorphic G_2 subgraph.

- First we would like to include all clique nodes in the subgraphs and match each clique in the first graph with some clique in the second. Observe that, besides the clique nodes, no node in the graph G_2 is in a clique of size five or more. This means that if five or more clique nodes in the same clique are included in the subgraph of G_1, then they must be matched with clique nodes in the other subgraph. In the same way we see that besides the clique nodes, no node in the graph G_1 is in a clique of size seven or more, so if seven or more clique nodes in the same clique are included in the subgraph of G_2, then they must be matched with clique nodes in the subgraph of G_1.

  In each clique in G_1 which has $5 \leq p < 18$ nodes included in the subgraph, we add the rest of the clique to the subgraph and remove every clause node which is connected to an added clique node. We have added $18 - p$ clique nodes and removed at most the same number of clause nodes.

  The matched clique in G_2 must also have p clique nodes in the subgraph. Add the remaining $18 - p$ clique nodes and remove the nodes which are matched with the removed clause nodes in G_1. Since the cliques now are separate connected components we can match them with each other without problems.

  Perform the same operation for each clique in G_1 which has at least five but not all nodes included in the subgraph. The resulting subgraphs are at least as large as the original subgraphs, and they are still isomorphic.

  Now every clique either has all nodes in the subgraph or at most four nodes in the subgraph. For each clique in G_1 with $0 \leq p \leq 4$ nodes we do the following.

  We add $18 - p$ clique nodes to the subgraph of G_1 and remove every clause node which is connected to a clique node in the current clique. We have added $18 - p$ nodes and removed at most 7. In G_2 we choose one clique with $0 \leq q \leq 6$ nodes in the subgraph and add the remaining $18 - q$ clique nodes. We remove the p nodes which are matched with the old clique nodes in the first subgraph and the at most 7 nodes which are matched with the removed nodes in the first subgraph. Furthermore we have to remove the q nodes in G_1 which are matched with the q old clique nodes in G_2. If the clause node in G_2 (which is a neighbour to the current clique in G_2) is included in the second subgraph we remove it and its matched node in G_1. We have now added $18 - p \geq 14$ nodes to the first subgraph and removed at most $7 + q + 1 \leq 14$. We have added $18 - q \geq 12$ nodes to the second subgraph and removed at most $7 + p + 1 \leq 12$. As before, since the cliques are now separate connected components we can match them with each other without problems.

  The two subgraphs are still isomorphic, and thus a solution to the problem. They are at least as large as before, but now all clique nodes are included in the subgraphs and are matched with each other (but they are not necessarily matched in order yet).

- We observe that in each 7-group of clause nodes in G_1 at most one node is in the subgraph. The explanation is that every clause node is in contact with an 18-clique, which is completely in the subgraph, but in G_2 only one node can be in contact with each 18-clique (namely the corresponding clause node). Hence a structure with two or more nodes connected to an 18-clique cannot be isomorphic with any structure in G_2. Furthermore we can see that clause nodes in one of the subgraphs can only be matched with clause nodes in the other subgraph and, since all clique nodes are matched with clique nodes, literal nodes must be matched with literal nodes.

- We would like to change the subgraphs so that each literal 3-clique is either totally included in a subgraph or is not included at all. Furthermore at most one of the two literal 3-cliques in G_1 corresponding to the same variable may be included in the subgraph. Suppose that there is (at least) one node in the literal 3-clique x which is included in the subgraph of G_1. Let y be the literal node in the subgraph of G_2 with which this node is matched. We examine two cases.

Case 1. At least one of the clause nodes connected to x is included in the subgraph of G_1. Let b be one of these clause nodes and let c be the clause node in G_2 with which b is matched. See Figure 3. First we shall see that no node in the $\overline{x}$ literal 3-clique can be in the subgraph. This is because the nodes in the $\overline{x}$ clique are connected to the nodes in the x clique but not to b (since we have assumed that x and $\overline{x}$ cannot be in the same clause), but in G_2 there are no literal nodes which are connected to y but not to c. Since all three nodes in the x clique have the same connections

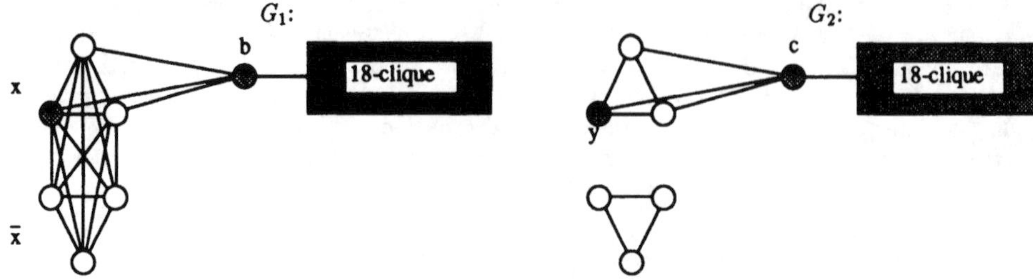

Figure 3: The structure in case 1.

to the environment in G_1 and all three nodes in the literal 3-clique containing y have the same environment in G_2 we still have isomorphic subgraphs if we include the whole x 3-clique in the subgraph of G_1 and the whole 3-clique containing y in the subgraph of G_2.

Case 2. None of the clause nodes connected to x are in the subgraph of G_1. If one or more nodes in $\overline{x}$ are in the subgraph of G_1 then none of the clause nodes which are connected to $\overline{x}$ can be in the subgraph, since we in that case would be in case 1 with the clause node as b and $\overline{x}$ as x. Thus we have a separate k-clique (with $1 \leq k \leq 6$) of literal nodes which by the above must be matched with a separate k-clique of literal nodes in G_2. In G_2 the largest possible clique of literal nodes is of size 3. Therefore the only possible cases are $1 \leq k \leq 3$. We remove those k nodes and instead include the whole x 3-clique in the subgraph of G_1 and the whole 3-clique containing one of the matched nodes in the subgraph of G_2.

In both cases we get a subgraph of G_1 where each literal 3-clique is either totally included in a subgraph or is not included at all and where both of the literal 3-cliques corresponding to the same variable are never included in the subgraph of G_1 at the same time.

- We now forget about the subgraph of G_2 and concentrate on the subgraph of G_1. It contains all clique nodes, at most one of each 7-group of clause nodes and at most one of each pair of literal 3-cliques. First we will include literal nodes so that every pair of literal 3-cliques has exactly one of the 3-cliques in the subgraph. We will have to remove some of the clause nodes, but the subgraph should contain at least as many nodes as before. Then we reorder the clause nodes to form a legal encoding of a MAX 3SAT -6 solution.

  1. Suppose there are k variables which do not have any of their literal 3-cliques in the subgraph and that there are j clauses which contain these variables. We know that each variable can occur in at most six clauses, thus $j \leq 6k$. Using a simple algorithm (see for example [13]) we can give values to the k variables so that at least half of the j clauses are satisfied. We first remove all of the j clause nodes which are in the subgraph from the subgraph and then include one clause node for each clause which was satisfied by the algorithm and the literal 3-cliques corresponding to the k variable values. We will then have removed at most j nodes and included at least $3k + j/2 \geq j/2 + j/2 = j$ nodes.
  2. In order to create a subgraph of G_1 which is a legal encoding of a MAX 3SAT -6 solution we may have to substitute some clause nodes in the subgraph for other clause nodes in the same 7-groups. Every clause node in the resulting subgraph should have connections to exactly those literal 3-cliques which are included in the subgraph of G_1 and correspond to literals in the corresponding clause. It is easy to see that this operation is always possible.

- As the subgraph of G_2 choose nodes as shown in the description of the encoding above. This is possible since the subgraph of G_1 is a legal encoding of a MAX 3SAT -6 solution. The isomorphic matching is then trivial.

We have now shown that every solution to the MAX CIS $-B$ problem can be transformed to an at least as large solution which is a legal encoding of a MAX 3SAT -6 solution. Moreover this transformation can be done in polynomial time.

If the optimal number of satisfied clauses is r and we do not have more variables than clauses then the optimal number of nodes in a subgraph is $3n + 18m + r \leq (3+18)m + r \leq 21 \cdot 2r + r = 43r$, since we can always satisfy at least half of the clauses. Thus the transformation f_1 of a MAX 3SAT -6 problem to a MAX CIS $-B$ problem, where $B \geq 25$, is an L-reduction with $\alpha = 43$ and $\beta = 1$. □

Theorem 6 MAX CIS *is at least as hard to approximate as* MAX CLIQUE.

PROOF. We will use an L-reduction f_2 from MAX CLIQUE to MAX CIS. The reduction is the same as the one Garey and Johnson use to prove that SUBGRAPH ISOMORPHISM is NP-complete. The MAX CLIQUE input is given as a graph G. Let the first of the MAX CIS graphs G_1 be this graph. Let G_2 be a $|V_1|$-clique, that is a complete graph with the same number of nodes as G_1.

Every induced subgraph in G_2 is a clique. Thus each common induced subgraph is a clique. The optimal solution of the MAX CLIQUE problem is a clique of size at most $|V|$, and this clique is also the largest clique in G_1 and is therefore the largest common induced subgraph.

In order to prove that f_2 is an L-reduction we also have to show that given a solution to the MAX CIS problem we in polynomial time can find an at least as good solution to the MAX CLIQUE problem. But since every common subgraph is a clique we can directly use the subgraph as a solution to the MAX CLIQUE problem. The solutions have the same size. □

The MAX CLIQUE problem is actually the same problem as MAX INDEPENDENT SET on the complementary graph. Berman and Schnitger have analyzed the approximability of this problem and have shown that if some MAX SNP-complete problem does not have a randomized polynomial time approximation scheme then for some constant $c > 0$ it is impossible to approximate MAX INDEPENDENT SET within n^c [2]. Thus the unbounded MAX CIS problem is at least that hard to approximate.

Another recent result showing that MAX CLIQUE is hard to approximate says that the problem cannot be approximated within $2^{(\log n)^{(1-\epsilon)}}$ in polynomial time, unless NP $\subset \tilde{\mathrm{P}}$, where $\tilde{\mathrm{P}}$ denotes the set of languages accepted in quasi polynomial time (i.e. time $2^{\log^c n}$) [10].

When we gave an approximation algorithm for the MAX CIS $-B$ problem in Theorem 1 we constructed a maximal independent set to use as the common subgraph. Somehow this feels like cheating, because an independent set of nodes is usually not the type of common subgraph we want. Perhaps we would rather like to find a big common *connected* subgraph. Unfortunately the MAX CIS problem, where we demand that the common subgraph is connected, is *provably* hard to approximate. We will prove that this problem is NPO PB-complete under L-reductions, which means that it cannot be approximated within $O(n^\epsilon)$ for any $\varepsilon > 0$, unless NP = P.

In general, for similar graph problems, the demand that the solution subgraph is connected seems to lead to harder problems [24].

Theorem 7 MAXIMUM COMMON INDUCED CONNECTED SUBGRAPH (MAX CICS) *is* NPO PB-*complete under L-reductions.*

PROOF. NPO PB consists of all NPO problems which are polynomially bounded, that is all problems F satisfying $opt_F(I) \leq p(|I|)$ for all problem instances $I \in \mathcal{I}_F$ where p is a polynomial. It is obvious that MAX CICS satisfies this and therefore is in NPO PB.

We know that LONGEST INDUCED PATH in a graph G is NPO PB-complete under L-reductions [2] so we will L-reduce this problem to MAX CICS in order to show that the latter problem is NPO PB-complete.

Choose G as the first graph G_1 and choose a simple path with $|V|$ nodes as the second graph G_2. We observe that every induced connected subgraph in G_2 must be a simple path. The maximum induced connected subgraph is the longest induced path that can be found in G_1, that is the optimal solution to the LIP problem with input G. Since every non-optimal solution of size c immediately gives a solution to the LIP problem of size c the transformation is an L-reduction. □

Finally we return to the edge subgraph problem and show that if the degrees of the graphs are bounded then the MAX CIS problem is at least as hard to approximate as the MAX CES problem.

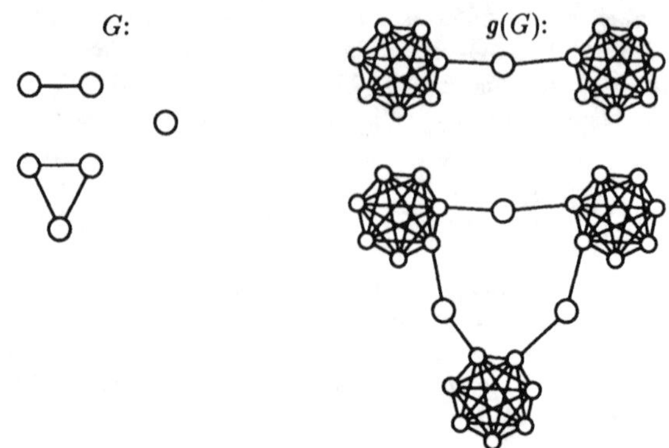

Figure 4: An example of the transformation g in Theorem 8.

Theorem 8 *There is an L-reduction from* MAX CES $-B$ *to* MAX CIS$-(2B+3)$.

PROOF. Let f_3 be the following transformation from MAX CES $-B$ to MAX CIS $-(2B+3)$. An input instance $\{G_1{}^E, G_2{}^E\}$ of MAX CES $-B$ shall be transformed to an instance $\{G_1{}^I, G_2{}^I\}$ of MAX CIS $-(2B+3)$. Let $G_1{}^I = g(G_1{}^E)$ and $G_2{}^I = g(G_2{}^E)$ where g transforms each node with degree greater than zero to a $(2B+3)$-clique and each edge to two edges connected with an *edge node*. The two edges are connected to one node in each of the two $(2B+3)$-cliques corresponding to the end points of the edge in the original graph. This shall be done so that every clique node is connected to at most one edge node. The constructed graphs have degree $2B+3$. See Figure 4.

Solutions $\{E_1{}', E_2{}'\}$ to the MAX CES $-B$ problem is encoded as solutions $\{V_1{}', V_2{}'\}$ to the MAX CIS problem where an edge node is in V_i' iff the corresponding edge is in E_i' and where $\min(|V_1'|, |V_2'|)$ $(2B+3)$-cliques, among them all cliques corresponding to nodes adjacent to edges in E_1' and E_2', are included in each subgraph.

In the same way as in the proof of Theorem 5 we will show that given a solution of the MAX CIS problem, which is d smaller than the optimal solution, we can in polynomial time find a solution to the MAX CES $-B$ problem which is at most d smaller than the optimal solution.

Given solutions V_1' and V_2' we first modify them so that all clique nodes are included. Observe that cliques of size three or more only can be found in the $(2B+3)$-cliques in $G_1{}^I$ and $G_2{}^I$ and thus that a $(2B+3)$-clique in $G_1{}^I$ with k nodes ($k \geq 3$) in V_1' only can be matched with a $(2B+3)$-clique in $G_2{}^I$ with exactly k nodes in V_2' (and vice versa). For each $(2B+3)$-clique in $G_1{}^I$ with k nodes ($3 \leq k < 2B+3$) in V_1' we can add the remaining $2B+3-k$ nodes if we remove all nodes in V_1' which are connected to added nodes and perform the same operations on the matched clique in the other graph.

Now every clique either has all nodes in the subgraph or at most two nodes in the subgraph. For each clique in $G_1{}^{I'}$ with $0 \leq p \leq 2$ nodes we do the following (until there are no nonfull cliques left in one of the subgraphs).

We add $2B+3-p$ clique nodes to the subgraph of $G_1{}^I$ and remove every edge node which is connected to a clique node in the current clique (at most B nodes). In $G_2{}^I$ we choose one clique with $0 \leq q \leq p$ nodes in the subgraph. If one of the p nodes in the first subgraph is matched with a clique node in the second subgraph (which is always the case if $p = 2$ because two edge nodes never are connected) we choose this clique. We add the remaining $2B+3-q$ clique nodes and remove every edge node which is connected to a clique node in the current clique (at most B nodes).

If one of the q nodes in the second subgraph is matched with an edge node in the first subgraph we have to remove this edge node from the first subgraph. If one of the p nodes in the first subgraph is matched with an edge node in the second subgraph we have to remove this edge node from the second subgraph.

Furthermore we have to remove the at most B nodes in the first subgraph which are matched with edge nodes which are connected with nodes in the current clique in the second subgraph. And symmetrically we have to remove the at most B nodes in the second subgraph which are matched with edge nodes which are connected with nodes in the current clique in the first subgraph.

We have now added $2B+3-p \geq 2B+1$ nodes to the first subgraph and $2B+3-q \geq 2B+1$ nodes from the second subgraph and removed at most $B+1+B=2B+1$ nodes from the first subgraph and $B+1+B=2B+1$ nodes from the second. If we match the cliques with each other the two subgraphs are still isomorphic.

Now every clique node in at least one of the graphs, say in ${G_1}^I$ are included in the corresponding subgraph and are matched with clique nodes in the other subgraph. Therefore the edge nodes in the second subgraph must be matched with edge nodes in the first subgraph. Every edge node in the first subgraph must be matched with an edge node in the second subgraph, because it is adjacent to a $(2B+3)$-clique in the first subgraph, an no clique node in the second subgraph is adjacent to a $(2B+3)$-clique. Thus we have subgraphs which are an encoding of a MAX CES $-B$ solution, where an edge node is in the MAX CIS subgraph if and only if the corresponding edge is in the MAX CES $-B$ subgraph.

If the subgraphs in an optimal solution to the MAX CES $-B$ problem contain k edges then the number of nodes in the subgraphs in the optimal solution to the MAX CIS problem is

$$k+(2B+3)\cdot 2\cdot \min(|{E_1}^E|,|{E_2}^E|) \leq k+(2B+3)\cdot 2\cdot (B+1)k \leq (4B^2+10B+7)k$$

Thus the reduction f_3 is an L-reduction with $\alpha = 4B^2+10B+7$ and $\beta = 1$. □

5 Discussion

We have studied the approximability of some NPO problems. The following figure illustrates the situation. Here $P_1 < P_2$ means that there is an L-reduction from problem P_1 to problem P_2, i.e. P_2 is at least as hard to approximate as P_1.

$$\begin{array}{ccccc} \text{MAX 3SAT } -B & < & \text{MAX CLIQUE} & < & \text{LIP} \\ \wedge & & \wedge\vee & & \wedge\vee \\ \text{MAX CIS } -B & < & \text{MAX CIS} & < & \text{MAX CICS} \\ \vee & & \vee & & \\ \text{MAX CES } -B & < & \text{MAX CES} & & \end{array}$$

Yannakakis observed in [25] that problems on edges tend to be easier to solve than their node-analogues. We have seen that this is valid for the approximability of the maximum common graph problem as well.

Acknowledgements

I would like to thank my advisor Johan Håstad for valuable support.

References

[1] H. Barrow and R. Burstall. Subgraph isomorphism, matching relational structures and maximal cliques. *Information Processing Letters*, 4:83–84, 1976.

[2] P. Berman and G. Schnitger. On the complexity of approximating the independent set problem. In *Proc. 6th Annual Symposium on Theoretical Aspects of Computer Science*, pages 256–268. Springer-Verlag, 1989. Lecture Notes in Computer Science 349.

[3] M. Bern and P. Plassmann. The Steiner problem with edge lengths 1 and 2. *Information Processing Letters*, 32:171–176, 1989.

[4] A. Blum, T. Jiang, M. Li, J. Tromp, and M. Yannakakis. Linear approximation of shortest superstrings. In *Proc. Twenty third Annual ACM symp. on Theory of Comp.*, pages 328–336. ACM, 1991.

[5] R. Boppana and M. M. Halldórsson. Approximating maximum independent sets by excluding subgraphs. In *Proc. SWAT 90*, pages 13–25. Springer-Verlag, 1990. Lecture Notes in Computer Science 447.

[6] D. Bruschi, D. Joseph, and P. Young. A structural overview of NP optimization problems. In *Proc. Optimal Algorithms*, pages 205–231. Springer-Verlag, 1989. Lecture Notes in Computer Science 401.

[7] N. Christofides. Worst-case analysis of a new heuristic for the travelling salesman problem. Technical report, Graduate School of Industrial Administration, Carnegie-Mellon University, Pittsburgh, 1976.

[8] P. Crescenzi and A. Panconesi. Completeness in approximation classes. In *Proc. FCT '89*, pages 116–126. Springer-Verlag, 1989. Lecture Notes in Computer Science 380.

[9] R. Fagin. Generalized first-order spectra, and polynomial-time recognizable sets. In R. Karp, editor, *Complexity and Computations*. AMS, 1974.

[10] U. Feige, S. Goldwasser, L. Lovász, S. Safra, and M. Szegedy. Approximating clique is almost NP-complete. In *Proc. of 32nd Annual IEEE Sympos. on Foundations of Computer Science*, pages 2–12, 1991.

[11] M. R. Garey and D. S. Johnson. *Computers and Intractability: a guide to the theory of NP-completeness.* W. H. Freeman and Company, San Fransisco, 1979.

[12] O. H. Ibarra and C. E. Kim. Fast approximation for the knapsack and sum of subset problems. *Journal of the ACM*, 22, 1975.

[13] D. S. Johnson. Approximation algorithms for combinatorial problems. *Journal of Computer and System Sciences*, 9:256–278, 1974.

[14] V. Kann. Maximum bounded 3-dimensional matching is MAX SNP-complete. *Information Processing Letters*, 37:27–35, 1991.

[15] V. Kann. Maximum bounded H-matching is MAX SNP-complete. Manuscript, submitted for publication, 1991.

[16] V. Kann. Which definition of MAX SNP is the best? Manuscript, submitted for publication, 1991.

[17] P. G. Kolaitis and M. N. Thakur. Logical definability of NP optimization problems. Technical Report UCSC-CRL-90-48, Board of Studies in Computer and Information Sciences, University of California at Santa Cruz, 1990.

[18] P. G. Kolaitis and M. N. Thakur. Approximation properties of NP minimization classes. In *Proc. 6th Annual Conf. on Structures in Computer Science*, pages 353–366, 1991.

[19] M. W. Krentel. The complexity of optimization problems. *Journal of Computer and System Sciences*, 36:490–509, 1988.

[20] L. Lovász and M. D. Plummer. *Matching Theory*, volume 121 of *North-Holland Mathematics studies*. North-Holland, Amsterdam, 1986.

[21] A. Panconesi and D. Ranjan. Quantifiers and approximation. In *Proc. Twenty second Annual ACM symp. on Theory of Comp.*, pages 446–456. ACM, 1990.

[22] C. Papadimitriou and M. Yannakakis. Optimization, approximation, and complexity classes. In *Proc. Twentieth Annual ACM symp. on Theory of Comp.*, pages 229–234. ACM, 1988.

[23] C. Papadimitriou and M. Yannakakis. The traveling salesman problem with distances one and two. *Mathematics of Operations Research*, 1991. To appear.

[24] M. Yannakakis. The effect of a connectivity requirement on the complexity of maximum subgraph problems. *Journal of the ACM*, 26:618–630, 1979.

[25] M. Yannakakis. Edge deletion problems. *SIAM Journal on Computing*, 10:297–309, 1981.

The Complexity of Colouring Circle Graphs

(extended abstract)

Walter Unger

Fachbereich 17
University of Paderborn
Warburger Straße 100
4790 Paderborn, Germany
Email: Walter.Unger@Uni-Paderborn.de

Abstract

We study the complexity of the colouring problem for circle graphs. We will solve the two open questions of [Un88], where first results were presented.

1. Here we will present an algorithm which solves the 3-colouring problem of circle graphs in time $O(n \log(n))$. In [Un88] we showed that the 4-colouring problem for circle graphs is *NP*-complete.
2. If the largest clique of a circle graph has size k then the $2 \cdot k - 1$-colouring is *NP*-complete. Such circle graphs are $2 \cdot k$-colourable [Un88].

Further results and improvements of [Un88] complete the knowledge of the complexity of the colouring problem of circle graphs.

Classification: algorithms and data structures, computational complexity

1 Introduction

A *circle graph* is an undirected graph which is an *intersection graph* of a set of chords in one circle. The vertex set of an intersection graph of a set of chords is that set of chords. Two vertices are joined by an edge iff the representing chords intersect each other. An example of such an intersection graph is given in Figure 1.

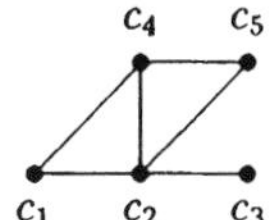

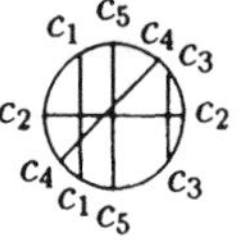

Figure 1: intersection graph of a circle graph

There are other graph classes defined as intersection graphs, such as interval graphs, path graphs and circular-arc graphs. For these and many other graph classes the complexity of colouring — and other — problems was investigated. For a survey of such graph classes see [Br90]. Before listing results for other problems of circle graphs we will describe the fields where the colouring problem for circle graphs arises.

- There is a relationship between chord colouring and the page-number of a graph. The page-number problem consists of arranging the vertices on the spline of a book so that edges are embedded without crossings on a minimum number of pages. If the vertices of the graph are already arranged on the spline of the book, then the remaining problem — the embedding of the edges on the pages — is identical to the k-colouring problem of the corresponding circle graph.

- The problem of realizing a given permutation using a minimum number of parallel stacks is modeled by the colouring of chords in a special circle graph, a permutation graph. This is presented in [EvIt71]. The number of stacks you have to use is the same as the number of colours you need to colour the corresponding permutation graph.

- The colouring of chords is closely related to some layer-assignment problems in VLSI. The layer-assignment to k layers of 2-point nets in a closed region is the same as the k-colouring of circle graphs.

There are several papers concerned with the *recognition problem* for circle graphs [Bo75, Bo85, GaHsSu85, Na85]. The algorithm with the best running time was given by Spinrad [Sp88]: Circle graphs are recognizable in time $O(n^2)$ where n is the number of vertices. The first algorithm for the *clique problem* for circle graphs was presented by Gavril [Ga75] with a running time of $O(n^3)$. The time was improved by Rotem and Urrutia [RoUr81] to $O(n^2)$. They also presented an algorithm for the k-clique problem with a running time of $O(n \log(n))$. Wessel and Pöschel [WePo85] showed that every outerplanar graph is a circle graph. Also, permutation graphs are circle graphs. Peter Damaschke [Da89] showed that the *Hamiltonian path* problem for circle graphs is *NP*-complete. Also the *dominating set* problem is *NP*-complete due to Keil [Ke90].
The first result about the complexity of the *colouring problem* for circle graphs was given by Garey et al. [GaJoMiPa80]. They considered circular-arc graphs: the k-colouring problem for circular-arc graphs is solvable in linear time — with a quite large constant factor — and the colouring problem for circular-arc graphs is *NP*-complete. Using the last result they were able to show that the colouring problem for circle graphs is also *NP*-complete. First results on the complexity of the k-colouring problem of circle graphs were given in [Un88]. It was shown that the 4-colouring of circle graphs is *NP*-complete. Furthermore: if the largest clique has size k and $k \geqslant 3$ then the $\lceil 3/2 \cdot k \rceil$ colouring is *NP*-complete and a $2 \cdot k$ colouring is always possible. There remained the gap between $\lceil 3/2 \cdot k \rceil$ and $2 \cdot k$. The complexity of the 3-colouring of circle graphs was also not solved.
This paper will solve these open questions and will give a full description of the complexity of the k-colouring problem of circle graphs. Due to the quite large complexity of the proofs of most Theorems, we will not present their full version here. We will describe in detail the algorithm for the 3-colouring problem for circle graphs.
Section 2 is concerned with the formal definition of circle graphs and their representation. In Section 3 we will present our algorithm which solves the 3-colouring problem for circle graphs in time $O(n \log(n))$. The full description of the complexity of the k-colouring problem of circle graphs is presented in Section 4. Section 5 is concerned with g-segment graphs, which are subclasses of the class of circle graphs. Polynomial time algorithm for these g-segment graphs were also presented in [St90].
Throughout this paper we assume that circle graphs are given by their representation as a set of chords. The construction of that representation is possible in time $O(n^2)$ (see [Sp88]). The full version of this paper is presented in [Un90].

2 Representation of circle graphs and definitions

We will define in this Section the type of representation which we will use in this paper. Without loss of generality we assume that no two chords have a common endpoint on the circle. We will not use the representation as shown in Figure 1. In our representation:

- The circle is broken up and straightened.

- The arcs become now arcs above the bottom line representing the circle.

This is a direct transformation from that one being presented in Figure 1. This *circle graph representation* is shown in Figure 2. It is also known as the overlap graph model.

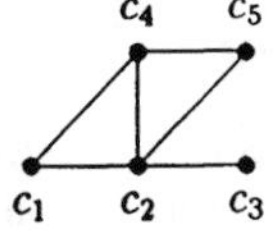

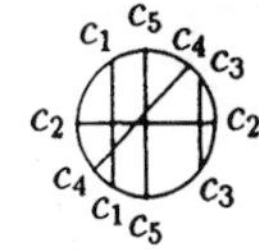

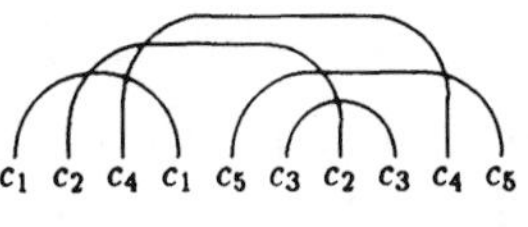

Figure 2: representation of a circle graph

In the following the arcs are still called chords. If a chord is named x we will denote its *left endpoint* reaching the bottom line with $\overleftarrow{x}$ and the right one with $\overrightarrow{x}$. With this notation we obtain the following formal definition of circle graphs.

Definition 1 (Circle graphs)
An undirected graph $G = (V, E)$ with $V = \{v_1, v_2, ..., v_n\}$ is a *circle graph* iff there exits a set of chords $C = \{(\overleftarrow{c_i}, \overrightarrow{c_i})\,;\, 1 \leqslant i \leqslant n \,\wedge\, \overleftarrow{c_i} < \overrightarrow{c_i}\,\}$ such that:

$$\{v_i, v_j\} = e \in E \;\text{ iff }\; c_i \otimes c_j$$

where the Boolean predicate $\otimes$ denotes the intersection of chords:

$$v \otimes w \;\text{ iff }\; \overleftarrow{v} < \overleftarrow{w} < \overrightarrow{v} < \overrightarrow{w} \;\text{ or }\; \overleftarrow{w} < \overleftarrow{v} < \overrightarrow{w} < \overrightarrow{v}$$

In the following we will only use the representation of a circle graph by a set of chords, thus a circle graph C is always of the form: $C = \{\,(\overleftarrow{c_i}, \overrightarrow{c_i})\,;\, 1 \leqslant i \leqslant n\,\}$.
Furthermore $\mathcal{N}_C(c)$ denotes the set of neighbours of c in C: $\mathcal{N}_C(c) = \{\,d \in C\,;\, d \otimes c\,\}$. □

In Figure 2 we see that some chords *cover* another and some chords are not covered by any other chord. From this we get the definition of the level structure of circle graphs.

Definition 2 (Level structure of circle graphs)
Let C be a circle graph. A chord x *covers* a chord y — denoted by: $x \rightsquigarrow y$ — iff

$$\overleftarrow{x} < \overleftarrow{y} < \overrightarrow{y} < \overrightarrow{x}$$

A chord x *directly covers* a chord y — denoted by: $x \rightarrow y$ — iff $x \rightsquigarrow y$ and there is no chord z with $x \rightsquigarrow z \wedge z \rightsquigarrow y$. The *level structure* is defined by recursion as follows:

$$\begin{aligned}
\texttt{level}^{\downarrow}_{C}(1) &= \{\, c \in C\,;\, \nexists x \in C : \; x \rightsquigarrow c\,\} \\
\texttt{level}^{\downarrow *}_{C}(i) &= \textstyle\bigcup_{j=1}^{i} \texttt{level}^{\downarrow}_{C}(j) \\
\texttt{level}^{\downarrow}_{C}(i+1) &= \texttt{level}^{\downarrow}_{C \setminus \texttt{level}^{\downarrow *}_{C}(i)}(1)
\end{aligned}$$

The number of levels is given by $\texttt{LEV}(C)$, and by $\texttt{index}^{\downarrow}_{C}(c)$ we denote the number of the level which c belongs to.

$$\begin{aligned}
\texttt{LEV}(C) &= \max\{\, i\,;\, \texttt{level}^{\downarrow}_{C}(i) \neq \emptyset\,\} \\
\texttt{index}^{\downarrow}_{C}(c) &= i \;\text{ iff }\; c \in \texttt{level}^{\downarrow}_{C}(i)
\end{aligned}$$

We call $\texttt{level}^{\downarrow}_{C}(1)$ the top level of C, $\texttt{level}^{\downarrow}_{C}(i)$ the i-th level of C and $\texttt{level}^{\downarrow}_{C}(\texttt{LEV}(C))$ the bottom level of C.
Furthermore let $\overset{\rightsquigarrow}{\mathcal{O}}_C\,(c) = \{\,d \in C\,;\, c \rightsquigarrow d\,\}$ be the set of chords covered by c. □

This level structure depends on the chosen representation.
A function $\texttt{col} : C \rightarrow \{1, 2, ..., k\}$ is a k-colouring iff $(a \otimes b \;\Rightarrow\; \texttt{col}(a) \neq \texttt{col}(b))$. With $\texttt{COL}(C)$ we denote the minimal number of colours to colour C and with $\texttt{CLI}(C)$ the size of the largest clique in C.

3 The 3-colouring problem for circle graphs

This Section contains the description of an algorithm for the 3-colouring problem of circle graphs. This algorithm will illustrate the fundamental difference between the 3-colouring problem and the 4-colouring problem of circle graphs. The latter one is *NP*-complete [Un88]. Thus this algorithm is "optimal", because there is — assuming $P \neq NP$ — no polynomial time algorithm for the 4-colouring of circle graphs. We will not describe the full algorithm which solves the 3-colouring problem of a circle graph C in time $O(n\log(n))$. That algorithm is too complicated to be presented here. But the "basic version" of that algorithm will be described.
The algorithm has three phases:

- The first phase computes a set of "important subgraphs" of C. Each important subgraph C_I is checked for conditions necessary for a correct 3-colouring of C_I. These conditions are expressed in a boolean formula $\mathcal{F}$.
- In the second phase an assignment is computed for the variables of $\mathcal{F}$ such that $\mathcal{F}$ is satisfied. If $\mathcal{F}$ is not satisfiable then C is not colourable with 3 colours and the algorithm will finish.
- Otherwise, the third phase uses the assignment of the variables of $\mathcal{F}$ to compute a valid 3-colouring of C. This colouring is done level by level. In round i level $\mathtt{level}^{\downarrow}_C(i)$ is coloured.

The three phases will be described in detail below. Phases one and two have to gather enough information for the third phase. This information will be given by a function $\mathtt{val}_3 : \mathcal{D} \to$ boolean, where $\mathcal{D}$ is a subset of the set of pairs of chords. A pair of chords $\{a, b\}$ is in $\mathcal{D}$ iff the following condition holds:

$$\begin{array}{ll} \neg(a \otimes b) & \text{and} \\ \mathcal{N}_C(a) \cap \mathcal{N}_C(b) \neq \emptyset & \text{and} \\ \mathtt{index}^{\downarrow}_C(a) = \mathtt{index}^{\downarrow}_C(b) \text{ or } a \to b & \end{array}$$

If $\mathtt{val}_3(a, b)$ is $\mathtt{true}$ then the two chords a and b have to be coloured with one colour, otherwise they have to be coloured by two different colours.
We will illustrate this gathering of information and the colouring using $\mathtt{val}_3$ with the example shown in Figure 3.

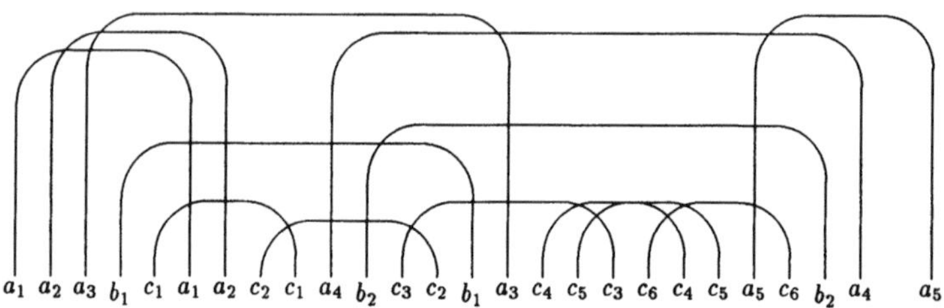

Figure 3: circle graph C_e

The circle graph C_e has the following three levels:

$$\begin{array}{lllllll} \mathtt{level}^{\downarrow}_{C_e}(1) & = & \{a_1, a_2, a_3, a_4, a_5\} & \quad & \mathtt{level}^{\downarrow}_{C_e}(2) & = & \{b_1, b_2\} \\ \mathtt{level}^{\downarrow}_{C_e}(3) & = & \{c_1, c_2, c_3, c_4, c_5, c_6\} & & & & \end{array}$$

The function $\mathtt{val}_3$ has to be defined for the following domain:

$$\mathcal{D} = \left\{ \begin{array}{c} \{a_1,a_4\},\{a_2,a_4\},\{a_3,a_5\},\{a_3,b_1\},\{a_4,b_2\},\{b_1,c_1\},\{b_1,c_2\}, \\ \{b_2,c_3\},\{b_2,c_6\},\{c_1,c_3\},\{c_2,c_4\},\{c_2,c_5\},\{c_3,c_6\} \end{array} \right\}$$

We will now look for important subgraphs from which we will get some conclusions for the function $\mathtt{val}_3$. This is done as follows: For every important subgraph there is a 3-colouring which obeys the function $\mathtt{val}_3$. A 3-colouring of an important subgraph C_I obeys $\mathtt{val}_3$ iff for all chords a,b in C_I $\mathtt{val}_3(a,b) = (\mathtt{col}(a) = \mathtt{col}(b))$ holds.
The first important subgraph is $\{a_1,a_2,a_3,b_1\}$, which is also shown in Figure 4.

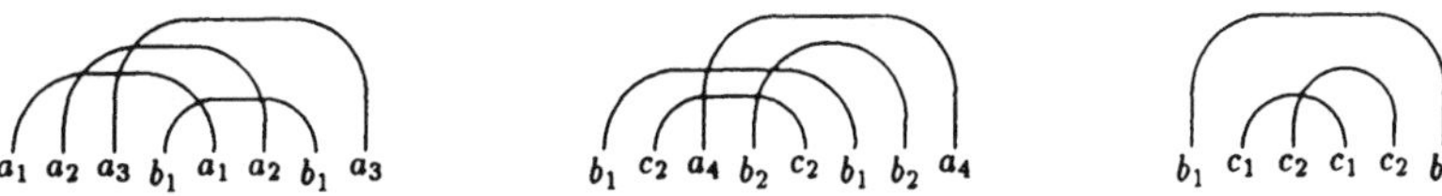

Figure 4: examples of important subgraphs

In a correct 3-colouring the chords a_3 and b_1 have to be coloured with the same colour, because of the two intersecting chords a_1,a_2. Thus we have to define: $\mathtt{val}_3(a_3,b_1) = \mathtt{true}$.
There are more values for $\mathtt{val}_3$ that we have to define like the above one:

$$\mathtt{val}_3(b_1,c_1) = \mathtt{true} \quad \text{by important subgraph: } \{a_1,a_2,b_1,c_1\}$$

$$\mathtt{val}_3(c_3,c_6) = \mathtt{true} \quad \text{by important subgraph: } \{c_3,c_4,c_5,c_6\}$$

Another type of important subgraph is $\{a_4,b_1,b_2,c_2\}$. This subgraph — also seen in Figure 4 — forms a cycle of length four. It is not possible to have $\mathtt{val}_3(a_4,b_2) = \mathtt{val}_3(b_1,c_2) = \mathtt{false}$, because then this subgraph must be coloured with four colours. We will express this by the following clause: $(\mathtt{val}_3(a_4,b_2) \vee \mathtt{val}_3(b_1,c_2))$. In C_e there are more of these cycles of length four. Therefore, we also get the following clauses:

$$(\mathtt{val}_3(b_2,c_3) \vee \mathtt{val}_3(b_1,c_3)) \qquad (\mathtt{val}_3(c_3,d_1) \vee \mathtt{val}_3(b_1,c_3))$$

$$(\mathtt{val}_3(a_3,b_1) \vee \mathtt{val}_3(a_4,b_2)) \qquad (\mathtt{val}_3(a_3,b_1) \vee \mathtt{val}_3(b_2,c_3))$$

$$(\mathtt{val}_3(a_4,b_2) \vee \mathtt{val}_3(a_3,a_5))$$

The next type of observation is made with the important subgraph $\{b_1,c_1,c_2\}$, see Figure 4.
For these chords the values of $\mathtt{val}_3(b_1,c_1)$ and $\mathtt{val}_3(b_1,c_2)$ have to be defined, but both of them cannot be $\mathtt{true}$. If both of them are true, we have to colour c_1 and c_2 with the same colour as b_1. But this is not possible, because c_1 and c_2 intersect each other. Thus our formula $\mathcal{F}$ will include the clause $(\neg\mathtt{val}_3(b_1,c_1) \vee \neg\mathtt{val}_3(b_1,c_2))$.
The last conditions we get for $\mathcal{F}$ result from the important subgraphs $\{a_3,a_5,b_2,c_3,c_4,c_5,c_6\}$ and $\{a_2,a_4,b_1,c_1,c_2\}$. From these we get some more complex clauses, which we will not state here. But we keep these subgraphs in mind for the evaluation of the formula $\mathcal{F}$.
In the next step we have to evaluate $\mathcal{F}$ to get a valid assignment to the function $\mathtt{val}_3$. For this evaluation we have listed "many" clauses of length two and "some" conditions from two important subgraphs which we did not list. Thus we will use in our example just the clauses of length two. The evaluation of $\mathcal{F}$ is started with the following conclusions:

$$\begin{array}{lll} \mathtt{val}_3(b_1,c_1) = \mathtt{true} & \Rightarrow & \neg\mathtt{val}_3(b_1,c_2) \\ & \Rightarrow & \mathtt{val}_3(a_4,b_2) \wedge \mathtt{val}_3(b_2,c_3) \wedge \mathtt{val}_3(c_3,d_1) \\ \mathtt{val}_3(b_2,c_3) \wedge \mathtt{val}_3(c_3,c_6) & \Rightarrow & \mathtt{val}_3(b_2,c_6) \end{array}$$

The second conclusion is made using the subgraph $\{a_3, a_5, b_2, c_3, c_4, c_5, c_6\}$. We may define some values of $\mathtt{val}_3$ without getting any conclusions:

$$\mathtt{val}_3(d_1, d_3) = \mathtt{true} \quad \text{and} \quad \mathtt{val}_3(a_3, a_5) = \mathtt{false}$$

From other values of *val* we get some conclusions:

$$\mathtt{val}_3(a_2, a_4) = \mathtt{false} \Rightarrow \mathtt{val}_3(a_1, a_4) \wedge \mathtt{val}_3(a_2, c_2) \wedge \neg\mathtt{val}_3(a_4, c_1)$$
$$\mathtt{val}_3(a_3, c_4) = \mathtt{false} \Rightarrow \mathtt{val}_3(a_3, c_5) \wedge \neg\mathtt{val}_3(a_5, c_5)$$

We will now use the information stored in function $\mathtt{val}_3$ to compute a valid 3-colouring of C_e. This colouring will be done level by level. At first the chords a_i $(1 \leqslant i \leqslant 3)$ are coloured with colour i. We will see that from now on the function $\mathtt{val}_3$ will define the colours of the other chords in a unique way.
Because of the value of $\mathtt{val}_3(a_1, a_4)$ the chord a_4 is coloured with 1 and because of $\mathtt{val}_3(a_3, a_5)$ a_5 is coloured with 2. Also the colouring of the next level b_1, b_2 is uniquely defined by $\mathtt{val}_3$: $\mathtt{col}(b_1) = 3$ and $\mathtt{col}(b_2) = 1$. The chord c_1 has to be coloured with 3, because $\mathtt{val}_3(b_1, c_1)$ holds. For the chord c_2 the only possible colour is 2, and this obeys function $\mathtt{val}_3$. This technique is continued until C_e is coloured. The final colouring is seen in Figure 5.

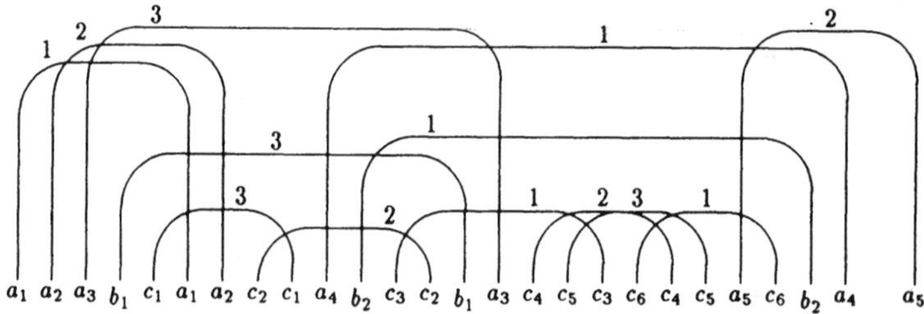

Figure 5: coloured example circle graph C_e

This example already gives quite a good description of the algorithm. There are three questions left:
How is it possible to get the claimed running time? Using the description from above this is not possible. We introduced the important subgraphs which form a cycle of length four. But there may be $\Omega(n^2)$ such important subgraphs and $\Omega(n^2)$ clauses from them. We have to be very careful in defining the clauses which have to be computed. But the full description of this will not fit into this paper. Hence we will describe a polynomial time algorithm.
How is it possible to evaluate the formula $\mathcal{F}$ in the claimed — or at least polynomial — time? For the part of $\mathcal{F}$ which consists of the clauses of length two, this is easy, but for the other clauses this seems to be hard.
We have to prove that the second phase stops iff the circle graph is not 3-colourable and that the third phase will always succeed.
We will answer the latter questions in the following part.

3.1 Phase 1: Computing formula $\mathcal{F}$

First we have to define what the important subgraphs of a circle graph C are. For the $O(n\log(n))$ time version of this algorithm this is rather complicated. Here we will give in Table 1 a "simplified" version of important subgraphs by listing them as undirected graphs.
We have to make the following remarks: The clauses will only be defined if the parameters of $\mathtt{val}_3$ are in $\mathcal{D}$. The latter important subgraphs — the cycles of length five or more, denoted as "type 6" subgraphs — are not defined in the correct way by the above picture. To be more precise: A cycle C_l of length $l \geqslant 5$ is an important subgraph iff the following is true:

	subgraphs	clauses		subgraphs	clauses
1)	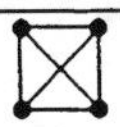	`false`	4)		$\texttt{val}_3(a,b) \vee \texttt{val}_3(a,c)$
2)	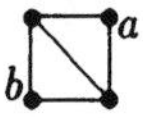	$\texttt{val}_3(a,b)$	5)		$\neg\texttt{val}_3(e,b) \vee \neg\texttt{val}_3(e,d)$
3)	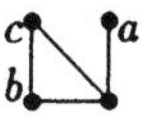	$\texttt{val}_3(a,b) \neq \texttt{val}_3(a,c)$	6)	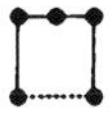	"Some large expression"

Table 1: Important subgraphs and clauses

- There exists an i such that $C_I \subseteq \texttt{level}_C^{\downarrow}(i) \cup \texttt{level}_C^{\downarrow}(i+1)$.
- It is not possible using an extra cord c to split C_I into two different cycles. We will illustrate this using the example from Figure 3. If we insert a chord x into C_e such that $\{x\} = \mathcal{N}_{C_e}(c_4) \cap \mathcal{N}_{C_e}(c_5) \cap \mathcal{N}_{C_e}(b_2)$ holds then $\{a_3, a_5, b_2, c_3, c_4, c_5, c_6\}$ will not be an important subgraph. There will be two other important subgraphs: $\{a_3, b_2, c_3, c_4, c_5, x\}$ and $\{a_5, b_2, c_4, c_5, c_6, x\}$.

The rather tedious work of defining the clauses for such an important subgraph is omitted.
At this stage we have computed a formula $\mathcal{F}$. This formula contains many clauses of length two and some clauses from these type 6 subgraphs.
We used the set of important subgraphs like a "fishing net" which we throw over C to extract the information for $\mathcal{F}$. The important subgraphs are the meshes of the fishing net. To gather enough information, we have to be sure that the meshes are small enough.
Let I be the set of important subgraphs. We will define the connectivity graph $\mathcal{G}$ as follows:

$$\begin{aligned} \mathcal{G} &= (I, E) \qquad\qquad \text{with} \\ \{I, J\} \in E &\text{ iff } \exists a, b \in I \cap J : a \otimes b \end{aligned}$$

Without loss of generality we may assume that C is 2-connected. Then — using careful observation — we prove that $\mathcal{G}$ is connected. This will ensure that we are gathering enough information. Furthermore, it will also ensure that the third phase will compute a correct colouring of C.

3.2 Phase 2: Computing a solution for $\mathcal{F}$

The formula $\mathcal{F}$ consists of two parts. $\mathcal{F}_1$ contains the clauses of length two and $\mathcal{F}_2$ the clauses from the important subgraphs of type 6. The evaluation of $\mathcal{F}_1$ is possible in linear time, it is an instance of 2-SAT. If there is no assignment to $\texttt{val}_3$ such that $\mathcal{F}_1$ is satisfied, then we are done. In this case C is not 3-colourable. From now on we assume that $\mathcal{F}_1$ is satisfiable.
The clauses of $\mathcal{F}_2$ are not so easy to evaluate. We have to use a backtracking technique. In a preprocessing step we will prepare the implications of $\mathcal{F}_1$ for later use.
Compute the dependencies of variables implied by $\mathcal{F}_1$ in $\mathcal{F}_2$. This preprocessing is done as follows: For each variable of the form $\texttt{val}_3(a,b)$ which is in $\mathcal{F}_2$, assign the value `true` to it and compute all conclusions we get by using $\mathcal{F}_1$. If we get a conclusion for another variable $\texttt{val}_3(c,d)$ which is also in $\mathcal{F}_2$, we will store this information in a formula $\mathcal{F}_3$ as:

$$\begin{aligned} \texttt{val}_3(a,b) &\Rightarrow \texttt{val}_3(c,d) \qquad \text{or as} \\ \texttt{val}_3(a,b) &\Rightarrow \neg\texttt{val}_3(c,d) \end{aligned}$$

The same is done for the assignment $\texttt{val}_3(a,b) = \texttt{false}$. We will use $\mathcal{F}_3$ together with $\mathcal{F}_2$ in the backtracking to compute a solution for $\mathcal{F}$.
We will use a modified backtracking, where the backtracking tree is stored. Thus our algorithm may jump from one node in the tree to another node to continue the computation at the second

node. This jumping is used if the assignment to the variables yields a contradiction. If a contradiction did occur, then we will try to resolve this contradiction. I.e. we will look for a variable $\mathtt{val}_3(a, b)$ with the following properties:

- This contradiction is broken up by changing the value $\mathtt{val}_3(a, b)$.
- The new value of $\mathtt{val}_3(a, b)$ is not already included in the backtracking tree.

If there is no such variable, then there is also no solution for $\mathcal{F}$, and C is not 3-colourable. If there is such a variable, our algorithm will jump in the backtracking tree.
Furthermore if the algorithm finds a solution to $\mathcal{F}$ it will continue with the third phase using this solution.
For this backtracking strategy we are able to prove the following Lemma:

Lemma 1 (Size of the backtracking tree)
The backtracking tree of the above algorithm has at most $O(\log(n))$ leaves. □

Because of the small number of leaves in the backtracking tree the algorithm for evaluating $\mathcal{F}$ will run in polynomial time. The algorithm is also correct, because we did just change the traversal of the backtracking tree, not the tree itself.

3.3 Phase 3: Using $\mathcal{F}$ for the colouring

At this stage we have computed a function $\mathtt{val}_3$ such that for each important subgraph of C there exists a colouring which obeys $\mathtt{val}_3$.
The colouring of the circle graph C using the function $\mathtt{val}_3$ is done level by level. In the i-th round level $\mathtt{level}_C^{\downarrow}(i)$ is coloured. Without loss of generality we assume that each level is connected and that the leftmost chord of a the i-th level has a neighbour in level $i-1$. Then each level is coloured in a left to right motion. We have to prove that this algorithm will never fail. This is done step by step with the following list of statements.

- If the value of $\mathtt{val}_3(a, b)$ is defined and the chords a and $c \in \mathcal{N}(a) \cap \mathcal{N}(b)$ are already coloured, then the value of $\mathtt{val}_3(a, b)$ determines a unique and correct colouring of the chord b.
- If two intersecting chords of an important subgraph C_I are coloured then the function $\mathtt{val}_3$ defines a correct and unique colouring of C_I.
- If the first two chords of the top level are coloured then the function $\mathtt{val}_3$ defines a correct and unique colouring of the top level.
- If $\mathtt{level}_C^{\downarrow *}(i)$ $(1 \leqslant i < \mathtt{LEV}(C))$ is coloured then the function $\mathtt{val}_3$ defines a correct and unique colouring of $\mathtt{level}_C^{\downarrow *}(i+1)$.

The proofs of these statements are done by a close observation of the structure of important subgraphs. This observation uses the connectivity graph defined above.
This also finishes the description of our algorithm and the proof of correctness.

4 The k-colouring of circle graphs: a full description

In this Section the complexity of the colouring problem of circle graphs is thoroughly investigated. We will investigate the following properties: 4-cycle free, 3-cycle free, bounded level size, bounded degree and bounded clique size. For all these properties we will describe the complexity of the corresponding colouring problem. Due to the bounded space we have to skip most of the proofs.
First we will consider circle graphs without cycles of length four. A cycle in a circle graph is the same as a cycle in the corresponding undirected graph.

Theorem 2 (Colouring circle graphs without cycles of length four [Un88])
A circle graph with no cycle of length four is colourable with three colours. □

In the next step we consider circle graphs without cycles of length three (clique of size three).

Theorem 3 (Colouring circle graphs without cycles of length three [Un88])
A circle graph with no cycle of length three is colourable with four colours. □

The above Theorems are tight, because there is a circle graph without cycles of length three which is not colourable with three colours and there is a circle graph without cycles of length four which is not colourable with two colours.
We will now consider the colouring of circle graphs with bounded level depth. First let us recall some simple definition: A proper interval graph is an interval graph which does not have two intervals I, J with: $I \subseteq J$. A graph G is perfect iff for each node induced subgraph G' of G $\mathtt{COL}(G') = \mathtt{CLI}(G')$ holds.

Lemma 4 (Circle graphs with $\mathsf{LEV}(C) = 1$)
The class of circle graphs with only one level is the same as the class of proper interval graphs. Thus this class is perfect and the colouring problem is solvable in time $O(n \log(n))$.

Proof:
A circle graph C with $\mathsf{LEV}(C) = 1$ does not contain two chords c, c' with $c \rightsquigarrow c'$. We will assign to each chord $c_i \in C$ the interval $I_i = [\overleftarrow{c_i}, \overrightarrow{c_i}]$. Then two chords c_i, c_j intersect iff the corresponding intervals I_i, I_j have a nonempty intersection. The interval graph defined in this way is proper, because there are no intervals I_i, I_j with $I_i \supseteq I_j$. Otherwise $c_i \rightsquigarrow c_j$ would be true. ■

Theorem 5 (k-colouring circle graphs with $\mathsf{LEV}(C) = d$)
The k-colouring problem for circle graphs with bounded level depth is solvable in time $O(n \log(n))$.

Proof:
Finding a clique of size $k+1$ is possible in time $O(n \log(n))$ [RoUr81]. Because of this we may assume without loss of generality: $\mathsf{LEV}(C) = d$ and $\mathtt{CLI}(C) \leqslant k$.
Let $\overleftrightarrow{T}_C\,(x)$ be the set of chords covering a point $x \in \mathbb{R}$: $\overleftrightarrow{T}_C\,(x) = \{c \in C\,;\, \overleftarrow{c} < x < \overrightarrow{c}\}$. The clique size is bounded by k, thus for any level $\mathtt{level}_C^{\downarrow}(i)$ and any $x \in \mathbb{R}$ $|\overleftrightarrow{T}_{\mathtt{level}_C^{\downarrow}(i)}\,(x)| \leqslant k$ will hold. Then $|\overleftrightarrow{T}_C\,(x)| \leqslant k \cdot d$ holds for all $x \in \mathbb{R}$. Thus a line-sweep algorithm has to store a finite amount of information and will solve the problem. ■

If k is not fixed — is a part of the input — the situation changes:

Theorem 6 (The colouring of circle graphs with $\mathsf{LEV}(C) \geqslant 7$)
The colouring of circle graphs with at least 7 levels is NP-complete. □

The next lemma — an improved version from [Un88] — is devoted to degree bounded circle graphs.

Lemma 7 (Circle graphs with bounded degree)
A circle graph C with degree at most d has a d-seperator[1]. □

In [Un88] the existence of a $2 \cdot d$-seperator was presented and the following Theorem was concluded.

Theorem 8 (Colouring circle graphs with bounded degree [Un88])
The k-colouring problem for circle graphs where the degree is at most d is solvable in time $O(n \log(n))$. □

Theorem 9 (The 3-colouring problem for circle graphs)
The 3-colouring of circle graphs is solvable in time $O(n \log(n))$.

---

[1] A graph has a d-seperator iff there exist d nodes such that the deletion of these d nodes slit the graph into two parts which are not connected and each part contains at least 1/3 of the nodes of the graph.

Proof:

See Section 3 ■

Theorem 10 (The 4-colouring problem for circle graphs [Un88])

The 4-colouring problem for circle graphs with $\texttt{CLI}(C) = 3$ is *NP*-complete. □

Using a circle graph C_k with clique-size k which contains an independent set which uses 2 colours in any $\lceil(2/3)\cdot k\rceil$-colouring we were able to prove in [Un88]: The $\lceil(2/3)\cdot k\rceil$-colouring problem is *NP*-complete for circle graph with clique-size k and for all $k \geqslant 3$.

We have now constructed a circle graph C'_k with clique-size k which contains an independent set which uses 2 colours in any $2\cdot k - 1$-colouring. Thus we are able to close the gap:

Theorem 11 (The $2\cdot k-1$-colouring problem for circle graphs with $\texttt{CLI}(C) = k$)

The $2\cdot k-1$-colouring problem for circle graphs with $\texttt{CLI}(C) = k$ is *NP*-complete. □

Theorem 12 (The $2\cdot k$-colouring problem for circle graphs with $\texttt{CLI}(C) = k$ [Un88])

Let C be a circle graph with $k = \texttt{CLI}(C)$. Then $\texttt{COL}(C) \leqslant 2\cdot k$ holds and a $2\cdot k$-colouring is computable in time $O(n\log(n))$. □

5 Segment graphs

Circle graphs were defined by a set of chords on one circle, but we may replace the circle by a g-polygon. In this g-polygon we will place chords and define an intersection graph in the same way as for circle graphs. By splitting the g-polygon at one corner and straightening the g-polygon we will get the following formal definition of these g-segment graphs.

Definition 3 (g-segment graphs)

A circle graph C is called a ***g*-segment graph** iff there exist $g-1$ points $e_1, e_2, \ldots e_{g-1}$ such that for all $c \in C$ there is one e_i such that $\overleftarrow{c} < e_i < \overrightarrow{c}$ holds. □

We start with a note on 2-segment graphs. The class of 2-segment graphs is the same as the class of circle graphs which have an equator. A circle graph C has an equator iff we are able to add a single chord which crosses each other chord of C. Using the well known result that these graph class is the same as the permutation graphs we have the following Lemma.

Lemma 13 (2-segment graphs)

The class of 2-segment graphs is the same as the class of permutation graphs. Thus this class is perfect. □

Lemma 14 (g-segment graphs with $\texttt{CLI}(C) = k'$)

There is a function $f(g,k')$ such that for each g-segment graph C with $\texttt{CLI}(C) = k'$ there is a g-segment graph $C' \subseteq C$ with: $\texttt{COL}(C') = \texttt{COL}(C)$ and $|C'| \leqslant f(g,k')$. This subgraph C' is computable in time $O(n\log(n))$. □

Using this we are able to solve the k-colouring problem for g-segment graphs.

Theorem 15 (k-colouring g-segment graphs with $\texttt{CLI}(C) = k'$)

The k-colouring of g-segment graphs with $\texttt{CLI}(C) = k'$ is solvable in time $O(n\log(n))$. □

6 Conclusions

We presented in this article a nearly complete description of the complexity of the colouring problems of circle graph. Table 2 shows this in a compact version.

There are two problems still open. One is a minor one: How many levels must a circle graph have, such that the colouring problem is *NP*-complete? The second open problem is the complexity of the colouring problem for g-segment graphs.

circle graph C	x-Colouring Problem $x=3$	$x=4$	$x=k$	$x=2k-1$	$x=2k$	$x=\epsilon$
C without restrictions	$\mathcal{P}$	$\mathcal{NPC}$	$\mathcal{NPC}$	$\mathcal{NPC}$	$\mathcal{NPC}$	$\mathcal{NPC}$-
$\mathrm{CLI}(C)=k\geqslant 3$	$\mathcal{P}$	$\mathcal{NPC}$	$\mathcal{NPC}$	$\mathcal{NPC}$	$\mathcal{P}+$	$\mathcal{NPC}$
degree$(C)=k$	$\mathcal{P}$	$\mathcal{P}$	$\mathcal{P}$	$\mathcal{P}$	$\mathcal{P}$	$\mathcal{P}$
C without 4-cycle	$\mathcal{P}+$	$\mathcal{P}+$	$\mathcal{P}$	$\mathcal{P}$	$\mathcal{P}$	$\mathcal{P}$
C without 3-cycle	$\mathcal{P}$	$\mathcal{P}+$	$\mathcal{P}$	$\mathcal{P}$	$\mathcal{P}$	$\mathcal{P}$
C is a 2-segment graph	$\mathcal{P}*$	$\mathcal{P}*$	$\mathcal{P}*$	$\mathcal{P}*$	$\mathcal{P}*$	$\mathcal{P}*$
C is a g-segment graph	$\mathcal{P}$	$\mathcal{P}$	$\mathcal{P}$	$\mathcal{P}$	$\mathcal{P}$	open
$\mathrm{LEV}(C)=1$	$\mathcal{P}*$	$\mathcal{P}*$	$\mathcal{P}*$	$\mathcal{P}*$	$\mathcal{P}*$	$\mathcal{P}*$
$2\leqslant \mathrm{LEV}(C)\leqslant 6$	$\mathcal{P}$	$\mathcal{P}$	$\mathcal{P}$	$\mathcal{P}$	$\mathcal{P}$	open
$\mathrm{LEV}(C)\geqslant 7$	$\mathcal{P}$	$\mathcal{P}$	$\mathcal{P}$	$\mathcal{P}$	$\mathcal{P}$	$\mathcal{NPC}$

Entry	Description of Entry
$\mathcal{P}$	Solvable in polynomial time
$\mathcal{P}+$	Colouring always possible and solvable in polynomial time
$\mathcal{P}*$	Solvable in polynomial time because that class is perfect
$\mathcal{NPC}$	$\mathcal{NP}$-complete
$\mathcal{NPC}$-	$\mathcal{NP}$-complete, not shown here or in [Un88]
open	open Problem

Table 2: The complexity of the colouring of circle graphs

Acknowledgment

Many discussions with B. Monien accompanied this work. His help is gratefully acknowledged.

References

[Bo75] *A. Bouchet*, **Reducing prime graphs and recognizing circle graphs**, Combinatorica, 1975

[Bo85] *A. Bouchet*, **Characterizing and recognizing circle graphs**, Proc. of the 6th Yugoslav Seminar on Graph Theory, Dubrovnik 1985

[Br90] *Andreas Brandstädt*, **Special graph classes — a survey (preliminary version)**, Forschungsergebnisse Friedrich-Schiller-Universität Jena, Sektion Mathematik, 1990

[Da89] *Peter Damaschke,* **The Hamilton Cycle Probelm is *NP*-Complete for Circle Graphs,** Technical Report, University of Jena, O-6900 Jena, Germany

[EvIt71] *S. Even and A. Itai*, **Queues, stacks and graphs**, Theory of Machines and Computations, 1971, 71-86

[Fr84] *H. de Fraysseix*, **A Characterization of Circle Graphs**, Europ. J. Combinatorics, Vol 5, 1984, 223-238

[GaHsSu85] *C.P. Gabor and W.-L. Hsu and K.J. Supowit*, **Recognizing Circle Graphs in Polynomial Time**, FOCS, 85

[GaJoMiPa80] *M.R. Garey and D.S. Johnson and G.L. Miller and Ch.H. Papadimitriou*, **The Complexity of Coloring Circular Arcs and Chords**, SIAM Alg. Disc. Meth., 1. No. 2., 1980, 216-227

[Ga75] *F. Gavril*, **Algorithms for a Maximum Clique and A Maximum Independent Set of a Circle Graph**, Networks 3, 1975, 261-273

[GuLeLe82] *U.I. Gupta and D.T. Lee and J.Y.T. Leung*, **Efficient Algorithms for Interval Graphs and Circular-Arc Graphs**, Networks, 12, 1982, 459-467

[Ke90] *J.M. Keil*, **The Dominating Set Problem for Circle Graphs is *NP*-Complete**, Personal commonications with A. Brandstädt

[Le84] *J.Y.-T. Leung*, **Fast Algorithms for Generating All Maximal Independent Sets of Interval, Circular-Arc and Chordal Graphs**, Journal of Algorithms 5,, 1984, 11, 22-35

[Na85] *W. Naji*, **Reconnaissance des graphes de cordes**, Discr. Math. 54, 1985, 329-337

[OrBoBo82] *J.B. Orlin and M.A. Bonuccelli and D.P. Bovet*, **An $O(n^2)$ Algorithm for Coloring Proper Circular Arc Graphs**, SIAM Alg. Disc. Meth., 2. No. 2., June 1982, 88-93

[RePoUr82] *R.C. Read and D. Rotem and J. Urrutia*, **Orientation of Circle Graphs**, Journal of Graph Theory, 6, 1982, 325-341

[RoUr81] *D. Rotyem and J. Urrutia*, **Finding Maximum Cliques in Circle Graphs**, Networks, 11, 1981, 269-278

[Sp88] *J.P. Spinrad*, **Recognition of Circle Graphs**, Manuscript 1988, Dept. of CS, Vanderbilt Univ., Nashville, TN

[St90] *L. Stewart and E. Elmallah and J. Culberson*, **Polynomial algorithms on *k*-polygon graphs**, Proc. 21th SE Conf. Combin. Graph Theory and Computing, Boca Raton, Florida 1990

[Tu80] *A. Tucker*, **An Efficient Test for Circular-Arc Graphs**, SIAM J. Comput. Vol. 9. No. 1., 1980, 1-24

[Un90] *W. Unger*, **Färbung von Kreissehnengraphen**, Ph.D. Thesis, University of Paderborn, Germany

[Un88] *W. Unger*, **On the k-colouring of Circle Graphs**, Lecture Notes in Computer Science, 294, Springer Verlag, proc. STACS 88 Bordeaux, pp 61-72

[WePo85] *W. Wessel and R. Pöschel*, **On Circle Graphs**, Graphs, Hypergraphs and Applications, 1985, 207-210

[Ya86] *M. Yannakakis*, **Four Pages are necessary and sufficient for Planar Graphs**, ACM, 1986, 104-108

Graph Isomorphism is low for PP*

Johannes Köbler
Theoretische Informatik
Universität Ulm
Oberer Eselsberg
D-7900 Ulm, Germany

Uwe Schöning
Theoretische Informatik
Universität Ulm
Oberer Eselsberg
D-7900 Ulm, Germany

Jacobo Torán†
Departamento L.S.I.
U. Politecnica de Catalunya
Pau Gargallo 5
E-08028 Barcelona, Spain

Abstract

We show that the graph isomorphism problem is low for PP and for C=P, i.e. it does not provide a PP or C=P computation with any additional power when used as oracle. Furthermore, we show that graph isomorphism belongs to the class LWPP (see Fenner, Fortnow, Kurtz [FeFoKu 91]). A similar result holds for the (apparently more difficult) problem Group Factorization. The problem of determining whether a given graph has a nontrivial automorphism, Graph Automorphism, is shown to be in SPP, and is therefore low for PP, C=P, and $\mathrm{Mod}_k\mathrm{P}$, $k \geq 2$.

1 Introduction

The problem of finding an efficient (i.e. polynomial time) algorithm for testing whether two given graphs are isomorphic has withstood all attempts for a solution up to date. The worst case running time of all known algorithms is of exponential order, and just for certain special types of graphs, polynomial time algorithms have been devised (for further reference see [Ba 81, Hof 82a, Jo 85]). Although the possibility that Graph Isomorphism (GI) is NP-complete has been discussed [GaJo 79], strong evidence against this possibility has been provided [Mat 79, BaMo 88, BoHaZa 87, GoMiWi 85, GoSi 86, Sch 88]. In the first place it was proved by Mathon [Mat 79] that the decision and counting versions of the problem are polynomial time equivalent (in the sense of a truth-table reduction) which does not seem to be the case for NP-complete problems. In the second place it has been shown that the assumption Graph Isomorphism is NP-complete implies that the polynomial time hierarchy collapses to AM $\subseteq \Pi_2^p$.

Still, a complete complexity classification (positive or negative) of Graph Isomporphism has not yet been achieved, and it has been conjectured that this problem might lie strictly between P and NP-complete.

Although GI is the best known example of a problem with this property, it is not an isolated case. There are many other graph and group theoretic problems related to Graph Isomorphism that lie between P and NP-complete and whose exact complexity is not known either (cf. [Hof 82a, Hof 82b, Luks 82]).

The present work makes a contribution towards a better (structural) complexity classification of Graph Isomporhism and other related problems. Using a group theoretic approach we study

*This research was supported by the DAAD (Acciones Integradas 1991, 313-AI-e-es/zk). A full version of this paper is available as *Ulmer Informatik-Bericht Nr. 91-04*.

†Research partially supported by ESPRIT-II Basic Research Actions Program of the EC under Contract No. 3075 (project ALCOM)

certain "counting properties" of these problems which show differences among them and which allow finer classifications than the existing ones.

Very recently the class Gap-P of functions that compute the difference between accepting and non-accepting paths of a nondeterministic polynomial time Turing machine has been considered [FeFoKu 91]. This class is basically a generalization of #P ("counting P" [Va 79]) with certain advantages. For example, it allows functions to have negative values and is closed under subtraction.

Some of our results can be best understood in the context of Gap-P functions. We show that for certain problems like Graph Automorphism (GA) there is a nondeterministic, polynomial time Turing machine M with the following properties:

- For every input graph G, the difference $gap_M(G)$ between the number of accepting and non-accepting computations of M is either 0 or 1,
- $G \in$ GA if and only $gap_M(G) = 1$.

From this fact follows immediately that the problem is in the classes $\oplus$P and $\mathrm{C_=P}$. Observe also that the accepting behaviour of machine M is similar to one for a language in UP (unambiguos NP [Va 76]). A machine with the above accepting mechanism appeared for the first time in [KöScToTo 89]. Recently the corresponding language class has been considered. This class is called XP "exact P" in [OgHe 91] and SPP "stoic PP" in [FeFoKu 91].

In the case of Graph Isomorphism and some other related problems we are able to construct a nondeterministic, polynomial time Turing machine M with the following properties:

- For every input pair of graphs (G_1, G_2) with n nodes, $gap_M(G_1, G_2) \in \{0, f(n)\}$,
- $(G_1, G_2) \in$ GI if and only if $gap_M(G_1, G_2) = f(n)$.

Here f is a polynomial time computable function. The class of languages for which there is a machine with such an accepting behaviour was also considered in [FeFoKu 91] and called LWPP "length dependent wide PP". In the mentioned paper the authors ask whether there are nontrivial examples of languages in SPP. GA seems to be the first such example, whereas GI is the first natural problem in LWPP which is not known to be in SPP.

The classes SPP and LWPP have the property that they are *low* for PP, i.e. they do not provide a PP machine with any additional help when used as oracle in a PP computation. This fact provides one of the main corollaries in this work: Graph Isomorphism and all the related problems are low for PP (in symbols: $\mathrm{PP}^{\mathrm{GI}} = \mathrm{PP}$). We use the lowness result for GI to provide further evidence for the problem not being NP-complete.

Note. *In this version most proofs have been omitted except those being directly concerned with the* PP*-lowness of the graph isomorphism problem.*

2 Notation and Preliminaries

2.1 Group Theory

We will denote groups by upper case Roman letters and elements of the groups by lower case Greek letters. The order of a group A is represented by $|A|$. The expression $B < A$ denotes that B is a subgroup of A. We consider only permutation groups. The group of permutations on $\{1, \ldots, n\}$ is denoted by S_n. We represent the identity permutation by id. The *orbit* of i in A is the set $\{j : \exists\varphi \in A,\ \varphi(i) = j\}$.

If X is a subset of $\{1, \ldots, n\}$, then the *pointwise stabilizer of* X *in* A is $A_{[X]} = \{\varphi \in A : \forall x \in X,\ \varphi(x) = x\}$. In case that $X = \{x\}$ is a singleton, $A_{[X]}$ is denoted by $A_{[x]}$. Let $X_i = \{1, \ldots, i\}$ and denote A by $A^{(0)}$ and $A_{[X_i]}$ by $A^{(i)}$, then one obtains the following "tower"

$$\{id\} = A^{(n)} < A^{(n-1)} < \ldots < A^{(1)} < A^{(0)} = A.$$

We will present nondeterministic algorithms for certain group problems in which the number of accepting computations can be only among two integers which are known beforehand. The following lemma is the key for such a behaviour.

Lemma 2.1 [Hall 59, Theorem 5.2.2] *Let $A = A_{[i]}\pi_1 + A_{[i]}\pi_2 + \ldots + A_{[i]}\pi_d$. Then d is the size of the orbit of i in A, and for all $\psi \in A_{[i]}\pi, \pi(i) = \psi(i)$.*

In other words, if $\{j_1, \ldots, j_d\}$ is the orbit of i in the permutation group A, and if we denote the set of all permutations in A which map i to j by $A_{[i \mapsto j]}$, then we can partition A into d sets of equal cardinality, i.e. $A = A_{[i \mapsto j_1]} + \cdots + A_{[i \mapsto j_d]}$. Thus, $|A| = d * |A_{[i]}|$ and for every $j \in \{1, \ldots, n\}$ the number of permutations in A which map i to j is either 0 or $|A_{[i]}|$. Since $A^{(i)}$ is the stabilizer of i in $A^{(i-1)}$, we can state the following two corollaries.

Corollary 2.2 *Let d_i be the size of the orbit of i in $A^{(i-1)}, 0 \le i \le n$. Then the order of A is equal to $\prod_{i=1}^{n} d_i$.*

Corollary 2.3 *Let $A < S_n$ and consider the pointwise stabilizers $A^{(i)}, 0 \le i \le n$. Then for every $j > i$ the number of permutations $\varphi \in A^{(i-1)}$ such that $\varphi(i) = j$ is either 0 or $|A^{(i)}|$.*

We will consider only simple undirected graphs $G = (V, E)$ without self-loops. We denote vertices in V by natural numbers, i.e. if G has n nodes then $V = \{1, \ldots, n\}$. Let $G = (V, E)$ be a graph ($|V| = n$). An *automorphism* of G is a permutation $\varphi \in S_n$ that preserves adjacency in G, i.e. for every pair of nodes i, j, $\{i, j\} \in E$ if and only if $\{\varphi(i), \varphi(j)\} \in E$. The set of automorphisms of G, $Aut(G)$, is a subgroup of S_n.

Two graphs $G_1 = (V_1, E_1)$, $G_2 = (V_2, E_2)$ are *isomorphic* if there is a bijection φ between V_1 and V_2 such that for every pair of nodes $i, j \in V_1$, $\{i, j\} \in E_1$ if and only if $\{\varphi(i), \varphi(j)\} \in E_2$. Let $Iso(G_1, G_2)$ denote the set of all isomorphisms between G_1 and G_2.

Lemma 2.4 *Let G_1 and G_2 be two graphs. If there is an isomorphism between G_1 and G_2, then $Iso(G_1, G_2)$ is a right coset of $Aut(G_1)$ and thus $|Iso(G_1, G_2)| = |Aut(G_1)|$.*

2.2 Complexity Theory

All sets considered here are languages over the alphabet $\Sigma = \{0, 1\}$. For a string $x \in \Sigma^*$, $|x|$ denotes the length of x. We assume the existence of a pairing function $\langle .,. \rangle : \Sigma^* \times \Sigma^* \mapsto \Sigma^*$ which is computable in polynomial time and has inverses also computable in polynomial time. $\langle y_1, y_2, \ldots, y_n \rangle$ stands for $\langle n, \langle y_1, \langle y_2, \ldots, y_n \rangle\rangle\rangle$. For a set A, $|A|$ is the cardinality of A.

We assume that the reader is familiar with (nondeterministic, polynomial time bounded, oracle) Turing machines and complexity classes (see [BaDiGa 87, Sch 86]). FP is the set of functions computable by a deterministic polynomial time bounded Turing machine.

For a nondeterministic machine M and a string $x \in \Sigma^*$, let $acc_M(x)$ be the number of accepting computation paths of M on input x. Analogously, for a nondeterministic oracle machine M, an oracle A, and a string $x \in \Sigma^*$, $acc_M^A(x)$ is the number of accepting paths of M^A with input x.

Next we define the complexity classes PP, C=P, ⊕P that are defined considering the number of computation paths of a nondeterministic machine. These classes were first introduced in [Gi 77, Wa 86, PaZa 83], respectively. The counting classes Mod$_k$P, $k \ge 2$, were independently defined in [BeGiHe 90] and [CaHe 89].

A language L is in the class PP if there is a nondeterministic polynomial time machine M and a function $f \in$ FP such that for every $x \in \Sigma^*$,

$$x \in L \iff acc_M(x) \ge f(x).$$

PP is called CP in the notation of [Wa 86]. A language L is in the class C=P if there is a nondeterministic polynomial time machine M and a function $f \in$ FP such that for every $x \in \Sigma^*$,

$$x \in L \Longleftrightarrow acc_M(x) = f(x).$$

The class of all languages whose complement is in C=P is denoted by C≠P. Note that C≠P is a generalization of NP. A language L is in the class Mod$_k$P, $k \geq 2$, if there is a nondeterministic polynomial time machine M such that for every $x \in \Sigma^*$,

$$x \in L \Longleftrightarrow acc_M(x) \not\equiv 0 \pmod{k}.$$

The class Mod$_2$P is also called ⊕P ("Parity P").

Closely related to the language class PP is the function class #P, defined by Valiant in [Va 79]. A function f is in #P if there is a nondeterministic polynomial time machine M such that for every x in Σ^*, $f(x) = acc_M(x)$. Fenner et al. [FeFoKu 91] defined the *gap* produced by M on input x as

$$gap_M(x) = acc_M(x) - acc_{\overline{M}}(x),$$

where $\overline{M}$ is the same machine as M but with the accepting and non-accepting states interchanged. A function f is in Gap-P if there is a nondeterministic polynomial time machine M such that for every x in Σ^*, $f(x) = gap_M(x)$.

For a language L and a complexity class K (which has a sensible relativized version $K^{()}$), we will say that L is *low* for K (L is K–low) if $K^L = K$. For a language class C, C is low for K if for every language L in C, $K^L = K$. The following class SPP which is denoted by XP in [OgHe 91] was also defined in [FeFoKu 91].

Definition 2.5 SPP *is the class of all languages L such that there exists a nondeterministic polynomial time machine M such that, for all x,*

$$\begin{aligned} x \in L &\implies gap_M(x) = 1, \\ x \notin L &\implies gap_M(x) = 0. \end{aligned}$$

Theorem 2.6 [FeFoKu 91, Theorem 5.5] SPP $= \{L \mid \text{Gap-P}^L = \text{Gap-P}\}$.

A consequence of the above theorem is the lowness of SPP for any uniformly gap-definable class like PP, C=P, Mod$_k$P, and SPP (see [FeFoKu 91]). In particular, it follows that SPP is closed under Turing reductions. Another interesting subclass of PP is WPP.

Definition 2.7 [FeFoKu 91] WPP *("wide" PP) is the class of all languages L such that there exists a nondeterministic polynomial time machine M and a function $f > 0$ in* FP *such that for all x,*

$$\begin{aligned} x \in L &\implies gap_M(x) = f(x), \\ x \notin L &\implies gap_M(x) = 0. \end{aligned}$$

It is clear that SPP ⊆ WPP ⊆ C=P ∩ C≠P. Fenner, et al. have also defined a restricted version LWPP of WPP, where the FP function f depends only on the length of x, and they showed that LWPP is low for PP, WPP, LWPP, and C=P.

3 Graph Automorphism

We consider the following problem:

Graph Automorphism (GA): Given a graph G, decide whether the automorphism group of G contains a non-trivial automorphism (i.e. an automorphism different from the identity).

Graph Automorphism is polynomial time truth-table reducible to Graph Isomorphism [Mat 79] but it seems to be an easier problem (a reduction in the opposite direction is not known). The special "counting properties" of this problem have been used by Boppana, Hastad and Zachos [BoHaZa 87] to show that Graph Automorphism is in the class co-AM. In the next theorem we show another property of this problem.

Theorem 3.1 GA *is in* SPP *and therefore is low for* SPP, Mod$_k$P, $\oplus$P, *and* PP.

Proof. Consider the function

$$f(G) = \prod_{1\le i<j\le n} (|Aut(G)_{[i]}| - |Aut(G)_{[i\mapsto j]}|).$$

First observe that this function is in the class Gap-P.

We show now that $f(G)$ is either 0 or 1 depending on whether $G \in$ GA. If $G \in$ GA then there is a non-trivial automorphism $\pi \in Aut(G)$ such that $\pi(i) = j$ for some pair of vertices $i, j,\ i < j$. This implies that $\pi \in Aut(G)_{[i\mapsto j]}$ and by lemma 2.1 $|Aut(G)_{[i]}| = |Aut(G)_{[i\mapsto j]}|$. Hence, at least one of the factors in the definition of f is 0 and this implies $f(G) = 0$.

On the other hand, if $G \notin$ GA, then for every pair $i, j,\ i < j$, $Aut(G)_{[i\mapsto j]}$ is empty, i.e. $|Aut(G)_{[i\mapsto j]}| = 0$. Moreover, for every i the only automorphism in $Aut(G)_{[i]}$ is the identity. All the factors of f have value 1 and therefore $f(G) = 1$. □

Remarkably, the following problem, which is not known to be in NP $\cup$ co-NP is polynomial time truth-table equivalent to Graph Automorphism.

Unique Graph Isomorphism (UGI): Given two graphs G_1 and G_2, decide whether there is a unique isomorphism between G_1 and G_2.

For the reduction from UGI to GA one has to make three independent queries asking whether G_1, G_2, and $G_1 \cup G_2$ belong to GA. It can be checked that $\langle G_1, G_2\rangle \in$ UGI if and only if the sequence of answers is "No,No,Yes". For the reduction in the other direction one query is enough since $G \in$ GA if and only if $\langle G, G\rangle \notin$ UGI.

Since the class SPP is closed under Turing reducibility, the reductions imply that Unique Graph Isomorphism is in SPP.

We present now a connection between Graph Automorphism and the area of promise problems (cf. [EvSeYa 84]).

Definition 3.2 *A promise problem is a pair of sets* (Q, R). *A set* L *is called a solution to the promise problem* (Q, R) *if*

$$\forall x(x \in Q \Rightarrow (x \in L \Leftrightarrow x \in R)).$$

Let 1SAT denote the set of formulas with at most one satisfying assignment. We will consider the promise problem (1SAT,SAT). Observe that a solution for this problem has to agree with SAT in the formulas with a unique satisfying assignment as well as in the unsatisfiable formulas.

Theorem 3.3 *Graph Automorphism is polynomial time disjunctively reducible to any solution* L *of the promise problem* (1SAT,SAT).

Valiant and Vazirani [VaVa 86] have proved that if the promise problem (1SAT,SAT) has a solution in P then NP is included in the probabilistic class R. Using the above result we can draw another conclusion from this hypothesis.

Corollary 3.4 *If the promise problem* (1SAT,SAT) *has a solution in* P, *then Graph Automorphism belongs to* P.

4 Graph Isomorphism

We show in this section that the Graph Isomorphism problem is low for the counting classes PP and GP. Our first step in this direction is done by the next theorem which shows that GI can be recognized by an oracle Turing machine M asking an NP oracle only about strings y for which the potential number of witnesses is known beforehand. This means that if y is a member of the oracle then the number of witnesses is equal to some specific value. Moreover, this value depends not on the actual query but can be controlled by an additional parameter m in the input to M. For this special purpose, we define the problem

$$\mathrm{GI}^* = \{\langle G,H,m\rangle \mid G \text{ and } H \text{ are isomorphic graphs and } m \geq n\},$$

where n is the number of vertices of G. Clearly, GI* is many-one equivalent to GI. The oracle machine M of Theorem 4.1 recognizes GI* asking its NP oracle only about strings y for which the number of witnesses is either 0 or $m!$. This machine will be used to construct for a given nondeterministic oracle machine M' which works under oracle GI another machine M'' working under some NP oracle A such that $gap_{M'}^{\mathrm{GI}} = gap_{M''}^{A}$, and for all positively answered queries y which are asked by M'' on input x there are exactly $f(0^{|x|})$ witnesses for membership in A ($f \in \mathrm{FP}$).

In a second step, we take the oracle queries out of the computation of M'', resulting in a nondeterministic machine M''' such that $gap_{M'''}(x) = g(0^{|x|}) * gap_{M''}^{A}(x)$ for a polynomial time computable function $g > 0$. From this, it can be easily seen that GI is low for PP and GP.

Theorem 4.1 *There is a deterministic polynomial time oracle Turing machine M and an oracle set $A \in \mathrm{NP}$ recognized by a nondeterministic polynomial time machine N s.t.*

i) $L(M,A) = \mathrm{GI}^*$,

ii) M^A *on input* $\langle G,H,m\rangle$ *only asks queries y for which* $acc_N(y) \in \{0, m!\}$.

Proof. Let A be the oracle set $B \oplus C$ where

$$B = \{\langle G,i,j,k\rangle : \text{there exists an automorphism in } Aut(G)^{(i-1)} \text{ which maps } i \text{ to } j\}$$

and

$$C = \{\langle G,H,k\rangle : G \text{ and } H \text{ are isomorphic}\}.$$

Note that the value of k does not affect membership of $\langle G,i,j,k\rangle$ in B and of $\langle G,H,k\rangle$ in C. k is only used by M to pass to the oracle the information about how many accepting path's should be produced for every automorphism (isomorphism, respectively). More specifically, there is a nondeterministic polynomial time machine N which on input y of the form $0\langle G,i,j,k\rangle$ has exactly $k * |Aut(G)^{(i-1)}_{[i\mapsto j]}|$ accepting paths, and if $y = 1\langle G,H,k\rangle$, then the number of accepting paths of N is equal to $k * |Iso(G,H)|$.

Using B as oracle, the following deterministic polynomial time oracle Turing machine M on input $\langle G,H,m\rangle$ first computes the size of $Aut(G)$ and afterwards asks oracle C whether G and H are isomorphic. The preceding computation of $|Aut(G)|$ is necessary because in order to fulfill condition *ii*) of the theorem, M has to know the potential number of isomorphisms between G and H.

```
input ⟨G, H, m⟩;
n := the number of vertices in G;
if m < n then reject end;
d := 1; /* d counts the number of automorphisms of G */
for i := n downto 1 do
/* determine the size d_i of the orbit of i in Aut(G)^(i-1) */
   d_i := 1;
   for j := i + 1 to n do
   /* test whether j lies in the orbit of i in Aut(G)^(i-1) */
     if ⟨G, i, j, ⌊m!/d⌋⟩ ∈ B then d_i := d_i + 1 end;
   end;
   d := d * d_i; /* now d is the order of Aut(G)^(i-1) */
end;
if ⟨G, H, ⌊m!/d⌋⟩ ∈ C then accept else reject end;
```

For every $i = n, \ldots, 1$, M computes in the inner for-loop the size d_i of the orbit of i in $Aut(G)^{i-1}$. It follows from Corollary 2.3 that whenever M^A makes a query $y = \langle G, i, j, \lfloor \frac{m!}{d} \rfloor \rangle$ to oracle B, then there are either 0 or exactly $d = |Aut(G)^{(i)}|$ automorphisms in $Aut(G)^{(i-1)}$ mapping i to j. Since for each such automorphism, N branches into $\lfloor \frac{m!}{d} \rfloor$ accepting computations, and since d as the order of a subgroup of S_n divides $n!$, $acc_N(0y) = m!$ for all $y \in B$.

As it is stated in Corollary 2.2, the order of $Aut(G)$ is equal to the product $\prod_{i=1}^{n} d_i$, which is computed in d. Thus, when M^A makes the last query $y = \langle G, H, \lfloor \frac{m!}{d} \rfloor \rangle$ to oracle C, then, according to Lemma 2.4, there are exactly $d = |Aut(G)|$ many isomorphisms between the two graphs G and H if they are isomorphic and 0 otherwise. Since for each isomorphism, N branches into $\lfloor \frac{m!}{d} \rfloor$ accepting computations, condition *ii*) of the theorem is also fulfilled for this last query. □

The determination of $|Aut(G)|$ in the above proof is based on Mathon's reduction of the Graph Isomorphism counting problem to GI [Mat 79]. But whereas his reduction is completely nonadaptive and therefore a truth-table reduction, our computation is highly adaptive. The adaptiveness seems to be necessary to obtain property *ii*) above which will be exploited in the next theorem.

In Theorem 4.2, it is shown that – modulo a multiplicative factor – a G≠P oracle A can be removed from a Gap-P^A computation that on input x only queries for A about strings y whose gap is in the set $\{0, f(x)\}$ where f is a given function in Gap-P.

Theorem 4.2 *Let M be a nondeterministic polynomial time oracle machine and let $A = \{y : gap_N(y) \neq 0\}$ be a set in* G≠P. *If there is a function $f \in$ Gap-P such that M^A on input x only asks queries y for which $gap_N(y) \in \{0, f(x)\}$, then there is a polynomial q such that the function $x \mapsto gap_M^A(x) * f(x)^{q(|x|)}$ is in* Gap-P.

The proof of Theorem 4.2 is similar in flavor to the proof of Theorem 5.5 in [FeFoKu 91] that SPP is low for Gap-P. The crucial observation here is that queries which are made in the sequel of a wrong oracle answer to a previous query don't need to possess the restricted counting properties imposed by PP-low classes like SPP and LWPP.

Now we are ready to state the main result of this section, namely that for every function g in Gap-P^{GI} there is a nondeterministic polynomial time machine whose gap h is a multiple of g, i.e. $h(x) = g(x) * f(0^{|x|})$ for a function f in FP. This can be easily derived from the last two theorems and is the key for the lowness properties of Graph Isomorphism with respect to the counting classes PP and C=P.

Theorem 4.3 *For every $g \in$ Gap-P^{GI} there is a function $f > 0$ in* FP *such that the function $x \mapsto g(x) * f(0^{|x|})$ is in* Gap-P.

Corollary 4.4 GI *is in* LWPP *and is low for* PP *and* C=P.

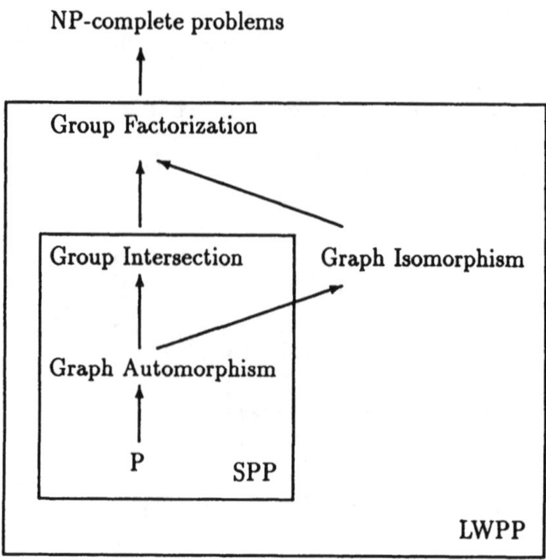

Figure 1: Results concerning membership to PP-low classes.

5 General Permutation Group Problems

In this section it is shown that Group Intersection and Group Factorization which can be seen as generalizations of the Graph Automorphism and the Graph Isomorphism problem [Hof 82a] are in SPP and LWPP, respectively.

Group Intersection: Given generating sets for two permutation groups A, B, decide whether $A \cap B$ contains a non-trivial permutation.

Group Factorization: Given generating sets for the groups $A, B < S_n$ and a permutation $\pi \in S_n$, decide whether $\pi \in AB$.

Theorem 5.1

i) Group Intersection *is in* SPP *and therefore is low for* Mod_kP, PP, *and* C=P.

ii) Group Factorization *is in* LWPP *and therefore is low for* PP *and* C=P.

A summary of our results concerning membership to the PP-low classes SPP and LWPP can be seen in Figure 1. The arrows indicate known polynomial time Turing reductions among the problems.

6 Some Consequences of the results

In this section, we show that if GI were NP-complete, then the polynomial time hierarchy would be low for the classes PP and C=P. The last one is especially unlikely since C=P seems to be a very weak class (for example in [Gre 91, Tar 91] relativizations are shown under which Σ_2^p or BPP are not even Turing reducible to C=P).

Theorem 6.1 *If the Graph Isomorphism problem is NP-complete, then the polynomial time hierarchy is low for PP.*

The following lemma is just a generalization of Theorem 4.4 from [Sch 88] to the class G=P. Its proof follows exactly the lines of the original one.

Lemma 6.2 NP ∩ co-AM *is low for* G=P^{NP}.

Theorem 6.3 *If the Graph Isomorphism problem is NP-complete, then the polynomial time hierarchy is low for* G=P.

At the end of this section we consider some problems related to the well known Graph Reconstruction Conjecture. Recently, these problems were defined by Kratsch and Hemachandra [KrHe 91] in order to study the complexity-theoretic aspects of graph reconstruction.

Let $G = (V, E)$ be a graph, $V = \{1, \ldots, n\}$. A sequence $\langle G_1, \ldots, G_n \rangle$ of graphs is a *deck* of G if there is a permutation $\pi \in S_n$ such that for every $i = 1, \ldots, n$, $G_{\pi(i)}$ is isomorphic to the one-vertex-deleted-subgraph $(V - \{i\}, E - \{\{i,j\} : j \in V\})$ of G. In this case, G is called a *preimage* of the deck $\langle G_1, \ldots, G_n \rangle$. The Reconstruction Conjecture says that for any legitimate deck, there is just one preimage of it, up to isomorphism.

Among other problems, Kratsch and Hemachandra investigated the following problems.

Deck Checking (DC): Given a graph G and a sequence of graphs G_i, $i = 1, \ldots, n$, decide whether G is a preimage of the deck $\langle G_1, \ldots, G_n \rangle$.

Legitimate Deck (LD): Given a sequence of graphs G_i, $i = 1, \ldots, n$, decide whether there is a preimage G for the deck $\langle G_1, \ldots, G_n \rangle$, i.e. the deck is legitimate.

Preimage Counting: Given a sequence of graphs G_i, $i = 1, \ldots, n$, compute the number PCount$(\langle G_1, \ldots, G_n \rangle)$ of all nonisomorphic preimages for the deck $\langle G_1, \ldots, G_n \rangle$.

Kratsch and Hemachandra showed DC $\leq_m^p$ GI $\leq_m^p$ LD. They left it as an open question whether there is also a reduction from LD to GI.

We show that if the Reconstruction Conjecture holds, then LD lies in LWPP. This follows immediately from the next theorem which shows without any assumption that the function PCount can be computed in Gap-P, modulo a polynomial time computable factor.

Theorem 6.4 *There is a function f in* FP *such that the following function is in* Gap-P

$$\langle G_1, \ldots, G_n \rangle \mapsto f(n) * \mathrm{PCount}(\langle G_1, \ldots, G_n \rangle).$$

Corollary 6.5 *If the Reconstruction Conjecture holds, then* LD *is in* LWPP *and therefore low for* G=P *and* PP.

References

[Ba 81] L. Babai, Moderately exponential bound for graph isomorphism. In *Proceedings Fundamentals of Computation Theory, Lecture Notes in Computer Science 117* (1981), 34-50.

[BaMo 88] L. Babai, S. Moran, Arthur-Merlin games: A randomized proof system, and a hierarchy of complexity classes. In *Journal of Computer and System Sciences 36* (1988), 254-276.

[BaDiGa 87] J.L. Balcázar, J. Díaz, J. Gabarró, *Structural Complexity I.* Springer, 1987.

[BeGiHe 90] R. Beigel, J. Gill, U. Hertrampf, Counting classes: Thresholds, parity, mods, and fewness. In *Proceedings 7th Symposium on Theoretical Aspects of Computer Science, Lecture Notes in Computer Science 415* (1990), 49-57.

[BeHeWe 89] R. BEIGEL, L. HEMACHANDRA, G. WECHSUNG, On the power of probabilistic polynomial time: $P^{NP[\log]} \subseteq$ PP. In *Proceedings 4th Structure in Complexity Theory Conference*, p. 225-227, IEEE Computer Society, 1989.

[BoHaZa 87] R. BOPPANA, J. HASTAD, AND S. ZACHOS, Does co-NP have short interactive proofs? In *Information Processing Letters 25* (1987), 127-132.

[CaHe 89] J. CAI, L.A. HEMACHANDRA, On the power of parity. In *Proceedings 6th Symposium on Theoretical Aspects of Computer Science, Lecture Notes in Computer Science 349* (1989), 229-240.

[Co 71] S.A. COOK, The complexity of theorem-proving procedures. In *Proceedings of the 3rd ACM Symposium on Theory of Computing* 1971, 151-158.

[EvSeYa 84] S. EVEN, A. SELMAN Y. YACOBI, The complexity of promise problems with applications to public-key cryptography. In *Information and Control 61* (1984), 114-133.

[FeFoKu 91] S. FENNER, L. FORTNOW, S. KURTZ, Gap-definable counting classes. In *Proceedings of the 6th Structure in Complexity Theory Conference* 1991, 30-42.

[FuHoLu 80] M. FURST, J. HOPCROFT, E. LUKS, Polynomial time algorithms for permutation groups. In *Proceedings of the 21st ACM Symposium on Theory of Computing* 1980, 36-41.

[GaJo 79] M.R. GAREY, D.S. JOHNSON, *Computers and Intractability: A Guide to the Theory of* NP*-Completeness*, Freeman, San Francisco, 1979.

[Gi 77] J. GILL, Computational complexity of probabilistic Turing machines. In *SIAM Journal on Computing 6* (1977), 675-695.

[GoMiWi 85] O. GOLDREICH, S. MICALI, AND A. WIGDERSON, Proofs that yield nothing but their validity and a methodology of cryptographic protocol design. In *Proceedings of the 27th Symposium on Foundations of Computer Science* 1986, 174-187.

[GoMiRa 85] S. GOLDWASSER, S. MICALI, AND C. RACKOFF, The knowledge complexity of interactive proofs. In *Proceedings of the 17th ACM Symposium on the Theory of Computing* 1985, 291-304.

[GoSi 86] S. GOLDWASSER, M. SIPSER, Private coins versus public coins in interactive proof systems. In *Proceedings of the 18th ACM Symposium on the Theory of Computing* 1986, 59-68.

[Gre 91] F. GREEN, On the Power of Deterministic Reductions to $C_=P$. Technical Report LSI-91-14 Universidad Politecnica de Catalunya, 1991.

[Hall 59] M. HALL, *The Theory of Groups*, Macmillan, New York, 1959.

[Hof 82a] C. HOFFMANN, *Group-Theoretic Algorithms and Graph Isomorphism*, Springer-Verlag *Lecture Notes in Computer Science #136*, 1982.

[Hof 82b] C. HOFFMANN, Subcomplete generalizations of graph isomorphism. In *Journal of Computer and System Sciences 25* (1982), 332-359.

[Jo 85] D.S. JOHNSON, The NP-completeness column: An ongoing guide. In *Journal of Algorithms 6* (1985), 434-451.

[KrHe 91] D. KRATSCH, L.A. HEMACHANDRA, On the complexity of graph reconstruction. In *Fundamentals of Computing Theory* 1991, to appear.

[KöScToTo 89] J. KÖBLER, U. SCHÖNING, J. TORÁN AND S. TODA, Turing Machines with few accepting computations and low sets for PP. In *Proceedings of the 4th Structure in Complexity Theory Conference* 1989, 208-216.

[Luks 82] E. LUKS, Isomorphism of Graphs of Bounded Valence can be tested in Polynomial Time. In *Journal of Computer and System Sciences 25* (1982), 42-65.

[Mat 79] R. MATHON, A note on the graph isomorphism counting problem. In *Inform. Process. Lett. 8* (1979), 131-132.

[OgHe 91] M. OGIWARA, L. HEMACHANDRA, A complexity theory for closure properties. In *Proceedings of the 6th Structure in Complexity Theory Conference* 1991, 16-29.

[PaZa 83] C. PAPADIMITRIOU, S. ZACHOS Two remarks on the power of counting. In *6th GI Conference on Theoretical Computer Science, Lecture Notes in Computer Science 145* (1983) 269-276.

[Sch 86] U. SCHÖNING, *Complexity and Structure.* Springer-Verlag *Lecture Notes in Computer Science #211,* 1986.

[Sch 88] U. SCHÖNING, Graph isomorphism is in the low hierarchy. In *Journal of Computer and System Sciences 37* (1988), 312-323.

[Si 75] J. SIMON, On some central problems in computational complexity. Ph.D. Thesis, Cornell University (1975).

[Sims 71] C. SIMS, Computation with permutation groups. In *Proceedings of the 2nd ACM Symposium on Symbolic and Algebraic Manipulations* 1971, 23-28.

[Tar 91] J. TARUI, Degree complexity of boolean functions and its applications to relativized separations. In *Proceedings of the 6th Structure in Complexity Theory Conference* 1991, 382-390.

[Tod 89] S. TODA, On the computational power of PP and $\oplus$P. In *Proceedings of the 30th Symposium on Foundations of Computer Science* 1989, 514-519.

[Tod 91] S. TODA, Private communication.

[Tor 88] J. TORÁN, An oracle characterization of the Counting Hierarchy. In *Proceedings of the 3rd Structure in Complexity Theory Conference* 1988, 213-224.

[Va 76] L.G. VALIANT, The relative complexity of checking and evaluating. In *Information Processing Letters 5* (1976), 20-23.

[Va 79] L.G. VALIANT, The complexity of computing the permanent. In *Theoretical Computer Science 8* (1979), 189-201.

[VaVa 86] L.G. VALIANT V.V VAZIRANI, NP is as easy as detecting unique solutions In *Theoretical Computer Science 47* (1986), 85-93.

[Wa 86] K.W. WAGNER, The complexity of combinatorial problems with succinct input representation. In *Acta Informatica 23* (1986), 325-356.

ALGORITHMS 1

A simple linear time algorithm for triangulating three-colored graphs*

Hans Bodlaender and Ton Kloks †

Abstract

In this paper we consider the problem of determining whether a given colored graph can be triangulated, such that no edges between vertices of the same color are added. This problem originated from the Perfect Phylogeny problem from molecular biology, and is strongly related with the problem of recognizing partial k-trees. In this paper we give a simple linear time algorithm that solves the problem when there are three colors. We do this by first giving a complete structural characterization of the class of partial 2-trees. We also give an algorithm that solves the problem for partial 2-trees.

1 Introduction

Consider a graph of which the vertices are colored such that no two adjacent vertices have the same color. In this paper we consider the problem to determine whether we can triangulate the graph (i.e. add edges, such that the resulting graph does not have an induced cycle of length at least 4), such that in the triangulated graph no two adjacent vertices have the same color. The problem of triangulating a colored graph such that no two adjacent vertices have the same color is polynomially equivalent to the *Perfect Phylogeny problem*, see e.g. [7]. This problem, which is concerned with the inference of evolutionary history from DNA sequences, is of major importance to molecular biologists. Very recently, this problem was proven to be NP-complete by Fellows and Warnow [4]. In [11] it was shown by Morris, Warnow and Wimer that the problem is solvable in $O(n^{k+1})$ time for k-colored graphs. Another special case was solved in [7] by Kannan and Warnow. In this paper we consider the problem, for the case that there are at most three colors. In [8] this problem was solved in $O(n\alpha(n))$ time by Kannan and Warnow. Very recently, Kannan and Warnow improved on their algorithm, and found a variant, that uses linear time. In this paper we give another, much simpler, linear time algorithm. This work was done more or less simultaneously with, and independent from this recent result of Kannan and Warnow.

This paper is organized as follows. In section 2 we give a number of important definitions and properties of triangulated graphs. In section 3 we give a complete characterization of partial 2-trees. In section 4, precise conditions are given when a 3-colored graph is c-triangulatable, and when a colored partial 2-tree is c-triangulatable. In section 5 we give a linear time algorithm, that solves the problem of triangulating a three-colored graph, using the characterization of section 4. In section 6 we discuss a simpler variant of the algorithm, that only decides whether a c-triangulation exists, but does not yield the triangulation itself, and give an algorithm for triangulating t-colored partial 2-trees.

*This research was supported in part by the Foundation for Computer Science (S.I.O.N.) of the Netherlands Organization for Scientific Research (N.W.O.), and in part by the ESPRIT II Basic Research Actions Program of the EC under contract no. 3075 (project ALCOM).

†Department of Computer Science, Utrecht University, P.O.Box 80.089, 3508 TB Utrecht, The Netherlands.

2 Definitions and basic properties

In this section we discuss some basic properties of triangulated graphs. Triangulated graphs are also known as *chordal graphs.* For further information we refer to [6]. The subgraph of graph $G = (V, E)$, induced by a set of vertices $W \subseteq V$, is denoted by $G[W] = (W, \{(v, w) \in E \mid v, w \in W\})$.

Definition 2.1
A graph is *triangulated* (or chordal) if it has no induced cycle of length strictly greater than three.

Definition 2.2
A vertex x of a graph G is *simplicial* if its adjacency set induces a complete subgraph (i.e. $G[\{v \in V \mid (v, x) \in E\}]$ is a complete graph.)

Definition 2.3
Let G be a graph and let $\sigma = [v_1, \ldots, v_n]$ be an ordering of the vertices. σ Is called a *perfect elimination scheme* for G if for all i, v_i is simplicial in $G[\{v_i, \ldots, v_n\}]$.

Fulkerson and Gross ([5]) gave the following characterization of triangulated graphs.

Theorem 2.1 *A graph G is triangulated if and only if it has a perfect elimination scheme. Furthermore, if a graph is triangulated then* any *simplicial vertex can start a perfect elimination scheme.*

This theorem gives us an easy algorithm for the recognition of triangulated graphs, namely repeatedly locate a simplicial vertex and remove it from the graph. The graph is triangulated if and only if this process ends with the empty graph. If the graph is not a clique then there are at least *two* nonadjacent simplicial vertices.

Definition 2.4
A subset S of vertices is called a vertex separator for nonadjacent vertices a and b, if a and b are in different connected components of $G[V - S]$.

The following characterization was found by Dirac ([3]).

Theorem 2.2 *A graph is triangulated if and only if every minimal vertex separator induces a complete subgraph.*

There is yet another characterization, which says that a graph is triangulated if and only if it is the intersection graph of a family of subtrees of a tree. Examples of triangulated graphs are interval graphs and k-trees. Triangulated graphs are *perfect* (i.e. for every induced subgraph the chromatic number is equal to the size of a maximum clique, or equivalently, for every induced subgraph the size of a clique cover is equal to the size of a maximum stable set). Triangulated graphs can be recognized in linear time. There exist linear time algorithms for many NP-complete problems when restricted to triangulated graphs, for example coloring, clique, stable set and clique-cover (see [6]).

Definition 2.5
A triangulation of a graph G is a graph H with the same number of vertices such that G is a subgraph of H and such that H is triangulated. We say that G is *triangulated into H*.

Clearly, every graph can be triangulated (into a clique). Triangulating a graph such that the number of added edges is minimum is called the minimum fill-in problem, and triangulating such that the maximum clique is minimum is called the treewidth problem (the treewidth of a graph is one less than the minimum size of a maximum clique in any triangulation). Both these problems (treewidth and minimum fill-in) are NP-complete.

In the next section we give a definition of a special type of triangulated graphs, called k-trees, and we characterize the biconnected partial 2-trees.

3 Characterization of biconnected partial 2-trees

Definition 3.1
A *k-tree* is a graph with at least k vertices for which there exists a perfect elimination scheme $\sigma = [v_1, \ldots, v_n]$ such that for all $i \leq n-k$, v_i is adjacent to all vertices of a k-clique (i.e. a complete graph with k vertices) in the subgraph $G[\{v_i, \ldots, v_n\}]$, and to no other vertices in $G[\{v_i, \ldots, v_n\}]$.

From this definition follows that every maximal clique in a k-tree has size $k+1$, and that every minimal vertex separator has size k. An equivalent way to obtain k-trees is by the following recursive definition: A k-clique is a k-tree; given a k-tree T_n with n vertices we construct a k-tree with $n+1$ vertices by taking a new vertex x_{n+1} which is made adjacent to every vertex of a k-clique in T_n and to no other vertex. Notice that a 1-tree is an ordinary tree.

Definition 3.2
A *partial k-tree* is a subgraph of a k-tree with at least k vertices, or equivalently, a partial k-tree is a graph that can be triangulated into a k-tree.

It is an easy exercise to show that every triangulated graph with a maximum clique of size at most $k+1$ is a partial k-tree. It turns out that many interesting classes of graphs are contained in a class of partial k-trees for some k (see e.g. [1]). A large number of NP-hard problems become solvable in polynomial or even linear time when restricted to the class of partial k-trees for some constant k. Partial k-trees are recognizable in $O(n \log^2 n)$ time, see [2].

In this section we address the problem of characterizing partial 2-trees. A graph is a partial 2-tree if and only if all its biconnected components are partial 2-trees, so when characterizing partial 2-trees we can restrict ourselves to biconnected partial 2-trees.

Lemma 3.1 *Let G be a biconnected partial 2-tree. Let $S = \{x, y\}$ be a separator such that $G[V-S]$ has at least three connected components. Then in any 2-tree embedding (x, y) is an edge.*

Proof:
Let x and y be vertices such that $G[V - \{x, y\}]$ has at least three connected components. Since the graph is biconnected, both x and y have at least one neighbor in every connected component and so there are paths between x and y with internal vertices in every connected component. Assume that G can be triangulated into a 2-tree H such that (x, y) is not an edge in H. In any 2-tree every minimal vertex separator is an edge. Every minimal vertex separator for x and y contains at least one vertex of every connected component of $G[V - \{x, y\}]$ (otherwise there would be a path between x and y). So the minimal separator contains at least three vertices, which is a contradiction. □

Definition 3.3
Let the *cell-completion* of G be the graph $\bar{G}$, obtained from G by adding an edge between all pairs $\{x, y\}$ for which $G[V - \{x, y\}]$ has at least three connected components. A *cell* of G is a set of vertices which form a chordless cycle in the cell-completion.

Note that, by lemma 3.1 the cell-completion $\bar{G}$ is a subgraph of any triangulation of G.

Definition 3.4
A *tree of cycles* is a graph which is defined recursively as follows: A chordless cycle is a tree of cycles; Given a tree of cycles with s chordless cycles T_s, we construct a tree of cycles with $s+1$ chordless cycles T_{s+1} as follows. Take a new chordless cycle and identify the endvertices of an edge of this cycle with the endvertices of an edge in T_s. So at each stage a new cycle is *glued to* exactly one edge of the part of the tree of cycles that has already been constructed.

For example, a 2-tree is a tree of triangles (i.e. a tree of cycles in which every cycle is a triangle). We show in the next theorem that a biconnected graph is a partial 2-tree if and only if its cell-completion is a tree of cycles. The following lemma will be useful.

Lemma 3.2 *Let G be a biconnected partial 2-tree and let G be triangulated into a 2-tree H. For any edge $e = (x, y)$ in H, the number of common neighbors of x and y in H is at least three if and only if the number of components in $G[V - \{x, y\}]$ is at least three.*

Proof:
Let $e = (x, y)$ be an edge in H such that x and y have at least three common neighbors in H. Two of these common neighbors can not be adjacent in H otherwise there would be a 4-clique. Also, two of these common neighbors can not be in the same connected component of $H[V - \{x, y\}]$, since e must be a minimal vertex separator for them. Hence the number of components in $G[V - \{x, y\}]$ is at least three. To prove the converse, let the number of components of $G[V - \{x, y\}]$ be at least three. Let C be a component of $G[V - \{x, y\}]$. G is biconnected so there is a path between x and y with internal vertices in C. Since (x, y) is an edge in H this implies that, in H, x and y must have a common neighbor in C. This shows that the number of common neighbors of x and y in H is at least three. □

The following characterization appears in [9], where it is used to enumerate all biconnected partial 2-trees. By triangulating a chordless cycle with m vertices, we mean adding $m - 3$ edges (i.e. the minimum number) such that the new graph is triangulated.

Theorem 3.3 *A biconnected graph G is a partial 2-tree if and only if its cell-completion $\bar{G}$ is a tree of cycles. By triangulating each chordless cycle of $\bar{G}$ in all possible ways we obtain all possible triangulations of G into a 2-tree.*

Proof:
The only if part can be seen as follows. We must show that $\bar{G}$ is a tree of cycles. Consider a triangulation of $\bar{G}$ into a 2-tree H. Consider the edges in H that are not in $\bar{G}$. Such an edge of H must be incident with at least two triangles since $\bar{G}$ is biconnected. If there are *three* triangles incident with an edge of H, the edge is also present in $\bar{G}$ (lemma 3.2). Since H is a tree of triangles, and the only edges that are not in $\bar{G}$ are edges that are incident with two triangles, it follows that $\bar{G}$ must be a tree of cycles.

Now let $\bar{G}$ be a tree of cycles. Then by triangulating each cycle into a 2-tree, the resulting graph is a tree of triangles, and hence a 2-tree. □

In the next section we focus on c-triangulating 3-colored graphs.

4 c-Triangulating 3-colored graphs

Recall that the problem that we consider in this paper is the following. Given a graph G with a vertex coloring $c : V \to C$, where C is a set of t colors, such that each vertex is colored with one of t colors and such that no two adjacent vertices have the same color. Triangulate G (if possible) without introducing edges between vertices of the same color. If such a triangulation exists we call it a *c-triangulation*. If each of the t colors is used at least once, we call the graph *t-colored*.

Note that a graph can be c-triangulated if and only if all the biconnected components can be c-triangulated, and a c-triangulation of a graph G can be obtained by c-triangulating all biconnected components. Hence, in the remainder we assume that G is a biconnected graph. We can also observe that the maximum clique size in a c-triangulation can be at most the number of different colors, since otherwise there would be an edge between vertices of the same color. The following lemma states an even stronger result.

Lemma 4.1 *Let $G = (V, E)$ be a $(k+1)$-colored graph. Then G can be c-triangulated if and only if G can be c-triangulated into a k-tree.*

Proof:
Consider a c-triangulation of G, say H. Clearly, every maximal clique in H has size at most $k+1$. Let n be the number of vertices of G. We show, by induction on n, that every $k+1$-colored graph which is c-triangulated can be c-triangulated into a k-tree. This proves the lemma since G can be c-triangulated into H and H can be c-triangulated into a k-tree. Notice that, since H is $k+1$-colored, $n \geq k+1$. If $n = k+1$, then the lemma is obviously true. Assume $n > k+1$. Since H is triangulated it has a perfect elimination scheme $\sigma = [v_1, \ldots, v_n]$. Consider the graph obtained from H by removing the simplicial vertex v_1. Let C be the set of neighbors of v_1 in H. So C is a clique and $|C| \leq k$. By induction, $H[V \setminus \{v_1\}]$ can be triangulated into a k-tree H'. Since H' is a k-tree and C is a clique in H', C is contained in a $k+1$-clique C'. Since C' is a clique with $k+1$ vertices it contains exactly one vertex x with the same color as v_1. Clearly x can not be contained in C. We make v_1 adjacent to every vertex of $C' \setminus \{x\}$ and obtain a c-triangulation of H into a k-tree. □

In particular, we have that a $(k+1)$-colored graph can be c-triangulated, only if it is a partial k-tree. Consider a biconnected three-colored graph $G = (V, E)$. To be c-triangulatable, G must be a biconnected partial 2-tree. The following results give a precise characterization of c-triangulatable partial 2-trees.

Theorem 4.2 *Let G be a biconnected t-colored partial 2-tree, and let $\bar{G}$ be the cell completion of G. G can be c-triangulated, if and only if:*

1. *No two adjacent vertices of $\bar{G}$ have the same color, and*

2. *Every cell has at least three vertices with different colors.*

Proof:
From theorem 3.3 it follows, that it is sufficient to show that a cycle can be c-triangulated, if and only if it contains three vertices of different colors. This fact was proved in [8]. For reasons of completeness we also present a proof here.

If a cycle has only two colors, then it is easy to see that it can not be c-triangulated, since cycles of length 3 cannot be made. Suppose a cycle S has three colors. If S is a triangle there is nothing to proof. Otherwise we can add a chord to S such that the two new cycles made by this chord each have three colors. To find such a chord, we consider two cases. In case that there is a color that appears only once, then take any chord containing this color at one of its end-vertices. If every color appears at least twice in S, then there is a vertex v such that the two neighbors have different colors. Then take the chord connecting the two neighbors of v. Recursively apply the argument to the shorter cycles formed by the chord. □

A slightly different characterization is obtained in the following theorem.

Theorem 4.3 *Let G be a biconnected t-colored partial 2-tree, and let $\bar{G}$ be the cell completion of G. G can be c-triangulated, if and only if:*

1. *No two adjacent vertices of $\bar{G}$ have the same color, and*
2. *Every cycle in $\bar{G}$ contains at least three vertices with different colors.*

Proof:
Use theorem 4.2. Clearly, if each cycle in $\bar{G}$ contains at least three different colors, then each cell in the cell completion does so. This shows the 'if'-part. The 'only if'-part follows from the fact, that if $\bar{G}$ contains a cycle with only two colors, then $\bar{G}$ contains a subgraph that cannot be triangulated, hence G cannot be triangulated. □

It follows that a t-colored partial 2-tree can be c-triangulated, if and only if $\bar{G}$ does not contain an edge between vertices with the same color, and for every pair of colors, the subgraph of $\bar{G}$ induced by the vertices with these colors is cycle-free, i.e., is a partial 1-tree. This partly generalizes: If a t-colored graph G can be c-triangulated, then for every subset of $s \leq t$ colors, the subgraph of G induced by vertices with a color in this set is a partial $(s-1)$-tree. This necessary condition is unfortunately not sufficient, not even for $t = 3$. The following result follows directly from lemma 4.1 and theorem 4.3.

Corollary 4.4 *Let G be a biconnected 3-colored graph, and let $\bar{G}$ be the cell completion of G. G can be c-triangulated, if and only if:*

1. *G is a partial 2-tree.*
2. *No two adjacent vertices of $\bar{G}$ have the same color, and*
3. *Every cycle in $\bar{G}$ contains at least three vertices with different colors.*

5 Algorithm for 3-colored graphs

In this section we describe an algorithm to c-triangulate a 3-colored graph, if possible. In section 6, we give an easier variant, that only tests whether it is possible to c-triangulate the graph, without actually yielding a c-triangulation. We also give a variant for t-colored partial 2-trees, for $t \geq 3$.

Suppose G is a biconnected 3-colored graph. Recall that a c-triangulation (if it exists) can be obtained by triangulating each chordless cycle of the cell-completion (theorem 3.3). Our algorithm to c-triangulate G, if possible, has the following structure:

1. Make *any* triangulation of G into a 2-tree H.
2. Given H we then can make the cell-completion of G, $\bar{G}$, using lemma 3.2. We check if no two adjacent vertices in $\bar{G}$ have the same color.
3. Make a list of all cells.
4. c-Triangulate each cell (if it has three different colors).

Notice that when step 1, 2 or 4 fails, the graph can not be c-triangulated, and otherwise the algorithm outputs a correct c-triangulation. Correctness of this method follows from theorem 4.2.

We now describe each step of this algorithm in more detail. As each step has a linear time implementation, we get the following result.

Theorem 5.1 *There exists a linear time algorithm, that given a 3-colored graph G, tests whether G can be c-triangulated, and if so, outputs a c-triangulation of G.*

5.1 Triangulating G into a 2-tree H

In this subsection we consider the problem to find a 2-tree H, that contains a given graph G as a subgraph, or output that such a graph does not exist. It is well known that this problem can be solved in linear time, see e.g. [10]. For reasons of completeness we describe the algorithm also here.

If G is a biconnected partial 2-tree, we can make a triangulation into a 2-tree H by successively choosing a vertex of degree two, making the neighbors of this vertex adjacent and removing the vertex from the graph (see for example [12].)

We can implement this as follows. Assume that we have for each vertex in the graph G a (linked) list of neighbors. Assume that with each edge (x, y) in the adjacency list of x, there is a pointer to the edge (y, x) in the adjacency list of y. We also keep a list of vertices of degree 2. Initialize $H := G$ (i.e. make a copy of the adjacency lists).

Choose a vertex of the list of vertices of degree 2, say x. Let y and z be the neighbors of x. We add z to the adjacency list of y and y to the adjacency list of z. It is possible, that we create a duplicate edge (y, z) in this way, i.e. in that case y appears twice in the adjacency list of z, and vice versa.

We now test, whether y, and z have degree 2 (and hence must be put on the list of vertices of degree 2). Look at the adjacency list of y. Scan this list until either we have encountered *three different* neighbors of y, or until the list becomes empty. When we encounter a duplicate edge while scanning the list, say (y, x'), we remove the second copy of it from the list of y, and remove its counterpart from the list of x'. If y has only two different neighbors, we put y in the list of vertices of degree two. We do the same for z.

Iterate the above until there are no vertices of degree 2 left. When the graph now has more than two vertices, then G was no partial 2-tree. Otherwise, H is a 2-tree, containing G as a subgraph. Correctness of the algorithm follows from [12]. The order in which the vertices have been removed, is a perfect elimination scheme for the 2-tree H.

To see that this algorithm runs in linear time notice the following. When scanning the adjacency lists, every step we either encounter a duplicate edge (which is then removed) or we find a new neighbor. Notice that the total number of duplicate edges is at most n (the number of vertices of G), since every time a vertex is removed at most one duplicate edge is created.

5.2 Making the cell-completion $\bar{G}$

Suppose that we now have the 2-tree embedding H with a perfect elimination scheme $\sigma = [v_1, \ldots, v_n]$ for H. Recall that $\bar{G}$ is a subgraph of H, so to find $\bar{G}$, we only have to test for each edge in H, whether it belongs to $\bar{G}$ or not. We use lemma 3.2 to make the cell-completion. The algorithms is as follows. First make a copy of G in $\bar{G}$. If vertices x and y have at least three common neighbors in H, add the edge (x, y) to $\bar{G}$. Testing this property can be done as follows.

From the adjacency list of a vertex v_i in H, remove the vertices v_j for all $j < i$. We do the same for $\bar{G}$, using the ordering of vertices of σ. Each adjacency list now has at most two elements, since σ is a perfect elimination scheme of a 2-tree. Number the edges of H in any order $1, \ldots, 2n-3$, and let a pointer point from the edges in the adjacency lists to its number and vice versa. Initialize an array $cn(1 \ldots 2n-3)$ to zero. Start with the triangle $\{v_{n-2}, v_{n-1}, v_n\}$. For each edge of this triangle, look up the number and increase the value in cn by 1.

Consider the other vertices one by one, in the reversed order of the perfect elimination scheme, i.e. in the order $v_{n-3}, v_{n-4}, \ldots, v_1$. For each vertex v_i $(i < n-2)$ we do the following. Suppose it is adjacent to vertices v_j and v_k (with $j > i$ and $k > i$). For the edges in this triangle increase the value in cn by one. It is straightforward to see (by induction) that for each edge in the induced subgraph $H[\{v_i, v_{i+1}, \ldots, v_n\}]$ the number of common neighbors is given by the value in cn. We use here that each triangle v_i, v_j, v_k is considered exactly once, namely when considering the lowest numbered vertex in the triangle. If a final value in cn is at least three, look up the edge in the adjacency list of H, and add this edge to $\bar{G}$ if it was not already present. Clearly, this procedure uses linear time. Correctness follows from lemma 3.2.

5.3 Making a list of the cells

Notice that the number of cells in $\bar{G}$ is at most $n-2$ (with equality if and only if $\bar{G}$ is already a 2-tree). For each cell we initialize an empty list. During the algorithm we keep a pointer from each edge in H we have encountered to the number of the last cell it is contained in and to its position in this cell. Again let $\sigma = [v_1, \ldots, v_n]$ be a perfect elimination scheme for H. Remember that for each vertex v_i we have removed all vertices v_j with $j < i$ from its adjacency list.

Put the vertices v_n, v_{n-1}, v_{n-2} in the first cell and for each edge in this triangle we make a pointer to the number of this cell and to its position in this cell. Consider the other vertices one by one, in the reversed order of the perfect elimination scheme. Suppose v_i has neighbors v_j and v_k in H with $j > i$ and $k > i$. Let $j < k$. Look in the adjacency list of v_j in $\bar{G}$ (which has at most two elements) if the vertices v_j and v_k are adjacent. If v_j and v_k are adjacent in $\bar{G}$ we make a new cell containing the three vertices v_i, v_j and v_k and for each edge of this triangle we update the number of the cell it is contained in. If v_j and v_k are not adjacent in $\bar{G}$, then they can be contained in at most one cell, since otherwise v_j and v_k would have at least three different neighbors in H (v_i included) and hence would be adjacent in $\bar{G}$. We have a pointer to this cell and to the position of v_j and v_k in this cell. We add the vertex v_i to this cell in the place between vertices v_j and v_k and we update the cell number for the edges (v_i, v_j) and (v_i, v_k). It is straightforward to see by induction that at each step each cell contains the vertices of the cell of $\bar{G}$, restricted to the vertices $v_i, v_{i+1}, \ldots, v_n$. So in this way we obtain a list of all cells in $\bar{G}$ in linear time.

5.4 Triangulating each cell

We have made a list of vertices for each cell such that consecutive vertices in this list are adjacent in $\bar{G}$. For each color we make a list of vertices in the cell that are of this color, and for every vertex we make a pointer to its position in the list. If there is a color with an empty list, the cell can not be triangulated. Suppose every color occurs in the cell. If there is only one vertex of a given color, we make this vertex adjacent to all other vertices. Otherwise, by going around the cycle clockwise, we find a vertex x such that the two neighbors are of a different color. We add the edge between the two neighbors and we delete x from the cell and from the color list. Move counterclockwise to the neighbor of x, and start again. It is easy to see that this algorithm runs in linear time.

6 Some variants

In this section we give some variants of the results and the algorithm of the previous section.

Theorem 6.1 *There exists a linear time algorithm, that given a colored partial 2-tree G, tests whether G can be c-triangulated, and if so, outputs a c-triangulation of G.*

Proof:
We can use the same algorithm as in section 5. □

Finally, we describe an algorithm, that given a 3-colored graph G, decides whether G is c-triangulatable. It does not produce a c-triangulation of G. However, it is slightly easier than the algorithm in section 5. The algorithm has the following structure:

1. Make *any* triangulation of G into a 2-tree H.

2. Given H we then can make the cell-completion of G, $\bar{G}$. Check if no two adjacent vertices in $\bar{G}$ have the same color.

3. For each pair of colors c_1, c_2, take the subgraph of G, induced by vertices of color c_1, or c_2, and check whether this graph does not contain a cycle.

G is c-triangulatable, if and only if step 1 succeeds (G is a partial 2-tree), in step 2 no pair of adjacent vertices with the same color is found, and in step 3 all three considered induced subgraphs are cycle-free. The correctness of this procedure follows from corollary 4.4. Implementation of steps 1 and 2 is done, as in section 5. It is easy to see, that step 3 has a linear time implementation.

Acknowledgements

We thank Mike Fellows, for drawing our attention to this subject, and the members of the Graph Algorithms Club of the Department of Computer Science in Utrecht for discussions and useful comments.

References

[1] H.L. Bodlaender, Classes of graphs with bounded tree-width, Tech. Rep. RUU-CS-86-22, Department of Computer Science, Utrecht University, Utrecht, 1986.

[2] H.L. Bodlaender and T. Kloks, Better algorithms for the pathwidth and treewidth of graphs, *Proceedings of the* 18*th International colloquium on Automata, Languages and Programming*, 544-555, Springer Verlag, Lecture Notes in Computer Science, vol. 510 (1991).

[3] G. Dirac, On rigid circuit graphs, *Abh. Math. Sem. Univ. Hamburg* **25**, 71 – 76 (1961).

[4] M.R. Fellows and T. Warnow, Personal communication, 1991.

[5] D. Fulkerson and O. Gross, Incidence matrices and interval graphs, *Pacific J. Math.* **15**, 835 – 855 (1965).

[6] M.C. Golumbic, *Algorithmic Graph Theory and Perfect Graphs*, Academic Press, New York, 1980.

[7] S. Kannan and T. Warnow, Inferring Evolutionary History from DNA Sequences, *in: Proceedings of the 31th Annual Symposium on Foundations of Computer Science*, pp. 362 – 371, 1990.

[8] S. Kannan and T. Warnow, Triangulating three-colored graphs, *in: Proceedings of the 1st Ann. ACM-SIAM Symposium on Discrete Algorithms*, pp. 337 – 343, 1990.

[9] T. Kloks, Enumeration of biconnected partial 2-trees, to appear.

[10] J. Matoušek and R. Thomas, Algorithms Finding tree-decompositions of graphs, *Journal of Algorithms* **12**, 1 – 22 (1991).

[11] C.F. McMorris, T. Warnow, and T. Wimer, Triangulating colored graphs, submitted to Inform. Proc. Letters.

[12] J.A. Wald and C.J. Colbourn, Steiner trees, partial 2-trees and minimum *IFI* networks, *Networks* **13** (1983), 159 – 167.

On locally optimal alignments in genetic sequences

Norbert Blum
Informatik IV, Universität Bonn
Römerstr. 164, W-5300 Bonn, F. R. Germany

Abstract

A substring $\tilde{x}$ of a text string x has c-locally minimal distance from a pattern string y, $c \in N \cup \{0\}$, if no other substring x' of x with smaller edit distance to y exists which overlaps $\tilde{x}$ by more than c characters. We show how to compute all substrings of x which have c-locally minimal distance from y and all corresponding alignments in $O(m \cdot n)$ time where n is the length of x and m is the length of y.

1 Introduction

Let $x = x_1x_2\ldots x_n$ and $y = y_1y_2\ldots y_m$ be two strings over a finite alphabet Σ. The *edit distance* $d(x,y)$ from string x to string y is the cost of the cheapest sequence of edit operations transforming the string x to the string y. We consider the following three types of edit operations

a) deleting a character from x,
b) inserting a character into x, and
c) replacing one character of x with another.

More formally, an *edit operation* is a pair $(a,b) \in (\Sigma \cup \{\lambda\})^2 \setminus \{(\lambda,\lambda)\}$ where

$$(a,b) = \begin{cases} \text{deletion of } a & \text{if } a \in \Sigma, b = \lambda \\ \text{insertion of } b & \text{if } a = \lambda, b \in \Sigma \\ \text{replacement of } a \text{ by } b & \text{if } a, b \in \Sigma \end{cases}$$

With each edit operation (a,b) we associate a positive cost $d(a,b)$, i.e. d is a *cost function*

$$d : (\Sigma \cup \{\lambda\})^2 \setminus \{(\lambda,\lambda)\} \mapsto R^+$$

If $a = b$ then $d(a,b) = 0$. The function d fulfills the metric axioms.

An *edit sequence* S is a sequence $S = s_1, s_2, \ldots, s_t$ of edit operations. S is a *derivation of x to y* if there is a sequence $z_0, z_1, z_2 \ldots z_t$ of strings such that

a) $x = z_0$, $y = z_t$,

b) we obtain z_i from z_{i-1} by one application of the edit operation s_i for all $1 \leq i \leq t$.

We say then that S *transforms x into y*. The *cost $d(S)$ of S* is defined by

$$d(S) = \sum_{i=1}^{t} d(s_i).$$

The *edit distance* $d(x,y)$ from string x to string y is defined by

$$d(x,y) = \min\{d(S) \mid S \text{ transforms } x \text{ into } y\}.$$

In genetic applications, x and y are macromolecular sequences, and $d(x,y)$ is also called the *evolutionary distance* between x and y. Overviews of sequence comparison with respect to genetic applications are given in [6, 9].

An *alignment* $A(x,y)$ between x and y consists of a matrix with two rows such that

a) the first row contains x, possibly interspersed with *null characters* λ,

b) the second row contains y, possibly interspersed with null characters,

c) there is no column which consists two null characters.

In the case $a \neq b$ the column $\left[\begin{smallmatrix} a \\ b \end{smallmatrix}\right]$ describes the edit operation (a,b) and in the case $a = b$ $\left[\begin{smallmatrix} a \\ b \end{smallmatrix}\right]$ describes a *continuation.*

The *cost* $d(A(x,y))$ of an alignment between x and y is defined by

$$d(A(x,y)) = \sum_{\left[\begin{smallmatrix} a \\ b \end{smallmatrix}\right] \in A} d(a,b).$$

By definition, a derivation of x to y can be described by an alignment between x and y.

Since the problem of computing the edit distance and a corresponding alignment of two given strings $x = x_1x_2\ldots x_n$ and $y = y_1y_2\ldots y_m$ has applications in different disciplines, such as speech processing, molecular biology, and computer science, the same algorithm for the solution of this problem has been discovered independently by many people [6, p. 23]. This algorithm uses dynamic programming and has time complexity $O(m \cdot n)$.

Using the "Four Russian's trick" Masek and Paterson [3] show how to solve the problem in $O(m \cdot n/\min\{\log n, m\})$ time where $m \leq n$. Since their algorithm beats the simple dynamic programming algorithm only for extremely long strings, their algorithm is only of theoretical interest.

Another problem in molecular biology which is closely related to the problem described above is finding substrings of a string which have minimal distance from a given pattern string.

Let $x = x_1x_2\ldots x_n$, $y = y_1y_2\ldots y_m$, $n \gg m$ be two strings over a finite alphabet Σ. A substring $\tilde{x} \subseteq x$ of x has *globally minimal distance* from y, if for all substrings x' of x $d(\tilde{x},y) \leq d(x',y)$. An alignment $A(\tilde{x},y)$ with $d(A(\tilde{x},y)) = d(\tilde{x},y)$ is *globally optimal.*

Essentially the same dynamic programming approach which solves the edit distance problem of two strings can be used for the solution of the following problem.

Problem 1 *Given two strings $x = x_1x_2\ldots x_n$, $y = y_1y_2\ldots y_m \in \Sigma^+$, $n \gg m$, and a cost function $d: (\Sigma \cup \{\lambda\})^2 \mapsto R^+$ compute all substrings $\tilde{x}$ of x with globally minimal distance from y and for each such a $\tilde{x}$ compute all corresponding globally optimal alignments.*

It is possible that there exists a substring x' of x which has no globally minimal distance from y, but the distance is only a little bit larger. In some cases, especially, if in a local region of x, x' has minimal distance from y it might be useful also to compute x' and the corresponding alignments $A(x',y)$. Sellers [7] defines a situation in which "y most resembles $\tilde{x} \subset x$ locally". In this situation we say that $\tilde{x}$ has weakly locally minimal distance from y.

Let $x = x_1x_2\ldots x_n$, $y = y_1y_2\ldots y_m \in \Sigma^+$, $n \gg m$. A substring $\tilde{x}$ of x has *weakly locally minimal distance* from y, if for all x' with $x' \subset \tilde{x}$ or $\tilde{x} \subset x'$ hold

$$d(\tilde{x},y) \leq d(x',y).$$

An alignment $A(\tilde{x},y)$ with $d(A(\tilde{x},y)) = d(\tilde{x},y)$ is *weakly locally optimal.*

Sellers [7] presents an $O(m \cdot n)$ algorithm for the computation of two matrices which make it possible to compute weakly locally optimal alignments. It is not clear how to solve the following problem in $O(m \cdot n)$ time with Sellers method.

Problem 2 *Given two strings $x = x_1x_2\ldots x_n$, $y = y_1y_2\ldots y_m \in \Sigma^+$, $n \gg m$ and a cost function $d : (\Sigma \cup \{\lambda\})^2 \mapsto R^+$ compute all substrings $\tilde{x}$ of x with weakly locally minimal distance from y and for each such a $\tilde{x}$ all corresponding weakly locally optimal alignments.*

A substring $\tilde{x}$ of x can have weakly locally minimal distance from y although there exists a substring x' of x such that

a) $\tilde{x} \cap x' \neq \emptyset$, i.e. $\tilde{x}$ and x' overlap
b) $d(x', y) < d(\tilde{x}, y)$.

It might be useless to compute such a substring $\tilde{x}$. Hence we define a stronger version of locally minimal distance.

Let $x = x_1x_2\ldots x_n$, $y = y_1y_2\ldots y_m \in \Sigma^+$, $n \gg m$. A substring $\tilde{x}$ of x has *locally minimal distance* from y, if for all substrings x' of x with $\tilde{x} \cap x' \neq \emptyset$ $d(\tilde{x}, y) \leq d(x', y)$. An alignment $A(\tilde{x}, y)$ with $d(A(\tilde{x}, y)) = d(\tilde{x}, y)$ is *locally optimal*.

Sometimes it might be useful to allow small overlaps.

A substring $\tilde{x}$ of x has *c-locally minimal distance* from y, $c \in N \cup \{0\}$ if $\tilde{x}$ has weakly locally minimal distance from y and if for all substrings x' of x with $|\tilde{x} \cap x'| > c$, i.e. $\tilde{x}$ and x' overlap by at least $c+1$ characters, $d(\tilde{x}, y) \leq d(x', y)$. An alignment $A(\tilde{x}, y)$ with $d(A(\tilde{x}, y)) = d(\tilde{x}, y)$ is *c-locally optimal*.

Problem 3 *Given two strings $x = x_1x_2\ldots x_n$, $y = y_1y_2\ldots y_m \in \Sigma^+$, $n \gg m$, $c \in N \cup \{0\}$ and a cost function $d : (\Sigma \cup \{\lambda\})^2 \mapsto R^+$ compute all substrings $\tilde{x}$ of x with c-locally minimal distance from y and for each such a $\tilde{x}$ compute all corresponding c-locally optimal alignments.*

Edit graphs are very useful for the solution of problem 1, problem 2 and some variations of these problems [7, 4, 8]. A careful examination of the structure of the edit graphs with respect to problem 2 leads to

1. a simpler and more efficient solution of problem 2 than Sellers solution,
2. a simple solution of problem 3.

All developed algorithms uses $O(m \cdot n)$ time and are practical.

2 Edit graphs, distance graphs, and their useful subgraphs

For $x = x_1x_2\ldots x_n$, $y = y_1y_2\ldots y_m \in \Sigma^+$ we can represent all alignments between y and substrings of x in a compact form by a directed acyclic graph.

A *directed graph* $G = (V, E)$ consists of a finite set of *nodes* V and a set of ordered pairs of distinct nodes the *edges* $E \subset V \times V$. A graph $H = (V', E')$ is a *subgraph* of $G = (V, E)$ if $V' \subseteq V$ and $E' \subseteq E$. For $v \in V$ we define $outdeg(v) = |\{(v, w) \in E\}|$ and $indeg(v) = |\{(w, v) \in E\}|$. A *path* P from $v \in V$ to $w \in V$ is a sequence of nodes $v = v_1, v_2, \ldots, v_k = w$ satisfying $(v_i, v_{i+1}) \in E$ for $0 \leq i < k$. The *length* of P is the number k of edges on P. Often we write $P = (v_1, v_2), (v_2, v_3), \ldots, (v_{k-1}, v_k)$. If there exists a path from v to w (of length 1) v is called a *(direct) predecessor* of w and w is called a *(direct) successor* of v. A graph $G = (V, E)$ is *connected* if we cannot partition the set V of nodes into two nonempty subsets V_1 and V_2 such that no edge $(v, w) \in E$ exists with $v \in V_1, w \in V_2$ or $v \in V_2, w \in V_1$. The *connected components* of G are the maximal connected subgraphs of G.

A common method for traversing a graph $G = (V, E)$ is performing a depth-first search of G. For a description of depth first search of a directed graph see [1, pp. 176–179, 187–189]. Another method for searching a graph is to consider in each step a node v for which all predecessors have been considered before. This kind of traversing a graph G is called a *topological search* of G. If we consider in each step a node v for which all successors have been considered before then we have a *backward topological search* of G. It is well known that all these graph search methods need only time $O(|V|+|E|)$.

The *edit graph* $E(x,y)=(V,E)$ with respect to x and y is a directed graph where

$$\begin{aligned} V = \ & \{[i,j] \mid 0 \leq i \leq m, 0 \leq j \leq n\}, \text{ and} \\ E = \ & \{([i,j],[i+1,j]),([i,j],[i,j+1]),([i,j],[i+1,j+1]) \mid 0 \leq i < m, \\ & 0 \leq j < n\} \\ & \cup\{([m,j],[m,j+1]) \mid 0 \leq j < n\} \\ & \cup\{([i,n],[i+1,n]) \mid 0 \leq i < m\} \end{aligned}$$

Each edge $e=([i,j],[i',j'])$ corresponds to an edit operation $op(e)$ in the following way.

$$op(e) = \begin{cases} \text{deletion of } x_{j+1} & \text{if } \ i'=i \text{ and } j'=j+1 \\ \text{insertion of } y_{i+1} & \text{if } \ i'=i+1 \text{ and } j'=j \\ \text{replacement of } x_{j+1} \text{ by } y_{i+1} & \text{if } \ i'=i+1 \text{ and } j'=j+1 \end{cases}$$

To the edge e we associate the cost $d(op(e))$. The *costs* $d(P)$ of a path P in $E(x,y)$ is defined by

$$d(P) = \sum_{e \in P} d(op(e)).$$

We can illustrate the edit graph by an $(m+1)\times(n+1)$-grid, where the grid point (i,j) corresponds to the node $[i,j]$.

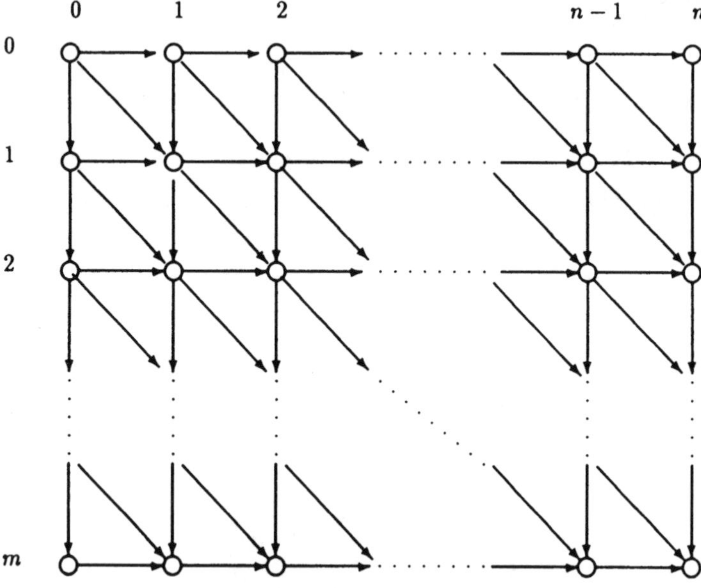

Figure 1:

It is easy to see that the paths from row 0 to row m with the property that the first edge is an edge to row 1 and the last edge is an edge from row $m-1$ correspond one-to-one to the alignments between y and substrings of x.

We prove that each of the three problems can be solved by deleting exactly these edges from E which are not on a path corresponding to an alignment with the desired property.

A path P in $E(x,y)$ is *globally optimal*, *weakly locally optimal* and *c-locally optimal*, respectively, if P corresponds to a globally optimal alignment, weakly locally optimal alignment and c-locally optimal alignment, respectively. Let

$$\begin{aligned} E_g &= \{e \in E \mid e \text{ is on a globally optimal path } P \text{ in } E(x,y)\}, \\ V_g &= \{v \in V \mid \exists w \in V : (v,w) \in E_g \text{ or } (w,v) \in E_g\}, \\ E_{wl} &= \{e \in E \mid e \text{ is on a weakly locally optimal path } P \text{ in } E(x,y)\}, \\ V_{wl} &= \{v \in V \mid \exists w \in V : (v,w) \in E_{wl} \text{ or } (w,v) \in E_{wl}\}, \\ E_{cl} &= \{e \in E \mid e \text{ is on a c-locally optimal path } P \text{ in } E(x,y), \\ V_{cl} &= \{v \in V \mid \exists w \in V : (v,w) \in E_{cl} \text{ or } (w,v) \in E_{cl}\}. \end{aligned}$$

Then we define

$$E_g(x,y) = (V_g, E_g),\ E_{wl}(x,y) = (V_{wl}, E_{wl}), \text{ and } E_{cl}(x,y) = (V_{cl}, E_{cl}).$$

$x(P)$ denotes the substring of x corresponding to the path P.

Our goal is to prove that for the solution of each of the three problems it suffices to construct the corresponding graph. Let

$$x^j = \begin{cases} \varepsilon & \text{if } j = 0 \\ x_1 x_2 \dots x_j & \text{if } j > 0 \end{cases}$$

where ε denotes the *empty word*. y^i is defined analogously. Let

$$e(x^j, y^i) = \begin{cases} 0 & \text{if } i = 0 \\ \min\{d(\tilde{x}, y^i) \mid \tilde{x} \text{ is suffix of } x^j\} & \text{if } i > 0 \end{cases}$$

The following lemma is very useful.

Lemma 1 *Let $P = P_1, [i,j], P_2$ be a globally optimal path, a weakly locally optimal path, or a c-locally optimal path. Then $d(P_1,[i,j]) = e(x^j, y^i)$.*

Proof: It is clear by definition that $d(P_1,[i,j]) \geq e(x^j,y^i)$. Assume that $d(P_1,[i,j]) > e(x^j,y^i)$. Then there exists a path P' such that $d(P',[i,j]) < d(P_1,[i,j])$. Hence the path $\tilde{P} = P',[i,j],P_2$ corresponds to an alignment between y and a suffix of $x_1 x_2 \dots x_s$, where $[m,s]$ is the last node on P. But then by construction, P cannot be globally, weakly locally or c-locally optimal. This contradicts our assumptions. □

Let

$$E_d = \{([i,j],[i',j']) \in E \mid e(x^{j'}, y^{i'}) = e(x^j, y^i) + d(op([i,j],[i',j']))\}$$

The *distance graph* $D(x,y)$ is defined by

$$D(x,y) = (V, E_d).$$

A immediate consequence of lemma 1 is the following lemma.

Lemma 2 $E_g(x,y) \subseteq D(x,y)$, $E_{wl}(x,y) \subseteq D(x,y)$, *and* $E_{cl}(x,y) \subseteq D(x,y)$.

Since each c-locally optimal path is also weakly locally optimal, we obtain:

Lemma 3 $E_{cl} \subseteq E_{wl}$.

The following theorem shows that we can solve the three problems by a computation of the graphs $E_g(x,y)$, $E_{wl}(x,y)$, and $E_{cl}(x,y)$, respectively.

Theorem 1 *There is an one-to-one correspondence between the paths from row 0 to row m in*

a) $E_g(x, y)$ *and the globally optimal alignments,*
b) $E_{wl}(x, y)$ *and the weakly locally optimal alignments,*
c) $E_{cl}(x, y)$ *and the c-locally optimal alignments.*

Proof: It is clear by construction that

a) $E_g(x, y)$ contains all globally optimal paths,
b) $E_{wl}(x, y)$ contains all weakly locally optimal paths,
c) $E_{cl}(x, y)$ contains all c-locally optimal paths.

Hence in every case, it suffices to prove that the graph contains no more paths from row 0 to row m.

Consider any path P from row 0 to row m in $E_g(x, y)$, $E_{wl}(x, y)$, and $E_{cl}(x, y)$, respectively. Let $e = ([m-1, s'], [m, s])$ be the last edge on P. Then by lemma 2, $d(P) = e(x^s, y^m)$.

If $P \in E_g(x, y)$ ($P \in E_{wl}(x, y)$) the assertion follows directly from the definition of globally (weakly locally) optimal path and the fact that the edge e is on such a path.

Assume that $P = e_1, e_2, \ldots, e_r \in E_{cl}(x, y)$ but P is not c-locally optimal.

Then there exists a path $\bar{P} \in E_{cl}(x, y)$ from row 0 to row m such that

i) $d(\bar{P}) < d(P)$
ii) $x(\bar{P})$ and $x(P)$ overlap at least $c + 1$ characters.

Since the edge e is on at least one c-locally optimal path, the left end of $x(P)$ must overlap the right end of $x(\bar{P})$. This situation is described by the following figure.

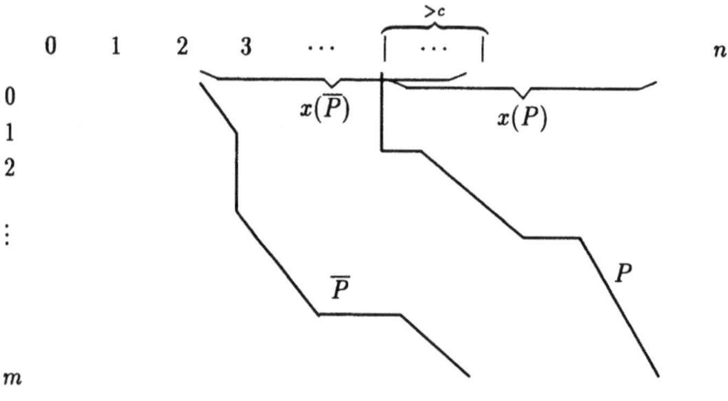

Figure 2:

From the facts that $x(P)$ and $x(\bar{P})$ overlap at least by $c + 1$ characters and $\bar{P}$ is c-locally optimal, we can conclude that for all c-locally optimal paths P' which contain e_1, $d(P') = d(\bar{P})$ holds.

Let e_t be the first edge on P which is on a c-locally optimal path $Q = Q_1, e_t, Q_2$ with $d(Q) > d(\bar{P})$. Since the edge e_r has this property, e_t exists.

Then there exists a c-locally optimal path $R = e_1, e_2, \ldots, e_{t-1}, R'$ with $d(R) <= d(\bar{P})$.

Consider the path $T = Q_1, R'$. Then $d(T) = d(R) <= d(\bar{P})$. By construction $x(T)$ is substring of $x(Q)$ or $x(Q)$ is substring of $x(T)$. Hence $d(T) >= d(Q)$. But this contradicts $d(Q) > d(\bar{P})$. Hence P is c-locally optimal. □

By lemma 2, the computation of the distance graph is very useful for the solution of each of the three problems. In the next section we review the well known method for the computation of the distance graph.

Implicitly Sellers [7] shows that a slight modified version of the Needleman-Wunsch method [5] for the computation of the evolutionary distance of two molecular sequences can be used for the computation of the distance graph. For this algorithm see [7] or the full paper [2]. This algorithm takes $O(mn)$ time.

3 The computation of the globally optimal alignments

Given the distance graph $D(x,y) = (V, E_d)$, our goal is to delete all edges from E_d which are not in E_g. The following lemma gives an exact characterization of E_g with respect to the distance graph $D(x,y)$.

Lemma 4 *Let* $e_g = \min_{1\leq j\leq n}\{e(x^j, y^m)\}$. *Then*

$$E_g = \{([i,j],[i',j']) \mid \exists P = [i,j],[i',j'],\dots[m,l] \in D(x,y) : e(x^l, y^m) = e_g\}$$

Proof: The assertion follows directly from the definition of a globally optimal path and the definition of the distance graph $D(x,y)$. □

It is easier to compute paths with given start nodes than to compute paths with given end nodes. Hence we give all edges in E_d the opposite direction and define the graph $D_{rev}(x,y) = (V, E_{d,rev})$ where

$$E_{d,rev} = \{([i',j'],[i,j]) \mid ([i,j],[i',j']) \in E_d\}.$$

Given $D_{rev}(x,y)$, we compute the graph $E_{g,rev}(x,y) = (V_g, E_{g,rev})$ where

$$\begin{aligned} E_{g,rev} = \; & \{([i',j'],[i,j]) \in E_{d,rev} \mid \exists P = [m,l],\dots,[i',j'],[i,j] \\ & \in D_{rev}(x,y) : e(x^l, y^m) = e_g\}. \end{aligned}$$

It is clear by construction that $E_g = \{([i,j],[i',j']) \mid ([i',j'],[i,j]) \in E_{g,rev}\}$.

We have reduced the problem of the computation of E_g to a reachability problem which can easily be solved using depth first search of $D_{rev}(x,y)$ with the start nodes $\{[m,l] \mid e(x^l,y^m) = e_g\}$. Thus the following algorithm computes the graph $E_g(x,y)$.

Algorithm 1

Input: distance graph $D(x,y) = (V, E_d)$ and the value $e(x^j, y^i)$ for all nodes $[i,j] \in V$

Output: the graph $E_g(x,y) = (V_g, E_g)$

Method:

(1) Compute $D_{rev}(x,y)$.

(2) Using depth first search with start nodes $\{[m,l] \mid e(x^l,y^m) = e_g\}$ compute $E_{g,rev} = \{([i',j'],[i,j]) \mid ([i',j'],[i,j])$ is considered during the depth first search $\}$.

(3) $E_g := \{([i,j],[i',j']) \mid ([i,j],[i',j']) \in E_{g,rev}\}$.

The correctness of algorithm 1 follows from lemma 5. Since depth first search needs only linear time in the size of the graph algorithm 1 has time complexity $O(m \cdot n)$. Hence we have obtained the following theorem.

Theorem 2 *The graph* $E_g(x,y) = (V_g, E_g)$ *can be computed in* $O(m \cdot n)$ *time.*

4 The computation of the weakly locally optimal alignments

For the solution of problem 2, Sellers [7] computes the "forward distance matrix" which corresponds to the distance graph. Analogously he computes the "reverse distance matrix" where he starts with the right end of x and y and terminates at the left end of the strings. Then he uses both distance matrices for the computation of weakly locally optimal alignments.

We show that the computation of the reverse distance matrix is superfluous. By a simple observation we show how to use merely the distance graph for the computation of $E_{wl}(x,y)$. This observation is formulated in the following lemma. We denote by $P_{l,k}$ a path from the node $[0,l]$ to the node $[m,k]$ in $D(x,y)$.

Lemma 5 *Let P_{l_1,k_1} and P_{l_2,k_2} be two paths in $D(x,y)$ such that $l_1 \leq l_2$ and $k_2 \leq k_1$. If P_{l_i,k_i}, $i \in \{1,2\}$ is weakly locally optimal then*

$$e(x^{k_i}, y^m) = \min\{e(x^{k_1}, y^m), e(x^{k_2}, y^m)\}.$$

Proof: By construction, the paths P_{l_1,k_1} and P_{l_2,k_2} have at least one node $[r,s]$ in common. Let $P_{l_1,k_1} = P_1, [r,s], P_1'$ and $P_{l_2,k_2} = P_2, [r,s], P_2'$.

Since the paths $P_{l_1,k_2} = P_1, [r,s], P_2'$ and $P_{l_2,k_1} = P_2, [r,s], P_1'$ correspond both to alignments between y and a substring of x, the assertion follows directly from the definition of weakly locally optimal path. □

Let

$$e_{l,k} = \min_{1\leq j\leq n} \{e(x^j, y^m) \mid \text{there is a path from } [l,k] \text{ to } [m,j] \text{ in } D(x,y)\}.$$

By the definition of $e_{l,k}$ we can conclude

$$e_{l,k} = \min\{e_{l',k'} \mid ([l,k],[l',k']) \in E_d\}.$$

The following lemma is a direct consequence of lemma 6.

Lemma 6

a) *If $([l,k],[l',k']) \in E_d$ is not on a path from $[l,k]$ to a node $[m,j]$ with $e(x^j, y^m) = e_{l,k}$ then $([l,k],[l',k']) \notin E_{wl}$.*

b) *Let $P_{l,k} \in D(x,y)$ be not weakly locally optimal. Then there exists an edge $([r,s],[r',s'])$ on $P_{l,k}$ with $e_{r,s} < e_{r',s'}$.*

The idea of the method is the following. In each step we choose an unconsidered node $[l,k]$ with the property that all successors of $[l,k]$ are considered before, i.e. we perform a backward topological search of $D(x,y)$. By lemma 7 a) an edge $([l,k],[\bar{l},\bar{k}]) \in E_d$ can only be in E_{wl} if $e_{\bar{l},\bar{k}} = e_{l,k}$. The set K of all these edges is computed. By lemma 7 b) we obtain E_{wl} from K by deleting the edges $([i,j],[i',j'])$ with $j > 0$ and $indeg([i,j]) = 0$ from K as long as such edges exists.

These considerations lead to the following algorithm. We use the variable $num([i,j])$ for storing the number of direct successors of $[i,j]$ which are not considered. After the consideration of the node $[i,j]$, $num([i,j])$ obtains the value 4, indicating that the node $[i,j]$ is considered.

Algorithm 2

Input: distance graph $D(x,y) = (V, E_d)$ and the value $e(x^i, y^j)$ for all $[i,j] \in V$

Output: $E_{wl}(x,y) = (V_{wl}, E_{wl})$

Method:

```
(1) K := ∅;
(2)for j from 1 to n
     do
        for i from 1 to m − 1
           do
              e_{i,j} := ∞;
              num([i, j]) := outdeg([i, j])
           od;
           e_{m,j} := e(x^j, y^m);
           num([m, j]) := 0
        od;
(3) while there exists a node [l, k] with num([l.k]) = 0
     do
        choose such a node [l, k];
        num([l, k]) := 4;
        for all ([l, k], [l̄, k̄]) ∈ E_d
           do
              if e_{l,k} = e_{l̄,k̄}
                 then K := K ∪ {([l, k], [l̄, k̄])}
              fi
           od;
        for all ([l', k'], [l, k]) ∈ E_d
           do
              e_{l',k'} := min{e_{l',k'}, e_{l,k}};
              num([l', k']) := num([l', k']) − 1
           od;
(4) while there exists ([i, j], [i', j']) ∈ K with j > 0 and indeg([i, j]) = 0
     do
        K := K \ {([i, j], [i', j'])}
     od;
(5) E_wl := K.
(6) V_wl := {v ∈ V | ∃w ∈ V : (v, w) ∈ E_wl or (w, v) ∈ E_wl}.
```

The correctness of algorithm 2 follows from lemma 7. Since the number of nodes and the number of edges in $D(x, y)$ is bounded by $O(m \cdot n)$, the time complexity of algorithm 2 is $O(m \cdot n)$. Altogether we have obtained.

Theorem 3 *The graph $E_{wl}(x, y) = (V_{wl}, E_{wl})$ can be computed in $O(m \cdot n)$ time.*

5 The computation of the c-locally optimal alignments

Given $E_{wl}(x, y) = (V_{wl}, E_{wl})$ we want to compute the graph $E_{cl}(x, y) = (V_{cl}, E_{cl})$ by deleting the nodes from V_{wl} and the edges from E_{wl} which are not contained in V_{cl} and in E_{cl}, respectively. For the exploration of this deletion process, a careful examination of the structure of the graph $E_{wl}(x, y)$ is very helpful. After knowing this structure, the solution of the problem 3 becomes simple.

First we characterize $E_{wl}(x, y)$ with respect to its connected components.

Lemma 7 *Let CC be a connected component of $E_{wl}(x, y)$. Then all paths $P_{l,k}$ in CC have the same costs.*

Proof: The assertion follows directly from lemma 6 and lemma 7. □

For a connected component CC of $E_{wl}(x,y)$, $d(CC)$ denotes the uniquely determined cost $d(P_{l,k})$ of any path $P_{l,k}$ in CC.

$Left(CC)$ denotes the leftmost position of CC with respect to the string x, i.e.

$$Left(CC) = \min_{1 \leq l \leq n} \{l \mid [0,l] \in CC\}.$$

$Right(CC)$ denotes the rightmost position of CC with respect to the string x, i.e.

$$Right(CC) = \max_{1 \leq k \leq n} \{k \mid [m,k] \in CC\}.$$

Assume that we have drawn the graph $E_{wl}(x,y)$ into the $(m+1) \times (n+1)$-grid in the obvious way. We call this drawing *griddrawing* of $E_{wl}(x,y)$.

Let us number the connected components of $E_{wl}(x,y)$ from left to right with respect to the griddrawing of $E_{wl}(x,y)$. CC_p denotes the p-th connected component of $E_{wl}(x,y)$.

The following lemma is the key of our method.

Lemma 8 *Let* $[0,l],[m,k] \in CC_p$ *for* $p \geq 1$. *Let* $p' = \max_{1 \leq q < p}\{q \mid d(CC_q) < d(CC_p)\}$ *and* $p'' = \min_{q > p}\{q \mid d(CC_q) < d(CC_p)\}$. *Then*

a) $Right(CC_{p'}) \geq l + c \Rightarrow [0,l] \notin V_{cl}$.

b) $Left(CC_{p''}) \leq k - c \Rightarrow [m,k] \notin V_{cl}$.

Proof: The assertion follows immediately from the definition of c-locally optimal path and the structure of $E_{wl}(x,y)$.

Note that $Right(CC_{p'}) \geq l + c$ implies that a path $P_{l,k'} \in CC_p$ with $k' \leq l + c$ would not be weakly locally simple. Hence such a path cannot exist in CC_p. □

Lemma 9 a) characterizes exactly these nodes $[0,l]$ in row 0 of the griddrawing of $E_{wl}(x,y)$ which cannot be in V_{cl} since all weakly locally optimal substrings $x_l x_{l+1} \ldots$ are overlapped from the left by at least $c+1$ characters of a substring of x which has smaller distance to y. We say then that $[0,l]$ is *excluded by the left.*

Lemma 9 b) characterizes exactly that nodes $[m,k]$ in row m of the griddrawing of $E_{wl}(x,y)$ which cannot be in V_{cl} since all weakly locally optimal substrings $\ldots x_{k-1}x_k$ are overlapped from the right by at least $c+1$ characters of a substring of x which has smaller distance to y. We say then that $[m,k]$ is *excluded by the right.*

Inductively we define that a node $[i,j] \in V_{wl}$ is *excluded by the left* (*excluded by the right*) if all its predecessors (successors) are excluded from the left (excluded by the right).

The following lemma characterizes exactly the other nodes of V_{wl} which cannot be in V_{cl}.

Lemma 9 *Let* $[i,j] \in V_{wl}$. *Then* $[i,j] \notin V_{cl}$ *if and only if* $[i,j]$ *is excluded by the left or* $[i,j]$ *is excluded by the right.*

Proof: It is clear that $[i,j] \notin V_{cl}$ if $[i,j]$ is excluded by the left or $[i,j]$ is excluded by the right.

For proving the other direction consider $[i,j] \notin V_{cl}$, such that i minimal and, for the chosen i, j is minimal such that $[i,j]$ is neither excluded by the left nor excluded by the right.

It is clear that no direct predecessor of $[i,j]$ is in V_{cl} or no direct successor is in V_{cl}. Otherwise also the node $[i,j]$ has to be in V_{cl}. We distinguish two cases.

case 1: $i = 0$ or no direct predecessors of $[i,j]$ is in V_{cl}.

If $i = 0$ then $[i,j]$ cannot be excluded with respect to lemma 9 a). Otherwise $[i,j]$ would be excluded by the left. Hence no direct successor of $[i,j]$ in $E_{wl}(x,y)$ is in V_{cl}.

If $i > 0$ then by assumption no direct predecessor of $[i,j]$ in $E_{wl}(x,y)$ is in V_{cl}. By the choice of $[i,j]$ all these direct predecessors must be excluded by the left. Hence by definition, $[i,j]$ must be excluded by the left. This contradicts the choice of $[i,j]$.

case 2: No direct successors of $[i,j]$ is in V_{cl}, $[i,j]$ is not excluded by Lemma 9 a) and some direct predecessors of $[i,j]$ is in V_{cl}.

It is easy to prove by induction that no successor $[i',j']$ of $[i,j]$ is in V_{cl}. It is also easy to see that all these successors cannot be excluded by the left.

Consider the successor $[i',j']$ with j' maximal and, for the chosen j', i' maximal such that $[i',j']$ is not excluded by the right.

If $j' = m$ then $[i',j']$ must be excluded by an application of lemma 9 b). But then $[i',j']$ is excluded by the right, a contradiction.

If $j' < m$ then by the choice of $[i',j']$ all direct successors are excluded by the right. But then by definition, $[i',j']$ is also excluded by the right, a contradiction.

This proves the lemma. □

According to lemma 10, the computation of $E_{cl}(x,y)$ is now simple. We obtain the following algorithm.

Algorithm 3

Input: $x = x_1x_2 \ldots x_n$, $y_1y_2 \ldots y_m$, cost function $d : (\Sigma \cup \{\lambda\})^2 \mapsto R^+$

Output: $E_{cl}(x,y)$

Method

(1) Simultaneously compute $E_{wl}(x,y)$, the connected components CC_1, CC_2, ..., and for each connected component CC_p the needed information $d(CC_p)$, $Left(CC_p)$ and $Right(CC_p)$.

(2) For each connected component CC_p compute p' and p'' as defined in lemma 9.

(3) Compute all nodes in row 0 which are excluded by the left according to lemma 9 a) and all nodes in row m which are excluded by the right according to lemma 9 b).

(4) By a topological search of $E_{wl}(x,y)$ where we consider in row 0 only the nodes excluded by the left, compute all nodes which are excluded by the left. Let K_1 be this set of nodes.

(5) By a backward topological search of $E_{wl}(x,y)$ where we consider in row m only the nodes excluded by the right, compute all nodes which are excluded by the right. Let K_2 be this set of nodes.

(6) $V_{cl} := V_{wl} \setminus (K_1 \cup K_2)$.

(7) $E_{cl} := E_{wl} \cap V_{cl} \times V_{cl}$.

The following theorem shows that the algorithm 3 computes $E_{cl}(x,y)$ in time $O(m \cdot n)$.

Theorem 4 *Algorithm 3 computes the graph* $E_{cl}(x,y) = (V_{cl}, E_{cl})$ *in time* $O(m \cdot n)$.

Proof: The correctness of the algorithm follows from lemma 9 and lemma 10.

It is not hard to extend algorithm 2 such that without increasing the time complexity by more than a small constant factor we compute $E_{wl}(x,y)$, the connected components in the sorted order, and for each connected component CC_p we compute the needed information $d(CC_p)$, $Left(CC_p)$ and $Right(CC_p)$. Hence step (1) can be performed in $O(m \cdot n)$ time.

By checking the connected components from left to right it is easy to perform step (2) in linear time in the number of connected components. Since there are at most n connected components, this time is bounded by $O(n)$.

Step (3) can also easily be performed in $O(n)$ time.

Since topological search and also backward topological search have linear time complexity in the size of the graph, step (4) and also step (5) need only $O(m \cdot n)$ time.

It is also trivial to perform the steps (6) and (7) in $O(m \cdot n)$ time. □

References

[1] Aho, A. V., Hopcroft, J. E., and Ullman, J. D. [1974]. *The Design and Analysis of Computer Algorithms*, Addison-Wesley, Reading, Mass.

[2] Blum, N. [1991]. *On locally optimal alignments in genetic sequences*, Research Report 8567–CS, Dept. of Computer Science, University of Bonn.

[3] Masek, W. J., and Paterson, M. S. [1980]. A faster algorithm for computing string-edit distances. *Journal of Computer and System Sciences* **20**, 18-31.

[4] Myers, E. W. [1986]. An $O(ND)$ difference algorithm and its variations. *Algorithmica* **1**, 251-266.

[5] Needleman, S. B., and Wunsch, C. D. [1970]. A general method applicable to the search for similarities in the amino-acid sequence of two proteins. *Journal of Molecular Biology* **48**, 443-453.

[6] Sankoff, D., and Kruskal, J. B. (eds.) [1983]. *Time Warps, String Edits, and Macromolecules: The Theory and Practice of Sequence Comparison*. Addison-Wesley, Reading, Mass.

[7] Sellers, P. H. [1980]. The theory and computation of evolutionary distances: Pattern recognition. *Journal of Algorithms* **1**, 359-373.

[8] Ukkonen, E. [1985]. Algorithms for approximate string matching. *Information and Control* **64**, 100-118.

[9] Waterman, M. S. [1984]. General methods of sequence comparison. *Bulletin of Mathematical Biology* **46**, 473-500.

CRYPTOGRAPHY

Secure Commitment Against A Powerful Adversary

A security primitive based on average intractability

(EXTENDED ABSTRACT)

Rafail Ostrovsky* Ramarathnam Venkatesan† Moti Yung‡

Abstract

Secure commitment is a primitive enabling information hiding, which is one of the most basic tools in cryptography. Specifically, it is a two-party partial-information game between a "committer" and a "receiver", in which a secure envelope is first implemented and later opened. The committer has a bit in mind which he commits to by putting it in a "secure envelope". The receiver cannot guess what the value is until the opening stage and the committer can not change his mind once committed.

In this paper, we investigate the feasibility of bit commitment when one of the participants (either committer or receiver) has an unfair computational advantage. That is, we consider commitment to a *strong receiver* with a large computational power (requiring that despite his power he can not "open" the secret commitment) or commitment by a *strong committer* (requiring that despite his power he can not change the value of the committed bit). We allow the strong party to use its computational resources and investigate the underlying complexity assumptions necessary for the feasibility of these primitives.

We show how to base commitment by a *strong committer* on *any* hard on the average problem. In fact, this is the first application of average case completeness to hiding information in a security primitive. We also show how commitment to a *strong receiver* with *information theoretic security* can be implemented based on *any* one-way function.

In addition, we show that commitment to a strong receiver is *complete* for *all* partial information games between weak and strong players. That is, given any implementation of the commitment protocol to a strong receiver, any partial-information game between a weak and a strong player can be implemented based solely on such a protocol.

* MIT Laboratory for Computer Science, 545 Technology Square, Cambridge MA 02139, USA. Supported by IBM Graduate Fellowship. Part of this work done while at IBM T.J. Watson Research Center.

† Bell Communications Research, 2M-344, 445 South St, Morristown NJ 07960, USA. Part of the work done at Boston University supported by NSF-CCR9015276.

‡ IBM Research, T.J. Watson Research Center, Yorktown Heights, NY 10598 USA.

1 Introduction

Secure protocols can be viewed as partial information games among mutually distrustful players (see, e.g., [GMW2, Co]). Many of these games can be based on a very simple game, called *bit-commitment* (BC) (see, e.g., [B1, B2, BM, BCC, BCY, BMO, EGL, GMW1, IY, SRA]). Here, we investigate the interplay between the computational power of the players in the commitment protocol and the complexity assumptions needed for its feasibility. A *strong* player has unlimited computing power; we often specify the exact needed power. A *weak* player is limited to polynomial time computations.

Different computational resources of the participants imply different notions of the security of the commitment. We say that bit commitment protocol is *computationally secure* if polynomially bounded receiver can not deduce the value of the committed bit before the reveal stage, however if receiver is given sufficient computational resources, he can discover the value of the committed bit. In contrast, we say that bit commitment protocol is *information-theoretically secure* if even with infinite resources, receiver can not gain any information about the bit before the reveal stage.

For commitment to a weak player, earlier Naor [N] exhibited a computationally secure bit-commitment protocol using any one-way function; when both players are weak (called the *symmetric* case), this is the best possible since such a protocol implies a one-way function [ImLu]. For the strong committer case, we relax this assumption much further, by basing it on any hard-on-average problems in PSPACE. This is the first application of Levin's theory of average case completeness to playing partial-information games. In fact, let C be any class inside PSPACE with a complete problem which (1) has an interactive proof whose prover is also in C (2) is hard-on-average. Then, assuming (1) and (2) the above (i.e. computationally secure) bit commitment protocol could be implemented from committer in C to receiver for whom complexity class C is hard on the average.

In the opposite direction, (i.e. for the commitment to a strong player) the goal is to construct an *information-theoretically* secure bit commitment protocol. (That is, to prevent the strong receiver from gaining any information about the committed secret despite his superior resources.) Previous implementations used a trapdoor permutation [GN], or a variety of specific algebraic assumptions, (e.g. [B2, BCY, BMO]). We improve this to *any* one-way function.

To get the later result we use another security primitive, the Oblivious Transfer (OT) protocol, introduced by Rabin [R]. This is a protocol by which one party sends a bit to a receiver, the bit gets there with probability 1/2 and the sender does not know the result of the transfer. We first show that the existence of the following three protocols is equivalent:

1. *BC* from weak to strong
2. *OT* from weak to strong
3. *OT* from strong to weak

That is, given an implementation of any one of these three protocols, we show how to implement the others *without any additional assumptions.* Thus, bit-commitment from a weak to strong player is "as hard" as any other protocol between weak and strong player (since OT is complete [K]). The corresponding result for the symmetric case is unknown and is unlikely to be proven using "black box" reductions [IR]. Finally, we use the above reduction and our recent result that "OT from weak to strong" can be based on any one-way function [OVY]) to get the bit commitment to a strong receiver based on *any* one-way function.

1.1 Preliminaries

The model we consider for two-party protocols is the standard system of communicating probabilistic machines [GMR]. In this section, we describe a few disclosure primitives and relations among them.

We start with an informal definition of Bit-commitment: *BC* may be thought of as a way for player S (the Sender) to commit a bit b to player R (the Receiver) in such a way that the bit may be revealed to R at a later point in time. Before b is revealed (but even after b has been committed), no information about b is revealed to R. When b is revealed, it is guaranteed to be the same as the value to which it was originally committed.

Oblivious Transfer (OT) is a two-party protocol introduced by Rabin [R]. Rabin's OT assumes that S possesses a value x, after the transfer R gets x with probability $\frac{1}{2}$ and it knows whether or not it got it (*equal-opportunity requirement*). A does not know whether B got the value (*oblivious-ness requirement*). A similar notion of 1-2-OT (one out of two OT) was introduced by [EGL]. In 1-2-OT, player S has two bits b_0 and b_1 and R has a selection bit i. After the transfer, R gets only b_i, while S does not know the value of i. Equivalently, R may get a random bit in $\{b_0, b_1\}$, or the game can be played on strings rather then bits. Further, there are many other flavors of OT [C, BCR, K, CK] all of which are information-theoretically equivalent. That is, given any one of these protocols, one can implement the other ones. Thus, by "OT" we can refer to any one of them.

The following notations will be used. By ($\texttt{weak}\xrightarrow{\text{BC}}\texttt{strong}$), we denote BC from a polynomially-bounded player to an infinitely-powerful one. We use ($\texttt{strong}\xrightarrow{\text{BC}}\texttt{weak}$), ($\texttt{strong}\xrightarrow{\text{OT}}\texttt{weak}$), ($\texttt{weak}\xrightarrow{\text{OT}}\texttt{strong}$) with similar meanings.

We must stress, that our results hold for the insecure communication environment. This should be contrasted with the work of [BGW, CCD, RB, BG, K, CK] where they assume right from the start that some form of OT already exists, or that secure channels exist. Instead, we concentrate on the two party scenario where secure channels do not help and investigate the required complexity assumptions for achieving BC.

1.2 Previous and related work

Our main primitive is *BC*, used as a basic building block in many different settings [B1, B2, BM, BCC, BCY, BMO, GMW1, K, N, Ost, SRA]. As was noted earlier, in the symmetric case BC and one-way functions are equivalent [BM, ILL, H, N]. We consider any hard on average problems (in PSPACE) as a base for the *BC* primitive.

The second primitive we apply is Oblivious Transfer. Rabin [R] defined and implemented OT for honest parties based on the intractability of factoring; Fischer, Micali and Rackoff [FMR] improved this result to be robust against cheaters. Other variations of OT were studied and shown to be information theoretically equivalent. Yao [Y] used OT (based on factoring) to construct secure circuit evaluation. Goldreich, Micali and Wigderson [GMW2] based OT for symmetric case (which also extends the asymmetric case of ($\texttt{strong}\xrightarrow{\text{OT}}\texttt{weak}$)) on the existence of any trapdoor permutation, and used it for multi-party circuit evaluation. Thus, secure circuit evaluation for poly-bounded players was made possible, assuming one-way trapdoor permutations exists. OT was also shown to be complete for secure circuit evaluation [K]. OT was also used to implement non-interactive and bounded-interaction zero-knowledge proof systems for NP [KMO]. This paper investigates the connection of asymmetric OT and asymmetric BC.

Since we deal with an asymmetric two-party model, let us point out what was considered in this model in addition to zero-knowledge proof systems of Goldwasser, Micali and Rackoff [GMR]. Note that this model represents naturally interaction between a small user and an all-powerful organization which may possess very large computational power. One such case is the context of zero-knowledge arguments of Brassard, Crépeau and Chaum [BCC], which assume an all-powerful verifier from which information has to be hidden. (Here we note that their protocols can be executed by polynomial time parties with cryptographic applications in mind while our results concentrate on allowing one party to have infinite power and use it in the computation. Recently, investigating the symmetric case, new results which reduce complexity assumptions in the practical context of [BCC] were also achieved [NOVY].) Another setting similar to ours is the model of using a powerful oracle to compute a value while keeping the real argument secret, [AFK, FO] where the oracle indeed uses its power.

2 Bit-commitment from strong to weak

In a $\texttt{strong}\xrightarrow{BC}\texttt{weak}$ protocol, if an infinitely-powerful "committer" (or Sender) tries to cheat by changing the value of the committed bit, the probabilistic polynomial-time "receiver" can catch this with overwhelming probability (over receiver's coin tosses). The actual work to be performed by the sender to execute the protocol is stated in the theorems below. Of course, if the receiver breaks the assumption, the value of the committed bit will be available before decomittal.

We first give a bit-commitment protocol based on an *average case complete* [L, VL, G, ImLe] problem. Randomized NP (RNP) consists of problems from NP under samplable distributions. For convenience we fix one such problem, namely Graph Coloration Problem (GCP) (see below). If there is any NP problem which is hard on average under any samplable (i.e., generatable in polynomial time) distribution, then so is this complete problem under random inputs. Thus, if a one-way function exists, then this complete problem is hard-on-average but the reverse implication that some complete (and thus hard-on-average) problem implies a one-way functions is open.

Let x be generated according to a distribution μ. An algorithm $A(x)$ is polynomial on average if it runs in time $(|x|r(x))^{O(1)}$, where $\mathbf{E}_\mu r(x) < 1$. Intuitively, $r(x)$ is a randomness test that takes small values on "typical" strings and large values on "rare" or "atypical" x. So, A can run longer on some rare inputs. Also, ignoring polynomial (in k) factors, an algorithm can take $2^{O(k)}$ time, with probability (over inputs) at most 2^{-k}. Let AP be the class of NP problems under samplable distributions which can be solved in polynomial on average time. A problem under a distribution μ is called *hard-on-average* if it is not in AP. In general, we may consider any complexity class instead of NP for defining AP. It is not hard to show (See the Corollary in [L] and [VL] for discussions) that a hard-on-average problem yields a problem with polynomial fraction of hard instances.

Lemma 1 *Unless $RNP = AP$, there is a protocol for committing a bit by a strong sender to a weak receiver, where the Sender needs only be a $(NP \cup co-NP)$ machine.*

Proof (Sketch): The following can be deduced from [N, GL]:

[N]: Assume there is pseudo-random generator (unpredictable for the receiver) that can be computed by the committer and which can be checked (given its seed) by the receiver. Then, there is a bit-commitment protocol from the committer to receiver.

[GL]: (List Decoding)Let $f(x) = y$ be polynomial time computable. Let $G(y,r) \in \{\pm 1\}$ be an algorithm that predicts the inner product $b(x,r)$ with a correlation $\mathbf{E}_r G(y,r)(-1)^{b(x,r)} = \varepsilon$. Then, there is an algorithm $A(y)$ that in $1/\varepsilon^{O(1)}$ time outputs a list L containing $1/\varepsilon^2$ strings such that $x \in L$.

Thus, if $|y| = n$ and $b(x,r)$ can be predicted with probability (over r) $1/2 \pm 1/n^c$, x can be computed in $n^{O(1)}$ time. Notice the absence of samplability requirement over x. This yields a hard-core bit based on a hard-on-average problem. Let f be the function checking the relation GCP which takes a edge-colored (with 4 colors) digraph and outputs the uncolored digraph, the number of edges of each color, and the list of all 3-node induced colored subgraphs with nodes relabeled 1,2,3 ; then $b(x,r)$ is hard-to-predict from y, r unless $RNP = AP$. Now, using the constructions of [H, ILL] the committer can generate pseudo-random bits. □.

Next we show the optimal conditions for commitment from strong to weak.

Theorem 1 *There exists a bit-commitment protocol from an infinitely-powerful sender to a weak receiver, based on any complete problem for any complexity class in PSPACE which is hard on the average.*

The proof has two steps described in the following proposition and lemma: first, we exhibit a complete problem in $RPSPACE$, second, we use analogous construction to Lemma 1, basing a generator on this complete problem. We also argue that this is the hardest language to base commitment on.

Let u be a machine with some fixed polynomial space bound, where $u(p,x,b) = (p,x)$ if the program p accepts x and $b = 1$ or p rejects x and $b = 0$. Otherwise $u(p,x,b) = 0000...00$. The problem of inverting u on an arbitrary input is equivalent to the halting problem for $PSPACE$. Let (μ, u) be the problem of inverting u when its inputs are randomly distributed under the distribution μ. By $RPSPACE$ we mean the class of all such pairs (μ, R) where $\mu \in P$ and $R \in PSPACE$. We define completeness similar to as in [L]. Let λ be the uniform distribution over all strings with $x \in \{0,1\}^n$, $\lambda(\{x\}) = \frac{2^{-n}}{n(n+1)}$.

Proposition 1 (λ, u) *is complete for* $RPSPACE$.

Proof (Sketch): Given an instance x of a problem (μ, R), the reduction in [L] produces an instance y for (λ, u). In our case u runs in polynomial space. □.

That is, (λ, u) is hard on the average unless every problem in $PSPACE$ under every polynomial time computable distribution has a polynomial on average algorithm. Note that this is weaker than the assertion that for example, Graph Coloration is hard-on average.

Let $\bar{x} = x_1 \circ x_2 \circ \cdots x_k$, $\bar{p} = p_1 \circ p_2 \circ \cdots p_k$, $\bar{b} = b_1 \circ b_2 \circ \cdots b_k$, and $u^*(\bar{p}, \bar{x}, \bar{b}) = \bar{p}, \bar{x}$. Then u^* is hard-to-invert for some $k = |x_i|^{O(1)}$ if u is.

If a bit $b(x)$ can not be predicted with probability p, one can amplify the unpredictability using independent $x_i, i := 1 \cdots n/p^2$ at random and taking the Xor of $b(x_i)$. We now obtain an unpredictable bit as follows. Let $e(x)$ be an encoding of x so that x can be uniquely decoded from any y in the Hamming Sphere of radius $0.05|e(x)|$ centered at $e(x)$. Then for $f(x) = y$, $b(x,i) = i$-th bit of $e(x)$ is hard to predict given y on constant fraction of i's, if x is hard-to-predict from y.

We note that assumption in the next lemma (a special case of the next lemma was independently shown in [K2]) can not be further weakened to any class larger than $PSPACE$ since any language provable by a prover to a polynomial-time verifier must be in $PSPACE$

as was first observed by P. Feldman; (in particular, proving "or opening" the language induced by the commitment protocol and value).

Lemma 2 *Unless RPSPACE=AP, there exists a bit commitment protocol from a (PSPACE) sender to a weak receiver.*

Generalizing the above lemma even further, we show that for any complexity class C inside PSPACE, if there is an interactive proof of membership for a complete language in C by the prover who is also in C, and if C is hard on the average, then a bit-commitment protocol can be constructed, in which the prover need not be more powerful then C.

3 Bit-commitment from weak to strong

Theorem 2 *The existence of the following three protocols is equivalent, provided that the* strong *player can perform $P^{\#P}$ (or stronger) computations:*

- (weak$\xrightarrow{BC}$strong)
- (weak$\xrightarrow{OT}$strong)
- (strong$\xrightarrow{OT}$weak)

Proof sketch:
(weak$\xrightarrow{BC}$strong) $\Longleftrightarrow$ (strong$\xrightarrow{OT}$weak):

($\Longrightarrow$) We are given a protocol (weak$\xrightarrow{BC}$strong) and we show how to execute (strong$\xrightarrow{OT}$weak) when strong player has b_0, b_1 as two input random bits to transmit via 1-2-OT.

Let ω_v denote the random tape of the weak player (wlog, we assume it's a string of a fixed (polynomial) size l). Let C denote the transcript of the messages exchanged when the weak player commits a bit in (weak$\xrightarrow{BC}$strong). Let $A_b(C) \leftarrow \{\omega_v :$ the conversation is C when weak player's random tape string is ω_v and weak player later decomits bit $b\}$. If we have a fixed C in context we just write A_0 and A_1. Note that these sets (i.e., $A_b(C)$) are disjoint and we may take C to be such that these are non-empty; otherwise the strong player can compute which value is being committed. Also, after the conversation, the weak player having a fixed C, and a (consistent) $\omega_v \in A_0(C)$, can not compute a string in $A_1(C)$; otherwise his committed bit and decommitted bit need not be the same. The protocol for 1-2-OT is as follows:

- strong and weak player execute (weak$\xrightarrow{BC}$strong) protocol. Let the conversation be C, the random tape of the weak player be $\omega_v \in \{0,1\}^l$, and the committed bit be b'.
- For $\beta \leftarrow 0$ to 1 do:

  Set $i \leftarrow 1$;
  [Repeat:]
  (strong): sends random $h_i^\beta \in \{0,1\}^l$
  (weak): sends $b_i^\beta := B(\omega_v, h_i^\beta)$ (the inner product) if $\beta = b'$ and a random bit otherwise.
  (strong): sends "stop" and exits loop if $\exists! \, \omega_v^\beta \in A_b \; \forall j \leq i \quad B(\omega_v^\beta, h_j^\beta) = b_j^\beta$.
  $i \leftarrow i+1$;
  [goto Repeat]

- End-For
- Then, the strong player chooses a random h so that $B(\omega_v^0, h) \neq B(\omega_v^1, h)$ and sends it to the weak player.

The above step is repeated thrice. The weak player randomly chooses two out of the three conversations and asks the strong player to convince him that the strong player acted according to the protocol (using the fact that this could be done in $P^{\#P}$ [LFKN]). If the strong player fails, the weak player aborts. Otherwise, the remaining conversation is used as follows: Let ω_0, ω_1 be the remaining "decomittal" strings of the third, unqueried conversation. The strong player selects a random string p, $|p| = l$ and sends to the weak player p, and two pairs $< \gamma_i, v_i >$, $i \in \{0,1\}$, where $v_i = b_i \oplus B(p, \omega_i)$, and $\gamma_i = B(h, \omega_i)$.

This can be shown to yield α-1-2-OT (where the sender can guess the result of the transfer with a slight advantage α), which is information-theoretically equivalent to OT [CK] using polynomial-time reductions.

($\Longleftarrow$): is straightforward: the strong player selects two random strings and plays 1-2-string-OT with the weak player. The "selection-bit" of a weak player serves as his committal. □

$(\texttt{weak} \xrightarrow{\text{BC}} \texttt{strong}) \iff (\texttt{weak} \xrightarrow{\text{OT}} \texttt{strong})$:

($\Longleftarrow$) : BC is known to follow from OT [C, K].

($\Longrightarrow$): Assume the weak player has two bits b_0 and b_1 and he wishes to execute 1-2-OT to the strong player. Since we assume $(\texttt{weak} \xrightarrow{\text{BC}} \texttt{strong})$, it follows that the strong player can do both OT and BC to the weak player. So the strong player can commit a bit by putting it in an "envelope". The strong player makes envelopes with names $e_1, \cdots, e_4$ and forms the pairs $P_0 = \{e_1, e_2\}$ and $P_1 = \{e_3, e_4\}$ satisfying:

1. the contents one pair are identical, while the contents in the other pair are different.
2. there is a label $l(e_i) \in \{0,1\}$ such that it is distinct for each envelope within a pair.

The above step is repeated $2k$ times, where k is the security parameter. Subsequently, for k-size randomly chosen subset, weak player requests to see the contents of both pairs. If the above constraints are not verified, weak player aborts the protocol. If not, then for the remaining k pairs $(P_0^1, P_1^1), \ldots, (P_0^k, P_1^k)$ the weak player chooses random bits $b_0^1, b_1^1, \ldots b_0^k, b_1^k$ and chooses (using appropriate OT protocol) the contents c_0^j (for j from 1 to k) of the envelope $e_i^j \in P_0^j$ with $l(e_i) = b_0^j$ and the content c_1^j of $e_i \in P_1$ with $l(e_i) = b_1^j$. Then the weak player sends c_0^j, c_1^j to the strong player. The strong player divides c_i^j into two equal size groups, (putting into one group bits which are pairwise distinct), and sends to the weak player indices of this two groups (without specifying which group is which, of course). The weak player takes an Xor of the first input bit (i.e. b_0) with the corresponding b_i^j bits of the first group and Xor of the second input bit (i.e. b_1) with the second group and sends this two bits back to the strong player. For the set for which the strong player knows *all* the b_i^j, he can compute the value of the input bit, while for the other bit, with overwhelming probability the value is hidden. (Alternatively, the strong player can ask which of groups to use with which input bit, first or second). □

We can conclude that:

Corollary 1 *Given a $(\texttt{weak} \xrightarrow{\text{BC}} \texttt{strong})$ protocol, then any partial information game of polynomial-size between a weak and a strong ($P^{\#P}$ or stronger) player is realizable.*

Bit commitment from weak to strong: In the bit-commitment protocol from the weak player to the strong one, recall that the goal is that even an infinitely-powerful "receiver" can not guess the committed bit with probability better then $\frac{1}{2}+\epsilon$, but such that a polynomially-bounded committer can not change a committed value, unless he breaks the assumption (which is explicitly) stated in the theorem.

In [OVY] we show how OT can be implemented in the asymmetric model under general complexity assumptions. For the sake of completeness, we explain briefly the technique behind this construction in the appendix. Using the results there and applying theorem 2, we get:

Corollary 2 *Given any one-way permutation, there exists a (weak-to-strong) bit-commitment protocol from a probabilistic poly-time "committer" to an (NP or stronger) "receiver".*

Corollary 3 *Given any one-way function, there exists a (weak-to-strong) bit-commitment protocol from a probabilistic poly-time "committer" to a ($P^{\#P}$ or stronger) "receiver".*

We stress that in the above two lemmas, once committed, the value of the committed bit is protected from the receiver *information-theoretically.*

Acknowledgments

We would like to thank Gilles Brassard, Shafi Goldwasser, Silvio Micali, Moni Naor, and Noam Nisan for helpful discussions.

References

[AFK] M. Abadi, J. Feigenbaum and J. Kilian. *On Hiding Information from an Oracle* J. Comput. System Sci. 39 (1989) 21-50.

[B1] Blum M., *Applications of Oblivious Transfer*, Unpublished manuscript.

[B2] Blum, M., "Coin Flipping over the Telephone," IEEE COMPCON 1982, pp. 133-137.

[BM] Blum, M. and S. Micali, "How To Generate Cryptographically Strong Sequences of Pseudorandom Bits," FOCS 82, (Also SIAM J. Comp. 84).

[BCC] G. Brassard, D. Chaum and C. Crepeau, *Minimum Disclosure Proofs of Knowledge*, JCSS, v. 37, pp 156-189.

[BCR] G. Brassard, C. Crépeau and J.-M. Robert, *"Information Theoretic Reductions among Disclosure Problems"*, FOCS 86 pp. 168-173.

[BCY] Brassard G., C. Crépeau, and M. Yung, "Everything in NP can be proven in Perfect Zero Knowledge in a bounded number of rounds," *ICALP* 89.

[BG] Beaver D., S. Goldwasser *Multiparty Computation with Faulty Majority* FOCS 89, pp 468-47.

[BMO] Bellare, M., S. Micali and R. Ostrovsky, "The (True) Complexity of Statistical Zero Knowledge" STOC 90.

[BGW] Ben-Or M., S. Goldwasser and A. Wigderson, *Completeness Theorem for Noncryptographic Fault-tolerant Distributed Computing*, STOC 88, pp 1-10.

[CCD] D. Chaum, C. Crepeau and I. Damgard, *Multiparty Unconditionally Secure Protocols*, STOC 88, pp 11-19.

[Co] A. Condon, *Computational Models of Games*, Ph.D. Thesis, University of Washington, Seattle 1987. (MIT Press, ACM Distinguished Dissertation Series).

[C] C. Crépeau, *Equivalence between Two Flavors of Oblivious Transfer*, Crypto 87.

[CK] C. Crépeau, J. Kilian *Achieving Oblivious Transfer Using Weakened Security Assumptions* , FOCS 88.

[EGL] S. Even, O. Goldreich and A. Lempel, *A Randomized Protocol for Signing Contracts*, CACM v. 28, 1985 pp. 637-647.

[FMR] Fischer M., S. Micali, C. Rackoff *An Oblivious Transfer Protocol Equivalent to Factoring*, Manuscript.

[GHY] Z. Galil, S. Haber and M. Yung, *Cryptographic Computations and the Public-Key Model*, Crypto 87.

[FO] J. Feigenbaum and R. Ostrovsky, *A Note On One-Prover, Instance-Hiding Zero-Knowledge Proof Systems* In Proceedings of the first international symposium in cryptology in Asia, (ASIACRYPT'91),

November 11-14, 1991, Fujsiyoshida, Yamanashi, Japan.

[GL] O. Goldreich and L. Levin, *Hard-core Predicate for ANY one-way function* , STOC 89.

[GMW1] O. Goldreich, S. Micali and A. Wigderson, *Proofs that Yields Nothing But their Validity*, FOCS 86, pp. 174-187.

[GMW2] O. Goldreich, S. Micali and A. Wigderson, *How to Play any Mental Poker* , STOC 87.

[GMR] S. Goldwasser, S. Micali and C. Rackoff, *The Knowledge Complexity of Interactive Proof-Systems*, STOC 85, pp. 291-304.

[GN] S. Goldwasser and N. Nisan, personal communication.

[G] Y. Gurevich, Average Case Completeness, Journ. of Comp Sys. Sci, 1991.

[H] Hastad, J., "Pseudo-Random Generators under Uniform Assumptions", *STOC 90*.

[ImLu] R. Impagliazzo and M. Luby, *One-way Functions are Essential for Complexity-Based Cryptography* FOCS 89.

[ILL] R. Impagliazzo, R., L. Levin, and M. Luby "Pseudo-Random Generation from One-Way Functions," *STOC* 89.

[ImLe] R. Impagliazzo, R., L. Levin, No better ways to generate hard NP instances than to choose uniformly at random, FOCS 90.

[IR] R. Impagliazzo and S. Rudich, *On the Limitations of certain One-Way Permutations* , STOC 89.

[IY] R. Impagliazzo and M. Yung, *Direct Minimum-Knowledge Computations*, Proc. of Crypto 87, Springer Verlag.

[K] J. Killian, *Basing Cryptography on Oblivious Transfer* , STOC 1988 pp 20-31.

[K2] J. Kilian *Interactive Proofs With Provable Security Against Honest Verifiers* CRYPTO 90, pp. 371-384.

[KMO] J. Killian, S. Micali and R. Ostrovsky *Minimum-Resource Zero-Knowledge Proofs*, FOCS 1989.

[L] L. Levin *Average Case Complete Problems* SIAM J. of Computing, 1986 VOL 15, pp. 285-286.

[LFKN] Lund, C., L. Fortnow, H. Karloff, and N. Nisan, "Algebraic Methods for Interactive Proof Systems" FOCS 90.

[N] M. Naor *"Bit Commitment Using Pseudo-Randomness"* Crypto-89 pp.123-132.

[NOVY] M. Naor, R. Ostrovsky, R. Venkatesan, M. Yung, Zero-Knowledge Arguments for NP can be Based on General Complexity Assumptions, manuscript.

[Ost] R. Ostrovsky *One-way Functions, Hard on Average Problems and Statistical Zero-knowledge Proofs* In Proceedings of 6'th Annual Structure in Complexity Theory Conference. June 30 - July 3, 1991, Chicago. pp. 51-59.

[OVY] R. Ostrovsky, R. Venkatesan, M. Yung, Fair Games Against an All-powerful Adversary, *Sequences 91*, July 1991, Positano, Italy, to appear in Springer Verlag. (Also presented at *DIMACS 1990 Cryptography Workshop*, 1-4 October 1990, Princeton.)

[R] M., Rabin *"How to exchange secrets by oblivious transfer"* TR-81 Aiken Computation Laboratory, Harvard, 1981.

[RB] T. Rabin and M. Ben-Or, *Verifiable Secret Sharing and Secure Protocols* , STOC 89.

[S] A. Shamir *IP=PSPACE* , FOCS 90.

[SRA] A. Shamir, R. Rivest and L. Adleman, *Mental Poker*, Technical Memo MIT (1979).

[VL] Venkatesan R., and L. Levin *Random Instances of a Graph Coloring Problem are Hard* STOC 88. Almost Journal version available.

[Y] A. C. Yao, *How to Generate and Exchange Secrets*, FOCS 86.

Appendix

We briefly recall our results from [OVY] on how $\texttt{strong}\xrightarrow{\text{OT}}\texttt{weak}$ protocols can be based on general complexity assumption. Assume that the strong player (the Sender S) has a secret random input bit b, which he wants the weak player (the Receiver R) to get with probability 1/2. R wants S not to know whether or not R received the bit.

For simplicity, let f be a strong one-way permutation (invertible in polynomial time only on a exponentially small fraction of the inputs). Below, S is given a secret input bit b at the beginning of the protocol, $B(x, y)$ denotes the dot-product mod 2 of x and y, and all $h_i \in \{0,1\}^n$ are linearly independent. The following is a "zooming" technique which can be described as gradually focusing on a value, while maintaining information-theoretic uncertainty.

- $\{R(0)\}$: R selects x' of length n at random and computes $x = f(x')$. He keeps both x' and x secret from S.

- **For i from 1 to $(n-1)$ do the following steps:**

$\{S(i)\}$: S selects at random h_i and sends it to R.

$\{R(i)\}$: R sends $c_i := B(h_i, x)$ to S.

- $\{S(n)\}$: Let x_0, x_1 be the ones which satisfy $\forall i, 1 \leq i < n, B(h_i, x_{\{0,1\}}) = c_i$. S flips a random coin j, selects a random string p, $|p| = l$ and sends to R a triple $< p, x_j, v >$, where $v = b \oplus B(p, f^{-1}(x_j))$.

- $\{R(n)\}$: R checks if for his x, $x = x_j$, and if so, computes $b' = v \oplus B(p, x')$ as the resulting bit he gets from S via an "OT" protocol and outputs (x, b').

We omit the proofs of the following theorems. (The proofs involve applying the basing zooming technique based on the power of the sender and what he can interactively prove.)

Theorem 3 *There exists a protocol implementing OT from a strong (at least probabilistic NP or stronger) player to a probabilistic polynomial-time player, based on any one-way permutation.*

Theorem 4 *There exists a protocol implementing OT protocol from an all-powerful (at least probabilistic $P^{\#P}$ or stronger) player to a probabilistic polynomial-time player, given any one-way function.*

Communication Efficient Zero-Knowledge Proofs of Knowledge (With Applications to Electronic Cash)

Extended Abstract

ALFREDO DE SANTIS* GIUSEPPE PERSIANO*†

Abstract

We show that, after a constant-round preprocessing stage, it is possible to give any polynomial number of *Non-Interactive Zero-Knowledge Proofs of Knowledge* for any NP language. Our proof-system is based on the sole assumption that one-way functions and Non-Interactive Zero-Knowledge Proof Systems of *Language Membership* exist.

The new tool has applications to multi-party protocols. We present the first protocol for *Electronic Cash* with the following properties.

- It is provably secure under general complexity assumptions. Its security is based on the existence of one-way functions and Non-Interactive Zero-Knowledge Proof Systems.
- It does not require the presence of a trusted center; not even the Bank is required to be trusted by the users.
- Each transaction requires only constant rounds of interaction. Actually, most of the transactions requires just one round of interaction.
- Each transaction can be performed by only the users that are interested; that is, it is not necessary for all the users to take part in each single transaction to guarantee privacy and security. Moreover, the transcript of each transaction can be used to prove that the transaction has actually taken place and to prove eventual frauds committed.

1 Introduction

A *Zero-Knowledge Proof of Knowledge* (in short, ZKPK) is a protocol between two polynomial time parties called the Prover and the Verifier. The Prover wants to convince the Verifier that he *knows* a witness w_x that an input string x belongs to a language L (for example x is a graph, L is the language of the Hamiltonian graphs and w_x is a Hamiltonian path in x) without revealing any additional information. The concept of ZKPK is closely related to the concept of a Zero-Knowledge Proof of Membership where an infinite-power Prover wants to convince a polynomial-time Verifier that an input string x belongs to the language L. In fact, it can be easily seen that the interactive Zero-Knowledge Proofs of Membership for NP

*Dipartimento di Informatica ed Applicazioni, Università di Salerno, 84081 Baronissi (Salerno), Italy. E-mail: {ads,giuper}@udsab.dia.unisa.it. This work was partially supported by the Italian Ministry of the University and Scientific Research and by CNR within the framework of the project "Crittografia e sicurezza nel trattamento dell'informazione".

†Aiken Computation Laboratory, Harvard University, Cambridge, MA 02138, USA. E-mail: persiano@endor.harvard.edu

of [GoMiWi] are also Proofs of Knowledge. However, in non-interactive settings, proofs of knowledge seem to be more difficult to obtain than proofs of membership; for example, it is not known whether the zero-knowledge proofs of membership of [BlFeMi, BlDeMiPe] are also proofs of knowledge.

In this paper we consider the problem of obtaining Non-Interactive ZKPK and propose the following model. The prover and the verifier perform a small interactive preprocessing stage after which the only communication allowed is from the Prover to the Verifier. We prove that in this model it is possible to implement Non-Interactive ZKPK. More specifically, we prove that, based solely on the minimal assumption of the existence of one-way functions and of Non-Interactive Zero-Knowledge Proofs of Membership, after a constant-round interactive preprocessing (seven rounds suffice) it is possible for the prover to give any polynomial number of Zero-Knowledge Proofs of Knowledge for any NP language.

Besides having its own interest, the problem of obtaining non-interactive ZKPK has also a dramatic impact on the design of cryptographic protocols. Indeed, the problem of reducing the communication needed for basic cryptographic primitives has recently emerged as a major line of research in the area of cryptographic protocols. [BlFeMi, BlDeMiPe] have proved that it is possible to dispose of interaction in zero-knowledge proofs of membership (see also [DeMiPe1, DeMiPe2, DePe, FeLaSh, DeYu]). [BrCrYu] showed how to obtain constant-round zero-knowledge proofs in the dual model and [FeSh] obtained constant-round zero-knowledge proofs of knowledge. The need of reducing the communication complexity becomes even more pressing when we consider multi-party protocols. [BaBe] were the first to investigate reducing the round complexity for secure function evaluation and exhibited a non-cryptographic method that saves a logarithmic number of rounds. In [BeMiRo], it is shown how any function can be securely computed using only a constant number of rounds of interactions, under the assumption that one-way functions exist.

In this paper, we consider the problem of designing a communication efficient Electronic Cash System. An Electronic Cash System is the digital counterpart of paper banknote and can be seen as a collection of protocols with one distinguished player called Bank. It supports transactions of four different types each of which is the digital equivalent of a real-life transaction: opening an account, issuing, spending and depositing a coin. Even though an Electronic Cash System is a special and important example of multi-party protocol, the result of [BeMiRo] is of no help. In fact, if we use the protocol of [BeMiRo] we would have an Electronic Cash System which requires each user, whenever someone spends, requests or deposits a coin, to perform some computations and to exchange messages with other users. This is clearly unacceptable, as, if such a system would be in use in a real-world situation, we would spend all of our time computing and exchanging messages to check that no one is cheating. In contrast, our solution only requires the two parties involved in a transaction to be active; for example, to spend a coin only the customer and the merchant need to perform some computations and to exchange messages. Nonetheless, our protocol is *minimal* in the number of rounds of communication required: only one-round of communication is required to perform each transaction once an account has been opened and a constant number of rounds (seven rounds suffice) are needed to open an account. Moreover, because of its generality, the protocol of [BeMiRo], requires a majority of honest users, whereas our protocol is secure even in the presence of an immoral majority. The study of this problem has been initiated by [ChFiNa] who presented protocols that, however, lacked a formal security proof (which was left as an open problem) and exploited the algebraic properties of the RSA function, a conjectured hard functions. In contrast, the security of our protocol can be proved based on general and weak complexity assumptions: the existence of any one-way functions and of Non-Interactive Zero-Knowledge Proofs. The protocol presented by [OkOh], even though is based on general assumptions, is highly inefficient from a communication point of view as each of the transactions requires an unbounded number of rounds.

2 Notation

Following the notation of [GoMiRi], if S is a probability space, then "$x \leftarrow S$" denotes the algorithm which assigns to x an element randomly selected according to S. If F is a finite set, then the notation "$x \leftarrow F$" denotes the algorithm which assigns to x an element selected according to the probability space whose sample space is F and uniform probability distribution on the sample points. If $p(\cdot,\cdot,\cdots)$ is a predicate, the notation $Pr(x \leftarrow S; y \leftarrow T; \ldots : p(x,y,\cdots))$ denotes the probability that $p(x,y,\cdots)$ will be true after the ordered execution of the algorithms $x \leftarrow S$, $y \leftarrow T, \ldots$. An *efficient non-uniform algorithm* $D = \{D_n\}$ is a family of expected polynomial-time algorithms with polynomial program-size. The symbol $a \oplus b$ denotes the bitwise xor of the binary strings a and b. If a is shorter than b, then we pad a with 0's so that the two strings have the same length.

2.1 Basic tools

In this section we will briefly review the concept of a pseudo-random function and of a secure commitment scheme that will play an important role in our construction of Non-Interactive Zero-knowledge Proofs of Knowledge.

The concept of pseudo-random function has been introduced by Goldreich, Goldwasser, and Micali [GoGoMi]. Intuitively, we can say that a collection of functions $f : \{0,1\}^n \rightarrow \{0,1\}^{n^c}$ is c-pseudo-random if the output of a function chosen at random from the collection on arguments chosen by a polynomial time algorithm cannot be distinguished from the output of a truly random function. In [GoGoMi] it is shown how to construct a c-pseudo-random collection of function for any constant c from any pseudo-random generator, which in turn can be constructed from any one-way function ([Ha, ImLeLu]). In the sequel, whenever the constant c is clear from the contest, we will just say pseudo-random function.

Another basic tool we use is the following.

Definition 1. A *secure commitment scheme* is an efficient Turing machine E that, on input a bit b, and coin tosses s, r, outputs a commitment $E(b,s,r)$, such that

1. {Uniqueness of decommitment: It is hard to compute a commitment which can be decommitted both as 0 and as 1.}

   $\forall c > 0$, and for all sufficiently large n,

   $$Pr(s \leftarrow \{0,1\}^n : \text{there are } r \neq r' \text{ such that } E(1,s,r) = E(0,s,r')) < n^{-c}.$$

2. {Indistinguishability of commitment: It is computationally hard to distinguish commitments of 0 from commitments of 1.}

   Let $E_n(b)$ be the probability space $\{s, r \leftarrow \{0,1\}^n : (s, E(b,s,r))\}$. Then, for all efficient non-uniform algorithms $C = \{C_n\}$, $\forall d > 0$, and for all sufficiently large n,

   $$\left| Pr(a \leftarrow E_n(0) : C_n(a) = 1) - Pr(a \leftarrow E_n(1) : C_n(a) = 1) \right| < n^{-d}.$$

If property 2 does not hold, we say that there is a "substantial advantage" in breaking the commitment scheme.

Remark 1. The secure commitment scheme can be easily extended to any string x. If $x = b_1...b_m$ is a m-bit string then $E(x,s,r)$ is intended to consist of the m different commitments $E(b_i,s,r_i)$, $i = 1,...,m$, where $r = r_1...r_m$ is a nm-bit string. In this case, it can be seen that property 2 can be extended to any two strings.

Remark 2. If $\alpha = E(x,s,r)$ and s and r are known, then the string x committed to by α can be easily computed.

Remark 3. Our definition of commitment is inspired by the bit-commitment protocol of Naor [Na] which is based on any pseudorandom number generator (and thus on any one-way function [Ha, ImLeLu]). The commitment is based on a random challenge given to the committer. For a random challenge and a random generator, there is no way to cheat in a commit phase. Whenever a random reference string σ is available, we can implement the commitment scheme by applying Naor's protocol by choosing the generator (one way function) and the long-enough challenging string *non-interactively* from some portion of the random string σ (used just for this purpose). All the commitments will, then, use this choice. For more details see [Na].

2.2 Zero-knowledge Proofs of Knowledge

Zero-knowledge proofs of membership were introduced by Goldwasser, Micali, and Rackoff [GoMiRa]. Roughly speaking, a Zero-knowledge proof of membership is a two-party protocol in which one party, called the Prover, convinces the second party, called the Verifier, that a certain input string x belongs to a language L without releasing any additional knowledge.

Subsequently, Tompa and Woll [ToWo] and Feige, Fiat, and Shamir [FeFiSh], considered the concept of a Zero-Knowledge Proof of Knowledge. Here, the Prover wants to convince the Verifier that he *knows* a witness w that the string x belongs to the language L (which needs to be in NP.) This is formalized by introducing the concept of an *extractor* of knowledge. An extractor is an efficient algorithm that, on input x, and by interacting with the prover (upon which he has complete control) tries to compute the witness whose knowledge the prover is claiming. If the extractor succeeds with very high probability, then we say that the Prover "knows" a witness that $x \in L$.

In this paper we address the problem of constructing Non-Interactive Zero-Knowledge Proofs of Knowledge. Although in the interactive setting it is easy to prove that the Zero-Knowledge Proofs (of membership) of [GoMiWi] are also proofs of Knowledge, it is not clear whether the non-interactive zero-knowledge proofs of [BlFeMi, BlDeMiPe] are also proofs of knowledge.

2.3 Non-Interactive Zero-Knowledge Proofs

The concept of a Non-Interactive Zero-Knowledge Proof has been introduced by [BlFeMi] where it is proved that, if a random string readable by both the prover and the verifier is available, it is possible for the prover to give any polynomial number of non-interactive zero-knowledge proofs. (See [BlDeMiPe] for formal definitions and proofs.) The property of zero-knowledge was proved by exhibiting a simulator which generates transcript of reference strings and proofs which a polynomial machine cannot tell apart from a real proof. The simulator was defined as a machine which first gets theorems to be proved and then starts the computation. For our construction we need a different kind of simulation called *on-line* simulation that has been introduced by [DeYu]. In an on-line simulation the simulator works in the following mode. In a preprocessing stage the simulator prepares a prefix of a simulation (the reference string); when given a theorem, the simulated proof with respect to the reference string is generated. In [DeYu], it is proved that, if one-way functions exist, any NIZK Proof System with polynomial-time prover can be transformed into NIZK Proof System with on-line simulator. The proof system of [FeLaSh] also has an on-line simulator.

3 Non-Interactive ZKPK with preprocessing

In this section we introduce the concept of non-interactive zero-knowledge proofs of knowledge with preprocessing.

In our model the prover and verifier perform a small interactive preprocessing stage during which the Prover interactively proves that he has some specific knowledge (in our implementation the string he has committed to.) Later, he can give to the Verifier (or to anyone else who trusts the correctness of the interactive stage) any polynomial (in the length of the preprocessing stage) number of Zero-Knowledge Proofs of Knowledge. This is similar to the model of [DeMiPe2], in which all the interaction needed for a zero-knowledge proof (of language membership) is squeezed to an interactive preprocessing stage. Our model, however, is stronger in the following respect. The length of the preprocessing stage does not bound the overall size of proofs that can be non-interactively proved, whereas in [DeMiPe2] after a preprocessing of size n only one theorem of size at most $n^{1/3}$ could be non-interactively proved. It is interesting to note that, even though the model is a stronger one, we are able to base our protocol on the very weak assumption of the existence of one-way functions and non-interactive Zero-Knowledge Proofs of Language Membership.

3.1 The proof system (P,V)

The proof system consists of two stages: the preprocessing stage and the proof stage.

The Preprocessing Stage. The preprocessing stage is executed only once and is completely independent from the particular choice of all subsequent theorems that will be later proved. Here is the protocol for the preprocessing stage.

Preprocessing Stage.

Let $\sigma = \sigma_1 \circ \sigma_2$ be a random string, with $|\sigma_1| = |\sigma_2| = n$.
(If not otherwise available, P and V can generate σ by executing a coin-flipping protocol.)

- P randomly chooses an n-bit string s and an n^2-bit string r.
- P computes $com = E(s, \sigma_1, r)$ and sends com to V.
- P interactively proves to V in zero-knowledge that he "knows s and r such that $com = E(s, \sigma_1, r)$."

The round complexity of the Preprocessing Stage. The interactive zero-knowledge proof of knowledge can be performed in five rounds by using the protocol of [FeSh]. If the string σ is not available, the coin-flipping protocol can be executed as follows and only requires two rounds. The Verifier commits to a $2n$-bit string v. The Prover sends to the Verifier a random string p. The Verifier decommits the string v and the output string is computed as $p \oplus v$.

The Proof Stage. Let the pair (A, B) be a Non-Interactive Zero-Knowledge Proof System of Membership for any NP language. We now describe the prover's program for a language $L \in NP$. On input a string $x \in L$ of length n (if x is longer we can use the techniques of [BlDeMiPe]) and a witness $w = w(x)$ that $x \in L$, P non-interactively proves to V that he "knows a witness for $x \in L$." As $L \in NP$, there is a constant $c > 0$, called the *expansion constant* of L, such that for all $x \in L$ the witness w has length at most $|x|^c$.

The Prover's program.

Let c be the expansion constant of L and let $\mathcal{F} = \{f_s\}$ be a pseudo-random collection of functions $f_s : \{0,1\}^n \rightarrow \{0,1\}^{n^c}$, where $n = |s|$.

Inputs to P: $x_1, x_2, \ldots \in L$ and witnesses $w_1, w_2, \ldots$; the reference string $\sigma = \sigma_1 \circ \sigma_2$, and com, r, s from the preprocessing stage.

P's computations: For $i = 1, 2, \ldots$

- Compute $f_s(x_i)$ and $\beta_i = w_i \oplus f_s(x_i)$.
- Execute A's program to obtain a NIZK proof $\mathtt{Proof}_i$ that "there exist strings r, s, w_i such that w_i is a witness for $x_i \in L$, $com = E(s, \sigma_1, r)$, and β_i is equal to $w_i \oplus f_s(x_i)$."
- Send V: x_i, com, β_i, and $\mathtt{Proof}_i$.

The Verifier's program.

Inputs to V: the reference string $\sigma = \sigma_1 \circ \sigma_2$, com from the preprocessing stage, and $(x_i, com, \beta_i, proof_i)$ for $i = 1, 2, \ldots$, sent by P.

V's computations: For $i = 1, 2, \ldots$

- Verify that com is the same string of the preprocessing stage.
- Execute B's program to verify that $\mathtt{Proof}_i$ has been correctly computed.

If all checks are successfully passed accept otherwise reject.

The choice of the NIZK proof of membership. Our protocol is quite flexible in the choice of the NIZK proof system which is used in the proof stage. If other NIZK proof systems are available, the above protocol could be modified to accommodate them. For example, if in the preprocessing step prover and verifier establish n oblivious channels, then they can use the protocol by [KiMiOs] for NIZK proofs. There are few differences if we use either of the two NIZK proof systems. First, the minimal assumption on which such systems have been shown to exist: one-way trapdoor permutation in one case and oblivious transfer in the other. Second, the transferability. Using the reference string the proofs are transferable to others who trust the correctness of the construction of the string com. This is a fundamental property for our protocol for electronic cash. Another possibility is the use of the non-interactive Zero-Knowledge Proof-Systems with Preprocessing of [DeMiPe2]. This has the advantage of being based on the sole existence of one-way functions, but as drawback only a small theorem (its size depends on the preprocessing stage) can be non-interactively proved.

The proof that the above protocol is indeed a Non-Interactive Zero-Knowledge Proof of Knowledge will be divided into two parts. First, we prove that it meets the completeness and soundness requirements and then we shall prove the more subtle property of Zero-Knowledge.

Theorem 1. If (A, B) is a Non-Interactive Zero-Knowledge Proof of Membership with efficient prover A, then the pair (P,V) is a Non-Interactive Proof of Knowledge with Preprocessing.

Proof. The *completeness* property holds since for all possible choices of $x_1, x_2, \ldots, x_m$, each of them in L, the prover P convinces the verifier V.

The *soundness* property holds as well. If the prover cannot compute a witness w for x, then he has negligible probability of convincing the verifier. More formally, we have to exhibit an extractor that interacting (only in the preprocessing stage) with the prover, is able to compute a witness of the theorems that are being proved. Since there is a proof of knowledge in the preprocessing stage in which P proves V that he "knows s and r such that $com = E(s, \sigma_1, r)$" then there is extractor that can compute such s and r. It is then an easy task to compute witnesses for all theorems $x \in L$ which are proved. Just compute $f_s(x)$ and extract the witness w as $\beta \oplus f_s(x)$. □

In the next theorem we prove that (P,V) is zero-knowledge. Intuitively (P,V) is zero-knowledge because all verifier sees is just commitments and zero-knowledge proofs on those commitments.

Theorem 2. If one-way functions exist and (A, B) is a Non-Interactive Zero-Knowledge Proof of Membership with efficient prover A, then the pair (P,V) is a Non-Interactive Zero-Knowledge Proof of Knowledge with Preprocessing.

Proof. We exhibit a simulator M that, for each verifier V', outputs a random variable that, under the assumption that one-way functions exist and that (A, B) is zero-knowledge, cannot be distinguished from the view of V' by any efficient non-uniform algorithm.

Here is a sketch of the simulator M. Let S be the (on line) simulator for the NIZK proof system used in the proof stage. To simulate the preprocessing stage, M duplicates what happens between the real prover and V' so to obtain the same random variable as seen by the verifier V' during the preprocessing stage. In the proof stage, the simulator does not compute β_i as $f_s(x_i) \oplus w_i$ (he does not know w_i!) but instead assigns β_i a truly random value and then runs the simulator S for the NIZK. Thus the only differences between the verifier's view and the simulator's output is that σ_2 is either a truly random string or S's output, the NIZK proof is either P's output or S's output, and the β_i's are either computed using a pseudo-random function or are truly random strings. Given the zero-knowledgeness of the NIZK proof system and properties the pseudo-random function, the simulator's output is undistinguishable from the verifier's view. In the final paper, we will prove that the existence of an efficient non-uniform algorithm that distinguishes the output of the simulator from the view of V' can be used to distinguish the output of the function f_s from the output of a truly random function or to violate the zero-knowledgeness of the NIZK (A, B). □

3.2 How to avoid the recycling of non-interactive proofs

Non-Interactive ZKPK can be used in multi-party computations. In certain applications it is desirable to avoid that the same proof is given to different verifiers; for example, in our protocol for Electronic Cash, a ZKPK is essentially a coin and thus we do not want the same coin to be given (i.e. to be spent) more than once. It is easy to detect recycling of the same proof but, in some cases, that might not be enough. We want to be able to punish the author of such a fraud by discovering his identity and thus obtaining a proof of his misbehavior. This can be achieved by suitably modifying the previous protocol and making the proof depending on the verifier's identity, denoted by ID. The preprocessing stage is the same, except that P commits to two strings s and s'. The resulting string *com* is the same for all verifiers. In the proof stage when proving the theorem $x \in L$ to the verifier with name ID, P will compute $f_{s'}(x)$ and break it into n-bit pieces $[f_{s'}(x)]_i$. Then, P sends $\gamma_i = [f_{s'}(x)]_i$ if the ith bit of ID is 0, $\gamma_i = r \oplus [f_{s'}(x)]_i$ otherwise. To prove that the γ_i's have been correctly computed, the prover sends a NIZK proof of correctness based on σ_2. If all names of verifiers are different, then any two names have a position k at which they differ. Then, the XOR of the two γ_k is equal to r, and the commitment *com* can be easily opened. (See Remark 2.)

4 An efficient protocol for Electronic Cash

In this section we show how to use the concepts developed in the previous sections to implement an *Electronic Cash System.* An Electronic Cash System is the digital counterpart of paper banknote and can be seen as a collection of multi-party protocols with one distinguished player called Bank. It supports transactions of four different types each of which is the digital equivalent of a real-life transaction.

1. *Opening of the Account.* During this phase a user A and the Bank establish some common information that will constitute A's credentials. The credentials will be later used by A to request, spend and deposit his electronic coins.
2. *Issuing a Coin.* User A requests an electronic coin from the Bank, which charges A's account for the appropriate amount.
3. *Spending a Coin.* User A passes an electronic coin to B.
4. *Depositing a Coin.* User A passes the coin to the Bank which credits A's bank account.

As we shall see later, the "knowledge" of the signature, produced using the Bank's key, of a certain message constitutes an electronic coin. This electronic coin is spent by giving a proof of knowledge of the signature.

Because paper banknotes are physical objects they do not constitute much of a problem in terms of security for both the Bank and the User. On the other hand, an electronic implementation of paper cash should provide at least the same level of security guaranteed by its physical counterpart. From the Bank's point of view, we would like to avoid the following problems.

1. *Forgery.* A User should not be able to generate an electronic coin; that is we want to make sure that only the Bank is able to issue coins.
2. *Multiple Spending.* Once an electronic coin has been properly issued to the User, it can only be spent once. If a coin is spent twice then the User is exposed.

We also would like to protect the User from possible cheating from the Bank. That is, we want to address the following.

1. *Framing.* If a coin has been spent only once, then the Bank (even if in cooperation with other users) should not be able to frame User as a multiple spender.
2. *Traceability.* When an electronic coin is deposited, it should not be possible for the Bank to trace it back to the user it was originally issued to. This not true for the paper banknote as this can be traced, at least in principle, from the serial number.

In this section we present a communication-efficient protocol for the electronic cash: each transaction requires only a constant number of rounds of communication. Actually, the operations of Issuing, Spending and Depositing a coin only require one round of interaction, which, as can be easily seen, is optimal. Moreover, only the parties involved in the transaction need to be active; e.g. the spending protocol needs only the customer and the merchant to be active and the other users need not to take part. The protocol we present requires the presence of a single randomly chosen reference string readable by all users and of a public file which contains the ID of each user. Moreover, we require the Bank to have published a public key for his signature scheme secure against adaptive chosen-message attacks (we refer the reader to [GoMiRi, NaYu, Ro] for a background on digital signature schemes) and a collection $\mathcal{F}$ of pseudo-random functions.

4.1 Opening the account

This protocol is executed only once, when the user U wants to open an account. We assume that U has a unique n-bit name ID. During this phase, U commits to two indices s_1 and s_2, of random functions from the family $\mathcal{F}$ and interactively proves in Zero-Knowledge that he knows the strings he has just committed to. In subsequent phases of the protocol, U will use the functions f_{s_1} and f_{s_2} to obtain random bits. The Bank, first, verifies the proof produced by U and then signs U's commitment.

Opening the account

Let $\sigma = \sigma_1 \circ \sigma_2$ be the random reference string with $|\sigma_1| = |\sigma_2| = n$.

U's computations:

- Randomly choose a $2n$-bit string $s = s_1 \circ s_2$, where $|s_1| = |s_2| = n$ and a $2n^2$-bit string r.
- Compute $com = E(s, \sigma_1, r)$. Send com to the Bank.

Interactive computation by U and the Bank:

- U interactively and in zero-knowledge proves to the Bank that "he knows s and r such that $com = E(s, \sigma_1, r)$".

Bank's computations:

- If convinced, continue. Otherwise, cancel the transaction.
- Send U a signature sig of com and ID.

Output: U's credentials (com, sig, ID).

4.2 Issuing a coin

Each user that has an account can ask for coins to the Bank. Below is the one-round (which is two-moves) protocol. First, the user U sends its request to the Bank and then the Bank sends its authorization to the user U. Two communication steps are the minimum possible to issue a coin. Let us now briefly describe the protocol used to issue a new coin; a formal description can be found in the picture below. The user U chooses a random n-bit string $coin$ and gets a signature from the bank. However, he cannot just have the Bank sign the string $coin$, for otherwise the coin would be easily traceable. Instead, he proceeds in the following way. The user U commits to $coin$ by computing $\alpha = E(coin, \sigma_1, f_{s_1}(coin))$ and (non-interactively) proves that he knows the string $coin$ he has just committed to. If satisfied by the proof of U, the Bank signs the string α. The coin consists of the triple $(coin, \alpha, sig_\alpha)$.

Issuing a coin

Let $\mathcal{F} = \{f_s\}$ be a collection of pseudo-random functions.

U's computations:

- Randomly choose an n-bit string $coin$.
- Send to the Bank:
  U's credentials (com, sig, ID) and $\alpha = E(coin, \sigma_1, f_{s_1}(coin))$.
- Send to the Bank a NIZK proof based on σ_2 that "there are s, r, and $coin$ such that $com = E(s, \sigma_1, r)$, $s = s_1 \circ s_2$, and $\alpha = E(coin, \sigma_1, f_{s_1}(coin))$.

The Bank's computations:

- If sig is a correct signature of ID and com by the Bank, and the NIZK proof is convincing, continue. Otherwise, cancel the transaction.
- Compute and send U a signature sig_α of α.

Output: The coin $(coin, \alpha, sig_\alpha)$.

4.3 Spending a coin

The user A want to pass a coin $(coin, \alpha, sig_\alpha)$ to the user B. Both of them have an account. We now briefly describe a protocol in which B need not to reply to A's message, he just accepts or rejects the coin. A more precise description can be found in the picture below. A cannot give α and sig_α to B otherwise the coin could be traced back by the Bank to A. Instead, A passes the string *coin* to B and proves that he "knows" a signature of the associated α. If the proof is convincing, B accepts the coin. To avoid multiple spending, we use the technique described in Section 3.2.

Spending a coin

A's computations:

- Send B: the credentials (com, ID, sig) and the string *coin*.
- Let T be the statement "there exist strings α and sig_α such that $\alpha = E(coin, \sigma_1, f_{s_1}(coin))$, $com = E(s_1 \circ s_2, \sigma_1, r)$ and sig_α is a correct signature of α".
- Send B the string $\beta = f_{s_1}(T) \oplus (\alpha \circ sig_\alpha)$ and $proof_1$, a NIZK proof based on σ_2 that β has been correctly computed.
- Compute $f_{s_2}(coin)$ and divide it into n blocks $[f_{s_2}(coin)]_1 ... [f_{s_2}(coin)]_n$, each consisting of n bits.
- For $i = 1, \ldots, n$
  Compute and send B the string $\gamma_i = [f_{s_2}(coin)]_i$ if the ith bit of B's identity ID_B is 0; otherwise $\gamma_i = r \oplus [f_{s_2}(coin)]_i$.
- Send B $proof_2$, a NIZK proof based on σ_2 that the γ_i's have been correctly computed.

B's computations:

- Receive from A: $(com, ID, sig), coin, \beta, proof_1, \gamma_1, \ldots, \gamma_n, proof_2$.
- If $proof_1, proof_2$ and the credentials are convincing accept *coin*. Otherwise, cancel the transaction.

4.4 Depositing a coin

The user B passes the coin he has received to the Bank. Recall that B has received from A the strings $(com, ID_A, sig_A), coin, \beta, proof_1, \gamma_1, \ldots, \gamma_n, proof_2$. Computing on those values, B sends a message to the Bank asking for the coin to be deposited on his account. The Bank accepts or rejects the transaction without having to reply to B's message.

If a double spending has been detected, then the Bank proceeds to compute the author of the tentative. The Bank knows two different entries $(coin, \gamma_1, ..., \gamma_n)$ and $(coin, \gamma'_1, ..., \gamma'_n)$. Since the same coin has been given to two different users, say B and C, then there is at least one position in which the two identities ID_B and ID_C differ. Let d be such a position. Then $\gamma_d \oplus \gamma'_d$ is equal to r. Using such r the Bank breaks the commitment *com* in the credentials of the author of the fraud. This is the only way for the Bank to break a secure commitment and thus can be used to prosecute the author of the fraud.

Depositing a coin

B's computations:

- Send to the Bank: his own identity ID_B, the strings *coin* and $\gamma_1, ..., \gamma_n$.
- Send to the Bank: a NIZK proof of knowledge that "he knows legal credentials (com, ID_A, sig_A), and two NIZK proofs $proof_1, proof_2$ based on σ_2 that would have convinced the Bank".

The *Bank*'s computations:

- If the NIZK proof of knowledge received from B is convincing, continue. Otherwise, cancel the transaction.
- Put *coin* on B's account ID_B.
- *Check for double spending.* Search in the coin database for an entry whose first field is *coin*. If the search is successful then a double spending has been detected.
- If the search is unsuccessful, record in the coin database the entry $(coin, \gamma_1, ..., \gamma_n)$.

4.5 Properties of the protocol

Let us briefly sketch the properties of the protocol. We refer the reader to the final version for formal proofs.

Unforgeability. To be able to forge a coin, any coalition user must be able to produce a signature of α, or to give proofs of knowledge without knowing the witness. However, this contradicts the properties of the signature scheme of the Bank (which is secure against chosen-ciphertext attacks) and of the Non-Interactive ZKPK.

Untraceability. This property can be is easily proved using the fact that all proofs of knowledge are zero-knowledge.

One-time usable. As argued before, if the same coin is spent twice, then the Bank can identify the author and charge him one extra coin, even if a coalition of users tried to cheat the Bank. In fact, if the same coin has been given to two different users, say B and C, then there is at least one position in which the two identities ID_B and ID_C differ. Let d be such a position. Then $\gamma_d \oplus \gamma_d'$ is equal to r. Using such r the Bank breaks the commitment *com* in the credentials of the author of the fraud. This is the only way for the Bank to break a secure commitment and thus can be used to prosecute the author of the fraud.

No need to trust the Bank. The Bank cannot withdraw money from an account and justify it as money used to issue a coin; that is, the Bank can issue a coin (and thus withdraw money from an account) only to a user who has expressly requested the coin. Moreover, if a user U spends a coin only once, then the Bank (even if in cooperation with other users) cannot frame U as a multiple spender.

Acknowledgments.

We would like to thank Maurizio Sereno for useful discussions.

References

[BaBe] J. Bar Ilan and D. Beaver, *Non-Cryptographic Fault-Tolerant Computation in a Constant Number of Rounds of Interaction*, in Proc. of the 8th PODC (1989) pp. 201–209.

[BeMiRo] D. Beaver, S. Micali, and P. Rogaway, *The Round Complexity of Secure Protocols*, Proceedings of the 22nd Annual Symposium on the Theory of Computing, 1990, pp. 503–513.

[BlDeMiPe] M. Blum, A. De Santis, S. Micali, and G. Persiano, *Non-Interactive Zero-Knowledge*, SIAM Journal on Computing, December 1991. Preliminary version: MIT Research Report MIT/LCS/TM-430, May 1990.

[BlFeMi] M. Blum, P. Feldman, and S. Micali, *Non-Interactive Zero-Knowledge Proof Systems and Applications*, Proceedings of the 20th Annual ACM Symposium on Theory of Computing, Chicago, Illinois, 1988.

[BrCrYu] G. Brassard, C. Crépeau, and M. Yung, *Everything in NP can be Proven in Perfect Zero-Knowledge in a Bounded Number of Rounds*, Proceedings of the 16th ICALP, July 1989.

[ChFiNa] D. Chaum, A. Fiat, and M. Naor, *Untraceable Electronic Cash*, in "Advances in Cryptology - CRYPTO 88", Ed. S. Goldwasser, vol. 403 of "Lecture Notes in Computer Science", Springer-Verlag, pp.319–327.

[DeMiPel] A. De Santis, S. Micali, and G. Persiano, *Non-Interactive Zero-Knowledge Proof-Systems*, in "Advances in Cryptology - CRYPTO 87", vol. 293 of "Lecture Notes in Computer Science", Springer Verlag.

[DeMiPe2] A. De Santis, S. Micali, and G. Persiano, *Non-Interactive Zero-Knowledge Proof-Systems with Preprocessing,* in "Advances in Cryptology - CRYPTO 88", Ed. S. Goldwasser, vol. 403 of "Lecture Notes in Computer Science", Springer-Verlag, pp. 269–282.

[DePe] A. De Santis and G. Persiano, *Public-Randomness in Public-key Cryptography,* in "Advances in Cryptology - EUROCRYPT 90", Ed. I.B.Damgård, vol. 473 of "Lecture Notes in Computer Science", Springer-Verlag, pp. 46–62.

[DeYu] A. De Santis and M. Yung, *Cryptographic Applications of the non-interactive Metaproof and Many-prover Systems,* CRYPTO 1990.

[FeFiSh] U. Feige, A. Fiat, and A. Shamir, *Zero-knowledge Proofs of Identity,* Journal of Cryptology, vol. 1, 1988, pp. 77–94. (Preliminary version in Proceedings of the 19th Annual ACM Symposium on Theory of Computing, New York, 1987, pp. 210–217.)

[FeLaSh] U. Feige, D. Lapidot, and A. Shamir, *Multiple Non-interactive Zero-knowledge Proofs Based on a Single Random String,* Proceedings of the 22nd Annual Symposium on the Theory of Computing, 1990, pp. 308–317.

[FeSh] U. Feige and A. Shamir, *Zero knowledge proof of knowledge in two rounds,* in "Advances in Cryptology - CRYPTO 89", vol. 435 of "Lecture Notes in Computer Science", Springer-Verlag, pp. 526–544.

[GoGoMi] O. Goldreich, S. Goldwasser, and S. Micali, *How to Construct Random Functions,* Journal of the Association for Computing Machinery, vol. 33, no. 4, 1986, pp. 792–807.

[GoMiRa] S. Goldwasser, S. Micali, and C. Rackoff, *The Knowledge Complexity of Interactive Proof-Systems,* SIAM Journal on Computing, vol. 18, n. 1, February 1989.

[GoMiRi] S. Goldwasser, S. Micali, and R. Rivest, *A Digital Signature Scheme Secure Against Adaptive Chosen-Message Attack,* SIAM Journal of Computing, vol. 17, n. 2, April 1988, pp. 281–308.

[KiMiOs] J. Kilian, S. Micali, and R. Ostrowsky, *Minimum-Resource Zero-Knowledge Proofs,* Proceedings of the 30th IEEE Symposium on Foundation of Computer Science, 1989, pp. 474–479.

[GoMiWi] O. Goldreich, S. Micali, and A. Wigderson, *Proofs that Yield Nothing but their Validity and a Methodology of Cryptographic Design,* Proceedings of 27th Annual Symposium on Foundations of Computer Science, 1986, pp. 174–187.

[Ha] J. Håstad, *Pseudorandom Generators under Uniform Assumptions,* Proceedings of the 22nd Annual ACM Symposium on Theory of Computing, 1990, pp. 395–404.

[ImLeLu] R. Impagliazzo, L. Levin, and M. Luby, *Pseudo-Random Generation from One-way Functions,* Proceedings of the 21st Annual ACM Symposium on Theory of Computing, May 1989.

[Na] M. Naor, *Bit Commitment using Pseudo-randomness,* in "Advances in Cryptology - CRYPTO 89", vol. 435 of "Lecture Notes in Computer Science", Springer-Verlag.

[NaYu] M. Naor and M. Yung, *Universal One-way Hash Functions and their Cryptographic Applications,* Proceedings of 21st Annual Symposium on the Theory of Computing, May 1989.

[OkOh] T. Okamoto and K. Ohta, *Disposable Zero-knowledge authentications and their Applications to Untraceable Electronic Cash,* in "Advances in Cryptology - CRYPTO 89", vol. 435 of "Lecture Notes in Computer Science", Springer-Verlag, pp. 481–496.

[Ro] J. Rompel, *One-way Functions are Necessary and Sufficient for Secure Signatures,* Proceedings of the 22nd Annual Symposium on the Theory of Computing, 1990, pp. 387–394.

[ToWo] M. Tompa and H. Woll, *Random Self-Reducibility and Zero-knowledge Interactive Proofs of Possession of Information,* Proceedings of 28th Symposium on Foundations of Computer Science, 1987, pp. 472–482.

ALGORITHMS 2

Four Results on Randomized Incremental Constructions*

Kenneth L. Clarkson[†] Kurt Mehlhorn[‡] Raimund Seidel[§]

Abstract

We prove four results on randomized incremental constructions (RICs):

- an analysis of the expected behavior under insertion and deletions,
- a fully dynamic data structure for convex hull maintenance in arbitrary dimensions,
- a tail estimate for the space complexity of RICs,
- a lower bound on the complexity of a game related to RICs.

1 Introduction

Randomized incremental construction (RIC) is a powerful paradigm for geometric algorithms [CS89, Mul88, BDS+]. It leads to simple and efficient algorithms for a wide range of geometric problems: line segment intersection [CS89, Mul88], convex hulls [CS89, Sei90], Voronoi diagrams [CS89, MMO91, GKS90, Dev], triangulation of simple polygons [Sei91], and many others. In this paper we make four contributions to the study of RICs.

- We give a simple analysis of the expected behavior of RICs; cf. § 2. We deal with insertions and deletions and derive bounds for the expected number of regions constructed and the expected number of conflicts encountered in the construction. In the case of deletions our bounds are new, but compare [DMT91, Mul91a, Mul91b, Mul91c, Sch91] for related results, in the case of insertions the results were known, but our proofs are simpler.
- We apply the general results on RIC to the problem of maintaining convex hulls in d-dimensional space; cf. § 3. We show that random insertions and deletions take expected time $O(\log n)$ for $d \leq 3$ and time $O(n^{\lfloor d/2 \rfloor - 1})$ otherwise. If the points are in convex position, which is, e.g., the case when Voronoi diagrams are transformed into convex hulls of one higher dimension, the deletion time becomes $\log \log n$ for $d \leq 3$. Schwarzkopf [Sch91] has obtained the same bounds for all $d \geq 6$. For $d < 6$ our bound is better.
- We derive a tail estimate for the number of regions constructed in RICs; cf. § 4.
- We study the complexity of a game related to the $O(n \log^* n)$ RICs of [Sei91] and [Dev] and show that the complexity of the game is $\Theta(n \log^* n)$; cf. § 5.

Sections 4 and 5 are omitted from this extended abstract.

*A full version of this extended abstract is available from the authors.

[†] AT&T Bell Laboratories

[‡] Max Planck Institut für Informatik and Universität des Saarlandes; supported in part by ESPRIT II Basic Research Actions Program of the EC under contract no. 3075(project ALCOM)

[§] Computer Science Division, University of California at Berkeley

2 Randomized Incremental Constructions: General Theorems

Let S be a set with $|S| = n$ elements, which we will sometimes call *objects*. Let $\mathcal{F}(S)$ be a multiset whose elements are nonempty subsets of S, and let b be the size of the largest element of $\mathcal{F}(S)$. We will call the elements of $\mathcal{F}(S)$ *regions* or *ranges*. If all the regions have size b, we will say that $\mathcal{F}(S)$ is *uniform*. For a region $F \in \mathcal{F}(S)$ and an object x, if $x \in F$ we say that F *relies on* x or x *supports* F. For $R \subseteq S$, define $\mathcal{F}(R) = \{F \in \mathcal{F}(S) \mid F \subset R\}$. (That is, the multiplicity of F in $\mathcal{F}(R)$ is the same as in $\mathcal{F}(S)$.) We also assume a *conflict relation* $C \subseteq S \times \mathcal{F}(S)$ between objects and regions. We postulate that for all $x \in S$ and $F \in \mathcal{F}(S)$, if $(x, F) \in C$ then F does not rely on x.

For a subset $R \subseteq S$, $\mathcal{F}_0(R)$ will denote the set of $F \in \mathcal{F}(R)$ having no $x \in R$ with $(x, F) \in C$; that is, $\mathcal{F}_0(R)$ is the set of regions over R which do not conflict with any object in R.

Clarkson and Shor [CS89] analyzed the incremental computation of $\mathcal{F}_0(S)$. In the general step, $\mathcal{F}_0(R)$ for some subset $R \subseteq S$ is already available, a random element $x \in S \setminus R$ is chosen, and $\mathcal{F}_0(R \cup \{x\})$ is constructed from $\mathcal{F}_0(R)$.

Let $(x_1, \ldots, x_j)$ be a sequence of pairwise distinct elements of S, and R_j the set $\{x_1, \ldots, x_j\}$. Let $R_0 = \{\}$, the empty set. The *history* $H = H(x_1, \ldots, x_r)$ for insertion sequence $(x_1, \ldots, x_r)$ is defined as $H = \bigcup_{1 \leq i \leq r} \mathcal{F}_0(R_i)$. Let Π_S be the set of permutations of S. For $\pi = (x_1, \ldots, x_n) \in \Pi_S$, $H_r(\pi)$ or simply H_r denotes the history $H(x_1, \ldots, x_r)$.

First, some simple facts about random permutations, whose proofs we leave to the reader:

Lemma 1 *If $\pi = (x_1, \ldots, x_n)$ is a random permutation of S, then R_j is a random subset of S of size j, $(x_1, \ldots, x_j)$ is a random permutation of R_j, x_j is a random element of R_j, and if δ is a (fixed) permutation, then $\pi\delta$ is a random permutation.*

We are now ready for an average case analysis of randomized incremental constructions. All expected values are computed with respect to a random ordering $(x_1, \cdots, x_n) \in \Pi_S$ of the objects in S.

For subset $R \subseteq S$, $r = |R|$, and distinct objects $x, y \in R$, let

$$\begin{aligned}
deg(x, R) &= |\{F \in \mathcal{F}_0(R); x \text{ supports } F\}| \\
pdeg(x, y, R) &= |\{F \in \mathcal{F}_0(R); x \text{ and } y \text{ support } F\}| \\
c(R) &= \frac{1}{r} \sum_{x \in R} deg(x, R) \\
p(R) &= \frac{1}{r(r-1)} \sum_{(x,y) \in R^2} pdeg(x, y, R)\,.
\end{aligned}$$

We call $deg(x, R)$ the *degree* of x in R, $pdeg(x, y, R)$ the *degree* of the ordered pair (x, y) in R, $c(R)$ the *average degree* of a random object in R and $p(R)$ the *average pair degree* of a random pair of objects in R.

For integer r, $1 \leq r \leq n$, let

$$c_r = E[c(R)] = \sum_{R \subseteq S, |R| = r} c(R) \Big/ \binom{n}{r}$$

and

$$p_r = E[p(R)] = \sum_{R \subseteq S, |R| = r} p(R) \Big/ \binom{n}{r}$$

be the expected average degree and pair degree for random $R_r \subset S$, and let

$$f_r = \sum_{R \subseteq S, |R| = r} |\mathcal{F}_0(R)| \Big/ \binom{n}{r}$$

be the expected number of conflict-free regions of $\mathcal{F}(R)$, with respect to random R_r. Note that $c_1 = f_1$. It will be convenient to adopt the convention that $c_j = p_j = f_j = 0$ for $j < 1$ or $j > n$, and (almost always) convenient to adopt the convention that $p_1 = f_1$.

Lemma 2 *The expectations c_r, p_r, and f_r satisfy $c_r \leq bf_r/r$, and for $r > 1$, $p_r \leq b(b-1)f_r/r(r-1)$, with equality if $\mathcal{F}(S)$ is uniform.*

Proof: For every region $F \in \mathcal{F}(S)$ there are at most b objects and at most $b(b-1)$ ordered pairs of objects which support F, and exactly as many if $\mathcal{F}(S)$ is uniform. ∎

Theorem 3 *Let C_r be the expected size of H_r. Then $C_r = \sum_{j\leq r} c_j$.*

Proof: H_0 is empty and hence $C_0 = 0$. For $r \geq 1$ the number of elements of H_r which are not already elements of H_{r-1} is equal to $deg(x_r, R_r)$. Since R_r is a random subset of S of size r and x_r is a random object in R, we have

$$E[deg(x_r, R_r)] = E[c(R)] = c_r.$$

∎

In §4 we will strengthen Theorem 3 and prove a tail estimate for $|H_n|$.

Theorem 4 *The expected number of regions in H_{r-1} which are in conflict with x_r is $-c_r + \sum_{j\leq r} p_j$.*

Proof: Let X be the number of regions $F \in H_{r-1}$ with $(x_r, F) \in C$. Let $H = H_{r-1} = H(x_1, \ldots, x_{r-1})$ and $H' = H(x_r, x_1, \ldots, x_{r-1})$, i.e., in H' we "pretend" that x_r was put in first. We have

$$|H| + |H' \backslash H| = |H'| + |H \backslash H'|,$$

which holds for any two finite sets. Now $X = |H\backslash H'|$ since $H\backslash H'$ is the set of regions in H which conflict with x_r. On the other hand, $H'\backslash H$ comprises regions incident to x_r; to count these regions, we count the number that appear when x_j is inserted. That is, letting $R'_j = R_j \cup \{x_r\}$, for each region there is exactly one $j \geq 1$ such that $F \in \mathcal{F}_0(R'_j)$ and x_j supports F. A region in $H'\backslash H$ is also supported by x_r, and so for given j the number of regions we count is $pdeg(x_r, x_j, R'_j)$. Putting these observations together,

$$X = |H| - |H'| + |\mathcal{F}_0(\{x_r\})| + \sum_{1\leq j\leq r-1} pdeg(x_r, x_j, R'_j),$$

and so

$$EX = E|H| - E|H'| + E|\mathcal{F}_0(\{x_r\})| + \sum_{1\leq j\leq r-1} E[pdeg(x_r, x_j, R'_j)]$$

We have $E|H| = C_{r-1}$ by Theorem 3, and $E|H'| = C_r$ by Theorem 3 and Lemma 1. Also $E|\mathcal{F}_0(\{x_r\})| = f_1 = p_1$ by convention, and $E[pdeg(x_r, x_j, R'_j)] = p_{j+1}$, since $R'_j = R_j \cup \{x_r\}$ is a random subset of S of size $j+1$ and x_r and x_j are random elements of this subset. ∎

The following estimates are also useful.

Lemma 5 *For $j \leq r$ the following holds:*

(a) *The expected number of regions in $\mathcal{F}_0(R_{j-1})$ in conflict with x_r is $f_{j-1} - f_j + c_j$.*

(b) *The expected number of regions in $\mathcal{F}_0(R_{j-1})$ supported by x_{j-1} and in conflict with x_r is at most $b(f_{j-1} - f_j + c_j)/(j-1)$, with equality if $\mathcal{F}(S)$ is uniform.*

Proof:

(a) We have

$$\begin{aligned}\mathcal{F}_0(R_{j-1} \cup \{x_r\}) &= \mathcal{F}_0(R_{j-1}) \backslash \{F \in \mathcal{F}_0(R_{j-1}); (x_r, F) \in C\} \\ &\cup \{F \in \mathcal{F}_0(R_{j-1} \cup \{x_r\}); x_r \text{ supports } F\}\end{aligned}$$

and hence the desired quantity is

$$\begin{aligned}E|\mathcal{F}_0(R_{j-1})| - E|\mathcal{F}_0(R_{j-1} \cup \{x_r\}| + E|\{F \in \mathcal{F}_0(R_{j-1} \cup \{x_r\}); x_r \text{ supports } F\}| \\ = f_{j-1} - f_j + c_j\end{aligned}$$

(b) x_{j-1} is a random element of R_{j-1}. Hence a region considered in part (a) is supported by x_{j-1} with probability at most $b/(j-1)$.

∎

Summation of the bound in Lemma 5b for j from 1 to $r-1$ gives an alternative bound on the expected number of regions in H_{r-1} which conflict with x_r.

The *conflict history* $G = G_n = G(\pi)$ for insertion sequence $\pi = (x_1, \cdots, x_n)$ is the relation $C \cap (S \times H_n)$. We may also describe this relation as a bipartite graph, with an edge between object $x \in S$ and region $F \in H_n$ when x and F conflict. The conflict history corresponds to the union (over time) of the conflict graphs in [CS89]. We use $|G|$ to denote the size of the conflict history, the number of pairs in it.

Theorem 6 *The expected size of the conflict history is*

$$E|G| = -C_n + \sum_j (n-j+1)p_j$$

Proof: Theorem 4 counts the expected number of edges incident to node $x_r \in S$. The claim follows by summation over r. ∎

We next turn to random deletions. For $\pi = (x_1, \ldots, x_n) \in \Pi_S$ and $r \in [1..n]$, let

$$\pi \backslash r = (x_1, \ldots, x_{r-1}, x_{r+1}, \ldots, x_n),$$

and $\pi \backslash r = \pi$ for $r \notin [1..n]$. We bound the expected size of the difference between $H(\pi)$ and $H(\pi \backslash r)$ and between $G(\pi)$ and $G(\pi \backslash r)$ for random $\pi \in \Pi_S$ and random $r \in [1..n]$.

Theorem 7

$$\frac{1}{n!n} \sum_{\pi \in \Pi_S} \sum_r |H(\pi) \oplus H(\pi \backslash r)| \leq 2b\frac{C_n}{n} - c_n,$$

with equality if $\mathcal{F}(S)$ is uniform.

Proof: For finite sets A and B,

$$|B \oplus A| = |A| - |B| + 2|B \backslash A|,$$

and so for $H = H(\pi)$ and $H(\pi \backslash r)$,

$$|H \oplus H(\pi \backslash r)| = |H(\pi \backslash r)| - |H| + 2|H \backslash H(\pi \backslash r)|.$$

The set $H \backslash H(\pi \backslash r)$ comprises regions in H supported by x_r. By Theorem 3, $E|H| = C_n$, and any $F \in H$ is supported by no more than b objects, with equality if $\mathcal{F}(S)$ is uniform. Therefore on average the random $x_r \in S$ supports no more than bC_n/n regions of H. By Theorem 3 and Lemma 1, we have $E|H(\pi \backslash r)| = C_{n-1}$, and the theorem follows since $C_{n-1} - C_n = -c_n$ by definition. ∎

Theorem 8

$$\begin{aligned} E|G(\pi\backslash i)\backslash G(\pi)| &= \frac{1}{n!n}\sum_{\pi\in\Pi_S}\sum_i |G(\pi\backslash i)\backslash G(\pi)| \\ &\le c_n - (b+1)C_n/n + \sum_j bp_j - \sum_j (b+1)(j-1)p_j/n, \end{aligned}$$

with equality if $\mathcal{F}(S)$ is uniform.

Proof: Letting $G = G(\pi)$, we have

$$|G(\pi\backslash i)\backslash G| = |G(\pi\backslash i)| - |G| + |G\backslash G(\pi\backslash i)|,$$

and by linearity of expectation,

$$E|G(\pi\backslash i)\backslash G| = E|G(\pi\backslash i)| - E|G| + E|G\backslash G(\pi\backslash i)|.$$

Theorem 6 gives an expression for $E|G|$, and that theorem with Lemma 1 gives a similar one for $E|G(\pi\backslash i)|$, yielding

$$E|G(\pi\backslash i)\backslash G| = E|G\backslash G(\pi\backslash i)| + c_n - \sum_j p_j.$$

(Alternatively, note that $E|G| - E|G(\pi\backslash i)|$ is the expected number of regions of H_n conflicting with x_n, and use Theorem 4.) We need to find $E|G\backslash G(\pi\backslash i)|$. A pair (x, F) is in $G\backslash G(\pi\backslash i)$ if it is in G and either $x_i = x$ or $x_i \in F$. At most $b+1$ choices of x_i allow this, for any $(x, F) \in G$, and so $E|G\backslash G(\pi\backslash i)| \le (b+1)E|G|/n$, with equality if $\mathcal{F}(S)$ is uniform. The result follows using Theorem 6 and easy manipulations. ∎

In the convex hull algorithm of §3, the conflicts of $G(\pi\backslash i)\backslash G(\pi)$ are not quite all those examined when deleting x_i. The following bound will also be useful.

Lemma 9 *Let I be the set of conflicts of the form (x_j, F) with $j > i$ and $F \in \mathcal{F}_0(R_{i-1}) \setminus \mathcal{F}_0(R_i)$. Then for random $\pi \in \Pi_S$ and random $i \in [1..n]$, $E|I| = (E|G| - E|H| + f_n)/n$.*

Proof: Let I_i denote the set I for x_i. Then $E|I| = \sum_i E|I_i|/n$, and since the I_i are disjoint, $E|I| = E|\cup_i I_i|/n$. For any conflict $(x_j, F) \in G$, either $F \in \mathcal{F}_0(R_{j-1})$, or there is exactly one $i < j$ such that $F \in \mathcal{F}_0(R_{i-1}) \setminus \mathcal{F}_0(R_i)$. In the latter case, $(x_j, F) \in I_i$. To count the conflicts (x_j, F) with $F \in \mathcal{F}_0(R_{j-1})$, note that each $F \in H\backslash\mathcal{F}_0(S)$ appears this way exactly once. ∎

3 Dynamic Convex Hulls

We apply the results of §2 to the problem of maintaining the convex hull in d-dimensional space under insertions and deletions of points. Let $X \subset \mathbb{R}^d$ be a set of points, which we assume to be in general position: no $d+1$ are coplanar. For $R \subseteq X$, let $\operatorname{conv} R$ denote the convex hull of R. We let $x_1, x_2, \ldots, x_n$ denote the points in X in the order of their insertion, and let R_i denote $\{x_1, \ldots, x_i\}$.

3.1 The Insertion Algorithm

To maintain the convex hull of R under insertions, we maintain a triangulation T of the hull: a simplicial complex whose union is $\operatorname{conv} R$. (A simplicial complex is a collection of simplices such that the intersection of any two is a face of each.) The vertices of the simplices of T are points of R. The triangulation is updated as follows when a point x is added to R: if $x \in \operatorname{conv} R$, and so is in some simplex S of T, leave T as it was. If $x \notin \operatorname{conv} R$, then for every facet F of the hull of R visible to x, add to T the simplex $S(F, x) = \operatorname{conv}(F \cup \{x\})$. Call F the *base* facet and x the *peak* vertex of the simplex. A facet is *visible* to x or x*-visible* just when $S(F, x)$ meets the hull only at F. We may also say, for x-visible F, that x is visible to F, and they *see* each other. Use T_r to denote the triangulation after the insertion of $x_1, x_2, \ldots, x_r$.

This process is called triangulation by "placing" [Ede87]. It should be clear that the stated conditions on the triangulation are preserved. (When $r \leq d+1$, we simply maintain a single $(r-1)$–dimensional simplex.) It will be convenient to extend the triangulation so that facets of the current hull are also base facets of simplices; this gives a uniform representation. The peak vertex of these simplices is a "dummy" that in effect is visible to all current facets. It will also be useful to put the origin O inside the first full-dimensional simplex created, when $r = d+1$. (Here we use the assumption of general position.) Call this the *origin simplex*. The dummy vertex is now an *anti-origin*, denoted $\overline{O}$: while the origin sees no facets of the current hull of R, the anti-origin sees all of them (These conventions can be made precise and rigorous in "two-sided space."[Sto87]). We use T to also denote the extended triangulation. To carry the uniformity even further, we designate an arbitrary vertex of the origin simplex as its peak and call its opposite facet the base of the origin simplex. In this way, there are $d+2$ simplices in the (extended) triangulation when the first full-dimensional simplex is created: the origin simplex and $d+1$ simplices with peak $\overline{O}$. One facet of the origin simplex (better: its two sides) is base facet of two simplices and all other facets of the origin simplex are base facet of one simplex.

Two simplices of T are *neighbors* if they share a facet. The neighbor relation defines the neighborhood graph on the set of simplices. Call a neighbor of some simplex S opposite to a vertex x of S, and vice versa, if the common facet does not contain x. In an implementation, we propose to store the directed version of the neighborhood graph augmented with the following information. For each simplex S store (a pointer to the edge to) the neighbor sharing the base facet of S, (a pointer to) the peak vertex of S, and for each edge (S, S') of the graph, i.e., for each facet of S, store (a pointer to) the vertex of S opposite to the facet. Finally, store for each simplex the equation of the hyperplane supporting the base facet of the simplex. The equation is normalized such that the peak lies in the positive half-space.

We discuss next two *search* methods for finding the x-visible current facets of conv R.

Here is one method: locate x in T by walking along the segment $\overline{Ox}$ beginning at O. If this walk enters a simplex whose peak vertex is the anti-origin, then an x-visible current facet has been found. Otherwise, a simplex of T containing x has been found, showing that $x \in \operatorname{conv} R$. In the former case, find all x-visible hull facets by a search of the simplices incident to the anti-origin. These simplices form a connected set in the neighborhood graph. We call this search method the *segment-walking* method.

Another search method is the following: starting at the origin simplex and the simplex sharing its base facet explore simplices according to the rule: if a simplex has an x-visible base facet, search its neighbors (not including the neighbor that shares the base facet). Here we say that a base facet F is x-visible if that was true (in the previous sense) at the time that F was a current hull facet. This search procedure reaches all x-visible current hull facets, i.e., all simplices $S(F, \overline{O})$ with x-visible base facet F, since the base facets of all simplices traversed in the segment-walking search method are x-visible. We call this search scheme the *all-visibilities* method.

We finally turn to the update procedure. At this point, we have found the current hull facets seeing x, in the form of the simplices whose base facets see x and with the anti-origin as their peak vertex. Let $\mathcal{V}$ be the set of such simplices. Now we update T by altering these simplices, and creating some others. The alteration is simply to replace the anti-origin with x in every simplex in $\mathcal{V}$.

The new simplices correspond to new hull facets. Such facets are the hull of x and a horizon ridge f; a *horizon ridge* is a $(d-2)$–dimensional face of conv R with the property that exactly one of the two incident hull facets sees x. Each horizon ridge f gives rise to a new simplex A_f with base facet $\operatorname{conv}(f \cup \{x\})$ and peak $\overline{O}$. For each horizon ridge of conv R there is a non-base facet of a simplex in $\mathcal{V}$ such that x does not see the base facet of the other simplex incident to the facet. Thus the set of horizon ridges is easily determined.

It remains to update the neighbor relationship. Let $A_f = S(\operatorname{conv}(f \cup \{x\}), \overline{O})$ be a new simplex corresponding to horizon ridge f. In the old triangulation (before adding x) there were two simplices V and N incident to the facet $\text{conv}(f \cup \{\overline{O}\})$; $V \in \mathcal{V}$ and $N \notin \mathcal{V}$. In the updated triangulation V has peak x. The neighbor of A_f opposite to x is N and the neighbor opposite to $\overline{O}$ is (the updated version) of V. Now consider any vertex $q \in f$ and let $\mathcal{S} = \mathcal{S}_{f,q}$ be the set of simplices with peak x and including $f \setminus \{q\} \cup \{x\}$ in their vertex set. We will show that the neighbor of A_f opposite to q can be determined by a simple walk through $\mathcal{S}$. Note first that $V \in \mathcal{S}$. Consider next any simplex $S = S(F, x) \in \mathcal{S}$. Then $F = \text{conv}(f \setminus \{q\} \cup \{y_1, y_2\})$ for some vertices y_1 and y_2. Thus S has at most two neighbors in $\mathcal{S}$, namely the neighbors opposite to y_1 and y_2 respectively. Also, V has at most one neighbor in $\mathcal{S}$, namely the neighbor opposite to q (Note that the neighbor opposite to y, where $\text{conv}(f \cup \{y\})$ is the base facet of V, is the simplex $A_f \notin \mathcal{S}$.). The neighbor relation thus induces a path on the set $\mathcal{S}$ with V being one end of the path. Let V' with base facet $\text{conv}(f \setminus \{q\} \cup \{y_1, y_2\})$ be the other end of the path. Assume that the neighbor of V' opposite to y_1, call it B, does not belong to $\mathcal{S}$ and that $y_1 = q$ if $V = V'$, i.e., the path has length zero. The simplex B includes $f \setminus \{q\} \cup \{y_2, x\}$ in its vertex set and does not have peak x. Thus B has peak $\overline{O}$ and hence B is the neighbor of A_f opposite to q. This completes the description of the update step.

3.2 Analysis of Insertions

The cost of adding a point to set R is the time needed to locate the point x in the triangulation T, plus the time needed to update the triangulation.

We need some additional notation. Let t_0 be the number of simplices visited by the walk along segment $\overline{Ox}$, let t_1 be the set of simplices with x-visible base facet, let t_2 be the set of simplices visited by the all-visibilites method, let t_3 be the number of simplices with peak x, and let t_4 be the number of new hull facets. Then $t_0 \leq t_1$, since the base facets of all simplices traversed by the segment-walking method see x, and $t_2 \leq (d+1) \cdot t_1$ since a simplex has $d+1$ neighbors.

In the segment-walking method the time spent on the walk is $O(d^2) \cdot t_0$, since given the entry point of segment $\overline{Ox}$ into a simplex S the exit point can be found in time $O(d^2)$; it takes time $O(d)$ per facet to compute the point of intersection, i.e., $O(d^2)$ altogether, and $O(d)$ time to select the first intersection following the entry point. The segment-walk determines the simplex containing x. All visible hull facets can then be determined in time $O(d^2) \cdot t_3$, since visiblity can be checked in time $O(d)$ per base facet and since a visible facet has at most d invisible neighbors. We define the search time of the segment-walking method to be $O(d^2) \cdot t_0 = O(d^2) \cdot t_1$ and include the $O(d^2) \cdot t_3$ term in the update time.

The search time for the all-visibilities method is $O(d) \cdot t_2 = O(d^2) \cdot t_1$, since $O(d)$ per simplex is needed for the visibility check and since the degree of the neighborhood graph is $d+1$.

Let's turn to the update time next. We need to alter t_3 simplices; this takes time $O(1) \cdot t_3$. For each new simplex we have to compute the equation of the hyperplane supporting the base facet. This takes time $O(d^3) \cdot t_4$, since solving the linear systems for the normal vectors requires $O(d^3)$ time per simplex (A factor of d can be removed using complicated rank-one updating techniques, if desired.). Finally, we need to update the neighbor relation. Let $\mathcal{S} = \mathcal{S}_{f,q}$ be defined as in the previous section. The walk through $\mathcal{S}$ takes time $O(|\mathcal{S}|)$. Next observe, that a simplex $S = S(F, x) \in \mathcal{V}$ can belong to at most $2d(d-1)$ different sets $\mathcal{S}_{f,q}$, since $f \setminus \{q\}$ can be obtained from F by deleting two vertices ($d(d-1)$ choices) and since there are only two choices for q once $f \setminus \{q\}$ is fixed (Note that there are only two horizon ridges containing $f \setminus \{q\}$.). Thus the time to update the neighbor relation is $O(d^2) \cdot t_3$ and total update time is $O(d^3) \cdot t_4 + O(d^2) \cdot t_3$.

We next establish the connection to § 2. Our regions are half spaces. More formally, we have $b = d$ and $\mathcal{F}(X)$ contains two copies of each subset $\{x_1, \ldots, x_d\} \subseteq X$ of cardinality d. These two copies are identified with the two open half-spaces defined by the hyperplane through points $x_1, x_2, \ldots, x_d$. A point x is said to conflict with a half-space if it is contained in the half-space. In this way, for $|R| \geq d+1$ the regions in $\mathcal{F}_0(R)$ correspond precisely to the facets of the convex hull of R (recall that we assume our points to be in general position) and a facet F of conv R is visible from $x \notin R$ if x conflicts with the half-space supporting the facet. Also $|\mathcal{F}_0(R)| = 2$ if $|R| = d$,

$\mathcal{F}_0(R) = \emptyset$ for $|R| < d$, and $\mathcal{F}(X)$ is uniform. Using the notation of §2, we therefore have $f_r = 0$ for $r < d$ and $f_d = 2$; for $r > d$, f_r is the expected number of facets of conv R for random subset $R \subseteq X$ with $|R| = r$.

Theorem 10 (a) *The expected number of simplices of T_r is $C_r = \sum_{j \leq r} df_j/j$.*

(b) *The expected search time for x_r, using either search method, is $O(d^2)$ times*

$$-c_r + 2\sum_{j \leq r} p_j = -\frac{d}{r}f_r + 2\sum_{j \leq r} \frac{d(d-1)}{j(j-1)} f_j.$$

(c) *The expected time to construct the convex hull of n points using either search method is*

$$O(d^3)\sum_j \frac{d}{j} f_j + O(d^2)\sum_j \frac{d(d-1)}{j(j-1)}(n-j+1)f_j = O(d^4)\sum_j \frac{nf_j}{j(j-1)}.$$

Proof:

(a) Each simplex has a base facet, and so the bound follows from Theorem 3 and Lemma 2.

(b) From the above discussion, we need to find t_1, the expected number of facets that either are x_r–visible. The expected number of visible facets is $-c_r + \sum_{j \leq r} p_j$, by Theorem 4.

(c) The work per simplex of T_n is $O(d^3)$, as discussed above. The bound follows, using (a) and summing the bound of (b) over r.

∎

Since $f_r = O(r^{\lfloor d/2 \rfloor})$ in the worst case, the running time is $O(n \log n)$ for $d \leq 3$, and $O(n^{\lfloor d/2 \rfloor})$ for $d \geq 4$. We note also that for many natural probability distributions, the expected complexity of the hull of random points satisfies $f_r = O(r)$ for fixed d. For such point sets, our algorithm requires $O(n \log n)$ expected time.

3.3 The Deletion Algorithm

The global plan is quite simple. When a point is deleted from R, we change the triangulation T so that in effect x was never added. This is in the spirit of § 2. The effect of the deletion of x on the triangulation T is easy to describe. All simplices having x as a vertex disappear (If x is not a vertex of T then T does not change). The new simplices of T resulting from the deletion of x all have base facets visible to x, with peak vertices inserted after x. These are the simplices that would have been included had x not been inserted into R. Let $R(x)$ be the set of points of R incident to simplices with vertex x, and also inserted after x. We will, in effect, reinsert the points of $R(x)$ in the order in which they were inserted into R, constructing only those simplices that have bases visible to x. On a superficial level, this describes the deletion process. The details follow.

Let $\pi = (x_1, \ldots, x_n)$ be the insertion order and assume that $x = x_i$ is deleted. We first characterize the triangulation $T(\pi \setminus i)$.

Lemma 11 *Assume that $x = x_i$ is a vertex of T. The triangulation $T(\pi \setminus i)$ can be obtained from the triangulation $T(\pi)$ as follows:*

1. *Set $T = T(\pi)$.*

2. *Remove all simplices having x_i as a vertex.*

3. $k \leftarrow i+1$;
while $k \leq n$
do *(* invariant A holds here *)*
for *all facets F of* $\text{conv}(R_{k-1} \setminus \{x_i\})$ *visible to* x_i *and* x_k
do *add* $S(F, x_k)$ *to T* **od**;
$k \leftarrow k+1$
od
(invariant A holds here with* $k = n+1$ **)*

Lemma 11 suggests that an algorithm akin to the insertion algorithm can be used to retriangulate the cavity created by the removal of x. In the insertion algorithm we first identified a single hull facet visible from the new point and then all visible hull facets by a search of the neighborhood graph. This strategy still works as the next lemma shows.

Lemma 12 *Let P be a polyhedron with vertices in X and let x and y be points in X. Then the set of facets of P visible to x as well as y is neighbor-connected.*

Lemma 11 characterizes the new simplices added to T. We show next that only points in $R(x)$ can contribute new simplices (Lemma 13) and that the search for facets visible from x_i and x_k can be streamlined (Lemma 14).

Lemma 13 *Let* $i < k$ *and assume that* $x = x_i$ *is a vertex of T. Then* $y = x_k \in R(x_i)$ *if some facet F of* $\text{conv}(R_{k-1} \setminus \{p\})$ *is visible to x and y.*

Lemma 14 *Step 3 in Lemma 11 may be replaced by:*

$B \leftarrow$ *set of facets of* $\text{conv}\, R_{i-1}$ *visible to* x;
for *all* $y \in R(x)$ *in ascending insertion order*
do for *all y–visible facets F of B*
do *add simplex* $S(F, y)$ *to T and remove F from B* **od**;
add all x–visible hull facets of new simplices $S(F, y)$ *to B*
od

We are now ready for the algorithmic details.

To handle deletions, we must augment our data structure slightly. For each point, store a pointer to a simplex incident to the point and for each simplex store the set of points incident to the simplex.

Check first, whether x is a vertex of the simplex pointed to by x. If not, x is removed and we are done. If so, construct the set $R(x)$ by inspection of all simplices incident to x. This takes time proportional to d times $|R(x)|$ plus the number of simplices with peak x. (Note that a simplex which has x in its base facet contributes its peak to $R(x)$ and that a simplex has at most $d+1$ neighboring simplices).

Next, sort the points in $R(x)$ by the time of insertion. This takes time $O(\min\{n, |R(x)| \log\log n\})$, where the former bound is obtained by bucket sort and the latter bound comes from the use of bounded ordered dictionaries ([vKZ77, MN90]). Also, collect for each point $y \in \{p\} \cup R(x)$ the set of simplices $\mathcal{S}(y)$ with peak y and also having x as a vertex. Next remove all simplices incident to x from T (cf. step 2 in Lemma 11). The set B (cf. Lemma 14) is initialized to the set of base facets of simplices in $\mathcal{S}(x)$. The neighborhood graph on the set B is given by the incidence information of the simplices in $\mathcal{S}(x)$.

The points in $R(x)$ are now processed in the insertion order. Consider $y \in R(x)$. First, a facet $y \in B$, if any, visible from y has to be determined. Then a graph search in the neighborhood graph of B determines all y–visible facets in B, the simplices in $S(F, y)$ for visible F are added, the neighborhood graph is updated as described in § 3.1, and the new hull facets visible to x

are determined. All of this (except for the search for the first visible facet in B) takes time proportional to d^3 times the number of new simplices constructed.

How can a y–visible facet in B be found? We distinguish cases. Assume first that y is a vertex. For each simplex $S \in \mathcal{S}(y)$ let $f(S)$ be the ridge with all the vertices of S but x and y. $f(S)$ is a ridge of conv R_{k-1} not incident to x and hence a ridge of $\text{conv}(R_{k-1} \setminus \{p\})$. Also, $f(S)$ is visible from x and y and hence a ridge of B. Moreover, if a facet in B is y–visible then a facet in B incident to a ridge $f(S)$ for some $S \in \mathcal{S}(y)$ is y–visible. Thus if we maintain the correspondence (via a dictionary for $(d-2)$–tuples representing ridges) between the ridges of removed simplices and ridges of B, we can find a y–visible facet in B in time proportional to the number of simplices in $\mathcal{S}(y)$.

For non–vertices y one has to work harder. Assume first that y is contained in a simplex $S \in \mathcal{S}(x)$. Let $S = (F, x)$, and let O be the intersection of F with the line through x and y. Locate y by a walk along $\overline{Oy}$ starting at O. Assume next that y is contained in a simplex $S \in \mathcal{S}(z)$ for some $z \in R(x)$. The ridge $f(S)$ of S with all vertices but x and y is a ridge of B when point z is reinserted and hence the facet spanned by $f(S)$ and z is added to T when point z is reinserted. Let O be the intersection of that facet with the line through x and y. Locate y by a walk along $\overline{Oy}$ starting at O.

Lemma 15 *The walk along $\overline{Oy}$ traverses only newly constructed simplices whose base facet is y–visible.*

If the points in X are in convex position, as when computing Voronoi diagrams in one less dimension [Ede87, section 1.8], all points y are vertices and hence the search time is well covered by the time to construct new simplices.

Having reinserted the points in $R(x)$ the cavity created by the removal of x is filled. A traversal of the new simplices and the boundary of the cavity allows to match the new simplices with the old simplices sharing a common facet. This completes the update step.

In summary, the cost of the removal of x is bounded by the sum of the following quantities:

(1) $\min(n, |R(x)| \log\log n)$,

(2) $O(d^3)$ times the number of removed and newly constructed simplices,

(3) $O(d^2)$ times the sum over all points $y \in R(x)$ of the number of y–visible facets ever contained in B.

If the points are in convex position, then the sum of the first two quantities suffices.

3.4 Analysis of Deletions

We analyze the cost of a deletion under the assumption that the points were inserted in random order and that a random point is deleted.

Lemma 16 *The expected number of removed simplices is bounded by*

$$\sum_{i \leq n} d(d+1) f_i / (i \cdot n)$$

and the expected number of new simplices is no larger.

Proof: The expected number of simplices in $T(\pi)$ is $C_n - f_n$ and the expected number of simplices in $T(\pi \backslash i)$ is $C_{n-1} - f_{n-1}$ according to Theorem 10 and Lemma 1. Also each simplex of $T(\pi)$ has $d+1$ vertices and therefore the expected number of removed simplices is $(d+1)(C_n - f_n)/n$. The expected number of new simplices is thus $(C_{n-1} - f_{n-1}) - (C_n - f_n - (d+1)(C_n - f_n)/n)$ which is no larger than the number of removes simplices. The bound now follows from Theorem 3 and Lemma 2.

∎

Lemma 17 *The expected size of $R(x)$ is bounded by*

$$(d+1)\left(2+d\sum_{i\le n} f_i/(i\cdot n)\right).$$

Proof: Let $R_1(x)$ be the set of points $y \in R(x)$ which are vertices of T and let $R_2(x) = R(x) \setminus R_1(x)$. To bound $|R_1(x)|$, observe that $|R_1(x)|$ is at most d plus the number of destroyed simplices. Thus

$$E\left[|R_1(x)|\right] \le d + \sum_{i\le n} d(d+1) f_i/(i\cdot n).$$

To bound $|R_2(x)|$, observe that each non-vertex y is incident to exactly one simplex (recall that our points are in general position) and that x is the vertex of such a simplex with probability $(d+1)/n$. Thus

$$E\left[|R_2(x)|\right] \le n(d+1)/n = (d+1).$$

∎

We next bound the sum over $y \in R(x)$ of the number of y-visible facets ever contained in B. Such a facet is either a y- and x-visible facet of $\operatorname{conv} R_{i-1}$ (recall that $x = x_i$) or a newly constructed base facet visible to y and x.

Lemma 18 (a) *The expected number of facets of* $\operatorname{conv} R_{i-1}$ *visible to x_i and x_j summed over $j > i$ and for random i is $(E|G| - E|H| + f_n)/n$.*

(b) *The expected number of new base facets visible to x_j summed over $j > i$ and for random i is $E|G(\pi\backslash i) \setminus G(\pi)|$.*

Proof: Part (a) follows from Lemma 9 and part (b) is obvious. ∎

Theorem 19 *The expected time to delete a random point from the convex hull of n points (constructed by random insertions) is*

$$O\left(\min\left\{n, \left(d + d^2 \sum_{i\le n} f_i/(i\cdot n)\right) \log\log n\right\} + d^5 \sum_{i\le n} f_i/(i\cdot n) + d^5 \sum_{2\le i\le n} f_i/(i(i-1))\right).$$

If the points are in convex position, then time

$$O\left(\min\left\{n, \left(d + d^2 \sum_{i\le n} f_i/(i\cdot n)\right) \log\log n\right\} + d^5 \sum_{i\le n} f_i/(i\cdot n)\right)$$

suffices.

Proof: This follows immediately from the summary at the end of § 3.3, Lemmas 16 to 18, and Theorems 3 and 8. ∎

We have $f_i = O(i^{\lfloor d/2\rfloor})$. A deletion from a convex hull in $\mathbb{R}^3$ therefore takes time $O(\log n)$ and a deletion from a Voronoi diagram in $\mathbb{R}^2$ takes time $O(\log\log n)$. For $d \ge 4$, a deletion from a convex hull in $\mathbb{R}^d$ and a Voronoi diagram in $\mathbb{R}^{d-1}$ takes time $O(n^{\lfloor d/2\rfloor - 1})$. We note also that for many natural probability distributions, the expected complexity of the hull of random points satisfies $f_r = O(r)$ for fixed d. For such point sets, a random deletion requires $O(\log n)$ expected time.

References

[BDS+] J.D. Boissonnat, O. Devillers, R. Schott, M. Teillaud, and M. Yvinec. Applications of random sampling to on-line algorithms in computational geometry. *Discrete and Computational Geometry*. To be published. Available as Technical Report INRIA 1285. Abstract published in IMACS 91 in Dublin.

[CEG+91] B. Chazelle, H. Edelsbrunner, L.J. Guibas, M. Sharir, and J. Snoeyink. Computing a face in an arrangement of line segments. *2nd Ann. ACM-SIAM Symp. on Discrete Algorithms*, pages 441 – 448, 1991.

[CS89] K. L. Clarkson and P. W. Shor. Applications of random sampling in computational geometry, II. *Journal of Discrete and Computational Geometry*, pages 387–421, 1989.

[Dev] O. Devillers. Randomization yields simple $o(n \log^* n)$ algorithms for difficult $\omega(n)$ problems. *International Journal on Computational Geometry and Aplications*. To be published. Full paper available as Technical Report INRIA 1412. Abstract published in the Third Canadian Conference on Computational Geometry 1991 in Vancouver.

[DMT91] O. Devillers, S. Meiser, and M. Teillaud. Fully dynamic Delaunay triangulation in logarithmic expected time per operation. In *WADS 91*, volume LNCS 519. Springer Verlag, 1991. Full version available as Technical Report INRIA 1349.

[Ede87] H. Edelsbrunner. *Algorithms in Combinatorial Geometry*. Springer Berlin-Heidelberg, 1987.

[GKS90] L.J. Guibas, D.E. Knuth, and M. Sharir. Randomized incremental construction of Delaunay and Voronoi diagrams. *Proc. of ICALP*, pages 414 – 431, 1990. also to appear in Algorithmica.

[MMO91] K. Mehlhorn, St. Meiser, and C. Ò'Dunlaing. On the construction of abstract voronoi diagrams. *Discrete Comput. Geom.*, 6:211 – 224, 1991.

[MN90] K. Mehlhorn and St. Näher. Bounded ordered dictionaries in $O(\log\log\ n)$ time and $O(n)$ space. *Information Processing Letters*, 35:183 – 189, 1990.

[Mul88] K. Mulmuley. A fast planar partition algorithm, I. *Proc. of the 29th FOCS*, pages 580–589, 1988.

[Mul91a] K. Mulmuley. Randomized multidimensional search trees: dynamic sampling. *Proc. ACM Symposium on Computational Geometry*, 1991.

[Mul91b] K. Mulmuley. Randomized multidimensional search trees: further results in dynamic sampling. *32nd IEEE FOCS*, pages 216 – 227, 1991.

[Mul91c] K. Mulmuley. Randomized multidimensional search trees: Lazy balancing and dynamic shuffling. *32nd IEEE FOCS*, pages 180 – 194, 1991.

[Sch91] O. Schwarzkopf. Dynamic maintenance of geometric structures made easy. *32nd IEEE FOCS*, pages 197 – 206, 1991.

[Sei90] R. Seidel. Linear programming and convex hulls made easy. *Proc. 6th Ann. ACM Symp. Computational Geometry*, pages 211 – 215, 1990.

[Sei91] R. Seidel. A simple and fast incremental randomized algorithm for computing trapezoidal decompositions and for triangulating polygons. *Computational Geometry: Theory and Applications*, 1:51 – 64, 1991.

[Sto87] A. J. Stolfi. Oriented projective geometry (extended abstract). *Proceedings of the 3rd Annual ACM Symp. on Computational Geometry*, pages 76–85, 1987.

[vKZ77] P. van Emde Boas, R. Kaas, and E. Zijlstra. Design and implementation of an efficient priority queue. *Math. Systems Theory*, 10:99 – 127, 1977.

Enclosing many boxes by an optimal pair of boxes *

Bruno Becker[†] Paolo Giulio Franciosa[‡] Stephan Gschwind[†]
Thomas Ohler[†] Gerald Thiemt[†] Peter Widmayer[†]

Keywords: computational geometry, covering problems, axis-parallel rectangles

Abstract

We look at the problem: Given a set M of n d-dimensional intervals, find two d-dimensional intervals S, T, such that all intervals in M are enclosed by S or by T, the distribution is balanced and the intervals S and T fulfill a geometric criterion, e.g. like minimum area sum. Up to now no polynomial time algorithm was known for that problem. We present an $O(dn \log n + d^2 n^{2d-1})$ algorithm for finding an optimal solution.

1 Introduction

Throughout the years, several fast heuristics have been proposed for a combinatorial optimization problem that is important in the area of spatial data structures. Given a set of (axis-parallel) rectangles in the plane, the problem is to find two rectangles, say S and T, such that each given rectangle is enclosed by S or by T (or both), each of S and T enclose at least a certain number of given rectangles, and S and T together minimize some measure, e.g. the sum of their areas. This problem must be solved whenever an overflowing page of an R-tree or its variants must be split into two pages [BKSS90, Gre89, Gut84]. The bound on the number of rectangles enclosed by S and by T serves to balance the split; S and T are the bounding boxes for the two pages after the split. There is a number of suggestions on what to minimize in order to achieve a good data structure performance, apart from the sum of areas, e.g. the area of the overlap, the sum of perimeters, and combinations of these. Surprisingly, quite a sorry state of affairs seems to have been accepted: almost all heuristics compute approximate solutions that can be arbitrarily bad as compared with the optimal solution, even though no lower bounds on the problem complexity have been established — the only comment in this direction says that an exhaustive search for an optimal solution takes time proportional to 2^n for n rectangles [Gut84].

In this paper, we show that the optimal solution for all of the variants of the problem investigated in the literature can be obtained in polynomial time. We propose a conceptually simple algorithm that is easy to implement and runs in optimal time for the one-dimensional case and in time $O(n^3)$ for two-dimensional rectangles; for comparison, the better (in terms of data structure performance)

*Partially supported by grant Wi810/2-5 of the Deutsche Forschungsgemeinschaft DFG, and by the ESPRIT II BRA Project 3075 ALCOM of the European Community.

[†]Institut für Informatik, Universität Freiburg, Rheinstraße 10–12, D-7800 Freiburg

[‡]Università "La Sapienza", Dipartimento di Informatica e Sistemistica, via Salaria 113, I-00198 Roma, Italy. Work done while the author was partially supported by Istituto di Analisi dei Sistemi ed Informatica, CNR, Viale Manzoni 30, I-00185 Roma, Italy

of the two originally proposed heuristics takes time $O(n^3)$. Our algorithm is general enough to work for any dimension and any objective function in a fairly large class. For $d \geq 2$ dimensions, it takes time $O(d^2 n^{2d-1})$ in the worst case, using linear space. Furthermore, we propose a variation of our algorithm that is faster in low-dimensional cases, if the value of the objective function is sufficiently low.

The problem attacked in this paper can be considered as a member of a family of *covering and approximation* problems. For instance, if we drop the requirement that each given rectangle be enclosed by S or by T, and instead merely request that each rectangle be enclosed by the union of S and T, the problem changes its nature considerably [BFG+91]. As a fringe benefit, our algorithm slightly improves on the best solution known for a special case of another problem in the family: for n input points instead of rectangles, a minimum perimeter rectangle that contains k points can be found with higher-order Voronoi diagrams in time $O(k^2 n \log n)$ [AIKS89], whereas our algorithm solves this problem in time $O(n^3)$, an improvement at least for $k = \Theta(n)$.

We describe the enclosing problem formally in chapter 2 where we also explain the most important ideas of our solution. In chapter 3 we propose an efficient algorithm for the problem and inspect its performance in detail. We show in chapter 4 that $O(n \log n)$ is a lower time bound and that our algorithm is optimal in the one–dimensional case. Chapter 5 describes a further development of the algorithm, which can in special cases result in an enhanced time performance for low dimensions.

2 The Minimum Measure Enclosing Problem

For any positive integer d, a *d-dimensional (closed) interval* I is written as a $2d$-tuple $(I_1, I_2, \ldots, I_{2d-1}, I_{2d})$, where $I_j \in \mathcal{R}$ for each $j \in \{1, \ldots, 2d\}$, and for each $k \in \{1, \ldots, d\}$, we have $I_{2k-1} \leq I_{2k}$. Let $\mathcal{I}^{2d} \subset \mathcal{R}^{2d}$ denote the set of all d-dimensional intervals. I_{2k-1} is the *lower*, I_{2k} the *upper boundary* of interval I in dimension k. In the sequel, we restrict ourselves to intervals of dimension d, unless the dimension is stated explicitly. An interval J *encloses* an interval I, denoted by $I \subseteq J$, if $I_{2k-1} \geq J_{2k-1}$ and $I_{2k} \leq J_{2k}$ for each $k \in \{1, \ldots, d\}$; if at least one of the inequalities is strict, J *strictly encloses* I, for short $I \subset J$. The *bounding box bbox(M)* of a set M of intervals is the unique interval for which $I \subseteq bbox(M)$ for all $I \in M$, and there is no other interval J with $J \subset bbox(M)$ and $I \subseteq J$ for all $I \in M$; in this sense, $bbox(M)$ is the smallest interval enclosing all intervals in M. For an interval S, let $M_S := \{I | I \in M, I \subseteq S\}$ be the set of intervals in M that are enclosed by S, and let $|M_S|$ be their number. For the set $\mathcal{I}^{2d}$ of all possible d-dimensional intervals, let a *combined measure* function of *two* d-dimensional intervals be any function in the set $F := \{f | f : \mathcal{I}^{2d} \times \mathcal{I}^{2d} \rightarrow \mathcal{R}_+; \forall I, I', J \in \mathcal{I}^{2d} : f(I,J) = f(J,I)$ and $(I \subseteq I' \Rightarrow f(I,J) \leq f(I',J))\}$. Note that all measure functions that are of interest for splitting pages in data structures [BKSS90, Gre89, Gut84] belong to set F.

The Minimum Measure Enclosing Problem now can be defined as follows:

Minimum Measure Enclosing Problem ME(M, b): Given a set M of n d-dimensional intervals and a positive integer b, with $b \leq \lfloor n/2 \rfloor$, compute two intervals S and T with minimum measure such that

1. every interval in M is enclosed by S or by T (*enclosing condition*).

2. $|M_S| \geq b, |M_T| \geq b$ (*balance condition*).

To solve this problem, first observe that the boundaries of the intervals S, T of an optimal solution must be chosen from the set of boundary values of the intervals in M, i.e., for each boundary S_k of S there is an interval $J \in M$ whose boundary J_k equals S_k. This can be shown as follows. If *bbox*(M_S) is strictly enclosed by S, a solution of smaller measure can be found by shrinking S to *bbox*(M_S). The same statement and argument holds for T. A pair of intervals is called an *admissible solution*, if it fulfills the enclosing condition, the balance condition, and its boundaries are chosen from the respective set of boundaries of the intervals in M. An interval pair is called *optimal solution*, if it is an admissible solution and its measure is minimum.

For simplicity of the presentation, we assume that the intervals are in general position, i.e., they have pairwise different boundaries in each dimension. This is no loss of generality, since identical boundary values can be made distinct by shrinking the i-th given interval by $i \cdot \varepsilon$ at each boundary, for $\varepsilon > 0$ and $i = 1, \ldots, n$. It can be shown that for sufficiently small ε, this transformation preserves all admissible solutions in such a way that the relative order of the measure is unchanged among different admissible solutions, and hence any optimal solution is preserved.

A polynomial time algorithm

A brute force algorithm creates at least all instances of S with boundaries matching boundaries of some $J \in M$ by a combination of all possible values for each of the $2d$ boundaries. This can be done in time $O(n^{2d})$. T is created analogously. For all pairs (S, T) the enclosing condition and balance condition can be checked in time $O(d \cdot n)$. Not every measure function in our class F can be computed fast, but many of the interesting ones, like sum of the areas of S and T, sum of the perimeters of S and T, area of the union of S and T, perimeter of the union of S and T, area of the intersection of S and T, perimeter of the intersection of S and T, and combinations of these, can be computed in time $O(d)$. Since the complexity of evaluating other functions in F is not our primary concern, we assume throughout the paper that $O(d)$ time is enough for computing the value of the objective function for given S and T. Hence, the brute force algorithm needs total time of $O(d \cdot n^{4d+1})$.

A first improvement

Obviously, many useless interval pairs are built by this algorithm, and many pairs of boundaries do not even define intervals at all. Clearly, after S has been created, T should enclose at least all those intervals in M which are not enclosed by S. For this purpose we define the *complement* $I^c(M)$ of an interval I with respect to a set of intervals M (I^c for short) as the bounding box of those intervals of M that are not enclosed by I. Then we can state the following:

Lemma 2.1 *For any optimal solution (S, T) of ME(M, b) there is $T = S^c$ or $S = T^c$.*

Proof. Assume (S, T) is an optimal solution with $T \neq S^c$ and $S \neq T^c$. Because both S^c and T^c are bounding boxes, $T \subset S^c$ or $S \subset T^c$ implies that not each interval of M is enclosed by S or by T. Therefore, we only need to consider the case $T \supset S^c$ and $S \supset T^c$. Then, S^c and T^c each enclose less than b intervals, because otherwise either (S, S^c) or (T^c, T) would be an admissible solution with smaller measure. Hence, S and T each enclose more than $n - b \geq b$ intervals. T strictly encloses S^c, i.e., at least one of its boundaries differs from the boundaries of S^c. Therefore, an interval i in M exists that is enclosed by T and not by S^c, and hence is enclosed by S. Because T encloses more than b intervals, it is possible to shrink T by modifying this boundary without enlarging S or violating the balance condition. Therefore, (S,T) cannot be an optimal solution. □

Now we can modify the brute force algorithm slightly. As before, all admissible instances of S are created by the combination of boundary values. But then, for each instance S we compute T as the complement of S. For each instance of S, each interval J in M is inspected; if J is not enclosed by S, T is enlarged so as to enclose J. In this way, time $O(d \cdot n)$ is sufficient to create the complement $T = S^c$. The balance condition can be checked by keeping track of the numbers of intervals in S and T. Because of symmetry it is not necessary to create all admissible intervals T and S as the complement of T. The overall time complexity of this algorithm is therefore $O(d \cdot n^{2d+1})$.

A second improvement

Another observation can be made: Let I denote the bounding box of M. If (S,T) is an optimal solution, then each boundary of I must be matched by a boundary of either S or T (but not both, because of the general position assumption and Lemma 2.1). If one interval of S and T, say T, equals I, then it must be the complement of an interval S that has no boundary in common with I. In this case, S must enclose exactly b intervals, because otherwise S could be shrunk without violating the enclosing or the balance condition.

This fact can be used for a second improvement of the algorithm. We distinguish whether S has at least one boundary in common with I, or no boundary of S matches a boundary of I. In the first case, the instances of S can be created by equating one boundary of S with the corresponding boundary of I and combining all other $(2d-1)$ boundaries of S as before. This results in no more than $2d \cdot n^{2d-1}$ different instances of S, a change in the number of instances by a factor of $O(d/n)$. In the second case, S is restricted to enclose exactly b intervals. In this case, the number of instances of S decreases, because $2d-1$ given boundaries for S uniquely determine the last boundary. We leave the details to the next section, where we also describe how to compute the complementary T efficiently.

3 An Efficient Algorithm

Let us now describe an algorithm for the d-dimensional $ME(M,b)$. Let n be the number of intervals in M and I the bounding box of M. We say an admissible solution (S,T) of the $ME(M,b)$ is of *type 1*, if one or more boundaries of S equal the corresponding boundaries of I and T is the complement of S. An admissible solution (S,T), where S has no common boundary with I, therefore the complementary T is equal to I and S encloses exactly b intervals, is said to be of *type 2*. The first algorithm we propose computes admissible solutions of *type 1* in a first and admissible solutions of *type 2* in a second part.

Algorithm 1: *Compute an optimal solution* (S,T) *for* $ME(M,b)$.

Part 1:

(1) **for** $j := 1$ **to** $2d$ **do**
  let $S_j := I_j$;
  { one boundary of S is fixed, $2d-1$ boundaries left to be fixed }
(2) choose an arbitrary $k \in \{1,..,2d\}, k \neq j$; $\{S_k$ will be fixed last$\}$
(3) **forall** combinations of admissible boundary values for
    all boundaries of S excluding S_j and S_k **do**
   { $2d-1$ boundaries of S fixed, boundary S_k left to be fixed }
(4)   **foreach** admissible boundary value for S_k **do**
    { all boundaries of S are fixed }
(5)   let T be S^c;
(6)   **if** both S and T enclose at least b intervals each **then**
    { (S,T) is an admissible solution }
(7)   keep track of the minimum measure of the (S,T)
      computed so far

Part 2:

(8) **foreach** candidate S that has no boundary in common
    with I and that encloses exactly b intervals **do**
  { T as complement of S is always equal to I }
  { $|M_T| = n$, $|M_S| = b$, balance condition is fulfilled }
(9) keep track of minimum measure interval S

Before filling in the details in this algorithm, let us describe the data structure on which we base our operations.

Data Structure, Space and Preprocessing Complexity

Altogether the n intervals in M have $2dn$ different lower and upper boundaries. The boundary values are stored in $2d$ linear, doubly linked boundary lists $B_1, B_2, \ldots, B_{2d}$. The kth boundary of an interval $J \in M$ is stored in the list B_k, for all $k \in \{1,\ldots,2d\}$. We need the values of each list to be sorted, lists with lower boundaries (lists with odd index) in ascending order, lists with upper boundaries (lists with even index) in descending order. The bounding interval $I = bbox(M)$ is then represented by the first elements of the boundary lists. For the purpose of simplifying the description of loops, we append to each boundary list one more element: At the end of each lower boundary list B_{2k-1} there is an element with the value of I_{2k}, and at the end of each upper boundary list B_{2k} there is an element with the value of I_{2k-1}. Elements of the same interval in different lists are said to *correspond* to each other. The set M is stored as a set L_M of n intervals, where each interval is an array containing $2d$ pointers to the corresponding elements in the boundary lists. In our algorithm, we want to access the boundary value of a given interval known from some interval list element and a given boundary index in constant time. To this end, we store with each element of each boundary list a pointer to the corresponding interval (array) in the interval set L_M. Because we need to mark an interval of M for different purposes, e.g.

to assign it to S or T, each element in L_M contains space for two binary markers. S and T are realized as arrays of $2d$ pointers to boundaries in the boundary lists, so called S- and T-pointers. Hence, there exist one set L_M with space $O(d)$ per element and $2d$ boundary lists with space $O(1)$ per element. In each of these lists, there are $n+1$ elements. The representation of S and T consumes space $O(d)$. Altogether, we need space $O(dn)$; that is, the required working space is linear in the length of the input.

Since the boundary lists must be sorted, we need the preprocessing time $O(dn \log n)$ for the $2d$ boundary lists.

Details of Algorithm 1

On the basis of precomputed interval boundary lists, let us now describe how the high-level operations of Algorithm 1 can be performed efficiently.

In Part 1, fixing one boundary S_j to be I_j amounts to setting the pointer of S_j to the beginning of list B_j, a constant time operation. In line (3), all requested combinations of boundary values can be obtained by repeatedly traversing all necessary $2d-2$ boundary lists; this amounts to a constant time operation for each combination of boundary values. In the innermost loop at line (4), the last boundary of S gets fixed. At the beginning of the loop, the pointer S_k is initialized to the last element of B_k; it is moved through list B_k to the first element during the loop. If w.l.o.g. k is an odd index, the last element of B_k equals the first element of B_{k+1}. This can be interpreted as S being flat in the dimension of k and therefore not enclosing any interval of M. As the pointer S_k moves to the beginning of B_k, S gets enlarged in each step of the loop.

The computation of $T := S^c$ in line (5) and the checking of the balance condition in line (6) can be done incrementally as follows. Because S does not enclose any interval at the beginning of the innermost loop, the complement T of S must enclose all intervals of M. Therefore, initially T equals I. At each step of this loop, S grows; by growing, S may or may not enclose a new interval J. Whenever S encloses a new interval in T that touches T's boundary, T shrinks; otherwise, T is unchanged. To control the shrinking of T use for each interval J in L_M a binary marker, which tells wether or not S covers J. If T shrinks, it may only shrink to boundaries of intervals, which are not enclosed by S, i.e. that are not marked.

The balance condition is checked on the fly by counting the intervals of M that are enclosed by S and those enclosed by T. The number of intervals enclosed by S is maintained in a counter. This counter needs to be incremented whenever S encloses an interval as a result of a growth step; initially, it is zero. Similarly, a counter for the number of intervals enclosed by T is decremented whenever, as a result of a shrink step, an interval is no longer enclosed by T. However, the test whether a given interval leaves a shrinking T costs $\theta(d)$ computation steps, if it is performed by comparing all $\theta(d)$ boundary values. To get the complexity of this test down to $\theta(1)$ – an essential step for the overall efficiency of our algorithm –, we make use of the fact that each interval leaves T only once: a second binary marker for each interval indicates whether or not the interval is enclosed by T.

In more detail, the computation of the complement of S and the verification of the balance condition can be initialized before line (4) as follows:

```
remove all marks assigning intervals to S;
Counter_S:=0;
initialize the T-pointers to the first elements of the boundary lists;
mark all intervals to be assigned to T;
Counter_T:=n;
```

The body of the innermost loop now can be refined as:

```
{ let J denote the interval corresponding to the new
      boundary value of S_k chosen for S in the innermost loop }
if   J is enclosed by S   then
      mark interval J to be assigned to S;
      increment Counter_S;
      foreach   boundary T_i that equals the corresponding boundary J_i  do
          { change the boundary T_i }
          repeat
              if   the T_i-pointer refers to an interval J' assigned to T   then
                  decrement Counter_T;
                  remove the mark that assigns J' to T;
              move the T_i-pointer to the next element in its boundary list
          until   the T_i-pointer refers to an interval not assigned to S
else
      { the boundaries of T do not change }
```

The assignment of an interval J to S may cause a T-pointer not only to move to the next element of its list, but to skip over several boundaries of intervals that were assigned to S before J. But every T-pointer moves through its list in only one direction and therefore visits each of the n elements at most once during the execution of the innermost loop. Therefore, each T-pointer is moved in $O(n)$ steps through its list. For $2d$ lists, altogether $O(dn)$ steps are sufficient.

To analyze the time that suffices for Part 1 of Algorithm 1, note that the body of the outer loop runs exactly $2d$ times. Since there are n^{2d-2} combinations of the $2d-2$ boundary values for each looping, $O(dn^{2d-2})$ steps altogether suffice for all of Part 1 except the innermost loop times. Under the assumption that the evaluation of the objective function takes time $O(d)$, the innermost loop needs time $O(nd)$. Hence, Part 1 needs time $O(d^2n^{2d-1})$.

In Part 2 of Algorithm 1, we generate all candidates of S that have no boundary in common with I and enclose exactly b intervals. Hence, all $2d$ boundaries of S need to be fixed. But if $2d-1$ boundaries are already fixed, the last variable boundary must be placed such that exactly b intervals of M are enclosed by S. Therefore, we need only $2d-1$ nested loops to fix all boundaries. Figure 1 shows different instances of S in a two-dimensional case with already fixed lower and left boundary. While the right boundary moves to the right, the upper boundary moves down.

Similar to Part 1, all combinations of boundaries except S_{2d-2} and S_{2d} can be fixed by repeatedly traversing the necessary $2d-2$ lists. For each of these combinations, the boundary values of S_{2d-2} and S_{2d} are fixed as described below.

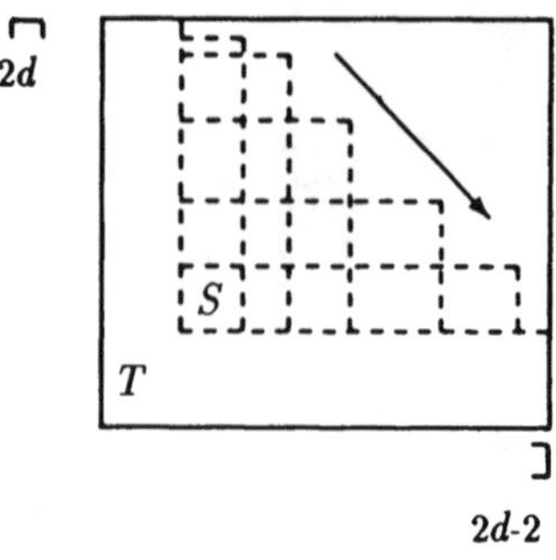

Figure 1

We define an interval or an element of a boundary list to be *active* with respect to some partially fixed S, if the already fixed boundaries of S enclose the corresponding boundaries of the interval. That is, active intervals are the only candidates that may be enclosed by the completed interval S. After having fixed all but two boundaries of S, all intervals can be marked as active or not active in time $O(dn)$. To do the marking, one of the two binary markers used in Part 1 for assigning intervals to S or T can be used here. If less than b intervals are active, S need not be completed. Otherwise, S is completed as follows:

let S_{2d} point to the first element of B_{2d}
{ i.e. highest upper boundary in dimension d };
let S_{2d-2} point to the next to last element of B_{2d-2}
{ i.e. lowest upper boundary in dimension $d-1$ };
move S_{2d-2} to the b-th active element of B_{2d-2},
assign the corresponding b active intervals to S, and
move S_{2d} to the boundary of the first interval in B_{2d} assigned to S;
keep track of the minimum measure interval S
while the first element of B_{2d-2} is not reached **do**
move S_{2d-2} to the next active element towards the front of the list
and assign the corresponding active interval to S;
move S_{2d} to the next element towards the rear of the list
whose corresponding active interval is assigned to S;
keep track of the minimum measure interval S

To analyze the time complexity of Part 2 of Algorithm 1, observe that n^{2d-2} combinations of boundary values are fixed first. The remaining two boundary values are fixed by traversing two lists in time $O(n)$. The complement T of S equals I, and the balance condition is fulfilled. To compute the measure for a given S we spend time $O(d)$, as explained for Part 1 (for quite a few functions, measure computation can even be performed incrementally, so time $O(1)$ suffices). Hence, Part 2 takes time $O(dn^{2d-1})$. Together with the time complexity of preprocessing and that of Part 1, we have:

Theorem 3.1 *The d-dimensional Minimum Measure Enclosing Problem ME(M, b) can be solved in time* $O(dn\log n + d^2n^{2d-1})$ *and space* $O(dn)$.

Corollary 3.2 *The one-dimensional Minimum Measure Enclosing Problem $ME(M,b)$ can be solved in time $O(n \log n)$ and space $O(n)$.*

Corollary 3.3 *The two-dimensional Minimum Measure Enclosing Problem $ME(M,b)$ can be solved in time $O(n^3)$ and space $O(n)$.*

4 A lower bound

To obtain a lower bound for the problem $ME(M,b)$ we look at the one-dimensional problem. Let I denote the bounding box of a set M of one-dimensional intervals. Because each of the boundaries of I has to be matched by either S or T, there are two different types of solutions (see Figure 2): Either each interval has one boundary in common with I, or one interval equals I and the other one encloses exactly b intervals.

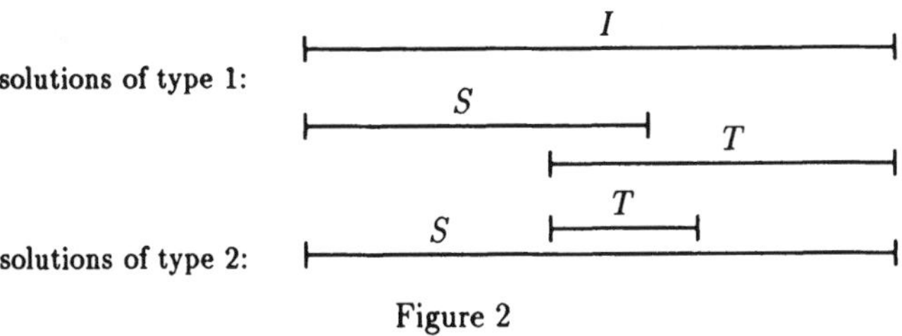

Figure 2

Lemma 4.1 *$ME(M,b)$ needs $\Omega(n \log n)$ operations in the algebraic decision tree model.*

Proof. We show that the *ε-Closeness Problem* [PS85] can be transformed in linear time to the Minimum Measure Enclosing Problem with balance constant $b = 2$. The ε-Closeness Problem is defined as follows: Given a set M of n real numbers $x_1, ..., x_n$ and a real number $\varepsilon > 0$, determine whether in M there exist x_i, x_j, $i \neq j$, such that $| x_i - x_j | < \varepsilon$. Let x_{max} denote the maximum and x_{min} denote the minimum in M. Transform each real number x_k into an interval $[x_k, x_k + \varepsilon]$, $k = 1, \ldots, n$, and build one interval $J = [x_{min}, x_{max} + \varepsilon]$. Finding the maximum and minimum value and the transformation of the given reals into intervals can be done in linear time.

An optimal solution (S,T) for the Minimum Measure Enclosing Problem with these $n+1$ intervals and balance constant $b = 2$ must be of type 2 (see Figure 2), because J has to be enclosed. Let S be the interval equal to I. Hence, T must enclose exactly two intervals. The measure of T is less than $2 \cdot \varepsilon$ if and only if there are two numbers in M with a distance less than ε. It follows that we can solve the ε-Closeness Problem, which needs $\Omega(n \log n)$ operations in the algebraic decision tree model, with the Minimum Measure Enclosing Problem. □

Together with Corollary 3.2 we get

Theorem 4.2 *The one-dimensional Minimum Measure Enclosing Problem $ME(M,b)$ can be solved in optimal time $\Theta(n \log n)$.*

5 A faster algorithm for low dimension

In the previous section, the bound on the runtime of Algorithm 1 has been shown to match the lower bound for the one-dimensional problem. In any dimension, Algorithm 1 always actually needs the runtime stated as an upper bound in Theorem 3.1. In this section, we propose an alternative way of solving the Minimum Measure Enclosing Problem, and we show that this second algorithm may sometimes be faster, especially for low dimension, a practically interesting case.

To see how a finer grained case analysis leads to a potentially more efficient algorithm, let us recall the working principle of Algorithm 1. There, we create various instances of the first enclosing interval S, and we compute the second enclosing interval T as the complement of S. By fixing one boundary of S to the corresponding boundary of $I = bbox(M)$, on the one hand we reduce the number of created instances of S by a factor of $1/n$; on the other hand, since there are $2d$ choices for the boundary to be fixed to I, the net effect on the number of created instances of S is only a factor of $2d/n$. In addition, if S must enclose exactly b intervals, the number of created instances of S is reduced by a factor of $1/n$, since there is just one correct position for the last free upper boundary, and this position can be determined incrementally at constant cost.

Based on these observations, Algorithm 1 distinguishes between instances of S with one fixed boundary and $2d-1$ free boundaries and instances with $2d$ free boundaries. This distinction is useful only because in the latter case it is sufficient to create instances of S with exactly b enclosed intervals, whereas in the former case, for each instance of S the complement must be computed. This immediately raises the crucial question of whether the number of created instances of S can be cut down further, similar to the case of exactly b enclosed intervals. To this end, let us distinguish the instances of S according to the number of fixed boundaries. Algorithmically, let us create all instances of S with exactly k free boundaries, for $k = 0, 1, \ldots, 2d$; as before, all combinations of values for free boundaries, together with the fixed ones, yield the instances of S.

To see which of these instances can be created efficiently or need not even be created, recall that the complement of S with all boundaries fixed is the empty interval, and that the complement of S with no boundary fixed is I. In the general case, let us look at an instance of S with exactly k free boundaries; that is, exactly $2d-k$ boundaries of $T = S^c$ are not boundaries of I. We can see that not all instances of S with exactly k free boundaries need to be created by distinguishing between $k \leq d$ and $k > d$: the complements of S for $k \leq d$ coincide with instances of S for $k > d$. Let us now restrict k to $k \leq d$, and consider an instance of S with $2d-k$ free boundaries, together with its complement $T = S^c$. Due to the restriction on k, T has already been created (as a former instance of S, so to speak), and if (T, T^c) is an admissible solution, its measure certainly cannot be worse than the measure of (T, S), since $T^c \subseteq S$. If (T, T^c) is not an admissible solution, T^c does not enclose b intervals, and therefore we can shrink S so as to enclose just b intervals. We conclude that for more than d free boundaries it is sufficient to create instances of S with exactly b enclosed intervals. More formally, as a refinement of Lemma 2.1, we have:

Lemma 5.1 *For any optimal solution (S,T) of $ME(M,b)$, either ($S \neq T^c$ and $T = S^c$ and $|M_S| = b$) or ($S = T^c$ and $T \neq S^c$ and $|M_T| = b$) or ($S = T^c$ and $T = S^c$)*

Proof. For any optimal solution (S,T) by Lemma 2.1 there is $S = T^c$ or $T = S^c$. This leads to three cases, namely ($S \neq T^c$ and $T = S^c$) or ($S = T^c$ and $T \neq S^c$) or ($S = T^c$ and $T = S^c$). For the first case we will show that $|M_S| = b$. Assume $|M_S| > b$. Because $S \neq T^c$, S encloses T^c and has at least one boundary different from T^c. Therefore, some interval J in M is enclosed by S and not by T^c, hence it is enclosed by T. Because $|M_S| > b$, it is possible to shrink S by modifying one boundary without increasing T, and the balance condition is still satisfied, a contradiction. A symmetric argument holds for the second case and the third case is the same as in Lemma 2.1. □

An algorithm making use of Lemma 5.1 can then be formulated as follows:

Algorithm 2: *Minimum Measure Enclosure with more fixed boundaries*

```
Part 1:
(1)  for k := 0 to d do
(2)      foreach combination of 2d - k boundaries of S fixed to I do
(3)          forall combinations of boundary values for the remaining
                 k boundaries of S do
(4)              let T be S^c;
(5)              if both S and T enclose at least b intervals then
(6)                  keep track of minimum measure;
Part 2:
(7)  for k := d + 1 to 2d - 1 do
(8)      foreach combination of 2d - k boundaries of S fixed to I do
(9)          forall combinations of boundary values for the remaining
                 k boundaries of S such that S encloses exactly
                 b intervals do
(10)             let T be S^c;
                 { balance condition is always fulfilled }
(11)             keep track of minimum measure;
Part 3:
(12) if minimum measure computed so far > measure(I,empty) then
(13)     forall instances S with 2d free boundaries
             that enclose exactly b intervals do
(14)         let T = I;
             { balance condition is always fulfilled }
(15)         keep track of minimum measure
```

For the efficiency of Algorithm 2, the fast computation of the complement in line (10) is crucial. Similar to Algorithm 1, all instances of S are created by combining free boundary values (line (9)). Within the innermost loop (see also Figure 1), the two last free boundary values are fixed simultaneously. To compute the complement incrementally, we make use of the fact that each interval may enter S once and may leave S once within each innermost looping. An interval that enters S may shrink the complement $T = S^c$, and an interval that leaves S may grow T. Since each interval can shrink and grow T at most once, an implementation similar to the one described for Algorithm 1 achieves $O(d \cdot n)$ runtime to compute the complements of n instances of S. By adding the costs for preprocessing and for Parts 1, 2 and 3 of Algorithm 2, we get:

Theorem 5.2 *The d-dimensional Minimum Measure Enclosing Problem ME(M,b) can be solved in time*

$$O\left(d \cdot n \log n + d \cdot \left(\sum_{k=0}^{d}\left(\binom{2d}{k} \cdot n^k\right) + \sum_{k=d+1}^{2d-1}\left(\binom{2d}{k} \cdot n^{k-1}\right) + n^{2d-1}\right)\right).$$

Because Part 3 of Algorithm 2 is the most costly part, we get for the special case $d = 2$:

Corollary 5.3 *The two-dimensional Minimum Measure Enclosing Problem can be solved in time* $O(n^2)$ *if the minimum measure is smaller than the measure of (bbox(M),empty).*

Even though in some situations, such as those characterized in Corollary 5.3, Algorithm 2 is faster than Algorithm 1, in others the reverse may be true. Especially in cases where d is large compared to n, Algorithm 1 is clearly superior. That is, the proper choice of the algorithm must be made on the basis of the problem input at hand.

References

[AIKS89] A. Aggarwal, H. Imai, N. Katoh, and S. Suri. Finding k points with minimum diameter and related problems. *Proc. of the 5th Annual ACM Symposium on Computational Geometry*, 5:283–291, 1989.

[BFG+91] B. Becker, P. G. Franciosa, S. Gschwind, T. Ohler, G. Thiemt, and P. Widmayer. An optimal algorithm for approximating a set of rectangles by two minimum area rectangles. In 7th *Workshop on Computational Geometry, Bern, Lecture Notes in Computer Science*, page to appear, 1991.

[BKSS90] N. Beckmann, H.P. Kriegel, R. Schneider, and B. Seeger. The r*-tree: An efficient and robust access method for points and rectangles. *ACM SIGMOD International Conf. on Management of Data*, 19:322–331, 1990.

[Gre89] D. Greene. An implementation and performance analysis of spatial access methods. *Fifth IEEE International Conference on Data Engineering, Los Angeles*, 5:606–615, 1989.

[Gut84] A. Guttman. R-trees: A dynamic index structure for spatial searching. *ACM SIGMOD International Conf. on Management of Data*, 12:47–57, 1984.

[PS85] F. P. Preparata and M. I. Shamos. *Computational Geometry, an Introduction*. Springer-Verlag, New York, 1985.

VLSI

Performance Driven k-Layer Wiring*

Michael Kaufmann Paul Molitor Wolfgang Vogelgesang

Max-Planck-Institut
für Informatik
Im Stadtwald
W-6600 Saarbrücken 11, FRG

Fachbereich Informatik
Universität des Saarlandes
Im Stadtwald
W-6600 Saarbrücken 11, FRG

Abstract

Given a grid based wire layout, the objective of the layer assignment problem we investigate in this paper is to minimize the interconnect delay by taking into account the conductivity of interconnection wires and vias. For MOS circuits with two or more layers for interconnections, the problem is shown to be NP-hard. It remains NP-hard for the case of two layers having the same conductivity ($PDW(=)_2$) and for the case of two layers with extremely different conductivities ($PDW(\neq)_2$). However, $PDW(\neq)_2$ can be reduced to a generalized flow problem which gives hope of good approximating heuristics not only for $PDW(\neq)_2$ but also for the general problem. $PDW(\neq)_2$ can be solved in polynomial time when the routing grid is sufficiently fine.

1 Introduction

A main feature of VLSI-design is the routing aspect. A routing problem is given by a routing region, a set of multiterminal nets (the demands) and the number of available layers. Usually, the routing region is a square grid. Square grid graphs model the popular constraint that wires can only run horizontally and vertically in a natural way. Cross-overs, junctions, knock-knees, bends and vias may only be placed on vertices of the routing region.

Typically, routing consists of two steps: *wire layout* and *wiring* (*layer assignment*). The first step determines the placement of the routing segments. A 'planar' wire layout is obtained. In the wiring step, this layout is converted into an actual three-dimensional configuration of wires. Formally, wiring is formulated as follows: Given a wire layout L, assign to each routing segment of L an unique layer such that the segments are electrically connected in the proper way. Adjacent edges of the same wire lying on different layers are connected with vertical connections, called *vias* or *contacts*.

A wire layout L is called *k-layer wirable* if there exists a wiring of L in k layers. The decision problem as to whether a given wire layout is k-layer wirable and the problem of minimally stretching a wire layout in one or two dimensions to obtain a k-layer wirable layout has been exhaustively investigated in the literature which is surveyed in [9].

Traditionally, the objective of the layer assignment has been to minimize the total number of vias in the routing. The motivation of this optimization problem consists of the fact that excess vias lead to decreased performance of the electrical circuit, decreased yield of the manufacturing process and increased amount of area required for interconnections. This problem is called *constrained via minimization problem* denoted by CVM_k when k layers are available. Results on this problem can be found in [5, 8, 9, 10, 11].

⋆ **Research supported by DFG, SFB124, TP B1/B2, VLSI Entwurfsmethoden und Parallelität.**

Unfortunately, the solutions of the CVM_k problem are global (cf. [11]), i.e., in the process of minimizing the total number of vias we may burden one signal net with an excessive number of vias and/or we may embed critical signal nets in layers with poor conductivity. Since the optimization of the electrical performance of interconnections in VLSI chips is more important than the minimization of the number of vias, layer assignments are needed which minimize signal delays through interconnection lines. This problem – we call it *performance driven k-layer wiring* (PDW_k) – was recently formulated by Ciesielski [1]. He investigates the case of two available layers and generalizes Pinter's model [11] in order to handle the problem. Unfortunately, his reduction results in a maximum cut problem for nonplanar graphs, which is known to be NP-hard. It remained open whether the problem itself is NP-hard.

This paper presents some new results on PDW_k. First, we handle the case that all the layers have the same conductivity and show that this restricted PDW_k problem is NP-hard for any $k \geq 2$. For $k \geq 3$, the result can be derived from the fact that the decision problem as to whether a k-layer wirable wire layout can be wired in k layers without using a via is known to be NP-hard. The NP-hardness for $k=2$ is proved by reducing 3SAT to it. In the third section, we handle the case of two layers which have extremely different conductivities. By reducing 3SAT to this problem, we show its NP-hardness. However, the problem can be reduced to a generalized flow problem [2] so that approximating flow algorithms may lead to good heuristics. For the special case that a via may be placed between every two adjacent cross-overs and/or knock-knees, the corresponding flow problem can be solved in polynomial time.

In both cases, we use a very simple model to determine the signal delays. However, the methods can easily be extended to the delay model of Rubinstein, Penfield, and Horowitz [13] by weighting the vias and the wire segments which are embedded in the poor layer, respectively, by load capacities. For the sake of simplicity, we omit these constructions.

Beneath the results presented by Ciesielski [1], this is the first paper handling performance driven wiring, so that many problems of this area are not investigated and remain open.

2 When the layers have same conductivity ...

Assume that all the layers have same conductivity. Because a via decreases the performance of the electrical circuit, i.e., increases the corresponding net delay, we have to investigate the following problem in order to solve this restricted version of PDW_k, which we denote by $PDW(=)_k$.

Instance: Let L be a k-layer wirable wire layout consisting of m nets $s_1, \ldots, s_m$, and let $v = (v_1, \ldots, v_m) \in \mathbb{N}_0^m$ be a vector consisting of m nonnegative numbers.

Question: Is there a k-layer wiring of L such that for any $i \in \{1, \ldots, m\}$ the number w_i of vias inserted on signal net s_i is less than or equal to v_i?

2.1 $PDW(=)_k$ is NP-hard for $k \geq 3$

First, let us concentrate on the decision problem $0-CVM_k$ as to whether a k-layer wirable wire layout ($k \geq 3$) can be wired in k layers without using one via (i.e. $v = (0, \ldots, 0)$). A straightforward reduction from Graph-3-Colorability of planar graphs, which is known to be NP-complete, shows that $0-CVM_k$ is NP-complete, too (cf. [8]). Thus, it is NP-complete to decide whether there is a k-layer wiring of a given k-wirable wire layout L with maximum number of vias inserted on a signal net equal to 0.

Theorem 2.1 *$PDW(=)_k$ is NP-hard for $k \geq 3$.*

2.2 $PDW(=)_2$ is NP-hard as well

For the case of two available layers of same conductivity, we have to recall the results proved in [8] (cf. also [3]).

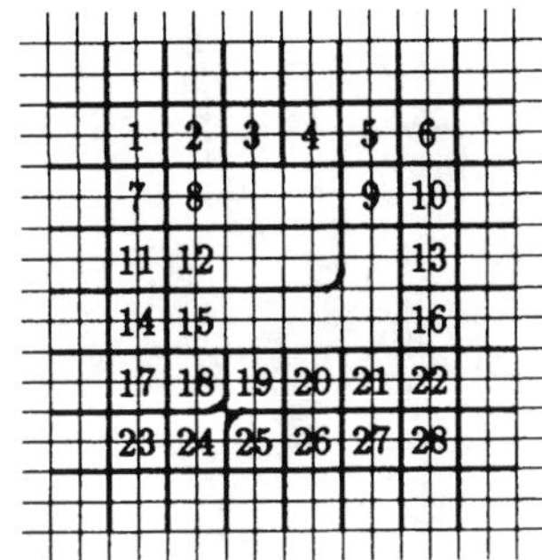

Figure 1: Wire layout L. The thin lines represent the square grid.

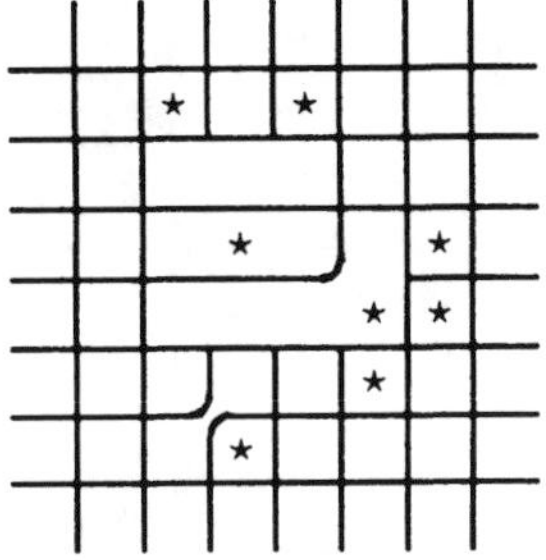

Figure 2: The odd faces of L are marked by $\star$.

2.2.1 A graph-theoretical approach to 2-layer wiring

An example shall illustrate the 2-layer wiring problem. Formal definitions of the notions can be found in [5, 8].
Let L be the wire layout shown in figure 1. L divides the plane into elementary regions (*faces*), 28 inner faces and the outer face. The inner faces are enumerated in figure 1. Each face r has a boundary. We call a face r *odd*, if its boundary cannot be embedded into two layers without using a via. (In the case of knock-knee mode layouts, this definition is equivalent to the following description: a face r is odd if there is an odd number of vertices on its boundary, where the two adjacent wire segments enclosing r cannot be placed in the same layer.) Otherwise the face is called *even*. For our example, in figure 2 all the odd faces of L are marked by $\star$.

Lemma 2.2 [8] *Let L be a wire layout. There is a 2-layer wiring of L needing no (further) via if and only if each face of L is even.*

Therefore, in order to obtain a 2-layer wiring, the odd faces have to be transformed into even ones by inserting vias on their boundaries. (We consider a via as a vertex where the layer has to be changed.) Obviously, a via may only be located on a grid vertex which is touched by at most one signal net.
In our example, if a via is placed, e.g., on the vertical wire segment (on the grid vertex) between face 2 and face 3, face 2 which was odd becomes even and face 3 which was even becomes odd. The insertion of a via between face 3 and face 4 transforms these two odd faces into even ones. To sum it up, we have transformed the two odd faces 2 and 4 into even ones by joining them by a path of vias. We have *married* face 2 and face 4.

Lemma 2.3 [8] *In any layout L, there is an even number of odd faces.*

So, marrying each odd face to exactly one other odd face results in a layout where each face is even. By lemma 2.2, this new layout is 2-layer wirable without further vias. Figure 3 shows a 2-layer wiring of L. Here, faces 2 and 4, 13 and 16, 12 and 15, and 21 and 25 are joined. Note that the faces 21 and 25 are joined by a path of vias which runs across the outer face.
In order to find the marriage paths, we need the notion of the dual graph of a wire layout.

Definition 2.1 *Let L be a wire layout. The* dual graph $G = (V, E)$ *of L is defined as follows. The set V of vertices consists of the faces of L, and for any grid vertex g adjacent to two different faces f and f' onto which a via may be located there is an edge e_g in E connecting f and f'.*

The dual graph of our example layout L which is shown in figure 4 is obviously connected (in the sense of graph theory [7]). Therefore, it is easy to find a perfect matching of the odd faces which

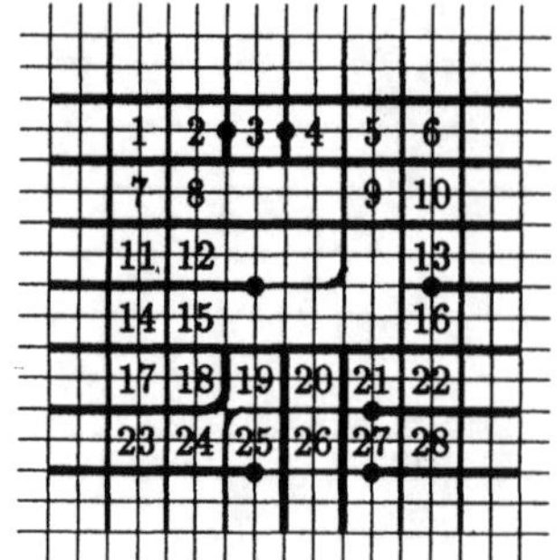

Figure 3: Two-layer wiring of L. The thick lines represent the one layer, the extra thick lines represent the other one.

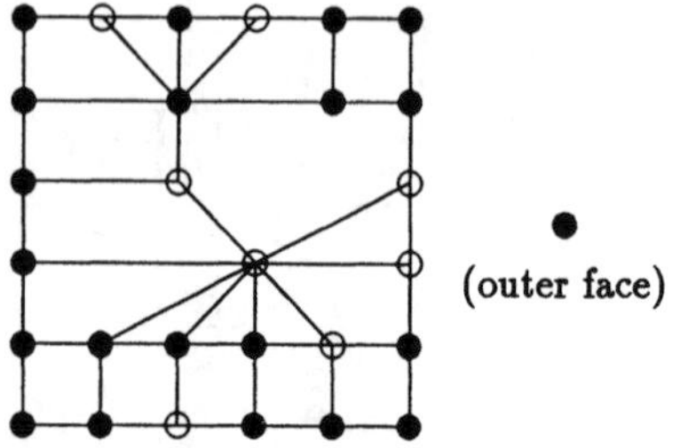

Figure 4: The dual graph of the wire layout of figure 1. The vertices marked by ∘ and • represent the odd and even faces, respectively. For the sake of clarity, multiple edges and edges adjacent to the outer face are omitted.

determines the marriages. The situation is more complex in the case of dual graphs consisting of more than one connected component.

Lemma 2.4 *Let L be a wire layout and let G be its dual graph. There is a 2-layer wiring of L if and only if each connected component of G contains an even number of odd faces.*

In the following, we confine our attention to 2-wirable wire layouts.

2.2.2 A formal approach to $PDW(=)_2$

Let $s_1, \ldots, s_m$ be the signal nets of wire layout L. We assign a vector $w(e)=(w(e)_1, \ldots, w(e)_m) \in \{0,1\}^m$ of dimension m to each edge e of the dual graph with $w(e)_j = 1$ if and only if edge e corresponds to a wire segment of signal net s_j (cf. definition 2.1). Obviously, one can now assign a vector $w(p)=(w(p)_1, \ldots, w(p)_m)$ to each path $p = (e_1, e_2, \ldots, e_r)$ in the dual graph by defining

$$w(p)_j = \sum_{k=1}^{r} w(e_k)_j.$$

Thus, setting $v=(\mu, \ldots, \mu)$ results in the following restricted version of $PDW(=)_2$:
Let L be a wire layout containing $2q$ odd faces. Find q simple paths $p_1, \ldots, p_q$ in the dual graph of L marrying the $2q$ odd faces of L such that

$$max\left\{\sum_{k=1}^{q} w(p_k)_j \mid j = 1, \ldots, m\right\} \leq \mu.$$

For the case '$q=1$', i.e., L only contains two odd faces, the problem can be reduced to the problem:
Let $G=(V,E)$ be a planar graph, $w : E \longrightarrow \mathbb{N}_0^m$ a function assigning a vector $w(e)$ of dimension m to each edge e, and let a, b be two distinguished vertices in V. Find a path p connecting a and b such that

$$max\left\{w(p)_j \mid j = 1, \ldots, m\right\} \leq \mu.$$

We denote this problem by MINMAX(1), because an algorithm with polynomial running time which solves the problem can easily be extended in a polynomial running time algorithm which determines the minimal value of μ, such that there is a k-layer wiring with the above property.

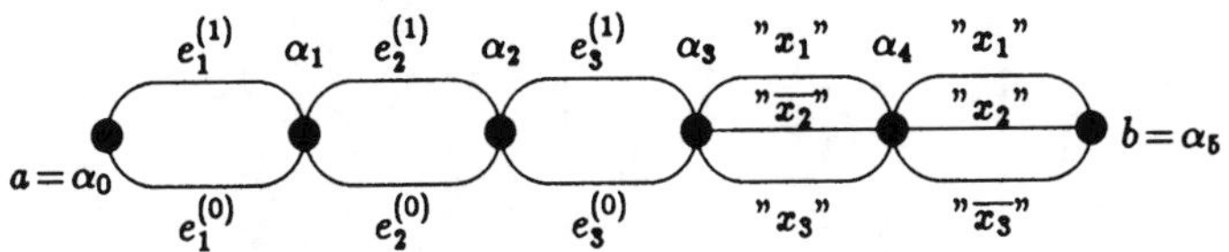

Figure 5: Reduction of 3SAT to MINMAX(1) illustrated by the example $X = \{x_1, x_2, x_3\}$ and $C=\{x_1 \vee \overline{x_2} \vee x_3, x_1 \vee x_2 \vee \overline{x_3}\}$.

2.2.3 MINMAX(1) is NP-hard

We transform 3SAT to MINMAX(1).
Let $X = \{x_1, \ldots, x_n\}$ be the set of variables and $C = \{c_1, \ldots, c_h\}$ the set of clauses defining any instance of 3SAT.
We will construct a MINMAX(1)-problem $P = (G, w, a, b)$ such that there is a path p in G connecting the two given vertices a and b with $w(p)_j \leq 3h$ for any component of the vector $w(p) \in \mathrm{N}^{2n}$ if and only if C is satisfiable.

The graph consists of two parts which correspond to the variables and the clauses, respectively. The set X of the variables is represented by the subgraph $G_{var} = (V_{var}, E_{var})$ with $V_{var} = \{\alpha_0, \alpha_1, \ldots, \alpha_n\}$. For any $i \in \{1, \ldots, n\}$, there are two edges $e_i^{(0)}$ and $e_i^{(1)}$ connecting the vertices α_{i-1} and α_i. To each edge e of E_{var} we assign a vector $w(e)$ of dimension $2n$ defined as follows:

$$w(e)_j = \begin{cases} 2h & , \text{ if } j \leq n \text{ and } e = e_j^{(0)} \\ 2h & , \text{ if } j > n \text{ and } e = e_{j-n}^{(1)} \\ 0 & , \text{ otherwise.} \end{cases}$$

The h clauses are represented by the subgraph $G_{cl} = (V_{cl}, E_{cl})$ with $V_{cl} = \{\alpha_n, \ldots, \alpha_{n+h}\}$. Clause c_i is represented by a *clause-bundle* consisting of the vertices α_{n+i-1} and α_{n+i} and the three edges f_i, f_i', f_i'' between these two vertices. We explain the construction by an example. Let $c_i = x_j \vee \overline{x_u} \vee x_s$ be the i-th clause. Edge f_i represents x_j and is assigned the vector $w(f_i) \in \{1,2\}^{2n}$ with $w(f_i)_k = 2$ if and only if $k = j$. (Marking the edges with vectors $\in \{1,2\}^{2n}$ instead of $\{0,1\}^{2n}$ is necessary because of the construction described in section 2.2.4.) Edge f_i' represents the item $\overline{x_u}$ and is assigned the vector $w(f_i') \in \{1,2\}^{2n}$ with $w(f_i')_k = 2$ if and only if $k = n + u$. Analogously, edge f_i'' represents the item x_s of c_i and is assigned the vector $w(f_i'') \in \{1,2\}^{2n}$ with $w(f_i'')_k = 2$ if and only if $k = s$.

Figure 5 illustrates this construction for the case of $X = \{x_1, x_2, x_3\}$ and $C = \{x_1 \vee \overline{x_2} \vee x_3, x_1 \vee x_2 \vee \overline{x_3}\}$.
Setting $a = \alpha_0$ and $b = \alpha_{n+h}$, the equivalence of the problems is easy to understand.

Lemma 2.5 *C is satisfiable if and only if there is a path p connecting a and b in G such that for the path vector $w(p)$ the inequations $w(p)_k \leq 3h$ hold for all $k \in \{1, \ldots, 2n\}$.*

Proof
Assume that p is a path connecting a and b with $w(p)_k \leq 3h$ for all $k \in \{1, \ldots, 2n\}$. The first n edges of the path determine an assignment $A: X \longrightarrow \{0,1\}$. If $e_i^{(1)}$ is on the path, x_i is set to 1. Otherwise, x_i is set to 0. Note that p contains either $e_i^{(0)}$ or $e_i^{(1)}$. In the first case ($x_i = 1$), the edge contributes the weight $2h$ to the $(n+i)$-th component of the path vector. Since in each of the h clause-bundles each component of the path vector is incremented, the path does not use an edge representing $\overline{x_i}$ in a clause-bundle. In the case '$x_i = 0$', the path does not use an edge representing x_i in a clause-bundle. Thus, each clause is satisfiable with respect to A and A is a satisfying assignment for C.

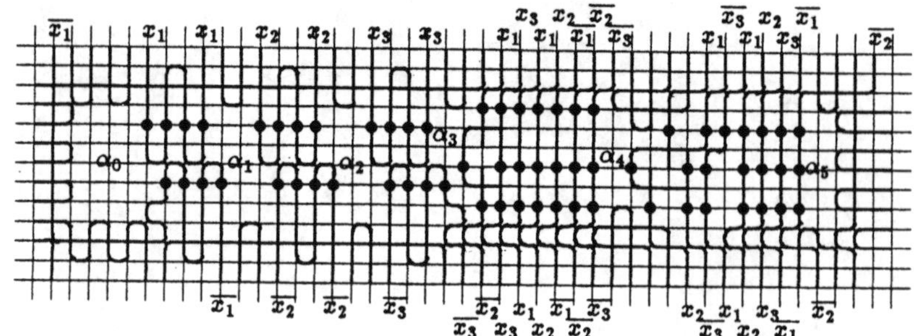

Figure 6: Reduction of 3SAT to $PDW(=)_2$ for knock-knee mode wire layouts illustrated by the example $X=\{x_1,x_2,x_3\}$ and $C=\{x_1\vee\overline{x_2}\vee x_3, x_1\vee x_2\vee\overline{x_3}\}$. The possible (interesting) locations for vias which can be reached from the faces $a=\alpha_0$ and $b=\alpha_5$ are marked by $\bullet$.

Now we prove the other direction. Assume that $A: X\longrightarrow\{0,1\}$ is a satisfying assignment for C. Then, we construct the following path $p=(e_1,\ldots,e_{n+h})$ where $e_i=e_i^{(A(x_i))}$ for $i\in\{1,\ldots,n\}$ and edge e_{n+j} ($j\in\{1,\ldots,h\}$) is taken from one of the true-setting items of clause c_j. Obviously, the path vector $w(p)$ is such that $w(p)_k\leq 3h$ for $k=1,\ldots,2n$. □

Corollary 2.6 *MINMAX(1) is NP-hard.*

2.2.4 $PDW(=)_2$ is NP-hard

To prove the NP-hardness of $PDW(=)_2$, we have to show that each graph used in the proof of lemma 2.5 can be interpreted as the dual graph of a wire layout L containing exactly 2 odd faces, namely those which are represented by the vertices α_0 and α_{n+h}.

The wire layout in knock-knee mode corresponding to the graph of figure 5 is shown in figure 6. This instance of 3SAT consists of 3 variables x_1, x_2, x_3 and 2 clauses, namely $(x_1\vee\overline{x_2}\vee x_3)$ and $(x_1\vee x_2\vee\overline{x_3})$, i.e., $h=2$. It is not hard to check that α_0 and α_5 are the only odd faces. The possible locations for vias which can be reached from $a=\alpha_0$ or $b=\alpha_5$ (and which are of interest) are marked by $\bullet$.

Obviously, α_0 has to be married to α_5 by a path of vias crossing $\alpha_1,\ldots,\alpha_4$. It is easy to see that for each i either the wire x_i or the wire $\overline{x_i}$ has to be cut $4(=2h)$ times by any marriage path. Connecting face α_3 to α_4, each signal net of $\{x_1,x_2,x_3,\overline{x_1},\overline{x_2},\overline{x_3}\}$ is cut at least once. One of these, namely one appearing in the corresponding clause $(x_1\vee\overline{x_2}\vee x_3)$, is cut twice. A similar statement holds for a subpath connecting α_4 and α_5. Thus, the 'weighted' dual graph of the wire layout of figure 6 is that shown in figure 5.

It is easy to see that this construction can be generalized to any instance of 3SAT. For the sake of brevity, we omit this general construction as well as the wire layout which is round this construction to make available the signal nets $x_1,\overline{x_1},\ldots,x_n,\overline{x_n}$ at the different positions.

Theorem 2.7 *$PDW(=)_2$ is NP-hard.*

In [4] we have shown that this result is valid for wire layouts in overlap mode, too. It remains as an open problem whether such a construction can also be given for Manhattan mode wire layouts.

3 When the layers have extremely different conductivities ...

Now, let us consider the case that two layers l_1 and l_2 are available to embed the wire layout which have the following property: the conductivity of l_1 is extremely better than the conductivity of l_2

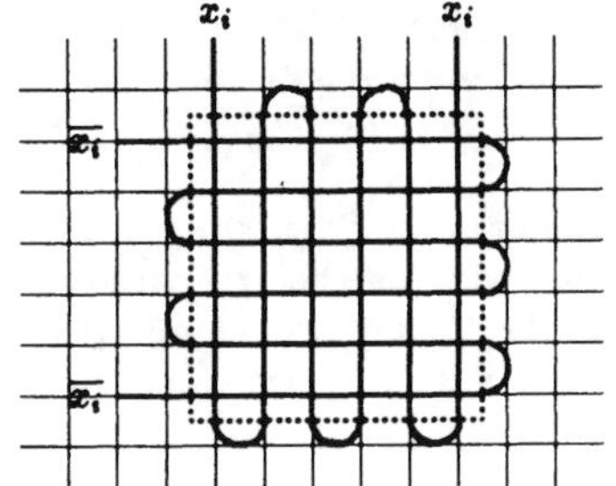

Figure 7: Variable-cluster belonging to variable x_i. It consists of two wires representing literal $\overline{x_i}$ and x_i, respectively. The cluster itself is marked by the dashed box.

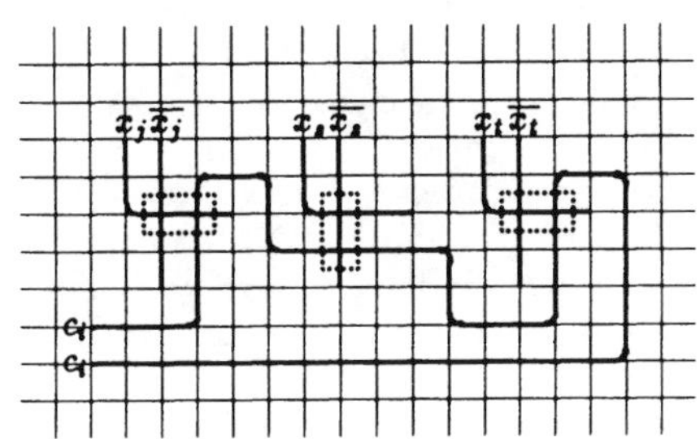

Figure 8: Clause-clusters belonging to clause $c_i = x_j \vee \overline{x_s} \vee x_t$. The three clusters are marked by dashed boxes.

(we will denote l_2 by 'the poor layer'), and the costs (with respect to signal delays) of a via are much lower than the costs caused by embedding one unit of the corresponding wire in the poor layer l_2. A unit length is defined to be the distance between two adjacent vertices of the routing grid.

Let f be a 2-layer wiring of a wire layout L and s_i a signal net of L. We denote by u_i the number of unit lengths of wire s_i which are embedded in the poor layer (by f). Then, our special case of PDW_2, which will be denoted by $PDW(\neq)_2$, is equivalent to the following problem.

Instance: Let L be a 2-layer wirable wire layout consisting of m nets $s_1, \ldots, s_m$, and let $v = (v_1, \ldots, v_m) \in \mathbb{N}_0^m$ be a vector consisting of m nonnegative numbers.

Question: Is there a 2-layer wiring of L such that for any $i \in \{1, \ldots, m\}$ the inequation $u_i \leq v_i$ holds?

3.1 $PDW(\neq)_2$ is NP-hard

We need some definitions (cf. [6]). Let L be a wire layout. A *via candidate* (of L) is a vertex of the routing grid, which is touched by at most one wire of L, i.e., which can accommodate one via. A *cluster* is a maximal connected (in the sense of graph theory) part of the routing grid, which does not contain a via candidate. Note that in each cluster, once a wire segment (a wire segment is defined to be a piece of wire connecting two via candidates) is assigned to a certain layer, 2-layer wiring of the rest of the cluster is fixed. Thus, there are only two possible ways to assign the wire segments of a cluster to two layers. Now, we will show the NP-hardness of $PDW(\neq)_2$.

Theorem 3.1 *$PDW(\neq)_2$ is NP-hard.*

Proof

Instead of giving the complete proof, which can be found in [4], we will only show the basic idea of our construction. We transform 3SAT to $PDW(\neq)_2$. Let $X = \{x_1, \ldots, x_n\}$ be the set of variables and $C = \{c_1, \ldots, c_h\}$ the set of clauses defining any instance of 3SAT. Essentially, our construction consists of the following two types of clusters:

1. To any variable x_i, a special cluster (see figure 7) is related, which we denote by x_i-*cluster*. It consists of two wires of length α which represent literal $\overline{x_i}$ and literal x_i, respectively. We denote these wires by $\overline{x_i}$-*wire* and x_i-*wire*.

2. Any clause $c_i \in C$ is represented by a wire layout as shown in figure 8 for the clause $c_i = x_j \vee \overline{x_s} \vee x_t$. It consists of three clusters which represent the appearances of the literals x_j, $\overline{x_s}$ and x_t in the clause. We denote them by (c_i, x_j)-*cluster*, $(c_i, \overline{x_s})$-*cluster* and (c_i, x_t)-*cluster*.

It is obvious that in each x_i-cluster either the x_i-wire or the $\overline{x_i}$-wire has to be put in the poor layer. By choosing α and the v-values appropriately, we can prevent this wire from being embedded in the poor layer in the rest of the wire layout. The embeddings in the x_i-clusters correspond to an assignment A to the variables of X in the following sense: putting x_i in the poor layer is equivalent to setting $A(x_i) = 0$.
In each clause-cluster (c_j, l_k) either the wire l_k or the wires c_j and $\overline{l_k}$ must be put in the poor layer. By choosing the v-values of the wires c_j accordingly, we can force that in at least one of the three clause-clusters (c_j, l_k) of clause c_j the l_k-wire has to be put in the poor layer. This is only possible if $A(l_k) = 1$, i.e., the clause c_j is satisfied by l_k.

3.2 A reduction of $PDW(\neq)_2$ to a generalized flow problem

3.2.1 Basic definitions

A *directed network* $N = (V, E, cap)$ is given by a directed graph $G = (V, E)$ and a function $cap: E \longrightarrow \mathbb{N}_0$. A *generalized flow problem* (cf. [2]) consists of a network $N = (V, E, cap)$, one designated vertex s (the source) and a *gain function* $g: E \longrightarrow \mathbb{R}^+$. A function $gf: E \longrightarrow \mathbb{R}^+$ is a *generalized flow* when it satisfies the capacity constraints

$$0 \le gf(e) \le cap(e)$$

for all $e \in E$ and the conservation laws

$$\sum_{e \in in(v)} gf(e) \cdot g(e) = \sum_{e \in out(v)} gf(e)$$

for all $v \in V \setminus \{s\}$, where $in(v)$ ($out(v)$) is the set of edges entering (leaving) v. The gain function can be interpreted as follows: let $e = (v, w)$ be any edge and gf a generalized flow; if $gf(e)$ units leave vertex v towards w then $gf(e) \cdot g(e)$ units arrive at w, i.e., the flow along the edge e is multiplied by the gain $g(e)$.
The *value* of the generalized flow is given by the flows on the edges adjacent to s, i.e.,

$$val(gf) = \sum_{e \in in(s)} gf(e) \cdot g(e) - \sum_{e \in out(s)} gf(e).$$

A positive flow value is obtained when there are *flow-generating cycles*, i.e., cycles with a gain greater than 1, where the gain of a path is defined as the product of the gains of the belonging edges.

Before reducing $PDW(\neq)_2$ to a generalized network flow problem, we introduce some further notations (see figure 9): As seen before, there are two possible 2-layer wirings for every cluster of a wire layout, i.e., for any cluster c_i, there are two states $c_i^{(0)}$ and $c_i^{(1)}$, denoted also by state 0 and 1, respectively. Each state is associated with a (weight-)vector $d(c_i^{(j)}) = (d_{i,1}^{(j)}, \ldots, d_{i,m}^{(j)})$ of dimension m (m is the number of nets), which is interpreted as follows: $d_{i,k}^{(j)}$ is the number of units of net s_k which are embedded in the poor layer if cluster c_i is in state j. We will denote

$$w(c_i^{(j)}) := \sum_{p=1}^{m} d_{i,p}^{(j)}$$

as the *weight* of cluster c_i in state j.

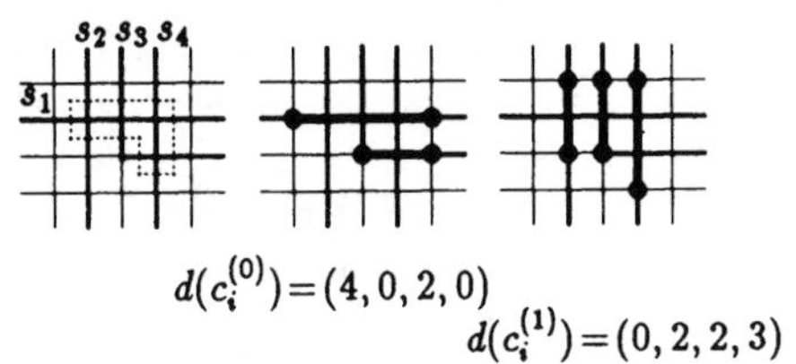

Figure 9: Example for a cluster, the two possible states and the corresponding weight-vectors.

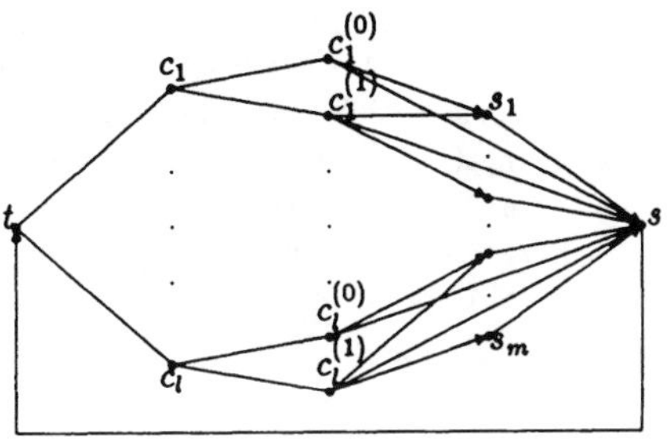

Figure 10: Generalized network flow problem to $PDW(\neq)_2$. The capacities and gains are omitted for the sake of clarity.

3.2.2 The reduction

Now, let us consider the reduction of $PDW(\neq)_2$ to the **integer** version of a generalized network flow problem (V, E, s, cap, g) (see figure 10). V is partitioned into the following sets:

- $V_{cl} = \{c_1, \ldots, c_l\}$ representing the l clusters of the layout ,
- $V'_{cl} = \{c_1^{(0)}, c_1^{(1)}, \ldots, c_l^{(0)}, c_l^{(1)}\}$ representing all states for the clusters ,
- $V_n = \{s_1, \ldots, s_m\}$ representing the signal nets, and
- two distinguished vertices s and t.

These vertices are connected by the following edges $(i = 1, \ldots, l\ ;\ j = 0, 1\ ;\ k = 1, \ldots, m)$:

- $e_i = (t, c_i)$ with $cap(e_i) = g(e_i) = 1$,
- $e_i^{(j)} = (c_i, c_i^{(j)})$ with $cap(e_i^{(j)}) = 1$ and $g(e_i^{(j)}) = \max\{w(c_i^{(0)}), w(c_i^{(1)})\}$,
- $e_{i,k}^{(j)} = (c_i^{(j)}, s_k)$ with $cap(e_{i,k}^{(j)}) = d_{i,k}^{(j)}$ and $g(e_{i,k}^{(j)}) = 1$,
- $e_{i,s}^{(j)} = (c_i^{(j)}, s)$ with $cap(e_{i,s}^{(j)}) = \max\{w(c_i^{(0)}), w(c_i^{(1)})\} - w(c_i^{(j)})$ and $g(e_{i,s}^{(j)}) = 1$,
- $e_{i,s} = (s_i, s)$ with $cap(e_{i,s}) = v_i$ and $g(e_{i,s}) = 1$, and
- $e_{s,t} = (s, t)$ with $cap(e_{s,t}) = l$ and $g(e_{s,t}) = 1$.

If we are only interested in integer flows, then the following statement holds:

Lemma 3.2 *An instance of $PDW(\neq)_2$ is solvable if and only if the corresponding generalized integer flow problem has a maximum flow value of*

$$gf_{max} := \sum_{p=1}^{l} \max\{w(c_p^{(0)}), w(c_p^{(1)})\} - l.$$

Proof

Assume that we have a solvable instance of our wiring problem, i.e., we have to show that there is a generalized flow with value gf_{max}. We send l units from s to t which are distributed to all c_i. According to the wiring, we send the unit of flow arriving at c_i either to $c_i^{(0)}$ or $c_i^{(1)}$. This selected vertex $c_i^{(j)}$ which corresponds to a state with weight $w(c_i^{(j)})$ can send $\max\{w(c_i^{(0)}), w(c_i^{(1)})\} - w(c_i^{(j)})$ units of the arriving $\max\{w(c_i^{(0)}), w(c_i^{(1)})\}$ via edge $e_{i,s}^{(j)}$ directly to s, the other $w(c_i^{(j)})$ units must be sent to the corresponding vertices s_p, which represent those wires, which are burdened by $c_i^{(j)}$. So, our construction guarantees, that each s_p obtains exactly $d(c_{i,p}^{(j)})$ units from the selected $c_i^{(j)}$. Therefore, if the wiring does not violate v, the total arriving flow at any s_p can be sent back to s. Now, it is easy to see that a generalized flow with value gf_{max} is obtained, because l units leave vertex s which are distributed to l flow-generating cycles that send $\sum_{p=1}^{l} \max\{w(c_p^{(0)}), w(c_p^{(1)})\}$ back to s. Because of the capacities of the edges e_i, this value is the maximal flow value.

Assume now that we have a generalized integer flow with value gf_{max}. Then we have l flow-generating cycles containing the vertices of V_{cl}. Because $cap(e_i)=1$ for all $i=1,\ldots,l$ and the flows are integers, for every cluster c_i exactly one of the edges $e_i^{(0)}$ and $e_i^{(1)}$ is selected. This selection induces a wiring of all the clusters. Because the flow satisfies the capacity constraints for the edges $e_{i,s}$, the corresponding wiring is a legal solution for our instance of $PDW(\neq)_2$. □

A fallout of the above reduction and the proof of the NP-hardness of $PDW(\neq)_2$ is

Corollary 3.3 *The integer version of the generalized flow problem is NP-hard.*

3.2.3 Applications of the reduction

If we are interested in minimal signal delays of wirings for a given routing, we could apply the following procedure (for simplicity, we assume that $v=(q,\ldots,q)$):

At first, we look for the minimal q such that the real–valued version of the corresponding generalized flow problem has a maximal flow value of gf_{max}. The maximum flow value can be computed by solving the corresponding dual linear program or by applying the combinatorial algorithm of Goldberg–Plotkin–Tardos [2].

For a flow with that maximal value, it is obvious that one unit of flow arrives at every vertex c_i. Now, we take a look at the flows on the edges $e_i^{(j)}$. For every c_i, the equation

$$gf_{max}(e_i^{(0)}) + gf_{max}(e_i^{(1)}) = 1$$

holds. In order to obtain the desired integer solution which induces a 2-layer wiring for all clusters, we use randomized rounding, i.e., we decide to put cluster c_i in state 0 with probability $gf_{max}(e_i^{(0)})$.

The results of Raghavan [12], who studied this method for *maximum multicommodity flow*, let us hope, that we get solutions close by the optimum. It remains open whether it is possible to specify upper bounds for the difference between approximated solution and optimum, as Raghavan did.

It is worth mentioning that this approximation method can be generalized to the case that there is an exact relation between the conductivities of the two available layers. The definition of the weight-vectors of the clusters has to be extended by taking into consideration not only the wire segments in layer l_2 but also those in layer l_1 and the vias.

A further application of the above reduction consists of a polynomial algorithm for a special case of $PDW(\neq)_2$, namely that the routing grid is so fine that it is possible to place a via between every two adjacent cross–overs and/or knock–knees.

Thus, in order to obtain a 2-layer wiring, it is sufficient to construct a 'bridge' for every cross-over and knock-knee by putting two units of one of the two wires in the poor layer l_2; all other wire segments can be embedded in the preferred layer l_1. Figure 11 illustrates this remark.

Obviously, this special case of $PDW(\neq)_2$ is equivalent to the problem of not burdening one signal net with an excessive number of bridges. If we regard such a bridge as one unit, then in our generalized flow problem, all gains are 1. Furthermore, if we delete the edge connecting s and t, we have a normal flow problem with source t and sink s.

Theorem 3.4 [7] *In a directed network $N=(V,E,cap)$ with $n=|V|$ and $e=|E|$, a maximum flow from a to b can be computed in time $O(n \cdot e)$.*

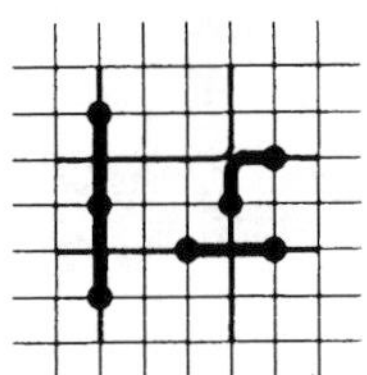

Figure 11: Illustration of *bridges*. The thick lines represent layer l_2.

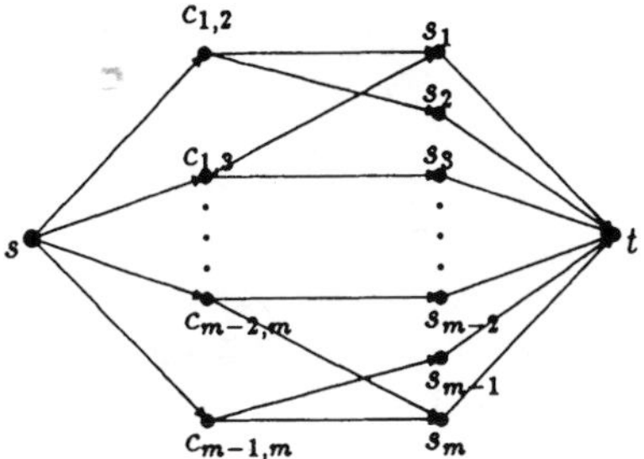

Figure 12: Maximum network flow problem to the special case of $PDW(\neq)_2$. The capacities are omitted because of clarity.

For our graph, it is obvious that $|V| = |E| = O(l + m)$, because in every cluster only two signal nets are involved. Thus, a solution can be computed in time $O((l + m)^2)$. When l exceeds m significantly, we can reduce the running time by combining all clusters where the two relevant wire segments belong to the same signal nets. Then, we get the following flow problem, where $c\{i, j\}$ denotes the number of cross-overs and knock-knees that are built by wire segments of signal net s_i and s_j (see figure 12):

- $V = \{s, t\} \cup \{c_{i,j} | 1 \leq i < j \leq m\} \cup \{s_1, \ldots, s_m\}$.
- there are the following edges:
  - $(s, c_{i,j})$ with capacity $c\{i, j\}$,
  - $(c_{i,j}, s_i)$ and $(c_{i,j}, s_j)$ with capacity $c\{i, j\}$ for both ,
  - (s_i, t) with capacity v_i.

In this case, $|V| = |E| = O(m^2)$. Thus, the following statement holds:

Theorem 3.5 *If the routing grid is sufficiently fine, $PDW(\neq)_2$ is solvable in running time $O((\min\{(l + m), m^2\})^2)$, where l denotes the number of cross-overs and knock-knees.*

4 Conclusions

We have investigated k-layer wiring for layouts taking into account the conductivity of interconnection wires and vias. Special cases have been approximated by some minmax-problems. For the case that all the layers have the same conductivity, the corresponding minmax-problem minimizes the maximum number of vias inserted on a signal net of the wire layout. The problem has been shown to be NP-hard. For the case of two layers with extremely different conductivities and the costs of a via being much lower (with respect to signal delays) than the costs of one unit wire length of the poor layer, the corresponding minmax-problem minimizes the maximum sum of the lengths of those wire segments of a signal net which are embedded in the poor layer. A polynomial running time algorithm is proposed if the wire layout is such that a via may be placed between every two adjacent cross-overs and/or knock-knees. In general, this problem is shown to be NP-hard. But note that the proposed network flow problem may lead to good heuristics for performance driven wiring in the general case.

References

[1] M.J. Ciesielski. Layer assignment for VLSI interconnect delay minimization. *IEEE Transactions on Computer Aided Design*, CAD-8(6):702–707, June 1989.

[2] A.V. Goldberg, E. Tardos, and R.E. Tarjan. Network flow algorithms. Technical Report 860, School of Operations Research and Industrial Engineering, Cornell University, Ithaca, NY 14853-7501, September 1989.

[3] M. Kaufmann and P. Molitor. Minimal stretching of a layout to ensure 2-layer wirability. *INTEGRATION, the VLSI Journal*, December 1991. 14 pages.

[4] M. Kaufmann, P. Molitor, and W. Vogelgesang. Performance driven k-layer wiring. Technical Report TR-14/1990, Sonderforschungsbereich 124 *VLSI Entwurfsmethoden und Parallelität*, Fachbereich Informatik, Universität des Saarlandes, Im Stadtwald, W–6600 Saarbrücken 11, FRG, 1990.

[5] R. Kolla, P. Molitor, and H.G. Osthof. *Einführung in den VLSI-Entwurf.* Leitfäden und Monographien der Informatik. B.G. Teubner Verlag, Stuttgart, 1989.

[6] Y.S. Kuo, T.C. Chern, and W. Shih. Fast algorithm for optimal layer assignment. In *Proceedings of the 25th ACM/IEEE Design Automation Conference (DAC88)*, pages 554–559, June 1988.

[7] K. Mehlhorn. *Data Structures and Algorithms 2: Graph Algorithms and NP-Completeness.* EATCS Monographs on Theoretical Computer Science. Springer-Verlag, Berlin Heidelberg New York Tokio, 1984.

[8] P. Molitor. On the contact minimization problem. In *Proceedings of the 4th Annual Symposium on Theoretical Aspects of Computer Science (STACS87)*, pages 420–431, February 1987.

[9] P. Molitor. A survey on wiring. *EIK Journal of Information Processing and Cybernetics*, EIK 27(1):3–19, 1991.

[10] N.J. Naclerio, S. Masuda, and K. Nakajima. Via minimization for gridless layouts. In *Proceedings of the 24th ACM/IEEE Design Automation Conference (DAC87)*, pages 159–165, June 1987.

[11] R. Pinter. Optimal layer assignment for interconnect. In *Proceedings of the IEEE International Conference on Circuits and Computers*, pages 398–401, September 1982.

[12] P. Raghavan. Probabilistic construction of deterministic algorithms: Approximating packing integer programs. In *Proceedings of the 27th Annual Symposium on Foundations of Computer Science*, pages 10–18, 1986.

[13] J. Rubinstein, P. Penfield, and M.A. Horowitz. Signal delay in rc networks. *IEEE Transactions on Computer–Aided Design*, CAD-2:202–210, July 1983.

Synthesis for Testability: Binary Decision Diagrams

(Extended Abstract)

Bernd Becker
Computer Science Department
J.W.G.-University
6000 Frankfurt
Germany

Abstract

We investigate the testability properties of Boolean circuits derived from (Reduced Ordered) Binary Decision Diagrams. It is shown that BDD-cirucits (or at least) BDD-like circuits are easily testable with respect to different fault models (cellular, stuck-at and path delay fault model). Furthermore the circuits and the test sets can be constructed efficiently.

Keywords VLSI structures, algorithms and data structures, synthesis, (complete, full) testability, fault model

1 Introduction

Binary Decision Diagrams as a data structure for Boolean functions were introduced by Lee [Lee59] in 1959. They have been intensively studied in several differing areas of computer science and electrical engineering: in theoretical computer science, especially in complexity theory (e.g. [Weg87,Mei89]), in electronic design automation, especially in design verification (e.g. [Ake78,Bry86,BRB90]). Here a restricted form of BDD's, namely reduced ordered BDD's, have gained a whitespread use. In this paper we point out that BDD's are not only well suited for design verification which is necessary to prove or at least give some evidence of the correctness of the design, but also may be used to synthesize circuits whose testability can be guaranteed in advance. At first, we want to emphasize the practical relevance of testing problems by making some general remarks. For a detailed treatment of the topic see e.g. [Wun91].

As a result of technological improvements, VLSI electronic circuitry may nowadays contain hundreds of thousands of transistors on a single silicon chip. Even if the chips are correctly designed, i.e., design verification has been done successfully, a non negligible fraction of them will have physical defects caused by imperfections occurring during the manufacturing process (e.g., open connections induced by dust particles). Therefore, there has to be a test phase in which "production" verification is performed, i.e., in which the "good" chips are sorted from the "bad" ones.

Because of the variety of possible defects restrictions on a subset of the possible faults are necessary; these simplifying assumptions based on the experience of many years are manifested in fault models. Since tests are generated to test for the fault mechanisms described by the assumptions the reliability of the chip is at least partly determined by the accuracy and effectiveness of the fault model (measured e.g., in detected physical failures). The classical stuck at fault model (SAFM) is well-known and used throughout the industry [BF76]. It assumes that a defect causes a basic cell input or ouput to be fixed to either 0 or 1. Thus all failures with this effect will be detected by tests for stuck-at faults. It has been observed by many authors that a large amount of defects typical for today's VLSI technologies are not covered by stuck-at faults [BART82,BA83,GCV80]. Other fault models are necessary to overcome this problem, e.g., flexible fault models which verify the correct static behaviour of a combinational circuit based on inductive fault analysis [Mal87,FC88] or dynamic fault models which allow to model stuck open faults or timing issues [Wad78,Smi85]. The strongest cell-based fault model to control the correct static behaviour of a combinational circuit is the cellular fault model (CFM) which tries to completely verify the function of each basic cell in the circuit [Fri73,BS88]. A fault model considering timing issues is the path delay fault model (PDFM), which checks the correct timing of any path from a primary input to a primary output [Smi85]. Correctness in PDFM is a very strong result, it especially implies the non existence of stuck-open faults and gate delay faults. Of course, it is desirable to prove the correctness of a chip under these fault models. However, due to the large costs of the test phase, in practice often only SAFM is considered. Furthermore the investigations are usually based on the single fault assumption, i.e., one assumes that there is at most one fault (according to the considered fault model) in the circuit.

With increasing complexity of VLSI circuits, the costs for the test phase have risen dramatically, at least 25% and up to 60-70% of the total product costs count on testing [Mil85,WDGS86]. Thus, the necessity of new methods for the test phase is evident. Specialists in the field of testing agree that testability issues have to be considered from the very beginning of the design process to control the test costs and to guarantee the testability of the circuit at the end of the manufacturing process. One step in this direction is the use of synthesis algorithms to generate logic designs that not only meet area and/or speed constraints but also are easily testable. For this, it is important to understand the relation between testability and structural properties of Boolean functions and circuit realizations. There has been a lot of work in this direction starting with investigations of testable realizations of Boolean functions in the seventies and leading to the incorporation of testability constraints in commercially available logic synthesis systems today. For an overview see e.g. [BF76,Fuj85,DMNS88].

In this paper we focus on the synthesis of circuits derived from BDD's and their testability aspects. As pointed out before BDD's are often used in design verification. In this case they are available for free in the subsequent phases of the design process. In addition, they can be easily transformed into multiplexer-based Boolean circuits which can for example be used as a first realization during the synthesis process for irregular logic like control circuitry. Thus, it is natural to investigate the testability of these circuits and to ask for synthesis algorithms which enhance testability, if necessary. The relation between BDD's and testability has so far only been studied in the context of functional testing based on BDD's [Ake79]. We show that BDD's lead to circuits with good testability properties. Criteria used in the following to measure testability are

- number of different fault models considered
- single or multiple fault assumption
- complexity of complete test set construction
- number of redundancies and complexity of their detection
- possibility and complexity of circuit modifications to obtain a fully testable circuit

A little bit more in detail the results of this paper can be sumarized as follows: The basis of our investigations is a characterization and (efficient) computation of the testability properties of BDD-circuits with respect to the cellular fault model. Testable and redundant faults can be easily classified. A sufficient condition is given, which either implies the non-existence of redundant faults or at least allows the removal of the redundancies. In this case the resulting BDD-like circuit is fully testable in the cellular fault model.

It follows easily that in a BDD-circuit any (single) stuck-at fault on an input- or output-line of a multiplexer is testable, i.e., a BDD-circuit as a circuit realized with multiplexers is fully testable in the (single) stuck-at model. This must not necessarily be the case, if the multiplexers are replaced by a standard *AND-*, *OR-*, *INVERTER*-based realization: there may be redundant stuck-at 1 faults at the inputs of the *AND* gates. However, it is shown that they correspond to redundant cellular faults, which can be removed in any BDD-circuit. Sufficient conditions are given, that guarantee the full stuck-at testability of the resulting BDD-like circuit. While the above results rely on the usual single fault assumption, for fully stuck-at testable BDD-circuits (or BDD-like circuits) we can even construct test sets, which detect any multiple stuck-at fault. Thus we obtain full testability with respect to multiple faults.

We then consider the path delay fault model to verify the correctness of the circuit with respect to timing constraints. The set of paths which is (robustly) testable with respect to path delay faults is determined. This implies a characterization of the BDD-circuits, which are fully path delay testable.

The remaining part of the paper is structured as follows: in section 2 we provide further definitions from the field of testing including our model for combinational logic circuits. Then we introduce BDD's and BDD-circuits in section 3. In section 4 the testability properties of BDD- and BDD-like circuits are derived. We finish with a discussion of the results and mention open problems.

2 Testing and Fault Models

In this section we give a short introduction to basic notions and concepts of testing which are important for the understanding of this paper.

A *combinational logic circuit* (CLC) over a fixed library is modeled as a directed acyclic graph $C = (V, E)$ with some additional properties: each vertex $v \in V$ is labeled with the name of a basic cell or with the name of a primary input (PI) or primary output (PO). The collection of basic cells available is given in advance by the fixed library. Very often the library STD consisting of the 2-input, 1-output *AND, OR* gate and the 1-input, 1-output inverter *NOT* is used. In this paper we consider circuits constructed with help of basic cells from STD $\cup$ MUXLIB $\cup$ $\{XOR\}$. MUXLIB is a library, which merely consists of several types of multiplexers with (at most) three inputs and one output, *XOR* is a cell with two inputs and one output. The inputs and outputs of each basic cell are ordered. There is an edge (u, v) in E from vertex u to v, if an output pin of the cell associated to u is connected to an input pin of the cell associated to v, i.e., edges contain additional information to specify the pins of the source and sink node they are connected to. Vertices have exactly one incoming edge per input pin. Nodes labeled as PI (PO) have no incoming (outcoming) edges. The *size* of the circuit $|C|$ is as usually given by $|E| + |V|$.

Now assume further, that for each basic cell (n inputs, m outputs) in the library the functional behaviour of the cell is defined by a Boolean function from $\mathbf{B}^n \rightarrow \mathbf{B}^m$. (For STD the functions are well-known, for the multiplexer cells in MUXLIB the functional behaviour is defined in section 3 (see figure 2 and 3). An *XOR* cell realizes the *exclusive or* operation.) A CLC with N PI's and M PO's then defines a Boolean function from $\mathbf{B}^N \rightarrow \mathbf{B}^M$ in the obvious way. For an example of a CLC over the libraries STD and MUXLIB see the circuit C_{MUXLIB} in figure 4 and the circuit C_{STD} in figure 5, respectively. These examples also demonstrate that a circuit over the library MUXLIB can easily be "expanded" into a circuit over STD with identical behaviour: the multiplexer cells have to be substituted by a standard cell realization of a multiplexer over STD. In general, let C_1 be a CLC over a fixed library L_1. Assume that some basic cells in C_1 can be substituted by some subcircuits over a library L_2 without changing the overall behaviour of the subcircuits, then C_2 is called an *expansion* of C_1. (For a detailed definition of this notion see e.g. [BHK*87,BBH*90].)

As mentioned above, even if a CLC is correctly designed, a fraction of them will behave faulty because of physical defects caused by imperfections during the manufacturing process. Fault models which cover a wide range of the possible defects are defined and tests for faults in the fault models are constructed. We give a short description of the fault models considered in this paper.

The Cellular Fault Model (CFM)

In the CFM ([Fri73,BS88] it is assumed that a fault modifies the behavior of exactly one node v in a given CLC C and that the modified behavior is still combinational. Since this fault can be detected by observing the incorrect output values of v for one suitable input combination, it suffices to test for faults of the following kind. A *cellular fault* in C is a tuple $(v, I, X/Y)$, where v is the faulty node (= fault location), I is the input for which v does not behave correctly, and X (Y) is the output of the correct (faulty) node on input I.

The Stuck-at Fault Model (SAFM)

A fault in the SAFM ([BF76]) causes exactly one input or output pin of a node in C to have a fixed constant value (0 or 1) independently of the values applied to the PI's of the circuit. Usually the stuck-at model is only considered on the gate level, i.e., for circuits over the library STD. We have given here an obvious generalization to circuits over arbitrary libraries.

We finish the discussion of CFM and SAFM with some general definitions and remarks on the relation between both fault models. For this let C be any CLC over a fixed library and FM a fault model as defined above. An input t to C is a test for a fault f in FM, iff the output values of C on applying t in the presence of f are different from the output values of C in the fault free case. A fault in FM is *testable*, if there exists a test for this fault. The goal of any test pattern generation process is a *complete* test set for the circuit under test in the considered fault model FM, i.e., a test set that contains a test for each testable fault. It follows easily from the definitions that, given a fixed circuit C, a complete test set in CFM is also complete in SAFM. Thus, the cellular fault model is stronger than the stuck-at fault model. In any case, the construction of complete test sets requires the determination of the faults which are not testable (= *redundant*), even though it is easily seen that in general the detection of redundancies is *co-NP complete*. Redundancies have further unpleasant properties: they may e.g. invalidate tests for testable faults and often correspond to locations of the circuit where area is wasted ([BF76]). All in all, synthesis procedures which result in non redundant circuits or at least circuits with known redundancies are desirable. A node v in C is called *fully testable* in FM, if there does not exist a redundant fault in FM with fault location v. If all nodes in C are fully testable in FM, then C is called *fully testable* in FM.

Now consider a circuit C_2 which results from a circuit C_1 by expansion (see figures 4 and 5). Then one can easily show that a complete test set for C_1 in CFM is also a complete test set for C_2 in CFM (and SAFM). Thus the CFM is more powerful, if the size of the basic cells increases. We call this property the *completeness property* of CFM. Notice, that in SAFM there is a trend in the opposite direction. This is the reason why in general the strongest version of SAFM, i.e., SAFM for circuits over STD is considered.

The Path Delay Fault Model (PDFM)

Whereas the CFM and SAFM are fault models to verify the static behaviour of a circuit, the purpose of delay testing is to ascertain that the circuit under test meets its timing specifications. In the PDFM ([Smi85,SFF89]) it is checked whether the propagation delays of all paths in a given CLC are less than the system clock interval.

For our discussion of the PDFM assume that the considered CLC C is defined over the library STD. A *path* π is given by an alternating sequence of nodes and edges $(v_0, e_0, v_1, \ldots, v_n, e_{n+1}, v_{n+1})$ starting in a PI v_0 and ending in a PO v_{n+1}. Inputs of nodes on the path where no edge e_i of the path ends are called *off-path inputs*.

A *transition* ($0 \to 1$ = *rising* or $1 \to 0$ = *falling*) *propagates along* π, if a sequence of transitions $t_0, t_1, \ldots, t_{n+1}$ occur at the nodes $v_0, v_1, \ldots, v_{n+1}$, such that t_i occurs as a result of t_{i-1}. π has a *path delay fault*, if the actual propagation delay of a (rising or falling) transition along π exceeds the system clock interval. For the detection of a path delay fault a pair of patterns (I_1, I_2) is required rather than a single pattern as in the CFM and SAFM: The *initialization vector* I_1 is applied and all signals of C are allowed to stabilize; then the *propagation vector* I_2 is applied and after the system clock interval the outputs of C are controlled. A two-pattern test is called a *robust test* for a path delay fault on π, if it detects that fault *independently* of all *other delays* in the circuit and all *other delay faults* not located on π. A two-pattern test is called a *non-robust test* for a path delay fault on π, if it detects that fault under the assumption that the off-path inputs of all nodes on π stabilize to their final (non-controlling) values prior to the time, at which the transition propagated along π. (A *controlling* value at the input of a node is the value that completely determines the value at the output, e.g., 1 (0) is the controlling value for *OR* (*AND*) and 0 (1) is the non-controlling value for *OR* (*AND*).)

In this paper we concentrate on the robust testing of path delay faults. It turns out that the construction of tests with the following property is possible: for each testable path delay fault there exists a robust test (I_1, I_2) which sets all off-path inputs to the non-controlling values on application of I_1 and remains stable during application of I_2, i.e., the values on the off-path inputs are not invalidated by hazards or races.

3 Binary Decision Diagrams

In this section we shortly review the essential definitions and properties of BDD's and introduce BDD-circuits. For a detailed treatment see [Bry86].

A *Binary Decision Diagram* (BDD) is a rooted directed acyclic graph $G = (V, E)$ with vertex set V containing two types of vertices, *nonterminal* and *terminal* vertices. A nonterminal vertex v has as label an argument index $ind(v) \in \{1, \ldots, n\}$ and exactly two outgoing edges with sink nodes $zero(v), one(v) \in V$. A terminal vertex v is labeled with a value $val(v) \in \{0, 1\}$ and has no outgoing edges. For an example see figure 1. There the edges $(v, zero(v))$ $((v, one(v)))$ are marked with a 0 (1).

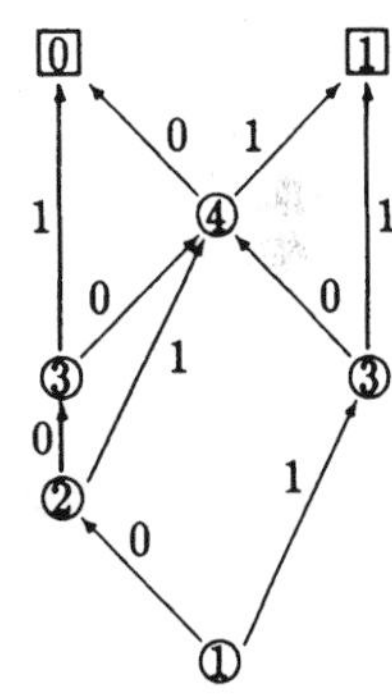

Figure 1: The reduced, ordered BDD for the Boolean function $f(x_1, x_2, x_3, x_4) = x_1x_3 + x_2x_4 + \bar{x}_3x_4$

There exists the following correspondence between BDD's and Boolean functions: A BDD having root vertex v denotes a function f_v defined recursively as:

> If v is a terminal vertex and $val(v) = 1$ ($val(v) = 0$), then $f_v = 1$ ($f_v = 0$). If v is a nonterminal vertex with $ind(v) = i$, then f_v is the function $f_v(x_1, \ldots, x_n) = \bar{x}_i \cdot f_{zero(v)}(x_1, \ldots, x_n) + x_i \cdot f_{one(v)}(x_1, \ldots, x_n)$.

Vice versa, it follows directly that for each Boolean function f there exists a BDD denoting f. BDD's that are used as data structure in design automation normally fulfil the following two additional properties:

- The BDD's are *ordered*, i.e., for any nonterminal vertex v, if $zero(v)$ is also nonterminal, then $ind(v) < ind(zero(v))$, and similarly, if $one(v)$ is also nonterminal, then $ind(v) < ind(one(v))$.
- The BDD's are *reduced*, i.e., there exists no $v \in V$ with $zero(v) = one(v)$ and there are no two vertices v and v' such that the sub-BDD's rooted by v and v' are isomorphic.

Henceforth we only consider reduced, ordered BDD's. The example BDD in figure 1 is reduced and ordered. Reduced, ordered BDD's have the following important properties:

- Let $G = (V, E)$ be an ordered BDD for f with size $|G| = |V| + |E|$. Then the satisfiability problem for f can be solved in time $O(|G|)$.
- The reduced ordered BDD of a given function f is uniquely determined and can be computed in time $O(|G| \log |G|)$, given an ordered BDD G for f.
- Let G_i be a (reduced, ordered) BDD for f_i ($i = 1, 2$) and $\diamond$ any binary operator. Then the BDD for $f_1 \diamond f_2$ can be computed in time $O(|G_1||G_2|)$.

It is well-known, that BDD's directly correspond to multiplexer based Boolean circuits called BDD-circuits in this paper. More exactly: BDD-circuits are CLC's over the libraries MUXLIB or STD. MUXLIB consists of a *multiplexer cell MUX* and *degenerated multiplexer cells DegMUXi*. A *MUX* is defined in figure 2 by its standard *AND-*, *OR-*, *INVERTER*-based realization. The western input is called *control input*, the left (right) northern input *0-input* (*1-input*). $DegMUX_i$

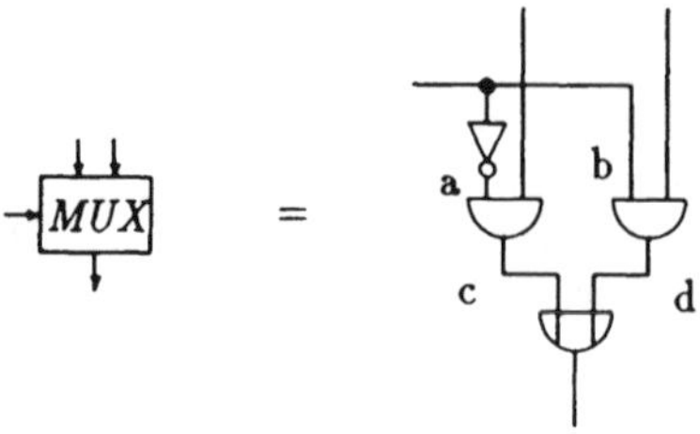

Figure 2: multiplexer cell *MUX*

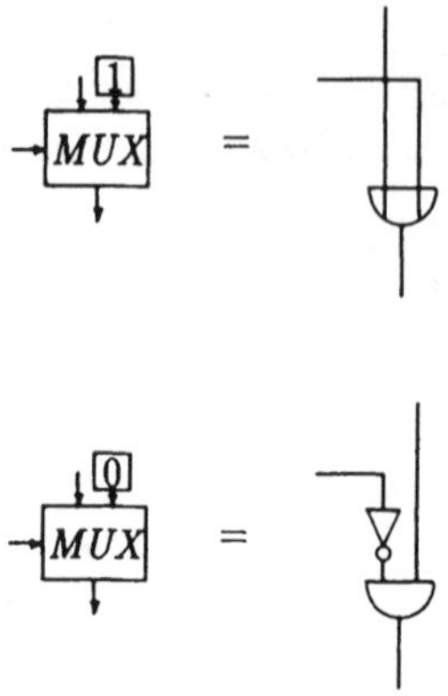

Figure 3: two degenerated multiplexer cells

cells result from a *MUX*, if one (both) of the northern input lines has a constant value 0 or 1 (have different constant values 0 and 1). It follows directly that there are 6 different cells $DegMUX_1, \ldots, DegMUX_6$. Two of them are given in figure 3. The remaining ones are defined analogously.

The *BDD-circuit* C_{MUXLIB} *of a BDD G over the library* MUXLIB is now obtained by the following construction: Traverse the BDD in topological order and replace each nonterminal node v in the BDD by a *MUX* cell, connect the control input with the PI $x_{ind(v)}$, connect the 0-input to $zero(v)$, the 1-input to $one(v)$. At the end replace *MUX* cells which are connected to a terminal node by a corresponding $DegMUX_i$ and connect the output of the multiplexer which replaced the root node with a PO. Figure 4 shows the BDD-circuit of the BDD given in figure 1. For simplicity we often identify a node in a BDD G with the corresponding multiplexer node in its BDD-circuit C_{MUXLIB}. Also $f_{zero(v)}$ ($f_{one(v)}$) is used to denote the function computed at the 0-input (1-input) of the multiplexer v. We now define BDD-circuits over STD. For this take the BDD-circuit C_{MUXLIB} of a BDD G and substitute the multiplexer cells by the standard realization of the *MUX* and $DegMUX_i$ cells over STD. The resulting CLC C_{STD} is called the *BDD-circuit of G defined over* STD. The BDD-circuit over STD for the example BDD is given in figure 5.

4 Results

In this section we derive the testability properties of BDD-circuits (over MUXLIB and STD) with respect to the fault models introduced in section 2. Furthermore we discuss possibilities to remove redundancies from BDD-circuits, if occurring, and thereby synthesize circuits with enhanced testability properties.

We start with some general remarks concerning the controllability of a multiplexer cell in a BDD-circuit over MUXLIB: The control input of any multiplexer (degenerated or non degenerated) is directly controllable. Furthermore, the values of the northern inputs of a multiplexer cell do not depend on the value of the control input, since we consider ordered BDD's. It follows that the control input can be set to 0 or 1 independently of the values of the northern inputs and that the controllability of the cell is completely determined, if we know which of the four possible input

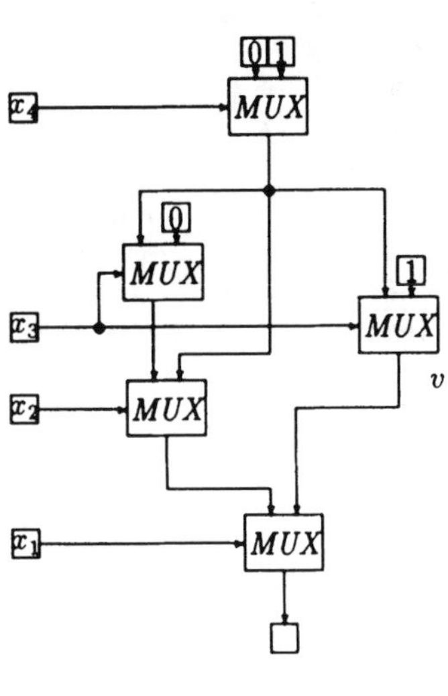

Figure 4: BDD-circuit C_{MUXLIB} over MUXLIB for the example BDD in figure 1

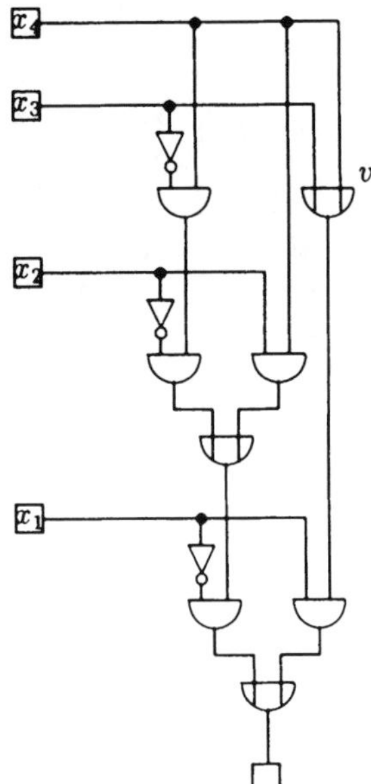

Figure 5: BDD-circuit C_{STD} over STD for the example BDD in figure 1

combinations $00, 01, 10, 11$ are applicable to the northern inputs. (The first (second) bit represents the value of the 0-input (1-input)).

In a first theorem we now show that the determination of redundancies and the computation of complete test sets with respect to CFM can be done in polynomial time for any BDD-circuit.

Theorem 1 (CFM testability) *Let C be a BDD-circuit over* MUXLIB. *The redundancies of C and a complete test set for C in CFM can be computed in time $O(|C|^3)$.*

Proof: Let v be any multiplexer node in the BDD-ciruцit C. The set of applicable inputs is determined as follows: Compute the BDD's of the functions $\bar{f}_{zero(v)} \cdot f_{one(v)}$, $f_{zero(v)} \cdot \bar{f}_{one(v)}$, $\bar{f}_{zero(v)} \cdot \bar{f}_{one(v)}$, $f_{zero(v)} \cdot f_{one(v)}$. This can be done in time $O(|C|^2)$, since the BDD corresponding to C is ordered and reduced. Then solve the satisfiability problem for the resulting BDD's (in linear time) to determine the set of applicable inputs and to obtain (partially defined) primary input combinations to C which generate the applicable input values at the northern inputs of v. (Notice that e.g. $\bar{f}_{zero(v)} \cdot f_{one(v)}$ is satisfiable, iff 01 is applicable to the northern inputs of v.)

Now consider any cellular fault $(v, I, X/Y)$ (with v a multiplexer node, a PI or a PO). The difference X/Y can be propagated to the PO by setting the control inputs on a path from v to the PO appropriately. (Notice that these control inputs have not to be set to generate any input at v and are set exactly once on the path.) So, propagation is no problem and we get: a cellular fault $(v, I, X/Y)$ is redundant, if and only if I is not applicable. This finishes the proof. ■

A test set constructed according to the above lemma is the strongest test set which can be obtained for a circuit derived by expansion from a given BDD-circuit over MUXLIB in any static combinational fault model. This is formalized in the following corollary whose proof follows directly from the completeness property of CFM and the relation between CFM and SAFM pointed out in section 2.

Corollary 1 *Let C be a BDD-circuit over* MUXLIB *and T a complete test set for C in CFM. Then T is a complete test set of any expansion of C for CFM and SAFM.*

Theorem 1 and the corollary give evidence to our claim that circuits synthesized from BDD's are easily testable. Nevertheless there may be redundancies, as indicated by the example in figure 4: it is easy to see that in the example exactly the cellular faults with northern input combination 10 at the non degenerated multiplexers are redundant. We now turn to the general question how redundancies in BDD-circuits can be classified and possibly removed.

Firstly, we show that degenerated multiplexers in BDD-circuits never have redundant faults. Then we partition the non degenerated multiplexer nodes of a BDD-circuit into disjoint *redundancy classes.* A redundancy class contains all multiplexer nodes in the BDD-circuit that have identical input combinations applicable to the northern inputs of the multiplexer. Following the proof of the above theorem the redundancies at a node v are given by the set of non applicable input combinations at the northern inputs and thus two nodes in the same class have the same types of redundant faults. At a first glance one might think that there are 15 different cases for the set of applicable input combinations at the northern inputs of a non degenerated multiplexer node resulting in 15 redundancy classes. The subsequent lemma shows that the number of classes can be reduced to 6. This turns out to be the key to many of the testability properties shown in the remainder of the paper.

Lemma 1 *Let C be a BDD-circuit over* MUXLIB, *v a node of C. Then the following holds:*

i) If v is a degenerated multiplexer with one northern input, then each value is applicable to this input, i.e., a degenerated multiplexer node has no redundancies.

ii) If v is a non degenerated multiplexer, then v belongs to one of the following 6 redundancy classes:

red. class	*input combinations applicable at northern inputs*			
1	*00*	*01*	*10*	*11*
2		*01*	*10*	*11*
3	*00*	*01*	*10*	
4		*01*	*10*	
5	*00*	*01*		*11*
6	*00*		*10*	*11*

Proof: Since the BDD is reduced we know that the function computed at a northern input of any node in the BDD-circuit is neither the 0-function nor the 1-function - this implies part i) of the lemma - and that one of the combinations 01 or 10 must be applicable to the northern inputs of any *MUX* cell. If 01 and 10 are both applicable, then v belongs to one of the classes 1-4. If 01 is and 10 is not applicable, we easily conclude that 00 and 11 can be generated at the northern inputs and thus, v belongs to class 5. The arguments for class 6 are analogous. This proves the lemma. ■

Nodes in class 1 are nodes with no redundant cellular faults. Nodes in class 2-6 have redundancies. We consider possibilities to remove these redundancies by a modification of the corresponding nodes. It is known that the removal of redundancies may create new redundancies. Therefore, in general it is not clear that the overall testability properties improve. It turns out that in this context the following *propagation property (PP)* is important: A BDD-circuit C has *(PP)*, iff

> the output edge of any multiplexer node is connected to the primary output via a path that avoids right input edges of nodes in redundancy classes 4,5 and left input edges of nodes in redundancy class 6.

Figure 4 gives an example of a BDD-circuit that does not fulfil *(PP)* (look at node v). The construction of the modified circuit, a *BDD-like circuit*, works as follows. At first, consider a node v in redundancy class 4. The function computed at the 0-input is the complement of the function computed at the 1-input, i.e., one of the northern inputs, say the 1-input, is superfluous. We therefore delete the edge connected to the 1-input. If the source node of this edge has no remaining outgoing edges, the source node is also deleted. The deletion process is continued until neither edges nor nodes are superfluous. The multiplexer at node v is now replaced by an *XOR* cell. This transformation can be executed for each node in class 4. The resulting circuit is a BDD-like circuit over MUXLIB $\cup$ $\{XOR\}$, denoted by $C_{\{4\}}$. The overall functional behaviour of the circuit has not changed and the redundancies at nodes in class 4 have been removed.

Furthermore, new redundancies have not been created. Now look at nodes in redundancy class 5, i.e., input combinations $00, 01, 11$ are applicable to the northern inputs. Consider the circuit over STD given by figure 6. It is not difficult to check that the function of this circuit is identical

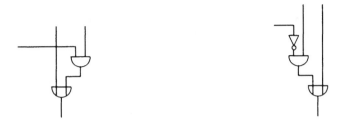

Figure 6: removal of the redundancies for nodes in class 5

Figure 7: removal of the redundancies for nodes in class 6

to the multiplexer function up to the redundant input values (= input combinations with 10 at the northern inputs). In addition, the circuit (as a circuit over STD) has no redundant cellular faults, even if we only allow input combinations unequal to 10 at the northern inputs. Therefore all multiplexer nodes with redundancy class 5 in a BDD-circuit C can be replaced by the circuit given in figure 6 to remove this type of redundancy without changing the functional behaviour. The resulting circuit is a BDD-like circuit over STD $\cup$ MUXLIB. Analogous arguments allow the removal of the redundancies for class 6. Thereby a node in class 6 is replaced by the circuit of figure 7.

It is easy to see that these transformations can be combined and the resulting synthesized BDD-like circuit denoted by $C_{\{4,5,6\}}$ is a well-defined CLC over STD $\cup$ MUXLIB $\cup$ $\{XOR\}$ which can be efficiently constructed from C. Again the functional behaviour has not changed compared to C. All redundancies for nodes in classes 4,5,6 have been removed. It can be shown that there unforunately is no way to remove the remaining redundancies for nodes in the classes 2 and 3 by using a strategy as indicated above. Furthermore, the removal of redundancies for nodes in classes 5 and 6 may create new redundancies, since the propagation properties of the circuit may change due to the modified multiplexers in classes 5 and 6: the BDD-like circuit $C_{\{4,5,6\}}$ of the example BDD has a redundant stuck-at 0 fault at the right input of the *OR* gate v (see figure 5). If a BDD-circuit has *(PP)*, we may conclude that new redundancies are not created by the construction of the BDD-like circuit. All in all, this leads to the following theorem.

Theorem 2 (Full CFM-testability) *Let C be a BDD-circuit over* MUXLIB *with propagation property (PP) and empty redundancy classes 2 and 3. Then the BDD-like circuit $C_{\{4,5,6\}}$ over* STD $\cup$ MUXLIB $\cup$ $\{XOR\}$ *is fully testable in CFM.*

We now come to the testability properties of BDD-circuits with respect to SAFM and PDFM. It turns out that most of the results can be easily deduced from the analysis of the CFM-testability. We start with the results for SAFM.

Theorem 3 (SAFM-testability) *Let G be a BDD, C_{MUXLIB} the BDD-circuit over* MUXLIB, *C_{STD} the BDD-circuit over* STD.

i) C_{MUXLIB} is fully testable in SAFM.

ii) C_{STD} is fully testable in SAFM, if and only if C_{MUXLIB} has empty redundancy classes 5 and 6.

iii) If C_{MUXLIB} has (PP), a BDD-like circuit over STD *which realizes the function of G and is fully testable in SAFM can be efficiently constructed.*

Proof: We give a sketch of the proof. Part i) follows easily from lemma 1 and the proof of theorem 1.

Consider a multiplexer node v in C_{MUXLIB} and the corresponding subcircuit in C_{STD}. If the multiplexer is degenerated, it follows directly that all stuck-at faults in the subcircuit are testable. If the multiplexer is non degenerated, we conclude from i) that only the testability of stuck-at faults on the internal lines a, b, c, d (compare figure 2) has to be checked. A detailed analysis shows that internal faults - with the exception of the stuck-at 1 faults on lines a and b - are testable independently of the redundancy class of v. Check figure 2 to see that a stuck-at 1 fault on a (b) is testable if and only if v is not in redundancy class 5 (6). This finishes the proof of ii).

Consider the the BDD-like circuit $C_{\{4,5,6\}}$ over STD $\cup$ MUXLIB $\cup$ $\{XOR\}$ and replace each XOR cell and each cell from MUXLIB in $C_{\{4,5,6\}}$ by a realization of this cell over STD to obtain a circuit over STD. The resulting circuit is a BDD-like circuit over STD which is fully testable in SAFM. (Make sure that full CFM-testability of an XOR cell implies full SAFM-testability of the equivalent realization over STD.) ■

Until now our investigations were based on the single fault assumption, that exactly one fault is in the circuit. In contrast to this the *multiple stuck-at fault model* (MSAFM) considers a collection of single stuck-at faults as a fault. Due to the large number of possible multiple faults (exponential in the size of the circuit) testing for multiple faults is usually not performed, even though the model is the more realistic one. We point out in the following theorem that synthesis for testability may yield circuits which are fully testable in MSAFM.

Theorem 4 (MSAFM Testability) *Let G be a BDD and C the corresponding BDD- or BDD-like circuit over* STD. *Then full SAFM-testability of C implies full MSAFM-testability.*

Proof: We outline the proof. Show that a multiplexer cell (as a circuit over STD) and the circuits of figures 6 and 7 are fully multiple testable in MSFAM. Then consider the root v in G. Let C_0 (C_1) be the part of C that computes $f_{zero(v)}$ ($f_{one(v)}$). Assume inductively that test sets T_0 (T_1) exist, which detect any multiple fault in C_0 (C_1). Then one can show that the following test set detects any multiple fault in C:

> Combine any test in C_0 (C_1) with a 0 (1) on the control input of v. Compute two inputs which generate 01 and 10, respectively, at the northern inputs of v and combine both inputs with 0 and 1 at the control input of v. (Notice that the existence of the inputs follows from the full SAFM-testability of C.)

■

We finish this section by a discussion of BDD-circuits under the path delay fault model (PDFM). Let G be a BDD, C_{STD} its BDD-circuit over STD and π a path in C_{STD} from a PI x_i to the PO. Consider the subcircuit of C_{STD} corresponding to the first multiplexer v met by π. Assume that v is degenerated. Then π is uniquely determined in the subcircuit and the off-path inputs in the subcircuit can be easily set to non-controlling values by setting PI's unequal x_i appropriately. Assume that v is non degenerated. Then x_i is connected to the control input of v. There exist exactly two paths to the output of the subcircuit. On each path two transitions have to be checked. A close look at the subcircuit shows that the off-path inputs in the subcircuit can be sensitized in all cases if and only if 10 and 01 are applicable at the northern inputs, i.e., if and only if v is not in redundancy class 5 or 6. The remaining part of the path is uniquely determined by the specification of a sequence of nodes in the BDD. The path is sensitized by setting the control inputs of these nodes appropriately. The propagation vector then differs from the initialization vector exactly at position x_i. (Notice that a change in the value of x_i does not influence the off-path inputs.) Thus, we have proved the following theorem.

Theorem 5 (PDFM-testability) *Let G be a BDD, C_{MUXLIB} the BDD-circuit over* MUXLIB, *C_{STD} the BDD-circuit over* STD. *Then C_{STD} is fully testable in PDFM with robust tests, if and only if C_{MUXLIB} has empty redundancy classes 5 and 6.*

5 Conclusions and Open Problems

We have characterized the testability of BDD-circuits with respect to the cellular fault model (CFM), the single stuck-at fault model (SAFM) and the path delay fault model (PDFM).

A detailed investigation for CFM turned out to be essential for the proof of the remaining results. We showed that the computation of a complete test set T in CFM is possible in time cubic in the size of the circuit. If properly chosen, T remains complete also for the modified circuits constructed for CFM and SAFM as well. For a classification of the redundancies of a BDD-circuit we defined 6 disjoint classes. This classification was used to determine redundancies which are removable by a simple linear time substitution process. We finally obtained a sufficient criterion for the existence of a fully testable BDD-like circuit in CFM. The redundancy classes were further used to characterize the full testability in SAFM and to finally obtain fully testable BDD-like circuits for SAFM and MSAFM. We finished by showing that the results for CFM can also be applied to characterize the BDD-circuits which are fully robustly testable. Our results demonstrate that BDD-circuits are easily testable and in many cases can be modified to be fully testable for different fault models which together cover a large variety of possible physical defects.

Nevertheless some questions are left open in this paper: No modification was given to obtain full testability in CFM at nodes in redundancy classes 2 and 3. Is there any possibility to obtain full testability with different methods of modification, even in the case that *(PP)* is not fulfilled? Notice, that for the example in figure 5 the removal of the redundant *OR* gate works. In the stuck-at model we proved the full testability in the presence of multiple faults. A similar result for cellular faults is not given. The existence of nodes in redundancy classes 5 and 6 implies the existence of non testable faults in PDFM. A modification that yields full robust testability in PDFM is not given. One might think that the BDD-like circuit over STD, which results from the BDD-circuit by modification of the nodes in classes 5 and 6 and which is fully testable in SAFM, is also fully testable in PDFM. A look at the circuit derived from the example BDD proves that this is not the case.

The circuits considered in this paper resulted from **reduced, ordered** BDD's. These restrictions may lead to BDD's of large size (possibly exponential in the number of variables). It is an interesting question if the results can be generalized (in modified form) to other classes of BDD's. It is clear for example that many arguments given in this paper remain valid if the BDD's are *read once only*, i.e., if the indices on any path in the BDD are pairwise disjoint.

References

[Ake78] S. B. Akers. Binary decision diagrams. *IEEE Transactions on Computers*, C-27(6):509–516, June 1978.

[Ake79] S. B. Akers. Functional testing with binary decision diagrams. In *9th Int. Symposium on Fault Tolerant Computing*, pages 75–82, 1979.

[BA83] P. Banerjee and J.A. Abraham. Generating tests for physical failures in MOS logic circuits. In *Proc. IEEE Int. Test Conference*, pages 554–559, 1983.

[BART82] C.C. Beh, K.H. Aray, C.E. Radke, and K.E. Torku. Do stuck-at fault models reflect manufacturing defects? In *Proc. IEEE Int. Test Conference*, pages 35–42, 1982.

[BBH*90] B. Becker, T. Burch, G. Hotz, D. Kiel, R. Kolla, P. Molitor, H.G. Osthof, G. Pitsch, and U. Sparmann. A graphical system for hierarchical specifications and checkups of VLSI circuits. In *Proc. of the 1st European Design Automation Conference (EDAC90)*, pages 174–179, 1990.

[BF76] M.A. Breuer and A.D. Friedman. *Diagnosis & reliable design of digital systems*. Computer Science Press, 1976.

[BHK*87] B. Becker, G. Hotz, R. Kolla, P. Molitor, and H.G. Osthof. Hierarchical design based on a calculus of nets. In *Proceedings of the 24th Design Automation Conference*, pages 649–653, June 1987.

[BRB90] K. S. Brace, R. L. Rudell, and R.E. Bryant. Efficient implementation of a BDD package. In *Proc. 25th Design Automation Conference*, pages 40–45, 1990.

[Bry86] R.E. Bryant. Graph-based algorithms for boolean function manipulation. *IEEE Transactions on Computers*, C-35(8):677–691, 1986.

[BS88] B. Becker and U. Sparmann. A uniform test approach for RCC-Adders. In *Proc. of the 3rd Aegean Workshop on Parallel Computation and VLSI Theory, Lecture Notes in Comput. Sci. 319*, pages 288–300, 1988.

[DMNS88] S. Devadas, H.K.T. Ma, R. Newton, and A. Sangiovanni-Vincentelli. Optimal logic synthesis and testability: two faces of the same coin. In *Proc. IEEE Int. Test Conference*, pages 4–12, 1988.

[FC88] F.J. Ferguson and J.P. Chen. Extraction and simulation of realistic CMOS faults using inductive fault analysis. In *Proc. IEEE Int. Test Conference*, pages 475–484, 1988.

[Fri73] A.D. Friedman. Easily testable iterative systems. *IEEE Transactions on Computers*, C-22:1061–1064, December 1973.

[Fuj85] H. Fujiwara. *Logic Testing and Design for Testabilty.* The MIT Press, 1985.

[GCV80] J. Galiay, Y. Crouzet, and M. Vergniault. Physical versus logical fault models MOS LSI circuits: impact on their testability. *IEEE Transactions on Computers*, C-29(6):527–531, June 1980.

[Lee59] C.Y. Lee. Representation of switching circuits by binary decision programs. *Bell Systems Technical Journal*, 38:985–999, 1959.

[Mal87] W. Maly. Realistic fault modeling for VLSI testing. In *Proc. 24th Design Automation Conference*, pages 173–180, 1987.

[Mei89] C. Meinel. *Modified Branching Programs and Their Computational Power. volume 370 of LNCS*, Springer Verlag, 1989.

[Mil85] B. Milne. Testability, 1985 technology forecast. *Electronic Design*, 10:143–166, 1985.

[SFF89] M.H. Schulz, F. Fink, and K. Fuchs. Parallel pattern fault simulation of path delay faults. In *Proc. of the 26th Design Automation Conference*, June 1989.

[Smi85] G.L. Smith. A model for delay faults based upon paths. In *Proc. of IEEE Int. Test Conference*, pages 342–349, September 1985.

[Wad78] R.L. Wadsack. Fault modeling and logic simulation of CMOS and MOS integrated circuits. *The Bell System Technical Journal*, 57, 1978.

[WDGS86] T.W. Williams, W. Daehn, M. Grützner, and C.W. Starke. Comparison of aliasing errors for primitive and non-primitive polynomials. In *Proc. IEEE Int. Test Conference*, pages 282–288, 1986.

[Weg87] I. Wegener. *The Complexity of Boolean functions.* Wiley Teubner, 1987.

[Wun91] H.J. Wunderlich. *Hochintegrierte Schaltungen: Prüfgerechter Entwurf und Test.* Springer Verlag, 1991.

INVITED LECTURE

Compression and Entropy*

Georges Hansel
Université de Rouen

Dominique Perrin
Université de Paris 7

Imre Simon†
Universidade de São Paulo

Abstract

The connection between text compression and the measure of entropy of a source seems to be well known but poorly documented. We try to partially remedy this situation by showing that the topological entropy is a lower bound for the compression ratio of any compressor. We show that for factorial sources the 1978 version of the Ziv-Lempel compression algorithm achieves this lower bound.

1 Introduction

The main objective of this paper is to elucidate the connection between text compression and the measure of entropy of a source. Such connection seems to be well known but poorly documented. We try to partially remedy this situation by showing that the topological entropy is a lower bound for the compression ratio of any compressor. We also show that for factorial sources the 1978 version of the Ziv-Lempel compression algorithm achieves this lower bound.

More precisely, we model compression as any injective function whose domain and co-domain are subsets of free monoids and define a measure of the compression obtained on any set of words. We model a source through the set of words it can emit, and measure its entropy by the so called topological entropy.

Among other results we show that

(1) the entropy is always a lower bound for the compression,

(2) ranking the words of a source is always a compression which achieves the entropy,

(3) the 1978 version of the Ziv-Lempel algorithm is a universal continuous compression which achieves the entropy on every factorial source,

*Authors addresses: G.H.: Université de Rouen, Mathématiques, BP 67, F-76130 Mont Saint-Aignan, France, <gh@litp.ibp.fr>; D.P.: Université de Paris VII, LITP, 2 place Jussieu, 75251 Paris Cedex 05, France, <dp@litp.ibp.fr>; I.S.: Universidade de São Paulo, IME, Caixa Postal 20570, 01498 São Paulo, SP, Brasil, <is@ime.usp.br>

†This work was done with partial support from FAPESP and from BID/USP (project 30.01).

(4) the factorial hypothesis is necessary in the previous item.

We also address briefly the problem of estimating the topological entropy of a source, both theoretically and practically. Finally, we remark that neither our results nor our arguments depend on any probabilistic concepts and in particular no explicit hypothesis of ergodicity of the source are assumed. On the other hand, an effort was made to make the paper self-contained (apart from some omitted proofs) and to state the concepts and results as simply and precisely as possible.

In the final section we consider the metric case, where both entropy and compression are computed according some invariant probability measure on the free monoid. In this case again entropy is a lower bound for compression.

2 Compression

In this section we define compression and propose measures of the compression rate achieved on a set of words. We note that our measure is geared to the model of a source we shall precise in the next section. Should we use a different model for the source we would need another measure of compression rate to maintain our results; another model will be seen in the last section.

Let A be any finite alphabet and let $\mathbf{2} = \{0,1\}$ be a fixed binary alphabet. A *compression* or a *compressor* for us is any injective partial function $\gamma\colon A^* \to \mathbf{2}^*$. Note that injectivity is essential in order to keep the ability of unique reconstruction of the original text. Note also that allowing γ to be partial is a concession to achieve better compression rates, since the compressor does not have to worry about possibly irrelevant words. Finally, note that we did not require any algorithmic property of γ. Indeed, we do allow functions which are not even recursive. However, we observe that from a practical point of view the efficiency of a compressor is one of its main virtues; any compressor which does not work in linear time is more or less useless in practice.

A few words now about the alphabet size. Our definition requires a 2-letter output alphabet. This is not a severe restriction, it is done to allow all logarithms in this paper to be taken on base 2 and to allow to normalize the measure of both entropy and compression in "bits per character". Later on we shall see what to modify in order to recover the general case. For the time being we only alert the reader that this convention might pre-empt the semantics of the word compression since it might force compression rates much higher than 1.

We shall need a special class of compressions called ranking which we now define. Consider an alphabet A and fix a total order on it. Corresponding to this order on A we define the *lexicographic order* on A^*, where s precedes t if either s is a prefix of t or there exist words $u, s', t' \in A^*$ and letters $a < b$ in A, such that $s = uas'$ and $t = ubt'$. Now we define the *military order* on A^*: for $s, t \in A^*$ we write $s \le t$ if either $|s| < |t|$ or if $|s| = |t|$ and s lexicographically precedes t. Clearly, both lexicographic and military orders are total over A^*. Let x be a word in a subset X of A^*. the *rank of x in X* is defined by $r(x, X) = |\{\, t \in X \mid t \le x \,\}|$. In other words, $r(x, X)$ is the index word x receives when the elements of X are enumerated in military order as $x_1 < x_2 < \cdots$. Given such an enumeration of X we shall say that x_p $(p \ge 1)$ is the p-th word of X in military order.

When $X = \mathbf{2}^*$, using the obvious order on $\mathbf{2}$, we denote $\mu(x) = r(x, \mathbf{2}^*)$, for $x \in \mathbf{2}^*$. We will also need the inverse of μ, denoted $\rho(n) = \mu^{-1}(n)$, for $n > 0$. Thus, if $\mathbb{N}_+$ denotes the set of strictly positive naturals then μ and ρ are bijections,

$$\mu\colon \mathbf{2}^* \to \mathbb{N}_+ \quad\text{and}\quad \rho\colon \mathbb{N}_+ \to \mathbf{2}^*.$$

Let $n \in \mathbb{N}$ be the canonical interpretation of word $x \in \mathbf{2}^*$. For instance, $n = 5$ for every $x \in 0^*101$. It is easy to see that $\mu(x) = n + 2^{|x|}$ and that $\rho(n)$ is the binary representation of $n - 2^{\lfloor \log n \rfloor}$ padded with zeros on the left to give a word of length $\lfloor \log n \rfloor$.

Let X be an infinite subset of A^*. Then $\gamma_X(x) = \rho(r(x, X))$ is a compression, called the *ranking of* X. Another way of looking at γ_X is by noting that it is the unique partial function

$$\gamma_X\colon A^* \to \mathbf{2}^*$$

with domain X which is monotonic with respect to military order and whose restriction to X is a bijection from X to $\mathbf{2}^*$. Note that $\gamma_{\mathbf{2}^*}$ is just the identity function on $\mathbf{2}^*$.

Now we measure the compression rate achieved by a given compressor $\gamma\colon A^* \to \mathbf{2}^*$. For $x \in A^*$ the *compression of* x is $\tau(x) = |\gamma(x)|/|x|$. For an infinite set $X \subseteq A^*$ some words might be compressed more than others. For $p > 0$, let $x_p \in X$ be the p-th word of X in military order. Assuming that $X \subseteq \operatorname{dom}\gamma$ we define the *compression of* γ *on* X by

$$\tau(X) = \limsup_{p\to\infty} \tau(x_p).$$

This is an asymptotic worst case measure of the compression. A more realistic measure is the (asymptotic worst case) *average compression rate* of γ on X:

$$\tau_{ave}(X) = \limsup_{n\to\infty} \frac{\sum_{x\in X\cap A^n} \tau(x)}{|X \cap A^n|}.$$

where we interpret $0/0$ as being 0. The measure τ always refers to a fixed compression γ; from the context it will always be clear to which γ does τ refer to. The two measures introduced are related by the next Proposition whose proof we omit.

Proposition 1 *For every infinite subset X of A^* and for every compression $\gamma\colon A^* \to \mathbf{2}^*$, such that $X \subseteq \operatorname{dom}\gamma$, we have that $\tau_{ave}(X) \leq \tau(X)$. There are infinite sets X for which the above inequality is strict.*

We shall say that compressor γ is *uniform* on X if equality holds in the above Proposition.

2.1 The Ziv-Lempel Compression Algorithm

In this section we specify and study some properties of a truly remarkable algorithm, namely the 1978 version of the Ziv-Lempel compression algorithm, which we will call ZL. This is one of the most succesfull algorithms in the area, both in theory and in practice; its numerous variants constitute perhaps the most popular compression method in use. In particular the UNIX[1] utility `compress` is based on a variant of this algorithm. More details on this family of compressors can be found in [17, 2].

[1]UNIX is a trademark of AT&T Bell Laboratories.

In the sequel we describe the Ziv-Lempel algorithm, i.e. the total function ZL: $A^* \to 2^*$. Let $x \in A^*$ be a word to be compressed. Word x is factorized as $x = x_0 x_1 x_2 \dots x_m x_{m+1}$ for some m subject to the following three conditions:

(1) x_0 is empty,

(2) for each i, $1 \leq i \leq m$, x_i is the longest prefix of $x_i x_{i+1} \dots x_{m+1}$ which can be written as $x_i = x_{j_i} a_i$, for some $j_i < i$ and $a_i \in A$,

(3) $x_{m+1} = x_{j_{m+1}}$, for some $j_{m+1} < m+1$.

In other words, the first factor is empty and each other factor but the last one is a new factor which consists of a previous factor followed by one letter. Clearly such a factorization is unique.

The words x_i, $1 \leq i \leq m$ are entered in a digital tree (or trie) which we will denote by $T(x)$ and this tree is usually called the dictionary. Each factor x_i corresponds to a vertex of $T(x)$ and if we label this vertex with label i then it is easy to see that the correspondence $x \mapsto (T(x), j_{m+1})$ is injective.

Now we describe the output of ZL. Factor x_i, $1 \leq i \leq m$, is transmitted as the pair (j_i, a_i). Factor x_{m+1} is transmitted just as j_{m+1}. To transmit j_i one usually uses $\lceil \log i \rceil$ bits which are necessary and sufficient to specify any of the numbers $0, 1, 2, \dots i-1$. The character a_i can be transmitted by itself and since A is fixed it takes some bounded number of bits to transmit it.

Next we note that the length of the output depends exclusively on the number m of edges of $T(x)$. Indeed, we have

$$|\mathrm{ZL}(x)| = \sum_{1}^{m+1} \lceil \log i \rceil + K''m,$$

for an apropriate constant K''. Since $\lceil \log i \rceil \leq 1 + \log m$ we can put

$$|\mathrm{ZL}(x)| \leq m \log m + Km,$$

for some constant K depending on A. The reason behind the last inequality is that whatever is the shape and labeling of $T(x)$ the output always consists of $m+1$ numbers of predetermined length (at most $\lceil \log m \rceil$) and m characters from A. Returning to the constant K we note that it might even be negative. We also observe that it will usually depend on the details of the protocol used to form the output and that clever protocols might lead to smaller constants. While this will change the length of the output it does not change $\tau(X)$ because the term $m \log m$ is dominant in the sum.

Let us now analyse the length of the input x in function of the tree $T(x)$. For vertex i the contribution of factor x_i to $|x|$ is exactly the height of vertex i, denoted height(i). Thus,

$$|x| = \mathrm{height}(j_{m+1}) + \sum_{i=1}^{m} \mathrm{height}(i).$$

The dominant term $\sum_{i=1}^{m} \mathrm{height}(i)$ depends only on the shape of the tree $T(x)$ and it is usually called *internal path length* of $T(x)$; see for instance [8].

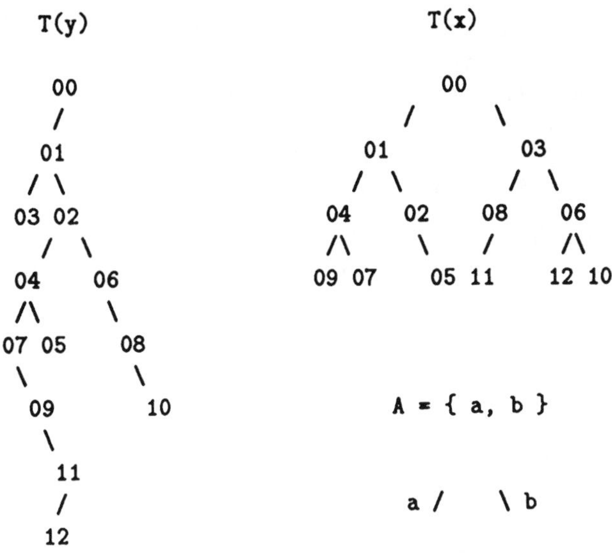

Figure 1: Trees $T(x)$ and $T(y)$.

¿From the above discussion one can conclude that if we fix m and consider any set of words x factored with m factors by ZL then the compression rate $\tau(x)$ is essentially inversely proportional to the internal path length of $T(x)$. In other words, if $T(x)$ is short and fat then τ is large and ZL is inefficient but if $T(x)$ is tall and slim then τ is small and ZL is efficient.

We illustrate this section with two examples. We consider $A = \{a, b\}$, and

$$x = a.ab.b.aa.abb.bb.aab.ba.aaa.bbb.baa.bba,$$

$$y = a.ab.aa.aba.abab.abb.abaa.abbb.abaab.abbbb.abaabb.abaabba, abaabba$$

$$\mathrm{ZL}(x) = a.1b.00b.01a.010b.011b.100b.011a.0100a.0110b.1000a.0110a.0000,$$

$$\mathrm{ZL}(y) = a.1b.01a.10a.100b.010b.100a.110b.0111b.1000b.1001b.1011a.1100,$$

$$m = 12, |x| = 28, |y| = 53, |\,\mathrm{ZL}(x)| = |\,\mathrm{ZL}(y)| = 49,$$

$$\tau(x) = 1.75, \tau(y) = 0.92.$$

On Figure 1 we represent the trees $T(x)$ and $T(y)$.

Observe that ZL is a continuous compression. By this we mean that function ZL is continuous with respect to the usual "longest common prefix" distance in A^*.

Observe also that to uncompress $\mathrm{ZL}(x)$ one has to reconstruct the tree $T(x)$ exactly as it was constructed during the compression and the factors x_i are obtained from this tree. The output protocol must be defined in order to allow this synchronized reconstruction of the tree $T(x)$.

For future reference we resume the properties of the Ziv-Lempel compression algorithm.

Proposition 2 *The Ziv-Lempel algorithm* ZL: $A^* \to 2^*$ *is a continuous compression. Let* $x \in A^*$ *be such that the tree* $T(x)$ *has* m *edges. Let* I *be the internal path length of* $T(x)$ *and let* τ *be the compression of* ZL. *Then, for some constant* K, $\tau(x) \leq (m \log m + Km)/I$.

3 Topological Entropy

Entropy is one of these words used for a number of related concepts in Mathematics and Physics. Usually it is expressed by a formula which through some magic mechanism leads one to unexpected properties. It seems that the word entropy was first used by R. Clausius in 1864 in a thermodynamical context. The classical version of the formula and the use of the letter H to denote entropy goes back to Boltzmann who used it in 1890 in the context of Statistical Mechanics:

$$h(S) = -\sum_{i=1}^{n} p_i \log p_i, \text{ with } \sum_{i=1}^{n} p_i = 1.$$

The modern use of entropy was pioneered by Claude Shannon in 1948 when he launched Information Theory [16]. Intuitively speaking Shannon showed that the above formula can be used to measure the information content, in bits per symbol, of a source which independently emits symbols i (over a fixed alphabet $\{1, 2, \ldots, n\}$) with probability p_i. Such a source is just a Bernoulli distribution. It is said that von Neumann suggested the term *entropy* to Shannon because of its former use by Boltzmann and because "no one knows what entropy really is, so in a debate you will always have the advantage" (Tribus and McIrvine, Energy and information, Scientific American, 1971). For more details on the misteries of entropy we refer the reader to the book of Rényi [15].

Shannon himself extended the formula to measure the entropy of sources modeled by finite Markov chains; we refer the reader to the books on Information Theory and especially to the little book of Khinchin [7] where we found the first explicit mention of the relationship between compression and entropy.

Perhaps the deepest use of entropy is encountered in modern Ergodic Theory, where entropy is one of the most successful invariants introduced by Kolmogorov and Sinai in 1958 and 1959. It is noteworthy to observe that even though entropy occupies a significant portion of any modern book on Ergodic Theory neither the concept nor the word occur in the classical book "Lectures on Ergodic Theory" of Paul Halmos which appeared in 1953. For a modern survey on entropy we refer the reader to the book of Martin and England [11].

Our interest here is to explore the relationship between entropy and compression. For the sources considered by Shannon such a relationship seems to be implicit in his work. For stationary ergodic sources the relationship has been known for some time but until recently it was not well documented. See the work of Hansel [6] and also the recent papers by Ornstein, Weiss, and Shields [13, 14]. ¿From our point of view there are some inconveniences with the approaches above: they are either too simplistic (the Shannon source) or too sophisticated (the stationary ergodic source) to be of easy use in Computer Science. This fact led us to the investigations reported in the sequel. Note that on the one hand our approach can be used to model a great variety of sources and on the other it does not depend on any probabilistic considerations (except some of the material in the last section).

The key concept we use is the topological entropy of a source. A *source* is any infinite subset X of A^*, the words in X constitute the messages which can be emitted by the source. The *topological entropy* of source X is defined by

$$H(X) = \limsup_{n\to\infty} \frac{1}{n} \log s_n,$$

where $s_n = |X \cap A^n|$. Here, by convention, $\log 0 = 0$. As far as we know the first explicit use of this formula goes back to Shannon who used it to compute the channel capacity. This motivated further investigations including its computation for recognizable and context free languages by Chomsky and Miller, de Luca, Kuich and others [3, 4, 9]. Later, Adler, Konheim and McAndrew introduced its modern use in 1965 [1] in Ergodic Theory which was recently surveyed by Newhouse [12]. We shall explore the relationship of this concept to text compression.

Intuitively speaking the topological entropy is a measure, in bits per character, of the asymptotic exponential growth rate of the numbers s_n; it is a collective property of X and not of any individual word $s \in A^*$. More precisely, assume we plot $\log s_n$ against n. Then, the entropy of X is the greatest angular coefficient of the lines through the origin which can be sustained infinitely often by the points of our graph.

We indicate an alternate view now. A subset X of A^* is *factorial* if every factor of every word in X also belongs to X.

Proposition 3 *Let $X \subseteq A^*$ be infinite and assume that* $\lim_{n\to\infty} \log(s_{n+1}/s_n)$ *exists. Then its value is exactly $H(X)$. The above limit might not exist even if we assume that X is factorial.*

If X is factorial then we can interpret s_{n+1}/s_n as being the average branching factor of X at words of length n (imagine the digital tree containing the words in X). Thus, for such a well behaved X, H is the asymptotic value of the number of bits to write the average branching factor of X. Another related view is obtained if we note that $(1/n)\log s_n = \log \sqrt[n]{s_n}$. Thus, the term $(1/n)\log s_n$ can be thought of as being the hypothetical branching factor at level n one would get if branching was uniform between levels 1 and n.

Next we compute $H(X)$ for some sources. If $X = A^*$ then $H(X) = \log|A|$. Now, clearly, if $Y \subseteq X$ then $H(Y) \leq H(X)$, hence $\log|A|$ is an upper bound for the entropy of any subset of A^*.

Now we give two examples of null entropy. Let $X = \{\, a^n b^n \mid n \geq 0 \,\}$, then $H(X) = 0$. Let now $A = \{\, a, b \,\}$ and let X be the set of factors of $\{\, \varphi^p(a) \mid p \geq 1 \,\}$, where φ is a morphism defined by $\varphi(a) = ab$ and $\varphi(b) = ba$. Thus, X is just the set of factors of the infinite Thue-Morse word (see Lothaire [10]). Then the reader can show that $H(X) = 0$. An interesting experience supporting this fact (and also the results in the next section) is to apply the UNIX `compress` to the words $\varphi^n(a)$ and observe the astonishing compression obtained which should be contrasted with the much poorer results obtained by `pack` or `compact` which use different algorithms (Huffman code and adaptive Huffman code respectively).

Finally, we leave it to the reader to show that for $X = \{\, a, b \,\}^* c \{\, a, b, d \,\}^*$, $H(X) = \log 3$; see also Section 5.

Regarding the breadth of our model we note that Shannon's First Theorem for the noiseless discrete channel states that the topological entropy H (there called channel capacity) of the set of most probable words is just $-\sum p_i \log p_i$. Hence, the Shannon source can be viewed as a particular case of the model we focus on.

4 Compression and Entropy

In this section we state our main results and prove the most important ones. Note that the proofs are based both on counting arguments and on approximations of the quantities in question.

We begin with two simple facts.

Lemma 4 *The length of the n-th word in* $\mathbf{2}^*$ *is* $\lfloor \log n \rfloor$; *i.e.* $|\rho(n)| = \lfloor \log n \rfloor$.

Lemma 5 *Let* $g(n)$, *for* $n \geq 1$, *denote the sum of lengths of the first n words of* $\mathbf{2}^*$ *in military order. Then,*

$$g(n) = \sum_{i=1}^{n} |\rho(i)| = (n+1)m - 2^{m+1} + 2,$$

where $m = \lfloor \log n \rfloor$. *In particular,*

$$n \log n - 3n \leq g(n) \leq \sum_{i=1}^{n} \log i \leq n \log n.$$

The next result shows in which sense is entropy a lower bound for compression.

Proposition 6 *For every compression* $\gamma\colon A^* \to \mathbf{2}^*$ *and for every infinite* $X \subseteq A^*$, *if* $X \subseteq \operatorname{dom}\gamma$ *then* $H(X) \leq \tau_{ave}(X)$.

Proof. Let $R \subseteq R' \subseteq B^*$ and, for some $n \geq 1$ let r_n and r'_n be the n-th word of R and R' in military order. Clearly, $|r_n| \geq |r'_n|$. Let then $R = \{\gamma(x) \mid x \in X \cap A^n\}$ and $R' = \mathbf{2}^*$. Since $X \subseteq \operatorname{dom}\gamma$ and γ is injective it follows that $|R| = s_n$, where $s_n = |X \cap A^n|$. Thus, in view of Lemma 5 it follows that

$$\sum_{x \in X \cap A^n} |\gamma(x)| \geq g(s_n) \geq s_n \log s_n - 3s_n.$$

Dividing by ns_n we get

$$\frac{\sum_{x\in X\cap A^n} |\gamma(x)|}{ns_n} \geq \frac{\log s_n}{n} - \frac{3}{n}.$$

Hence, taking the lim sup of the sequence and observing that the term $-3/n$ tends to zero we have

$$\tau_{ave}(X) = \limsup_{n\to\infty} \frac{\sum_{x\in X\cap A^n} \tau(x)}{|X \cap A^n|} = \limsup_{n\to\infty} \frac{\sum_{x\in X\cap A^n} |\gamma(x)|}{ns_n} \geq$$
$$\limsup_{n\to\infty} \frac{\log s_n}{n} - \frac{3}{n} = H(X),$$

as required. ∎

Now we show that ranking is a uniform compression which is optimal for every source X. Initially we state a technical Lemma without proof. For a sequence $(s_n)_{n>0}$ let H_s denote its topological entropy, i.e., $H_s = \limsup_{n\to\infty}(1/n) log s_n$.

Lemma 7 *Let $(s_n)_{n>0}$ be a sequence of real numbers such that, for some constant $k \geq 1$, either $s_n = 0$ or $1 \leq s_n \leq k^n$. Let $t_n = \sum_{i=1}^{n} s_i$. Then, $H_s = H_t$.*

The reader can verify that Lemma 7 is equivalent to stating that for every infinite $X \subseteq A^*$ and every $a \in A$, $H(X) = H(Xa^*)$.

Proposition 8 *For every infinite $X \subseteq A^*$, the compression rate $\tau(X)$ of γ_X satisfies $\tau(X) \leq H(X)$. Hence, γ_X is always a uniform optimal compression.*

Proof. Let $s_n = |X \cap A^n|$ and $t_n = \sum_{i=1}^{n} s_i$. For $p \geq 1$ let x_p be the p-th word of X in military order and assume that $|x_p| = n$. Then, $p \leq t_n$. Since γ_X is the ranking of X, in view of Lemma 4, we have that $\tau(x_p) = |\gamma_X(x_p)|/n \leq \lfloor \log t_n \rfloor / n$. It follows that the sequence $(\tau(x_p))_{p\geq 1}$ is majorized, term by term, by the sequence which consists of s_n terms equal to $(\log t_n)/n$, for $n = 1, 2, \ldots$ successively. Using Lemma 7 it follows that

$$\tau(X) = \limsup_{p\to\infty} \tau(x_p) \leq \limsup_{n\to\infty} \frac{\log t_n}{n} = \limsup_{n\to\infty} \frac{\log s_n}{n} = H(X).$$

This concludes the proof of Proposition 8. ∎

This results allows an alternative interpretation of entropy, a variant of which will play an important role in the sequel.

Corollary 9 *Let $X \subseteq A^*$ be an infinite set and let x_p denote the p-th word of X in military order. Then,*

$$H(X) = \limsup_{p\to\infty} \frac{\log p}{|x_p|}.$$

As noted earlier, usually γ_X is not continuous, though one can construct examples of sets where γ_X is a continuous compression. Does there exist continuous optimal compressions for every X? In the sequel we show that the Ziv-Lempel algorithm satisfies this requirement for certain sets X. First we need the promised variant of Corollary 9.

Proposition 10 *Let $X \subseteq A^*$ be an infinite set and let x_p denote the p-th word of X in military order, for $p \geq 1$. Then,*

$$H(X) = \limsup_{m\to\infty} \frac{\sum_{p=1}^{m} \log p}{\sum_{p=1}^{m} |x_p|}.$$

Proof. Let $z_m = \sum_{p=1}^{m} \log p / \sum_{p=1}^{m} |x_p|$, and let $H = H(X)$. Using Lemma 5 we have that $\sum_{p=1}^{m} \log p \geq m \log m - 3m$. Since $|x_p|$ is monotonic increasing with p we have that $\sum_{p=1}^{m} |x_p| \leq m|x_m|$. Then,

$$z_m \geq \frac{m \log m - 3m}{m|x_m|} = \frac{\log m}{|x_m|} - \frac{3}{|x_m|}.$$

Since, by Corollary 9, $H = \limsup(\log m)/|x_m|$ we conclude that,

$$\limsup_{m\to\infty} z_m \geq \limsup_{m\to\infty}\left(\frac{\log m}{|x_m|} - \frac{3}{|x_m|}\right) = \limsup_{m\to\infty} \frac{\log m}{|x_m|} = H.$$

In the next paragraph we will show that

$$\forall \epsilon > 0 \exists m_0 \forall m \geq m_0 : \quad z_m \leq H + \epsilon.$$

It will then follow that $\limsup z_m \leq H$, as required.

Since, by Corollary 9, $H = \limsup(\log p)/|x_p|$, we can conclude that, for $\delta = \epsilon/2$,

$$\exists p_0 \forall p \geq p_0 : \quad \frac{\log p}{|x_p|} \leq H + \delta.$$

Hence, $|x_p| \geq (\log p)/(H+\delta)$. Choose m_0 such that

$$\sum_{p=1}^{m_0} \log p \geq (2 + H/\delta)[\sum_{p=1}^{p_0} \log p - (H+\delta) \sum_{p=1}^{p_0} |x_p|].$$

Such an m_0 clearly exists since the left hand side tends to infinity while the right hand side is a constant. Let now $m \geq m_0$. We have,

$$z_m = \frac{\sum_{p=1}^{m} \log p}{\sum_{p=1}^{m} |x_p|} = \frac{\sum_{p=1}^{m} \log p}{\sum_{p=1}^{p_0} |x_p| + \sum_{p=p_0+1}^{m} |x_p|} \leq$$

$$\frac{\sum_{p=1}^{m} \log p}{\sum_{p=1}^{p_0} |x_p| + 1/(H+\delta) \sum_{p=p_0+1}^{m} \log p} = \frac{(H+\delta) \sum_{p=1}^{m} \log p}{(H+\delta) \sum_{p=1}^{p_0} |x_p| + \sum_{p=p_0+1}^{m} \log p} =$$

$$\frac{(H+\delta) \sum_{p=1}^{m} \log p}{(H+\delta) \sum_{p=1}^{p_0} |x_p| - \sum_{p=1}^{p_0} \log p + \sum_{p=1}^{m} \log p} = \frac{H+\delta}{1-Q},$$

where

$$Q = \frac{\sum_{p=1}^{p_0} \log p - (H+\delta) \sum_{p=1}^{p_0} |x_p|}{\sum_{p=1}^{m} \log p}.$$

Using now the choice of m_0 and the fact that $m \geq m_0$ we have that

$$\sum_{p=1}^{m} \log p \geq \sum_{p=1}^{m_0} \log p \geq (2 + H/\delta)[\sum_{p=1}^{p_0} \log p - (H+\delta) \sum_{p=1}^{p_0} |x_p|];$$

hence,

$$Q \leq \frac{\delta}{H + 2\delta},$$

and, consequently,

$$\frac{H+\delta}{1-Q} \leq \frac{H+\delta}{1 - \delta/(H+2\delta)} = H + \epsilon.$$

Altogether, $z_m \leq H + \epsilon$, as claimed. This concludes the proof of Proposition 10. ∎

Finally we have the main result of this paper.

Theorem 11 *Let X be an infinite factorial subset of A^* and let τ be the compression of the Ziv-Lempel algorithm,* ZL: $A^* \to \mathbf{2}^*$. *Then, $\tau(X) \leq H(X)$, i.e.* ZL *is a universal continuous compression which is optimal on every factorial source X.*

Proof. For $n \geq 1$, let x_n be the n-th word of X in military order. Let us consider the tree $T(x_n)$ and let m_n and I_n denote its number of edges and its internal length. Since X is factorial, every one of the m_n words contributing to I_n lies in X; hence

$$I_n \geq \sum_{p=1}^{m_n} |x_p|.$$

Using then Proposition 2 and Lemma 5

$$\tau(x_n) \leq \frac{m_n \log m_n + K m_n}{I_n} \leq \frac{\sum_{p=1}^{m_n} \log p + K' m_n}{\sum_{p=1}^{m_n} |x_p|},$$

for appropriate constants K and K'. Since $\sum \log p$ is the dominant term in the numerator we have

$$\tau(X) = \limsup_{n\to\infty} \tau(x_n) \leq \limsup_{n\to\infty} \frac{\sum_{p=1}^{m_n} \log p}{\sum_{p=1}^{m_n} |x_p|}.$$

Since there are only finitely many trees $T(x)$ with any given number m of edges, we have the fiirst equality below. The second one comes from Proposition 10.

$$\limsup_{n\to\infty} \frac{\sum_{p=1}^{m_n} \log p}{\sum_{p=1}^{m_n} |x_p|} = \limsup_{m\to\infty} \frac{\sum_{p=1}^{m} \log p}{\sum_{p=1}^{m} |x_p|} = H(X).$$

Altogether, $\tau(X) \leq H(X)$; concluding the proof of Theorem 11. ∎

Let A be a fixed set now. Consider, for every $n \geq 1$, the word x_n of length n such that $\tau(x_n) = \max\{ \tau(x) \mid x \in A^n \}$. The reader can verify that X is an infinite set of entropy $H(X) = 0$, for which $\tau(X) = \log |A|$. Of course if Y is the set of factors of X then $H(Y) = \tau(Y) = \log |A|$. It follows that Theorem 11 does not hold for arbitrary infinite subsets X of A^*.

5 Some Remarks

Throughout the paper we considered a binary output alphabet but now we wish to argue that the results do not depend on this particularity. Let B be an output alphabet, with $|B| = k$, and let $\gamma \colon A^* \to B^*$ be a compression. In order to maintain our results, we would have to define the compression coefficient as

$$\tau(x) = \frac{|\gamma(x)|}{|x|} \log k.$$

This is necessary in order to normalize the measure of the compression rate in "bits per character", and keep it compatible with the entropy.

An equivalent solution, perhaps even preferable, would be to change all the logarithms to base k and continue to use $\tau(x) = |\gamma(x)|/|x|$. This corresponds to measure entropy and compression in k-ary digits per character. This is preferable because the compression coefficients will be less than one exactly when there is a compression in the ordinary sense of the word.

We leave it as an interesting exercise to verify that all the results are indeed maintained after the obvious changes are made.

Another question relates to expansions. Let $\gamma\colon A^* \to A^*$ be a compression with identical input and output alphabets. If γ is a total function then clearly $|\gamma(x)| < |x|$ for some x if and only if $|\gamma(y)| > |y|$ for some y. In other words, every total proper compression must expand some words. Can one limit the amount of expansion? The answer is "yes" because we can use the first output character to show whether or not there is compression and in case the compression would expand we just transmit the input. With this protocol one can see that expansion can be limited to one character, without compromising the compression rate. We make two observations. First, something similar happens in real life. Indeed, the UNIX utility `compress` refuses to compress a file if it would expand. The corresponding bit indicating whether or not compression was performed is put in the name of the file which will or not have the extension ".Z". In second place we note that limiting expansions in the above way destroys continuity. Is it possible to limit expansion while maintaining continuity? This might lead to interesting questions. In particular one can verify that ZL uses unlimited expansion and thus it does not settle the question.

Now we make some rèmarks about the estimation of the topological entropy. A consequence of the proof of Theorem 11 is that if one can enumerate in military order the words of a set X then the compression rate $\tau(x_1x_2\ldots x_m)$ of the concatenation of the first m words of X will give a numerical estimation of the entropy of X. Here, τ refers to the Ziv-Lempel algorithm and we have to assume that X is factorial. Of course $H(X)$ is the $\limsup$ of the above numbers and some preliminary experiments indicate that convergence might be (very) slow. Nevertheless, one has a method of estimating the entropy of X.

Another way of estimating the entropy would be to consider the values of $(\log p)/|x_p|$. According Corollary 9 the $\limsup$ of such a sequence is always $H(X)$. This again assumes that we can compute $|x_p|$ for a given p.

The entropy of a rational language can be easily esimated using classical methods of numerical computation instead of performing tests on certain sequences as in the general case. Let indeed X be a rational language and let A be a deterministic (or, more generally, unambiguous) automaton recognizing X. We assume A to be trim. Let M be the adjacency matrix of the graph of the automaton. Then it is a classical result that

$$H(X) = \log \rho,$$

where ρ is the maximal eigenenvalue of M, also called its Perron eigenvalue. The computation of $H(X)$ therefore reduces to the computation of the maximal eigenvalue of a positive matrix. This is a well-known computational problem since ρ is also the maximal root of the characteristic polynomial $p(\lambda)$ of the matrix M.

In practice, one may also use the following more direct and elementary method. Let M be an $n \times n$ positive matrix. We assume M to be irreducible and aperiodic. Let x be a vector with all its components strictly positive and let $R_{max}(x) = \max\{\, (xM)_i/x_i \mid 1 \le i \le n\,\}$ and $R_{min}(x) = \min\{\, (xM)_i/x_i \mid 1 \le i \le n\,\}$. It is well known that for any such vector x one has the inequalities $R_{min}(x) \le \rho \le R_{max}(x)$, hence giving an estimate of the Perron eigenvalue of M. This estimate is especially visible in the case where x is the vector with all its components equal to one. In this case, R_{min} (resp. R_{max}) is the minimum (resp. the maximum) of the indegrees of the vertices. Also, the estimate can be made as precise as necessary since for any such vector x, the sequence starting at $x_0 = x$ and defined by $x_{n+1} = x_n M$ is such that R_{max} and R_{min} converge to ρ.

Ranking functions have received some attention recently, both as a way of compression and also from the computational point of view. See, for instance [5] and the references therein, where another theory of compression is delineated.

We close this paper by considering very briefly a probabilistic model of a source and indicate that in this case also entropy is a lower bound for compression. The results in this case however are not as complete as for the topological model of a source.

We consider an invariant probability measure ν on A^*. By this we mean a function $\nu \colon A^* \to [0,1]$ such that for every $u \in A^*$ one has

$$\nu(u) = \sum_{a \in A} \nu(au) = \sum_{a \in A} \nu(ua)$$

and such that $\nu(1) = 1$. The (metric) entropy of ν is given by

$$h(\nu) = \lim_{n \to \infty} \frac{1}{n} \sum_{u \in A^n} -\nu(u) \log \nu(u).$$

Let X be the support of ν, i.e. $X = \{ u \in A^* \mid \nu(u) > 0 \}$. Then $h(\nu) \leq H(X)$, since one can verify that

$$-\sum_{u \in A^n} \nu(u) \log \nu(u) \leq \log |X \cap A^n|.$$

Let γ be a total compression. We define the metric compression rate $\tau(\nu)$ of γ on ν by

$$\tau(\nu) = \limsup_{n \to \infty} \sum_{u \in A^n} \nu(u) \frac{|\gamma(u)|}{|u|}.$$

By exploiting the convexity of the logarithmic function one can prove the following result.

Proposition 12 *For every ν and γ as above, $h(\nu) \leq \tau(\nu)$.*

References

[1] R. Adler, A. Konheim, and M. McAndrew. Topological entropy. *Trans. Amer. Math. Soc.*, 114:309–319, 1965.

[2] T. C. Bell, J. G. Cleary, and I. H. Witten. *Text Compression.* Prentice Hall, Englewood Cliffs, N.J., 1990.

[3] N. Chomsky and G. A. Miller. Finite state languages. *Information and Control*, 1:91–112, 1958.

[4] A. de Luca. On the entropy of a formal language. In H. Brakhage, editor, *Automata Theory and Formal Languages*, pages 103–109, Springer-Verlag, Berlin, 1975. Lecture Notes in Computer Science, 33.

[5] A. V. Goldberg and M. Sipser. Compression and ranking. *SIAM J. Comput.*, 20:524–536, 1991.

[6] G. Hansel. Estimation of the entropy by the Lempel-Ziv method. In M. Gross and D. Perrin, editors, *Electronic Dictionaries and Automata in Computational Linguistics*, pages 51–65, Springer-Verlag, Berlin, 1989. Lecture Notes in Computer Science, 377.

[7] A. I. Khinchin. *Mathematical Foundations of Information Theory*. Dover, New York, 1957.

[8] D. E. Knuth. *The Art of Computer Programming, Vol. 1, Fundamental Algorithms*. Addison-Wesley Pu. Co., Reading, MA, 1968.

[9] W. Kuich. On the entropy of context-free languages. *Information and Control*, 16:173–200, 1970.

[10] M. Lothaire. *Combinatorics on Words*. Volume 17 of *Encyclopedia of Mathematics and its Applications*, Addison-Wesley Pu. Co., Reading, MA, 1983.

[11] N. F. G. Martin and J. W. England. *Mathematical Theory of Entropy*. Volume 12 of *Encyclopedia of Mathematics and its Applications*, Addison-Wesley Pu. Co., Reading, MA, 1981.

[12] S. E. Newhouse. Entropy in smooth dynamical systems. 1990. To appear in *Proceedings of the 1990 World Congress of Mathematicians*.

[13] D. Ornstein and B. Weiss. Entropy and data compression schemes. 1990. manuscript.

[14] D. S. Ornstein and P. C. Shields. Universal almost sure data compression. *The Annals of Probability*, 18:441–452, 1990.

[15] A. Rényi. *A Diary on Information Theory*. John Wiley & Sons, New York, NY, 1984.

[16] C. E. Shannon. A mathematical theory of communication. *Bell System Technical J.*, 27:398–403, 1948.

[17] J. A. Storer. *Data Compression - Methods and Theory*. Computer Science Press, Rockville,MD, 1988.

WORDS AND REWRITING

Iterative devices generating infinite words *

Karel Culik II
Dept.of Computer Science
University of South Carolina
Columbia, SC 29208, USA

Juhani Karhumäki
Department of Mathematics
University of Turku
Turku, Finland

Abstract

We consider various TAG-like devices that generate one-way infinite words in real time. The simplest types of these devices are equivalent to iterative morphisms (also called substitutions), automatic sequences and iterative DGSM's. We consider also a few new types. Mainly we study the comparative power of these mechanisms and develop some techniques for proving that certain devices cannot produce a particular infinite word.

1 Introduction

We study simple mechanisms that in real time generate (one-way) infinite sequences (ω-words) over a finite alphabet Σ. Formally, an ω-word w over alphabet Σ is a function f_w : $\boldsymbol{N} \rightarrow \Sigma$. If the function f_w can be specified by a finite automaton (more precisely a sequential machine) for a natural numbers represented in m-ary notation, then w is called m-automatic or simply automatic [7].

Simple mechanisms, called limited and general TAG-systems, that have the power to generate aperiodic sequences, have been also considered in [7]. We introduce them in Section 4 under the names D0L and CD0L TAG-systems, respectively. Cobham has shown that CD0L TAG-system generate exactly the automatic sequences. The definitions in Section 3 make clear that (C)D0L TAG-systems are equivalent to ω-(C)D0L systems considered in [13, 10, 2, 14]. In [14] and elsewhere an ω-D0L system is called a substitution.

An ω-word is effectively computable if it is produced by a deterministic Turing Generator [17, 19]. Note, that every deterministic generator with finite memory produces an ultimately periodic sequence. The (C)D0L TAG-systems constitute the simplest models that generate ω-words in real time, the real time generation is assured by allowing only nonerasing morphisms, therefore at least one symbol is written at each step. Viewed as a special case of a Turing generator, a TAG-system are not finite memory devices since there can be an unbounded gap between the reading and the writing head; however, it is space-efficient in the sense that it does not use any extra memory just rereads the generated sequence.

The other real time generators considered here are D0L TAG-systems with periodic or D0L outside control, DGSM TAG-systems and double D0L or triple D0L TAG-systems. The latter

*This work was done during the first author's stay at the University of Turku, Finland, supported by the Academy of Finland.

generate one or two auxiliary tapes that interact with the generating tape. In the case of a noninteractive control a finite number of morphisms is applied in a fixed sequence. The well known Kolakovski sequence [20, 22, 23, 14, 15, 30] is produced by a D0L TAG-system with periodic control, see Section 3.

In Section 3 we describe our real time iterative models and start to consider their relative power. In Section 4 we give a number of examples, some of them will be useful later when proving proper inclusion between classes of sequences produced by our devices. The last section contains some rather difficult technical lemmas that make it possible to prove negative criteria of the form: "certain ω-words cannot be generated by a device of certain type". We use these criteria to compare the power of the studied models. The results are summarized in Figure 2.

D0L words, DGSM words and other words generated by our devices are not only of interest to mathematicians but recently are increasingly used in other areas as well. D0L words interpreted as line drawings have been extensively used to generate various fractal images, especially of plants [31, 26]. The other more powerful mechanisms, e.g. DGSM, can be used as a more powerful tool to specify more complicated images, see [12]. An important property of our iterative mechanisms is that by using very simple rules they define some (apparently) complex sequences. This can be used when it is desired to obtain some mixture of order and chaos (disorder, complexity). One application is again in applied fractal geometry [9]. Regular sequences of affine transformation (MRFS) might produce a perfectly symmetric image, for example, an ideal tree. By using a DGSM-word to control the sequence of affine transformation a "naturally" looking trees are obtained. Here the "order" requires that every tree produced is of the same type, the "disorder" that every tree produced is different and within the same tree the "subtrees" also differ. DGSM-words and their two-dimensional generalization have been also used to obtain the proper mixture of order and chaos in crystalography, see [5, 3, 1].

2 Preliminaries

Here we fix our notation and recall some basic notions needed in this paper. Specific definitions of this paper are given in the next section. Also we note that several notions are defined, or proper references are given, in the connection with their actual use.

The central object of this paper is an infinite word, often referred to as an ω-word, over a finite alphabet Σ. Formally, such a word is a mapping from $\boldsymbol{N}$ into Σ, but we prefer to view it as a sequence of symbols of $\Sigma : a_0a_1a_2\ldots$. The set of all ω-words over Σ is denoted by Σ^ω. Sometimes we talk only about words when meaning infinite words. Since our topic is to generate an infinite word we identify, by convention, words w and $\$w$, where $w \in \Sigma^\omega$ and $\$$ is a symbol not in Σ.

We shall frequently need several special types of ω-words. We call an infinite word w *marked* over Σ iff it is of the form

$$w = w_1\#w_2\#w_3\#\ldots \tag{1}$$

where w_i's are finite words over Σ, that is in Σ^*, and $\#$ is not in Σ. We call a marked word *binary* (resp. *unary*) iff Σ is a binary (resp. unary) alphabet. Consequently marked unary words are actually over a binary alphabet. Further we call a marked word (1) *sparse* iff

$$\liminf_{i\to\infty} |w_i| = \infty, \tag{2}$$

where $|w_i|$ denotes the length of the word w_i.

Let w be an infinite word over Σ. We say that an infinite word w' *simulates* w if

(i) w' is over $\Sigma' = \Sigma \cup \{\#\}$, where $\# \notin \Sigma$, and

(ii) $\pi(w') = w$, where π is the projection of Σ' into Σ,

that is it erases the markers # and is the identity on Σ. Moreover, we say that w' *weakly simulates* w iff Σ' above is an arbitrary superset of Σ.

Our methods of generating infinite words are based on iteration or repeated application of simple mappings, namely morphisms of free monoids or as their generalization DGSM-mappings. However, we strongly concentrate on the real time generation of infinite words, so that, if not otherwise stated, our mappings are ε-free, that is nonerasing. Consequently, by a morphism h we mean a morphism between two finitely generated free semigroups, in symbols $h : \Sigma^+ \to \Delta^+$. A special type of such a morphism is a *coding* which maps each letter to another letter. Similarly, by a DGSM we mean an ε-free DGSM.

One important technical notion needed several times later is that of a *marker-to-marker* generation of a marked ω-word. The precise definition (before Lemmas 6 and 8), depends on the model used to generate such words. However, intuitively it means that the markers of the word are (ultimately) generated from previous markers in one-to-one way.

The theories of D0L systems, D0L functions and $\boldsymbol{N}$-rational functions are needed in our technical lemmas. Consequently, we need here some basic terminology, for more details cf. [27] and [28]. A D0L *system* consists of an alphabet Σ, an element w_0 of Σ^+ and a morphism h from Σ^* into Σ^*. It defines a D0L sequence $\{w_i\}_{i \geq 0}$, where $w_i = h(w_{i-1})$, and a D0L *function* $f : f(i) = |w_i|$. D0L functions are closely connected to $\boldsymbol{N}$-*rational functions*. Indeed, each $\boldsymbol{N}$-rational function $g : \boldsymbol{N} \to \boldsymbol{N}$ can be decomposed to a finite number of D0L functions, say $f_0, \ldots, f_{p-1}$, as follows:

$$f_j(i) = g(j + ip) \quad \text{for all } i \geq 0, 0 \leq j \leq p-1 \,, \tag{3}$$

and conversely, for any sequence $f_0, \ldots, f_{p-1}$ of D0L functions the function g defined by (3) is $\boldsymbol{N}$-rational. We shall need this important result, cf. [28], several times.

3 Models

Here we define our basic models to generate infinite words. First, however, we explain our approach. Our main interest is to generate infinite words by different but related mechanisms. We are interested in defining infinite words directly as limits of certain procedures rather than, for example, by using the adherence of finite words. Neither do we allow erasing in our models, thus emphasizing the real time generation. Several of our mechanisms have already been studied in a number of papers, but a few of them seem to be new. One of our goals is to introduce a unified model for these mechanisms.

Perhaps the simplest and most natural mechanism to define infinite words is to iterate a morphism $h : \Sigma^+ \to \Sigma^+$ on a letter a satisfying

$$a \prec h(a), \tag{4}$$

or equivalently, $h(a) = az$ for some z in Σ^+. Then, clearly

$$w = h^\omega(a) = \lim_{i \to \infty} h^i(a) \tag{5}$$

is a well defined element of Σ^ω. We call $h^\omega(a)$ an infinite word obtained by iterating h on a. Further if (4) is true, we say that h is *prolongable* on a, cf. [6], or satisfies the *prefix condition* on a, cf. [13]. Of course, instead of a letter a in (4) we could use an arbitrary word x, and results would not change.

The above infinite word w was obtained by *iterating* a morphism h, as illustrated in Figure 3: The same sequence can be obtained using the same objects h and a in a different way as follows. We consider a potentially infinite queue with one reading head on the left and one writing head on the right. At the beginning the contents of the queue is $h(a)$. Now, we proceed indefinitely by reading the current symbol b on the left and writing its image $h(b)$ on the right. This is illustrated in Fig. 4.

We say that an infinite word determined in this way starting from any word x (instead of $h(a)$) is TAG-*generated* by a morphism h. This mechanism was introduced by Post in 1920's and was made well-known in [24]. Actually, our model is a TAG-system with deletion number 1. We call the system shown in Fig. 4 a D0L TAG-*system*. The connection between the iteration and the TAG-generation is completely straightforward.

Lemma 1 *Each word $w \in \Sigma^\omega$ obtained by iterating a morphism is also obtained by a* D0L TAG-*system. Conversely, for each word $w \in \Sigma^\omega$ obtained by a* D0L TAG-*system, the word Sw, with $S \notin \Sigma$, is obtained by iterating a morphism.*

Although the above two mechanisms define the same class of infinite words, they define such words in quite a different way. Indeed, after n steps of the generation the length of the word obtained by iterating h is $|h^n(a)|$, that is typically of exponential size in terms of n, while the length of the word obtained by the TAG-generation is only of linear size.

We call an infinite word w a D0L *word* if it, or Sw, is defined by either of the above mechanisms. Indeed, the iterative mechanism determines such a word as the unique limit of a D0L sequence. Although, we have chosen the "D0L-terminology", the TAG-mechanism suits much better our purposes when defining more general devices producing infinite words.

There exist at least two natural ways to generalize the notion of a D0L TAG-system. First, the word obtained is processed after its generation by some mechanism. Second, mappings more general than morphisms are used.

A CD0L *word*, that is to say, a morphic image of a D0L word under a coding, provides a simple and natural extension of the first type. Such a sequence can be generated as follows. We use either of our earlier mechanisms with a modification that in the generation it is allowed to index letters with an invisible ink which is not noticed when the result is actually recorded. Another way to describe such sequences formally is to use two tapes: The second tape is a *control tape* and generates a D0L sequence by h, which is then translated by using a coding c to the contents of the first tape which is the actual *generating tape*. The situation is depicted in Fig. 5.

Of course, instead of codings we could use more general mappings such as ε-free morphisms (recall that we do not allow erasing), but this does not seem to provide anything essentially new.

A natural example of an extension of the second type is to consider an ε-free deterministic gsm M instead of a morphism h. Clearly, our earlier considerations on morphisms including Lemma 1 are valid for gsm's as well. Here, of course, an analog of (4) is assumed. To generate such an infinite word, a DGSM *word*, we prefer to use the following two-tape model which we call a DGSM TAG- *system*. As in the CD0L case the first tape is the generating tape and the second is the control tape. Here, of course, the control tape records only the current state of the DGSM M. The whole mechanism is illustrated in Figure 6, where q' is the new state and $\lambda(q,a)$ denotes the current output.

The reason we wanted to define both CD0L and DGSM mechanisms as two-tape TAG-type systems is that now both of them are obtained in a natural way as restricted cases of a more general system. A *double* D0L TAG-*system* consists of two D0L TAG-systems which are interconnected. Consequently, the rules are of the form

$$\begin{pmatrix} a \\ b \end{pmatrix} \rightarrow \begin{pmatrix} \alpha \\ \beta \end{pmatrix} \tag{6}$$

and the generation starts as follows: At the beginning the contents of the tapes are empty and after the first application of a rule of form 6 they are α and β, respectively. Consequently, we assume, although this is not important, that a is a prefix of α and b is a prefix of β like in iterative morphisms. The model can be illustrated as in Fig. 7.

By definition the two reading heads progress synchronously while the writing heads may run at different speeds. We say that an infinite word w is a *double* D0L *word* if it is obtained as a component, i.e. as the contents of the 1-st tape, of a double D0L TAG-system. This is so because we use double D0L TAG's to generate infinite words rather than pairs of infinite words. Therefore it is natural to refer to the first tape as the *generating* tape and to the second as the *control* tape. This terminology becomes even better motivated by the restricted cases we will be considering next.

Before defining these restricted versions of double D0L TAG-systems let us consider the control mechanisms occurring therein. In a TAG-type system the generation of a word is always controlled by the word itself, i.e. by the generating tape itself. Such a control can be called *self-control* or *inside* control. This occurs most typically in a D0L TAG-system. On the other hand, the control tape can be used as further control of the generation. We call such a control *outside* control which can be either *interactive* or *noninteractive*. In a CD0L TAG we have noninteractive outside control and no inside control, while in a DGSM TAG we have an inside control together with a restricted interactive outside control. The latter outside control, i.e. the sequence of states of a DGSM, is not a "pure" outside control because it depends on the generated sequence as well, hence the name interactive outside control. After these remarks it is natural to define some models using only noninteractive outside control.

A D0L TAG-*system with* X *control* is a double D0L TAG-system where the control tape is independent of the generating tape and itself generates a word of type X. Informally, such a model can be illustrated as in Fig. 8.

We restrict our considerations to the cases where X is as simple as possible, that is when X is either "periodic" or "D0L". Words obtained in this way are called D0L *words with periodic control* and D0L *words with* D0L *control*, respectively. The rewriting rules (6) for the latter case split into the following

$$\begin{pmatrix} a \\ b \end{pmatrix} \rightarrow \alpha \text{ and } b \rightarrow \beta \tag{7}$$

as is immediate by Fig. 8. Consequently, a D0L TAG-system with D0L control is determined by $n+1$ morphisms, where n is the size of the alphabet of the second tape. In the second tape only the $(n+1)$st morphism is used in the TAG way, while in the generating tape the choice of the morphism used in the TAG way is determined by the current control symbol.

We make the following simple observations.

Lemma 2 *Each* CD0L *word is a* D0L *word with* D0L *control.*

Lemma 3 *Each* D0L *word with periodic control is a* D0L *word with* D0L *control.*

Lemma 4 *Each* D0L *word with periodic control is a* DGSM *word.*

We introduced double D0L TAG-systems in order to capture both CD0L and DGSM extensions of D0L words. If we want to capture both of these extensions simultaneously, i.e. to generate codings of DGSM words, we can do it by introducing a third tape, i.e. *another control* tape, and it is intuitively pretty clear that this third tape is really necessary. Based on that we formalize a *triple* D0L TAG-*system* as three ordinary D0L TAG-systems, where tapes are completely interconnected. Consequently, the rules of a triple D0L TAG-system are of the form

$$\begin{pmatrix} a \\ b \\ c \end{pmatrix} \to \begin{pmatrix} \alpha \\ \beta \\ \gamma \end{pmatrix} \tag{8}$$

In this framework a CDGSM TAG-*system* can be depicted as in Fig. 10. Consequently, (8) splits in this case to

$$\begin{pmatrix} a \\ b \end{pmatrix} \to \begin{pmatrix} b' \\ \gamma \end{pmatrix} \quad \text{and} \quad c \to a'$$

where, moreover no queue is needed on the second tape. This means that CDGSM TAG-systems are very special cases of triple D0L TAG's. There are, of course, many other special cases of triple D0L TAG-systems but we do not see enough motivation to formalize those here.

According to our goals we use also triple D0L TAG's to generate infinite sequences, so that we can talk about *triple* D0L *words* and CDGSM *words.* Therefore it is again proper to call the first tape the generating tape and the other two the control tapes.

4 Examples

This section is devoted to examples illustrating the power of our different mechanisms to generate infinite words. A number of such examples has been introduced for the most basic mechanisms such as D0L- and DGSM-mechanisms, cf. [6]. More examples are given in the full version of this paper [11].

Example 1: CD0L sequences are obtained from D0L sequences by merging some letters. Therefore, for any morphism $h : \Sigma^+ \to \Sigma^+$, the sequence of the form

$$Sa^{|x|}\#a^{|h(x)|}\#a^{|h^2(x)|}\#\ldots$$

is a CD0L sequence, specific examples being

$$Sa\#a^4\#a^9\#\ldots\#a^{n^2}\#\ldots$$

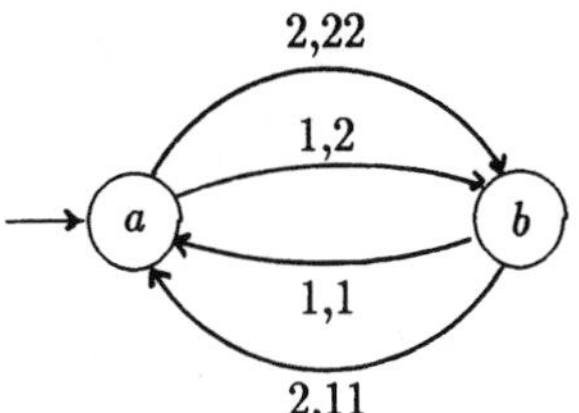

Figure 1: A DGSM for the Kolakovski sequence

and

$$Sa\#a^2\#a^3\#a^5\#\ldots\#a^{t_n}\#\ldots,$$

where t_n is the nth Fibonacci number.

Example 2: Even a periodic noninteractive control can produce really complicated sequences as shown by the Kolakovski sequence, cf. [15]. It is obtained by using morphisms

$$h_a : \begin{array}{ccc} 1 & \rightarrow & 2 \\ 2 & \rightarrow & 22 \end{array} \quad \text{and} \quad h_b : \begin{array}{ccc} 1 & \rightarrow & 1 \\ 2 & \rightarrow & 11 \end{array}$$

in the TAG way starting at 2 and with the control word $(ab)^\omega$:

$$\begin{array}{rcccccccccccccc} & 2 & 2 & 1 & 1 & 2 & 1 & 2 & 2 & 1 & 2 & 2 & 1 & 1 & \ldots \\ \mathit{control}: & a & b & a & b & a & b & a & b & a & b & a & b & a & \ldots \end{array}$$

The Kolakovski sequence possesses many interesting properties, cf. [15], and at the same time is a source of several intriguing open problems, cf. [15, 29, 12]. By Lemma 4, the Kolakovski sequence can be obtained by iterating a DGSM M on a word α. Indeed, we can choose $\alpha = 2$ and M as shown in Fig. 1.

Example 3: The morphism h given by

$$\begin{array}{ccc} \$ & \rightarrow & \$a\# \\ a & \rightarrow & abbc \\ b & \rightarrow & bc \\ c & \rightarrow & c \\ \# & \rightarrow & c\#, \end{array}$$

generates the D0L sequence

$$w = \$a\#abbcc\#abbcbcbcccc\#\ldots . \tag{9}$$

Here the lengths of the words in between markers (including the right-hand side marker) grows as $2 = 2 \times 1, 6 = 3 \times 2, 12 = 4 \times 3, \ldots (n+1) \times n \ldots$. Let this sequence be the D0L control sequence for the morphisms

$$\begin{array}{rcl} h_\$ & : & \begin{array}{lll} \# & \rightarrow & \#a\# \\ x & \rightarrow & x \quad \text{for } x \neq \#, \end{array} \\ h_a = h_b = h_c & : & x \rightarrow x \quad \text{for all } x, \\ h_\# & : & \begin{array}{lll} \# & \rightarrow & a\# \\ x & \rightarrow & x \quad \text{for all } x \neq \#. \end{array} \end{array}$$

It follows from this construction, that the sequence generated repeats $\#a$ once, then $\#aa$ twice, then $\#aaa$ three times and, in general, $\#a^n$ n times. In other words the sequence is

$$\#a(\#aa)^2(\#aa)^3\dots(\#a^n)^n\dots \tag{10}$$

Consequently, the n-th word in between markers is of the order $\sqrt{n}$. It is not difficult to see that the above can easily be modified to satisfy $|w_n| = [\sqrt{n}]$ where w_n is the n-th word in between markers.

We also have, cf. e.g. [4],

Example 4: The ω-word

$$0\#1\#01\#11\#\dots\#\text{r-bin}(n)\#\dots$$

is a DGSM word, where r-bin(n) is the reverse binary representation of the number n.

Example 5: For a given Turing machine M and its initial configuration w_0, let w_n be the configuration of the machine after the n-th step. Then

$$w_0\#w_1\#w_2\#\dots\#w_n\#\dots$$

is a DGSM word. This simple observation has several interesting consequences. Let us recall that an ω-word $a_1a_2\dots$ over Σ, or a real number $0.a_1a_2\dots$ at the base card(Σ), is *computable*, cf. [17], if it can be generated by a Turing machine. In order to generate words by Turing machines it is natural to use a specific output tape, but as it is well known, this is not necessary. For our purposes it is better to consider Turing machines having only one tape.

Now, let

$$w = a_1a_2a_3\dots \qquad , a_i \in \Sigma, \tag{11}$$

be a computable ω-word. Then, by above, w can be weakly simulated by a DGSM word, i.e. a DGSM word in $(\Sigma\cup\Sigma')^\omega$ where $\Sigma'\cap\Sigma = \emptyset$. Moreover, w can be simulated by a CDGSM word, i.e. a DGSM word in $(\Sigma\cup\#)^\omega$, where $\# \notin \Sigma$. Finally, if erasing is allowed, then w can be generated as a morphic image of DGSM word, even under the morphism which maps each letter to a letter or the empty word. Observe also that given w, that is an M defining w, the above mentioned DGSM and CDGSM words can be effectively generated.

On the other hand, not all ω-words of the form (11) are either CDGSM words, or triple D0L TAG words, or even their natural generalizations to the n-tape case. A simple argument for that is the standard diagonalization. Indeed, any of our TAG-mechanisms produces always an infinite word, and consequently *the next symbol in the generated sequence is always defined.* Hence, the diagonalization works here.

5 Results

The goal of this section is to search for methods to prove that certain infinite words are not generated by certain of our systems, for detailed proofs of our results cf. [11]. It has to be emphasized that there seems to be very few results of this nature. Indeed, although it is clear by a diagonalization argument that there exist computable ω-words which are not CDGSM words, or even triple D0L words, we do not know any concrete example of a word for which this can be proved.

Our first two criteria are based on so called subword complexity. We formulate it for infinite words, and not for languages of finite words as it is usually done. The *subword complexity* of an ω-word w is the function $f_w : \mathbf{N} \rightarrow \mathbf{N}$ satisfying

$$f_w(n) = \text{ the number of subwords of } w \text{ with the length } n.$$

By example 10, a DGSM word can have an exponential subword complexity. On the other hand it has been show in [16] that the subword complexity of a D0L word is always at most quadratic. Moreover, since codings do not increase the subword complexity, the same result holds for CD0L words as well.

Lemma 5 *For any* D0L *or* CD0L *word w, the subword complexity $f_w(n)$ is in $O(n^2)$.*

Next we turn to consider subword complexity in connection with interactive control. Our result, although stated in general terms, is really intended to show that the word of Example 10 cannot be obtained by a D0L TAG-system with periodic control.

We recall the definition of a marked word over Σ and define a *marked subword complexity* of such a word w to be the function $m_w : \mathbf{N} \rightarrow \mathbf{N}$ defined by

$$m_w(n) = \text{ the number of subwords of } w \text{ in } \Sigma^+ \text{ having the length } n.$$

Finally, we say that a marked infinite word w is *marker-to-marker* generated by a D0L TAG-system with or without control iff

(i) almost all occurrences of letters a from $\Sigma \cup \{\#\}$ produce at most one marker;

(ii) almost all occurrences of letters from Σ produce no markers;

(iii) almost all occurrences of markers produce exactly one marker.

As usual, the notion "almost all" means all but a finite number of cases. Observe also that (i) is actually superficial, and is stated here only for clarity.

With the above terminology we can prove

Lemma 6 *Let w be a marked binary word which is marker-to-marker generated by a* D0L TAG-*system with periodic control. Then the marked subword complexity $m_w(n)$ is in $0(2^{(1-\varepsilon)n})$ for some* $\varepsilon > 0$.

In our second criterion we deal with marked ω-words of the form

$$w = Sa^{n_0}\#a^{n_1}\#a^{n_2}\#\ldots \tag{12}$$

which, in addition are sparse, that is satisfy

$$\liminf_{i \rightarrow \infty} n_i = \infty. \tag{13}$$

Lemma 7 *Let w satisfy (12) and (13). Then w is a CD0L word iff the function*

$$f_w(i) = n_i \tag{14}$$

is $\mathbf{N}$-rational.

In our last lemma we again consider sparse marked unary ω-words, i.e. words satisfying (12) and (13). Our goal is to give a necessary condition for such a word to be a DGSM word. We need a preliminary lemma describing how such sequences can be generated. We redefine the notation of marker-to-marker generation for DGSM's as follows. We say that a DGSM *marker-to-marker* generates the sequence (12) if for some constant K the following conditions hold:

(i) almost all subwords of (12) of the length K generate at most one marker;

(ii) only a finite number of markers in (12) are generated from the letter a preceded by at least $K-1$ letters a;

(iii) almost all markers in (12) are generated from subwords of the form $\#a^{K-1}$, and almost all such occurrences generate exactly one marker.

Lemma 8 *Let w be a sparse marked unary ω-word. If w is a DGSM word, then it is marker-to-marker generated by a DGSM TAG-system.*

Lemma 9 *For each sparse marked unary* DGSM-*word*

$$w = Sa^{n_0}\#a^{n_1}\#a^{n_2}\#\ldots$$

there exist constants i_0 and k such that, for all $i \geq i_0$, one of the following conditions holds

(i) $n_i - \alpha < n_{i+k} \leq n_i + \alpha$, *for some* α,

(ii) $\alpha_1 n_i \leq n_{i+k} \leq \alpha_2 n_i$, *for some* $\alpha_1, \alpha_2 > 1$.

We have established criteria to show that some infinite words are not generated by certain devices. Now we will use them to show that particular infinite words are not generated by certain of our devices. Indeed, we are going to establish the diagram of Figure 2 showing relations between the families of infinite words considered in this paper.

In Figure 2 solid line denotes the inclusion (the lower family is included in the upper one), the arrow denotes a proper inclusion, and a dotted arrow from X to Y denotes that X is not included in Y. Consequently, two-directional dotted arrow denotes the incomparability. If there is no direct connection (nor an ascending chain of solid lines or arrows) between two families, then their relation is open.

To establish Figure 2 we consider the following four infinite words, cf. Examples 1, 4, 3, respectively:

$$\text{PER} = Sa\#aa\#aa\#aaa\#aaa\#\ldots,$$

where the number of a's in between markers is increased by one in every second time,

$$\text{SQUARE} = Sa\#a^4\#a^9\#a^{16}\#\ldots\#a^{n^2}\#\ldots,$$

$$\text{BIN} = S0\#1\#01\#11\#001\#\ldots,$$

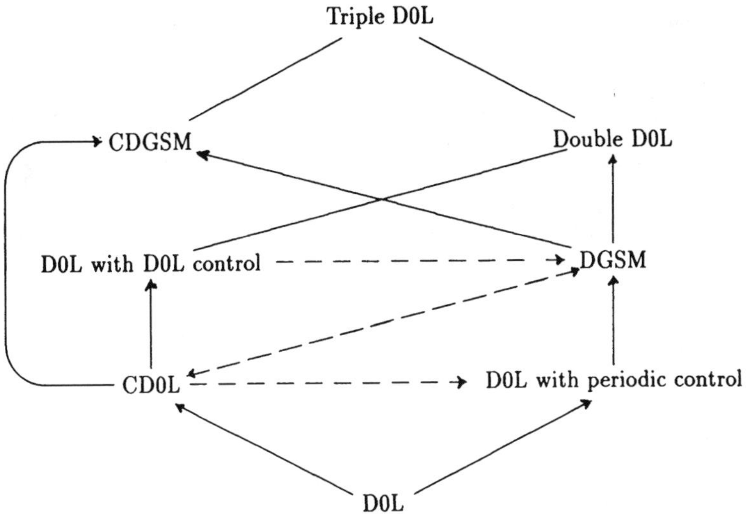

Figure 2: Inclusions of classes of ω-words

$$\text{SQRT} \;=\; Sa^{n_1}\#a^{n_2}\#a^{n_3}\#\ldots\,,$$

where $n_1 = 1, n_2 = n_3 = 2$, and, in general, the number of consecutive n_i's of equal size increases always by one. The details can be found in [11].

Consequently, there remain a number of open problems. However, we conjecture that the three inclusions are proper, and moreover all unsettled relations are incomparabilities. Indeed, we have candidates to solve these conjectures, for details cf. [11].

References

[1] J. P. Allouche, Finite Automata in 1-D and 2-D Physics, *Number Theory & Physics*, ed. by J. M. Luck, P. Moussa and M. Walschmidt, in: *Springer Proceedings in Physics* 47, Springer-Verlag (1990).

[2] J. P. Allouche, J. Betrema and J. O. Shallit, Sur des Points Fixes de Morphismes d'un Monoide Libre, R.A.I.R.O., *Informatique theorique et Applications* 23, 3, 235 - 249 (1989).

[3] J. P. Allouche and M. Mendes-France, Quasi-Crystal Using Chain and Automata Theory, *J. Stat. Phys* 42, 5/6 (1986).

[4] J. M. Autebert and J. Gabarro, Iterated GSM's and Co-CFL, *Acta Informatica* 26, 749 - 769 (1989).

[5] F. Axel, J. P. Allouche, M. Kleman, M. Mendes-France and J. Peyriere, Vibrational Modes in a One Dimensional "Quasi-Alloy": The Morse Case, *Journal de Physique*, Colloque C3, Suppl. 7, Tome 47 (1986).

[6] J. Berstel, Properties of Infinite Words: Recent results, in: *Lecture Notes in Computer Science* 349, Springer-Verlag, New York (1989).

[7] A. Cobham, Uniform tag Sequences, *Math. Systems Theory* 6, 164 - 192 (1972).

[8] K. Culik II, Homomorphisms: Decidability, Equality and Test Sets, in: *R. Book (ed.), Formal Language Theory, Perspectives and Open Problems*, Academic Press, New York (1980).

[9] K. Culik II and S. Dube, Balancing Order and Chaos in Image Generation, to appear in: *ICALP Proceedings*, Madrid (1991).

[10] K. Culik II and T. Harju, The ω-Sequence Equivalence Problem for D0L systems is Decidable, *JACM* 31, 277 - 298 (1984).

[11] K. Culik II and J. Karhumäki, Iterative Devices Generating Infinite words, Tech. Report TR 9106, Univ. of South Carolina (1991).

[12] K. Culik II, J. Karhumäki and A. Lepistö, Alternating Iteration of Morphisms and the Kolakovski Sequence, in: G. Rozenberg and A. Salomaa (eds.), *A memorial volume for A. Lindenmayer*, Springer-Verlarg (to appear).

[13] K. Culik II and A. Salomaa, On Infinite Words Obtained by Iterating Morphisms, *Theoret. Comput. Sci.* 19, 29 - 38 (1982).

[14] F. M. Dekking, Regularity and Irregularity of Sequences Generated by Automata, *Sem. Th. de Nombres de Bordeaux*, exp. 9 (1979-1980).

[15] F. M. Dekking, On the Structure of Selfgenerating Sequences, *Sem. Th. de Nombres de Bordeaux*, (1980-1981).

[16] A. Ehrenfeucht, K. P. Lee and G. Rozenberg, Subword Complexities of Various Classes of Deterministic Developmental Languages Without Interactions, *Theoret. Comput. Sci.* 1, 59 - 76 (1975).

[17] P. C. Fischer, A. A. Meyer and A. L. Rosenberg, Time-restricted Sequence Generation, *J. Comput. System Sci.* 4, 50 - 73 (1970).

[18] T. Harju and M. Linna, On the Periodicity of Morphisms in Free Monoids, R.A.I.R.O., *Theoret. Inform. Appl.* 20, 47 - 54 (1986).

[19] J. E. Hopcroft and J. D. Ullman, *Introduction to Automata Theory, Languages, and Computation*, Addison-Wesley, Reading, MA (1979).

[20] W. Kolakovski, Self Generating Runs, problem 5304, *American Math. Monthly* 71 (1965), solution by N. Ucoluk, same journal 73, 681 - 682 (1966).

[21] J. Karhumäki, Two Theorems Concerning Recognizable $\mathbb{N}$-subsets of δ^*. *Theoretical Computer Science* 1, 317 - 323 (1976).

[22] C. Kimberling, Problem 6281*, *Amer. Math. Monthly* 86, 793 (1979).

[23] D. Knuth, Solution to Problem E 2307, *Amer. Math. Monthly* 79, 773 - 774 (1972).

[24] M. L. Minsky, *Computation: Finite and Infinite Machines*, Prentice-Hall, Englewood Cliffs, N.J. (1967).

[25] J. J. Pansiot, Decidability of Periodicity for Infinite Words, R.A.I.R.O., *Theoret. Inform. Appl.* 20, 43 - 46 (1986).

[26] P. Prusinkiewicz and A. Lindenmayer, *The Algorithmic Beauty of Plants*, Springer-Verlag, Berlin (1990).

[27] G. Rozenberg and A. Salomaa, *The Mathematical Theory of L-Systems*, Academic press, New York (1980).

[28] A. Salomaa and M. Soittola, *Automata-Theoretic Aspects of Formal Power Series*, Springer-Verlag, New York (1978).

[29] J. Shallit, A Generalization of Automatic Sequences, *Theoretical Computer Science* 61, 1 - 16 (1988).

[30] J. Shallit, Open Problem on the Kolakovski Sequence, (private communication).

[31] A. R. Smith III, Plants, Fractals, and Formal Languages, *Computer Graphics* 18, 1 - 10 (1984).

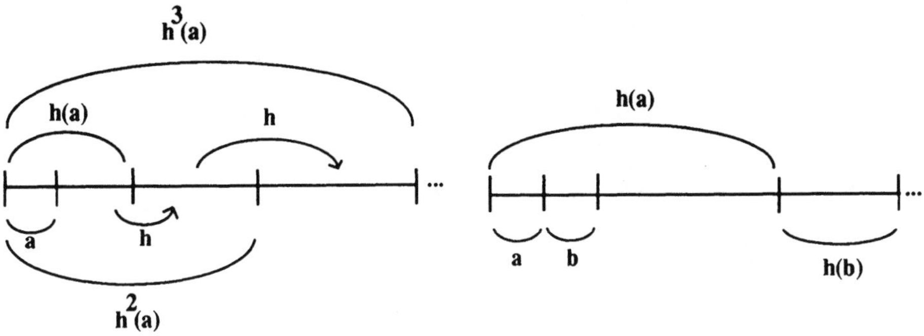

Fig. 3 : An iterative morphism

Fig. 4 : A DOL TAG-system

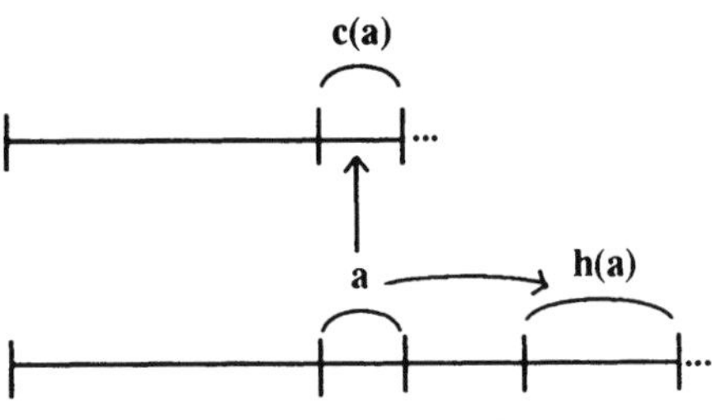

Fig. 5 : A CDOL TAG-system

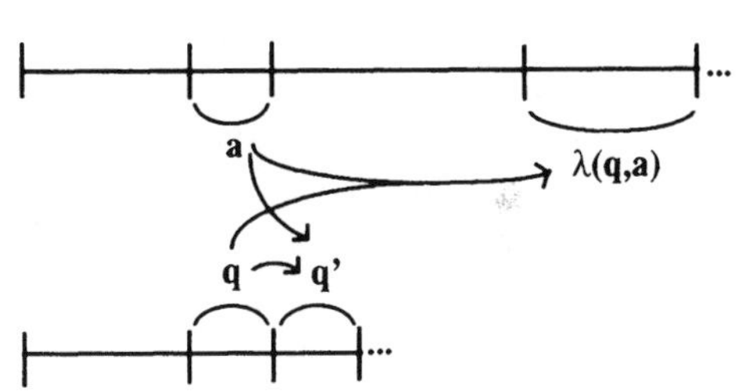

Fig. 6 : A DGSM TAG-system

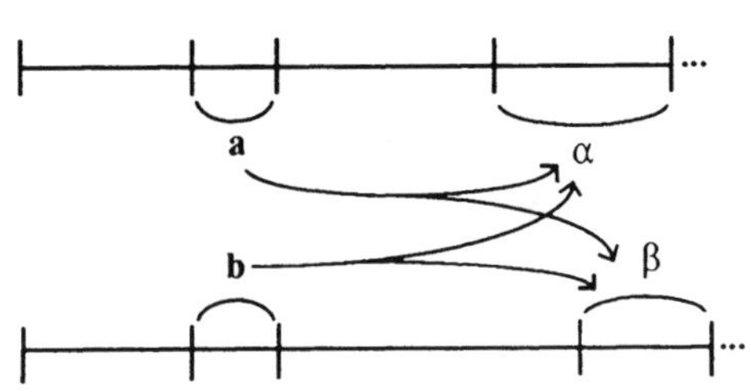

Fig. 7 : A double DOL TAG-system

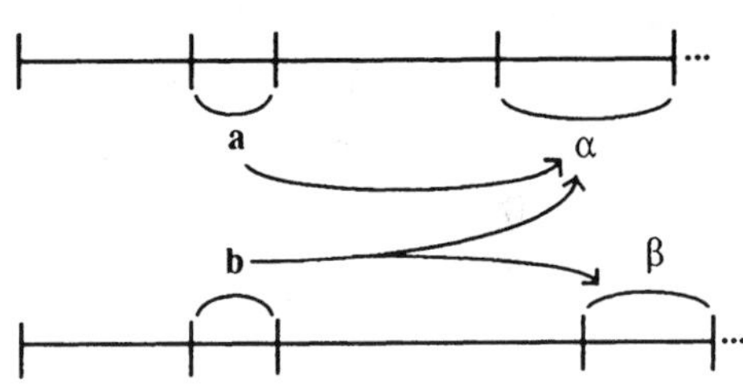

Fig. 8 : A DOL TAG-system with control

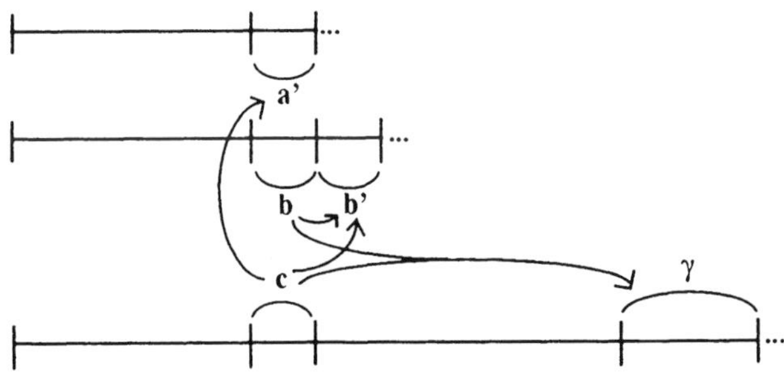

Fig. 9 : A CDGSM TAG-system

On the factorization conjecture

CLELIA DE FELICE
Dipartimento di Informatica ed Applicazioni
Università di Salerno
84081 Baronissi (SA), Italy

Abstract

We construct a family of finite maximal codes over the alphabet $\{a,b\}$ which verify the factorization conjecture on codes proposed by Schützenberger. This family contains any finite maximal code with at most three occurrences of the letter b by word.

1 Introduction

The theory of variable-length codes draws its origin from the theory of information devised by Shannon in the 1950s. This theory was subsequently developed in an algebraic way by M.P.Schützenberger and numerous researchers; it is now clearly a part of theoretical computer science, strongly related to automata theory, formal languages, combinatorics on words and the theory of semigroups [2, 24].

Codes C are naturally defined as subsets of A^* such that any word w in A^* has at most one factorization into words of C. One important aim of the theory of codes is to give a structural description of the codes in a way that allows their construction. This is easily accomplished for *prefix (suffix) codes*, i.e. codes such that none of their words is a left (right) factor of another one. This is also verified for *finite maximal biprefix codes* [8, 9] i.e. finite maximal codes both prefix and suffix. However no systematic method is known to construct all finite codes and particularly all finite maximal codes.

In this context, M.P.Schützenberger has proposed several conjectures [25]. A first one, still open today, is inspired by a problem of information theory and states that *any finite maximal code is commutatively equivalent to a prefix code* (i.e. they have the same commutative image) [18].In the case of the particular family of codes with bounded synchronization delay, Perrin and Schützenberger [18] gave a positive answer to this conjecture. On the other hand, it is known that the conjecture is false for finite codes which are not maximal: P. Shor constructed a code not commutatively equivalent to a prefix code [23].

A more general conjecture has been formulated in terms of non commutative polynomials of $\boldsymbol{Z}\langle A\rangle$: any finite maximal code $C \subseteq A^*$ is *factorizing*, i.e. there exist finite subsets P,S of A^* such that $\underline{C}-1=\underline{P}(\underline{A}-1)\underline{S}$ ($\underline{C}$ denotes the characteristic polynomial of C).It links properties of the code with factorization properties of its characteristic polynomial. This subject was deeply investigated [2, 14, 24, 25] but this study is far from being complete at the present time. The major contribution to this conjecture is due to C. Reutenauer [20, 21]. He proved that for any finite maximal code C equality $\underline{C}-1 = P(\underline{A}-1)S$ holds with P,$S \in \boldsymbol{Z}\langle A\rangle$. We call (P,S) a factorization of C. Then, if the factorization conjecture were true, there would exist some factorization (P,S) of C with P,$S \in \boldsymbol{N}\langle A\rangle$. In a natural way codes having multiple factorizations were investigated by several authors: J.M. Boë created a particular family of factorizing codes C with a unique factorization (see [4, 5]); families of factorizing codes having multiple factorizations were constructed by D. Perrin and M.Vincent in the framework of another Schützenberger's conjecture (see [17, 26]);recently V.Bruyére and C. De Felice characterized in a algebraic way codes with multiple factorizations [6, 7].

On the other hand algorithms exist to construct some families of codes, finite maximal codes over the alphabet $\{a, b\}$ which have few letters b by word: they are 1 - codes or 2 - codes, a *n-code* being a finite maximal code with n letters b or less by word. This shares some properties with a special class of binomial trees corresponding to particularly efficient algorithms [1].

1- and 2 - codes are factorizing and algorithms to construct them are all obtained by using a class of *factorizations* of the cyclic group $\boldsymbol{Z}_n$ [11, 19]. A factorization of $\boldsymbol{Z}_n$ is a pair (I, J) of subsets of $\boldsymbol{N}$ such that any $m \in \boldsymbol{Z}_n$ can be uniquely written as $m = i + j \pmod n$ with $i \in I, j \in J$. The structure of the factorizations of $\boldsymbol{Z}_n$ is unknown except for the class used for 1- and 2- codes. This class was also described by Hajós [12, 15].

The same properties hold for 3-codes: we prove in this paper that any 3- code is factorizing and can be constructed using Hajós factorizations.In order to establish this statement we prove a general result linking the factorization conjecture and a recursive construction of a family of n-codes.

This paper is organized as follows. In Section 2 we enunciate our main result and we recall the definitions and results to be used later. In Section 3 we prove a theorem which is interesting by itself and which is also used later in the proof of the main result. In Section 4 we solve the factorization conjecture for 3 - codes and we construct them in Section 5.

2 Preliminaries

In this section we first introduce the definitions and basic properties of codes.Then we enunciate the main results of the paper.

We denote A a finite alphabet and A^* the free monoid generated by A. We denote $\mid w \mid$ the length of the word $w \in A^*$ and $\mid w \mid_a$ the number of occurrences of the letter $a \in A$ in w. 1 is the empty word and A^+ is the set $A^* \setminus 1$.

$C \subseteq A^+$ is a *code* over the finite alphabet A if for any $c_1, \ldots, c_n, c'_1, \ldots, c'_m \in C$ we have

$$c_1 \ldots c_n = c'_1 \ldots c'_m \Rightarrow n = m \text{ and } \forall\, i\ c_i = c'_i$$

C is a *prefix code* if $C \cap CA^+ = \emptyset$. *Suffixes codes* C are defined symmetrically such that $C \cap A^+C = \emptyset$. A *biprefix code* is a code both prefix and suffix.

A code $C \subseteq A^+$ is said *maximal* (over A) if it cannot be strictly included in another code over A. Given a finite maximal code $C \subseteq A^+$, each letter $a \in A$ has a unique power a^n in C.

A systematic exposition of the theory of codes can be found in [2].

As usual $\mathbf{Z}$ denote the ring of the integer numbers, $\mathbf{N}$ the semiring of nonnegative integers. For any semiring K, $K\langle\langle A\rangle\rangle$ (resp. $K\langle A\rangle$) is the semiring of *series* (resp. *polynomials*) with noncommutative variables $a \in A$ and coefficients in K (see [3, 13] for a complete survey of this theory). For a series S, (S, w) denotes the coefficient of the word w in S. Any (finite) subset X of A^* will be identified with its *characteristic (polynomial) series* $\underline{X} = \sum_{x \in X} x$.

The *support supp*(P) of a series S is equal to $\{w \in A^* \mid (S, w) \neq O\}$. For any P in $K\langle A\rangle$ we will denote P_r $(r = O, \ldots, h)$ polynomials such that for any $w \in supp(P_r)$ we have $\mid w \mid_b = r$ and

$P = P_O + \cdots + P_h$. In the definition it is supposed that $P \neq 0$ implies $P_h \neq O$. For any P in $\boldsymbol{N}\langle A\rangle$, we will set $P \geq O$.

If M is a finite multiset of $\mathbf{N}$, we denote a^M the polynomial $\sum_{n\in \boldsymbol{N}}(M,n)a^n \in \boldsymbol{N}\langle A\rangle$. The computation rules are

$$a^{\emptyset} = O, a^O = 1, a^{M\cup N} = a^M + a^N, a^{M+N} = a^M a^N$$

The symbols $\cup$ and $+$ mean union and addition for multisets.

The notion of factorizing code shows the interplay between codes and polynomials. A code C over A is *factorizing* if there exist two subsets P, S of A^* such that $\underline{C} - 1 = \underline{P}(\underline{A} - 1)\underline{S}$.

Finite maximal prefix (resp. suffix) codes $C \subseteq A^*$ are factorizing since $\underline{C} - 1 = \underline{P}(\underline{A} - 1)$ (resp.$\underline{C} - 1 = (\underline{A} - 1)\underline{S}$) with P (resp.S) be the set of proper left (resp.right) factors of words in C.

As a special case, note that if C is a factorizing code with P, S *finite*, then C is a finite maximal code; conversely, if C is a finite maximal factorizing code, then P, S are finite sets [2]. However it is not known whether any finite maximal code is always factorizing:

Conjecture 2.1 *[18] Any finite maximal code is factorizing.*

The major contribution to this conjecture is due to C. Reutenauer

Theorem 2.1 *[20, 21].Let C be a finite subset of A^+. Then C is a finite maximal code iff there exist polynomials P, S in $\boldsymbol{Z}\langle A\rangle$ such that $\underline{C} - 1 = P(\underline{A} - 1)S$.*

Remark 2.1 *[20, 21].If $P, S \in \boldsymbol{N}\langle A\rangle$ are such that $P(\underline{A} - 1)S + 1 \geq 0$, then $P(\underline{A} - 1)S + 1$ is the characteristic polynomial of a finite maximal code and P,S have coefficients $0, 1$.*

For finite maximal codes over the alphabet $\{a, b\}$ having a few times the letter b inside their words, conjecture 2.1 is true. We call *n-code* any finite maximal code $C \subseteq \{a, b\}^+$ with n letters b or less by word. In the definition, it is supposed that at least one word of C exactly contains n letters b.

Theorem 2.2 *[10, 19]. Any 1- or 2- code $C \subseteq \{a, b\}^+$ is factorizing. Moreover, for any factorization (P, S) of C with $P, S \in \boldsymbol{Z}\langle A\rangle$, then P, S are characteristic polynomials.*

The second statement of this result does not hold in general [20].

In this paper we suppose $A = \{a, b\}$ as in Theorem 2.2. However we can generalize these results to arbitrary alphabets A containing at least two distinct letters a, b.

In Section 4 we prove the main result of this paper:

Theorem 2.3 *Any 3-code $C \subseteq \{a, b\}^+$ is factorizing.Moreover,for any factorization (P, S) of C with $P, S \in \boldsymbol{Z}\langle A\rangle$, then P, S are characteristic polynomials.*

Again,Hajós factorizations are essential in the construction of 3-codes (see Section 5).Moreover we use the following theorem by which we get a class of finite factorizing n-codes.

Theorem 2.4 *Let C be a finite maximal code and P, S a factorization of C.Let $S = \sum_{i=O}^{h} S_i$. Suppose that for any $r \in \{O, \ldots, h\}$ we have $P(\underline{A} - 1)(S_O + \cdots + S_r) \geq O$.Then P, S have coefficients $O, 1$.*

A characterization of the degree and the decomposability of n-codes with $1 \leq n \leq 3$ can be found in [6, 7]. It is obtained by an algebraic characterization of respectively, codes with multiple factorizations and decomposable codes (over arbitrary alphabets) (see [6, 7]).

From now on, we suppose all the codes to be *finite* and *maximal*, we briefly call them *codes*.

3 A family of factorizing codes

In this section we prove some technical preliminary results. Let C be a finite maximal code and (P, S) a factorization of C. We start with Remark 3.1 which states a factorization of the characteristic polynomial C_r in terms of the polynomials P, S.

In particular we will use this remark in the proof of Theorem 3.3. This theorem links the maximal number of b's in the words of C and the maximal number of b's in the words of $supp(P)$ and $supp(S)$. We will use Theorem 3.3 in Section 4, it enable us to study the different factorizations of a 3-code C in terms of the maximal number of b's in $supp(P)$ and $supp(S)$.

We will use Lemma 3.1 in the proof of Theorem 3.1. This theorem states that for any factorization (P, S) of a code C some words in $supp(P)$ and $supp(S)$ have nonnegative coefficients.

Let $\mathfrak{R}$ be the family of finite maximal code C such that C has a factorization (P, S) with $S \in \boldsymbol{Z}\langle a\rangle$. Theorem 3.2 (i) states that if $C \in \mathfrak{R}$ then C is factorizing. The proof use an argument that enable us to construct a code $C \in \mathfrak{R}$: if C is a n-code in $\mathfrak{R}$ and (P, S) is a factorization of C, then $C_{(n-1)} \in \mathfrak{R}$, where $C_{(n-1)}$ is the $(n-1)$-code defined by the factorization (P', S) and with $P' = P - P_{n-1}$. In other words C can be constructed by $C_{(n-1)}$, by a transformation of the words in $C_{(n-1)}$ having $(n-1)$ letters b. This construction is also given in [11].

Theorem 3.2 (ii) extends (i). Let us consider the family $\wp$ of codes inductively defined in the following way. If $C \in \mathfrak{R}$ then $C \in \wp$. Let C be a n-code. Suppose that there exists a $(n-1)$-code $D \in \wp$ with a factorization $(P, (S_O + \cdots + S_h))$ such that $(P, (S_O + \cdots + S_h + S_{h+1}))$ is a factorization of C.Then $C \in \wp$. Theorem 3.2 prove that any code in $\wp$ is factorizing.

We conclude this section with some examples of codes in $\mathfrak{R}$ and in $\wp$.

Remark 3.1 *Let $C \in \boldsymbol{N}\langle A\rangle, P, S \in \boldsymbol{Z}\langle A\rangle$ be polynomials such that $\underline{C} - 1 = P(\underline{A} - 1)S$. Then we have*

$$C_O = P_O(a-1)S_O + 1$$

$$\forall r > O\ C_r = \{w \in supp(C) \mid\mid w\mid_b = r\} = \sum_{i+j=r} P_i(a-1)S_j + \sum_{i+j=r-1} P_i b S_j$$

Lemma 3.1 *Let X, Y, Z, T be polynomials in $\boldsymbol{Z}\langle A\rangle$. Suppose that for any $w \in A^*$ we have*

$$w \in supp(X) \Rightarrow \mid w\mid_b = k; w \in supp(Y) \Rightarrow \mid w\mid_b = h;$$

$$w \in supp(Z) \Rightarrow \mid w\mid_b = k-1; w \in supp(T) \Rightarrow \mid w\mid_b = h-1$$

(i) If $x, y \in A^*$ are such that $\mid x\mid_b = k$ and $\mid y\mid_b = h$ then we have $(XbY, xby) = (X, x)(Y, y)$.

(ii) If XbY has coefficients $O, 1$ and $X, Y \neq O$, then X and Y (or $-X$ and $-Y$) have coefficients $O, 1$.

(iii) Suppose $X \neq O, Y \neq O$ have coefficients $O, 1$. If $X(a-1)Y + TbY + XbZ \in \boldsymbol{N}\langle A\rangle$ then $T \in \boldsymbol{N}\langle A\rangle \setminus O$ and $X = \sum_{t\in supp(T)} tba^{L_t}$ $(L_t \subseteq \boldsymbol{N})$ or $Z \in \boldsymbol{N}\langle A\rangle \setminus O$ and $Y = \sum_{z\in supp(Z)} a^{M_z}bz$ $(M_z \subseteq \boldsymbol{N})$.

PROOF :
(i) By definition we have

$$(XbY, xby) = \sum_{xby=x'by'} (X, x')(Y, y')$$

The nonzero terms in this sum are such that $(X, x') \neq O$ and $(Y, y') \neq O$. In this case, by the hypotheses we have $|x'|_b = k = |x|_b$ and $|y'|_b = h = |y|_b$. So, by $xby = x'by'$ we get $x = x'$ and $y = y'$.

(ii) It suffices to prove that X, Y (or $-X$ and $-Y$) are in $\boldsymbol{N}\langle A\rangle$. Assume the contrary, X or Y verifies one of the following two properties

$$1. \exists y \in supp(Y) : (Y, y) < O,\ \exists x \in supp(X) : (X, x) > O$$

$$2. \exists y \in supp(Y) : (Y, y) > O,\ \exists x \in supp(X) : (X, x) < O$$

Suppose that the first case holds. By (i) we have

$$(XbY, xby) = (X, x)(Y, y) < O$$

which is a contradiction. By a similar argument we get a contradiction in the other case.

(iii) Let $R, G, V, Q \in \boldsymbol{N}\langle A\rangle$ be polynomials such that $Z = V - Q$ and $T = R - G$ with $supp(R) \cap supp(G) = \emptyset = supp(V) \cap supp(Q)$. Moreover let X', Y' be subsets of A^* such that $X = \sum_{x \in X'} xba^{L_x}$, $Y = \sum_{y \in Y'} a^{M_y}by$, with M_y, L_x finite subsets of $\boldsymbol{N}$.

Let us prove that $X' \subseteq supp(R)$ or $Y' \subseteq supp(V)$. Assume the contrary, let $x \in X' \setminus supp(R)$, $y \in Y' \setminus supp(V)$, $\alpha = min L_x$, $\beta = min M_y$. Then we have

$$(X(a-1)Y + TbY + XbZ, xba^{\alpha+\beta}by) < O$$

which is a contradiction.

So $X' \subseteq supp(R)$ or $Y' \subseteq supp(V)$. Let us suppose that the first case occurs, the argument is similar in the other one. Let $g \in supp(G)$, $y = a^m by' \in Y$. By (i) we have

$$(XbV, gby) = (X, gba^m)(V, y')$$

and by definition we get

$$(XaY, gby) = \sum_{t+n=m-1} (X, gba^t)(Y, a^n by')$$

Then, by $X' \subseteq supp(R)$ we get $(XbV, gby) = (XaY, gby) = O$. Thus $(X(a-1)Y + TbY + XbZ, gby) < O$. Consequently $G = O$ and the conclusion follows.

□

Theorem 3.1 *Let C be a finite maximal code. Let $P = \sum_{i=0}^{k} P_i$, $S = \sum_{i=0}^{h} S_i$ be polynomials in $\boldsymbol{Z}\langle A\rangle$ such that $\underline{C} - 1 = P(\underline{A} - 1)S$. Then the following statements hold either for (P, S) or for $(-P, -S)$*

(i) P_k and S_h have coefficients $O, 1$

(ii) either $P_{k-1} \in \boldsymbol{N}\langle A\rangle \setminus O$ and $P_k = \sum_{p \in supp(P_{k-1})} pba^{L_p}$ or $S_{h-1} \in \boldsymbol{N}\langle A\rangle \setminus O$ and $S_h = \sum_{s \in supp(S_{k-1})} a^{M_s} bs$

Proof :
By Remark 3.1 the polynomials $C_{k+h+1} = P_k b S_h$ and $C_{k+h} = P_k(a-1)S_h + P_{k-1}bS_h + P_k b S_{h-1}$ have coefficients $O, 1$. We get the result by Lemma 3.1 (ii) and (iii) applied with $X = P_k$, $Y = S_h$, $Z = S_{h-1}$, $T = P_{k-1}$.
□

Theorem 3.2 *Let C be a finite maximal code. Let $P = \sum_{i=0}^{k} P_i$, $S = \sum_{i=O}^{h} S_i$ be polynomials in $\mathbf{Z}\langle A\rangle$ such that $\underline{C} - 1 = P(\underline{A}-1)S$.*

(i) Suppose P (resp.S) $\in \mathbf{Z}\langle a\rangle$. Then for any $r \in \{1,\ldots,h\}$ (resp. $\{1,\ldots,k\}$) $C_{(r)} = P_O(\underline{A}-1)(S_O+\cdots+S_r)+1 = \sum_{i\in\{O,\ldots,r-1\}} C_i + P_O b S_r$ (resp. $C_{(r)} = (P_O+\cdots+P_r)(\underline{A}-1)S_O+1 = C_O+\cdots+C_{r-1}+P_r b S_O$) is a code.

Moreover P, S have coefficients $O, 1$.

(ii) Suppose that for any $r \in \{O,\ldots,h\}$ we have $P(\underline{A}-1)(S_O+\cdots+S_r)+1 \geq O$. Then P, S have coefficients $O, 1$.

Proof :
(i) By Theorem 3.1 we can suppose P_O, S_k, $S_{k-1} \in \mathbf{N}\langle A\rangle$ (the argument is similar in the other cases). Then the result follows by induction over h.

(ii) If $r = O$ the statement holds thanks to (i). Then the conclusion follows by induction over h and by using Theorem 3.1 (i).
□

Theorem 3.3 *Let C be a code and let P, S be a factorization of C. C is a n-code, where $n > 1$, if and only if there exist $h, k \geq O$ such that $S = S_O+\cdots+S_h$, $P = P_O+\cdots+P_k$ and $h+k+1 = n$.*

Proof :
Straightforward,by Remark 3.1.
□

We conclude this section with some examples of codes in $\Re$ and $\wp$.

Example 3.1 *$C_2 = a^6 + a^{\{O,2,4\}}ba^{\{1,2\}} + a^{\{O,2,4\}}b^2a^{\{1,2\}} + a^{\{O,2,4\}}b^3a^{\{O,1\}} = 1 + (a^{\{O,2,4\}}(1+b+b^2))(\underline{A}-1)a^{\{O,1\}}$ is in $\Re$.*

There exists an algorithm (see [11]) which enable us to construct C_2 by C_1 and C_1 by C_O, where C_O, C_1 are the codes defined by

$$C_O = 1 + a^{\{O,2,4\}}(\underline{A}-1)a^{\{O,1\}}$$

$$C_1 = 1 + (a^{\{O,2,4\}}(1+b))(\underline{A}-1)a^{\{O,1\}}$$

Finally we give an example of a code C_3 in $\wp$ which can be constructed starting by C_2

$C_3 = 1+(a^{\{O,2,4\}}(1+b+b^2))(\underline{A}-1)(a^{\{O,1\}}+ba^{\{1,2\}}) = a^6+a^{\{1,3,5\}}ba^{\{1,2\}}+a^{\{O,2,4\}}ba^{\{O,1\}}ba^{\{1,2\}}+a^{\{O,2,4\}}b^3a^{\{O,1\}}+a^{\{O,2,4\}}b^2aba^{\{1,2\}}+a^{\{O,2,4\}}b^4a^{\{1,2\}}$

4 3-codes

In this section we prove that any 3-code is factorizing. First we need a lemma. In this lemma we consider some inequations related to the factorizations of the cyclic groups. Two parameters are associated with them: a polynomial $a^H \in N\langle a\rangle$ and a nonnegative integer α. we prove that α is a bound for the coefficient (a^H, a^t) of the word a^t on a^H, where $t \in N$. In Theorem 3.1 we will prove our result about 3-codes. The proof is by contradiction. We will get a contradiction by showing that (a^H, a^t) is greater than α, for a particular $t \in$ N.

Lemma 4.1 *(i) Let $P, a^H \in N\langle a\rangle$ be polynomials. Suppose that there exists $n \in N$ such that $P + a^H(a^n - 1) \geq O$. Then for any $t \in N$ we have*

$$(a^H, a^t) \leq \sum_{j \equiv t \pmod n} (P, a^j)$$

(ii) Let $I, H \subseteq N$. Suppose that there exists $n \in N$ such that $I = \{O, \ldots, n-1\} \pmod n$ and $a^{I'} = a^I + a^H(a^n - 1) \geq O$. Then $I' = \{O, \ldots, n-1\} \pmod n$.

(iii) Let $H, I \subseteq N$; $n, \alpha \in N$. Then $\alpha a^I + a^H(a^n - 1) \geq O$ implies $a^I + a^H(a^n - 1) \geq O$.

(iv) Let $a^H \in N\langle a\rangle$. Suppose that there exists $\alpha, n \in N$ such that

$$a^{I'} = \alpha\left(\frac{a^n - 1}{a - 1}\right) + a^H(a^n - 1) = \frac{a^n - 1}{a - 1}(\alpha + a^H(a - 1)) \geq O$$

Then for any $t \in N$ we have $(a^H, a^t) \leq \alpha$. If $\alpha = 1$ then a^H has coefficients $O, 1$ and $I' = \{O, \ldots, n-1\} \pmod n$. If $H \subseteq N$ then $((a^n - 1)/(a - 1))(1 + a^H(a - 1)) \geq O$ and it has coefficients $O, 1$.

PROOF :
(i) The proof is by induction on $card(supp(a^H))$. If $H = \emptyset$ then the conclusion follows. Otherwise let $h = \min(supp(a^H))$. Since $P + a^H(a^n - 1) \geq O$ and $(a^{H+n}, a^h) = O$ then $(a^H, a^h) \leq (P, a^h)$.

Let $a^{H'} = a^H - (a^H, a^h)a^h$ and $P' = P + (a^H, a^h)a^h(a^n - 1)$. We have $a^{H'}, P' \in N\langle a\rangle$. Moreover $P + a^H(a^n - 1) = P' + a^{H'}(a^n - 1) \geq O$ and $\sum_{j \equiv t \pmod n}(P, a^j) = \sum_{j \equiv t \pmod n}(P', a^j)$, for any $t \in N$.

Since $card(supp(a^{H'})) < card(supp(a^H))$ we can apply to $a^{H'}$ and P' the induction hypothesis. We obtain

$$\forall t \in N, t \neq h \; (a^H, a^t) = (a^{H'}, a^t) \leq \sum_{j \equiv t \pmod n} (P', a^j) = \sum_{j \equiv t \pmod n} (P, a^j)$$

Then the conclusion follows.

(ii) The proof is by induction on $card(H)$ and similar to (i).

(iii) By contradiction, suppose that there exists $t \in N$ such that $(a^I + a^H(a^n - 1), a^t) < O$. Then $t \in H, t \notin I, t \notin H + n$. This contradicts $(\alpha a^I + a^H(a^n - 1), a^t) \geq O$.

(iv) The conclusion follows straightforwardly by (i), (ii), (iii).

□

Remark 4.1 *It is known that a pair (T, R) of subsets of* N *is a factorization of Z_n if and only if $a^T a^R = ((a^n - 1)/(a - 1))(1 + a^H(a - 1)) \geq O$. H is the set of the holes of (T, R) [12].*

Theorem 4.1 *Let C be a 3-code and (P, S) a factorization of C. Then P, S have coefficients $O, 1$.*

Proof :
Let C be a 3-code and let (P,S) be a factorization of C. By Theorem 3.3, (P,S) verifies one of the following three properties: (i) $P\in \mathbf{Z}\langle a\rangle$, (ii) $S\in \mathbf{Z}\langle a\rangle$, (iii) $P=P_O+P_1, S=S_O+S_1$. By Theorem 3.2 (i), in case (i) or (ii) the conclusion follows.

Suppose that case (iii) holds. By Theorem 3.1 we can suppose that there exist finite subsets I,J,L_i $(i\in I)$, M_t $(t\in T)$ of $\mathbf{N}$, with $P_O\in \mathbf{N}\langle a\rangle$ and $a^I = supp(P_O)$ such that

$$P_1=\sum_{i\in I} a^iba^{L_i}, S_1=\sum_{t\in T} a^{M_t}ba^t$$

and with at least an $i\in I$ such that $L_i\neq\emptyset$ (otherwise C would be a 2-code).

By Remark 2.1 it suffices to prove that $S\in \mathbf{N}\langle A\rangle$, i.e. that $S_O\in \mathbf{N}\langle a\rangle$. Assume the contrary, let $Q=\{t\in \mathbf{N}\mid \exists\alpha_t>O:(S_O,a^t)=-\alpha_t\}\neq\emptyset$.

By Remark 3.1 we have the three inequations $C_r\geq O$, where $r\in\{O,1,2\}$. By every inequation we will infer some facts which will lead to a contradiction.

Let n be the positive integer such that $a^n\in C$. We have

$$C_O=a^n=P_OS_O(a-1)+1\geq O$$

So $1=(P_O,1)=(a^I,1)$.

Let us consider $C_2\geq O$:

$$C_2=\sum_{i\in I,t\in T} a^iba^{L_i}(a-1)a^{M_t}ba^t+\sum_{t\in T}P_Oba^{M_t}ba^t+\sum_{i\in I}a^iba^{L_i}bS_O\geq O$$

Let $i\in I$ be such that $L_i\neq\emptyset$ and let $t\in Q$, $v=\min L_i$. By $O\leq (C_2,a^iba^vba^t)\leq (\sum_{t\in T}P_Oba^{M_t}ba^t,a^iba^vba^t)+(S_O,a^t)\leq (\sum_{t\in T}P_Oba^{M_t}ba^t,a^iba^vba^t)$ we have

$$Q\subseteq T,\forall t\in Q\ M_t\neq\emptyset \tag{1}$$

Remember that $1\in supp(P_O)=a^I$. Let us consider $C_1\geq O$

$$C_1=\sum_{i\in I}a^iba^{L_i}(a-1)S_O+P_ObS_O+\sum_{t\in T}a^{M_t}(a-1)P_Oba^t\geq O$$

By this inequation we have $b(a^{L_O}(a-1)S_O+S_O)\geq ba^{L_O}(a-1)S_O+bS_O-\sum_{t\in T}a^{M_t}ba^t\geq O$. This inequation implies $L_O\neq\emptyset$ since $S_O\notin \mathbf{N}\langle a\rangle$ .So

$$L_O\neq\emptyset, a^{L_O}(a-1)S_O+S_O\geq O \tag{2}$$

Now $P_OS_O=(a^n-1)/(a-1)$. Then $P_O, S_O(1+a^{L_O}(a-1))\in \mathbf{N}\langle a\rangle$ imply $P_OS_O(1+a^{L_O}(a-1))=((a^n-1)/(a-1))(1+a^{L_O}(a-1))\in \mathbf{N}\langle a\rangle$. By Lemma 4.1 (iv), $P_OS_O(1+a^{L_O}(a-1))$ has coefficients $O,1$. So

$$P_O \text{ has coefficients } O,1 \tag{3}$$

In the following we will set $P_O=a^I$ and $-(S_O,a^t)=\alpha_t$, for any $t\in Q$.

By $C_2\geq O$ we get $O\leq ba^{L_O}(a-1)a^{M_t}ba^t+ba^{M_t}ba^t-\alpha_tba^{L_O}ba^t$. So we have

$$\forall t \in Q\ a^{L_O+M_t}(a-1) + a^{M_t} - \alpha_t a^{L_O} \geq O \text{ i.e. } \alpha_t a^{L_O} + a^{L_O+M_t} \leq a^{L_O+M_t+1} + a^{M_t} \tag{4}$$

Consequently we have

$$\forall \alpha \in \boldsymbol{N}\ a^{L_O+M_t}(a-1) + a^{M_t} + \alpha a^{L_O} \geq O \tag{5}$$

Moreover, let $v = \min L_O$. By (4) we have

$$1 \leq \alpha_t = \alpha_t(a^{L_O}, a^v) \leq (\alpha_t a^{L_O} + a^{L_O+M_t}, a^v) \leq (a^{L_O+M_t+1}, a^v) + (a^{M_t}, a^v) = (a^{M_t}, a^v) \leq 1$$

This inequation implies

$$\forall t \in Q\ \alpha_t = 1\ ,\ min\ M_t > O \text{ and } v \in M_t \tag{6}$$

Let $t \in Q, \alpha \in \boldsymbol{N}$. Define $a^H = a^{M_t} + \alpha a^{L_O} + a^{L_O+M_t}(a-1)$. By (5) we have that a^H is in $\boldsymbol{N}\langle a \rangle$.

Moreover by (6) we get

$$(a^H, a^v) = (a^{M_t}, a^v) + \alpha(a^{L_O}, a^v) = 1 + \alpha \tag{7}$$

Now, by $C_1 \geq O$ we get

$$O \leq a^{M_t}(a-1)a^I - a^I + \sum_{i \in I}(a^{L_i}(a-1)S_O, a^t)a^i \leq a^{M_t}(a-1)a^I + \alpha a^I$$

for any positive integer $\alpha > max_{i \in I}(a^{L_i}(a-1)S_O, a^t) - 1$.

So, by (2) for any $t \in Q$ there exists $\alpha \in \boldsymbol{N}$, $\alpha > O$ such that

$$O \leq \frac{a^n - 1}{a-1}(1 + a^{L_O}(a-1))(\alpha + a^{M_t}(a-1)) = \alpha(\frac{a^n - 1}{a-1}) + a^H(a^n - 1) \tag{8}$$

where $a^H = a^{M_t} + a^{L_O+M_t}(a-1) + \alpha a^{L_O}$.

By (5) we have $a^H \in \boldsymbol{N}\langle a \rangle$. By (7) we have $(a^H, a^v) = 1 + \alpha$. On the other hand, by Lemma 4.1 (iv) (applied to (8) with $t = v$) we have $(a^H, a^v) \leq \alpha$. This contradiction shows that $S_O \in \boldsymbol{N}\langle a \rangle$ and concludes the proof.

□

5 Structure of 3-codes

In this section we characterize the structure of 3-codes. Let C be a 3-code and let $P, S \in \boldsymbol{Z}\langle A \rangle$ such that $\underline{C} - 1 = P(\underline{A} - 1)S$. By Theorem 4.1, (P, S) verifies one of the following three properties

$$(i)\ P \in \boldsymbol{N}\langle a \rangle, S \in \boldsymbol{N}\langle A \rangle$$

$$(ii)\ P \in \boldsymbol{N}\langle A \rangle, S \in \boldsymbol{N}\langle a \rangle$$

$$(iii)\, P = P_O + P_1, S = S_O + S_1$$

In case (i) or (ii), C belongs to the class $\Re$ introduced in Section 3. These cases were described in [11]. In case (iii), by Theorem 3.1, we can suppose that there exist finite subsets I, J, T, L_i $(i \in I)$, M_t $(t \in T)$ of $\mathbf{N}$ such that

$$P_O = a^I,\ S_O = a^J,\ a^I a^J = \frac{a^n - 1}{a - 1},\ P_1 = \sum_{i \in I} a^i b a^{L_i},\ S_1 = \sum_{t \in T} a^{M_t} b a^t$$

(otherwise we turn C into its reverse).

The following theorem characterizes the factorizations (P, S) of a 3-code which verifies (iii). More precisely it characterizes the pairs (P_1, S_1) in a factorization of these codes. Indeed we can see that (P_O, S_O) is a Krasner factorization of $\mathbf{Z}_n$. Consequently we have a description of the structure of the corresponding codes.

Theorem 5.1 *Let (P, S) be a pair of polynomials in $\mathbf{N}\langle A\rangle$ such that*

$$P = a^I + \sum_{i \in I} a^i b a^{L_i},\ S = a^J + \sum_{t \in T} a^{M_t} b a^t$$

where (I, J) is a Krasner factorization of $\mathbf{Z}_n$ and T, L_i $(i \in I)$, M_t $(t \in T)$ are subsets of $\mathbf{N}$. Then, we have $P(\underline{A} - 1)S + 1 \in \mathbf{N}\langle A\rangle$ if and only if (P, S) verify the following conditions

$$(1)\ \forall i \in I\ a^{R_i} = a^{L_i}(a-1)a^J + a^J \geq O$$

$$(2)\ T \subseteq \cup_{i \in I} R_i$$

$$(3)\ \forall t \in T \text{ set } I_t = \{i \in I \mid t \in R_i\}. \text{Then we have}$$

$$a^{M_t}(a-1)a^I + a^I \geq a^{M_t}(a-1)a^I + a^{I_t} \geq O$$

$$(4)\ \forall i \in I, t \in T \text{ if } t \notin J \text{ then } a^{L_i}(a-1)a^{M_t} + a^{M_t} \geq O,\ \text{if } t \in J \text{ then } a^{L_i}(a-1)a^{M_t} + a^{M_t} + a^{L_i} \geq O$$

PROOF :

Let (P, S) be a pair of polynomials in $\mathbf{N}\langle A\rangle$ such that

$$P = a^I + \sum_{i \in I} a^i b a^{L_i},\ S = a^J + \sum_{t \in T} a^{M_t} b a^t$$

where (I, J) is a Krasner factorization of $\mathbf{Z}_n$ and T, L_i $(i \in I)$, M_t $(t \in T)$ are subsets of $\mathbf{N}$.

By Theorems 2.1,3.3 and 4.1, $\underline{C} = P(\underline{A} - 1)S + 1 \geq O$ if and only if C is a 3-code that verifies (iii). Then $C_i \geq O$, where $i \in \{O, 1, 2, 3\}$.

Suppose $\underline{C} = P(\underline{A} - 1)S + 1 \geq O$. Let us prove that $a^I a^J(1 + a^{M_t}(a-1)) \geq O$, for any $t \in T$. As in Theorem 4.1 we can prove that there exists $\alpha \in \mathbf{N}, \alpha > O$ such that $a^I(\alpha + a^{M_t}(a-1)) \geq O$.Then, by Lemma 4.1 (iv), we have that $a^I a^J(1 + a^{M_t}(a-1))$ has coefficients $O, 1$.

Let us prove (i). Assume the contrary, there exist $i \in I, h \in \mathbf{N}$ such that $(a^{L_i}(a-1)a^J + a^J, a^h) < O$. Consequently we have

$$O \leq (C_1, a^i b a^h) \Rightarrow h \in T \text{ and } (a^{M_h}(a-1)a^I, a^i) > O \Rightarrow (a^I(1 + a^{M_h}(a-1)), a^i) > 1$$

By this inequation, since $a^I a^J(1 + a^{M_h}(a-1))$ has coefficients $O, 1$, we get

$$\exists j \in J \setminus O, q \in \mathbf{N} : q + j = i,\ (a^I(1 + a^{M_h}(a-1), a^q) < O$$

Since $(a^I(\alpha + a^{M_h}(a-1)), a^q) \geq O$ with $\alpha > O$, we get $(a^I(\alpha - 1), a^q) > O$, i.e. $q \in I$. So we have $a^I a^J = (a^n - 1)/(a-1)$ and $q + j = i + O \in I + J$ with $j \neq O$, $i \neq q$ i.e. a contradiction.

Let $t \in T$ be such that $M_t \neq \emptyset$ and $m = \min M_t$. Then we have

$$O \leq (C_1, a^m ba^t) = (\sum_{i \in I} a^i ba^{R_i}, a^m ba^t) - (a^{M_t + I} ba^t, a^m ba^t) < \sum_{i \in I} (a^i ba^{R_i}, a^m ba^t)$$

Then (2) follows. By fixing $t \in T$ in $C_1 \geq O$ we get (3). By fixing $i \in I$ and $t \in T$ in $C_2 \geq O$ we get (4).

Conversely, suppose that (P, S) verifies (1),(2),(3) and (4). Set $C = P(\underline{A} - 1)S + 1$. One can prove that $C_i \geq O$, where $i \in \{O, 1, 2, 3\}$. Then $C \in \boldsymbol{N}\langle A \rangle$. By Remark 2.1 and Theorem 3.3, C is a 3-code and case (iii) holds.

□

We conclude this section with some remarks and open problems.

Pairs (I, J) such that $a^I a^J = (a^n - 1)/(a-1)$ are described in [16]. Subsets R_i, L_i and M_i which appear in Theorem 5.1 are described in [11, 15]. They are related to the Hajós factorizations (see [12, 15]).

Some natural questions arise in this framework. Can any factorizing code be constructed by using only Hajós factorizations? Does there exist a similar relation between a finite maximal code and a factorization of a cyclic group? Does there exist a recursive transformation which turns a finite maximal code into a factorizing code?. We conjecture that the first question has a positive answer.

ACKNOWLEDGEMENTS

I would like to thank C. Reutenauer for much encouragements throughout the course of this work. Many thanks are due to V. Bruyére,who read carefully the manuscript. I wish to thank also V. Auletta,A. De Santis and many colleagues which helped me in the Latex redaction of this paper.

The first part of this work was supported by the Istituto Nazionale di Alta Matematica (Roma,Italy) and by contract ESPRIT-EBRA ASMICS n.3166.

References

[1] J.L.Bentley,D.J.Brown,A general class of resource tradeoffs, J. Comp. System Sc. 25 (1982) 214-238 and Proc. 21th FOCS (1980) 217-228

[2] J.Berstel,D.Perrin,"Theory of codes" (Academic Press,New York, 1985)

[3] J.Berstel,C.Reutenauer,"Rational series and their languages" (EATCS Monographs 12, Springer-Verlag, 1988)

[4] J.M.Boë,Sur les codes synchronisants coupants,in "Non Commutative Structures in Algebra and Geometric Combinatorics" (A. de Luca ed.), Quaderni della Ric. Sc. del C.N.R. 109 (1981) 7-10

[5] J.M.Boë,Sur les codes factorisants,in "Théorie des codes" (D. Perrin ed.),LITP (1979) 1-8

[6] V.Bruyère,C.De Felice,Degree and decomposability of variable- length codes,Proc. ICALP'91,LNCS n.510 (1991) 342-350

[7] V.Bruyère,C.De Felice,Synchronization and decomposability for a family of codes (part 1 and part 2),submitted (1991)

[8] Y.Césari,Sur un algorithme donnant les codes bipréfixes finis,Math.Syst.Theory 6 (1972) 221-225

[9] Y.Césari,Propriétés combinatoires des codes bipréfixes finis,in "Théorie des codes" (D.Perrin ed.),LITP (1979) 29-46

[10] C.De Felice,C.Reutenauer,Solution partielle de la conjecture de factorisation des codes,C.R.Acad. Sc. Paris 302 (1986) 169-170

[11] C.De Felice,Construction of a family of finite maximal codes, in LNCS 294 (1988) 159-169 and Theor. Comp. Science 63 (1989) 157-184

[12] C.De Felice,Codes factorisants and factorisations of finite cyclic groups,in preparation (1991)

[13] S.Eilenberg,"Automata,languages and machines" (Academic Press, New York,1974)

[14] G.Hansel,D.Perrin,C.Reutenauer,Factorizing the commutative polynomial of a code,Trans.Amer.Math.Soc. 285 (1984) 91-105

[15] G.Hajós,Sur le problème de factorisation des groupes cycliques, Acta Math. Acad. Sc. Hungaricae 1 (1950) 189-195

[16] M.Krasner,B.Ranulac,Sur une propriété des polynômes de la division du cercle, C.R.Acad. Sc. Paris 240 (1937) 397-399

[17] D.Perrin,Codes asynchrones,Bull.Soc.Math.France 105 (1977) 385-404

[18] D.Perrin,M.P.Schützenberger, Un problème élémentaire de la théorie de l'information,"Théorie de l'Information", Colloques Internat. CNRS 276,Cachan (1977) 249-260

[19] A.Restivo,On codes having no finite completions,Discrete Math. 17 (1977) 309-316

[20] C.Reutenauer,Sulla fattorizzazione dei codici,Ricerche di Matematica 32 (1985) 115-130

[21] C.Reutenauer,Non commutative factorization of variable- length codes,J.Pure and Applied Algebra 36 (1985) 167-186

[22] C.Reutenauer,private communication

[23] P.Shor,A counterexample to the triangle conjecture, J.Comb.Theory A 38 (1985) 110-112

[24] M.P.Schützenberger,Une théorie algébrique du codage,Séminaire Dubreil-Pisot 1955-56,Exposé n.15

[25] M.P.Schützenberger,Codes à longueur variable, manuscript (1965),reprinted in "Théorie des codes" (D.Perrin ed.) LITP (1979) 247-271

[26] M.Vincent,Construction de codes indécomposables,RAIRO Inform. Théor. 19 (1985) 165-178

Conditional Semi-Thue Systems for Presenting Monoids

Thomas Deiß
Fachbereich Informatik
Universität Kaiserslautern
Erwin Schrödinger Straße
W-6750 Kaiserslautern
Germany
deiss@informatik.uni-kl.de

Abstract

There are well known examples of monoids in literature which do not admit a finite and canonical presentation by a semi-Thue system over a fixed alphabet, not even over an arbitrary alphabet. We introduce conditional Thue and semi-Thue systems similar to conditional term rewriting systems as defined by Kaplan. Using these conditional semi-Thue systems we give finite and canonical presentations of the examples mentioned above. Furthermore we show, that every finitely generated monoid with decidable word problem is embeddable in a monoid which has a finite canonical conditional presentation.

1 Introduction

Thue and semi-Thue systems [Boo85, Jan88] can be used to examine questions concerning monoids and groups. A Thue system R over an alphabet Σ induces a congruence on Σ^*, the congruence classes form the monoid M_R. A monoid M is finitely presented by (Σ, R) if M is isomorphic to M_R and both Σ and R are finite, it is finitely generated, if only Σ is finite. If R viewed as a semi-Thue system induces a canonical, i.e. confluent and noetherian, relation, it can be used to decide the word problem of M: Two strings u and v are congruent if and only if u and v have the same common irreducible descendant.

Therefore no monoid with an undecidable word problem admits a finite and canonical presentation. It has been shown by Narendran and Squier that there exist finitely presented monoids with decidable word problem which do not have a finite and canonical presentation using a fixed alphabet, see e.g. [KN85], resp. using an arbitrary but finite alphabet [Squ87].

To overcome this gap between decidability of the word problem and the existence of finite and canonical presentations we introduce in this paper conditional Thue and semi-Thue systems. They are defined similar to conditional term rewriting systems, see e.g. [Kap84, Kap87, JW86, Gan87]. We show, that using conditional semi-Thue systems we get finite and canonical presentations for the examples of Narendran and Squier. Furthermore we are able to strengthen a result of Bauer [Bau81]: Each finitely generated monoid with decidable word problem can be embedded in a monoid, presented by a finite, canonical, and conditional semi-Thue system.

Different conditional string rewriting systems have been used already by Siekmann and Szabo [SS82] to give a finite and canonical presentation of idempotent monoids. They use variables within their rules and a different system to evaluate the premises of a conditional rule and therefore are not a conditional semi-Thue system according to our definition.

2 Conditional Semi-Thue Systems

The form of the conditional rules we use follows that of conditional term rewriting systems as defined e.g. by Kaplan [Kap84]. Therefore the induced congruences are more difficult to handle as for unconditional systems. For example, the congruences are not decidable in general. These problems can be solved by introducing reductive systems analogously to simplifying and reductive conditional term rewriting systems [Kap87, JW86].

A *conditional Thue system* R is a set of conditional equations. Each equation consists of a *conclusion* $u_0 = v_0$ and a finite set $\{u_i = v_i | 1 \leq i \leq n\}$ of *premises*, all u_i, v_i are strings over an alphabet Σ. We write such a conditional equation as

$$\bigvee_{i=1}^{n} u_i = v_i :: u_0 = v_0$$

The relation $\Leftrightarrow_R$ is defined as follows: $u \Leftrightarrow_R v$ if and only if there exist $x, y \in \Sigma^*$ and an equation $\forall_{i=1}^{n} u_i = v_i :: u_0 = v_0$ in R such that $u = xu_0y$ and $v = xv_0y$ or $u = xv_0y$ and $v = xu_0y$ and for all $1 \leq i \leq n$ we have $xu_iy \Leftrightarrow_R^* xv_iy$. $\Leftrightarrow_R^*$ is the *Thue congruence* induced by R. x and y are the left resp. right context of the occurrence of u_0 resp. v_0 in u. In this case R is called a *left-right* conditional Thue system. If only the right context y is used in evaluating the premises, i.e. $u_iy \Leftrightarrow_R^* v_iy$, we call R a *right* conditional system.

To define conditional *semi-Thue* systems we restrict the application of the equations. Let $\forall_{i=1}^{n} u_i = v_i :: u_0 \rightarrow v_0$ be a rule of a conditional semi-Thue system R. u_0 is the left-hand side of this rule, v_0 is the right-hand side. Now $u \rightarrow_R v$ if and only if $u = xu_0y$ and $v = xv_0y$, where $x, y \in \Sigma^*$, and xu_iy and xv_iy have each a common descendant modulo $\rightarrow_R$ for $1 \leq i \leq n$. As for conditional Thue systems we distinguish left-right and right conditional semi-Thue systems.

Notice that the premises must have a common descendant in the context of rule application instead of being congruent as it was in the definition of conditional Thue systems. This causes the first difference to unconditional systems: The Thue congruence and the symmetric, transitive, and reflexive closure of the reduction relation may not coincide anymore. To recover this property we need in addition confluence of R.

Lemma 1 cf. [Kap84, theorem 3.2.]
a) There exists a conditional Thue system R with $\Leftrightarrow_R^* \neq \leftrightarrow_R^*$.
b) If R is confluent, we have $\Leftrightarrow_R^* = \leftrightarrow_R^*$.

Both $\Leftrightarrow_R^*$ and $\leftrightarrow_R^*$ are compatible with concatenation, hence the congruence classes modulo $\Leftrightarrow_R^*$ resp. $\leftrightarrow_R^*$ form a monoid. As a direct consequence of the lemma above these monoids are the same if R is confluent.

For a finite unconditional system R the relations $\Leftrightarrow_R$ and $\rightarrow_R$ are decidable. This changes, too, when considering conditional systems, cf. [Kap84, theorem 3.3.].

Lemma 2 *There exists a finite conditional Thue system R such that $\Leftrightarrow_R$ and $\rightarrow_R$ are undecidable.*

Similar to the case of conditional term rewriting systems there is a sufficient criterion such that the reduction relation becomes decidable: We call a conditional semi-Thue system *reductive* if for all rules in R the strings in the premises and the right-hand side are smaller than the left-hand side wrt. to a wellfounded ordering which is compatible with concatenation. In analogy to [Kap87, theorem 1.6.] we have

Lemma 3 *Let R be a finite reductive conditional semi-Thue system, then $\rightarrow_R$ is noetherian and decidable.*

The results of the lemmata 1 and 3 can be combined: If R is finite, confluent, and reductive, then we have $\Leftrightarrow_R^* = \leftrightarrow_R^*$ and $\Leftrightarrow_R^*$ resp. $\leftrightarrow_R^*$ are decidable. Hence R can be used to decide the word problem by means of string rewriting.

Notice that conditional Thue systems can only be more expressive in representing non-cancellative monoids. Let us assume that (Σ, R) is a finite presentation of a group or a cancellative monoid M. An equation e of R can be applied if its premises are congruent in the corresponding context, i.e. $xu_iy \Leftrightarrow_R^* xv_iy$. Since M is cancellative we have $xu_iy \Leftrightarrow_R^* xv_iy$ if and only if $u_i \Leftrightarrow_R^* v_i$. Now, if $u_i \Leftrightarrow_R^* v_i$ for all $1 \leq i \leq n$ the equation e can be applied in all contexts and the premises are superfluous. On the other hand if $u_i \not\Leftrightarrow_R^* v_i$ for an $i \in 1 \ldots n$ we cannot apply e at all and the equation itself is superfluous. Hence there is a unconditional system R' corresponding to R, R' is canonical if R is canonical. Now if M does not admit a finite, canonical, and unconditional presentation, it does not admit a conditional one either. As an example consider the system $\{aba \rightarrow bab\}$ [KN85] which is cancellative and does not have a finite and canonical presentation over the alphabet $\Sigma = \{a, b\}$.

3 Confluence and Equivalence

The previous results are valid for left-right as well as for right conditional systems. The following characterization of confluence using critical pairs can be shown for right conditional semi-Thue systems only. Since right conditional systems correspond to conditional term rewriting systems with unary function symbols, the definition of critical pairs and the proof of confluence can be carried over from term rewriting to semi-Thue systems, see e.g. [Kap87, JW86].

To show confluence and equivalence of conditional systems we have to know which contexts can be used to make the premises joinable. These contexts are called *solutions*, but we only need a subset of them. This set of *minimal* solutions is defined as follows.

Definition 1 *Let (r)* $\forall_{i=1}^n u_i = v_i :: u_0 \rightarrow v_0$ *be a conditional rule of a right conditional semi-Thue system R. Then* $sol(r) = \{x \in \Sigma^* | \forall_{i=1}^n u_i x \downarrow_R v_i x, x$ *is irreducible and there is no proper prefix* x' *of* x *such that* $x' \in sol(r)\}$. *We have* $sol(r) = \{\lambda\}$ *if* r *is an unconditional rule* .

To show confluence of a system R we take the usual approach. We show that R is reductive and therefore noetherian and then show that R is locally confluent, i.e. all critical pairs are joinable. Then R is confluent too [Hue80].

Definition 2 *Let* $(r)\forall_{i=1}^n u_i = v_i :: u_0 \rightarrow v_0$ *and* $(r')\forall_{i=1}^{n'} u_i' = v_i' :: u_0' \rightarrow v_0'$ *be two (not necessarily different) rules of a right conditional semi-Thue system R.*
If u_0' *is a substring of* u_0, *i.e.* $u_0 = xu_0'y, x, y \in \Sigma^*$ *then* $\forall_{i=1}^n u_i = v_i \wedge \forall_{i=1}^{n'} u_i'y = v_i'y :: v_0 = xv_0'y$ *is a critical pair of r and r'.*
If u_0 *and* u_0' *have an overlap, i.e.* $u_0x = yu_0', |x| < |u_0'|$, *then*
$\forall_{i=1}^n u_ix = v_ix \wedge \forall_{i=1}^{n'} u_i' = v_i' :: v_0x = yv_0'$ *is a critical pair of r and r'.*
CP(R) is the set of all critical pairs of R.
We say a critical pair (p) $\forall_{i=1}^n u_i = v_i :: u_0 = v_0$ is joinable *in R if and only if for all* $x \in sol(p)$ *we have* $u_0x \downarrow_R v_0x$.

Lemma 4 *Let R be a finite reductive right-conditional semi-Thue system. If all critical pairs in CP(R) are joinable in R, then R is locally confluent and thereby confluent.*
Sketch of proof: Follows the proof of confluence for conditional term rewriting systems in [Kap87]. Let us assume that we use the rules (r) $\forall_{i=1}^n u_i = v_i :: u_0 \rightarrow v_0$ and (r') $\forall_{i=1}^{n'} u_i' = v_i' :: u_0' \rightarrow v_0'$ to reduce a string u to v_1 resp. v_2. We have 3 cases: The occurrences of u_0 and u_0' are disjoint, u_0' is a substring of u_0 or u_0' and u_0 have an overlap. The first case corresponds to a variable overlap in a term rewriting system. The second and the third case both correspond to a critical pair overlap in a term rewriting system. □

In proving the equivalence of two conditional semi-Thue systems R_1 and R_2 we have to show that both sides of a rule of R_1 are congruent modulo R_2 and vice versa. If the rule is a conditional rule we have to show this in all contexts which are solutions of the premises of the rule. If we only want to check this for minimal solutions we need the additional requirement that both systems are confluent.

4 Examples

Now we are able to study the examples mentioned in the introduction. We first examine the example of Narendran. After that we briefly investigate an example due to Kirchner and Hermann [KH89] and finally we present the example of Squier.

4.1 A representation with no equivalent finite canonical presentation

We consider the one rule system $R = \{aba \to ba\}$ over the alphabet $\Sigma = \{a, b\}$. R itself is not confluent, by completing R we get the infinite system $R_{KB} = \{ab^n a \to b^n a | n \geq 1\}$. It is easy to see that R_{KB} is canonical and equivalent to R. But it has been shown, see e.g. [KN85]

Lemma 5 *There is no finite canonical unconditional semi-Thue system R' equivalent to R.*

As announced there is a conditional semi-Thue system which is finite, canonical, and equivalent to R.

Example 1 *The right conditional system $R_c = \{aba \to ba; ab = b :: abb \to bb\}$ is a finite canonical system equivalent to $R = \{aba \to ba\}$.*

We will use r_u and r_c as abbreviations of the unconditional resp. conditional rule of R_c. At first we determine the set of minimal solutions of the premise of r_c.

Lemma 6 $sol(r_c) = \{b^n a | n \geq 0\}$.
Proof: Let $S = \{b^n a | n \geq 0\}$

$S \subseteq sol(r_c)$: Induction on n shows that $abb^n a \downarrow bb^n a$.

$sol(r_c) \subseteq S$: Let $x \in sol(r_c)$, by definition x is irreducible. We show by induction on length l of x, that $x \in S$, i.e. $x = b^n a$.

$l = 0$: that means $x = \lambda$, but no left or right-hand side of r_u or r_c is a substring of ab resp. b. Therefore they are the only elements in their equivalence classes and they are not congruent. This contradicts $x \in sol(r_c)$.

$l = 1$: $x = a$, by definition of S we have $x \in S$ and $aba \downarrow ba$.
$x = b$, this again contradicts $x \in sol(r_c)$, since we would have to show $abb \downarrow bb$. Only r_c could be applied on abb, but the premise of r_c is not joinable within the context λ.

$l > 1$: $x = ax'$, this is a contradiction to $x \in sol(r_c)$ since a is a prefix of x and $a \in sol(r_c)$.
$x = bx'$, by assumption $x \in sol(r_c)$ we have $abx \downarrow bx$, i.e. $abbx' \downarrow bbx'$. x' is irreducible and therefore the only possibility to reduce $abbx'$ is $abbx' \to_{r_c} bbx'$. This rule may be applied if $x' \in sol(r_c)$. By induction hypothesis we have $x' = b^{l-2}a$, hence $x = b^{l-1}a$.

□

Lemma 7 *The system R_c is canonical.*
Proof: Since both rules of R_c are length reducing and the premises are shorter than the left-hand sides R_c is reductive. To show confluence we have to consider two overlaps. The critical pair $baba = abba$ corresponding to the overlap $ababa$ is joinable in bba. The other overlap $ababb$ results in the critical pair $ab = a :: babb = abbb$. It has the same premise as r_c, hence it has the same set $S = \{b^n a | n \geq 0\}$ of minimal solutions. Hence we have to show $babbb^n a \downarrow abbbb^n a$ for all $n \geq 0$. These can be reduced to $bbbb^n a$ using r_c. Therefore all critical pairs are joinable and we conclude that R_c is canonical. □

Concluding this example we show that R and R_c are equivalent.

Lemma 8 *R and R_c are equivalent.*
Proof: Since R is a subset of R_c it suffices to show for all $x \in sol(r_c)$ that $abbx \leftrightarrow_R^* bbx$. R and R_{KB} are equivalent, hence we can use R_{KB} to show this. By Lemma 6 we have $x = b^n a, n \geq 0$. But now $abbb^n a \rightarrow_{R_{KB}} bbb^n a$, which completes the proof. □

Kirchner and Hermann presented in [KH89] a term rewriting system which can be converted into the semi-Thue system $R' = \{fgf \rightarrow fh\}$. We use the reversed system $R = \{fgf \rightarrow hf\}$ as an example. Completing R gives the infinite canonical system $R_{KB} = \{fgh^n f \rightarrow h^{n+1} f \mid n \geq 0\}$. Again the monoid M_R can be presented by a finite canonical conditional semi-Thue system, the proof is left to the reader.

Example 2 *Let $R = \{fgf \rightarrow hf\}$, then $R_c = \{fgf \rightarrow hf; fg = h :: fgh \rightarrow hh\}$ is a finite canonical right conditional semi-Thue system equivalent to R.*

4.2 A monoid without a finite canonical presentation

Squier [Squ87, Squ88] defined a family of monoids $S_k, k \geq 1$, which cannot be presented by finite unconditional semi-Thue systems. We show that each of these monoids can be presented by a finite canonical right conditional system. We will study only the monoid S_1 in detail, but the results can be easily generalized to $S_k, k > 1$.

Let $\Sigma_1 = \{a, b, t, x, y\}, R_1' = \{xa \rightarrow atx; xt \rightarrow tx; xb \rightarrow bx; xy \rightarrow \lambda\}$ and $R_1 = R_1' \cup \{at^n b \rightarrow \lambda | n \geq 0\}$. S_1 is the monoid presented by (Σ_1, R_1), it is finitely presented by the system $R_1'' = R_1' \cup \{ab \rightarrow \lambda\}$, [Squ87]. Using a syllable or collected ordering as defined in [Sim87], based on the precedence $a > t > b > x > y$ we can show that R_1 is noetherian. Furthermore R_1 is confluent, hence R_1 is canonical. Therefore we are able to compute unique normal forms modulo R_1, hence the word problem of S_1 is decidable. Though we have an infinite semi-Thue system to solve the word problem of S_1, there is no finite system with this property.

Lemma 9 [Squ88]. *S_1 has no finite canonical unconditional presentation.*

To show that S_1 can be presented by a finite canonical conditional system, we need a canonical unconditional system $\overline{R}_1$ equivalent to R_1. It is similar to R_1, except that the rules $\{at^n b \rightarrow \lambda \; |n \geq 0\}$ are replaced by a set of rules $\{at^{n+1}b \rightarrow at^n b | n \geq 0\} \cup \{ab \rightarrow \lambda\}$. The rules of the conditional system will resemble the rules in $\overline{R}_1$. Especially the rules $at^{n+1}b \rightarrow at^n b$ will be replaced by a conditional rule.

Example 3 *The right conditional system $R_c = R_1' \cup \{ab \rightarrow \lambda; atb \rightarrow \lambda; at = a :: att \rightarrow at\}$ is a canonical presentation of the monoid S_1.*

The last 3 rules will be abbreviated with r_1, r_2 and r_c. Similar to lemma 6 we get $sol(r_c) = \{t^n b | n \geq 0\}$. Then we can show that R_c is canonical.

Lemma 10 *R_c is canonical.*
Proof: Since the rules of R_c can be ordered with the ordering defined above, R_c is reductive and therefore noetherian. To prove that R_c is confluent we have to consider the overlaps $xab, xatb$ and $xatt$. It is easy to see that the critical pairs $atxb = x$ and $atxtb = x$ corresponding to the first two overlaps are joinable. The overlap $xatt$ gives the critical pair $(p)\; at = a :: atxtt = xat$. Since p and r_c have the same premises, $sol(r_c) = sol(p) = \{t^n b | n \geq 0\}$. For all $n \geq 0, atxttt^n b$ and $xatt^n b$ are joinable in $attt^n bx$, hence R_c is confluent. □

We have equivalence of R_c and $\overline{R}_1$, the proof is similar to that of lemma 8. Since $\overline{R}_1$ and R_1 are equivalent, R_c and R_1 are equivalent too. Combining these results we get

Theorem 1 *There exist finitely presented monoids without a finite canonical unconditional presentation, but which can be presented by a finite canonical right conditional semi-Thue system.*

5 Embeddability of monoids

Throughout section 4 we used right conditional systems only and concentrated on single examples. To achieve a general result we resume to the use of left-right conditional semi-Thue systems.

In his dissertation Bauer used another approach to close the gap between decidability of the word problem and the existence of finite and canonical presentations [Bau81]. He showed that each finitely generated monoid with decidable word problem can be embedded in a monoid such that the following holds:

Lemma 11 cf. [Bau81]
Let (Σ, S) be a presentation of a finitely generated monoid M with decidable word problem. Then M can be embedded identically in a monoid M', which is finitely presented by (Σ', S'), such that

1. $\rightarrow^*_{S'}$ *is noetherian.*
2. *each word $w \in \Sigma^*$ has an unique irreducible normalform modulo $\rightarrow_{S'}$.*

Using left-right conditional semi-Thue systems we are able to strengthen this result:

Theorem 2 *Let (Σ, S) be a presentation of a finitely generated monoid M with decidable word problem, then M is embeddable in a monoid M' which has a finite canonical left-right conditional representation (Δ, C).*

Throughout the rest of this section we sketch the proof of this theorem. A technical report containing the full version of the proof is in preparation. The basic idea of the proof follows the proof of the result of Bauer: We have an injective homomorphism between M and M', induced by the function $\varphi : \Sigma \rightarrow \Delta^*$ which is defined as $\varphi(a_i) \rightarrow (a_i)$. The conditional system C will be used to compute an unique representant $\hat{w}$ of a word $w = a_{i_1} \dots a_{i_n} \in \Sigma^*$ in the following sense:

$$\varphi(a_{i_1}) \dots \varphi(a_{i_n}) = (a_{i_1}) \dots (a_{i_n}) \rightarrow^*_C \begin{cases} (\hat{w}) & \text{if } \hat{w} \neq \lambda \\ \lambda & \text{if } \hat{w} = \lambda \end{cases}$$

In the following we call a word $w \in \Sigma^*$ within delimiting parentheses a *configuration.* Computing the representant of a word w will be performed in two alternating steps. First, given a configuration (w), simulating a Turing machine, compute the representant of w: $(\hat{w})$. Second, given two configurations $(w_1), (w_2)$ concatenated as words, concatenate them as configurations: $(w_1 w_2)$.

Besides the choice of the representants there are some more problems. If $w = \hat{w}$ then (w) must be irreducible, otherwise we would have an infinite reduction. Second, in general the reduction $(w) \rightarrow (\hat{w})$ cannot be performed in one step, since an arbitrary large amount of information can be changed. Hence there are intermediate steps which should not interfere with other reductions. To solve both problems we use conditional rules. A third problem forces us to use left-right conditional rules. Otherwise, using right conditional ones to concatenate two configurations, we need some rules with left-hand side $(a_i$ and the solutions of the corresponding premises are $w_1)(w_2), w_1, w_2 \in \Sigma^*$. These rules insert a new letter e which goes to the right and deletes one occurrence of $)($. But (w_2) could be reduced to λ before this pair of parentheses is removed, this gives us two distinct normalforms $(a_i \widehat{w_1 w_2})$ resp. $(a_i w_1 e)$ of $(a_i w_1)(w_2)$. Our system would be no longer confluent.

It seems possible to avoid the use of left-right conditional rules by computing the representant of a word in a different way. Instead of a Turing machine one could use a Post machine, see e. g. [Man74, SS63], but the simulation of a Post machine by a semi-Thue system uses the same notion of a configuration: A string containing the contents of the tape with markers at the left and right end. Therefore we have the same problems as before when concatenating two configurations. Another possibility is the use of semi-Thue systems as defined by Sattler-Klein [Sat91] to compute representants without simulating a machine. This approach is part of current research.

The choice of the representants of the equivalence classes modulo S is the most obvious one. We define a length-lexicographical ordering on Σ^*. Then the representant $\hat{w}$ of $[w]_S$ is the smallest element of $[w]_S$ with respect to this ordering. Since the word problem of M is decidable, there is a computable function f, which computes for each word $w \in \Sigma^*$ the corresponding representant. It is defined as

$$f(w) = \mu_{z \leq w} (z \leftrightarrow_S^* w)$$

f is well-defined, hence we have $[u]_S = [v]_S$ if and only if $f(u) = f(v)$ for $u, v \in \Sigma^*$. Furthermore, it is decidable whether $w = \hat{w}$ and we define a function p as

$$p(w) = \begin{cases} w & \text{if } w \neq \hat{w} \\ \$w & \text{if } w = \hat{w}, \text{ where \$ is a new letter.} \end{cases}$$

It is easy to see that p is a computable function. Therefore there are two deterministic Turing machines M_f and M_p computing the functions f resp. p. According to Davis [Dav56] M_f and M_p can be constructed such that they halt when started from an arbitrary configuration. Hence we are able to construct two finite canonical unconditional semi-Thue systems R_f and R_p simulating M_f resp. M_p, see e.g. [DW83]. R_f resp. R_p are extended to perform the following tasks: Before starting the simulation of the Turing machine, R_f scans the configuration and copies it to another alphabet. Thereby it also determines whether the input is a correct configuration. After the computation the result is copied back to the original alphabet and superfluous blanks are eliminated. Let us have a look at a sample reduction:

$$\begin{array}{lll} [q^1 a_{i_1} \dots a_{i_n}) & \to_{R_f}^* [c'_{i_1} \dots c'_{i_n} q^2) & \text{scan to the right and copy} \\ & \to_{R_f}^* [q_0 c_{i_1} \dots c_{i_n}] & \text{go left and copy again} \\ & \to_{R_f}^* [b'^*_c q_f \widehat{c_{i_1} \dots c_{i_n}} b^*_c] & \text{simulate } M_f \\ & \to_{R_f}^* (\widehat{a_{i_1} \dots a_{i_n}}) & \text{clean up and copy back to } \Sigma \end{array}$$

R_p proceeds the same way on a configuration $\{p^1 w)$, but uses other copies of the alphabet and the parentheses. It stops with the word $|w)$ if $w \neq \hat{w}$, resp. $|\$w)$ if $w = \hat{w}$. Thereby $|$ is a new letter, which does not occur in the left-hand side of a rule, hence $|w)$, resp. $|\$w)$, are irreducible. We get the following behaviour of R_f resp. R_p.

Lemma 12 *Let $w \in \Sigma^+$, then*

1. $\{p^1 w) \to_{R_p}^* \begin{cases} |w) & \textit{if } w \neq \hat{w} \\ |\$w) & \textit{if } w = \hat{w} \end{cases}$

2. $[q^1 w) \to_{R_f}^* (\hat{w})$

Now we connect both systems using a set of conditional rules (one rule for each letter $a_i \in \Sigma$).

$$\{p^1 a_i \;=\; |a_i \;::\; (a_i \;\to\; [q^1 a_i \quad (1)$$

R_p is used to evaluate the premise, the right hand side $[q^1 a_i$ invokes a reduction with R_f. For $w_1, w_2 \in \Delta^*$ we have $w_1(a_i w_2 \to_1 w_1[q^1 a_i w_2$ if and only if $w_2 = w_3)w_4$, such that $w_3 \in \Sigma^*$ and $a_i w_3 \neq \widehat{a_i w_3}$, then $w_1[q^1 a_i w_3)w_4 \to_{R_f}^* w_1(\widehat{a_i w_3})w_4$. Thus we are able to compute the representant $(\hat{w})$ of a configuration (w), $w \in \Sigma^*$ using the noetherian system $R = R_f \cup R_p \cup \{(1)\}$.

To concatenate two configurations we use the following system T of conditional rules, $a_i, a_j, a_k, a_l \in \Sigma$:

$$\begin{array}{rclcrcll}
 & & & & () & \rightarrow & \lambda & (2) \\
 & & & & (a_i)(a_j) & \rightarrow & (a_i a_j) & (3) \\
(a_i)(a_k & = & (a_i a_k & :: & (a_i)(a_j a_k & \rightarrow & (a_i a_j a_k & (4) \\
a_i)(a_k) & = & a_i a_k) & :: & a_i a_j)(a_k) & \rightarrow & a_i a_j a_k) & (5) \\
a_i)(a_l & = & a_i a_l & :: & a_i a_j)(a_k a_l & \rightarrow & a_i a_j a_k a_l & (6)
\end{array}$$

The overall system C is defined as $C := R \cup T$. It is easy to see (induction on n) that C reduces some concatenated configurations to the corresponding representant.

Lemma 13 *for $w_1, \ldots, w_n \in \Sigma^*$ we have*
$(w_1)\ldots(w_n) \rightarrow_C^* w = \begin{cases} \lambda & \text{if } w_1 \widehat{\ldots} w_n = \lambda \\ (w_1 \widehat{\ldots} w_n) & \text{if } w_1 \widehat{\ldots} w_n \neq \lambda \end{cases}$
and w is irreducible modulo $\rightarrow_C$.

Since it is very difficult to determine the solutions of the conditional rules we proceed indirectly to show that $\rightarrow_C$ is canonical. At first we define a restricted relation $\rightharpoonup_C$ and show that $\rightharpoonup_C$ is canonical. Then we compare the relations $\rightharpoonup_C, \rightarrow_C$ and $\Leftrightarrow_C$.

$\rightharpoonup_C$ arises from $\rightarrow_C$ by further restricting the application of the conditional rules. Let $u = v :: l \rightarrow r$ be a conditional rule, then

$$xly \rightharpoonup xry \text{ iff } xuy \rightharpoonup_C^* xvy$$

That is, when evaluating the premise, the right part of it is fixed within its context, thus the solutions of the conditional rules are of a relatively simple form.

Lemma 14

- *solutions of rules (1)*
  $w_1(a_i w_2 \rightharpoonup_C w_1[q_1 a_i w_2$ *iff* $w_2 = w_3)w_4$, *with* $w_3 \in \Sigma^*$, $a_i w_3 \neq \widehat{a_i w_3}$,
  then $w_1(a_i w_2 \rightharpoonup_C^* w_1(\widehat{a_i w_3})w_4$
- *solutions of rules (3)-(6)*
  $w_1 a_i)(a_j w_2 \rightharpoonup_C w_1 a_i a_j w_2$ *iff* $w_1 = w_3(w_4,\ w_2 = w_5)w_6$, *with* $w_4, w_5 \in \Sigma^*$.

We were not able to find an ordering on strings, such that $\rightharpoonup_C$ is reductive with respect to this ordering. Hence we had to prove directly that $\rightharpoonup_C$ is decidable and noetherian. Confluence is an easy consequence of the form of the solutions and of Lemma 13.

Lemma 15 $\rightharpoonup_C$ *is canonical.*

Decidability and Noetherianity of $\rightarrow_C$ can be proven similar to that of $\rightharpoonup_C$. By noetherian induction we show that the reflexive, symmetric, and transitive closures of both relations coincide. Furthermore both relations have the same set of irreducible elements. Hence we conclude

Lemma 16 $\rightarrow_C$ *is canonical.*

Refering to the lemmata 1,15,16 and 13 we are able to conclude the final part of the proof.

Lemma 17 *for $w_1, w_2 \in \Sigma^*$ we have*
$w_1 \Leftrightarrow_S^* w_2$ *iff* $\varphi(w_1) \downharpoonleft_C \varphi(w_2)$ *iff* $\varphi(w_1) \downarrow_C \varphi(w_2)$ *iff* $\varphi(w_1) \Leftrightarrow_C^* \varphi(w_2)$

Thus, M is embeddable in M' using the injective homomorphism induced by φ, and M' has a finite canonical conditional representation (Δ, C).

Acknowledgements

I would like to thank Birgit Reinert, Andrea Sattler-Klein and Bernhard Gramlich for valuable discussion and Professor Madlener for initiating these investigations.

References

[Bau81] Günther Bauer. *Zur Darstellung von Monoiden durch konfluente Regelsysteme*. PhD thesis, Fachbereich Informatik, Universität Kaiserslautern, 1981. in German.

[Boo85] Ronald V. Book. Thue systems as rewriting systems. In *Proc. of 1st Rewriting Techniques and Applications*, pages 63–94. Springer, 1985. LNCS 202.

[Dav56] Martin D. Davis. A note on universal turing machines. In C. E. Shannon and J. McCarthy, editors, *Automata Studies*, pages 167–175. Princeton Press, 1956.

[DW83] Martin D. Davis and Elaine J. Weyuker. *Computability, Complexity, and Languages*. Academic Press, 1983.

[Gan87] Harald Ganzinger. A completion procedure for conditional equations. Technical Report 234, Fachbereich Informatik, Universität Dortmund, 1987.

[Hue80] Gérard Huet. Confluent reductions: Abstract properties and applications to term rewriting systems. *Journal of the ACM*, 27(4):797–821, oct 1980.

[Jan88] Matthias Jantzen. *Confluent String Rewriting*. Springer, Berlin, Heidelberg, New York, 1988.

[JW86] Jean Pierre Jouannaud and Bernard Waldmann. Reductive conditional term rewriting systems. In *Proceedings of the 3rd IFIP Working Conference on Formal Description of Programming Concepts*. North-Holland, 1986.

[Kap84] Stéphane Kaplan. Conditional rewrite rules. *Theoretical Computer Science*, 33:175–193, 1984.

[Kap87] Stéphane Kaplan. Simplifying conditional term rewriting systems: Unification, termination and confluence. *Journal of Symbolic Computation*, 4:295–334, 1987.

[KH89] Hélène Kirchner and Miki Hermann. Computing meta-rules from crossed rewrite systems. Technical report, CRIN, Nancy, 1989.

[KN85] Deepak Kapur and Paliath Narendran. A finite Thue system with decidable word problem and without equivalent finite canonical system. *Theoretical Computer Science*, 35:337–344, 1985.

[Man74] Zohar Manna. *Mathematical Theory of Computation*. Computer Science Series. McGraw-Hill, 1974.

[Sat91] Andrea Sattler-Klein. Divergence phenomena during completion. In Ronald V. Book, editor, *Proc. of 4th Rewriting Techniques and Applications*, pages 374–385. Springer, 1991. LNCS 488.

[Sim87] C. C. Sims. Verifying nilpotence. *Journal of Symbolic Computation*, 3:231–247, 1987.

[Squ87] Craig Squier. Word problems and a homological finiteness condition for monoids. *Journal of pure and applied algebra*, 49:201–217, 1987.

[Squ88] Craig Squier. A finiteness condition for rewriting systems. Department of Mathematical Sciences, SUNY–Binghamton, Binghamton, NY 13901, 1988.

[SS63] J. C. Shepherdson and H. E. Sturgis. Computability of recursive functions. *Journal of the ACM*, 10:217–255, 1963.

[SS82] Jörg Siekmann and P. Szabo. A noetherian and confluent rewrite system for idempotent semigroups. *semigroup forum*, 25:83–110, 1982.

ALGORITHMS 3

A Combinatorial Bound for Linear Programming and Related Problems*

Micha Sharir[†] Emo Welzl[‡]

Abstract

We present a simple randomized algorithm which solves linear programs with n constraints and d variables in expected $O(d^3 2^d n)$ time. The expectation is over the internal randomizations performed by the algorithm, and holds for any input.

The algorithm is presented in an abstract framework, which facilitates its application to a large class of problems, including computing smallest enclosing balls (or ellipsoids) of finite point sets in d-space, computing largest balls (ellipsoids) in convex polytopes, convex programming in general, etc.

KEYWORDS: computational geometry, combinatorial optimization, linear programming, randomized incremental algorithms.

1 Introduction

Linear programming is one of the basic problems in combinatorial optimization. The goal of this paper is two-fold: (i) to describe a simple randomized algorithm for solving linear programs; (ii) to provide an abstract framework in which the algorithm and its analysis can be presented, which facilitates the application of the algorithm to a large class of problems, including the computation of smallest enclosing balls

*Work by both authors has been supported by the German-Israeli Foundation for Scientific Research and Development (G.I.F.). Work by the first author has been supported by Office of Naval Research Grant N00014-90-J-1284, by National Science Foundation Grant CCR-89-01484, and by grants from the U.S.-Israeli Binational Science Foundation, and the Fund for Basic Research administered by the Israeli Academy of Sciences. Work by the second author has been supported by the ESPRIT II Basic Research Action Program of the EC under contract no. 3075 (project ALCOM).

†School of Mathematical Sciences, Tel Aviv University, Tel Aviv 69978, Israel, and Courant Institute of Mathematical Sciences, New York University, New York, NY 10012, USA, e-mail: sharir@math.tau.ac.il

‡Institut für Informatik, Freie Universität Berlin, Arnimallee 2-6, W 1000 Berlin 33, Germany e-mail: emo@tcs.fu-berlin.de

(or ellipsoids) of finite point sets in d-space, computing largest balls (ellipsoids) in convex polytopes in d-space, convex programming in general, etc.

The expected running time of the algorithm can be bounded by $O(d^3 2^d n)$ for a linear program with n constraints and d variables. (This expectation is with respect to the internal randomizations performed by the algorithm, and holds for any input.) Thus for constant d the algorithm has linear (expected) complexity in n. That algorithms with running time $O(C(d)n)$ exist for linear programming was first proved by Megiddo [Meg2], even for deterministic algorithms, but the factor $C(d)$ in his algorithm is 2^{2^d}; an improvement to $C(d) = 3^{d^2}$ can be found in [Dye1] and [Cla1]. Recently, a number of randomized algorithms have been presented for the problem, see [DyF], [Cla2], [Sei1], with a better dependence on d, where the best expected running time is given by Clarkson [Cla2]: $O(d^2 n + d^3\sqrt{n}\log n + d^{d/2+O(1)}\log n)$ This complexity is better than the one offered by our algorithm except when d is large (i.e. close to n). A closer look at Clarkson's algorithm shows that it reduces the problem to solving $O(d^2 \log n)$ small linear programs, each with $O(d^2)$ constraints and d variables. Clarkson suggests to solve these problems by the Simplex Algorithm, bounding the time by the maximum number of vertices of the polyhedron determining the feasible region of the program. Instead, we can plug our algorithm into Clarkson's algorithm, replacing the third term in its complexity by $O(d^5 2^d \log n)$. This gives a better bound and has the advantage that the whole algorithm works in the formal framework we consider (Clarkson's algorithm works also in this framework except for the base case of small linear programs).

Remark. After submission of this paper, Gil Kalai [Kal] announced a randomized simplex algorithm with expected running time bounded by $O(d^3 n \log n\, 16^{\sqrt{d}\log n})$ — thus providing the first subexponential bound for Linear Programming in terms of arithmetic operations. This bound is better than ours for large d (in particular, in the range of the small linear programs in Clarkson's algorithm). After that we reconsidered our analysis and proved a similar bound for our algorithm (as presented here), which, for linear programming, should read now as $O(\min\{d^3 2^d n, (d^3 + dn)e^{4\sqrt{d\ln(n+1)}}\})$, see [MSW].

The abstract framework we present considers the set H of n constraints, and a function w which maps every subset of H to its optimal solution, where w satisfies two simple conditions; as it turns out, this is all which is needed to prove the correctness of the algorithm, and to analyze its expected running time in terms of two primitive operations — violation tests and basis computations; see below for details. It facilitates the application of the algorithm to several problems as indicated above. As mentioned above, Clarkson's algorithm [Cla2] can also be shown to work in this framework, while the algorithm of Seidel [Sei1] (and its generalization in [Wel]) needs to make a more explicit use of the geometry of the problems. A different framework has been recently developed by Dyer [Dye2], which yields linear deterministic algorithms (with larger constants in d). His framework requires more reference to the geometry of the problem; still the smallest volume enclosing ellipsoid fits into the framework, while the problem of computing the largest volume ellipsoid in the intersection of n halfspaces has so far resisted a deterministic linear time solution (see [Dye2]).

Before we plunge into technicalities, we briefly describe the algorithm for the problem of computing the smallest enclosing circle of a planar point set. This problem has a long history going back to Sylvester [Syl] in 1857; see [Meg1] for the first linear time algorithm to solve this problem (and for more references).

For a set H of n points in the plane, let md(H) denote the smallest closed disk covering H. Such a smallest disk is unique, and it is determined by at most 3 of the points in H; that is, there is a set B of at most 3 points in H such that md(B) = md(H). We call B a *basis* of H, and we assume the availability of a (constant-time) procedure `basis` which computes a basis for any set of at most 4 points.

Now the algorithm proceeds as follows. It starts with a random ordering of H and computes a basis B of the first 3 elements, h_1, h_2, h_3, in this order, together with D = md(B). Then it goes through the list of remaining points in H, testing each point h whether it is contained in D. If $h \in D$, then D is also the smallest enclosing disk of the set F of points we have seen so far (including h). If $h \notin D$, then we compute D = md(F) recursively, by the same process, but we start now with the set `basis`$(B \cup \{h\})$ as the basis. After we have computed md(F) and a new basis B for F, we continue in the list of H. When we reach the end of the list, we have computed md(H) and a basis for H. This can be formulated as a recursive procedure `SMALLDISK`, which computes a basis of a set G of points, using a basis $T \subseteq G$ (not necessarily a basis for G) as an auxiliary information for its computations. (The procedure is initiated with the call `SMALLDISK`$(H, \emptyset)$.)

```
function procedure SMALLDISK(G,T);
    F := T;  B := T;  D := md(T);
    for all h ∈ G − T in random order do
        F := F ∪ {h};
        if h ∉ D then
            B := SMALLDISK(F, basis(B ∪ {h}));
            D := md(B);
        end if;
    end do;
    return B;
```

The intuition behind the algorithm is that whenever we make a recursive call, we start our computation with a 'better' set T, so that we expect to converge faster on the basis of the augmented point set. The algorithm resembles Seidel's linear programming algorithm [Sei1], which has expected running time $O(d!n)$, combined with the move-to-front heuristic suggested in [Wel]. Of course, many known algorithms follow the general line of computing a 'potential solution' and checking if a point (or several points) lie outside this potential solution (similar for linear programming). The 'subtlety' lies in the different ways of finding points outside the potential solution, and how to obtain a new potential solution.

2 An Abstract Framework

For a set X and an element x, we write $X + x$ short for $X \cup \{x\}$, and $X - x$ short for $X - \{x\}$.

A linear program with n constraints and d variables can be formulated as follows: We are given a set H of n halfspaces bounded by hyperplanes in $\mathbf{R}^d$, and we want to find the lexicographically smallest point s in $\bigcap_{h \in H} h$. (We use a 'backward' lexicographical order in which the last coordinate is the most significant; thus $x < y$ if for some $j \leq d$ we have $x_j < y_j$ and $x_k = y_k$ for all $j < k \leq d$.) If the intersection is empty, we define s to be a special symbol Ω, and if the intersection is nonempty, but contains no minimum, then we adopt a standard convention, defining s to be the direction of a (lexicographically) steepest ray which lies in this intersection. (All this assumes that the optimizing vector is $(0, 0, \ldots, 0, -1)$.)

The framework. Let w be the function which maps every subset G of H to its optimal solution $w(G)$ (in the context of linear programming, this is the s just defined above); that is we have a mapping

$$w : 2^H \to \mathcal{W}, \quad (\mathcal{W}, <),$$

where $\mathcal{W}$ is a set with a linear order $<$. It is easily seen that this mapping satisfies the following two conditions:

$$\text{(C1)} \quad F \subseteq G \subseteq H \quad \Rightarrow \quad w(F) \leq w(G)$$

$$\text{(C2)} \quad \left.\begin{array}{c} F \subseteq G \subseteq H \\ w(F) = w(G) \\ h \in H \end{array}\right\} \Rightarrow \left\{\begin{array}{c} w(F + h) > w(F) \\ \Leftrightarrow \\ w(G + h) > w(G) \end{array}\right.$$

We call a pair (H, w) satisfying conditions (C1) and (C2) an *LP-type problem.* The elements in H are called the *constraints*, and for $G \subseteq H$, we call $w(G)$ the *value* of G. A *basis* B is a set of constraints with $w(B') < w(B)$ for all proper subsets B' of B. A *basis for* a subset G of H is a basis B with $B \subseteq G$ and $w(B) = w(G)$. So a basis of G is a minimal subset of G with the same value as G. We write $\mathtt{basis}(G)$ or B_G for a basis of G (although in general such a basis in not unique).

The maximum cardinality of any basis is called the *combinatorial dimension of* (H, w), and it is denoted by $\delta = \delta_{(H,w)}$. Note that the combinatorial dimension is a monotone function: if $F \subseteq G$ then $\delta_{(F,w)} \leq \delta_{(G,w)}$.

For $G \subseteq H$, we say that a constraint h *violates* G if $w(G+h) > w(G)$ (which implies $h \notin G$), and we say h is *extreme in* G if $w(G-h) < w(G)$ (which implies $h \in G$).

Note that a constraint which is extreme in G has to lie in every basis of G; consequently, every subset $G \subseteq H$ has at most δ extreme constraints.

Examples of LP-type problems. Linear programs are easily seen to be LP-type problems. The set $\mathcal{W}$ is defined to be $\mathbf{R}^d$, plus the set of all directions in $\mathbf{R}^d$ (viewed as points on the $(d-1)$-sphere) whose last non-zero coordinate is negative, plus the nominal symbol Ω. We order $\mathcal{W}$ so that all directions precede all points in $\mathbf{R}^d$, which all precede Ω. The points of $\mathbf{R}^d$ are ordered by the backward lexicographical order defined above, and directions are ordered by an analogous backward lexicographical order. The function w maps, as described above, each subset G of H to its optimal solution, steepest ray, or Ω, according to whether the linear program for G has a solution, is unbounded, or has no feasible solution, respectively. The conditions (C1) and (C2) are satisfied, as has already been observed. The combinatorial dimension of a linear program with d variables is at most $d-1$ if the program is unbounded, it is d if there is a solution, and it may be as large as $d+1$ if the feasible region is empty.

Another example of an LP-type problem is the computation of the smallest enclosing ball of a set H of points in $\mathbf{R}^d$. Here $w(G)$ is the radius of the smallest ball containing G, which is known to be unique ($\mathcal{W}$ is just the set of nonnegative real numbers). Axiom (C1) is obviously satisfied. To see that condition (C2) also holds, we observe that a constraint (point) h violates G if and only if it lies outside the smallest ball containing G — this follows from the uniqueness of such a ball. Moreover, if $F \subseteq G$ and $w(F) = w(G)$, then the smallest enclosing balls of F and G are the same, and hence (C2) holds. The basis of a point set can have at most $d+1$ points (these points have to lie on the boundary of the smallest enclosing ball), and so the combinatorial dimension is $d+1$. Details can be found, e.g., in [Jun] (or [Wel]), where these claims are established.

Analogously, we can convince ourselves that the problem of computing smallest volume enclosing ellipsoids for point sets in $\mathbf{R}^d$ gives rise to an LP-type problem; here the combinatorial dimension is $(d+3)d/2$; see [Beh], [DLL], [Pos], [Juh].

Other examples of LP-type problems with constant combinatorial dimension are convex programming in d-space (of which the smallest enclosing ball problem is a special case), largest balls (ellipsoids) in the intersection of halfspaces, smallest enclosing ball of a set of balls, etc.

3 The Algorithm in the Abstract Setting

In this section we present our algorithm in the abstract setting, and analyze its expected running time. Let (H, w) be an LP-type problem with H a set of n

constraints and with $w : 2^H \rightarrow \mathcal{W}$ a function satisfying conditions (C1) and (C2). Let δ be the combinatorial dimension of (H, w), i.e. the maximum cardinality of a basis.

As the SMALLDISK procedure, our algorithm is a recursive procedure lptype(G, T) with two arguments, a set $G \subseteq H$ of constraints, and a basis $T \subseteq G$ (which is not necessarily a basis for G). Note also that the test '$h \notin D$' translates now to '$w(B) < w(B + h)$' in the abstract setting.

The procedure lptype(G, T) computes a basis B of G and uses the set T as an auxiliary parameter for guiding its computations.

```
function procedure lptype(G,T);
    F := T;  B := T;
    for all h ∈ G − T in random order do
        F := F + h;
        if w(B) < w(B + h) then
            B := lptype(F, basis(B + h));
        end if;
    end do;
    return B;
```

A basis for H can be computed by the call lptype$(H, \emptyset)$.

Let us first argue that the algorithm terminates. If the input basis T is already a basis for G, then no further recursive calls are necessary, and the procedure will correctly return T after $|G - T|$ steps.

Whenever there is a recursive call, then the second argument consists of a new basis with value larger than the previous one. Hence, there are at most $\sum_{i=0}^{\delta} \binom{n}{i}$ such calls, and each requires at most n tests of the form '$w(B) < w(B + h)$'. It follows that there are at most $O(n^{\delta+1})$ such tests — that is, the algorithm terminates after a finite number of steps. (Note that these bounds are independent of the order in which the elements of $G - T$ are being processed.)

The primitive operations required are *violation tests* '$w(B) < w(B + h)$?', and *basis computations* for sets $B + h$, where B itself is a basis, and h is a constraint violating B.

(In the context of linear programming in $\mathbf{R}^d$, a violation test is easy to implement — just substitute the current optimal solution into the equation of the new hyperplane h. To perform basis computations efficiently, we propose to maintain the d edges of the feasible region incident to the current optimum. This allows us to compute basis$(B + h)$ in time $O(d^2)$, by intersecting all these edges with h; the edges incident to $w(B + h)$ are then computed by a single Gaussian elimination step. In this manner, a basis computation takes $O(d^3)$ time. Note that the naive approach that computes the intersection points of h and each subset of $d - 1$ hyperplanes from B would have taken time $O(d^4)$. All this assumes that $w(B)$ is a point in $\mathbf{R}^d$; a similar technique can be devised when the current feasible region is unbounded.)

Returning to the abstract framework, we have already seen that there are at most

$O(n^\delta)$ basis computations, and $O(n^{\delta+1})$ violation tests; our goal is now to determine better bounds on the expected number of these operations.

As a warm-up exercise, let us analyze the probability that '$w(B) < w(B+h)$' is satisfied, when F has cardinality i for $|T| < i \leq |G|$. So we have a random set $F \supset T$ of i constraints (more precisely, $F - T$ is a random subset of $G - T$ with $i - |T|$ elements), a random $h \in F - T$, and a basis B_{F-h} of $F - h$, and we ask for the probability that $w(B_{F-h}) < w(B_{F-h} + h)$ under these assumptions.

Lemma 1

$$w(B_{F-h}) < w(B_{F-h} + h) \quad \text{iff} \quad w(F-h) < w(F) .$$

Proof. We can rewrite the statement as

$$w(B_{F-h}) < w(B_{F-h} + h) \quad \text{iff} \quad w(F-h) < w((F-h)+h),$$

which is an immediate consequence of condition (C2), since $w(B_{F-h}) = w(F-h)$ and $B_{F-h} \subseteq F - h$ by definition of a basis. □

We conclude that

$$\text{Prob}(w(B_{F-h}) < w(B_{F-h} + h)) \leq \frac{\delta}{i - |T|}, \tag{1}$$

because of Lemma 1, and the fact that there are at most δ extreme constraints among the $i - |T|$ constraints in $F - T$ we can choose from. This type of analysis is called 'backward analysis', since we reverse the process of adding a random h to a set of $i - 1$ elements (as it happens in the algorithm), and instead analyze the process of removing a random h from a set F of i elements (this follows the analysis of [Sei1] — more examples of backward analysis can be found in [Sei2]).

Note that (1) holds for every $F \supset T$ of i elements, as long as h is chosen randomly in $F - T$. The simple but crucial observation for our further analysis is that the probability estimated in (1) decreases, if some of the extreme constraints in F are already in T; for example, if all extreme constraints of F are in T, then this probability is 0. So a set T which contains many extreme constraints for all sets F with $T \subset F \subseteq G$ will be favorable for our purposes. This leads to the following definitions.

Given a pair (G,T), $T \subseteq G \subseteq H$, we call a constraint $h \in G$ *enforcing in* (G,T), if

$$w(G-h) < w(T) .$$

We get

Lemma 2 *If h is enforcing in (G,T), then* (i) $h \in T$, *and* (ii) *h is extreme in all F with $T \subseteq F \subseteq G$.*

Proof. (i) If $h \notin T$, then $T \subseteq G - h$, and so

$$w(T) \leq w(G - h) < w(T) ;$$

a contradiction.

(ii) If h is not extreme in F, then

$$w(F) = w(F - h) \leq w(G - h) < w(T) \leq w(F) ;$$

again, a contradiction. $\square$

Let $\Delta = \Delta_{(G,T)}$ denote δ minus the number of enforcing constraints in (G,T), and call Δ the *hidden dimension* of (G,T). The discussion above implies that the bound in (1) can be improved to $\frac{\Delta}{i-|T|}$.

The following lemma demonstrates that whenever we make a recursive call, then the hidden dimension decreases at least by 1. Since $\Delta = 0$ entails that T is a basis of G, we conclude that the recursion bottoms out after at most δ levels.

Lemma 3 *Let $T \subseteq F \subseteq G \subseteq H$, and let $h \in F - T$ be an extreme constraint in F. Then, for S a basis of $B_{F-h} + h$,*
(i) all constraints e enforcing in (G,T) are also enforcing in (F,S),
(ii) h is enforcing in (F,S), and
(iii) $\Delta_{(F,S)} \leq \Delta_{(G,T)} - 1$.

Proof. (i) holds because

$$w(F - e) \leq w(G - e) < w(T) \leq w(F - h) = w(B_{F-h}) < w(B_{F-h} + h) = w(S).$$

This derivation includes also $w(F - h) < w(S)$, which yields (ii). Finally, (iii) follows from (i) and (ii), observing that h is not enforcing in (G,T) (or else we would have $h \in T$). $\square$

We now show that the hidden dimension actually decreases much faster 'on the average'. In terms of backward analysis, we ask: given F, $T \subset F \subseteq G$, and a random $h \in F - T$, what is the probability that the addition of h to $F - h$ causes a recursive call, and if so, what is the expected hidden dimension of the arguments $(F, \texttt{basis}(B_{F-h} + h))$ of such a call.

Lemma 4 *Let $\{e_1, e_2, \ldots\}$ be the extreme constraints in F which are not in T, enumerated in such a way that*

$$w(F - e_1) \leq w(F - e_2) \leq \ldots .$$

Then for $1 \leq j \leq \ell$, e_j is enforcing in $(F, \texttt{basis}(B_{F-e_\ell} + e_\ell))$.

Proof. Using the fact that e_ℓ is extreme in F, and condition (C2), we obtain:

$$w(F - e_j) \leq w(F - e_\ell) = w(B_{F-e_\ell}) < w(B_{F-e_\ell} + e_\ell) = w(\mathtt{basis}(B_{F-e_\ell} + e_\ell)) \, .$$

□

Hence if $h \notin \{e_1, e_2, \ldots\}$, then there is no recursive call caused by the addition of h to $F-h$. If $h = e_\ell$, then the arguments of the recursive call have hidden dimension at most $\Delta_{(G,T)} - \ell$, since $\{e_1, e_2, \ldots, e_\ell\}$ will be enforcing in $(F, \mathtt{basis}(B_{F-h} + h))$, in addition to the enforcing constraints of (G,T). Since each e_ℓ is equally likely to be chosen as the constraint h, it follows that the decrease in Δ is rather rapid on the average.

Let $t_k(m)$ denote the maximum expected number of violation tests for a call to `lptype` (and all its subsequent recursive calls) with a pair (G,T) with $|G| = m$ and $\Delta_{(G,T)} = k$. Then $t_0(m) = m - \delta$, since in this case T must contain δ enforcing constraints, and we have to make only one violation test for each of the constraints in $G - T$ (all yielding negative results). For $m \geq \delta + 1$ and $k \geq 1$, the previous analysis yields

$$t_k(m) \leq t_k(m-1) + 1 + \frac{t_0(m) + t_1(m) + \cdots + t_{k-1}(m)}{m - \delta} \, . \qquad (2)$$

Lemma 5 $t_k(\delta + 1) \leq 2^{k+2}$.

Proof. Consider a pair (G,T) with $|G| = \delta + 1$ and $\Delta_{(G,T)} = k$, and let $E \subset G$ be the $\delta - k$ enforcing constraints. We look at the sequence of all violation tests '$w(B_i) < w(B_i + h_i)$?', $i = 1, 2, \ldots$ in the chronological order that they occur during the execution of the algorithm. Note first that there are at most 2^{k+1} distinct sets in the sequence of $(B_i + h_i)$'s, since all these sets have to contain E, so we may choose only $\leq k+1$ elements in $G-E$ to obtain such a set, and $|G-E| = k+1$. Furthermore, the values $w(B_i)$'s form a nondecreasing sequence, and whenever $w(B_i) < w(B_i + h_i)$ actually holds, then $w(B_i) < w(B_{i+1}) = w(B_i + h_i)$. Assume now that for $i < j$, $B_i + h_i = B_j + h_j$. Then $B_i \neq B_j$ and $w(B_i) < w(B_j) \leq w(B_j + h_j) = w(B_i + h_i)$. That is, $w(B_i) < w(B_i + h_i)$ and so $B_{i+1} = \mathtt{basis}(B_i + h_i)$ which yields $w(B_i + h_i) \leq w(B_j)$; hence, we have $w(B_j) = w(B_j + h_j)$. Consequently, there is no third index ℓ with $B_i + h_i = B_\ell + h_\ell$, and so the sequence has length at most $2 \cdot 2^{k+1}$. □

Lemma 6 $t_k(m) \leq 2^{k+2}(m - \delta)$ *for* $m \geq \delta + 1$.

Proof. We proceed by double induction on k and m. The case $k = 0$ is obvious, and the case $m = \delta + 1$ is proved in Lemma 5. Suppose the lemma has been established for all pairs k', m' with $k' < k$, or $k' = k$ and $m' < m$. Then (2) implies

$$t_k(m) \leq 2^{k+2}(m - \delta - 1) + 1 + \frac{(2^2 + 2^3 + \cdots 2^{k+1})(m - \delta)}{m - \delta} \leq 2^{k+2}(m - \delta)$$

□

Since every basis computation is preceded by a violation test, this bounds also the number of basis computations. We summarize the result of this section:

Theorem 7 *Procedure* `lptype` *solves an LP-type problem of combinatorial dimension δ with $n \geq \delta + 1$ constraints using an expected number of at most $2^{\delta+3}(n - \delta)$ primitive operations.* □

Since each primitive operation can be performed in $O(d^3)$ time in the case of linear programming, we have as an explicit corollary (representative for the class of LP-type problems):

Corollary 8 *A linear program with n constraints and d variables can be solved in expected time $O(d^3 2^d n)$.* □

4 Discussion

We have presented a simple randomized algorithm for linear programming and many related problems which has expected running time linear in the number of constraints (for d constant). As indicated in the introduction, the analysis of the algorithm can be fine-tuned to obtain a subexponential bound for linear programming in n *and* d (see [MSW]) similar to the one in [Kal]. It would be interesting to consider also the issue of lower bounds in the formal framework we presented.

Another line of research we are pursuing is to even further simplify the implementation of the algorithm — without recursion and with the constraints simply stored in a linear array (in random order). We hope to present progress in this direction in [MSW].

References

[Beh] F. Behrend, Über die kleinste umbeschriebene und die größte einbeschriebene Ellipse eines konvexen Bereiches, *Math. Ann.* **115** (1938), 379–411.

[Cla1] K. L. Clarkson, Linear Programming in $O(n3^{d^2})$ time, *Inform. Process. Lett.* **22** (1986), 21–24.

[Cla2] K. L. Clarkson, Las Vegas algorithms for linear and integer programming when the dimension is small, manuscript, 1989.

[DLL] L. Danzer, D. Laugwitz and H. Lenz, Über das Löwnersche Ellipsoid und sein Analogon unter den einem Eikörper eingeschriebenen Ellipsoiden, *Arch. Math.* **8** (1957), 214–219.

[Dye1] M. E. Dyer, On a multidimensional search technique and its application to the Euclidean one-center problem, *SIAM J. Comput* **15** (1986), 725–738.

[Dye2] M. E. Dyer, A class of convex programs with applications to computational geometry, manuscript (1991).

[DyF] M. E. Dyer and A. M. Frieze, A randomized algorithm for fixed-dimensional linear programming, manuscript, 1987.

[Juh] F. Juhnke, Löwner ellipsoids via semiinfinite optimization and (quasi-) convexity theory, Technische Universität Magdeburg, Sektion Mathematik, Report 4/90 (1990).

[Jun] H. Jung, Über die kleinste Kugel, die eine räumliche Figur einschließt, *J. Reine Angew. Math.* **123** (1901), 241–257.

[Kal] G. Kalai, A subexponential randomized simplex algorithm, manuscript (1991).

[MSW] J. Matoušek, M. Sharir and E. Welzl, A subexponential bound for linear programming, in preparation (1991).

[Meg1] N. Megiddo, Linear-time algorithms for linear programming in $\mathbf{R}^3$ and related problems, *SIAM J. Comput* **12** (1983) 759-776.

[Meg2] N. Megiddo, Linear programming in linear time when the dimension is fixed, *J. Assoc. Comput. Mach.* **31** (1984), 114-127.

[Pos] M. J. Post, Minimum spanning ellipsoids, *in* "Proc. 16th Annual ACM Symposium on Theory of Computing" (1984), 108–116.

[Sei1] R. Seidel, Low dimensional Linear Programming and convex hulls made easy, *Discrete Comput. Geom.* **6** (1991), 423-434.

[Sei2] R. Seidel, Backwards analysis of randomized geometric algorithms, manuscript (1991).

[Syl] J. J. Sylvester, A question in the geometry of situation, *Quart. J. Math.* **1** (1857) 79-79.

[Wel] E. Welzl, Smallest enclosing disks (balls and ellipsoids), *in* "New Results and New Trends in Computer Science", (H. Maurer, Ed.), *Lecture Notes in Computer Science* (1991) to appear.

In-place Linear Probing Sort

Svante Carlsson* Jyrki Katajainen† Jukka Teuhola‡

Abstract

We introduce the first sorting algorithm that is proven to sort n randomly drawn uniformly distributed elements in $\Theta(n)$ time *in situ*. The constants in this algorithm are small, and simulations have shown it competitive with other sorting algorithms. It is, furthermore, conceptually simple and easy to code, which makes it a practical distributive sorting algorithm.

Keywords: Analysis of algorithm, Sorting, Average-case analysis, Distributive sorting.

1 Introduction

The time to sort n elements using comparisons between elements as the basic operation is $\Omega(n \lg n)$ both in the worst case and in the average case[1]. In standard text books there are usually three algorithms described with a running time that matches the lower bound, namely Quicksort, Mergesort, and Heapsort. Although Quicksort suffers from a worst-case behavior of $\Theta(n^2)$ while Mergesort requires linear extra storage to sort the n elements. There are versions of these, however less efficient, that can handle the drawbacks of the two algorithms. Heapsort, on the other hand, was designed to be an in-place algorithm with a worst case of $O(n \lg n)$.

In this paper we will use the terms implicit, in-place, and *in situ* interchangeably. By this we mean that we have only a constant number of pointers or indices and one or two extra data locations, apart from the array storing the elements.

If we allow ourselves to perform address calculation depending on the elements value, and if the element are drawn from some known distribution, we can sort in $O(n)$ expected time. There are several algorithms that achieve this. Most of them are based on bucketing, where the domain of the data is divided into several *buckets*, and each element is placed in the corresponding bucket. If there are more than one element in a bucket these elements are sorted using some comparison based algorithm. There have been some attempts to reduce the extra storage of n pointers that is used in the straight forward implementation of such an algorithm. Knuth [8] gave an improvement of MacLaren's [10] two-levels bucketing algorithm that only requires $\Theta(\sqrt{n})$ extra pointers. Handley [7] claims that his algorithm, Clocksort, is an *in situ* distributive sorting algorithm. However, if he is only using a constant number of pointers the algorithm will not run in linear expected time. Gonzalez and Johnson [6] came close to our goals when they gave an algorithm where parts of the data keys are modified and then restored during the sorting. This made it possible to encode the extra storage needed within the keys. We would, however, like to regard the elements as being atomic so that the elements can be accessed at any time during the algorithm, and more important so that the algorithm works independently of the representation of the data.

One could note that there is an analogy between the type of sorting described above and direct chaining hashing [4]. Another linear time sorting scheme is based on the same ideas as in linear probing hashing. The sorting method, called Linear Probing Sort, uses a table of size n/α, $\alpha < 1$,

*Institutionen för datateknik, Tekniska högskolan i Luleå , S-951 87 Luleå , Sweden

†Datalogisk Institut, Københavns Universitet, Universitetsparken 1, 2100 København O, Denmark

‡Department of Computer Science, University of Turku, Lemminkäisenkatu 14 A, SF-20520 Turku, Finland

[1]We will use $\lg n$ to denote $\log_2 n$

to sort n elements. The minimal number of probes on the average is $(2+\sqrt{2})n = 3.412\cdots n$ when $\alpha = 2-\sqrt{2} = 0.5857\ldots$ [5]. The extra storage necessary is a linear number of data locations, and not pointers. We will use this to reduce the extra storage to a minimum.

We will give two in-place variants of linear probing sort. One that is conceptually easy and with an easy analysis, but that uses subroutines that are inefficient and difficult to code. The other one is a modification that is only slightly more complicated to analyze. It is, however, much easier to code and it is much more efficient.

2 Linear Probing Sort

For the understanding of our in-place version of Linear Probing Sort we present the underlying algorithm that we will use as a subroutine. It is based on the standard linear probing hashing method by Peterson [11] but moving colliding element as in Amble and Knuth's ordered hashing [1].

The algorithm works by processing one element at the time. Divide the domain of the data into m buckets, each corresponding to an entry in a sorting table. Find the table entry that corresponds to the processed element. If the entry is occupied let the smallest of the two elements stay in that position. The larger element will repeatedly be tried at the next location until we have found an empty place to put the element in. It is possible that we need to store elements outside the m addresses that we map into. These locations is called the overlow area. This will be referred to as the sorting phase of the algorithm.

The next phase is the compaction phase. The table will be scanned from left to right moving each element to the first free location of the table, leaving the n elements sorted in the first n locations of the array.

If we let $m = n/\alpha$ where α is a positive constant less than one this procedure will sort uniformly distributed random elements in linear time. The expected size of the overflow area is only constant. For more details on Linear Probing Sort we would like to refer to Gonnet and Munro [5].

3 In-place Linear Probing Sort

To be able to make Linear Probing Sort sort in linear time we need to use linear amount of extra storage. Since we do not have access to extra storage we must solve this problem differently. The idea of our algorithm is to use the whole array to sort only part of the data. If we can sort at least a constant fraction of the data in each pass in linear time, and if we do not introduce any dependencies between the passes we will have a linear-time sorting algorithm that is not too difficult to analyze.

We could, in each pass, sort only the pn smallest elements, where n is the number of elements left to be sorted and $0 < p < 1/2$. These elements can be found by an in-place selection algorithm, either an average-case efficient one [3] or one with a guaranteed worst-case behavior [9]. These elements should be stored in the last pn positions of the array leaving the $(1-p)n$ largest elements in the first part of the array. At this point we can sort the small elements (belonging to the pn smallest) into the first $(1-p)n$ positions of the array using Linear Probing Sort. The positions containing large elements (belonging to the $(1-p)n$ largest) will be regarded as empty by the Linear Probing Sort subroutine, except that when a small element is inserted into an "empty" position the large element occupying that place will be moved to the origin of the small element that has just been inserted. The small elements are moved stably to the first pn locations, leaving the rest of the elements to be sorted in the same way.

More structured we could describe the algorithm as follows:

1. Find the element x which rank is $\lceil pn \rceil$
2. Partition the elements around x with smaller elements at larger indices and larger or equal elements at smaller indices

3. Insert the elements smaller than x, in order of appearance in the array, into the first $n - \lceil pn \rceil$ locations, using Linear Probing Sort. If an element larger than or equal to x is encountered the location is regarded as empty. If an assignment has to be done at such a location the element present can be moved to the origin of the inserted small element.

4. Move stably the elements smaller than x to the first $\lceil pn \rceil$ positions of the array.

5. Repeat from the beginning with $n = n - \lceil pn \rceil$ until the array is completely sorted.

The correctness of the algorithm depends heavily on the correctness of Linear Probing Sort. The first two steps only ensure that there are no small elements in the first part of the array. The third step inserts the small elements in the first part of the array, leaving large elements immediately after the sorted elements. The small elements are at this point in the correct order among themselves. In the forth step the elements are moved so that if the larger elements are sorted correctly the table will be sorted correctly. If there is a need for an overflow area for Linear Probing Sort there will be as many large elements in the right part as elements inserted at that time, leaving enough overlow area.

Theorem 1 *The algorithm above sorts n independently drawn uniformly distributed elements in* $\Theta(n)$ *expected time.*

Proof: We can observe that all steps except Step 3 can be performed in $O(n)$ time in the worst case. Note also that if we have n elements drawn from a uniform(a,b) distribution and that the k th smallest element has a value x then the $k-1$ smallest elements can be regarded as $k-1$ elements drawn from uniform(a,x) distribution, and similar for the $n-k$ largest elements. This means that we do not have any dependencies between different passes of the algorithm, and that we can use the same analysis recursively.

Let $T(n)$ denote the running time of our in-place algorithm and $T_{LPS}(n, \alpha)$ the running time for regular Linear Probing Sort, where n is the number of elements to be sorted and α the ratio between n and the table size. In this case $\alpha = \frac{p}{1-p}$. Both T and T_{LPS} are stochastic variables. We have

$$T(n) \leq \begin{cases} C & \text{if } n \leq \frac{1}{p} \\ cn + T_{LPS}(\lceil pn \rceil, \frac{p}{1-p}) + T(n - \lceil pn \rceil) & \text{otherwise} \end{cases}$$

To find the expected running time of the algorithm we can use the expected time for the different parts of the algorithm. Since the expected running time of Linear Probing Sort is linear for any fixed α the solution to this is

$$E[T(n)] \leq \frac{cn}{p} + E[T_{LPS}(n, \frac{p}{1-p})]$$

the total expected running time of the algorithm is thus also linear. □

The algorithm above mostly has a theoretical value since it involves linear time implicit selection which can be a quite time consuming operation. Since, in the model of our analysis, the elements are drawn from a uniform(a,b) distribution we can estimate the pivot value that will split the elements into two parts of sizes pn and $n - pn$, respectively. By choosing the pivot to be $a + p(b-a)$ we will have a very good estimate of the value of the pn'th smallest element.

The probability that a random element is less than the pivot value is p. Let X be the total number of elements less than the pivot. Since we have n independent random elements, $X \in \text{Bin}(n,p)$. The expected number of elements less than p, $E[X]$, is np, and the variance, $V[X]$, is $np(1-p)$. Using the approximations by Angluin and Valiant [2] (using Chernoff bounds) we can show that the probability that X is more than a small fraction from its expected value is very small. More specific, we have

$$P(X \leq E[X](1-\epsilon)) \leq e^{\frac{-\epsilon^2 np}{2}}$$

and

$$P(X \leq E[X](1+\epsilon)) \leq e^{\frac{-\epsilon^2 np}{3}}$$

To keep the algorithm simple we have to get a slightly more complicated, or at least more complex, analysis. We will measure the complexity of the algorithm by the number of probes that we perform in the table. In order to perform this analysis we will need some lemmas on the way. First, however, we would like to recall the simple algorithm that we would like to call *In-place Linear Probing Sort.* Below we give the pseudo code. We assume that the elements are in the interval (a,b).

1. Partition the elements around the pivot value $x = a + p(b-a)$ with smaller elements at larger indices and larger or equal elements at smaller indices
2. Insert the elements smaller than x, in order of appearance in the array, into the first locations, using Linear Probing Sort. If an element larger than or equal to x is encountered the location is regarded as empty. If an assignment has to be done at such a location the element present can be moved to the origin of the inserted small element.
3. Move stably the elements smaller than x to the first positions of the array.
4. Repeat from the beginning with the rest of the elements until the array is sorted.

Lemma 1 *For any value of n and α, $T_{LPS}(n,\alpha) = O(n^2)$ even in the worst case.*

Proof: The worst case of the algorithm, which can easily be verified, is when all elements will be mapped into the same location. This means that all elements will be compared with all the others, possibly in the overflow area. This is clearly $O(n^2)$. □

The lemma above shows that Linear Probing Sort has a worst case and thus also an average case of $O(n^2)$ for any value of α. This is true even for the table that only consists in the overflow area. The next lemma also gives an upper limit on the expected running time of our algorithm. Both of these results will later be used as approximations in the more careful analysis of the number of probes performed by the algorithm.

Lemma 2 *$E[T(n)]$ is polynomial in n.*

Proof: First we note that Step 1 and Step 3 of the algorithm take linear time. Thus, for some constant, c, they take at most cn time.

Let P_i denote the probability that we get i elements less than the pivot value. Let also ϵ be a small value between 0 and 1. For n larger than $\frac{1}{p}$ we have

$$\begin{aligned} E[T(n)] &\leq cn + \sum_{i=0}^{n} P_i E[T_{LPS}(i, \frac{i}{n-i}) + T(n-i)] \\ &= cn + \sum_{i \leq np(1+\epsilon)} P_i E[T_{LPS}(i, \frac{i}{n-i})] + \sum_{i > np(1+\epsilon)} P_i E[T_{LPS}(i, \frac{i}{n-i})] \\ &+ \sum_{i \leq np(1-\epsilon)} P_i E[T(n-i)] + \sum_{i > np(1-\epsilon)} P_i E[(T(n-i)] \end{aligned}$$

By using the previous lemma and the fact that the expected running time of Linear Probing Sort is linear if the load factor α is constant, and also that $E[T(n)]$ is increasing with n we get

$$\begin{aligned} E[T(n)] &\leq cn + dn \left(\sum_{i \leq np(1+\epsilon)} P_i \right) + O(n^2) \left(\sum_{i > np(1+\epsilon)} P_i \right) \\ &+ E[T(n)] \left(\sum_{i \leq np(1-\epsilon)} P_i \right) + E[T(n(1-p+p\epsilon))] \left(\sum_{i > np(1-\epsilon)} P_i \right) \end{aligned}$$

Using the approximations of the binomial distribution from above, and the fact that the sum of probabilities are at most one we get

$$\leq \quad cn + dn + O(n^2)e^{\frac{-\epsilon^2 np}{3}} + E[T(n)]e^{\frac{-\epsilon^2 np}{2}} + E[T(n(1-p+p\epsilon))]$$

Subtracting the E[T(n)] term of the right side from both sides, and normalizing, gives

$$E[T(n)] \quad \leq \quad \frac{c+d}{1-e^{\frac{-\epsilon^2 np}{2}}} n + o(1) + \frac{E[T(n(1-p+p\epsilon))]}{e^{\frac{-\epsilon^2 np}{2}}}$$

After observing that $T(\frac{1}{p})$ is constant the solution to this recurrence is

$$E[T(n)] \leq nC \sum_{i=0}^{\log_{1/\beta} n} \left(\frac{\beta}{e^{\frac{-\epsilon^2 pn\beta^{i-1}}{2}}} \right)^i$$

where $\beta = 1 - p + p\epsilon$ and C is some constant. Since $n\beta^i$, the number of elements we sort in the i th pass, is larger than one for all i's in the sum, the quote between the terms is bounded by some constant. This gives a polynomial bound on the expected runing time of the algorithm. □

We now have an approximation of $E[T(n)]$ that we will use to get a sharper bound of itself.

Theorem 2 *Implicit Linear Probing Sort uses at most* $8.2n + o(n)$ *probes on the average to sort* n *uniformly distributed elements if* $0.39 \leq p \leq 0.40$.

Proof: As in the proof of the previous lemma we get

$$\begin{aligned} E[T(n)] \quad &\leq \quad cn + E[T_{LPS}(np(1+\epsilon), \alpha)] + O(n^2)e^{\frac{-\epsilon^2 np}{3}} \\ &+ \quad E[T(n)]e^{\frac{-\epsilon^2 np}{2}} + E[T(n(1-p+p\epsilon))] \end{aligned}$$

where cn in this case is the expected number of probes by Step 1 and Step 3 of the algorithm, and $\alpha = \frac{p(1+\epsilon)}{1-p(1+\epsilon)}$.

Using the result of the previous lemma we get

$$E[T(n)] \quad \leq \quad cn + E[T_{LPS}(np(1+\epsilon), \alpha)] + o(1) + o(1) + E[T(n(1-p+p\epsilon))]$$

The number of probes in the table during the partition is one for each element (to inspect it) and one for each element that is moved, if the partition is carefully implemented. The expected number of elements moved from the first part of the array is equal to the expected number of small elements in that part. This is $np(1-p)$. For each element that is moved from the left part to the right one element has to be moved the other way, giving a total of $(1 + 2p(1-p))n$ probes for Step 1.

During the compaction we need to look at all the elements until we have found all small elements and using one extra probe when a small element is encountered. The expected place of the last small element is equal to the table size plus the overflow area. This gives a total of $(1-p)n + pn = n$ probes for the compaction on the average. Putting everything together we get that $c = 2 + 2p - 2p^2$.

From [5] we get that the expected number of probes for Linear Probing Sort to sort $np(1+\epsilon)$ elements is

$$\frac{np(1+\epsilon)}{2\alpha} \left(\frac{1}{1-\alpha} - 1 + \alpha \right) + O(1)$$

Number of elements	ILPS	LPS	Quicksort	Heapsort
1 000	0.0365 ± 0.0206	0.0267 ± 0.0165	0.0284 ± 0.0155	0.0684 ± 0.0182
5 000	0.195 ± 0.022	0.125 ± 0.017	0.203 ± 0.014	0.400 ± 0.022
10 000	0.395 ± 0.027	0.249 ± 0.007	0.438 ± 0.022	0.848 ± 0.018
50 000	2.04 ± 0.02	1.26 ± 0.02	2.64 ± 0.02	5.06 ± 0.02
100 000	4.11 ± 0.03	2.50 ± 0.02	5.67 ± 0.03	10.79 ± 0.05
500 000	20.7 ± 0.1	12.9 ± 0.3	34.0 ± 0.1	61.9 ± 0.1
1 000 000	41.5 ± 0.1	no result	80.0 ± 0.1	130.8 ± 0.2

Table 1: 95% intervals of confidence for the running time in seconds on a SUN SPARC station SLC, for our algorithm (ILPS), Linear Probing Sort (LPS), Quicksort, and Heapsort.

Inserting this and solving the recurrence gives us

$$E[T(n)] \leq \left(2 + 2p - 2p^2 + \frac{p(1+\epsilon)}{2\alpha}\left(\frac{1}{1-\alpha} - 1 + \alpha\right)\right)\frac{1}{p - p\epsilon}n + O(1)$$

Letting ϵ tend to zero we get a minimum for $E[T(n)]$ of $8.1939\ldots n$ for $p = .4060\ldots$. □

For $p = 0.4$, $E[T(n)] \leq 8.2n$, which suggests that for simplicity we chose to use $p = 0.4$ in our algorithm.

In an analysis of the number of probes of an algorithm there is always a question of what is ment by a probe. We have counted all accesses in the table. We could have chosen to count only one for each element we have to examine, which may be more conform with the analysis of Gonnet and Munro. If we did we would have had only $6n$ probes for $p = 0.4$. This would in fact make the comparison with Gonnet and Munro's analysis of Linear Probing Sort more fair, since they are not even counting the initial access of the elements to sort. Compared to $3.412n$ probes our extra cost up to $6n$ probes is reasonable to get rid of all extra storage.

4 Conclusion

We have given a simple in-place algorithm that sorts n uniformly distributed elements using less than $8.2n$ probes in the table that we use for sorting. All other operations can be performed very fast. As cán be seen in Table 1 our algorithm competes well with the other in-place algorithms, such as Quicksort and Heapsort, but also with fast distributive algorithms such as Linear Probing Sort. We have used, to our knowledge, the fastest implementations of the sorting algorithms in the simulation. Our algorithm is easy to code. It can be done in less than 40 PASCAL lines.

We would also like to point out that if we find another linear time sorting algorithm that uses a linear amount of extra data locations (not pointers) we could use that algorithm instead of Linear Probing Sort. This will also give us a linear time in-place sorting algorithm. This could be very important if we find such a sorting algorithm that is more robust against non-uniform distributions.

We would be interested in seeing a natural algorithm that, together with the properties of Implicit Linear Probing Sort, also has a worst case of $O(n \lg\ n)$. We could, of course, use a counter of the number of probes, and when the counter is too large we could use Heapsort on the whole array. We do not, however, consider this to be a natural algorithm.

5 Acknowledgements

We would like to thank Ola Petersson for valuable discussions on the problem. We would also like to thank Christer Mattsson for helpful comments on Linear Probing Sort and on the analysis of the algorithm.

References

[1] O. Amble and D. E. Knuth, *Ordered hash tables*, Computer Journal, 17(2):135-142, (May 1974)

[2] D. Angluin and L. G. Valiant, *Fast probabilistic algorithms for Hamiltonian circuits and matchings*, Journal of Computer System Science, 18:155-193, (1979)

[3] R. W. Floyd and R. L. Rivest, *Expected time bounds for selection*, Communications of the ACM, 18(3):165-172, (Mar. 1975)

[4] G. H. Gonnet, *Handbook of Algorithms and Data Structures*, Addison-Wesley, Reading Mass., (1984)

[5] G. H. Gonnet and J. I. Munro, *The analysis of Linear Probing Sort by the use of a new mathematical transform*, Journal of Algorithms 5,(1984), 451-470.

[6] T. F. Gonzalez and D. B. Johnson, *Sorting numbers in linear expected time and optimal extra space*, Information Processing Letters, 15(3):119-124, (Oct. 1982)

[7] C. C. Handley, *An in situ distributive sort*, Information Processing Letters, 23(1986), 265-270

[8] D. E. Knuth, *The Art of Computer Programming, vol. III: Sorting and Searching* , Addison-Wesley, Reading Mass. (1973)

[9] T. W. Lai and D. Wood, *Implicit selection*, In proceedings SWAT 88, Halmstad, Sweden, Lecture Notes in Computer Science 318, 14-23, Springer Verlag, (Jul. 1988)

[10] M. D. MacLaren, *Internal Sorting by radix plus shifting*, Journal of the ACM, 13(3):404-411, (July 1966)

[11] W. W. Peterson, *Addressing for random-access storage*, IBM Journal on Research and Development, 1(4):130-146 (Apr. 1957)

SPEEDING UP TWO STRING-MATCHING ALGORITHMS

Maxime CROCHEMORE[1,2], Thierry LECROQ[1]
LITP, Institut Blaise Pascal, Université Paris 7, 2 Place Jussieu, 75251 Paris Cedex 05, France

Artur CZUMAJ, Leszek GASIENIEC, Stefan JAROMINEK, Wojciech PLANDOWSKI, Wojciech RYTTER
Institute of Informatics, Warsaw University, ul. Banacha 2, 00-913 Warsaw 59, Poland

Abstract

We show how to speed up two string-matching algorithms : the Boyer-Moore algorithm (BM algorithm) and its version called here the reversed-factor algorithm (the RF algorithm). The RF algorithm is based on factor graphs for the reverse of the pattern. The main feature of both algorithms is that they scan the text right-to-left from the supposed right position of the pattern, BM algorithm goes as far as the scanned segment is a suffix of the pattern, while the RF algorithm is scanning while it is a factor of the pattern. Then they make a shift of the pattern, forget the history and start again. The RF algorithm usually makes bigger shifts than BM, but is quadratic in the worst case. We show that it is enough to remember the last matched segment to speed up considerably the RF algorithm (to make linear number of comparisons with small coefficient) and to speed up BM algorithm with match-shifts (to make at most $2.n$ comparisons). Only a constant additional memory is needed for the search phase. We give alternative versions of an accelerated algorithm RF: the first one is based on combinatorial properties of primitive words, and two others use extensively the power of suffix trees.

1. INTRODUCTION

The Boyer-Moore algorithm [BM 77] is one of the string-matching algorithms very fast on average. However it is successful mainly for the case of big alphabets. For small alphabets its average complexity is $\Omega(n)$, see [BR 91]. We discuss a version of this algorithm, named here the RF algorithm, which is much faster on average, also for small alphabets. If the alphabet is of size at least 2 then the average complexity of the new is $O(n \log(m)/m)$, and reaches the lower bound given in [Yao 79]. The main feature of both algorithms is that they scan the text right-to-left from the supposed right position of the pattern. BM algorithm goes as far as the scanned segment (also called a factor) is a suffix of the pattern, while the RF algorithm matches the text against any factor of the pattern, traversing the factor graph or the suffix tree of the reversed pattern. Afterwards, both algorithms make a shift of the pattern to the right, forget the history and start again. We show that it is enough to remember the last matched segment to speed up the algorithm: an additional constant memory is sufficient.

We derive a version of BM algorithm named here the algorithm Turbo_BM. One of the advantages of this algorithm with respect to the original BM algorithm is the simplicity of the complexity analysis. At the same time the algorithm Turbo_BM looks as a superficial modification of BM algorithm. Only few additional lines are inserted inside the search phase of the original algorithm and two registers (constant memory for the last match) are added. The preprocessing phase is left unchanged. An algorithm remembering a linear number of previous matches has been given before by Apostolico and Giancarlo [AG 86] as a version of BM algorithm. The algorithm Turbo_BM given here seems to be an efficient compromise between the recording of a linear size history as in the Apostolico-Giancarlo algorithm, and no recording of any history about previous matches in the original BM algorithm.

Our method to speed up the BM and RF algorithms is an example of a general technique called in [BKR 91] the dynamic simulation - for a given algorithm A construct the algorithm A' which works in the same way as A but remembering a part of the information A is wasting; during the process such an information is used to save on a part of the computation the original algorithm A does. In our case the additional information is the constant size information about the last match. The transformation of the Boyer-Moore algorithm gives an algorithm of the same simplicity as the original Boyer-Moore algorithm and with the upper bound of $2.n$ on the number of comparisons, which improves slightly on the bound

---

[1] Work by these authors is partially supported by PRC "Mathématiques-Informatique".

[2] Work by this author is partially supported by NATO Grant CRG 900293

3.n of the original algorithm. The derivation of this bound is also much simpler than the 3.n bound in [Co 89]. The transformations of the RF algorithm show the applicability of data structures representing succinctly the set of all subwords of a pattern p of length m. We denote this set by $FACT(p)$. The set of all suffixes of p is denoted by $SUF(p)$. For simplicity of presentation we assume that the size of the alphabet is constant.

The general structure of BM and RF algorithms looks as in Figure 1.

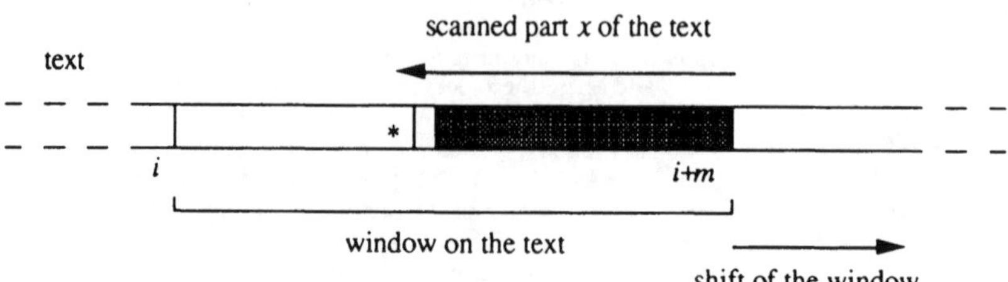

Figure 1. One iteration of Algorithm 1.
The algorithm scans right-to-left a segment (factor) x of the text.

Algorithm 1 /* common schema for algorithms BM and RF */;
i:=0;
while $i \leq n-m$ **do**
{ align pattern with positions $t[i+1..i+m]$ of the text;
scan the text right-to-left from the position $i+m$;
let x be the scanned part of the text;
if match found (x = pattern) **then** report it;
compute the shift;
i:=i+ *shift*; }
end.

In Algorithms BM and RF we use the synonym x for the lastly scanned segment $t[i+j..i+m]$ of the text t. This will shorten the presentation. In one algorithm we check if x is a suffix, and in the second algorithm we check if it is a factor of t. The shift uses a pre-computed function on x. In fact in BM algorithm x is identified with a position j on the pattern, while in the RF algorithm x is identified with a node corresponding to x^R in a data structure representing $FACT(p^R)$. We use the reversed pattern because we scan right-to-left, while most data structures for the set of factors are oriented to left-to-right scanning of the pattern. These orders are equivalent after reversing the pattern. In both cases a constant size memory is sufficient to identify x.

Both algorithms BM and RF can be viewed as instances of Algorithm 1. For a suffix x of length j denote here by *BM_shift*[x] the match-shift $d2[j-1]$ defined in [KMP 77] (see also [Ry 80] and [Ah 90]). The value of $d2[j-1]$ is, roughly speaking, the minimal shift (>0) of the pattern on itself such that the symbols aligned with the suffix x, except the first letter of x, agree. The symbol at the position aligned with the first letter of x , denoted by * in Figure 2, is distinct, if there is any symbol aligned (see Figure 2).

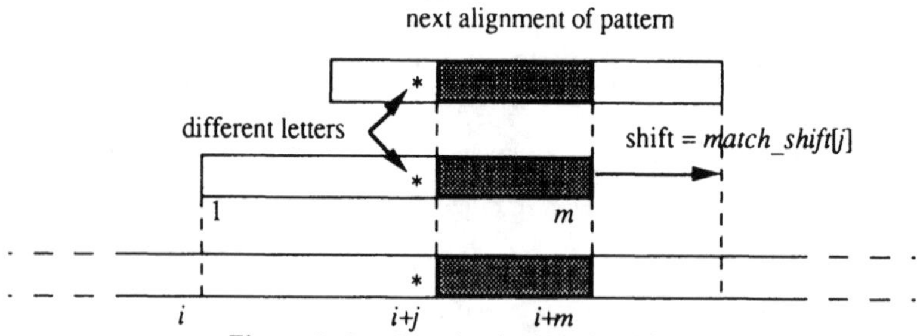

Figure 2. One iteration in BM algorithm.

Algorithm BM; /* reversed-suffix string matching */
i:=0; /* denote $t[i+j..i+m]$ by x, it is the lastly scanned part of the text */
while $i \le n-m$ **do**
{ j:=m; **while** j>1 & $x \in SUF(p)$ **do** j:=j-1;
if x = pattern **then** report match;
shift:= *BM_shift*[j];
i:=i+*shift*; }
end.

The work which Algorithm 1 spends at one iteration is denoted here by *cost*, the shift is denoted by *shift*. In BM algorithm a small cost gives usually a small shift. The strategy of the RF algorithm is more optimal: the smaller is the cost the bigger is the shift. The bigger shifts speed up the algorithm better. In practice $cost_i$ and the match at a given iteration is usually very small, hence the algorithm whose shifts are reversely proportional to the local matches is closer to optimal. The straightforward application of this strategy gives algorithm RF that is very successful on average, unfortunately it is quadratic in the worst case.

Algorithm RF; /* reversed-factor string-matching */
i:=0; /* denote $t[i+j..i+m]$ by x; it is the lastly scanned part of the text */
while $i \le n-m$ **do**
{ j:=m; **while** j>1 & $x \in FACT(p)$ **do** j:=j-1;
/* in fact, we check the equivalent condition $x^R \in FACT(p^R)$ */
$x = t[i+j..i+m]$ is the scanned part of the text;
if x = pattern **then** report match;
shift:= *RF_shift*[x]; i:=i+*shift*; }
end.

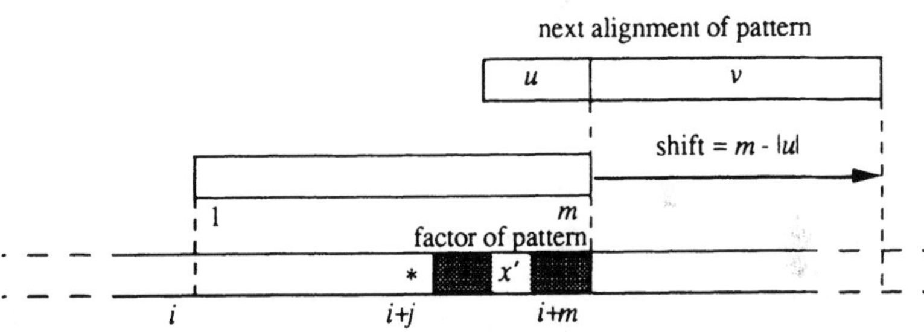

Figure 3. One iteration of Algorithm RF.
Word u is the longest prefix of the pattern that is a suffix of x'.

The algorithm RF makes an essential use of a data structure representing the set $FACT(p)$, see [BBEHCS 85] for the definition of directed acyclic word graphs (dawg's), [Cr 86] for the definition of suffix automata and see [Ap 85] for details on suffix trees. The graph $G = dawg(p^R)$ represents all subwords of p^R as labelled paths starting from the root of G. The factor z corresponds in a many-to-one fashion to a node $vert(z)$ such that the path from the root to that node "spells" z. Additionally we add to each node an information telling whether all paths corresponding to that node are suffixes of the reversed pattern p^R (prefixes of p). We traverse this graph when scanning the text right-to-left in the algorithm RF. Let x' be the longest word which is a factor of p found in a given iteration. When x=p x'=x; otherwise x' is obtained by cutting off the first letter of x (the mismatch symbol).

We define the shift *RF_shift*, and describe how to compute it easily. Let u be the longest suffix of x' which is a prefix of the pattern. It should be a proper suffix of x' iff x'=x=p. We can assume that we always know the actual value of u, it is the last node on the scanned path in G corresponding to a suffix of p^R. Then shift $RF\_shift[x] = m$-$|u|$ (see Figure 3).

The use of information about the previous match in the next iteration is the key to an improvement. However this application can be realized in many ways: we discuss three alternative transformations for RF. This leads to three versions Turbo_RF, Turbo_RF', and Turbo_RF" of the RF algorithm that are presented in Section 3.

Algorithms Turbo_BM and Turbo_RF, Turbo_RF', Turbo_RF" can be viewed as instances of Algorithm 2 presented below.

Algorithm 2 /* general schema for algorithms Turbo_RF, Turbo_RF', Turbo_RF" and Turbo_BM: a version of Algorithm 1 with an additional memory */;
i:=0; memory:=nil;
while $i \le n-m$ **do**
{ align pattern with positions $t[i+1..i+m]$ of the text;
scan the text right-to-left from the position $i+m$,
use memory to reduce number of inspections;
let x' be the part of scanned text; /* x' is here usually smaller than x in Algorithm 1 */
if match found **then** report it;
compute the shift $shift_i$ depending on x and memory; $i:=i+shift_i$;
update memory using the information about x; }
end.

2. SPEEDING UP THE REVERSED-FACTOR ALGORITHM

To speed up algorithm RF we memorize the prefix u of size m-$shift$ of the pattern. We have a situation depicted in Figure 3. We then scan the part of the text align with the part v of the pattern right-to-left. When we arrive at the boundary between u and v in a successful scan (all comparisons positive) then we are in a *decision point*. Now instead of scanning u until a mismatch is found, we just can scan a part of u, due to combinatorial properties of *primitive words*. A word is primitive iff it is not a proper power of a smaller word. We denote by $per(u)$ the length of the smallest period of u. Primitive words have the following useful properties:

a) the prefix of u of size $per(u)$ is primitive;

b) a cyclic shift of a primitive word is also primitive, hence the suffix z of u of size $per(u)$ is primitive;

c) if z is primitive then the situation presented in the Figure 4 is impossible.

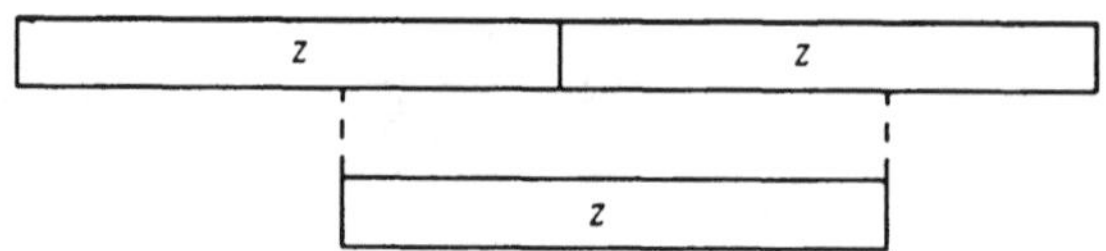

Figure 4. If z is primitive then such an overlap is impossible.

If $y \in FACT(p)$ we denote by $displ(y)$ the least integer d such that $y = p[m-d-|y|+1..m-d]$ (see Figure 5).

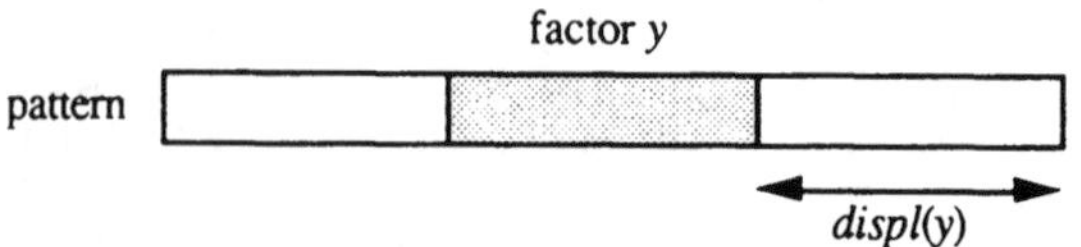

Figure 5. Displacement of factor y in the pattern.

The crucial point is that if we scan successfully v and the suffix of u of size $per(u)$ then we know the shift without further calculations: many comparisons are saved and the algorithm RF can be sped up in this moment. In terms of the next lemma we save $|x|-|zv|$, in the situation when $|x| \ge |zv|$.

```
Algorithm Turbo_RF ;
/* denote t[i+j..i+m] by x; it is the lastly scanned part of the text; we memorize the last prefix u of the
pattern; initially u is the empty word; */
i:=0; u:=empty;
while i≤n-m do
{     j:=m; while j>|u| & x ∈ FACT(p) do j:=j-1;
      if j=|u| then
/* we are at the decision point between u and v, after v has been successfully scanned */
            if v ∈ SUF(p) then report a match;
            else { scan right-to-left up at most per(u) symbols stopping at a mismatch;
                   let x be the successfully scanned text;
                   if |x|=m-|u|+per(u) then shift:=displ(x) else shift:=RF_shift(x); };
      else shift:=RF_shift(x);
      i:=i+shift; u:=prefix of pattern of length m-shift; }
end.
```

Lemma 1. (key lemma)
Let u, v be as in Figure 3. Assume that u is periodic ($per(u) \leq |u|/2$). Let z be the suffix of u of length $per(u)$ and let x be the longest suffix of uv that belongs to $FACT(p)$. Then

$$zv \in FACT(p) \text{ implies } RF\_shift(x) = displ(zv).$$

Proof.
It follows from the definition of $per(u)$, as the smallest period of u, and periodicity of u that z is a primitive word . The primitivity of z implies that occurrences of z can appear from the end of u only at distances which are multiples of $per(u)$. Hence $displ(zv)$ should be a multiple of $per(u)$, and this easily implies that the smallest proper suffix of uv which is a prefix of p has size $|uv|$-$displ(zv)$. Hence the next shift in the original algorithm RF is $shift$=$displ(zv)$.♦

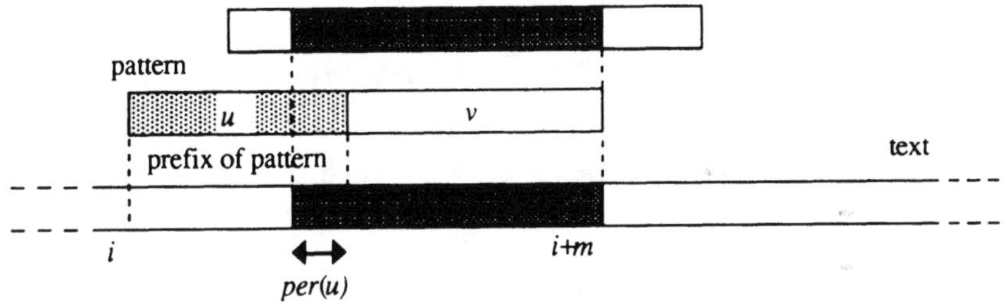

Figure 6. We keep in memory the prefix u of pattern. If u is periodic then at most $per(u)$ symbols of u are scanned again. However, this extra work is amortized by next shift. The symbols of v are scanned for the first time.

We need to explain how we add only a constant memory to the algorithm RF. The variables u and x need only pointers to the text. The values of displacements are in the data structure which represents $FACT(p)$, similarly the values of periods $per(u)$ for all prefixes of the pattern are pre-computed with the representation of $FACT(p)$ and read-only in the algorithm. In fact it is possible to remove the table of periods and values of $per(u)$ can be computed dynamically inside the algorithm Turbo_RF using constant additional memory.

Theorem 2.
The algorithm Turbo_RF makes at most 2.n symbol comparisons.
Proof.
If u is periodic then there are scanned again at most $per(u)$ symbols of u. If u is not periodic then we scan again at most half of u. Let extra_cost be the numbers of symbols inside u scanned in the actual stage. In each case extra_cost ≤ next_shift. Hence it is amortized by the next shift, this gives together at most n comparisons. The symbols in parts v are scanned for the first time in a given stage. They are disjoint in distinct phases. Hence they give together at most n comparisons. The work spent inside segments u and inside segments v is thus bounded by 2.n. This completes the proof. ♦

3. TWO OTHER VARIATIONS OF THE ALGORITHM TURBO_RF

Assume that we are at a decision point, when we just finished scanning v. At this moment we know that the part of text immediately to the left is a prefix u of p of size m-|v|. Denote by *nextpref*(v) the biggest suffix of uv that is a prefix of uv, and that is longer than v. If there is no such suffix then denote the corresponding value by *nil*. If we know this value then we could take as the next *RF_shift* the shift by distance m-|*nextpref*(v)|. The next value of u will be *nextpref*(v). All that is determined uniquely by v, hence after a suitable preprocessing of the pattern no symbols of u are to be read. The algorithm will make at most n comparisons of the pattern versus the text. However the complexity is affected by the computation of *nextpref*(v).

There are at least two possible approaches. One is to pre-compute a data structure which allows to compute at the k-th iteration the value of *nextpref*(v) in time $cost'_k$, such that the sum of all $cost'_k$'s is linear. The second solution is to preprocess the pattern in such a way that the value of *nextpref*(v) can be computed in constant time.

Technically it will be convenient to deal with suffixes, denote $p'=p^R$. We look at the computation of *nextpref* from a "reverse" perspective. Let *nextsuf*(v) = (*nextpref*(v))^R. In other words *nextsuf*(v) is the biggest prefix of v^Ru^R, which is a suffix of p', where u^R is the suffix of p' of length m-|v|.

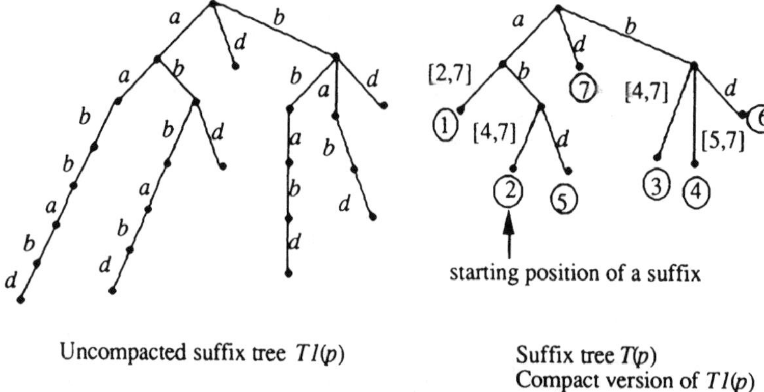

Uncompacted suffix tree $T1(p)$

Suffix tree $T(p)$
Compact version of $T1(p)$

Figure 7. The uncompacted tree and suffix tree T for p'=*aabbabd*. Each factor corresponds to a node in the first tree. The factors corresponding to the nodes in the suffix tree are *main* nodes. The representative *repr*(v) of the word v is the first descendant of v in the first uncompacted tree which is a node in the suffix tree.

In both approaches we represent the set *FACT*(p') by the suffix tree T. The edges of this tree are labelled by factors of the text represented by pairs (start-position, end-position). We take the compacted suffix tree for p'\$, in the sense of [Ap 85], then we cut off all edges labelled by \$. Afterwards each suffix of p' is represented by a node of T. In Figure 7 is presented an uncompacted suffix tree and a (compacted) suffix tree. The term "compacted" will be omitted later. Call factors of p' which correspond to nodes of T the main factors. The tree T has only linear number of nodes, hence not all factors of p' are main. Non-main factors correspond to a point on an edge of T. For a word v denote by *repr*(v) the node of T corresponding to the shortest word v' which is an extension of v (possibly $v=v'$). For example the whole string p' is main and for p'=*aabbabd* we have *repr*(*aa*)=p'.

The first approach.

Let P be the *failure function* of p, see [KMP 77]. $P(j)$ is the length of the longest proper suffix of the pattern which is also its prefix (a border). Assume that P is pre-computed.

Let $j = m$-|v|. Let *suf*(k) denotes here the node corresponding to the suffix of size k. Then it is easy to prove the following fact:

$$|nextsuf(v)| = \mathrm{MIN}\{\ k\ /\ k = |v|+P^h(j) \text{ and } suf(k) \text{ is a descendant of } repr(v^R)\ ,\ \text{for } h\geq 0\}.$$

If the set on the right side is empty then *nextsuf*(v) = *nil*.

We can check whether $suf(k)$ is a descendant of $repr(v^R)$ in a constant time, after preprocessing the suffix tree T. We can number the nodes of the tree in a DFS order. Then the nodes which are descendants of a given node form an interval of consecutively numbered nodes. Associate with each node such an interval. Then the question about descendants is reduced to an inclusion of an integer in a known interval. This can be answered in a constant time. Altogether this gives the algorithm Turbo_RF'.

Theorem 3.
The algorithm Turbo_RF' makes at most n symbol comparisons of pattern versus text t. The total number of iterations $P^h(j)$ done by the algorithm does not exceed n. The preprocessing time is also linear.

Proof.
We have already discussed the preprocessing phase. Each time we make an iteration of type $P^h(j)$ the pattern is shifted to the right of the text by at least one position, hence there are at most n such iterations. This completes the proof. ♦

The second approach

Here we improve the complexity of the search phase of the algorithm considerably. This increases the cost of the preprocessing phase that however is still linear. In the algorithm Turbo_RF' at a decision point we have sometimes to spend a linear time to make many iterations of type $P^h(j)$. In this new version, we compute the shift in constant time. It is enough to show how to preprocess the suffix tree T for p' to compute $nextsuf(v)$ for any factor v of p' in a constant time whenever it is needed.

First we show how to compute $nextsuf(v)$ for main factors, i.e. factors corresponding to nodes of T. Let us identify main factors with their corresponding nodes. The computation is in a bottom-up manner on a tree T.

Case of a bottom node: v is a leaf.
  $nextsuf(v) = nil$;
Case of an internal node:
  assume v has sons $v_1, v_2, ..., v_q$, then there exists a son v_j such that
  $nextsuf(v) = nextsuf(v_j)$ or, if v_j is a leaf then $nextsuf(v) = v_j$

We scan the sons v_i of v, and for each of them check if $nextsuf(v_j)$ or v_j is a *good* candidate for $nextsuf(v)$. We choose the longest good candidate, if there is any. Otherwise the result is *nil.*
The word v is prefix of each of the candidates. What does it exactly mean for a word y, whose prefix is v, to be a good candidate ? Let u be the prefix of the pattern p of length m-|v|. The candidate y is good iff the prefix of y of length |y|-|v| is a suffix of u, see the Figure 8. This means that the prefix of the pattern which starts at position |u|-(|y|-|v|) continues to the end of |u|. We have to be able to check it in constant time.

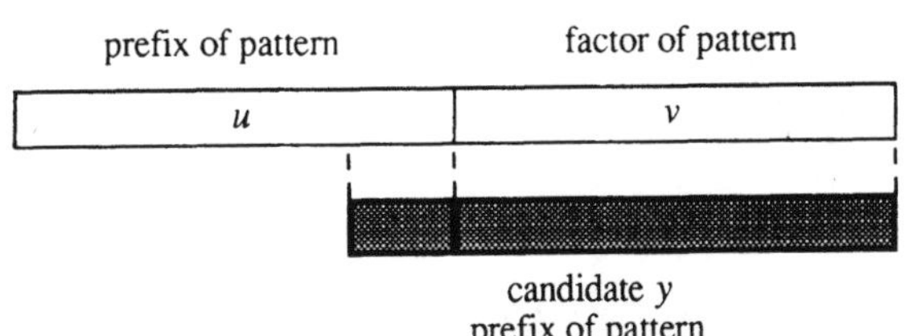

Figure 8. The candidate y is *good* iff $PREF[k] \geq |u|-k$, where $k=|u|-(|y|-|v|)$.

It is enough to have a table $PREF$, such that for each position k in the pattern the value $PREF[k]$ is the length of the longest prefix of p which starts at position k in p. This table can be easily computed in linear time as a side effect of the Knuth-Morris-Pratt algorithm for string-matching. After the table is computed we can check for each candidate in $O(1)$ time if it is good.
For nodes which are not main we set $nextpref(v)=y$, where $y=nextpref(repr(v^R))$, if y is a good candidate for v, i.e. $PREF[k] \geq |u|-k$, where $k=|u|-(|y|-|v|)$.

Hence after preprocessing we keep a certain amount of additional data: suffix tree, the table of *nextpref*(*v*) for all main nodes of this tree and the table *PREF*. Anyway, altogether this needs only linear size memory and is later accessed in a read-only way.

Theorem 4.
The pattern can be preprocessed in linear time in such a way that the computation of the *RF_shift* in the algorithm Turbo_RF can be accomplished in a constant time whenever it is needed. The data structure used in the preprocessing phase is read-only at search phase. Only a constant read-write memory is used at search phase.

Denote by Turbo_RF" the version of the algorithm Turbo_RF in which the computation of the RF_shift is computed at decision points according to Theorem 4. The resulting algorithm Turbo_RF" can be viewed as an automata-oriented string-matching. We scan the text backwards and change the state of the automaton. The shift is then specified by state of the automaton where the scanning stops. This idea applied directly gives a kind of Boyer-Moore automaton of polynomial (but not linear) size [Le 91]. However it is enough to keep in memory only a linear part of such an automaton (implied by the preprocessing referred in the theorem).

4. THE ANALYSIS OF THE AVERAGE CASE COMPLEXITY OF THE ALGORITHM RF AND ITS VERSIONS.

Denote by A the alphabet and by r the size of A. Assume $r>1$.
Assume that the input alphabet has $r>1$ letters. We consider the situation when the text is random. The probability of the occurrence of a specified letter on the i-th position is $1/r$, and these probabilities are independent (for distinct positions).

Theorem 5.
The expected time of the algorithm RF is $O(n\log_r(m)/m)$.
Proof.
Let $cost_i$ be the number of comparisons made in the i-th iteration of the algorithm. The shift computed in this iteration is denoted by $shift_i$. If $cost_i \leq 4\log_r m$ then $shift_i \geq m-4\log_r m$ and $shift_i$ is called *long*. It is called *short* otherwise. For a m which is big enough each long shift satisfies : $shift_i \geq m/2$.
The proof relies on two claims that we first establish :

Claim -i-.
The probability that $shift_i$ is short is less than $1/m^2$.
Proof of Claim -i-.
There exists less than m^2 different factors of length $4\log_r m$ in the pattern and there may be $r^{4\log_r m} = m^4$ different strings of this length in the scanned text. If we divide these numbers we obtain the required result. This completes the proof of the claim.

Let us partition the text t into disjoint segments of length $m/2$. The sequence of the iterations of the algorithm is called the k-th *phase* iff it consists of all iterations of the algorithm with the supposed end of the pattern placed in the k-th segment. Hence there are at most $2n/m$ phases in the algorithm. Now we study the expected cost of one phase. Let X_k be the random variable, whose value is the cost spent by the algorithm in the k-th phase.

Claim -ii-.
The expected cost of X_k is $\text{ave}(X_k) = O(\log_r m)$.
Proof of Claim -ii-.
Let us estimate separately the cost of the first iteration and the expected cost of other iterations in the k-th phase. The probability that the first shift in the k-th phase is short is less than $1/m^2$, due to Claim -i-. If this shift is long then the cost is logarithmic. Otherwise, the cost of all other iterations in the k-th phase does not exceed m^2. However the probability that we start the second iteration in the phase is less than $1/m^2$. Hence all these iterations contribute together $O(m^2.1/m^2)$ average cost. Altogether we have $O(\log m + 1)$, which is of a logarithmic order. This completes the proof of the claim.

The cost of the whole algorithm is :
$$\text{ave}(X_1+X_2+...+X_{2n/m}) = \text{ave}(X_1)+\text{ave}(X_2)+...+\text{ave}(X_{2n/m}) = O(n.\log(m)/m).$$
This completes the proof of the theorem. ♦

5. SPEEDING UP THE BOYER-MOORE ALGORITHM

The linear time complexity of the Boyer-Moore algorithm is quite nontrivial. The first proof of the linearity of the algorithm appeared in [KMP 77]. Other authors have work on it (see [Ga 79] and [GO 80]), however it was needed more than a decade for the full analysis. R.Cole has proved that the algorithm makes at most $3.n$ comparisons, see [Co 90], and that this bound is tight. The "mysterious" behavior of the algorithm is due to the fact that it forgets the history and the same part of the text can be scanned an unbounded (at most logarithmic) number of times. The whole "mystery" disappears when the whole history is memorized and additional $O(m)$ size memory is used. Then in successful comparisons each position of the text is inspected at most once. The resulting algorithm is an elegant string-matching algorithm (see [AG 86]) with a straightforward analysis of the text searching phase. However it requires more preprocessing and more tables than the original BM algorithm. In our approach no extra preprocessing is needed and the only table we keep is the original table of shifts used in BM algorithm. Hence all extra memory is of a constant size (two integers). The resulting algorithm Turbo_BM forgets all its history except the most recent one and its behavior has again a "mysterious" character. Despite that, the complexity is improved and the analysis is simple.

The main feature of the algorithm Turbo_BM is that during the process the factor of the pattern that matched the text during the last attempt is memorized. This has two advantages: it can lead to both a jump over the factor during the scanning phase and to, what is called, a *Turbo_shift*.

We now explain what is a *Turbo_shift*. Let x be the longest suffix of p that matches the text at a given position. Let also *fact* be the stored factor that matches the text at the same position. For different letters a and b, ax is a suffix of p aligned with bx in the text (see Figure 9). A *Turbo_shift* can occur when x is shorter than *fact*. In this situation the suffix ax of the pattern is aligned with the factor bx of the text and a, b are different letters. Since x is shorter than *fact*, ax is a suffix of *fact*. Thus a and b occur at distance m-g in the text. But, since suffix $p[f+1..m]$ of p has period m-g it cannot overlap both letters a and b. As a consequence the smallest shift of the pattern is then $(m-f)-(m-j)-(m-g)$, that is, $j+g-f-m$ (see Figure 10). This proves that if Algorithm BM is correct so is Algorithm Turbo_BM.

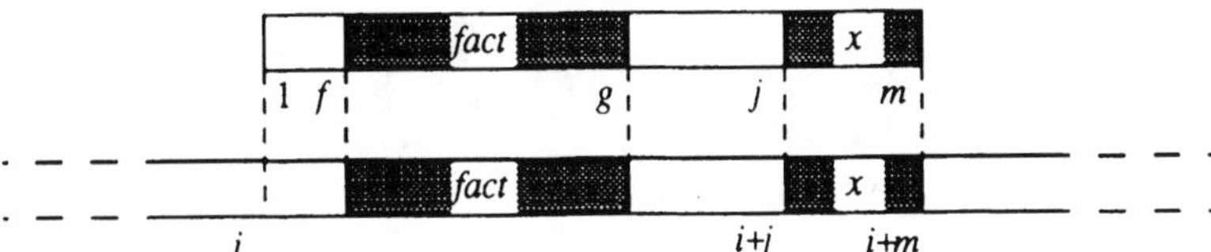

Figure 9. Variables i, j, f and g in Turbo_BM (matching parts in grey).

```
Algorithm Turbo_BM; /* reversed-suffix string matching */
i:=0; /* denote t[i+j..i+m] by x, it is the lastly scanned part of the text */
f:=0; g:=0; /* factor p[f+1..g] of p matches the text t[i+f+1..i+g] */
while i≤n-m do
{     j:=m; while j>0 & x ∈ SUF(p) do if j=g+1 then j:=f /* jump */ else j:=j-1;
      if x = pattern then report match;
      if g-f>m-j then Turbo_shift :=g+j-m-f else Turbo_shift :=0;
      if BM_shift[j] >= Turbo_shift then
      {     i = i+BM_shift[j];
            f = max(0, j-BM_shift[j]; g = m-BM_shift[j]; }
      else
      {     i = i+Turbo_shift;
            f = 0; g= 0; }
end.
```

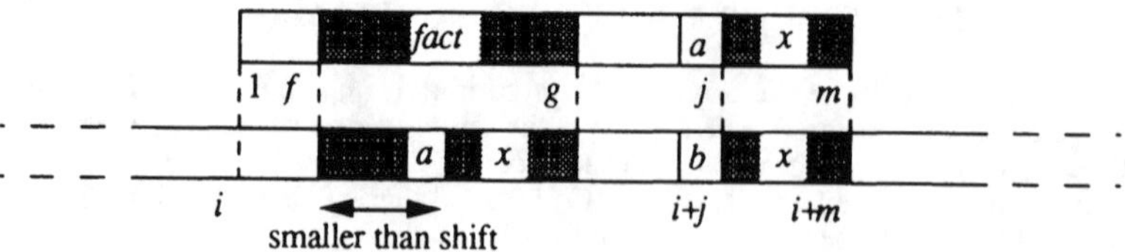

Figure 10. *Turbo_shift.*

Theorem 6.
The algorithm Turbo_BM makes at most 2.n comparisons.
Proof.
We decompose the searching phase into stages. Each stage is itself divided into the two operations: scan and shift. At stage k we call Suf_k the suffix of the pattern that matches the text and suf_k its length. It is preceded by a letter that does not match the aligned letter in the text. At the end of stage k, suffix Suf_k is stored to avoid scanning it again. At next stage, we say that there is a (true) jump if the letter of the text which precedes the occurrence of Suf_k is tested again. We also call $shift_k$ the length of the shift done at stage k. Let $cost_k$ be the number of comparisons done at this stage.

Consider three types of stages according to the natures of the scan and of the shift:
(i) stage followed by a stage with jump,
(ii) no type (i) stage with long shift.
(iii) no type (i) stage with short shift,
We say that a shift is short if $2.shift_k<(suf_k+1)$.

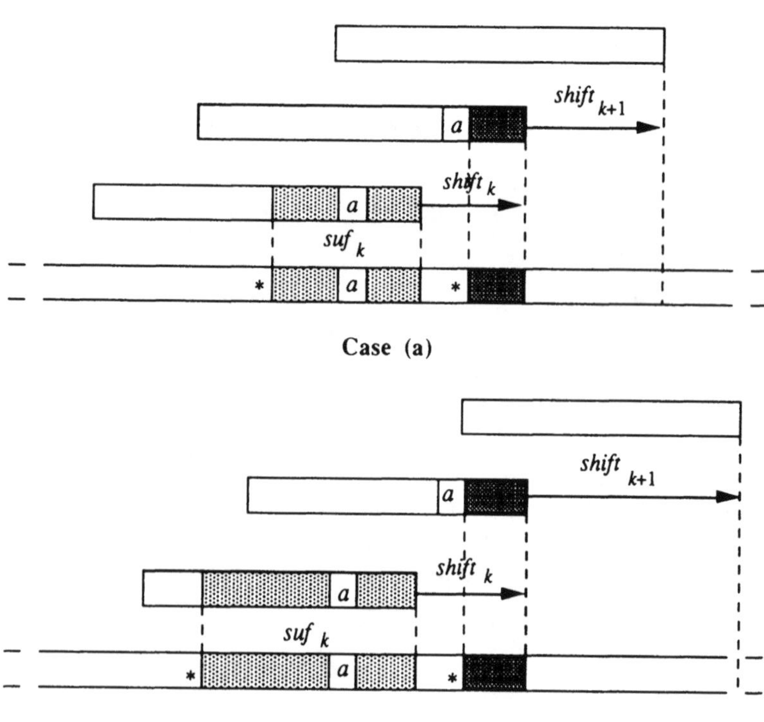

Figure 11. The costs of stages k and k+1 correspond to shadowed areas plus mismatches denoted by stars (together at most $suf_k+shift_k+1$ comparisons). If $shift_k$ is small then $shift_{k+1}$ is big enough (together with $shift_k$) to amortize the costs.

To prove the theorem it is enough to show that the sum of all costs is less than twice the sum of all shifts: $\Sigma cost_k \leq 2.\Sigma shift_k$.

We count separately the number of comparisons done during stages. We consider a stage k.

If stage k is of type (i) with no jump its cost is suf_k+1. The comparison with mismatch is one, obviously less than $2.shift_k$. The rest of the cost, suf_k, is accounted to the next stage. If stage k is of type (i) with a jump its cost is also suf_k+1 by the previous rule. Then the same applies.

If stage k is of type (ii), $cost_k = suf_k+1 \leq 2.shift_k$.

It remains to consider stages of type (iii). The situation is displayed in the Figure 11. This case is itself divided into two subcases. In case (a), $suf_k+shift_k \leq m$ and the contrary holds in case (b).

Case (a). $suf_k+shift_k \leq m$

Since stage $k+1$ makes no jump, a mismatch should occur inside the suffix of p of length $shift_k$. The *Turbo_shift* then implies $shift_{k+1} \geq suf_k - cost_{k+1}+1$.
Therefore, $cost_k+cost_{k+1} \leq suf_k+1+shift_k \leq 2.shift_k+shift_{k+1}+1 \leq 2.(shift_k+shift_{k+1})$.

Case (b). $suf_k+shift_k > m$

If no mismatch happens at stage $k+1$, $cost_k+cost_{k+1} \leq m+shift_k$. An occurrence of p is found in the text. Then we count $cost_k$ as $shift_k$ and $cost_{k+1}$ as m. In other words, $m-shift_k$ comparisons are reported from stage k to stage $k+1$.
If a mismatch happens at stage $k+1$, as in case (a), the *Turbo_shift* implies $shift_{k+1} \geq m-2.shift_k+1$. Then $cost_k+cost_{k+1} \leq m+cost_{k+1} \leq shift_{k+1}+2.shift_k-1+cost_{k+1}$.
If $cost_{k+1} \leq shift_{k+1}$ we get the result. The assumption holds in particular if stage $k+1$ is of type (i) because then its cost is 1.
It remains to look at the case where $shift_{k+1} < cost_{k+1}$ and stage $k+1$ is not of type (i). The *Turbo_shift* at stage $k+2$ implies that $shift_k+shift_{k+1}+shift_{k+2} > m$. On the other hand, $cost_k+cost_{k+1}+cost_{k+2} \leq m+shift_k+shift_{k+1} < 2.m$ (because $shift_k+shift_{k+1} < m$). This shows that $cost_k+cost_{k+1}+cost_{k+2} \leq 2.(shift_k+shift_{k+1}+shift_{k+2})$.
This ends case (b) and the whole proof. ♦

6. FINAL REMARKS

Remark 1 (on the algorithm Turbo_BM)

In the algorithm Turbo_BM we deal only with match shifts of the Boyer-Moore algorithm. If the alphabet is binary then occurrence shifts in Boyer-Moore are useless. Generally for small alphabets the occurrence heuristics have little effect. For bigger alphabets we can include in the algorithm the occurrence shifts in many possible ways. The simplest way to do it is to choose the match_shift if both match_shift and occ_shift are small (at most $suf_k/2$). In other words, the occurrence shift is considered only when it is reasonably big. Then the same analysis applies.

Another alternative is to consider all occurrence shifts, but this leads to an algorithm which uses only additional memory dynamically updated during the text-search phase. However, it seemingly requires additional linear preprocessing and a linear size table which after preprocessing will be read only.

Remark 2 (on the fast multi-pattern string-matching)

One can easily extend the RF algorithm to the multi-pattern string matching. Let $p_1, p_2, ..., p_k$ be a set of patterns and $FACT(p_1, p_2, ..., p_k)$ be the set of all factors of the pattern.

Algorithm multi-RF; /*reversed-factor string-matching for patterns $p_1, p_2, ..., p_k$ */
m:=length of the shortest pattern
i:=0; /* denote $t[i+j..i+m]$ by x, it is the lastly scanned part of the text */
while $i \leq n-m$ **do**
{ j:=m; **while** $j>1$ & $x \in FACT(p_1, p_2, ..., p_k)$ **do** j:=j-1;
$x = t[i+j..i+m]$ is the scanned part of the text;
if $x = p_r$ for some r in [1..k] **then** report match;
shift:= *multi-RF_shift*[x]; i:=i+*shift* ; }
end.

The definition of the table *RF_shift*[x] can be extended to many patterns in a natural way. When defining shift *RF_shift*[x] = m-|u|, we allow uv to be a prefix of one of $p_1, p_2, ..., p_k$, see Figure 3.

The algorithm multi-RF is also fast on average, however similarly as RF it takes quadratic time in pessimistic case. We are able to make an accelerated version of this algorithm similar to Turbo-RF'. The accelerated algorithm Turbo-multi-RF has $O(n \log(m))$ time complexity, or it can have $O(n)$ time complexity if we use a table of $O(m^2)$ size. This table does not need to be initialized and only a linear sized part of it is used.

REFERENCES

[Ah 90] A.V. AHO, Algorithms for finding patterns in strings, in: (J. VAN LEEUWEN, editor, *Handbook of Theoretical Computer Science*, vol A, *Algorithms and complexity*, Elsevier, Amsterdam, 1990) 255-300.

[Ap 85] A. APOSTOLICO, The myriad virtues of suffix trees, in: (A. APOSTOLICO, Z. GALIL, editors, *Combinatorial Algorithms on Words*, NATO Advanced Science Institutes, Series F, vol. 12, Springer-Verlag, Berlin, 1985) 85-96.

[AG 86] A. APOSTOLICO, R. GIANCARLO, The Boyer-Moore-Galil string searching strategies revisited, *SIAM J.Comput.* 15 (1986) 98-105.

[BR 91] R.A. BAEZA-YATES, M. RÉGNIER, Average running time of the Boyer-Moore-Horspool algorithm, *Theoret. Comput. Sci.* (1991) to appear.

[BBEHCS 85] A. BLUMER, J. BLUMER, A. EHRENFEUCHT, D. HAUSSLER, M.T. CHEN, J. SEIFERAS, The smallest automaton recognizing the subwords of a text, *Theoret. Comput. Sci.* 40 (1985) 31-55.

[BKR 91] L. BANACHOWSKI, A. KRECZMAR, W. RYTTER, *Analysis of algorithms and data structures*, Addison Wesley, 1991.

[BM 77] R.S. BOYER, J.S. MOORE, A fast string searching algorithm, *Comm. ACM* 20 (1977) 762-772.

[Co 90] R. COLE, Tight bounds on the complexity of the Boyer-Moore pattern matching algorithm, in: (*2nd annual ACM Symp. on Discrete Algorithms*, 1991) 224-233

[Cr 86] M. CROCHEMORE, Transducers and repetitions, *Theoret. Comput. Sci.* 45 (1986) 63-86.

[Ga 79] Z. GALIL, On improving the worst case running time of the Boyer-Moore string searching algorithm, *Comm. ACM* 22 (1979) 505-508.

[GO 80] L.J. GUIBAS, A.M. ODLYZKO, A new proof of the linearity of the Boyer-Moore string searching algorithm, *SIAM J.Comput.* 9 (1980) 672-682.

[KMP 77] D.E. KNUTH, J.H. MORRIS Jr, V.R. PRATT, Fast pattern matching in strings, *SIAM J.Comput.* 6 (1977) 323-350.

[Le 91] T. LECROQ, A variation on Boyer-Moore algorithm, *Theoret. Comput. Sci.* (1991) to appear.

[Ry 80] W. RYTTER, A correct preprocessing algorithm for Boyer-Moore string searching, *SIAM J.Comput.* 9 (1980) 509-512.

[Ya 79] A.C. YAO, The complexity of pattern matching for a random string, *SIAM J.Comput.* 8 (1979) 368-387.

SYSTEMS

The ANIGRAF System*

Michel Billaud
LaBRI† - Université Bordeaux I
351, Cours de la Libération
33405 Talence Cedex (France)

billaud@geocub.greco-prog.fr

Graph Rewriting (or Relabeling) Systems with Priorities (PGRS for short) have been introduced in [3] as a formalism for designing distributed algorithms on graphs. A PGRS is a set of rules telling how to change the labels attached to edges or vertices if some pattern is recognized. When two occurrences of rules overlap in a graph, a priority relation may solve the conflict by excluding one rule.

In [1, 5, 7] various algorithms are given under the form of PGRS (computing trees, election, finding an hamiltonian path, etc.) together with their formal proofs.

People design a PGRS with pencil and paper, and they use small examples in order to grasp more intuition on the problem. The ANIGRAF system was developped as a simple interactive tool to help them: running a collection of concrete examples sometimes reveals that some rules are never used, or that the PGRS doesn't work at all! By hand, the task quickly becomes tedious, as computations on relatively small graphs can involve huge numbers of steps.

Using ANIGRAF

The user has to provide a source file containing a set of rules, priority declarations, and some sample graphs. During the session, the screen of ANIGRAF looks like:

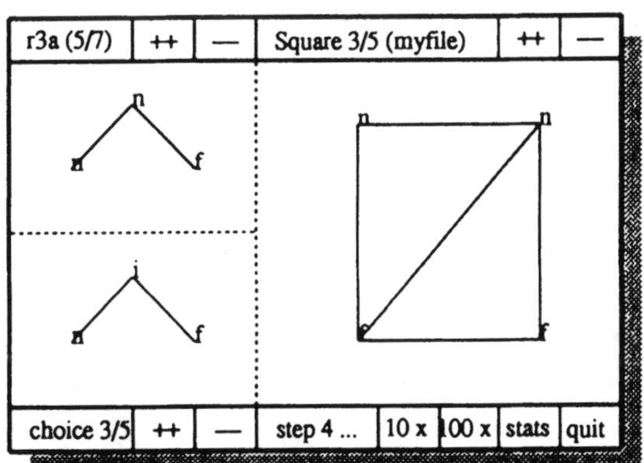

*This work has been partially supported by the GraGra project

†Laboratoire Bordelais de Recherche en Informatique, Unité Associée 1304 du C.N.R.S.

The upper menu bar shows the name of the rule "r3a" (5th of a set of 7) drawn on the left part of the screen, and the name of the current sample graph "Square" from the source file "myfile". Clicking [1] on `++` or `--` browse through rules and graphs.

The lower bar indicates that there are 5 possible rewritings on the graph, the third being proposed as a candidate for the 4th step. Buttons `10x` and `100x` start random sequences. `Stats` displays how many times each rule was used, and `quit` works as expected.

Implementation

ANIGRAF is written in Turbo-Pascal 4.0 under MS-DOS. It runs on IBM-PC/XT compatible computers with very limited harware requirements: 512 KB of memory, one floppy disk, any reasonable graphics card (Hercules, CGA, EGA, ATT ...) supported by TP 4.0, and a Microsoft-compatible mouse.

A previous version for X-Windows was written in C on a SUN 3/80 (with 8 MB) by Jean Luc Lafaye as a student's project [4].

References

[1] M. Billaud, *Some Backtracking Graph Algorithms expressed by Graph Rewriting Systems with Priorities,* Report LABRI 8989.

[2] M. Billaud, *Un Interpréteur pour les Systèmes de Réécriture de Graphes avec Priorités,* Report LABRI 9040 (in french).

[3] M. Billaud, P. Lafon, Y. Métivier, E. Sopena, *Graph Rewriting Systems with Priorities, in Graph-theoretic Concepts in Computer Science,* 15th Workshop on Graphs'89, LNCS 411, pp. 94-106.

[4] J.L. Lafaye, *Mémoire pour l'Obtention du D.U.T.,* Département Informatique de l'IUT 'A', Université Bordeaux I (1990).

[5] I. Litovsky, Y. Métivier, *Graph Rewriting Systems as a Tool to design and to prove Graph and Network Algorithms,* Report LABRI 9070.

[6] I. Litovsky, Y. Métivier, *Computing Trees with Graph Rewriting Systems with Priorities,* Report LABRI 9085.

[7] I. Litovsky, Y. Métivier, Computing with Graph Rewriting Systems with Priorities, *Fourth International Workshop on Graph Grammars and Their Applications to Computer Science, Bremen (1990).* Also Report LABRI 9087.

[1]The left mouse button starts the action, the right button gives context dependent help.

A programming language for symbolic computation of regular languages, automata and semigroups

J.M. CHAMPARNAUD

I.B.P., L.I.T.P., Université Paris 7, Tour 55-56, 2 Place Jussieu, 75251 Paris Cedex 05, France
e-mail : jmc@litp.ibp.fr

We describe a programming language designed for symbolic computation of regular languages, extended regular expressions, finite automata, and finite semigroups. An interpreter has been written in C and implemented on the VAX/780 of the L.I.T.P. running under UNIX. It has been constructed with the help of lexical analyser generator Lex and LALR parser generator Yacc. It makes use of the functions defined in the libraries of the AUTOMATE package [1].

1 Functionalities

- **Syntax.** We have chosen a C-like syntax.
- **Data types.** The basic types are the following : *char*, *integer*, *float*, *automaton*[1]and *monoid*. Derived types may be constructed from these fundamental types, with the help of the *array* and *structure* type constructors. The type-checking is rather strong.
- **Expressions.** Integral and floating arithmetics are implemented in a classical way. Automaton type arithmetic provides the following operations : union, concatenation product, Kleene star, positive closure, power, intersection, set difference, left quotient, right quotient, and shuffle product. Relational expressions make it possible to compare two expressions so long as they have the same type (for instance two integers or two automata).
- **Control-flow constructions.** The most common control-flow statements are provided : statement grouping, selection (`if else`), iteration (`while, for`), exit from a function (`return`).
- **Functions.** It is possible to define recursive functions which return a basic type. Call-by-value and call-by-address are implemented. As in C, functions are external (nested function definitions are not allowed). Unlike C, parameters number and types are checked.
- **Predefined functions.** The functions of the libraries on automaton and monoid types defined inside AUTOMATE package may be called. These libraries contain in particular the functions which allow to make an automaton be trim, deterministic, or minimal, and to compute the transition monoid of a deterministic automaton.
- **Programmation environment.** Programmation environment is made up of commands which, in particular, allow to list defined variables and their values, to free a variable, to save an automaton in a file, or to load a function from a file.

[1]The type *automaton* is rather a type "regular language representation", since access to the elements of an automaton (alphabet and sets of states, initial states, final states and edges) is not implemented.

2 An example

The following session has been worked out on L.I.T.P. VAX/780. The interpreter is runned by typing the command automate. We first define a function power. This function has two arguments : an automaton a and an integer n, and it returns the nth power of the automaton a, computed by logarithmic method. This function is then called to compute the minimal automaton of the expression "a"^10. Of course this result can be computed directly.

```
% automate
**********************   automate  (version 91)  *****************************
> aut power(aut a; int n;)              /* a, n : parametres                 */
aut b, c;                               /* b, c : locales a la fonction      */
{
        if (n == 0) return UN;          /* UN : automate predefini           */
        if (a == VIDE || a == UN)       /* VIDE : automate predefini         */
             return a;
        if (n == 2) return a a;         /* a.a ou a a : concatenation        */
        if (n % 2 == 0)
             c = UN;
        else
             c = a;
        b = a;
        n = n / 2;
        while (n > 0) {
                b = b b;
                if (n % 2 == 1) c = c b;
                n = n / 2;
        }
        return c;                       /* power retourne un automate        */
}
> aut v;
> v = power("a", 10);                   /* appel par valeur                  */
> v;                                    /* affichage de l'automate v         */

There is 1 final state : 11

    1  2  3  4  5  6  7  8  9 10 11
 a  9 11 10  3  4  5  6  7  8  2  0
>
```

Reference

[1] Champarnaud J.M. and G. Hansel, AUTOMATE, a computing package for automata and finite semigroups, *J. Symbolic Computation*, **12**, 1991, 197-220.

μSPEED: a System for the Specification and Verification of Microprocessors

H. COLLAVIZZA

Université de Provence, 3 Place Victor Hugo, 13331 Marseille France
e-mail: BCL@FRCCUP51.BITNET

μSPPED is a system for the specification of micro-processor behaviour at the assembly language level and micro-sequence level. Starting from a user-friendly interface, the description is built according to a functional model inspired from the denotational semantics of programming languages. The verification consists in the correspodence between the functions obtained at the two levels.

Framework

With the increasing complexity of VLSI circuits it becomes necessary to provide CAD tools in order to ensure a correct circuit design. Among the several verification methods such as simulation and synthesis, we prone the use of formal verification methods. These methods consist in verifying the correctness of the circuit behaviour for all inputs and state variables, using formal model and formal proof techniques (see [6] for a complete survey of formal hardware verification).

To our knowledge, specifying and verifying the abstract levels of processors, has been done only using specific formalisms, related to particular theorem provers [3], [4], [7]. On the opposite, we believe that formal verification can be handled with reasonable effort only if it is performed step by step, and integrated in the design process. Therefore, we intend to define a complete verification system that would get the descriptions in a user-friendly way, and perform the proofs rather automatically, without any knowledge of the theorem prover used.

Objectives

To be fully automatic, the verification must be done as far as possible using simple proof techniques such as rewriting techniques. This can be achieved by decomposing the proof process into successive steps of description and verification between adjacent levels. In our initial study of processor verification [1], the specification levels have been identified as: 1-Assembly language level, 2-Micro-sequence level (takes into account the memory and peripheral exchanges), 3-Micro-instruction level, 4-Micro-operation level and 5-Logic level. A functional model of description has been defined for all specification levels; for abstract levels (1 and 2) it is inspired from the denotational semantics of programming languages. The proof consists in the correspondence between state transition functions describing the processor behaviour at two adjacent levels (see [5] and [8]).

Another objective is that the specification and the proof process would be integrated in a CAD tool. The functional specification of a processor (state definition, addressing mechanisms, and instruction execution results), amounts to a rather large and complex set of lambda expressions. If well adapted to symbolic computations and proofs, such a specification is hard to write for the electronic engineer. Our approach consists in automatically producing the functional specification from a user-friendly input syntax. At levels 3 to 5, a functional model can be automatically obtained from the compilation of a description in an appropriate HDL [2]. At levels 1 and 2, we think that classical HDL are not well adapted to this type of description task. So it was necessary to construct a semantic-directed interactive tool called "μSPEED" (for μprocessor SPEcification EDitor), that would obtain the necessary informations from the designer, and automatically construct the functional expressions.

Global structure of μSPEED

The specification editor is divided into three main parts:

The **user interface** is a multi-window hierarchy of environments; each environment will be seen

---

... CNET #883 D09 10 790 0245 CNS and "CHARME" BRA #3216

by the user as a menu allowing a particular characteristic or element of a processor to be defined. The user dialogue is based on mouse-driven selection of actions displayed as buttons and scrolling windows, together with specialized editors for the interpretation of textual information. The specification is segmented into two parts: the state definition and the instruction definition; in turn, the instruction definition is decomposed into three sub-parts which define: the instruction set (to name and class the instructions and define their syntax), the operand computation functions and the instruction functions. This interface maintains an intermediate data structure, in one to one correspondence with the informations given by the user, under the form of "property lists" (P-lists in Lisp jargon). The user interface is written in AIDA, a tool box based on Le-Lisp which provides a set of primitives to build editing windows, scroll bars, buttons...

The **application interface** takes as input the set of P-lists for a particular processor specification, and builds the functional model of the processor. It automatically introduces the main concepts of our formalism, such as the state and instruction attribute selection functions, environment function (definition of operands) and state updating functions [5], [8]. In particular, the instructions are specified as operand assignments in the user interface while they are translated as state transition functions expressed in terms of updating functions applied to operand functions.
The application interface is written in Le-Lisp.

The **proof supervisor** allows to select a particular instruction to be verified and to enable the proof process; it calls on a specific rewriting tool. The proof is performed by showing the equivalence between expressions obtained at two adjacent levels, after evaluation of the specification functions built by the application interface.

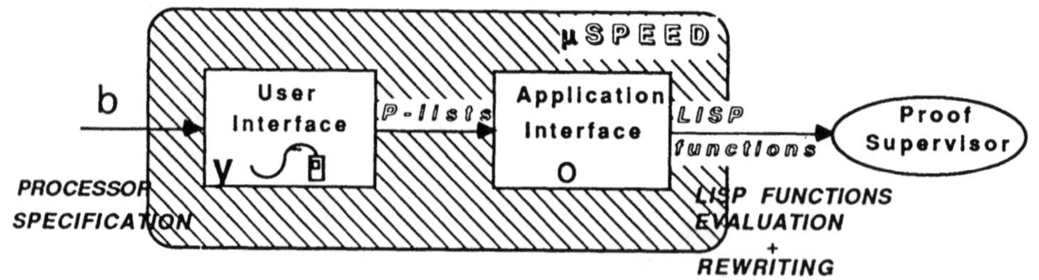

Conclusion

A menu-driven environment is suitable for an efficient implementation of prototypes, as well as it is easy and pleasant to use. Since the informations are well organized, and since we keep the functions used at the two levels in an abstract form, the correspondence proof is quite easy. The description of the MTI processor [1] at level 1 taked less than one hour and a portion of the verification was done by rewriting only (using rules based on bit vector operations).

References

[1] Borrione, Camurati, Paillet, Prinetto: "*A functional approach to formal hardware verification: The MTI experience*", Proc. IEEE ICCD'88, Port Chester, New York. October 1988.

[2] Borrione, Pierre, Salem: "*Formal verification of VHDL descriptions in Boyer-Moore: first results*", 1st European Conference on VHDL Methods, Marseille, France, 4-7 Sept. 1990.

[3] Bickford, Srivas: "*Verification of a Pipelined Microprocessor Using Clio*", Proc. Work. on "Hardware Specification, Verification and Synthesis: Mathematical Aspects", Cornell University, Ithaca, USA, July 1989

[4] Cohn: "*A Proof of Correctness of the Viper Microprocessor: The First Level* ", in Proc. "VLSI Specification, Verification and Synthesis", Calgary, Canada, Jan. 1987.

[5] Collavizza: "*Functional Semantics of Microprocessors at the Micro-program level and Correspondence with the Machine Instruction level*", Proc IEEE EDAC Glasgow, Scotland, 12-15 March, 1990.

[6] Camurati, Prinetto: "*Formal verification of hardware correctness: an introduction* ", Proc. IFIP 8th Int. Conf. CHDL, Amsterdam, April 1987 (North Holland)

[7] Hunt: "*FM8501 : A Verified Microprocessor* ", Technical Report 47, Institute for Computing Science. University of Texas at Austin, Feb. 1986.

[8] Paillet: "*Functional Semantics of Microprocessors at the Machine Instruction Level*", Proc. 9th IFIP Int. Conf. CHDL, Washington D.C., USA, June 89 (North Holland)

A DISCRETE EVENT SIMULATOR OF COMMUNICATION ALGORITHMS IN INTERCONNECTION NETWORKS

Miltos D. Grammatikakis
Laboratoire de l' Informatique du Parallelisme
Ecole Normale Superieure de Lyon
Lyon 69364, France

Jung-Sing Jwo
Department of Information Science
Providence University
Shalu 43309, Taiwan R.O.C.

EXTENDED ABSTRACT:

We demonstrate a discrete-event simulator for evaluating the performance of various routing algorithms (probabilistic and deterministic) in both multicomputers, and multistage parallel interconnection networks. This simulator can route packets in a mesh [1], a generalized hypercube [2], or a star graph [3]. It can also realize connection requirements for a given permutation, on an omega network [4], an augmented data manipulator [5], or a multistage circuit-switched hypercube [6].

Current multicomputer, and multiprocessor designs tend to become increasingly complex. Therefore, it is very difficult to analytically evaluate their performance, and simulator tools become very useful [7]. In this study, we are motivated to design a flexible and efficient simulator of various routing schemes, on both multicomputer, and multiprocessor architectures.

Such parallel systems usually incorporate a highly symmetric interconnection network. By exploiting this symmetry, we avoid an explicit representation (adjacency list, or matrix) of the underlying network. Instead, the communication algorithm is "smart" to route messages along adjacent nodes. The implementation is also event-driven, which is faster and easier to parallelize, as reported in previous studies [8]. Next, we will briefly describe the available options, and provide a glimpse at current results and future extensions of this tool.

• Simulation of Multicomputers

In simulating multicomputer architectures we assume full duplex communication. Several architecturally, and algorithmically induced options are provided.

(1) The communication mode may be chosen, as *1-port*, or *d-port*. This option affects the probability of congestion: a) In *1-port* communication a conflict arises when some (≥2) packets attempt to move simultaneously to the same node. b) In *d-port* communication a conflict occurs when some (≥2) packets attempt to follow the same link at the same time..

(2) Various *queuing disciplines*, such as random, or furthest-first have being implemented. Note, that options (1), and (2) are transparent to architectural and algorithmic changes.

(3) *Randomized protocols* (of various types), as well as *deterministic* (greedy, minimal path routing). The implementation of these protocols can be performed either in synchronous, or asynchronous manner.

• Simulation of Multiprocessors

In multiprocessors, we study the characteristics of permutations realizable by various blocking multistage interconnection networks. The usually studied permutations include the commonly encountered classes *BPC* (Bit, Permute, and Complement) [10], Omega [4], and inverse omega permutations [9]. A backtracking algorithm for routing permutations, based on Waksman's algorithm (for the Benes network) [11], is available. Furthermore, by simulating a large number of random permutations, we can approximate the network's blocking probability, and thus we can empirically evaluate its permutation capability [6].

• A Sample of Simulation Results

In Table 1, we show a simulation-based comparison of greedy routing on the star graph, and the generalized hypercube architecture. For both graphs we assume, full duplex d-port communication, and a furthest-first queuing discipline. From this table, we can see that the star graph architecture, is better than the generalized hypercube; for approximately the same number of nodes, as the star offers shorter delay times, and also smaller expected queue lengths, and number of links— which implies a smaller design cost.

Table 1. Comparison of Star and Generalized Hypercube Architecture (based on d-port duplex communication)

Architecture	Nodes(N)	Edges	Diameter	Avg. Distance	Degree	Time Delay	Queue Length
Star Network	5040	30240	9	5.88	6	9.04	1.93
Binary Hypercube	4096	49152	12	6	12	11.94	3.02
Generalized Hypercube ($b = 4$)	4096	73728	6	4.5	18	9.32	3.08
Generalized Hypercube ($b = 8$)	4096	114688	4	3.5	28	7.55	3.03

• Further Refinements and Extensions

An extension to general h-relations, or partial permutations [12], as well as to adaptive (heuristic) algorithms, is straightforward. Currently, simulations of promising multicomputer architectures based on Cayley graphs, such as the alternating group sequence graph [13], as well as parallel implementations using C, and MPL on a MasPar MP I architecture, are underway.

Another interesting extension is in simulating local and large area networks. In such systems (e.g. ARPANET [14]), packets are created dynamically, and routed to their final destinations. Performance studies, under various conditions of fault, are challenging.

Finally, this tool, in an appropriate refined form, could be used in teaching courses related to parallel communication algorithms, interconnection networks, queuing theory and simulation. The design and validation techniques, as well as more simulation data, are described in [15].

REFERENCES

[1] F. T. Leighton, "Average Case of Greedy routing Algorithms", 1990, preprint.

[2] L. N. Bhuyan and D. P. Agrawal, "Generalized Hypercube and Hyperbus Structures for a Computer Network", *IEEE Transactions on Computers,* C-33 (4), April 1984, pp. 323-334.

[3] S. B. Akers, D. Harel, and B. Krishnamurthy, "The Star Graph, an attractive alternative to the n-cube", *1987 International Conference on Parallel Processing,* pp. 393-400.

[4] Duncan H. Lawrie, "Access and Alignment of Data in an Array Processor", *IEEE Transactions on Computers,* C-24 (12), December 1975, pp. 1145-1155.

[5] M-D. P. Leland, "On the Power of the Augmented Data Manipulator Network", *1985 International Conference on Parallel Processing,* pp. 74-78.

[6] Ted Szymanski, "On the permutation capability of a circuit-switched hypercube", *1989 International Conference on Parallel Processing,* Volume I, pp. 103-110.

[7] M. T. Raghunath, A. Ranade, "A Simulation-Based comparison of Interconnection Networks", *Annual IEEE Conference on Parallel and Distributed Processing,* 1990, pp. 98-103.

[8] F. Wieland, D. Jefferson, "Case Studies in Serial and Parallel Simulation", *1989 International Conference on Parallel Processing,* Vol. III, pp. 255-258.

[9] J. Lenfant, "Permutations of Data: A Benes Network Control Algorithm for frequently used permutations", *IEEE Transactions on Computers,* Vol. C-27 (7), July 1978, pp. 637-647.

[10] D. Steinberg, "Notes Invariant Properties of the Shuffle/Exchange and a Simplified Cost-Effective Version of the Omega Network, *IEEE Transactions on Computers,* C-32 (5), March 1983, pp. 444-450.

[11] A. Waksman. "A Permutation Network", *Journal of the ACM,* 15 (1), January 1968, pp. 159-163.

[12] L. G. Valiant, "A Bridging Model for Parallel Computation", *Communication of the ACM,* August 1990, pp. 103-11.

[13] J. Jwo, S. Lakshmivarahan, S. K. Dhall, " A New Class of Interconnection Networks Based on The Alternating Group", Networks 1991 (submitted).

[14] L. Kleinrock, *Communication Networks,* MC Graw Hill 1964.

[15] M. Grammatikakis, J. Jwo, "A Simulation-Based Analysis of Communication in Multicomputers " , International Parallel Processing Symposium 1992 (submitted).

ALPHA DU CENTAUR: An environment for the design of systolic arrays*

Hervé LE VERGE
IRISA, Campus de Beaulieu
35042 Rennes Cedex
France

November 22, 1991

Abstract

ALPHA DU CENTAUR is an interactive environment for the design of regular VLSI parallel integrated circuits. It is based on ALPHA, a restricted functional language whose definition is well-suited to express the specification of an algorithm and also its execution on a synchronous parallel architecture. Programs, written in ALPHA, are rewritten using formal transformations (space-time reindexing, pipelining, etc.), and finally translated into a form suited to conventional VLSI design tools. Our demonstration will show the current state of development of ALPHA DU CENTAUR, and will illustrate its use on classical systolic algorithms.

ALPHA DU CENTAUR is a prototype environment for the interactive design of systolic algorithms, built on top of CENTAUR. It results from research on formalism for synthesizing regular parallel problems, such as systolic algorithms. ALPHA DU CENTAUR is based on the ALPHA language [LMQ91], a restricted functional language for modeling systems of linear recurrence equation [QD89] which allows one to derive synchronous hardware, starting with a high-level description of a problem. The synthesis process consists in a sequence of rewriting, mostly implemented using inference rules. Another attempt in the same direction is the language Crystal[Che86] designed at Yale University.

Figure 1 shows the ALPHA specification of the unidimensional convolution $y_i = \sum_{k=1}^{4} w_k \times x_{i-k+1}$, as seen in a CENTAURwindow. An ALPHA program is a function from a set of input variables to a set of output variables defined by means of recurrence equations. ALPHA variables and expressions are functions from an index set (a subset of Z^n included in a finite union of disjoint convex polyedra) to a set of values. In addition to point-wise arithmetical and boolean

*This work was partially funded by the French Coordinated Research Program C^3 and by the Esprit BRA project NANA.

operators (+, −, …, or, not, …), ALPHAexpressions are built using *restriction*, *case* operators and *dependence* functions (affine integral functions). The transformations are based on the use of substitution together with axiomatic rules which make it possible to simplify expressions. Any ALPHAexpression can be reduced to a so called *case-restriction-dependence* normal form. Using substitution and normalization, one can get the expression of output variables only in term of input variables by removing recursively all occurrences of local variables. This is very helpful to compare two descriptions of the same algorithm.

```
File Display Edit Selections ALPHA GENERALISATION
system convolution (w : {i|4>=i;i>=1} of integer;
                    x : {i|i>=1} of integer)
       returns (y : {i|i>=4} of integer);
  var
    Y : {i,k|i>=1;4>=k;k>=0} of integer;
  let
    Y = case
        {i,k|k=0} : 0.(i,k->);
        {i,k|4>=k;k>=1} : Y.(i,k->i,k-1) + w.(i,k->k) * x.(i,k->i-k+1);
        esac;
    y = Y.(i->i,4);
  tel;
```

Figure 1: An ALPHAprogram (Convolution)

An unimodular *change of basis* can be applied to any local variable, in order to place all computations in a time/processor space. This way, some indexes can be interpreted in term of time and others in term of processor number. Moreover, timing (respectively processors) dependencies represents delays (respectively interconnection paths). ALPHA DU CENTAURalso provides useful transformations, for removing data broadcasting, handling special kinds of initializations, etc. The goal of all these transformations is to derive a program in which all equations can be interpreted as synchronous hardware. This can then easely be translated in VLSI, or used a as starting point for code generation.

References

[Che86] M.C. Chen. Transformations of parallel programs in crystal. *Information Processing*, 455–462, 1986.

[LMQ91] H. Le Verge, C. Mauras, and P. Quinton. The ALPHA language and its use for the design of systolic arrays. *Journal of VLSI Signal Processing*, 3:173–182, 1991.

[QD89] P. Quinton and V. Van Dongen. The mapping of linear recurrence equations on regular arrays. *The Journal of VLSI Signal Processing*, 1:95–113, 1989.

Verification of Communicating Processes by means of Automata Reduction and Abstraction

Eric Madelaine[1] and Didier Vergamini[2]

The semantics of communicating processes can be expressed as labelled transition systems (cf [9] for instance). Thus, verification on a communicating system may consist in verification on a corresponding automaton, derived from its algebraic specification. Nowadays, two main complementary verification techniques are used:

- model checking: the user verifies if the obtained automaton verify a set of properties given as modal logic formulas [3,1],
- equivalence/preorder proofs: the user verifies if the obtained automaton is in relation (in bisimulation for instance) with another one, obtained by derivation of another specification [4,7,6].

AUTO accepts as inputs terms of a process algebra called MEIJE[2], and since [8], of any process algebra that can be described in terms of structural operational semantics rules. The finiteness property is imposed by restriction of the syntax of terms. The main goal of this system is the computation of automata as reductions of abstractions of terms. Abstractions are parameterised by user-defined abstract actions, while reduction is defined using the classical notion of bisimulation. AUTO uses weak and strong bisimulation congruence properties to reduce structurally its components before combining them rather than building first a huge global automaton and reducing it afterwards. This allows to deal with potentially enormous systems. AUTO now uses improved algorithms to compute strong, weak and branching bisimulation grace to a new tool called FCTOOL: the interface between the two systems is a so-called *common format* for transition systems. This format is used by several other verification systems such as the CWB [4], and ALDEBARAN [6]; each of these tools has its own particularities, and through the common format interfaces, the user can use all of them in conjunction. AUTO also provides *debugging* functionalities: a sequence of actions computed on an abstract automaton can be projected on the initial term, thus the user can see which concrete sequence of actions and synchronisations leads to a given abstract behaviour.

AUTO is interfaced with a graphical editor named AUTOGRAPH for the display of the produced automata, and for edition of networks of communicating automata. The user's manuals of AUTO and AUTOGRAPH are available in [5,10].

References

[1] A. Arnold. Construction et analyse des systèmes de transitions: le système MEC. In *Actes du Premier Colloque* C^3, 1985.

[2] G. Boudol. Notes on algebraic calculi of processes. In K. Apt, editor, *Logics and Models for Concurrent Systems*. Springer-Verlag, 1985.

[1] INRIA, Route des Lucioles, Sophia Antipolis, 06565 Valbonne Cedex (France), email: madelain@sophia.inria.fr

[2] CERICS, Avenue Albert Einstein, Sophia Antipolis, 06565 Valbonne Cedex (France), email: dvergami@sophia.inria.fr

[3] M. Browne, E. Clarke, et al. Automatic verification of sequential circuits using temporal logic. Technical Report CMU-CS-100, Carnegie Mellon University, 1984.

[4] R. Cleaveland, J. Parrow, and B. Steffen. The concurrency workbench: A semantics-based tool for the verification of finite-state systems. *ACM Transactions on Programming Languages and Systems*, To appear.

[5] R. de Simone and D. Vergamini. Aboard AUTO. Technical Report RT111, INRIA, October 1989.

[6] Jean-Claude Fernandez. Aldebaran: A tool for verification of communicating processes. Rapport technique SPECTRE C14, Laboratoire de Génie Informatique — Institut IMAG, Grenoble, September 1989.

[7] H. Korver. The current state of bisimulation tools. Technical Report P9101, CWI, January 1991.

[8] E. Madelaine and D. Vergamini. Finiteness conditions and structural construction of automata for all process algebras. In R. Kurshan, editor, *Proceedings of Workshop on Computer Aided Verification*, New-Brunswick, June 1990.

[9] R. Milner. *A Calculus of Communicating Systems*, volume 92 of *Lectures Notes in Computer Science*. Springer-Verlag, 1980.

[10] V. Roy and R. de Simone. An AUTOGRAPH primer. Technical Report RT112, INRIA, October 1989.

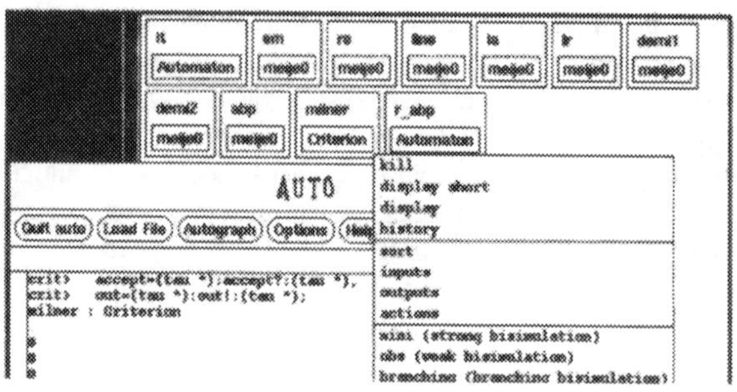

DISTRIBUTED SYSTEM SIMULATOR (DSS)

P. Spirakis B. Tampakas M. Papatriantafillou K. Konstantoulis
K. Vlaxodimitropoulos V. Antonopoulos P. Kazazis T. Metallidou
D. Spartiotis

Computer Technology Institute
and
Computer Science and Engineering Department
Patras University-Greece

November 1991

Abstract

DSS ([1],[2],[3]) is a powerful simulation tool (event driven simulator+simulation environment) that provides efficient simulation of distributed algorithms by modelling the specs of a distributed algorithm as a set of communicating automata with countable number of states. **DSS** has been implemented under the BRA ALCOM project. The user has to provide the specs of the distributed algorithms and the topology of the interconnections in the distributed system. The degree of asynchrony (process rates, message delays) can also be specified in a flexible way.
keywords: Simulation, Distributed Algorithms, Communicating Automata, Simulation Environment.

GENERAL DESCRIPTION

Each processor is modeled as an automaton that behaves as follows. At each time when an event occurs the corresponding processor examines its state and decides whether to send a message to each of its neighbours, and what message to send. These messages will be received after some time corresponding to the communication delay of the transmittion line. Each processor uses its current state and the above messages to update its state. The ordering of events of the simulated system is probabilistic. For every node and for every link of the distributed system the simulator needs some parameters about the step and communication delays such as upper and lower bounds, mean values etc. depending on the desired time distribution.

The simulation is event driven, i.e. the simulator schedules the events of the future in time ordering and when the next event causes a simulator call to the protocol, the time of the simulation model advances by a variable amount. To implement this concept, the simulator actually proceeds by keeping track of an event list. This list has all the kinds of events, and has the structure of a heap which means that every new event that is created during the execution of an action of the protocol is inserted in this list at the proper place to keep the time ordering of events.

In order to make a more realistic handling of the time that it takes each node to execute an action of the protocol, the notion of **idle time** is introduced. This means that after the execution of an action the corresponding node will remain idle, i.e. unable to execute a next event, for an amount of time equal to the currently chosen step , even though such an event could be scheduled for an earlier time.

DSS is able to simulate completely **asynchronous**, **synchronous**, **archimedean** and **ABD** networks. All the information concerning the timing parameters of a specific simulation run can be found in a corresponding *input file.* This input file is created and modified by the user.

As an example we give a short description of the implementation of the asynchronous systems. Archimedean and synchronous systems are implemented in a similar way. In the asynchronous timing approach, the steps and the communication delays are randomly chosen from a time distribution. More specificaly, every time a node executes an action of the protocol, selects a new step (and remains idle for this amount of time) depending on time distribution selected and the time distribution parameters for its step. Also, every time a message is sent, a delay is chosen depending on the time distribution selected and on the delay parameters for the communication line used. A constraint in this concept is that, on purpose to keep the simulation steady, the step of a node can be changed only when no timer is running. This constraint seems hard but in all the simulation experiments with protocols with many timers the steps changed lots of times and no problem arised.

DSS is also able to simulate networks with **faulty links**. Real systems always have a possibility of suffering from link failures, so a realistic simulation must allow for links to fail. Each time and before a specific protocol execution, one of the above types of failures can be selected and an upper bound of the number of links to fail can be determined. The number of links to fail is a fundamental parameter for all the available solutions. A **synchronizer** is also offered to the DSS user, allowing the execution of synchronous algorithms on ABD networks. The user is also able to simulate networks with clock drifts.

Measurments are obtained through either tracing files or a flexible user-friedly visual interface with on-line debug facilities. Simulator supports user by a flexible Visual Interface that provides the ability of tracing the flow of the protocol execution. This interface prints on the display, of terminal VT220 or of a SUN workstation terminal, windows that contain information about the current protocol execution and also provides to the user the flexibility of interactive changing several parameters. For each process there is such a window that contains information concerning the specific process, like, the process id, the old and the new state, the more recent scheduled event, the receiving and sending messages and other user-defined information. Besides the main Visual Interface there is another option for debugging by means of a debug file that is created for every execution of simulator. In this file some information about the run is print and is available for off-line checking.

A version of DSS has alredy been used in the industrial application area (ESPRIT 2080) for the simulation of a telecommunication switch whose specifications are written in SDL specification language. Under this goal, DICOMP tool ([4]) has been created, accepting files containing the SDL specifications and producing files containing data and C code. These files can be used as input for the above version of DSS tool. DICOMP may be viewed as a compiler for SDL. As any other compiler, it does error detection and produces some sort of output which is similar with the files DSS accepts for the simulation of a distributed protocol. It is worth noting that DICOMP, in compination with DSS, is a general tool that can be used for the simulation of the function of any telecommunications switch.

REFERENCES

[1] *"The DSS tool"*, K. Konstantoulis, P. Spirakis, B. Tampakas and K. Vlahodimitropoulos, T. R. 90.10.24, Computer Technology Institute, Patras 1990, Project ALCOM.

[2] *"The DSS tool(version 2)"*, V. Antonopoulos, P. Kazazis, T. Metallidou, P. Spirakis and B. Tampakas, T. R. 90.12.35, Computer Technology Institute, Patras 1990, Project ALCOM.

[3] *"Distributed System Simulator (DSS):An Overview"*, P. Spirakis and B. Tampakas, T. R. 91.06.20, Computer Technology Institute, Patras 1991, Project ALCOM.

[4] *"DICOMP, a Compiler for SDL into Distributed System Simulator Specs"*, D. Spartiotis, P. Spirakis and B. Tampakas, April 1991, Internal Report (ESPRIT 2080) REX-WP4.

An Interactive Proof Tool for Process Algebras *

Huimin Lin
School of Cognitive and Computing Sciences
University of Sussex
Brighton BN1 9QH
England

Abstract

A proof assistant for process algebras is described. The main novelty of this proof system is that it is parameterised by process calculi: the users can define their own calculi by providing suitable signatures and axioms. Proofs are constructed by directly manipulating process terms, and some forms of induction has been built-in to handle recursion.

PAM (for *Process Algebra Manipulator*) is a general proof tool for process algebras. Unlike operational semantics based tools such as the Concurrency Workbench [CPS 89], TAV [GLZ 89] and Auto [BRSV 89], PAM is an interactive proof assistant. It makes no attempt at fully automatic verification – it is the users who conduct the proofs; But it does provide the users with great amount of help – the most tiresome parts of algebraic reasoning, e.g. applying axioms (including expansion laws) to transform a term from one form into another and folding/unfolding definitions, can be done automatically by PAM. As proofs are constructed by directly manipulating process terms without resort to any internal representation, PAM does not suffer from "state explosion" problem – in fact even infinite state problems can be dealt with. Open term reasoning is also easy.

PAM differs from other algebraic tools such as [NIN 89] (which is only for CCS) in that it does not have any calculi built-in. Instead it allows the users to define their own favourite process algebras and then perform proofs in them.

The logic which PAM implements is essentially equational logic plus recursion, with some features tailored to the particular requirements of process algebras. At the core of PAM is a rewrite machine which is capable of handling associative and commutative operators. Infinite processes can be defined by mutual recursion, and some forms of induction (notable the unique fixpoint induction) have been built into the system to cope with such recursively defined processes.

Calculi are defined in a format similar to what people write on paper. A calculus definition basically consists of a signature and a set of axioms which are equations (possibly with side conditions). A scheme is provided for defining interleaving (or expansion) laws in various calculi, and applications of these laws are treated as special forms of rewriting.

*This work has been supported by the Science and Engineering Research Council of UK and the ESPRIT II BRA project CONCUR.

To modify a calculus, simply change the definition file and recompile it which takes only a few minutes. This feature makes PAM a handy tool for process algebra designers as a test-bed for their languages.

Problems to prove are described in separated files. For each problem a proof window is created. So at the same time one can work on several problems in one calculus, or try out proofs of the same problem described in different calculi. With the help of the window interface, proofs are driven by mouse-buttons – one selects a process term to work with by clicking the mouse-button at it, then invoke an appropriate proof step by clicking on the suitable command button.

It is possible to prove a conjecture, enter it as a named theorem, and use it in other proofs. In fact theorems can be generated "on the fly": when a theorem is needed during a proof, one can create a separate proof window for the theorem, prove it, and then go back to the original proof and use it.

Presented below is a definition of counter problem (in CCS) and a proof window for it with a completed proof. Although this is a infinite state problem, the proof consists of only a few steps.

```
conjecture

 C = P

where

 C = up.(down.NIL | C)
 P = up.(~s.P | B)\{s}
 B = s.down.NIL

need sort computation

end
```

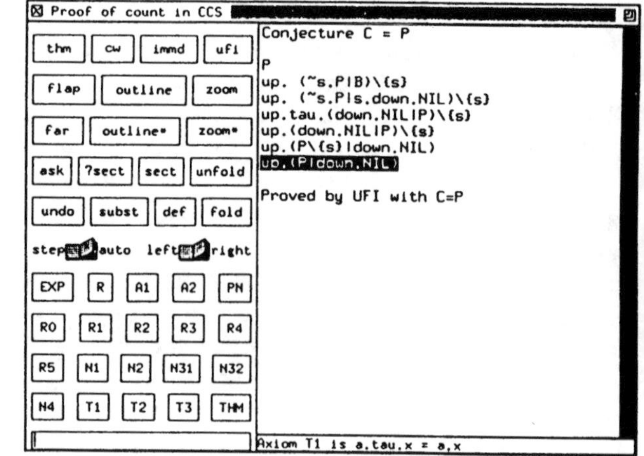

References

[BRSV 89] Boudol, G., Roy, V., de Simone, R., Vergamini, D., *Process Calculi, From Theory to Practice: Verification Tools*, INRIA Report No 1098, 1989.

[CPS 89] Cleaveland, R., Parrow, J. and Steffen, B., "The Concurrency Workbench", *Proc. of the Workshop on Automated Verification Methods for Finite State Systems*, LNCS 407, 1989.

[GLZ 89] Godskesen, J.C., Larsen, K.G., Zeeberg, M., *TAV Users Manual*, Internal Report, Aalborg University Centre, Denmark, 1989.

[NIN 89] De Nicola, R., Inverardi, P., Nesi, M., "Using the Axiomatic Presentation of Behavioural Equivalences for Manipulating CCS Specifications", *Proc. Workshop on Automatic Verification Methods for finite State Systems*, LNCS 407, 1989.

SPECI90

A term rewriting and narrowing system

by

Peter Bachmann, Thomas Drescher, Sabine Nieke
Technische Universität Dresden
Institut für Algebra

Overview

SPECI90 is a term rewriting and narrowing system which allows a user:

- to create, store and change specifications and tasks with a built-in editor,
- to compile specifications and tasks,
- to work with compiled specifications in three ways:
  ① derivation of normal forms of terms
  ② proof of equations
  ③ solving equations

Specification Language

The *specification language* allows a user to define the signature and the axioms of the specified theory and to write down tasks.
For every part, i.e. for sorts, operators, variables and axioms a corresponding phrase exists in specification language. The sequence of the several phrases is not important, the only restriction is that an object has to be introduced before it can be used.
A specification can be based on other specifications, i.e. it can use predefined sorts and operators. Identifiers for sorts and operators are any strings. The identifiers of local and global objects can coincide. In this case a qualification of identifiers is possible.
Axioms are conditional equations. The order of axioms is important.
The rewriting of terms, the proof or the solving of equational systems is specified in tasks.
A task is connected to a specification by a based-on phrase.

Editor

SPECI90 works with its own built-in editor for editing specifications, tasks and protocols. Its facilities can be compared with standard editors. Its special features include a macrohandler and search for missing parathesis.

Compiler

The compiler translates specifications and tasks into an internal form. At the same time, it checks the syntax. Errors are indicated immediately in the corresponding window. If a specification is based on another one, then this specification is automatically loaded and translated.

Windowing

SPECI90 works with windows, i.e. all input, output, and menus are shown in windows. The current active window is shadowed. In general all commands refer to this window on the screen, whereas other windows can be overlapped. Windows can be moved and zoomed. The *library* window contains a list of all complete paths where SPC- and TSK-windows may be created. *SPC*-windows display all specification names in the current directories. In *TSK*-windows all tasks are shown. In *Edit*-windows texts for specifications or tasks can be edited. The *trace*- and *result*-windows show log respectively result of rewriting or narrowing processes. Logs and result expressions can be edited and stored as files.

Find

In SPECI90 it is possible to find a certain specification, task, operator definition, or operator occurence. The search expression can include wildcards. It will be executed in all paths contained in the library window, the user path, and the system path. The result is a list of all occurences of the searched expression. When an item is selected, related edit window will be opened.

Term rewriting

A conditional term rewriting is implemented according to the use of conditional equations as axioms. The axioms are taken sequence in which they are written where local axioms are are prefered over global ones. In the current term, at first a matching on the rightmost-innermost position is searched for all axioms. If such a axiom is not found then the next node according the rightmost-innermost strategie is taken. If there is such a redex then the derivation mechanism tries to prove all premisses of the axiom used.

Narrowing

The conditional narrowing works analogously to the SLC of Fribourg. Since only a small amount of storage is available, a deep search strategy was implemented. Some optimizations help cut infinite and useless paths. Also, another kind of narrowing which uses E-equality can be used.

Implementation

SPECI90 was implemented in ModulaII for MS-DOS. It consists of more than 30 modules. The overall size of all modules amounts to about 650 kbytes. SPECI90.exe is about 164 kbytes.

Index of Authors

Lecture Notes in Computer Science

For information about Vols. 1–481
please contact your bookseller or Springer-Verlag

Vol. 482: Y. Kodratoff (Ed.), Machine Learning – EWSL-91. Proceedings, 1991. XI, 537 pages. 1991. (Subseries LNAI).

Vol. 483: G. Rozenberg (Ed.), Advances in Petri Nets 1990. VI, 515 pages. 1991.

Vol. 484: R. H. Möhring (Ed.), Graph-Theoretic Concepts in Computer Science. Proceedings, 1990. IX, 360 pages. 1991.

Vol. 485: K. Furukawa, H. Tanaka, T. Fuijsaki (Eds.), Logic Programming '89. Proceedings, 1989. IX, 183 pages. 1991. (Subseries LNAI).

Vol. 486: J. van Leeuwen, N. Santoro (Eds.), Distributed Algorithms. Proceedings, 1990. VI, 433 pages. 1991.

Vol. 487: A. Bode (Ed.), Distributed Memory Computing. Proceedings, 1991. XI, 506 pages. 1991.

Vol. 488: R. V. Book (Ed.), Rewriting Techniques and Applications. Proceedings, 1991. VII, 458 pages. 1991.

Vol. 489: J. W. de Bakker, W. P. de Roever, G. Rozenberg (Eds.), Foundations of Object-Oriented Languages. Proceedings, 1990. VIII, 442 pages. 1991.

Vol. 490: J. A. Bergstra, L. M. G. Feijs (Eds.), Algebraic Methods II: Theory, Tools and Applicatlons. VI, 434 pages. 1991.

Vol. 491: A. Yonezawa, T. Ito (Eds.), Concurrency: Theory, Language, and Architecture. Proceedings, 1989. VIII, 339 pages. 1991.

Vol. 492: D. Sriram, R. Logcher, S. Fukuda (Eds.), Computer-Aided Cooperative Product Development. Proceedings, 1989 VII, 630 pages. 1991.

Vol. 493: S. Abramsky, T. S. E. Maibaum (Eds.), TAPSOFT '91. Volume 1. Proceedings, 1991. VIII, 455 pages. 1991.

Vol. 494: S. Abramsky, T. S. E. Maibaum (Eds.), TAPSOFT '91. Volume 2. Proceedings, 1991. VIII, 482 pages. 1991.

Vol. 495: 9. Thalheim, J. Demetrovics, H.-D. Gerhardt (Eds.), MFDBS '91. Proceedings, 1991. VI, 395 pages. 1991.

Vol. 496: H.-P. Schwefel, R. Männer (Eds.), Parallel Problem Solving from Nature. Proceedings, 1990. XI, 485 pages. 1991.

Vol. 497: F. Dehne, F. Fiala. W.W. Koczkodaj (Eds.), Advances in Computing and Information - ICCI '91. Proceedings, 1991. VIII, 745 pages. 1991.

Vol. 498: R. Andersen, J. A. Bubenko jr., A. Sølvberg (Eds.), Advanced Information Systems Engineering. Proceedings, 1991. VI, 579 pages. 1991.

Vol. 499: D. Christodoulakis (Ed.), Ada: The Choice for '92. Proceedings, 1991. VI, 411 pages. 1991.

Vol. 500: M. Held, On the Computational Geometry of Pocket Machining. XII, 179 pages. 1991.

Vol. 501: M. Bidoit, H.-J. Kreowski, P. Lescanne, F. Orejas, D. Sannella (Eds.), Algebraic System Specification and Development. VIII, 98 pages. 1991.

Vol. 502: J. Bārzdiņž, D. Bjørner (Eds.), Baltic Computer Science. X, 619 pages. 1991.

Vol. 503: P. America (Ed.), Parallel Database Systems. Proceedings, 1990. VIII, 433 pages. 1991.

Vol. 504: J. W. Schmidt, A. A. Stogny (Eds.), Next Generation Information System Technology. Proceedings, 1990. IX, 450 pages. 1991.

Vol. 505: E. H. L. Aarts, J. van Leeuwen, M. Rem (Eds.), PARLE '91. Parallel Architectures and Languages Europe, Volume I. Proceedings, 1991. XV, 423 pages. 1991.

Vol. 506: E. H. L. Aarts, J. van Leeuwen, M. Rem (Eds.), PARLE '91. Parallel Architectures and Languages Europe, Volume II. Proceedings, 1991. XV, 489 pages. 1991.

Vol. 507: N. A. Sherwani, E. de Doncker, J. A. Kapenga (Eds.), Computing in the 90's. Proceedings, 1989. XIII, 441 pages. 1991.

Vol. 508: S. Sakata (Ed.), Applied Algebra, Algebraic Algorithms and Error-Correcting Codes. Proceedings, 1990. IX, 390 pages. 1991.

Vol. 509: A. Endres, H. Weber (Eds.), Software Development Environments and CASE Technology. Proceedings, 1991. VIII, 286 pages. 1991.

Vol. 510: J. Leach Albert, B. Monien, M. Rodríguez (Eds.), Automata, Languages and Programming. Proceedings, 1991. XII, 763 pages. 1991.

Vol. 511: A. C. F. Colchester, D.J. Hawkes (Eds.), Information Processing in Medical Imaging. Proceedings, 1991. XI, 512 pages. 1991.

Vol. 512: P. America (Ed.), ECOOP '91. European Conference on Object-Oriented Programming. Proceedings, 1991. X, 396 pages. 1991.

Vol. 513: N. M. Mattos, An Approach to Knowledge Base Management. IX, 247 pages. 1991. (Subseries LNAI).

Vol. 514: G. Cohen, P. Charpin (Eds.), EUROCODE '90. Proceedings, 1990. XI, 392 pages. 1991.

Vol. 515: J. P. Martins, M. Reinfrank (Eds.), Truth Maintenance Systems. Proceedings, 1990. VII, 177 pages. 1991. (Subseries LNAI).

Vol. 516: S. Kaplan, M. Okada (Eds.), Conditional and Typed Rewriting Systems. Proceedings, 1990. IX, 461 pages. 1991.

Vol. 517: K. Nökel, Temporally Distributed Symptoms in Technical Diagnosis. IX, 164 pages. 1991. (Subseries LNAI).

Vol. 518: J. G. Williams, Instantiation Theory. VIII, 133 pages. 1991. (Subseries LNAI).

Vol. 519: F. Dehne, J.-R. Sack, N. Santoro (Eds.), Algorithms and Data Structures. Proceedings, 1991. X, 496 pages. 1991.

Vol. 520: A. Tarlecki (Ed.), Mathematical Foundations of Computer Science 1991. Proceedings, 1991. XI, 435 pages. 1991.

Vol. 521: B. Bouchon-Meunier, R. R. Yager, L. A. Zadek (Eds.), Uncertainty in Knowledge-Bases. Proceedings, 1990. X, 609 pages. 1991.

Vol. 522: J. Hertzberg (Ed.), European Workshop on Planning. Proceedings, 1991. VII, 121 pages. 1991. (Subseries LNAI).

Vol. 523: J. Hughes (Ed.), Functional Programming Languages and Computer Architecture. Proceedings, 1991. VIII, 666 pages. 1991.

Vol. 524: G. Rozenberg (Ed.), Advances in Petri Nets 1991. VIII, 572 pages. 1991.

Vol. 525: O. Günther, H.-J. Schek (Eds.), Advances in Spatial Databases. Proceedings, 1991. XI, 471 pages. 1991.

Vol. 526: T. Ito, A. R. Meyer (Eds.), Theoretical Aspects of Computer Software. Proceedings, 1991. X, 772 pages. 1991.

Vol. 527: J.C.M. Baeten, J. F. Groote (Eds.), CONCUR '91. Proceedings, 1991. VIII, 541 pages. 1991.

Vol. 528: J. Maluszynski, M. Wirsing (Eds.), Programming Language Implementation and Logic Programming. Proceedings, 1991. XI, 433 pages. 1991.

Vol. 529: L. Budach (Ed.), Fundamentals of Computation Theory. Proceedings, 1991. XII, 426 pages. 1991.

Vol. 530: D. H. Pitt, P.-L. Curien, S. Abramsky, A. M. Pitts, A. Poigné, D. E. Rydeheard (Eds.), Category Theory and Computer Science. Proceedings, 1991. VII, 301 pages. 1991.

Vol. 531: E. M. Clarke, R. P. Kurshan (Eds.), Computer-Aided Verification. Proceedings, 1990. XIII, 372 pages. 1991.

Vol. 532: H. Ehrig, H.-J. Kreowski, G. Rozenberg (Eds.), Graph Grammars and Their Application to Computer Science. Proceedings, 1990. X, 703 pages. 1991.

Vol. 533: E. Börger, H. Kleine Büning, M. M. Richter, W. Schönfeld (Eds.), Computer Science Logic. Proceedings, 1990. VIII, 399 pages. 1991.

Vol. 534: H. Ehrig, K. P. Jantke, F. Orejas, H. Reichel (Eds.), Recent Trends in Data Type Specification. Proceedings, 1990. VIII, 379 pages. 1991.

Vol. 535: P. Jorrand, J. Kelemen (Eds.), Fundamentals of Artificial Intelligence Research. Proceedings, 1991. VIII, 255 pages. 1991. (Subseries LNAI).

Vol. 536: J. E. Tomayko, Software Engineering Education. Proceedings, 1991. VIII, 296 pages. 1991.

Vol. 537: A. J. Menezes, S. A. Vanstone (Eds.), Advances in Cryptology – CRYPTO '90. Proceedings. XIII, 644 pages. 1991.

Vol. 538: M. Kojima, N. Megiddo, T. Noma, A. Yoshise, A Unified Approach to Interior Point Algorithms for Linear Complementarity Problems. VIII, 108 pages. 1991.

Vol. 539: H. F. Mattson, T. Mora, T. R. N. Rao (Eds.), Applied Algebra, Algebraic Algorithms and Error-Correcting Codes. Proceedings, 1991. XI, 489 pages. 1991.

Vol. 540: A. Prieto (Ed.), Artificial Neural Networks. Proceedings, 1991. XIII, 476 pages. 1991.

Vol. 541: P. Barahona, L. Moniz Pereira, A. Porto (Eds.), EPIA '91. Proceedings, 1991. VIII, 292 pages. 1991. (Subseries LNAI).

Vol. 543: J. Dix, K. P. Jantke, P. H. Schmitt (Eds.), Nonmonotonic and Inductive Logic. Proceedings, 1990. X, 243 pages. 1991. (Subseries LNAI).

Vol. 544: M. Broy, M. Wirsing (Eds.), Methods of Programming. XII, 268 pages. 1991.

Vol. 545: H. Alblas, B. Melichar (Eds.), Attribute Grammars, Applications and Systems. Proceedings, 1991. IX, 513 pages. 1991.

Vol. 547: D. W. Davies (Ed.), Advances in Cryptology – EUROCRYPT '91. Proceedings, 1991. XII, 556 pages. 1991.

Vol. 548: R. Kruse, P. Siegel (Eds.), Symbolic and Quantitative Approaches to Uncertainty. Proceedings, 1991. XI, 362 pages. 1991.

Vol. 550: A. van Lamsweerde, A. Fugetta (Eds.), ESEC '91. Proceedings, 1991. XII, 515 pages. 1991.

Vol. 551:S. Prehn, W. J. Toetenel (Eds.), VDM '91. Formal Software Development Methods. Volume 1. Proceedings, 1991. XIII, 699 pages. 1991.

Vol. 552: S. Prehn, W. J. Toetenel (Eds.), VDM '91. Forma Software Development Methods. Volume 2. Proceedings, 1991 XIV, 430 pages. 1991.

Vol. 553: H. Bieri, H. Noltemeier (Eds.), Computational Geometry - Methods, Algorithms and Applications '91. Proceedings, 1991. VIII, 320 pages. 1991.

Vol. 554: G. Grahne, The Problem of Incomplete Informatior in Relational Databases. VIII, 156 pages. 1991.

Vol. 555: H. Maurer (Ed.), New Results and New Trends ir Computer Science. Proceedings, 1991. VIII, 403 pages. 1991.

Vol. 556: J.-M. Jacquet, Conclog: A Methodological Approach to Concurrent Logic Programming. XII, 781 pages. 1991.

Vol. 557: W. L. Hsu, R. C. T. Lee (Eds.), ISA '91 Algorithms Proceedings, 1991. X, 396 pages. 1991.

Vol. 558: J. Hooman, Specification and Compositional Verification of Real-Time Systems. VIII, 235 pages. 1991.

Vol. 559: G. Butler, Fundamental Algorithms for Permutatior Groups. XII, 238 pages. 1991.

Vol. 560: S. Biswas, K. V. Nori (Eds.), Foundations of Software Technology and Theoretical Computer Science. Proceedings, 1991. X, 420 pages. 1991.

Vol. 561: C. Ding, G. Xiao, W. Shan, The Stability Theory of Stream Ciphers. IX, 187 pages. 1991.

Vol. 562: R. Breu, Algebraic Specification Techniques in Object Oriented Programming Environments. XI, 228 pages. 1991.

Vol. 563: A. Karshmer, J. Nehmer (Eds.), Operating Systems of the 90s and Beyond. Proceedings, 1991. X, 285 pages. 1991.

Vol. 564: I. Herman, The Use of Projective Geometry in Computer Graphics. VIII, 146 pages. 1992.

Vol. 565: J. D. Becker, I. Eisele, F. W. Mündemann (Eds.), Parallelism, Learning, Evolution. Proceedings, 1989. VIII, 525 pages. 1991. (Subseries LNAI).

Vol. 566: C. Delobel, M. Kifer, Y. Masunaga (Eds.), Deductive and Object-Oriented Databases. Proceedings, 1991. XV, 581 pages. 1991.

Vol. 567: H. Boley, M. M. Richter (Eds.), Processing Declarative Kowledge. Proceedings, 1991. XII, 427 pages. 1991. (Subseries LNAI).

Vol. 568: H.-J. Bürckert, A Resolution Principle for a Logic with Restricted Quantifiers. X, 116 pages. 1991. (Subseries LNAI).

Vol. 569: A. Beaumont, G. Gupta (Eds.), Parallel Execution of Logic Programs. Proceedings, 1991. VII, 195 pages. 1991.

Vol. 570: R. Berghammer, G. Schmidt (Eds.), Graph-Theoretic Concepts in Computer Science. Proceedings, 1991. VIII, 253 pages. 1992.

Vol. 571: J. Vytopil (Ed.), Formal Techniques in Real-Time and Fault-Tolerant Systems. Proceedings, 1992. IX, 620 pages. 1991.

Vol. 572: K. U. Schulz (Ed.), Word Equations and Related Topics. Proceedings, 1990. VII, 256 pages. 1992.

Vol. 573: G. Cohen, S. N. Litsyn, A. Lobstein, G. Zémor (Eds.), Algebraic Coding. Proceedings, 1991. X, 158 pages. 1992.

Vol. 574: J. P. Banâtre, D. Le Métayer (Eds.), Research Directions in High-Level Parallel Programming Languages. Proceedings, 1991. VIII, 387 pages. 1992.

Vol. 576: J. Feigenbaum (Ed.), Advances in Cryptology - CRYPTO '91. Proceedings. X, 485 pages. 1992.

Vol. 577: A. Finkel, M. Jantzen (Eds.), STACS 92. Proceedings, 1992. XIV, 621 pages. 1992.